HANDBOOK OF

INDUSTRIAL and SYSTEMS ENGINEERING

HANDBOOK OF

INDUSTRIAL and SYSTEMS ENGINEERING

Adedeji B. Badiru

Taylor & Francis
Taylor & Francis Group
Boca Raton   London   New York

A CRC title, part of the Taylor & Francis imprint, a member of the
Taylor & Francis Group, the academic division of T&F Informa plc.

FIRST INDIAN REPRINT 2010

Published in 2006 by
CRC Press
Taylor & Francis Group
6000 Broken Sound Parkway NW, Suite 300
Boca Raton, FL 33487-2742

© 2006 by Taylor & Francis Group, LLC
CRC Press is an imprint of Taylor & Francis Group

No claim to original U.S. Government works
Printed and bound in India by Replika Press Pvt. Ltd.

International Standard Book Number-10: 0-8493-2719-9 (Hardcover)
International Standard Book Number-13: 978-0-8493-2719-3 (Hardcover)

Library of Congress Cataloging-in-Publication Data

Catalog record is available from the Library of Congress

Visit the Taylor & Francis Web site at
http://www.taylorandfrancis.com

and the CRC Press Web site at
http://www.crcpress.com

FOR SALE IN SOUTH ASIA ONLY

Dedication

This book is dedicated to all those who are committed to oiling the wheels of any system.

Foreword

Handbook of Industrial and Systems Engineering

While any handbook on a major topic such as Industrial Engineering and Systems will have inclusions and omissions, this handbook covers virtually all areas of industrial engineering as viewed from a system perspective. More than 40 authors have contributed state-of-the-art coverage of the most important modern industrial and systems engineering topics. Most of the authors are rising stars in our field, and are at the cutting edge of both theory and practice. This handbook will be invaluable, whether as a reference guide, a classroom text, or as support for the professional.

The premise of this new handbook is to incorporate more of the systems engineering aspects of industrial engineering than the two existing handbooks in our field. The material is presented by the newer authors of our field who have fresh (and possibly revolutionary) ideas. The objective of this handbook is to provide students, researchers, and practitioners with a comprehensive yet concise and easy-to-use guide to a wide range of industrial and systems tools and techniques.

The editor set himself the following goals:

1. To provide a one-stop reference for industrial and systems engineering
2. To use a comprehensive yet concise format
3. To include an up-to-date treatment of topics
4. To introduce new technologies for industrial and systems engineering
5. To use the systems integration approach
6. To provide coverage of information engineering
7. To offer a diversity of contributions from industry as well as academia
8. To provide up-to-date material for teaching, research and practice.

These have, in my opinion, been successfully met.

The editor, Dr. Adedeji B. Badiru, is a well-recognized and respected authority and leader in the fields of industrial and systems engineering, with numerous academic and professional publications to his credit. I know and respect Dr. Badiru, and I am proud to have been his dissertation advisor at the University of Central Florida.

Gary E. Whitehouse
Provost Emeritus
Distinguished University Professor
University of Central Florida
Orlando, FL

Preface

The *Handbook of Industrial and Systems Engineering* is the latest addition in the collation of the body of knowledge of industrial and systems engineering. The premise of the handbook is the incorporation of more of the systems engineering aspects, and expansion of the breadth of contributing authors beyond the traditional handbooks on industrial engineering.

There is a growing need for a handbook on the expanding field of industrial and systems engineering. The objective of this book is to fulfill this need and to provide students, researchers, and practitioners with comprehensive and easy access to a wide range of industrial and systems engineering tools and techniques in a concise format. The handbook:

1. Is a one-stop reference for industrial and systems engineering
2. Is comprehensive, yet concise
3. Provides an up-to-date treatment of topics
4. Is an introduction of new technologies for industrial and systems engineering
5. Has a systems integration approach
6. Provides coverage of information engineering
7. Has a diversity of contributions
8. Has up-to-date information for teaching, research, and practice

This handbook is expected to fill the gap that exists between the traditional and modern practice of industrial and systems engineering. The overall organization of the book is integrative with respect to quantitative models, qualitative principles, and computer techniques. Where applicable, the book incorporates a project approach in end-of-chapter exercises, rather than typical textbook problems. This is to provide open-ended problem exercises for readers, since most systems issues are open-ended challenges that are best handled from an integrated project perspective.

Part I of the book covers a general introduction with specific reference to the origin of industrial engineering and the ties to the Industrial Revolution. Part II covers the fundamentals of industrial engineering. Part III covers the fundamentals of systems engineering. Part IV contains chapters on manufacturing and production systems. Part V presents chapters on new technologies. Part VI contains general applications in industrial and systems engineering. Readers will find useful general information in the Appendix, which contains systems conversion factors and formulae.

Adedeji Bodunde Badiru
Knoxville, Tennessee

Acknowledgments

I acknowledge the contributions of all those who helped bring the idea of this book to fruition. It has been a long road and it could not have been done without the constant support of so many people. I thank my wife, Iswat, who as usual created an environment that is conducive for me to think and write. I appreciate all those around me who had to put up with temporary neglect while I focused on the demands of compiling this handbook. I thank my former professors and mentors, Dr. Sid Gilbreath and Dr. Gary Whitehouse, for continuing to be sources of inspiration for my intellectual pursuits.

Several individuals participated directly in putting together this edited work. I could not have done it without the administrative, editorial, typing, and copy-editing support of K. Christine Tidwell, Jeanette Myers, and Em Turner Chitty. They all worked tirelessly to fine-tune the manuscript.

I thank Cindy Renee Carelli, Senior Acquisitions Editor of CRC Press, for being a benevolent "pest," who frequently checked on my progress with the book. She jolted me back into action at the times that I thought stopping work was the right thing to do.

About the Editor

Professor Adedeji "Deji" B. Badiru is the department head in the Department of Industrial & Information Engineering at the University of Tennessee, Knoxville. He was previously professor of industrial engineering and Dean of University College at the University of Oklahoma. He is a registered professional engineer and a fellow of the Institute of Industrial Engineers. He holds a B.S. in Industrial Engineering, an M.S. in Mathematics, and an M.S. in Industrial Engineering from Tennessee Technological University, as well as a Ph.D. in Industrial Engineering from the University of Central Florida.

His areas of expertise cover mathematical modeling, project management, expert systems, economic analysis, industrial development projects, quality and productivity improvement, and computer applications. He is the author of several technical papers and books. His books include *Project Management in Manufacturing and High Technology Operation*; *Computer Tools, Models, and Techniques for Project Management*; *Project Management Tools for Engineering and Management Professionals*; *Industry's Guide to ISO 9000*; *Project Management for Research*; *Practitioner's Guide to Quality and Process Improvement*; *Comprehensive Project Management*; *Managing Industrial Development Projects*; *Expert Systems Applications in Engineering and Manufacturing*; and *Fuzzy Engineering Expert Systems with Neural Network Applications*.

Dr. Badiru is a member of several professional associations including the Institute of Industrial Engineers, the Society of Manufacturing Engineers, the Institute for Operations Research and Management Science, and the Project Management Institute.

He has served as a consultant to several organizations around the world including Russia, Mexico, Taiwan, Nigeria, and Ghana. He has conducted customized training workshops for numerous organizations including Sony, AT&T, Seagate Technology, U.S. Air Force, Oklahoma Gas & Electric, Oklahoma Asphalt Pavement Association, Hitachi, Nigeria National Petroleum Corporation, and ExxonMobil. He has won several awards including IIE Outstanding publication Award, University of Oklahoma Regents' Award for Superior Teaching, School of Industrial Engineering Outstanding Professor of the Year award, Eugene L. Grant Award for Best Paper in Volume 38 of *The Engineering Economist Journal*, University of Oklahoma College of Engineering Outstanding Professor of the Year award, Ralph R. Teetor Educational Award from the Society of Automotive Engineers, Award of Excellence as Chapter President from the Institute of Industrial Engineers, UPS Professional Excellence Award, Distinguished Alumni Award from Saint Finbarr's College (Lagos, Nigeria), and Distinguished Alumni Award from Department of Industrial and Systems Engineering, Tennessee Tech University. He holds a leadership certificate from the University Tennessee Leadership Institute.

He has served as a Technical Project Reviewer for The Third-World Network of Scientific Organizations, Italy. He has also served as a proposal review panelist for the National Science Foundation and the National Research Council, and a Curriculum reviewer for the American Council on Education. He is on the editorial and review boards of several technical journals and book publishers. Professor Badiru has served as an Industrial Development Consultant to the United Nations Development Program.

About the Editor

Contributors

Navin Acharya

H. Aiyappan

Mehmet Aktan

Sunday Asaolu

M. Affan Badar

Adedeji B. Badiru

Jonathan F. Bard

David Ben-Arieh

Richard E. Billo

Jennifer Blackhurst

Shing-Chi Chang

Zhifeng Chen

Garry D. Coleman

Lesia L. Crumpton-Young

Taren Daigle

Ashok Darisipudi

Gail W. DePuy

Ike C. Ehie

John V. Farr

Bobbie Leon Foote

Lincoln H. Forbes

Craig M. Harvey

Oye Ibidapo-Obe

Mohamad Y. Jaber

Wei Jiang

Kyoung-Yun Kim

Seong Kim

Dongjoon Kong

Tzong-Ru Lee

Pamela R. McCauley-Bell

Reinaldo J. Moraga

Larry Nabatilan

David A. Nembhard

Harriet Black Nembhard

Bartholomew O. Nnaji

Bryan A. Norman

S.A. Oke

Olufemi A. Omitaomu

J. David Porter

Richard J. Puerzer

Shivakumar Raman

Paul S. Ray

Marc L. Resnick

Ling Rothrock

Rupy Sawhney

Evangelos Triantaphyllou

Aashish Wadke

Xiaoting Wang

Yan Wang

Gary E. Whitehouse

Teresa Wu

Nong Ye

Contents

PART IV   MANUFACTURING AND PRODUCTION SYSTEMS

PART V   NEW TECHNOLOGIES

PART VI GENERAL APPLICATIONS

PART I

GENERAL INTRODUCTIONS

Origins of Industrial and Systems Engineering

Industrial engineering (IE) thrives on systems perspectives just as systems thrive on IE approaches. One cannot treat topics of IE effectively without recognizing systems perspectives and vice versa. Thus, it makes sense to have a handbook that integrates industrial and systems engineering (ISE) principles. A generic definition of IE, adopted by the Institute of Industrial Engineers (IIE) states:

"Industrial Engineer — *One who is concerned with the design, installation, and improvement of integrated systems of people, materials, information, equipment, and energy by drawing upon specialized knowledge and skills in the mathematical, physical, and social sciences, together with the principles and methods of engineering analysis and design to specify, predict, and evaluate the results to be obtained from such systems.*"

The above definition embodies the various aspects of what an industrial engineer does. Although some practitioners find the definition to be too convoluted, it nonetheless describes an industrial engineer. As can be seen, the profession is very versatile, flexible, and diverse. It can also be seen from the definition that a systems orientation permeates the work of industrial engineers. Some of the major functions of industrial engineers involve the following:

- Designing integrated systems of people, technology, process, and methods
- Developing performance modeling, measurement, and evaluation for systems
- Developing and maintaining quality standards for industry and business
- Applying production principles to pursue improvements in service organizations
- Incorporating technology effectively into work processes
- Developing cost mitigation, avoidance, or containment strategies
- Improving overall productivity of integrated systems of people, materials, and processes
- Recognizing and incorporate factors affecting performance of a composite system
- Planning, organizing, scheduling, and controlling production and service projects
- Organizing teams to improve efficiency and effectiveness of an organization
- Installing technology to facilitate work flow
- Enhancing information flow to facilitate smooth operations of systems
- Coordinating materials and equipment for effective systems performance

What is Industrial Engineering?

Industrial engineering can be described as the practical application of the combination of engineering fields, together with the principles of scientific management. It is the engineering of work processes and the application of engineering methods, practices, and knowledge to production and service enterprises. Industrial engineering places a strong emphasis on an understanding of workers and their needs in order to increase and improve production and service activities. Industrial engineering activities and techniques include the following:

1. Designing jobs (determining the most economic way to perform work)
2. Setting performance standards and benchmarks for quality, quantity, and cost
3. Designing and installing facilities

An important aspect of industrial engineering is its concern with the human element in industrial processes. The classical industrial engineering of the late 19th and early 20th centuries emphasized time studies, work sampling, methods engineering, costing methods, and employee incentives to make human interaction with industrial processes cost effective and reliable. Modern industrial engineering, in addition to the classical methods, deals with mathematical process modeling, management science methods, automation, and robotics. The use of advanced mathematical methods has become possible with the advent of computers. Mathematical process modeling allows the consideration of all available information on a process and the prediction of outcomes for given inputs and process parameters. The work of industrial engineers is varied and ranges from practical aspects of data gathering and analysis to the use of advanced mathematical methods of process simulation and optimization, as firms seek to reduce costs and increase productivity. Industrial engineers are in demand in all industries, ranging from manufacturing to service enterprises.

What is Systems Engineering?

Systems engineering involves a recognition, appreciation, and integration of all aspects of an organization or a facility. A system is defined as a collection of interrelated elements working together in synergy to produce a composite output that is greater than the sum of the individual outputs of the components. A systems view of a process facilitates a comprehensive inclusion of all the factors involved in the process.

Systems engineering is the application of engineering to solutions of a multi-faceted problem through a systematic collection and integration of parts of the problem with respect to the lifecycle of the problem. It is the branch of engineering concerned with the development, implementation, and use of large or complex systems. It focuses on specific goals of a system considering the specifications, prevailing constraints, expected services, possible behaviors, and structure of the system. It also involves a consideration of the activities required to assure that the system's performance matches the stated goals. Systems engineering addresses the integration of tools, people, and processes required to achieve a cost-effective and timely operation of the system.

Ties to the Industrial Revolution

Industrial engineering has a proud heritage with a link that can be traced back to the *Industrial Revolution*. Although the practice of IE has been in existence for centuries, the work of Frederick Taylor in the early 20th century was the first formal emergence of the profession. It has been referred to with different names and connotations. Scientific management was one of the original names used to describe what industrial engineers do.

Industry, the root of the profession's name, clearly explains what the profession is about. The dictionary defines industry generally as the ability to produce and deliver goods and services. The "industry" in IE can be viewed as the application of skills and cleverness to achieve work objectives. This relates to how human effort is harnessed innovatively to carry out work. Thus, any activity can be defined as "industry" if it generates a product, be it service or physical product. A systems view of IE encompasses all the details and aspects necessary for applying skills and accuracy to produce work efficiently. Hence the academic curriculum of IE must change, evolve, and adapt to the changing systems environment of the profession.

It is widely recognized that the occupational discipline that has contributed the most to the development of modern society is *engineering*, through its various segments of focus. Engineers design and build the infrastructures that sustains the society. This includes roads, residential and commercial buildings, bridges, canals, tunnels, communication systems, healthcare facilities, schools, habitats, transportation systems, and factories. The IE process of systems integration facilitates the success of these infrastructures. In this sense, the scope of ISE spans all the levels of activity, task, job, project, program, process, system, enterprise, and society. This handbook of ISE presents essential tools for the levels embodied by this hierarchy of functions. From the age of horse-drawn carriages and steam engines, to the present age of intelligent automobiles and aircraft, the impacts of ISE cannot be mistaken, even though the contributions may not be recognized in the context of the ISE disciplinary identification.

It is essential to recognize the alliance between "industry" and IE as the core basis for the profession. The profession has branched off on too many different tangents over the years. Hence, it has witnessed the emergence of IE professionals who claim sole allegiance to some narrow line of practice, focus, or specialization rather than the core profession itself. Industry is the original basis of IE and it should be preserved as the core focus. This should be supported by the different areas of specialization. While it is essential that we extend the scope of IE to other domains, it should be realized that over-divergence of practice will not sustain the profession. The continuing fragmentation of IE is a major reason for compiling a handbook such as this. A fragmented profession cannot survive for long. The incorporation of systems can help to bind everything together.

Notable industrial developments that fall under the purview of the practice of IE range from the invention of the typewriter to the invention of the automobile. Some examples are presented below.

Typewriter History

Writing is a basic means of communication and preservation of records. It is one of the most basic accomplishments of society. The course of history might have taken a different path if early writing instruments had not been invented at the time that they were. Below is the chronological history of events leading up to the development of the typewriter:

1714: Henry Mill obtains British patent for a writing machine.

1833: Xavier Progin creates a machine that uses separate levers for each letter.

1843: American inventor, Charles Grover Thurber, developes a machine that moves paper horizontally to produce spacing between lines.

1873: E. Remington & Sons of Ilion, New York, manufacturers of rifles and sewing machines, develop a typewriter patented by Carlos Glidden, Samuel W. Soule and Christopher Latham Sholes, who designed the modern keyboard. This class of typewriters wrote in only uppercase letters, but contained most of the characters on the modern machines.

1912: Portable typewriters were first introduced.

1925: Electric typewriters became popular. This led to typefaces becoming more uniform. International Business Machines Corporation (IBM) was a major distributor for this product.

In each case of product development, engineers demonstrate the ability to design, develop, manufacture, implement, and improve integrated systems that include people, materials, information, equipment, energy, and other resources. Thus, product development must include in-depth understanding of appropriate analytical, computational, experimental, implementation, and management processes.

The Heritage of Industrial and Systems Engineering

Going further back in history, several developments helped form the foundation for what later became known as IE. In America, George Washington was said to have been fascinated by the design of farm implements on his farm in Mt Vermon. He had an English manufacturer send him a plow built to his specifications that included a mold on which to form new irons when old ones were worn out or would needed repairs. This can be described as one of the early attempts to create a process of achieving a system of interchangeable parts. Thomas Jefferson invented a wooden mold board which, when fastened to a plow, minimized the force required to pull the plow at various working depths. This is an example of early agricultural industry innovation. Jefferson also invented a device that allowed a farmer to seed four rows at a time. In pursuit of higher productivity, he invented a horse-drawn threshing machine that did the work of ten men.

Meanwhile, in Europe, the Industrial Revolution was taking place at a rapid pace. Productivity growth, through reductions in manpower, marked the technological innovations of 1769–1800 Europe. Sir Richard Arkwright developed a practical code of factory discipline. In their foundry, Matthew Boulton and James Watt, developed a complete and integrated engineering plant to manufacture steam engines. They developed extensive methods of market research, forecasting, plant location planning, machine layout, work flow, machine operating standards, standardization of product components, worker training, division of labor, work study, and other creative approaches to increasing productivity. Charles Babbage, who is credited with the idea of the first computer, documented ideas on scientific methods of managing industry in his book entitled "*On the Economy of Machinery and Manufacturers*," which was first published in 1832. The book contained ideas on division of labor, paying less for less important tasks, organization charts, and labor relations. These were all forerunners of modern IE.

Back in America, several efforts emerged to form the future of the IE. Eli Whitney used mass production techniques to produce muskets for the U.S. Army. In 1798, Whitney developed the idea of having machines make each musket part so that it could be interchangeable with other similar parts. By 1850, the principle of interchangeable parts was widely adopted. It eventually became the basis for modern mass production for assembly lines. It is believed that Eli Whitney's principle of interchangeable parts contributed significantly to the Union victory during the U.S. Civil War.

Attempts of managements to improve productivity prior to 1880 did not consider the human element as an intrinsic factor. However, from 1880 through the first quarter of the 20th century, the works of Frederick W. Taylor, Frank and Lillian Gilbreth, and Henry L. Gantt created a long-lasting impact on productivity growth through consideration of the worker, and his or her environment.

Frederick Winslow Taylor (1856–1915) was born in the German town section of Philadelphia in a well-to-do family. At the age of 18, he entered the labor force, having foregone his admission to Harvard University due to an impaired vision. He became an apprentice machinist and pattern-maker in a local machine shop. In 1878, when he was 22, he went to work at the Midvale Steel Works. The economy was in a depressed state at the time. Frederick was employed as a laborer. His superior intellect was very quickly recognized. He was soon advanced to the positions of time clerk, journeyman, lathe operator, gang boss, and foreman of the machine shop. By the age of 31, he was made chief engineer of the company. He attended night school and earned a degree in mechanical engineering in 1883 from the Stevens Institute. As a work leader, Taylor often faced the following questions:

"Which is the best way to do this job?"

"What should constitute a day's work?"

These questions are still faced today by industrial and systems engineers. Taylor set about the task of finding the proper method for doing a given piece of work, instructing the worker in following the method, maintaining standard conditions surrounding the work so that the task could be properly accomplished, and setting a definite time standard, and payment of extra wages for doing the task as specified. Taylor later documented his industry management techniques in his book *The Principles of Scientific Management.*

The work of Frank and Lillian Gilbreth coincided with the work of Frederick Taylor. In 1895, on his first day on the job as a bricklayer, Frank Gilbreth noticed that the worker assigned to teach him how to lay bricks did his work in three different ways (when training someone on the job, when performing the job himself, and when speeding up). The bricklayer was insulted when Frank tried to tell him of his work inconsistencies. Frank thought it was essential to find one best way to do work. Many of Frank Gilbreth's ideas were similar to Taylor's. However, Gilbreth outlined procedures for analyzing each step of the work flow. Gilbreth made it possible to apply science more precisely in the analysis and design of the work place. Developing *therblig* (from Gilbreth spelled backwards) as elemental predetermined time units, Frank and Lillian Gilbreth were able to analyze the motions of a worker in performing most factory operations. They found there were a maximum of 18 steps. Working as a team, they developed techniques that later became known as work design, methods improvement, work simplification, value engineering, and optimization. Lillian (1878–1972) brought to the engineering profession the concern for human relations. The foundation for establishing the profession of IE was originated by Frederick Taylor and Frank and Lillian Gilbreth.

Henry Gantt's work advanced the management movement from an industrial management perspective. He expanded the scope of managing industrial operations. His concepts emphasized the unique needs of the worker by recommending the following considerations for managing work:

(a) Define his task, after a careful study
(b) Teach him how to do it
(c) Provide an incentive in terms of adequate pay or reduced hours
(d) Provide an incentive to surpass it

Henry Gantt's major contribution is the Gantt chart, which went beyond the works of Frederick Taylor or the Gilbreths. The Gantt chart related every activity in the plant to the factor of time. This was a revolutionary concept for its time. It led to better production planning control and better production control. It involved visualizing the plant as a whole as one big system made up of inter-related subsystems.

Major Chronological Events Marking the Origin of Industrial and Systems Engineering

Year	Major Publications and Events
1440	Venetian ships are reconditioned and refitted on an assembly line
1474	Venetian senate passes the first patent law and other industrial laws
1568	Jacques Besson publishes illustrated book on iron machinery as replacement for wooden machines.
1622	William Oughtred invents the slide rule.
1722	Rene de Reaunur publishes the first handbook on iron technology.
1733	John Kay patents the flying shuttle for textile manufacture — a landmark in textile mass production.
1747	Jean Rodolphe Perronet establishes the first engineering school.
1765	Watt invents the separate condenser, which made the steam engine the power source.
1770	James Hargreaves patents his "Spinning Jenny". Jesse Ramsden devises a practical screw-cutting lathe.
1774	John Wilkinson builds the first horizontal boring machine.
1775	Richard Arkwright patents a mechanized mill in which raw cotton is worked into thread.
1776	James Watt builds the first successful steam engine, which became a practical power source.
1776	Adam Smith discusses the division of labor in *The Wealth of Nations*.
1785	Edmund Cartwright patents a power loom.
1793	Eli Whitney invents the "cotton gin" to separate cotton from its seeds.
1797	Robert Owen uses modern labor and personnel management techniques in a spinning plant in the New Lanark Mills in Manchester, England.
1798	Eli Whitney designs muskets with interchangeable parts.
1801	Joseph Marie Jacquard designs automatic control for pattern-weaving looms using punched cards.
1802	"Health and Morals Apprentices Act" in Britain aims at improving standards for young factory workers. Marc Isambard Brunel, Samuel Benton, and Henry Maudsey designed an integrated series of 43 machines to mass produce pulley blocks for ships.
1818	Institution of Civil Engineers founded in Britain.
1824	The repeal of the Combination Act in Britain legalizes trade unions.
1829	Mathematician Charles Babbage designs "analytical engine," a forerunner of the modern digital computer.
1831	Charles Babbage publishes *On the Economy of Machines and Manufacturers*.
1832	The Sadler report exposes the exploitation of workers and the brutality practiced within factories.
1833	Factory law enacted in United Kingdom. The Factory Act regulates British children's working hours. A general Trades Union is formed in New York.
1835	Andrew Ure publishes *Philosophy of Manfacturers*. Samuel Morse invents the telegraph.
1845	Friederich Engels publishes *Condition of the Working Classes in England*.
1847	Factory Act in Britain reduces the working hours of women and children to 10 h per day. George Stephenson founds the Institution of Mechanical Engineers.
1856	Henry Bessemer revolutionizes the steel industry through a novel design for a converter.
1869	Trans-continental railroad completed in United States.
1871	British Trade Unions are legalized by Act of Parliament.
1876	Alexander Graham Bell invents a usable telephone.
1877	Thomas Edison invents the phonograph.
1878	Frederick W. Taylor joins Midvale Steel Company.
1880	American Society of Mechanical Engineers (ASME) is organized.
1881	Frederick Taylor begins time study experiments.
1885	Frank B. Gilbreth begins motion study research.
1886	Henry R. Towne presents the paper, *The Engineer as Economist*.

American Federation of Labor (AFL) is organized.

Vilfredo Pareto publishes *Course in Political Economy*.

Charles M. Hall and Paul L. Herault independently invent an inexpensive method of making aluminum.

1888   Nikola Tesla invents the alternating current induction motor, enabling electricity to take over from steam as the main provider of power for industrial machines.

Dr. Herman Hollerith invents the electric tabulator machine, the first successful data processing machine.

1890   Sherman Anti-Trust Act is enacted in the United States.

1892   Gilbreth completes motion study of bricklaying.

1893   Taylor begins work as consulting engineer.

1895   Taylor presents paper entitled *A Piece-Rate System* to ASME.

1898   Taylor begins time study at Bethlehem Steel.

Taylor and Maunsel White develop process for heat-treating high-speed tool steels.

1899   Carl G. Barth invents a slide rule for calculating metal cutting speed as part of Taylor system of management.

1901   American national standards are established.

Yawata Steel begins operation in Japan.

1903   Taylor presents paper entitled *Shop Management* to ASME.

H.L. Gantt develops the "Gantt chart".

Hugo Diemers writes *Factory Organization and Administration*.

Ford Motor Company is established.

1904   Harrington Emerson implements Santa Fe Railroad improvement.

Thorstein B. Veblen publishes *The Theory of Business Enterprise*.

1906   Taylor establishes metal-cutting theory for machine tools.

Vilfredo Pareto publishes *Manual of Political Economy*.

1907   Gilbreth uses time study for construction

1908   Model T Ford is built

Pennsylvania State College introduces the first university course in industrial engineering.

1911   Taylor publishes *The Principles of Scientific Management*.

Gilbreth publishes *Motion Study*.

Factory laws are enacted in Japan.

1912   Harrington Emerson publishes *The Twelve Principles of Efficiency*.

Frank and Lillian Gilbreth present the concept of "therblig".

Yokokawa translates into Japanese Taylor's *Shop Management* and *The Principles of Scientific Management*.

1913   Henry Ford establishes a plant at Highland Park, Michigan, which utilizes the principles of uniformity and interchangeability of parts: and of the moving assembly line by means of conveyor belt.

Hugo Munstenberg publishes *Psychology of Industrial Efficiency*.

1914   World War I.

Clarence B. Thompson edits *Scientific Management*, a collection of articles on Taylor's system of management.

1915   Taylor's system is used at Niigata Engineering's Kamata plant in Japan.

Robert Hoxie publishes *Scientific Management and Labour*.

1916   Lillian Gilbreth publishes *The Psychology of Management*.

Taylor Society established in United States.

1917   The Gilbreths publishes *Applied Motion Study*.

The Society of Industrial Engineers is formed in the Unites States.

1918   Mary P. Follet publishes *The New State: Group Organization, the Solution of Popular Government*.

1919   Henry L. Gantt published *Organization for Work*.

1920   Merrick Hathaway presents paper: *Time Study as a Basis for Rate Setting*.

General Electric establishes divisional organization.

Karel Capek publishes *Rossum's Universal Robots*. This play gave rise to the word "robot".

1921   The Gilbreths introduce process-analysis symbols to ASME.

1922   Toyoda Sakiichi's automatic loom is developed.

Henry Ford publishes *My Life and Work*.

1924   The Gilbreths announce results of micromotion study using therbligs.

Elton Mayo conducts illumination experiments at Western Electric.

1926   Henry Ford publishes *Today and Tomorrow*.

1927   Elton Mayo and others begin relay-assembly test room study at the Hawthorne plant.

1929   Great Depression.

International Scientific Management Conference held in France.

1930   Hathaway publishes *Machining and Standard Times*.

Allan H. Mogensen discusses 11 principles for work simplification in *Work Simplification*.

Henry Ford publishes *Moving Forward*.

1931   Dr. Walter Shewhart publishes *Economic Control of the Quality of Manufactured Product*.

1932   Aldous Huxley publishes *Brave New World*, a satire which prophesies a horrifying future ruled by industry.

1934   General Electric performs micromotion studies.

1936   The word "automation" is first used by D.S. Harder of General Motors. It is used to signify the use of transfer machines which carry parts automatically from one machine to the next, thereby linking the tools into an integrated production line.

Charlie Chaplin produces *Modern Times*, a film showing an assembly line worker driven insane by the routine and unrelenting pressure of his job.

1937   Ralph M. Barnes publishes *Motion and Time Study*.

1941   R.L. Morrow publishes on *Ratio Delay Study*, in *Mechanical Engineering* journal.

Fritz J. Roethlisberger publishes *Management and Morale*.

1943   ASME work standardization committee publishes glossary of Industrial Engineering terms.

1945   Marvin E. Mundel devises "memo-motion" study, a form of work measurement using time-lapse photography.

Joseph H. Quick devises work factors (WF) method.

1945   At a technical meeting of the Japan Management Association, Shigeo Shingo presents concept of production as a network of processes and operations and identifies lot delays as source of delay between processes.

1946   The first all-electronic digital computer Electronic Numerical Integrator and Computer (ENIAC) is built at Pennsylvania University.

The first fully automatic system of assembly is applied at the Ford Motor Plant.

1947   American mathematician, Norbert Wiener publishes *Cybernetics*.

1948   H.B. Maynard and others introduce methods time measurement (MTM) method.

Larry T. Miles develops value analysis (VA) at General Electric.

Shigeo Shingo announces process-based machine layout.

American Institute of Industrial Engineers is formed.

1950   Marvin E. Mundel publishes *Motion and Time Study, Improving Productivity*.

1951   Inductive statistical quality control is introduced to Japan from the United States.

1952   Role and sampling study of Industrial Engineering conducted at ASME.

1953   B.F. Skinner publishes *Science of Human Behaviour*.

1956   New definition of industria engineering is presented at the American Institute of Industrial Engineering Convention.

1957   Chris Argyris publishes *Personality and Organization*.

Herbert A. Simon publishes *Organizations*.

R.L. Morrow publishes *Motion and Time Study*.

1957   Shigeo Shingo introduces scientific thinking mechanism (STM) for improvements.

The Treaty of Rome establishes the European Economic Community.

1960 Douglas M. McGregor publishes *The Human Side of Enterprise.*

1961 Rensis Lickert publishes *New Patterns of Management.*

1961 Shigeo Shingo devises ZQC (source inspection and poka-yoke systems).

1961 Texas Instruments patents the silicon chip integrated circuit.

1963 H.B. Maynard publishes *Industrial Engineering Handbook.*

Gerald Nadler publishes *Work Design.*

1964 Abraham Maslow publishes *Motivation and Personality.*

1965 Transistors are fitted into miniaturized "integrated circuits".

1966 Frederick Hertzberg publishes *Work and the Nature of Man.*

1968 Roethlisberger publishes *Man in Organization.*

U.S. Department of Defense publishes *Principles and Applications of Value Engineering*

1969 Shigeo Shingo develops Single-Minute Exchange of Dies (SMED).

Shigeo Shingo introduces pre-automation.

Wickham Skinner publishes *Manufacturing — missing link in corporate strategy,* article in *Harvard Business Review.*

1971 Taiichi Ohno completes the Toyota production system.

1971 Intel Corporation develops the micro-processor chip.

1973 First annual Systems Engineering Conference of AIIE.

1975 Shigeo Shingo extols NSP-SS (non-stock production) system.

Joseph Orlicky publishes "*MRP: Material Requirements Planning*".

1976 IBM markets the first personal computer.

1980 Matsushita Electric used Mikuni method for washing machine production.

Shigeo Shingo publishes *Study of the Toyota Production System from an Industrial Engineering Viewpoint.*

1981 Oliver Wight publishes *Manufacturing Resource Planning: MRP II.*

1982 Gavriel Salvendy publishes *Handbook of Industrial Engineering*

1984 Shigeo Shingo publishes *A Revolution in Manufacturing: The SMED System.*

As can be seen from the historical details above, industry has undergone a hierarchical transformation over the past several decades. The picture below shows how industry has been transformed from one focus level to the next, ranging from efficiency of the 1960s to the present-day nanoscience trend.

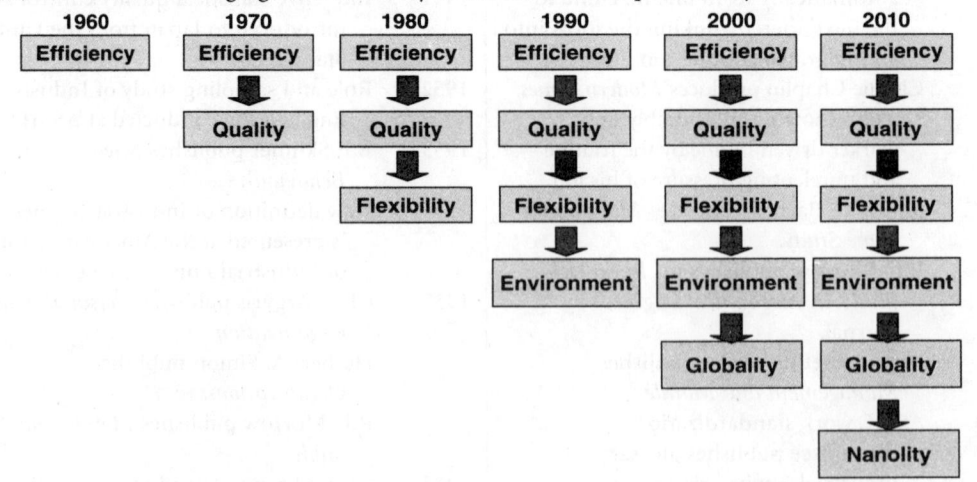

PART II

FUNDAMENTALS OF INDUSTRIAL ENGINEERING

1

Human Factors

Marc Resnick
Florida International University

1.1 Human Factors in Industrial and Systems Engineering

Human Factors is a science that investigates human behavioral, cognitive, and physical abilities and limitations in order to understand how individuals and teams will interact with products and systems. Human Factors engineering is the discipline that takes this knowledge and uses it to specify, design, and test systems to optimize safety, productivity, effectiveness, and satisfaction.

Human Factors is important to industrial and systems engineering because of the prevalence of humans within industrial systems. It is humans who, for the most part, are called on to design, manufacture, operate, monitor, maintain, and repair industrial systems. In each of these cases, Human Factors should be used to ensure that the design will meet system requirements in performance, productivity, quality, reliability, and safety. This chapter presents an overview of Human Factors, how it should be integrated into the systems engineering process, and some examples from a variety of industries.

The importance of including Human Factors in systems design cannot be overemphasized. There are countless examples that illustrate its importance for system performance. Mackenzie (1994) found that in a survey of 1100 computer-related fatalities between 1979 and 1992, 92% could be attributed to failures in the interaction between a human and a computer. The extent of the 1979 accident at the Three Mile Island nuclear power plant was largely due to Human Factors challenges (Bailey, 1996), almost resulting in a disastrous nuclear catastrophe. The infamous butterfly ballot problem in Florida in the 2000 U.S. presidential election is a clear example of an inadequate system interface yielding remarkably poor performance (Resnick, 2001). Web sites such as www.baddesigns.com and www.thisisbroken.com provide extensive listings of designs from everyday life that suffer from poor consideration of Human Factors. Neophytes often refer to Human Factors as common sense. However, the prevalence of poor design suggests that Human Factors sense is not as common as one might think. The consequences of poor Human Factors design can be inadequate system performance, reduced product sales, significant product damage, and human injury.

This chapter provides an overview of Human Factors and is intended to support the effective design of systems in a variety of work domains, including manufacturing, process control, transportation, medical care, and others. Section 1.2 presents some of the principal components of Human Factors analysis that must be addressed in any systems design, as well as the benefits of effectively integrated Human Factors. Section 1.3 describes a conceptual model of human information processing and outlines how each aspect affects performance. An example is provided for each one that illustrates the design challenges for Human Factors and how these challenges can be overcome. Section 1.4 describes two important consequences of design: the ability of humans to learn from their experience, and the likelihood of errors during system use.

1.2 Elements of Human Factors

In order to facilitate the design of effective systems, Human Factors must adopt a holistic perspective on human–system interaction. Systems engineers need to understand how people think, how these thoughts lead them to act, the results of these actions, and the reliability of the results of these actions. Thus, the following four elements should be considered: cognition, behavior, performance, and reliability.

1.2.1 Cognition

A considerable body of Human Factors research has been dedicated to human cognition. It is critical for systems engineers to understand and predict how users will perceive the information that they receive during system use, how this information will be processed, and the nature of the resulting behavior and decisions of users. Situation awareness (SA) (Endsley, 2000a) refers to the extent to which a user has perceived and integrated the important information in the world and can project that information into the future to make predictions about system performance.

Consider the case of an air-traffic controller who needs to monitor and communicate simultaneously with several aircraft to ensure that they all land safely. This job requires the controller to develop a composite mental model of the location and direction of each aircraft so that when new information appears, he or she can quickly decide on an appropriate response. The design of the system interface must anticipate this model so that information can be presented in a way that allows the controller to perceive it quickly and effectively integrate it into the mental model.

1.2.2 Behavior

The actions taken by the human components of a system are often more difficult to predict than the mechanical or electrical components. Unlike machines, people behave on the basis of experiences and beliefs that transcend the system, including factors such as corporate culture, personal goals, and past experience. It is critical for systems engineers to investigate the effects of these sources on behavior to ensure that the system will be successful.

For example, the Columbia Accident Investigation Board (CAIB) concluded that the accident causing the destruction of the space shuttle Columbia in 2003 was as much caused by NASA's organizational culture as it was by the foam that struck the orbiter. The CAIB report stated that systems were approved despite deviations in performance because of a past history of success (CAIB, 2003). At the consumer level, various factors may also be important. For instance, Internet retailers are interested in the factors that determine whether a consumer will purchase a product on the company's website. In addition to design factors such as the site's menu design and information architecture, the user's past history at other web sites can also affect his or her behavior on this site (Nielsen, 1999).

1.2.3 Performance

Most systems depend not only on whether an action is completed, but also on the speed and accuracy with which the action is completed. Many factors affect user performance, such as the number of information sources that must be considered, the complexity of the response required, the user's motivation for performing well, and others (Sanders and McCormick, 1993).

Call-center operations are a clear example of the need to include Human Factors in design to achieve optimal performance (Kemp, 2001). Call-center software must complement the way that operators think about the task, or performance may be significantly delayed. The cost structure of call centers relies on most customer service calls being completed within seconds. Early versions of some customer relationship management (CRM) software required operators to drill down through ten screens to add a customer record. This design slowed the task considerably. However, labels that lead to strong path recognition can have as great an effect as path length on performance (Katz and Byrne, 2003). Trade-offs between path recognition strength and path length must be resolved in the information architecture of the system. It is critical for systems that rely on speed and accuracy of performance to thoroughly integrate Human Factors into their designs.

1.2.4   Reliability

Human Factors is also important in the prediction of system reliability. Human error is often cited as the cause of system failures (Federal Aviation Administration, 1990; MacKenzie, 1994). However, the root cause is often traceable to an incompatibility between the system interface and human information processing. An understanding of human failure modes, the root causes of human error, and the performance and contextual factors that affect error probability and severity can lead to more reliable systems design.

Much of the research in human error has been in the domain of aerospace systems (Eurocontrol, 2002) and control center operations (Swain and Guttmann, 1983). Eurocontrol (2002) describes models that predict human errors in behavior, cognition, communication, perception, and others. Integrating human reliability models into the systems engineering process is essential.

1.2.5   The Benefits of Human Factors

There are many benefits that result from considering each of these four elements of Human Factors in systems design. The primary benefit is that the resulting system will be more effective. By accommodating the information-processing needs of the users, the system will better match the system requirements and will thus be more productive. Systems that incorporate Human Factors are also more reliable. Since human error is often the proximate cause of system failure, reducing the likelihood of human error will increase the reliability of the system as a whole.

Consideration of Human Factors also leads to cost reductions in system design, development, and production. When Human Factors are considered early in the design process, flaws are avoided and early versions are closer to the final system design. Rework is avoided, and extraneous features can be eliminated before resources are expended on developing them.

Human Factors also lead to reduced testing and quality assurance requirements. Problems are caught earlier and it becomes easier to prioritize what system components to modify. Systems that exhibit good Human Factors design reduce sales time and costs because they are easier to demonstrate, train, and set up in the field.

Finally, consideration of Human Factors leads to reduced costs for service and support. When systems are easy to use, there are fewer service calls and less need for ongoing training. The occurrence of fewer errors leads to reduced maintenance costs, fewer safety violations, and less frequent need for mishap/injury investigation.

1.3   A Human Factors Conceptual Model

Behavior and performance emerge from the way that humans process information. Human information processing is generally conceptualized using a series of processing stages (see Figure 1.1). It is important to keep in mind that these stages are not completely separate; they can work in parallel, and they are linked bi-directionally. A detailed discussion of the neurophysiology of the brain is beyond the scope of this chapter. But there is one underlying trait that is often overlooked. The

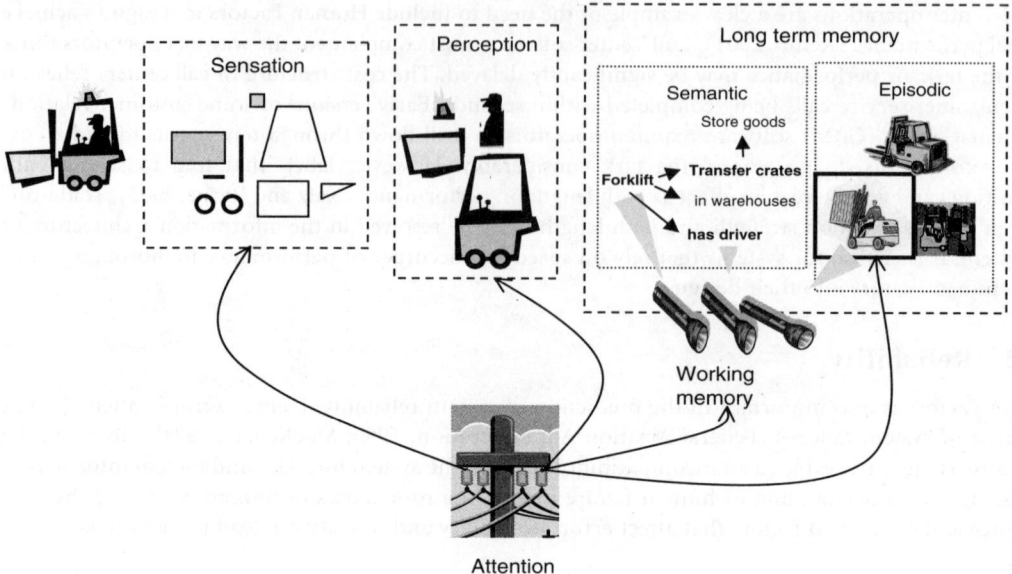

FIGURE 1.1   A conceptual model of human information processing.

human information-processing system (the brain) is noisy, a fact that can lead to errors even when a particular fact or behavior is well known. On the positive side, this noise also enables greater creativity and problem-solving ability.

1.3.1   Long-Term Memory

"Long-term memory" refers to the composite of information that is stored in an individual's information-processing system, the brain. It is composed of a vast network of interconnected nodes, a network that is largely hierarchical but that has many cross-unit connections as well. For example, a dog is an animal, but also lives (usually) in a house.

The basic unit of memory has been given many names, but for the purposes of this chapter will be called the *cell assembly* (Hebb, 1955). A cell assembly is a combination of basic attributes (lines, colors, sounds, etc.) that become associated because of a history of being activated simultaneously. Cell assemblies are combined into composites called schema. The size and complexity of the schema depend on the experience of the individual. The schema of an elephant will be very simple for a 4-year-old child who sees one for the first time in a storybook. On the other hand, a zoologist may have a complex schema composed of physical, behavioral, historical, ecological, and perhaps other elements. The child's elephant schema may be connected only to the other characters of the story. The zoologist will have connections between the elephant and many schemas throughout his or her long-term memory network.

Another important characteristic of memory is the strength of the connections between the units. Memory strength is developed through repetition, salience, or elaboration. Each time a memory is experienced, the ease with which that memory can be recalled in the future increases (Hebb, 1976). Thus, rote memorization increases memory strength by increasing the number of times the memory was activated. Similarly, experiences that have strong sensory or emotional elements have a disproportionate gain in memory strength. A workplace error that has significant consequences will be remembered much better than one that has none. Elaboration involves relating the new information to existing schema and incorporating it in an organized way. Memory strength has a substantial impact on cognition. Well-learned schema can be recalled faster and with less effort because less energy is required to activate the stronger connections.

1.3.1.1 Types of Long-Term Memory

Long-term memory can be partitioned into categories such as episodic, semantic, and procedural components (Tulving, 1989). Episodic memory refers to traces that remain from the individual's personal experiences. Events from the past are stored as sub-sensory cell assemblies that are connected to maintain important features of the event but generalize less important features. Thus, an athlete's episodic memory of the championship game may include very specific and detailed visual traces of significant actions during the game. But less important actions may really be represented using statistical aggregates of similar actions that occurred over the athlete's total experience of games. These aggregates would be reconstructed during recall to provide the recall experience.

Semantic memory is composed of conceptual information, including knowledge of concept definitions, object relationships, physical laws, and similar non-sensory information (Lachman et al., 1979). Connections within semantic memory link concepts that are related or are used together. While the composition of semantic memory is not as structured or organized as an explicit semantic network (see Figure 1.2), this is a reasonable simplification for the purposes of this chapter. There are also links between semantic memory and episodic memory. For example, the semantic memory of the meaning of "quality" may be linked to the episodic memory of a high-quality product.

Procedural memory refers to the combination of muscle movements that compose a unitary action (Johnson, 2003). Procedural memories are often unconscious and stored as one complete unit. For example, a pianist may be able to play a complex piano concerto, but cannot verbally report what the fourth note is without first imagining the initial three notes. These automatic processes can take over in emergency situations when there is not enough time to think consciously about required actions.

Memories are not separated into distinct units that are clearly demarcated within the human information-processing system. There are overlaps and interconnections within the memory structure that have both advantages and disadvantages for human performance. Creative problem-solving is enhanced when rarely used connections are activated to brainstorm for solution ideas. But this can lead to errors when random connections are assumed to represent fact or statistical associations are assumed to apply to inappropriate specific cases.

1.3.1.2 Implications for Design

The structure of long-term memory has significant implications for the design of industrial systems. Workers can master work activities faster when new processes match prior learning and overlap with existing schema. This will also reduce the probability that inappropriate schema will become activated in emergency situations, perhaps leading to errors. Similarly, terminology for labels, instructions, and information displays should be unambiguous to maximize speed of processing.

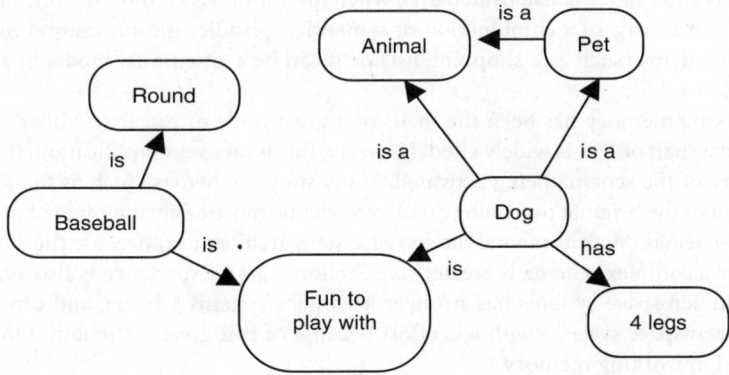

FIGURE 1.2   A semantic network.

Simulator and field training can focus on developing episodic memories that bolster semantic memory for system architecture and underlying physical laws. Training breadth can expand the semantic schema to include a variety of possible scenarios, while repetition can be used to solidify the connections for critical and frequent activities. Scielzo et al. (2002) found that training protocols that supported the development of accurate schema allowed the performance of new learners to approach that of experts.

1.3.1.3 Case: Power-Plant Control

Control rooms are often composed of a large set of monitors, controls, and displays that show the status of processes in graphical, tabular, and digital readouts. Operators are trained to recognize problems as soon as they occur, diagnose the problem, and initiate steps to correct the problem. The design of this training is critical so that operators develop schema that effectively support problem solving.

To maximize the ability of operators to identify major emergencies, training should include repeated simulations of these emergencies. According to Wickens et al. (2004), initial training leads to an accurate response for a given situation. But additional training is still necessary to increase the speed of the response and to reduce the amount of attention that is necessary for the emergency to be noticed and recognized. Thus, overlearning is important for emergency response tasks. With sufficient repetition, operators will have strong long-term memories for each emergency and can recognize them more quickly and accurately (Klein, 1993). They will know which combination of displays will be affected for each problem and what steps to take. Accurate feedback is critical to ensure that workers associate the correct responses with each emergency (Wickens et al., 2004). When errors are made, corrective feedback is particularly important so that the error does not become part of the learned response.

But problems do not always occur in the same way. In order for training to cover the variability in problem appearance, variation must be included in the training. Employees must be trained to recognize the diversity of situations that can occur. This will develop broader schema that reflect a deeper conceptual understanding of the problem states and lead to a better ability to implement solutions. Semantic knowledge is also important because procedures can be context-specific (Gordon, 1994). Semantic knowledge helps employees to adapt existing procedures to new situations.

Training fidelity is also an important consideration. The ecological validity of training environments has been shown to increase training transfer, but Swezey and Llaneras (1997) have shown that not all features of the real environment are necessary. Training design should include an evaluation of what aspects of the real environment contribute to the development of effective problem schema.

1.3.2 Working Memory

While the entire network of schemas stored in long-term memory is extensive, it is impossible for an individual to simultaneously recall more than a limited set of schemas. Working memory refers to the set of schemas that is activated at one point in time. A schema stored in long-term memory reverberates due to some input stimulus and can remain activated even when the stimulus is removed (Jones and Polk, 2002). Working memory can consist of a combination of semantic, episodic, and procedural memories. It can be a list of unrelated items (such as a shopping list) or it can be a situational model of a complex environment (see Section 1.3.5).

The size of working memory has been the focus of a great body of research. Miller's (1956) famous study that reported a span of 7±2 is widely cited. However, this is an oversimplification. It depends on the size and complexity of the schema being activated. Many simple schemas (such as the single digits and letters used in much of the original psychology research) can be more easily maintained in working memory than complex schemas (such as mental models of system architecture) because the amount of energy required to activate a complex schema is greater (see Section 1.3.4). Experience is also a factor. Someone who is an expert in aerospace systems has stronger aerospace systems schema, and can therefore recall schema related to aerospace systems with less effort because of this greater strength. More schemas can thus be maintained in working memory.

The size of working memory also depends on the ability of the worker to combine information into chunks (Wickens et al., 2004). Chunks are sets of working memory units that are combined into single units based on perceptual or semantic similarity. For example, mnemonics enhance memory by allowing workers to remember a single acronym, such as SEARCH for Simplify, Eliminate, Alter sequence, Requirements, Combine operations, and How often for process improvement brainstorming (from Konz and Johnson, 2000), more easily than a list of items.

How long information can be retained in working memory depends on the opportunity for workers to subvocally rehearse the information (Wickens et al., 2004). Without rehearsal, information in working memory is lost rapidly. Thus, when working memory must be heavily used, distractions must be minimized and ancillary tasks that also draw on this subvocalization resource must be avoided.

The similarity of competing information in working memory also affects the reliability of recall. Because working memory exists in the auditory modality, information that sounds alike is most likely to be confused (Wickens et al., 2004). The working memory requirements for any concurrent activities must be considered to minimize the risk of interference.

1.3.2.1 Implications for Design

A better understanding of working memory can support the development of more reliable industrial systems. The amount of information required to complete work activities should be considered in relation to the working memory capacities of the workers. Norman (1988) describes two categories of information storage. Information in the head refers to memory and information in the world refers to labels, instructions, displays, and other physical devices. When the amount of information required to complete a task exceeds the capacity of working memory, it must be made available in the physical world through computer displays, manuals, labels, and help systems. Of course, accessing information in the world takes longer than recalling it from memory, so this time must be considered when evaluating system performance requirements. This trade-off can also affect accuracy, as workers may be tempted to use unreliable information in working memory to avoid having to search through manuals or displays for the correct information (Gray and Fu, 2004).

Similarly, when information must be maintained in working memory for a long period of time, the intensity can fall below the threshold required for reliable recall. Here too, important information should be placed in the physical world. Interfaces that allow workers to store preliminary hypotheses and rough ideas can alleviate the working memory requirements and reduce the risk of memory-related errors. When information must be maintained in working memory for extended periods, the worker must be allowed to focus on rehearsal. Any other tasks that require the use of working memory must be avoided. Distractions that can interfere with working memory must be eliminated.

Training can also be used to enhance working memory. Training modules can be used to strengthen workers' conceptions of complex processes, and can thus reduce the working memory required to maintain it during work activities. This would allow additional information to be considered in parallel.

1.3.2.2 Case: Cockpit Checklists

Degani and Wiener (1993) describe cockpit checklists as a way to provide redundancy in configuring an aircraft and reduce the risk of missing a step in the configuration process. Without checklists, aircraft crews would have to retrieve dozens of configuration steps from long-term memory and maintain in working memory whether each step had been completed for the current flight. Checklists reduce this memory load by transferring the information into the world. Especially in environments with frequent interruptions and distractions, the physical embodiment of a procedure can ensure that no steps are omitted because of lapses in working memory.

1.3.3 Sensation

Sensation is the process through which information about the world is transferred to the brain's perceptual system through sensory organs such as the eyes and ears (Bailey, 1996). From a systems design perspective, there are three important parameters for each dimension that must be considered: sensory threshold, difference threshold, and stimulus/response ratio.

The sensory threshold is the level of stimulus intensity below which the signal cannot be sensed reliably. The threshold must be considered in relation to the work environment. In the visual modality, there are several important stimulus thresholds. For example, in systems that use lights as warnings, indicators and displays need to have a size and brightness that can be seen by workers at the appropriate distance. These thresholds were determined in ideal environments. When environments are degraded because of dust or smoke, or workers are concentrating on other tasks, the thresholds may be much higher. In environments with glare or airborne contaminants, the visual requirements will change.

Auditory signals must have a frequency and intensity that workers can hear, again at the appropriate distance. Workplaces that are loud or where workers will be wearing hearing protection must be considered. Olfactory, vestibular, gustatory, and kinesthetic senses have similar threshold requirements that must be considered in system design.

The difference threshold is the minimum change in stimulus intensity that can be differentiated; this is also called the "just noticeable difference" (Snodgrass et al., 1985). This difference is expressed as a percent change. For example, a light must be at least 1% brighter than a comparison for a person to be able to tell that they are different. On the other hand, a sound must be 20% louder than a comparison for a person to perceive the difference.

The difference threshold is critical for the design of systems when multiple signals must be differentiated. When different alarms are used to signal different events, it is critical that workers be able to recognize the difference. When different-sized connectors are used for different parts of an assembly, workers need to be able to distinguish which connector is the correct one. Although there has been little research in this area, it is likely that there is a speed/accuracy trade-off with respect to difference thresholds. When workers are forced to act quickly, either because of productivity standards or in an emergency situation, even higher difference thresholds may be required for accurate selection.

The third dimension is the stimulus/response ratio. The relationship between the increase in intensity in a sensory stimulus and the corresponding increase in the sensation of that intensity is an exponential function (Stevens, 1975). For example, the exponent for perception of load heaviness is 1.45, so a load that is 1.61 times as heavy as another load will be perceived as twice as heavy. Similarly, the exponent for the brightness of a light is 0.33, so a light has to be eight times as bright to be perceived as twice as bright. Predicting these differences in perception is critical so that systems can be designed to minimize human error in identifying and responding to events.

1.3.3.1 Implications for Design

To maximize the reliability with which important information will reach workers, work environment design must consider sensation. The work environment must be designed to maximize the clarity with which workers can sense important sources of information. Lighting must be maintained to allow workers to see at requisite accuracy levels. Effective choice of color for signs and displays can maximize contrast with backgrounds and the accuracy of interpretation. Background noise can be controlled to allow workers to hear important signals and maintain verbal communication. The frequency and loudness of auditory signals and warnings can be selected to maximize comprehension. Location is also important. Key sources of visual information should be placed within the worker's natural line of sight.

1.3.3.2 Case: Industrial Dashboards

Designing system interfaces to support complex decision making, such as with supply chain, enterprise, and executive dashboards, requires a focus on human sensory capabilities (Resnick, 2003). Display design requires selecting among digital, analog, historical, and other display types (Hansen, 1995; Overbye et al., 2002). The optimal design depends on how often the data change and how quickly they must be read.

The salience of each interface unit is also critical to ensure that the relevant ones attract attention from among the many others on the display (Bennett and Flach, 1992). A variety of techniques can be used in industrial dashboards to create salience, such as brightness, size, auditory signals, or visual animation. The design should depend on the kinds of hardware on which the system will be implemented. For example, when systems will be accessed through handheld or notebook computers, the display size and color capabilities will be limited and these limitations must be considered in the display design.

1.3.4 Perception

As these basic sensory dimensions are resolved, the perceptual system tries to put them together into identifiable units. This is where each sensation is assigned to either an object or the background. If there are several objects, the sensations that compose each of them must be separated (recall Figure 1.2). There is a strong interaction here with long-term memory (see Section 1.3.1). Objects with strong long-term memory representations can be recognized faster and more reliably because less energy is required to activate the corresponding schema. On the other hand, objects that have similar features to different but well-known objects are easily misidentified as these objects. This is called a "capture error" because of the way the stronger schema "captures" the perception and becomes active first.

There is also an interaction with working memory (see Section 1.3.2). Objects that are expected to appear are also recognized faster and more reliably. Expectations can be described as the priming of the schema for the object that is expected. Energy is introduced into the schema before the object is perceived. Thus, less actual physical evidence is needed for this schema to reach its activation threshold. This can lead to errors when the experienced object is not the one that was expected but has some similarities.

1.3.4.1 Implications for Design

The implications of perception for industrial systems design are clear. When designing work objects, processes, and situations, there is a trade-off between the costs and benefits of similarity and overlap. When it is important that workers are able to distinguish objects immediately, particularly in emergency situations, overlap should be minimized. Design efforts should focus on the attributes that workers primarily use to distinguish similar objects. Workers can be trained to focus on features that are different. When object similarity cannot be eliminated, workers can be trained to recognize subtle differences that reliably denote the object's identity.

It is also important to control workers' expectations. Because expectations can influence object recognition, it is important that they reflect the true likelihood of object presence. This can be accomplished through situational training. If workers know what to expect, they can more quickly and accurately recognize objects when they appear. For those situations where there is too much variability for expectations to be reliable, work procedures can include explicit rechecking of the identity of objects where correct identification is critical.

1.3.4.2 Case: In-Vehicle Navigation Systems

In-vehicle navigation systems help drivers find their way by showing information on how to travel to a programed destination. These systems can vary greatly in the types of information that they provide and the way in which the information is presented. For example, current systems can present turn-by-turn directions in the visual and auditory modalities, often adjusted according to real-time traffic information. These systems can also show maps that highlight the recommended route and traffic congestion in different colors.

There are many advantages provided by these systems. In a delivery application, optimization software can consider all of the driver's remaining deliveries and current traffic congestion to compute the optimal order to deliver the packages. For many multistop routes, this computation would exceed the driver's ability to process the information. Including real-time traffic information also enhances the capabilities of the system to select the optimal route to the next destination (Khattak et al., 1994).

A challenge for these systems is to provide this information in a format that can be quickly perceived by the driver. Otherwise, there is a risk that the time required for the driver to perceive the relevant information will require an extended gaze duration, increasing the likelihood of a traffic accident. Persaud and Resnick (2001) found that the display modality had a significant effect on decision-making time. Graphical displays, although the most common design, required the most time to parse. Recall scores were also lowest for graphical displays, possibly requiring the driver to look back at the display more often. This decrease in recognition speed can lead to the greater risk of a traffic accident.

1.3.5   Attention and Mental Workload

Because only a small subset of long-term memory can be activated at any one time, it is important to consider how this subset is determined. Ideally, attention will be focused on the most important activities and the most relevant components of each activity, but the prevalence of errors that are the result of inappropriately focused attention clearly indicates that this is not always the case.

There are many channels of information, both internal and external, upon which attention can be focused. In most industrial settings, there can be visual and auditory displays that are designed specifically to present information to workers to direct their job activities. There are also informal channels in the various sights, sounds, smells, vibrations, and other sensory emanations around the workplace. Communication with other workers is also a common source of information. Additionally, there are internal sources of information in the memory of the individual. Both episodic and semantic memories can be a focus of attention. But it is impossible for workers to focus their attention on all of these channels at once.

It is also important to consider that attention can be drawn to channels that are relevant to the intended activities, but also to those that are irrelevant. Daydreaming is a common example of attention being focused on unessential information channels. Attention is driven in large part by the salience of each existing information channel. Salience can be defined as the attention-attracting properties of an object. It can be derived based on the intensity of a channel's output in various sensory modalities (Wickens and Hollands, 2000). For example, a loud alarm is more likely to draw attention than a quiet alarm. Salience can also be based on the semantic importance of the channel. An alarm that indicates a nuclear accident is more likely to draw attention than an alarm signaling lunchtime. Salience is the reason that workers tend to daydream when work intensity is low, such as long-duration monitoring of displays (control center operators, air travel, and security). When nothing is happening on the display, daydreams are more interesting and draw away the worker's attention. If something important happens, the worker may not notice.

If humans had unlimited attention, then we could focus on all possible information sources, internal and external. However, this is not the case; there is a limited amount of attention available. The number of channels on which attention can be focused depends on the complexity of each channel. One complex channel, such as a multifunction display, may require the same amount of attention as several simple channels, such as warning indicators.

Another important consideration is the total amount of attention that is focused on an activity and how this amount varies over time. This *mental workload* can be used to measure how busy a worker is at any given time, to determine if any additional tasks can be assigned without degrading performance, and to predict whether a worker could respond to unexpected events. Mental workload can be measured in several ways, including the use of subjective scales rated by the individual doing the activity, or physiologically by measuring the individual's heart rate and brain function. A great deal of research has shown that mental workload must be maintained within the worker's capability, or job performance will suffer in domains such as air-traffic control (Lamoreux, 1997), driving a car (Hancock et al., 1990), and others.

1.3.5.1   Implications for Design

There are many ways to design the work environment to facilitate the ability of the worker to pay attention to the most appropriate information sources. Channels that are rarely diagnostic should be designed to have low salience. On the other hand, important channels can be designed to have high sensory salience through bright colors or loud auditory signals. Salient auditory alerts can be used to direct workers' attention toward key visual channels. Workers should be trained to recognize diagnostic channels so that they evoke high semantic salience.

Mental workload should also be considered in systems design. Activities should be investigated to ensure that peak levels of workload are within workers' capabilities. The average workload should not be so high as to create cumulative mental fatigue. It should also not be so low that workers are bored and may miss important signals when they do occur.

1.3.5.2 Case: Warnings

A warning is more than a sign conveying specific safety information. It is any communication that reduces risk by influencing behavior (Laughery and Hammond, 1999). One of the most overlooked aspects of warning design is the importance of attention. In a structured recall environment, a worker may be able to recall accurately the contents of a warning. However, if a warning is not encountered during the activity in which it is needed, it may not affect behavior because the worker may not think of it at the time when it is needed. When the worker is focusing on required work activities, the contents of the warning may not be sufficiently salient to direct safe behavior (Wogalter and Leonard, 1999). Attention can be attracted with salient designs, such as bright lights, sharp contrasts, auditory signals, large sizes, and other visualization enhancements.

Frantz and Rhoades (1993) reported that placing warnings in locations that physically interfered with the task could increase compliance even further. The key is to ensure that the warning is part of the attentional focus of the employee at the time it is needed and that it does not increase mental workload past the employees' capacity.

1.3.6 Situation Awareness

Situation awareness is essentially a state in which an observer understands what is going on in his or her environment (Endsley, 2000a). There are three levels of SA: perception, comprehension, and projection. Perceptional SA requires that the observer know to which information sources attention should be focused and how to perceive these sources. In most complex environments, many information sources can draw attention. Dividing one's attention among all of them reduces the time that one can spend on critical cues and increases the chance that one may miss important events. This is Level 1 SA, which can be lacking when relevant information sources have low salience, are physically obstructed, are not available at needed times, when there are distractions, or when the observer lacks an adequate sampling strategy (Eurocontrol, 2003). The observer must be able to distinguish three types of information sources: those that must be examined and updated constantly, those that can be searched only when needed, and those that can be ignored.

Comprehension is the process of integrating the relevant information that is received into a cohesive understanding of the environment and retaining this understanding in memory for as long as it is needed. This includes both objective analysis and subjective interpretation (Flach, 1995). Comprehension can be compromised when the observer has an inadequate schema of the work environment or is overreliant on default information or default responses (Eurocontrol, 2003).

Projection is when an observer can anticipate how the situation will evolve over time, can anticipate future events, and comprehends the implications of these changes. The ability to project supports timely and effective decision-making (Endsley, 2000a). One key aspect of projection is the ability to predict *when* and *where* events will occur. Projection errors can occur when current trends are under- or overprojected (Eurocontrol, 2003).

Endsley (2000a) cautions that SA does not develop only from official system interface sources. Workers can garner information from informal communication, world knowledge, and other unintended sources. SA is also limited by attention demands. When mental workload exceeds the observer's capacity, either because of an unexpected increase in the flow of information or because of incremental mental fatigue, SA will decline.

1.3.6.1 Implications for Design

Designing systems to maximize SA relies on a comprehensive task analysis. Designers should understand each goal of the work activity, the decisions that will be required, and the best diagnostic information sources for these decisions (Endsley, 2002).

It is critical to predict the data needs of the worker in order to ensure that these are available when they are needed (Endsley, 2001). However, overload is also a risk because data must be absorbed and assimilated in the time available. To avoid overload, designers can focus on creating information sources that perform some of the analysis in advance and present integrated results to the worker. Displays can also be goal-oriented, and information can be hidden at times when it is not needed.

It is also possible to design set sequences for the sampling of information sources into the work processes themselves. This can enhance SA because the mental workload due to task overhead is reduced. Workers should also be informed of the diagnosticity of each information source.

1.3.6.2   Case: Air-Traffic Systems

Situation awareness has been used in the investigation of air-traffic incidents and to identify design modifications to air traffic control systems that reduce the likelihood of future incidents (Rodgers et al., 2000). In this study, inadequate SA was linked to poor decision-making quality, leading both to minor incidents and major aircraft accidents. When air-traffic controllers are aware of developing error situations, the severity of the incident is reduced. The study identified several hypotheses to explain the loss of SA in both high- and low-workload situations. In high-workload conditions, operators had difficulty in maintaining a mental picture of the air traffic. As the workload shifts down from high to low, sustained periods can lead to fatigue-induced loss of SA. The evaluation of air-traffic controller SA led to insights into the design of the radar display, communication systems, team coordination protocols, and data-entry requirements.

Situation awareness has also been used in the design stage to evaluate competing design alternatives. For example, Endsley (2000b) compared sensor hardware, avionics systems, free flight implementations, and levels of automation for pilots. These tests were sensitive, reliable, and able to predict the design alternative that achieved the best performance.

1.3.7   Decision-Making

Decision-making is the process of selecting an option based on a set of information under conditions of uncertainty (Wickens et al., 2004). In contrast to the systematic way that deliberate decisions are made or programmed into computers, human decision making is often unconscious, and the specific mechanisms are unavailable for contemplation or analysis by the person who made them. Environments with many interacting components, degrees of freedom, and unclear data sources challenge the decision-making process. Decision-making processes are affected by neurophysiological characteristics that are influenced by the structure of long-term memory and the psychological environment in which the decision is made. For experienced decision makers, decisions are situational discriminations (Dreyfus, 1997) where the answer is obvious without comparison of alternatives (Klein, 2000). There are two major types of decision-making situations: diagnosis and choice.

1.3.7.1   Diagnosis

Diagnosis decisions involve evaluating a situation to understand its nature, and can be modeled as a pattern recognition process (Klein, 2000). Diagnosis describes decisions made in troubleshooting, medical diagnosis, accident investigation, safety behavior, and many other domains. The information that is available about the situation is compared with the existing schema in long-term memory, subject to the biasing effects of expectations in working memory. If there is a match, the corresponding schema becomes the diagnosis. For experts, this matching process can be modeled as a recognition-primed decision (Klein, 1993) whereby the environment is recognized as matching one particular pattern and the corresponding action is implemented.

The minimum degree to which the current situation must match an existing schema depends on the importance of the decision, the consequences of error, and the amount of time available. When the cost of searching for more information exceeds the expected benefits of that information, the search process stops (Marble et al., 2002). For important decisions, this match threshold will be higher so that more evidence can be sought before a decision is made. This leads to more reliable and accurate decisions. However, it may still be the case that an observed pattern matches an existing schema immediately and a decision is made regardless of how important the decision is.

Under conditions of time pressure, there may not be sufficient time to sample enough information channels to reach the appropriate threshold. In these cases, the threshold must be lowered and decisions will be made only on the basis of the information available (Ordonez and Benson, 1997). In these cases, individuals focus on the most salient source(s) of information (Wickens and Hollands, 2000) and select the closest match on the basis of whatever evidence has been collected at that point (Klein, 1993).

When the decision-maker is an expert in the domain, this process is largely unconscious. The matched schema may be immediately apparent with no one-by-one evaluation of alternatives. Novices may have less well-structured schema and so the match will not be clear. More explicit evaluation may be required.

1.3.7.2 Choice

In choice decisions, an individual chooses from a set of options that differ in the degree to which they satisfy competing goals. For example, when one is choosing a car, one model may have a better safety record and another may be less expensive. Neither is necessarily incorrect, although one may be more appropriate according to a specific set of optimization criteria.

When a person makes a choice decision, it is often based on an unconscious hybrid of several decision-making strategies (Campbell and Bolton, 2003). In the weighted-adding strategy, the score on each attribute is multiplied by the importance of the attribute, and the option with the highest total score is selected (Jedetski et al., 2002). However, this strategy generally requires too much information processing for most situations and often does not match the desired solution (Campbell and Bolton, 2003). In the satisficing strategy, a minimum score is set for each attribute. The first option that meets all of these minima is selected (Simon, 1955). If none do, then the minimum of the least important attribute is relaxed and so on until an option is acceptable. In the lexicographic strategy, the option with the highest score on the single most important attribute is selected without regard for other attributes (Campbell and Bolton, 2003).

Table 1.1 illustrates a decision matrix that depicts the differences among these strategies. With use of the weighted-adding strategy, option 1 would receive 173 points ($7^*8 + 3^*5 + 8^*9 + 5^*6$). Options 2 and 3 would receive 168 and 142, respectively. So option 1 would be selected. On the other hand, the company may have satisficing constraints for attributes such as safety and value. A safety score <5 and a value score <4 may be considered unacceptable regardless of the other attribute scores (eliminating options 1 and 2 from consideration), resulting in the selection of option 3. Finally, the company may choose to use a lexicographic strategy on value, selecting the option with the highest value regardless of all other attribute scores. In this case, option 2 would be selected.

While the weighted-adding strategy is often considered the most optimal, this is not necessarily the case. Some attributes, such as safety, should not be compensatory. Regardless of how fast, capable, reliable, or cost-effective a machine may be, risk to workers' safety should not be compromised. Lexicographic strategies may be justified when one attribute dominates the others, or the company does not have the time or resources to evaluate other attributes. For example, in an emergency situation, preventing the loss of life may dominate consideration of cost or equipment damage. The use of these strategies can be quite effective (Gigerenzer and Todd, 1999), and according to Schwartz (2004), benefits gained from making optimal decisions are often not worth the time and effort required.

Contrary to the systematic way that companies make official decisions, day-to-day decisions are often made with little conscious evaluation of the strategy. As with diagnosis decisions, time pressure and decision importance influence the decision-making process. When faced with limited time, workers may be forced to use faster and simpler strategies such as the lexicographic strategy (Ordonez and Benson, 1997).

TABLE 1.1 Example Decision-Making Matrix

Attributes	Weights (out of 10)	Option 1 (Scores)	Option 2 (Scores)	Option 3 (Scores)
Quality of output	8	7	6	5
Value	5	3	9	5
Safety	9	8	4	7
Durability	6	5	7	4

1.3.7.3 Decision-Making Heuristics

There are several decision-making heuristics that can reduce the information-processing requirements and often reduce the time required to make a decision. However, these shortcuts can also bias the eventual outcome (Brown and Ramesh, 2002). These are often not consciously applied, so they can be difficult to overcome when they degrade decision-making accuracy and reliability.

- *Anchoring*: When an individual develops an initial hypothesis in either a diagnosis or choice decision, it is very difficult to switch to an alternative. Contrary evidence may be discounted.
- *Confirmation*: When an individual develops an initial hypothesis in either a diagnosis or choice decision, he or she will have a tendency to search for information that supports this hypothesis even when other channels may be more diagnostic.
- *Availability*: When searching for additional information, sources that are more easily accessed or brought to mind will be considered first, even when other sources are more diagnostic.
- *Reliability*: The reliability of information sources is hard to integrate into the decision-making process. Differences in reliability are often ignored or discounted.
- *Memory limitations*: Because of the higher mental workload required to keep many information sources in working memory simultaneously, the decision-making process will often be confined to a limited number of information sources, hypotheses, and attributes.
- *Feedback*: Similar to the confirmation bias, decision makers often focus on feedback that supports a past decision and discount feedback that contradicts past decisions.

1.3.7.4 Implications for Design

Human factors can have a tremendous impact on the accuracy of decision-making. It is often assumed that normative decision-making strategies are optimal and that workers will use them when possible. However, neither of these is the case in many human decision-making situations. Limitations in information-processing capability often force workers to use heuristics and focus on a reduced number of information sources. Competing and vague goals can reduce the applicability of normative decision criteria.

Workers can be trained to focus on the most diagnostic sources in each decision domain. If they are only going to use a limited number of sources, they should at least be using the most effective ones. Diagnostic sources can also be given prominent locations in displays or be the focus of established procedures.

The reliability of various information sources should be clearly visible either during the decision-making process or during training. Workers can be trained to recognize source reliability or to verify it in real time. Similarly, workers can be trained to recognize the best sources of feedback. In design, feedback can be given a more prominent position or provided more quickly.

To avoid anchoring and confirmation biases, decision support systems (DSS) can be included that suggest (or require) workers to consider alternatives, seek information from all information sources, and include these sources in the decision-making process. At the least, a system for workers to externalize their hypotheses will increase the chance that they recognize these biases when they occur. However, the most successful expert systems are those that complement the human decision process rather than those that act as stand-alone advisors that replace humans (Roth et al., 1987).

In cases where decision criteria are established in advance, systems can be designed to support the most effective strategies. Where minimum levels of performance for particular criteria are important, the DSS can assist the worker in establishing the level and eliminating options that do not reach this threshold. The information-processing requirements of weighted-adding strategies can be offloaded to DSS entirely to free the worker for information-collecting tasks for which he or she may be more suited.

1.3.7.5 Case: Accident Investigation

Accident investigation and the associated root cause analysis can be fraught with decision-making challenges. Human error is often the proximate cause (Mullen, 2004) of accidents, but is much less often the root cause. During the accident investigation process, it is critical for investigators to explore the factors that led to the error and identify the design changes that will eliminate future risk (Doggett, 2004).

However, this process engenders many opportunities for decision-making errors. Availability is usually the first obstacle. When an accident occurs, there is often high-visibility evidence that may or may not lead directly to the root cause. The CAIB report found that the root cause that ultimately led to the Columbia accident was not a technical error related to the foam shielding, which was the early focus of the investigation, but rather, the organizational culture of NASA (CAIB, 2003). The confirmation bias can also challenge the investigation process. When investigators develop an initial hypothesis that a particular system component led to an accident, they may focus exclusively on evidence to confirm this component as the root cause rather than on general criteria that could rule out other likely causes. This appeared in the investigations of the USS Vincennes incident in the Persian Gulf and the Three Mile Island nuclear power incident (Wickens and Hollands, 2000).

Decision support systems can be used to remove a lot of the bias and assist in the pursuit of root causes. By creating a structure around the investigation, they can lead investigators to diagnostic criteria and ensure that factors such as base rates are considered. Roth et al. (2002) provide an overview of how DSS can reduce bias in decision-making. For example, DSS can inform users when the value for a particular piece of evidence falls outside a specified range. They can make confirming and disconfirming dimensions explicit and facilitate switching between them. But they warn that these systems can also introduce errors, such as by allowing drill-down into large data sources so that many data in one area are sampled without looking elsewhere. DSS can also exacerbate the availability bias by providing easy access to recent investigations.

1.4 Cognitive Consequences of Design

1.4.1 Learning

Every time an object is perceived, an event is experienced, a memory is recalled, or a decision is made, there are small, incremental changes in the structure of the human information-processing system. Learning is very difficult when there is no prior experience to provide a framework (Hebb, 1955). This explains the power of analogies in early training. Later learning is a recombination of familiar patterns through the transition of general rules into automatic procedures and the generalization and specialization of these procedures as experience develops (Taatgen and Lee, 2003). The magnitude of the change depends on the salience of the experience and how well it matches existing schema.

When a human–system interaction is exactly the same as past experiences, there is very little learning because no new information is gained. The only result is a small strengthening of the existing schema. It is unlikely that the worker will develop a strong episodic memory of the event at all. When a human–system interaction is radically different from anything that has been experienced before, a strong episodic memory may be created because of the inherent salience of confusion and possible danger. But the event will not be integrated into the semantic network because it does not correspond to any of the existing schema — there is nowhere to "put" it. Maximum learning occurs when a human–system interaction mirrors past experience but has new attributes that make sense, i.e., the experience can be integrated into the conceptual understanding of the system.

1.4.1.1 Implications for Design

Training programs should always be designed on the basis of an analysis of the workers' existing knowledge. Training of rote procedures where there will be no variability in workers' actions should be approached differently than training for situations where workers will be required to recognize and solve problems. In a study of novice pilot training, Fiore et al. (2003) found that diagrams and analogical representations of text content facilitate learning of procedures that require knowledge elaboration, but not on recognition or declarative rote memorization.

A better understanding of human learning mechanisms can also facilitate the development of experiential learning that workers gain on the job. System interfaces can be structured to maximize experiential learning by providing details that help employees develop accurate schema of the problem space. Over time, repeated exposure to this information can lead to more detailed and complex schema that can facilitate more elaborate problem solving. A cognitive analysis of the task requirements and possible situations can lead to a human–system interface that promotes long-term learning.

1.4.2 Error

Human behavior is often divided into three categories: skill-, rule-, and knowledge-based (Rasmussen, 1993). In skill-based behavior, familiar situations automatically induce well-practiced responses with very little attention. In deterministic situations with a known set of effective responses, simple IF-THEN decision criteria lead to rule-based behaviors. Knowledge-based behaviors are required in unfamiliar or uncertain environments where problem solving and mental simulation is required.

Each of these behavior types is associated with different kinds of errors (Reason, 1990). With skill-based behavior, the most common errors are related to competing response schema. Skill-based behavior results from a strong schema that is associated repeatedly with the same response. When a new situation shares key attributes with this strong schema, the old response may be activated in error. Because skill-based behavior requires little attention, the response is often completed before the error is noticed. In these cases, expertise can actually hurt performance accuracy. Unless there is salient feedback, the error may not be noticed and there will be no near-term recovery from the error.

Rule-based behavior can lead to error when a rule is erroneously applied, either because the situation was incorrectly recognized or because the rule is inappropriately generalized to similar situations. Rule-based behavior is common with novices who are attempting to apply principles acquired in training. Because rule-based behavior involves conscious attention, the error is likely to be noticed, but the employee may not know of a correct response to implement.

Knowledge-based errors occur when the employee's knowledge is insufficient to solve a problem. Knowledge-based behavior is the most likely to result in error because it is the type of behavior most often used in uncertain environments. When an employee is aware that his or her schema is not sophisticated enough to predict how a system will respond, he or she may anticipate a high likelihood of error and specifically look for one. This increases the chance that errors will be noticed and addressed.

1.4.2.1 Implications for Design

If system designers can anticipate the type(s) of behavior that are likely to be used with each employee–system interaction, steps can be taken to minimize the probability and severity of errors that can occur. For example, when skill-based behavior is anticipated, salient feedback must be designed into the display interface to ensure that employees will be aware when an error is made. To prevent skill-based errors from occurring, designers can make key attributes salient so that the inappropriate response will not be initiated.

To prevent rule-based errors, designers should ensure that the rules taught during training match the situations that employees will encounter when they are interacting with the system later. The triggers that indicate when to apply each rule should be made explicit in the system interface design. Signals that indicate when existing rules are not appropriate should be integrated into the interface design.

For complex systems or troubleshooting scenarios, when knowledge-based behavior is likely, errors can only be minimized when employees develop effective schema of system operators or when problem-solving activities are supported by comprehensive documentation and expert systems. Training should ensure that employees are aware of what they know and what they do not know. Employee actions should be easy to reverse when they are found to be incorrect.

1.5 Summary

Humans interact with industrial systems throughout the system lifecycle. By integrating human factors into each stage, the effectiveness, quality, reliability, efficiency, and usability of the system can be enhanced. At the requirements stage, it is critical for management to appreciate the complexity of human–system interaction and allocate sufficient resources to ensure that human factors requirements are emphasized. During design, human factors should be considered with the earliest design concepts to maximize the match between human capabilities and system operations. As the system develops, human factors must be applied to control and display design and the development of instructions and training programs. Maintenance operations should also consider human factors to ensure that systems can be preserved and repaired effectively. Human factors are also critical for human error analysis and accident investigation.

This chapter has presented a model of human information processing that addresses most of the relevant components of human cognition. Of course, one chapter is not sufficient to communicate all of the relevant human factors concepts that relate to the system lifecycle. But it does provide a starting point for including human factors in the process.

In addition to describing the critical components of human cognition, this chapter has described some of the implications of human cognition on system design. These guidelines can be applied throughout systems design. The specific cases are intended to illustrate this implementation in a variety of domains. As technology advances and the nature of human–system interaction changes, research will be needed to investigate specific human–system interaction effects. But an understanding of the fundamental nature of human cognition and its implications for system performance can be a useful tool for the design and operation of systems in any domain.

References

Bailey, R.W., *Human Performance Engineering*, 3rd ed., Prentice-Hall, Englewood Cliffs, NJ, 1996.

Bennett, K.B. and Flach, J.M., Graphical displays: implications for divided attention, focused attention, and problem solving, *Hum. Factors*, 34 (5), 513–533, 1992.

Browne, G.J. and Ramesh, V., Improving information requirements determination: a cognitive perspective, *Inform. Manage.*, 39, 625–645, 2002.

CAIB, The Columbia Accident Investigation Board Final Report, NASA, 2003. (Retrieved on 08/04/04 at http://www.caib.us).

Campbell, G.E. and Bolton, A.E., Fitting human data with fast, frugal, and computable models of decision making, *Proceedings of the Human Factors and Ergonomics Society 47th Annual Meeting*, Human Factors and Ergonomics Society, Santa Monica, CA, 2003, pp. 325–329.

Degani, A. and Wiener, E.L., Cockpit checklists: concepts, design, and use, *Hum. Factors*, 35 (2), 345–359, 1993.

Doggett, A.M., A statistical comparison of three root cause analysis tools, *J. Ind. Technol.*, 20 (2), 2–9, 2004.

Dreyfus, H.L., Intuitive, deliberative, and calculative models of expert performance, in *Naturalistic Decision Making*, Zsambok, C.E. and Klein, G., Eds., Lawrence Erlbaum Associates, Mahwah, NJ, 1997.

Endsley, M., *From Cognitive Task Analysis to System Design*, CTAResource.com Tutorial, 2002.

Endsley, M.R., Theoretical underpinnings of situation awareness: a critical review, in *Situation Awareness Analysis and Measurement*, Endsley, M.R. and Garland, D.J., Eds., Lawrence Erlbaum, Mahwah, NJ, 2000a.

Endsley, M.R., Direct measurement of situation awareness: validity and use of SAGAT, in *Situation Awareness Analysis and Measurement*, Endsley, M.R. and Garland, D.J., Eds., Lawrence Erlbaum, Mahwah, NJ, 2000b.

Endsley, M.R., Designing for situation awareness in complex systems, *Proceedings of the Second International Workshop of Symbiosis of Humans, Artifacts, and Environments*, Kyoto, Japan, 2001.

Eurocontrol, Technical Review of Human Performance Models and Taxonomies of Human Error in ATM (HERA), 2002 (Retrieved on 6/18/2004 at http://www.eurocontrol.int/humanfactors/gallery/content/public/docs/DELIVERABLES/HF26%20(HRS-HSP-002-REP-01)%20Released.pdf).

Eurocontrol, The Development of Situation Awareness Measures in ATM Systems, European Organisation for the Safety of Air Navigation, 2003 (Retrieved on 6/18/2004 at www.eurocontrol.int/humanfactors/docs/HF35-HRS-HSP-005-REP-01withsig.pdf).

Federal Aviation Administration, Profile of operational errors in the national aerospace system, Technical Report, Washington, DC, 1990.

Fiore, S.M., Cuevas, H.M., and Oser, R.L., A picture is worth a thousand connections: the facilitative effects of diagrams on mental model development, *Comput. Hum. Behav.*, 19, 185–199, 2003.

Flach, J.M., Situation awareness: proceed with caution, *Hum. Factors*, 37 (1), 149–157, 1995.

Frantz, J.P. and Rhoades, T.P., A task-analytic approach to the temporal and spatial placement of product warnings, *Hum. Factors*, 35, 719–730, 1993.

Giggerenzer, G. and Todd, P., *Simple Heuristics That Make us Smart*, Oxford University Press, Oxford, 1999.

Gordon, S.E., *Systematic Training Program Design*, Prentice-Hall, Englewood Cliffs, NJ, 1994.

Gray, W.D. and Fu, W.T., Soft constraints in interactive behavior: the case of ignoring perfect knowledge in-the-world for imperfect knowledge in-the-head. *Cognit. Sci.*, 28, 359–382, 2004.

Hancock, P.A., Wulf, G., Thom, D., and Fassnacht, P., Driver workload during differing driving maneuvers, *Accident Anal. Prev.* 22 (3), 281–290, 1990.

Hansen, J.P., An experimental investigation of configural, digital, and temporal information on process displays, *Hum. Factors*, 37 (3), 539–552, 1995.

Hebb, D.O., *The Organization of Behavior*, Wiley, New York, NY, 1955.

Hebb, D.O., Physiological learning theory, *J. Abnormal Child Psychol.*, 4 (4), 309–314, 1976.

Jedetski, J., Adelman, L., and Yeo, C., How web site decision technology affects consumers, *IEEE Internet Comput.*, 6 (2), 72–79, 2002.

Johnson, A., Procedural memory and skill acquisition, in *Handbook of Psychology, Experimental Psychology*, Healy, A.F., Proctor, R.W., and Weiner, I.B., Eds., Wiley, New York, 2003.

Jones, M. and Polk, T.A., An attractor network model of serial recall, *Cognit. Syst. Res.*, 3, 45–55, 2002.

Katz, M.A. and Byrne, M.D., Effects of scent and breadth on use of site-specific search on e-commerce web sites, *ACM Trans. Computer–Human Interaction*, 10 (3), 198–220, 2003.

Kemp, T., CRM stumbles amid usability shortcomings. *Internet Week Online*, 4/6/01, 2001 (Retrieved on 08/04/04 at www.internetweek.com/newslead01/lead040601.htm).

Khattak, A., Kanafani, A., and Le Colletter, E., Stated and Reported Route Diversion Behavior: Implications of Benefits of Advanced Traveler Information Systems in Transportation Research Record No 1464, 28, 1994.

Klein, G., *Sources of Power*, MIT Press, Cambridge, MA, 2000.

Klein, G.A., A recognition-primed decision (RPD) model of rapid decision making, in *Decision Making in Action: Models and Methods*, Klein, G.A., Orasanu, J., Calderwood, J., and MacGregor, D., Eds., Ablex Publishing, Norwood, NJ.

Konz, S. and Johnson, S., *Work Design: Industrial Ergonomics*, Holcomb Hathaway, Scotsdale, AZ, 2000.

Lachman, R., Lachman, J.L., and Butterfield, E.C., *Cognitive Psychology and Information Processing. Chapter 9 Semantic Memory*, Wiley, New York, 1979.

Lamoreux, T., The influence of aircraft proximity data on the subjective mental workload of controllers on the air traffic control task, *Ergonomics*, 42 (11), 1482–14591, 1997.

Laughery, K.R. and Hammond, A., Overview, in *Warnings and Risk Communication*, Wogalter, M.S., DeJoy, D.M., and Laughery, K.R., Eds., Taylor and Francis, London, 1999.

MacKenzie, D., *Science and Public Policy*, 21 (4), 233–248, 1994.

Marble, J.L., Medema, H.D., and Hill, S.G., Examining decision-making strategies based on information acquisition and information search time, *Proceedings of the Human Factors and Ergonomics Society 46th Annual Meeting*, Human Factors and Ergonomics Society, Santa Monica, CA, 2002.

Miller, G.A., The magical number seven, plus or minus two, *Psychol. Rev.*, 63, 81–97, 1956.

Mullen, J., Investigating factors that influence individual safety behavior at work, *J. Saf. Res.*, 35, 275–285, 2004.

Nielsen, J., When bad designs become the standard, *Alertbox*, 11/14/99, 1999 (Retrieved on 08/04/04 at www.useit.com/alertbox/991114.html).

Norman, D.A., *The Design of Everyday Things*, Basic Books, New York, 1988.

Ordonez, L. and Benson, L., Decisions under time pressure: how time constraint affects risky decision making, *Organ. Behav. Hum. Performance*, 71 (2), 121–140, 1997.

Overbye, T.J., Sun, Y., Wiegmann, D.A., and Rich, A.M., Human factors aspects of power systems visualizations: an empirical investigation, *Electric Power Components Syst.*, 30, 877–888, 2002.

Persaud, C.H. and Resnick, M.L., The usability of intelligent vehicle information systems with small screen interfaces, *Proceedings of the Industrial Engineering and Management Systems Conference*, Institute of Industrial Engineers, Norcross, GA, 2001.

Rasmussen, J., Deciding and doing: decision making in natural contexts, in *Decision Making in Action: Models and Methods*, Klein, G.A., Orasanu, J., Calderwood, J., and MacGregor, D., Eds., Ablex Publishing, Norwood, NJ, 1993.

Reason, J., *Human Error*, Cambridge University Press, Cambridge, UK, 1990.

Resnick, M.L., Task based evaluation in error analysis and accident prevention, *Proceedings of the Human Factors and Ergonomics Society 45th Annual Conference*, Human Factors and Ergonomics Society, Santa Monica, CA, 2001.

Resnick, M.L., Building the executive dashboard, *Proceedings of the Human Factors and Ergonomics Society 47th Annual Conference*, Human Factors and Ergonomics Society, Santa Monica, CA, 2003.

Roth, E.M., Bennett, K.B., and Woods, D.D., Human interaction with an intelligent machine, *Int. J. Man–Machine Stud.*, 27, 479–525, 1987.

Roth, E.M., Gualtieri, J.W., Elm, W.C., and Potter, S.S., Scenario development for decision support system evaluation, *Proceedings of the Human Factors and Ergonomics Society 46th Annual Meeting*, Human Factors and Ergonomics Society, Santa Monica, CA, 2002.

Sanders, M.S. and McCormick, E.J., *Human Factors in Engineering and Design*, 7th ed., McGraw-Hill, New York, 1993.

Schwartz, B., *The Paradox of Choice*, HarperCollins, London, 2004.

Scielzo, S., Fiore, S.M., Cuevas, H.M., and Salas, E., The utility of mental model assessment in diagnosing cognitive and metacognitive processes for complex training, *Proceedings of the Human Factors and Ergonomics Society 46th Annual Meeting*, Human Factors and Ergonomics Society, Santa Monica, CA, 2002.

Simon, H.A., A behavioral model of rational choice, *Q. J. Econ.*, 69, 99–118, 1955.

Snodgrass, J.G., Levy-Berger, G., and Haydon, M., *Human Experimental Psychology*, Oxford University Press, Oxford, 1985.

Stevens, S.S., *Psychophysics*, Wiley, New York, 1975.

Swain, A.D. and Guttman, H.E., A handbook of human reliability analysis with emphasis on nuclear power plant applications, NUREG/CR-1278, USNRC, Washington, DC, 1983.

Swezey, R.W. and Llaneras, R.E., Models in training and instruction, in *Handbook of Human Factors and Ergonomics*, Salvendy, G., Ed., 2nd ed., Wiley, New York, 1997.

Taatgen, N.A. and Lee, F.J., Production compilation: a simple mechanism to model complex skill acquisition, *Hum. Factors*, 45 (1), 61–76, 2003.

Tulving, E., Remembering and knowing the past, *Am. Scientist*, 77, 361–367, 1989.

Wickens, C.D. and Hollands, J.G., *Engineering Psychology and Human Performance*, Prentice-Hall, New York, 2000.

Wickens, C.D., Lee, J.D., Liu, Y., and Gordon Becker, S.E., *An Introduction to Human Factors Engineering*, 2nd ed., Prentice-Hall, New York, 2004.

Wogalter, M.S. and Leonard, S.D., Attention capture and maintenance, in *Warnings and Risk Communication*, Wogalter, M.S., DeJoy, D.M., and Laughery, K.R., Eds., Taylor and Francis, London, 1999.

Minsky, M. Society of Mind. In *The Society of Mind*. Simon and Schuster, New York, 1985.

Nielsen, J. When Bad Design Becomes the Standard. *Alertbox*, Feb 11, 1999. Retrieved from http://www.useit.com/alertbox/991121.html.

Norman, D.A. *The Design of Everyday Things*. Basic Books, New York, 1988.

Ortony, A. and Turner, T. J. What's basic about basic emotions? *Psychological Review*, 97 (3), 315–331, 1990.

Overbye, D., Smith, W. J., and Wingrate, D. A. P.A.M. Human factor aspects of power process plant control room investigations. *Ergonomics*, 1981, 50, 327–349, 1991.

Preece, J., Rogers, Y., and Sharp, H. *Interaction Design: Beyond Human-Computer Interaction*. John Wiley & Sons, New York, 2002.

Proceeding of the 2nd International Conference on Augmented Cognition. Strategic Analysis, Inc., Arlington, VA, 2006.

Rasmussen, J. *Information Processing and Human-Machine Interaction*. North-Holland, Amsterdam, 1986.

Reason, J. *Managing the Risks of Organizational Accidents*. Ashgate Publishing, Aldershot, UK, 1997.

Reason, J. *Human Error*. Cambridge University Press, Cambridge, 1990.

Reising, D.V. and Sanderson, P. M. Work domain analysis and sensors: An example of situation awareness. In *Proceedings of the Human Factors and Ergonomics Society 44th Annual Meeting*, 2000.

Reising, D.V. and Sanderson, P. M. Work domain analysis and sensors: An example of situation awareness. *Human Factors and Ergonomics Society Annual Meeting*, Santa Monica, CA, 2001.

Rohn, E.M., Bennett, K.B. and Woods, D. D. Human interaction with an intelligent machine. *Int. J. Man-Machine Studies*, 27, 479–525, 1992.

Roth, E. M., Gualtieri, J. W., Elm, W. C., and Potter, S. S. Scenario development for decision support system evaluation. *Proceedings of the Human Factors and Ergonomics Society 45th Annual Meeting*, Human Factors and Ergonomics Society, Santa Monica, CA, 2002.

Sanders, M. S. and McCormick, E. J. *Human Factors in Engineering and Design*, 7th ed. McGraw-Hill, New York, 1993.

Schiffrin, R. M. and Schneider, W. Controlled and automatic human information processing. *Psychological Review*, 84, 127–190, 1977.

Selby, R. *The Paradox of Choice*. HarperCollins, London, 2004.

Stanton, N. A., McIlroy, R. C., Harvey, C., Blainey, S., Hickford, A., Preston, J. M. and Ryan, B. Following the line: An investigation into the human factor dimension. *Ergonomics*, 2014.

Stanton, N. A. A behavioral model of rational choice. *J. Psych.*, 69, 99–118, 1955.

Stanton, N. A., Chambers, P. R. G., and Piggott, J. *Systems Reliability: Applications*. Oxford University Press, Oxford, 1957.

Stevens, S. S. *Psychophysics*. Wiley, New York, 1975.

Swain, A. D. and Guttmann, H. E. *A Handbook of Human Reliability Analysis with Emphasis on Nuclear Power Plant Applications, NUREG/CR-1278*. USNRC, Washington, DC, 1983.

Sweller, R. W. and Chandler, P. Models in learning and instruction. In *Handbook of Human Factors and Ergonomics*, Salvendy, G., Ed., 2nd ed. Wiley, New York, 1997.

Taatgen, N. A. and Lee, F. J. Production compilation: a simple mechanism to model complex skill acquisition. *Hum. Factors*, 15 (1), 61–76, 2003.

Tulving, E. Remembering and knowing the past. *Am. Scientist*, 77, 361–367, 1989.

Wickens, C. D. and Hollands, J. G. *Engineering Psychology and Human Performance*. Prentice-Hall, New York, 2000.

Wickens, C. D., Lee, J. D., Liu, Y. and Gordon Becker, S. E. *An Introduction to Human Factors Engineering*, 2nd ed. Prentice-Hall, New York, 2004.

Wickens, C. D. and Leonard, S. D. Attention capture and maintenance. In *Warnings and Risk Communication*, Wogalter, M. S., DeJoy, D. M., and Laughery, K. R., Eds. Taylor and Francis, London, 1999.

2

Human Factors and Ergonomics: How to Fit into the New Era

Dongjoon Kong
The University of Tennessee

2.1 Introduction

The purpose of Industrial Engineering (IE) is, in general, to seek an optimal solution under given conditions. As one of the main areas of IE, Human Factors (HF) engineering pursues the same goal. The only difference is the target, which is *people*. "Human Factors" and "ergonomics" have become familiar terms, as can be seen from the use of ergonomics in applications from simple tools to very sophisticated airplanes. The study of HF has as its goal maximizing human capacity, usability, and comfort while minimizing human errors, accidents, and injury. Therefore, in ergonomics, we design a system that takes into account human capabilities and skills while optimizing technology and human interactions.

The design of products without the use of HF input can cause loss of productivity and sometimes the loss of lives. Everyday life is also affected by this unsatisfactory approach, as is evidenced by the complexity of VCR programming tasks on remote controllers, automobile diagnostic repair problems, or even the setup and use of business or personal computers. A well-known example frequently mentioned with regard to a lack of HF input is the nuclear power plant accident at Three Mile Island on March 28, 1979. After the accident, the U.S. Nuclear Regulatory Commission (NRC) published a summary report. The NRC stated in the report:

> The accident at the Three Mile Island Unit 2 (TMI-2) nuclear power plant near Middletown, Pennsylvania, on March 28, 1979, was the most serious in U.S. commercial nuclear power plant operating history, even though it led to no deaths or injuries to plant workers or members of the nearby community. But it brought about sweeping changes involving emergency response planning,

reactor operator training, Human Factors engineering, radiation protection, and many other areas of nuclear power plant operations. The accident began about 4:00 a.m. on March 28, 1979, when the plant experienced a failure in the secondary, non-nuclear section of the plant. The main feedwater pumps stopped running, caused by either a mechanical or electrical failure, which prevented the steam generators from removing heat. First the turbine, then the reactor automatically shut down. Immediately, the pressure in the primary system (the nuclear portion of the plant) began to increase. In order to prevent that pressure from becoming excessive, the pilot-operated relief valve (a valve located at the top of the pressurizer) opened. The valve should have closed when the pressure decreased by a certain amount, but it did not. Signals available to the operator failed to show that the valve was still open. As a result, cooling water poured out of the stuck-open valve and caused the core of the reactor to overheat. As coolant flowed from the core through the pressurizer, the instruments available to reactor operators provided confusing information. There was no instrument that showed the level of coolant in the core. Instead, the operators judged the level of water in the core by the level in the pressurizer, and since it was high, they assumed that the core was properly covered with coolant. In addition, there was no clear signal that the pilot-operated relief valve was open. As a result, as alarms rang and warning lights flashed, the operators did not realize that the plant was experiencing a loss-of-coolant accident. They took a series of actions that made conditions worse by simply reducing the flow of coolant through the core. (See http://www.nrc.gov/reading-rm/doc-collections/fact-sheets/3mile-isle.html for details.)

Since the accident at Three Mile Island, an awareness of the importance of good design as well as product design has been acknowledged in many areas. Multiple terms are used to describe the skills applied to the design and development of systems and their products so that the results are user-centered and not equipment-centered. In addition to these design issues, the study of HF has made many contributions to safety issues. In the early era of HF research, productivity improvement was the main focus, but as technology has advanced, especially computer technology, the usability of tools, machines, and computer software has moved to the forefront of research topics.

The purpose of this chapter is to focus on recent issues in the study of HF. There are several HF handbooks available: *Handbook of Human Factors and Ergonomics* (Salvendy, 1997), *Human Factors Design Handbook* (Woodson et al., 1981), and *The Occupational Ergonomics Handbook* (Karwowski and Marras, 1999).

2.2   History of Human Factors

The history of HF is less than one century old if we consider the beginning of that history to date back only to when the first HF specialists wrote papers presenting their methods and results. However, if we regard any tool designed to be convenient to use as an indication of the study of HF, the history must go back to when the first man-made tool appeared (Strom, 2003). The history of HF and notable activities is summarized in Table 2.1. The development of the HF field has been inextricably intertwined with developments in technology (Sanders and McCormick, 1992).

2.3   Anthropometry

The term "anthropometry" from the Greek words *anthropos* (man) and *metrein* (to measure) explains how the physical dimensions of people vary (Konz and Johnson, 2004). Anthropometry is well defined by Sanders and McCormick (1992): *measurement of the dimensions and certain other physical characteristics of the body such as volumes, centers of gravity, inertial properties, and masses of body segments.*

When a task is given to a worker, there are two alternatives: selection or job modification. In selection, the worker is selected from the population of workers based on criteria such as strength, height, weight, age, or even gender. This selection strategy has been called *fitting the man to the task.* The other alternative is to modify the job so that almost anyone can do it. This job modification is called *fitting the task to the man.* Job modification has been widely implemented since the Americans with Disabilities Act (ADA)

TABLE 2.1 The History of Human Factors and Important Activities

Period	Notable Activities
Prehistory to eighteenth Century	• Prehistoric period ○ Beginning of HF: the first man-made tool • 1950 to 1900 BC ○ Realistic description of work conditions in different professions, probably from Egypt's Middle Kingdom • 960 AD: commonsense ergonomics ○ The Danish king, Harald Bluetooth, built four armed bases having symmetrical barracks ○ However, the exits from the barracks into the passageways between them broke the symmetry such that none of the exits was directly opposite any other exit ○ It appears that the architect wanted to prevent collisions between soldiers who were running out of two barracks at the same time ○ The displacement of the exits is a sign that some ergonomic thinking was applied • 1759 AD: comparative test ○ Danish proprietor, Borreschmith, introduced a new plough ○ He understood the concept of ergonomics. The new plough was tested under realistic conditions and it came with a set of user instructions
Nineteenth Century to World War II	• 1887: ergonomics with scientific base ○ The Danish government wanted to limit the amount of color added to margarine ○ The only question was where to set a limit to ensure that margarine was visibly different from butter ○ Alfred Lehmann published an article describing the human ability to discriminate among different colors, and designed a series of color cards showing different shades of yellow as a reference to the colors of margarine. The color tables came with detailed instructions • Early 1900s: real start of HF ○ Frank and Lillian Gilbreth began the first motion study ○ Their work included the study of skilled performance and fatigue, the design of workstations and equipment for the handicapped, and the analysis of hospital surgical teams • World War II: fitting the person to the job ○ The major emphasis of behavioral scientists was on the use of tests for selecting the proper people for jobs and on the development of improved training procedures ○ During World War II, however, it became clear that even with the best selection and training, the operation of some of the complex equipment still exceeded the capabilities of the people who had to operate it
1945 to 1960	• The HF profession was born; HF study in the United States was essentially concentrated in the military-industrial complex • At the end of the war in 1945, engineering psychology laboratories were established in the United States and Britain. In 1949, the Ergonomics Research Society was formed in Britain, and the first book on HF was published • In 1957, the journal *Ergonomics* from the Ergonomics Research Society appeared; the Human Factors Society was formed. Russia launched Sputnik and the race for space was on • In 1959, the International Ergonomics Association was formed

continued

TABLE 2.1   (*Continued*)

Period	Notable Activities
1960 to 1980	• Rapid growth and expansion of HF • Human Factors in the United States expanded beyond military and space applications • With the race for space and manned space flights, HF quickly became an important part of the space program • Human Factors considerations were incorporated into many industries, including those dealing in pharmaceuticals, computers, automobiles, and other consumer products • Industry began to recognize the importance and contribution of HF for both the design of workplaces and the products manufactured there

1980 to 1990	• Computer technology provided new challenges for the HF profession
	  ○ New control devices, information presentation via computer screen, and the impact of new technology on people were new areas for the HF profession
	• Several disasters related to HF:
	  ○ The incidents at Three Mile Island nuclear power station in 1979. The incident came very close to resulting in a nuclear meltdown
	  ○ In 1984, a leak of methylisocyanate (MIC) at the Union Carbide pesticide plant in Bhopal, India, claimed the lives of nearly 4000 people and injured another 200,000
	  ○ In 1986, an explosion and fire at the Chernobyl nuclear power station in Soviet Union resulted in more than 300 dead, widespread human exposure to harmful radiation, and millions of acres of radioactive contamination
	  ○ Three years later, in 1989, an explosion ripped through a Phillips Petroleum plant in Texas. It killed 23 people, injured another 100 workers, and resulted in the largest single U.S. business insurance loss in history ($1.5 billion)
	• Human Factors involvement increased dramatically in forensics and particularly product liability and personal injury litigations

1990 and beyond	• Intensive involvement of HF study in building a permanent space station
	• HCI: computers and the application of computer technology
	• Safety
	  ○ Workplace safety: the U.S. Occupational Safety and Health Administration (OSHA) regulations
	  ○ Aviation safety: the Federal Aviation Administration (FAA) expands its HF research efforts
	• Medicine
	  ○ The design of medical devices
	  ○ The design of products and facilities for the elderly
	• Security
	  ○ Human interaction with security technology
	  ○ Intelligence analysis

Sources: Strom G., *Ergon. Des.*, 11, 5–6, 2003; Sanders, M.S. and McCormick, E.J., *Human Factors in Engineering and Design* (7th ed.), McGraw-Hill, Singapore, 1992.

of 1990, which took effect in 1992. Since then, industries have incorporated this *fitting the task to the man* concept in consumer product designs.

Table 2.2 shows useful dimensions that apply to the particular postures needed for workplace design for adult males and females in the United States. Some of these data have been incorporated into mannequins that can be manipulated, or computer programs that enable product designers to simulate product usability at the planning stage of the design.

TABLE 2.2　Selected U.S. Civilian Body Dimensions (in cm with Bare Feet; add 3 cm to Correct for Shoes) of Industrial Relevance

Body Dimensions	Female			Male		
	5th	50th	95th	5th	50th	95th
Standing						
1. Tibial height	38.1	52.0	46.0	41.0	45.6	50.2
2. Knuckle height	64.3	70.2	75.9	69.8	75.4	80.4
3. Elbow height	93.6	101.9	108.8	100.0	709.9	119.0
4. Shoulder (acromion) height	121.1	131.1	141.9	132.3	142.8	152.4
5. Stature	149.5	160.5	171.3	161.8	173.6	184.4
6. Functional overhead reach	185.0	199.2	213.4	195.6	209.6	223.6
Sitting						
7. Functional forward reach	64.0	71.0	79.0	76.3	82.5	88.3
8. Buttock-knee depth	51.8	56.9	62.5	54.0	59.4	64.2
9. Buttock-popliteal depth	43.0	48.1	53.5	44.2	49.5	54.8
10. Popliteal height	35.5	39.8	44.3	39.2	44.2	48.8
11. Thigh clearance	10.6	13.7	17.5	11.4	14.4	17.7
12. Sitting elbow height	18.1	23.3	28.1	19.0	24.3	29.4
13. Sitting eye height	67.5	73.7	78.5	72.69	78.6	84.4
14. Sitting height	78.2	85.0	90.7	84.2	90.6	96.7
15. Hip breadth	31.2	36.4	43.7	30.8	35.4	40.6
16. Elbow-to-elbow breadth	31.5	38.4	49.1	35.0	41.7	50.6
Other dimensions						
17. Grip breadth, inside diameter	4.0	4.3	4.6	4.2	4.8	5.2
18. Interpupillary distance	5.1	5.8	6.5	5.5	6.2	6.8

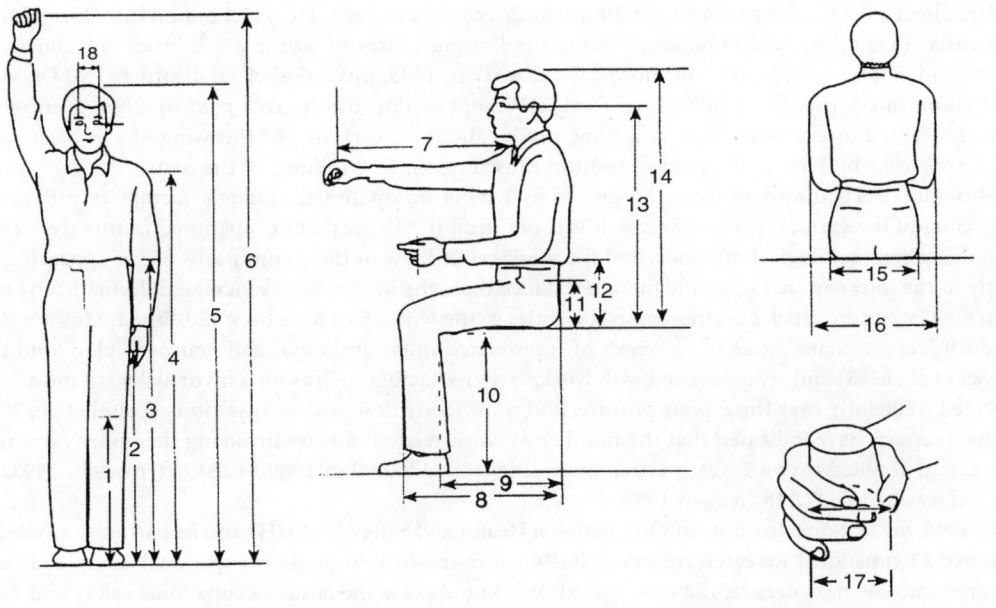

Source: Helander, 1995.

There are three design principles when designing for most individuals: *design for extremes, design for the average,* and *design for adjustability* (Niebel and Freivalds, 2003). Design for extremes implies that a specific design feature is a limiting factor in determining either the maximum or minimum value of a population variable that will be accommodated. For example, reach distances should be designed for the minimum individual, that is, a 5th percentile female arm length. Then, 95% of all females and almost all males will be able to reach. On the other hand, clearances, such as the height of an entry opening to a storage room, should be designed for the maximum individual, that is, a 95th percentile male stature, so that 95% of all males and almost all females will be able to enter the opening.

Design for the average is the cheapest but least preferred approach. Even though there is no individual with all average dimensions, there are certain situations where it would be impractical or too costly to include adjustability for all features. For example, most office desks have fixed dimensions and the design for extreme principle is not appropriate in this case. Therefore, the desk height is determined at the 50th percentile of the elbow height for the combined female and male populations (roughly the average of the male and female 50th percentile values) so that most individuals will not be unduly inconvenienced. However, the exceptionally tall male or very short female may experience some postural discomfort.

Design for adjustability is typically used for equipment or facilities that can be adjusted to fit a wide range of individuals. Chairs, vehicle seats, steering columns, and tool supports are devices that are typically adjusted to accommodate the worker population ranging from 5th percentile females to 95th percentile males. Obviously, designing for adjustability is the preferred method of design, but there is a trade-off with the cost of implementation.

2.4 Work-Related Injuries: Musculoskeletal Disorders

Each day, an average of 9000 U.S. workers sustain disabling injuries on the job, while 16 workers die from an injury suffered at work, and 137 workers die from work-related diseases. The Liberty Mutual 2002 Workplace Safety Index estimates that direct costs for occupational injuries rose to $38.7 billion in 1988, with indirect costs reaching over $150 billion; the direct costs increased to $40.1 billion in 1999, with indirect costs reaching over $200 billion. Among the leading causes of workplace injuries and illness that resulted in employees missing 5 or more days of work in 1998, musculoskeletal disorders (MSDs) occupied about one third ($12.1 billion) of workers' compensation direct costs paid in 1998; overexertion (injuries caused by excessive lifting, pushing, pulling, holding, carrying, or throwing of an object) stood at 25.57% ($9.8 billion), and repetitive motion caused 6.1% ($2.3 billion) of the costs.

Musculoskeletal disorders were recognized as having occupational etiologic factors as early as the beginning of the 18th century. However, it was not until the 1970s that occupational factors were examined using epidemiological methods, and the work-relatedness of these conditions began appearing regularly in the international scientific literature. Since then, the literature has increased dramatically; more than 6000 scientific articles addressing ergonomics in the workplace have been published. Musculoskeletal disorders are characterized by a series of interrelated musculoskeletal and neurovascular conditions (Keller et al., 1998) and are associated with workplace risk factors such as forceful or awkward movements repeated frequently over time, poor posture, and improperly designed workstations (Loisel et al., 2001). Various reports have indicated that the number of work-related injuries involving the upper extremities more than doubled over a 5-year period, increasing from 140,000 in 1989 to 281,000 cases in 1993, further increasing to 397,118 cases in 1999.

In 1996, the National Institute of Occupational Health and Safety (NIOSH) and its partners unveiled the National Occupational Research Agenda (NORA), a framework to guide occupational safety and health research into the next decade, not only for NIOSH but also for the entire occupational safety and health community. Approximately 500 organizations and individuals outside NIOSH provided input into the development of NORA. Before NORA, no national research agenda existed in the field of occupational safety and health, and no research agenda in any field had captured such broad input and consensus. The NORA process resulted in a remarkable consensus about the top 21 research priorities, as shown in Table 2.3. The areas highlighted in bold font are the areas in which HF specialists make major contributions to NORA.

TABLE 2.3 NORA Priority Research Areas

Disease and Injury	Work Environment and Workforce	Research Tools and Approaches
Allergic and irritant dermatitis	Emerging technologies	Cancer research methods
Asthma and chronic obstructive	Indoor environment	**Control technology and**
Pulmonary disease	Mixed exposures	**personal protective equipment**
Fertility and pregnancy abnormalities	Organization of work	**Exposure assessment methods**
Hearing loss	**Special populations at risk**	Health services research
Infectious diseases		Intervention effectiveness research
Low back disorders		**Risk assessment methods**
Musculoskeletal disorders		Social and economic consequences
Traumatic injuries		of workplace illness and injury
		Surveillance research methods

Source: NIOSH website (http://www2a.cdc.gov/NORA/NORAabout.html)
Note: Areas printed in bold type are those in which **HF** can make a major contribution to NORA.

TABLE 2.4 Number of Occupational Injuries and Illnesses (in 1000s) Involving Time Away from Work by Selected Nature of Injury and Illness, 1995 to 2001

Year	1995	1996	1997	1998	1999	2000	2001
Total cases	2040.9	1880.5	1833.4	1730.5	1702.5	1664.0	1537.6
Sprains, strains	876.8	819.7	799.0	760.0	739.7	728.2	669.9
Bruises, contusions	192.1	174.9	165.8	153.1	156.0	151.7	136.4
Cuts, lacerations	153.2	133.2	133.6	137.6	132.4	121.3	114.8
Fractures	124.6	120.5	119.5	115.4	113.7	116.7	108.1
Back pain	59.0	52.0	48.7	42.4	43.2	46.1	42.7
Carpal tunnel syndrome	31.5	29.9	29.2	26.3	27.8	27.7	26.8
Heat burns	36.1	29.0	30.0	28.4	27.1	24.3	25.1
Tendonitis	22.1	17.4	18.0	16.9	16.6	14.4	14.1

Table 2.4 and Figure 2.1 show the number of occupational injuries and illness involving time away from work by selected nature of injury and illness reported by the Bureau of Labor Statistics (BLS) (2003). Carpal tunnel syndrome (CTS; entrapment of the median nerve within the carpal tunnel) ranks 7th in the list of injuries and illnesses. In 1994, 92,576 repetitive motion cases of the upper extremity were reported as "days away from work"; 37,803 (40.8%) of them were CTS cases (Szabo, 1998). In 2002, 58,576 repetitive motion cases of the upper extremity and 22,478 (38.4%) CTS cases were reported (BLS, 2004). Since 1993, the BLS has reported that repetitive motion cases typically resulted in the lengthiest absence from work of any work-related injuries and that CTS cases required the longest median days away from work (27 days in 2000, 25 days in 2001, and 30 days in 2002). The total direct medical treatment expense ranged from $20,000 to $100,000 per case (Szabo, 1998), and would increase if indirect costs regarding production loss and rehabilitation expenses were included.

2.5 Epidemiological Research Methods

The relationship between MSDs and work-related factors remains the subject of considerable debate because some claim that MSDs are caused by other non-work-related factors, such as housekeeping, gardening, or sports activities. Musculoskeletal disorders are different from other occupational injuries and illnesses because they develop over a long period of time. For example, many studies have shown that CTS is a work-related disorder caused by extreme hand/wrist postures, highly repetitive motions, forceful exertion, use of hand-held vibration tools, or cold working environments (Armstrong et al., 1993; Bernard et al., 1994; Bovenzi, 1998; Delgrosso and Boillat, 1991; de Krom et al., 1990; Greening and Lynn, 1998; Hagberg et al., 1992; Latko et al., 1997; Masear et al., 1986; Roquelaure et al., 1997; Silverstein and Hughes, 1996; Szabo, 1998). This work-relatedness is supported by a higher industry-wide rate (2.0 cases per 10,000 full-time workers) than the prevalent rate of 1.25 per 10,000 general individuals in 1999. Also,

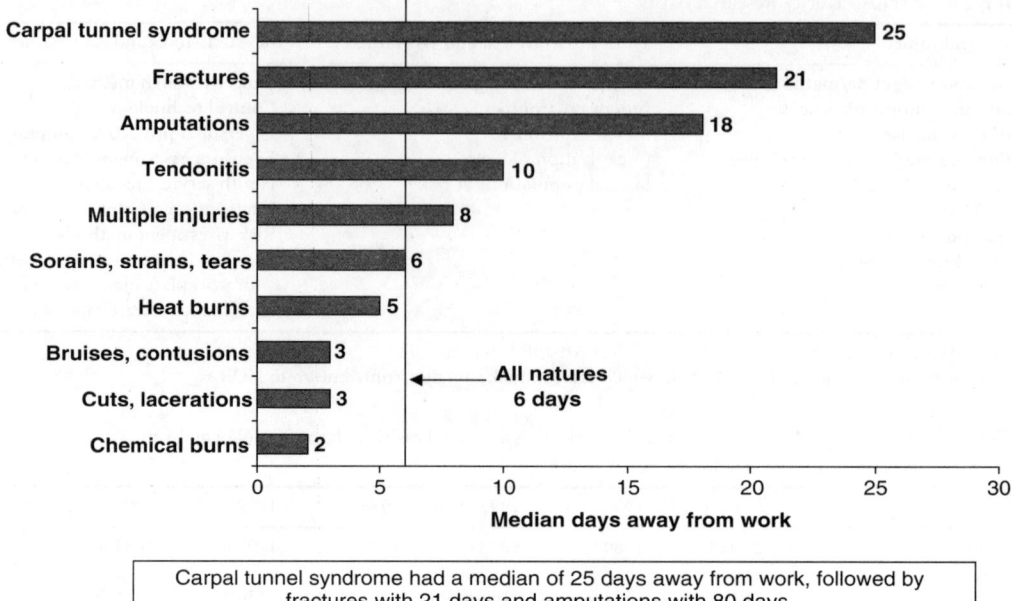

Carpal tunnel syndrome had a median of 25 days away from work, followed by
fractures with 21 days and amputations with 80 days.

Source: Bureau of Labor Statistics, U.S Departments of Labor, survey of occupational injuries and illnesses

FIGURE 2.1 Median days away from work due to non-fatal occupational injury or illness by nature, 2001 (Bureau of Labor Statistics, *News: Lost-Worktime Injuries and Illnesses: Characteristics and Resulting Time Away from Work*, Washington, DC, 2003).

TABLE 2.5 Notation of an Unpaired 2×2 Contingency Table Comparing Two Groups

	Outcome		Total	Observed	Anticipated
	Exposed (E)	Unexposed (U)		Proportion of Exposure	Proportion of Exposure
Group					
Case (CA)	$X_{CA/E}$	$X_{CA/U}$	N_{CA}	X_{CA}/N_{CA}	π_1
Control (CO)	$X_{CO/E}$	$X_{CO/U}$	N_{CO}	X_{CO}/N_{CO}	π_2
Total	N_E	N_u	N		

the number of cases varies largely depending on occupations. Owing to the high incidence rate and cost, determining the origin of CTS development — i.e., whether CTS is caused by the work or not — is a sensitive issue in the context of workers' compensation. A few researchers claim that CTS is not an occupational injury but is mainly caused by personal factors (Hadler, 1998; Zetterberg and Öfverholm, 1999). However, in the majority of studies, CTS is seen as a work-related disorder.

In epidemiological methodology, odds ratios (ORs) are used to prove that a certain group has a different proportion of illness or symptoms compared with a control group. Odds ratios are an outcome difference measure (Δ) to test the equivalence of attributes being investigated between groups defined. Table 2.5 illustrates a 2×2 contingency table consisting of two binomial distributions. In a case-control study, the odds of exposure in each group indicate a proportion of the probability of an outcome (p) to its complementary probability ($1-p$), as formulated in Equations (2.1) and (2.2), the odds of case-group exposure and the odds of control-group exposure, respectively. Then, the OR of exposure between the groups is defined as a ratio of these two odds' values, as shown in Equation (2.3) (Donald and Donner, 1987; Hauck, 1987).

We can derive the conclusion that there is a significant difference between a control group and a case group, if the OR is not 1, with 95% confidence:

$$\text{Odds of exposed case (CA)} = \frac{X_{CA/E}/N_{CA}}{X_{CA/U}/N_{CA}} = \frac{X_{CA/E}}{X_{CA/U}} \tag{2.1}$$

$$\text{Odds of exposed control (CO)} = \frac{X_{CO/E}/N_{CO}}{X_{CO/U}/N_{CO}} = \frac{X_{CO/E}}{X_{CO/U}} \tag{2.2}$$

$$\text{Odds ratio of risk} = \frac{X_{CA/E}/X_{CA/U}}{X_{CO/E}/X_{CO/U}} = \frac{X_{CA/E} \times X_{CO/U}}{X_{CA/U} \times X_{CO/E}} \tag{2.3}$$

Epidemiological study results regarding CTS were collected from various sources. On the basis of the literature survey, the risk factors for CTS were grouped into three major categories: (1) personal, (2) psychosocial, and (3) physical factors. Personal factors studied in previous research works included gender, age, marital status, gynecological status, medical conditions, and individual factors such as wrist depth/width ratio, general physical condition, obesity, alcohol intake, and smoking habits. Psychosocial factors included nonphysical factors at work, such as job demands, social support from management or coworkers, and workload. These stressors at work are related to MSDs (Bernard et al., 1994; Estryn-Behar et al., 1990; Feyer et al., 1992; Houtman et al., 1994), but some studies reported that there was no conclusive evidence for the relationship between them (Armstrong et al., 1993; Bongers et al., 1993). Physical factors for CTS include highly repetitive or forceful exertions of the hand and wrist, repetitive or forceful pinching, repeated flexion or extension of the wrist, ulnar deviation of the wrist, segmental vibration, and mechanical stress on the base of the palm (Greening and Lynn, 1998; Miller et al., 1994; Putz-Anderson, 1988; Szabo, 1998; Viikari-Juntura, 1998). An increased risk has been reported in workers with a high number of repetitive hand/wrist movements, such as butchers, grocery store workers, watch assemblers in Switzerland, and automobile assembly workers (Barnhart et al., 1991; Hagberg et al., 1992; Delgrosso and Boillat, 1991; Zetterberg and Öfverholm, 1999).

Many studies provided ORs and 95% confidence intervals (CIs) but not standard deviations (SDs). Table 2.6 shows descriptive statistics data of ORs and CIs found in 27 studies. A reciprocal of OR was used for avoiding misinterpretation and maintaining consistency with other values if OR < 1. Figure 2.2 shows a histogram with frequency data for OR of each factor and the total. Seventy-five percent (Q3) of OR are 2.4, 2.225, 2.125, and 2.49 for total, personal, psychosocial, and physical factors, respectively. For SDs, 0.3536, 0.3429, 0.3360, and 0.3536 are the third quartiles for total, personal, psychosocial, and physical factors, respectively. From the SD data, 57 out of 58 (98.3%) are within a range of [0, 0.5] and the quartile is close to 0.975, i.e., Q(0.975)=57.05, or 97.5% of SD data are less than 0.474, the 57th data point of ordered ranks.

TABLE 2.6 Descriptive Statistics for Odds Ratio (OR) and Standard Deviation (SD) Data from 27 Studies

Statistic	Factor	N	Mean	Median	Min	Max	Q1	Q3
Odds ratio (OR)	Personal	13	1.952	1.82	1.17	4.20	1.445	2.250
	Psychosocial	9	1.628	1.35	1.03	2.86	1.195	2.125
	Physical	36	2.226	1.83	1.00	7.00	1.225	2.490
	Total	58	2.072	1.76	1.00	7.00	1.285	2.400
Standard deviation (SD)	Personal	13	0.2573	0.2269	0.0810	0.4047	0.2069	0.3429
	Psychosocial	9	0.2522	0.2183	0.1029	0.4675	0.1738	0.3360
	Physical	36	0.2668	0.2409	0.0249	0.6519	0.1722	0.3536
	Total	58	0.2624	0.2306	0.0249	0.6519	0.1936	0.3536

Source: Kong, D., Ph.D. dissertation, Pennsylvania State University, University Park, PA, 2002, (unpublished).

(a)

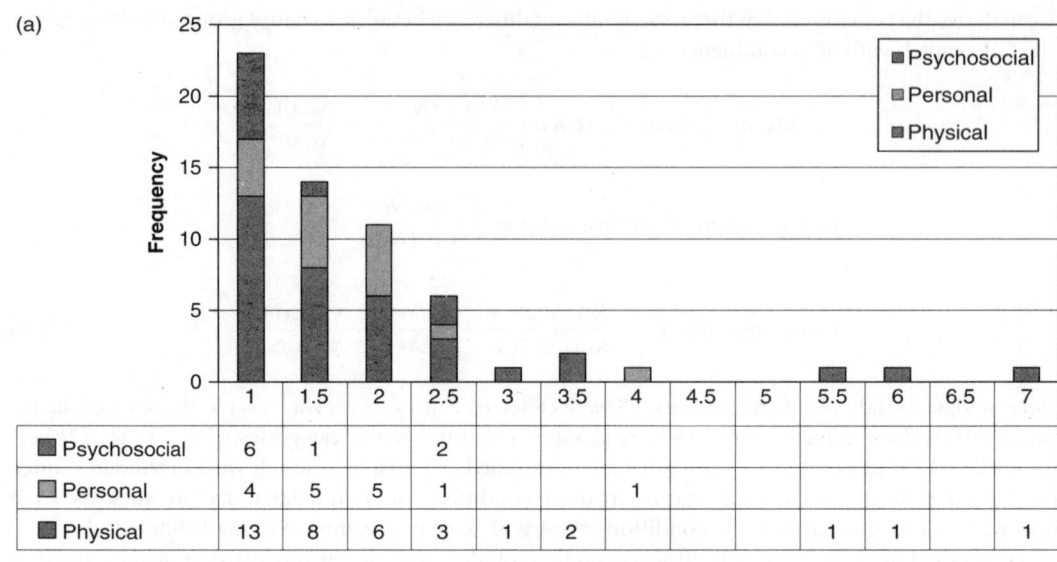

	1	1.5	2	2.5	3	3.5	4	4.5	5	5.5	6	6.5	7
■ Psychosocial	6	1		2									
□ Personal	4	5	5	1			1						
■ Physical	13	8	6	3	1	2				1	1		1

Odds ratio

(b)

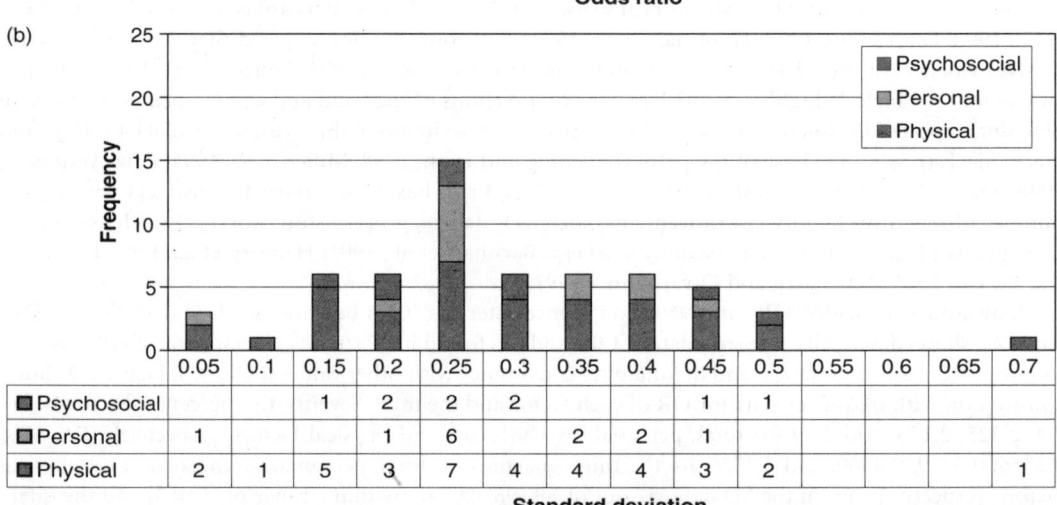

	0.05	0.1	0.15	0.2	0.25	0.3	0.35	0.4	0.45	0.5	0.55	0.6	0.65	0.7
■ Psychosocial			1	2	2	2			1	1				
□ Personal	1			1	6		2	2	1					
■ Physical	2	1	5	3	7	4	4	4	3	2				1

Standard deviation

FIGURE 2.2 Histograms and frequencies of odds ratio (OR) and standard deviation (SD). (a) OR distribution: about 40% (23/58) of OR values are less than 1.5 and overall trend of the distribution is exponentially decreasing. (b) SD distribution: about 98% (57/58) of SD values are less than 0.5.

2.6 Human–Computer Interaction

In 2001, 72.3 million workers used a computer at work in the United States (BLS, 2002). These workers accounted for 53.5% of total employment in 2001. Computer systems, including both hardware and software, have become more and more sophisticated. End users have different levels of skills and knowledge. But regardless of their skill and knowledge levels, end users have a single goal in common: to complete the given tasks as soon as possible or find what they want in time when surfing the Internet. For example, consider a voice recognition system. Many companies are now providing voice response systems rather than a touch-tone response system. Some customers will find it convenient, but others will find it difficult. The voice recognition technology may well be advanced enough not to worry about the

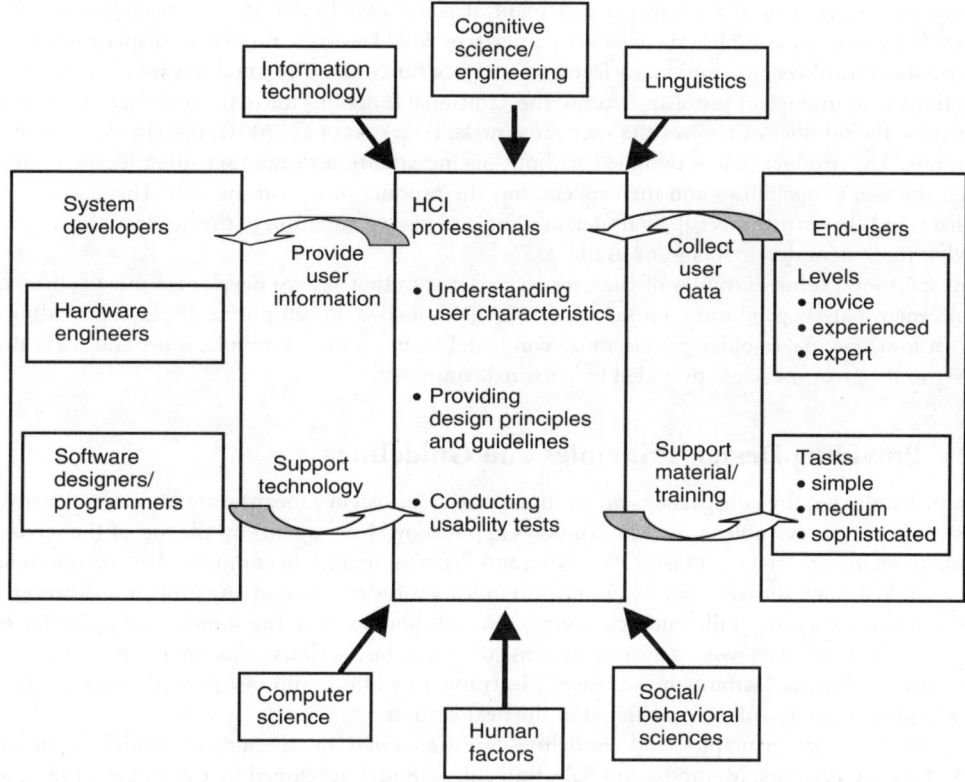

FIGURE 2.3 Conceptual diagram of relationships among system developers, HCI professionals, and end users. Not all of the related fields in HCI are present.

question, "How can my speech be perfectly recognized through the phone?" But for those who have a strong accent or who are beginning to learn English, the voice recognition system may not be favorable.

Human–computer interaction (HCI) is a relatively young field, still developing compared with other research fields. A wide range of groups are interested in HCI, including researchers in linguistics, social and behavioral sciences, computer engineering, information technology, and so on.

There are three major topics in HCI:

(1) Understanding user characteristics
(2) Providing design principles and guidelines
(3) Conducting usability tests

These topics require the collaborative work of experts from a variety of disciplines. Figure 2.3 illustrates a conceptual map for the relationships. HCI professionals, including HF specialists, collect user data and give user-characteristics information to the system developers so that the developers know more about what the user wants and how the user behaves. The HCI professionals also provide the user more user-friendly support materials with help from system developers. The related disciplines shown in Figure 2.3 are examples.

2.6.1 Understanding User Characteristics

"User-friendly" is a very familiar term in HCI. It indicates that consumers now prefer software packages or devices having features that are easy to use or manipulate. However, considering the nature of software, which ranges from performing very simple functions such as basic calculations to extremely complex functions such as control of a nuclear power plant, it is not easy to define "user-friendliness." Also, as products become more complex, the necessary interfaces will likely have numerous displays, menus, display formats, control systems, and many levels of interface functions. The trend toward a greater number of functions is an important problem because the additional functions make the interface more complex and increase the number of choices the user must make (Wickens et al., 2004), thereby decreasing user-friendliness. Any product that is designed without paying attention to the user often leaves a huge gap between the user's capabilities and the expectations the product places on the user. The goal of the HF specialist is to help narrow this gap by understanding and paying attention to the needs of the user rather than what the system developers want to make.

Table 2.7 shows some examples of the user characteristics that system developers must consider. For example, many elderly people are not satisfied with the font sizes on cell-phone displays; providing flexibility on font size makes older people more comfortable. In another example, most end users do not understand the error messages provided by personal computers.

2.6.2   Providing Design Principles and Guidelines

Design principles require consistency and good usability. Consistency means providing new systems that are easy to learn: for example, most MS Windows applications have menus on the top of the screen and a *File* menu on the top left containing *Open*, *Save*, and *Print* commands in common. Anyone who is familiar with any Windows software can use another Windows software without any problem. However, consider the menu structures built into cell phones. No cell phones have the same or even similar menu structures to indicate "start over." This lack of consistency can be a serious problem for consumers as they switch between systems. Usability means ease of learning, ease of use, and ease of recovering from errors. The concept of usability will be considered in the next section.

In general, design principles and guidelines are generated by theoretical models such as the GOMS (Goals, Operators, Methods, and Selection rules) model developed by Card et al. (1983). Many systems developers have the strong belief that people will become skilled in their systems' use and will want to have efficient methods for accomplishing routine tasks. This GOMS model can predict the impact of design decisions on this important measure of success. For example, consider the goal of sending an e-mail to a friend. The user will open a blank message screen, select a recipient or type an e-mail address, write a message, and click on the send button. The GOMS model assumes that the user sets a goal (sending an e-mail) and subgoals, if necessary, that the user achieves by way of methods and selection rules. A method is a sequence of steps that the user should follow, while selection rules are the choice of one or another method, such as "select a recipient" or "type an e-mail address," for example.

TABLE 2.7   Examples of User Characteristics and HCI Considerations

User Characteristics	Considerations for System Developers
Age	• Flexibility on font size for older people • Selection of words for children • Sensitivity of mouse and keyboard
Knowledge of computer	• Understandable error messages • Instruction manual • Help menus
Disability factor	• Use of color for the color-blind • Voice recognition and response system • Keyboard size
Preference	• Dominant hand when using a mouse • Flexibility of color selection • Option of auditory feedback

2.6.3 Conducting Usability Tests

Even a carefully designed system that uses the best theories must be evaluated in usability tests. Usability tests involve typical users using the system in realistic situations. All details of difficulties and frustrations the testers encounter should be recorded for the purpose of upgrading software quality. Most usability specialists use a variety of prototypes, from low- to high-fidelity methods. Low-fidelity methods, generally used early in the design process, include index cards, paper stickers, paper-and-pen drawings, and storyboards. Storyboards are a graphical depiction of the outward appearance of the software system, without any actual system functioning. High-fidelity methods include fully interactive screens with the look and feel of the final software (Wickens et al., 2004).

When designers are conducting usability testing, whether early in the low-fidelity prototyping stages or late in the design lifecycle, they must identify what they are going to measure, often called usability metrics. Usability metrics tend to change in nature and scope as the project moves forward. In early conceptual design phases, usability can be evaluated with a few users and focuses on the qualitative assessment of general usability (whether the task can even be accomplished using the system) and user satisfaction. Low-fidelity prototypes are given to users, who then imagine performing a very limited subset of tasks with the materials or screens (Carroll, 1995). At this point, there is usually little to no quantitative data collection; simply talking with a small number of users can yield a large amount of valuable information. As the design takes on more specific form, usability testing becomes more formalized and often quantitative. Several versions of usability questionnaires are available, and some companies have developed their own usability testing metrics. In general, effectiveness, efficiency, and subjective satisfaction are the main usability measures.

Usability testing is not limited to product designs or software development. As computers become more popular, companies develop their websites, trying to attract more visitors by providing more services. In many cases, it is really difficult for a user to find what he or she wants from websites, especially when the website contains tremendous amounts of data. For example, Yu et al. (1998) reported that the Kodak website (www.Kodak.com) contained over 25,000 pages and 74,000 files in 1998. In terms of traffic size, Kodak.com averaged around one million hits daily, including roughly a quarter of a million pages viewed and accessed by some 24,000 unique visitors each day. However, the main top-level design remained unchanged, and the visitors had a hard time finding what they wanted as a result of broken links and unmatched menu structures. The effort to develop a new design was started from the guestbook of Kodak.com, which provided important information on what the visitors wanted most (*user's needs*) and which problems the visitors had (*user's behavior*). As seen in this Kodak.com case, users are the core factor of design. No products will be favored without considering the user. In response to the growing demands for usability, the U.S. Department of Health and Human Services developed a website about usability (http://www.usability.gov/). This website provides useful information such as federal guidelines, Internet statistics, and lists of usability-related events.

2.7 Summary

Human Factors is a very broad concept. Only a few areas in HF are introduced in this chapter but the main idea is simple: *How can we provide employees safe and comfortable working conditions without interrupting productivity?* This can be accomplished only when government, employers, researchers, and employees collaborate together. The slightly old but still relevant slogan, "Fitting the task to the man," becomes even more important when technology advances at light speed.

2.8 Exercises

1. You may find any tool, desk, equipment, or even TV remote that doesn't match your hand size. Think about why it doesn't and how to fit it to you. If it fits you, what about your family members or your friends? How do you make a proposal that it accommodate more people?

2. What software program do you frequently use? Do you like the software? Why? What feature of the program do you like the least? Why? How do you want to change it? Do you think it is acceptable to most users?

3. Do you or anyone you know have any work-related illness or injury? How do you know it is work related? If it is work related, has the employer made a modification of the workplace? If so, has the incidence rate decreased?

References

Armstrong, T.J., Buckle, P., Fine, L.J., Hagberg, M., Jonsson, B., Kilbom, A., Kuorinka, I.A.A., Silverstein, B.A., Sjogaard, G., and Viikari-Juntura, E.R.A., A conceptual model for work-related neck and upper-limb musculoskeletal disorders, *Scand. J. Work, Environ., Health*, 19, 73–84, 1993.

Barnhart, S., Demers, P., Miller, M., Longstreth, W., and Rosenstock, L., Carpal Tunnel syndrome among ski manufacturing workers. *Scand. J. Work, Envir., Health*, 17, 46–52, 1991.

Bernard, B., Sauter, S., Fine, L., Peterson, M., and Hales, T., Job task and psychosocial risk factors for work-related musculoskeletal disorders among newspaper employees, *Scand. J. Work, Environ., Health*, 20, 417–426, 1994.

Bongers, P.M., Winter, C.R., Kompier, M.A.J., and Hilderbrandt, V.H., Psychosocial factors at work and musculoskeletal disease, *Scand. J. Work, Environ., Health*, 19, 297–312, 1993.

Bovenzi, M., Review: exposure-response relationship in the hand-arm vibration syndrome: an overview of current epidemiology research, *Int. Arch. Occup. Environ. Health*, 71, 509–519, 1998.

Bureau of Labor Statistics, *News: Lost-Worktime Injuries and Illnesses: Characteristics and Resulting Time Away From Work*, Washington, DC, 2002, 2003, 2004.

Card, S.K., Moran, T.P., and Newell, A., *The Psychology of Human-Computer Interaction*, Erlbaum, Hillsdale, NJ, 1983.

Carroll, J.M., *Senario-Based Design: Envisioning Work and Technology in System Development*, Wiley, New York, NY, 1995.

Delgrosso, I. and Boillat, M-A., Carpal tunnel syndrome: role of occupation, *Int. Arch. Occup. Environ. Health*, 63, 267–270, 1991.

de Krom, M.C.T.F.M., Kester, A.D.M., Knipschild, P.G., and Spaans, F., Risk factors for carpal tunnel syndrome, *Am. J. Epidemiol.*, 132, 1102–1110, 1990.

Donald, A. and Donner, A., The effect of clustering on the analysis of sets of 2x2 contingency tables, in *Biostatistics*, Vol. 38, MacNeill, I. and Umphrey, G.J., Eds., University of Western Ontario Series in Philosophy of Science, 1987, pp. 151–167.

Estryn-Behar, M., Kaminski, M., Peigne, E., Bonnet, N., Vaichere, E., Gozlan, C., Azoulay, S., and Giorgi, M., Stress at work and mental health status among female hospital workers, *Br. J. Ind. Med.*, 47, 20–28, 1990.

Feyer, A-M., Williamson, A., Mandryk, J., de Silva, I., and Healy, S., Role of psychosocial risk factors in work-related low-back pain, *Scand. J. Work, Environ., Health*, 18, 368–375, 1992.

Greening, J. and Lynn, B., Vibration sense in the upper limb in patients with repetitive strain injury and a group of at-risk office workers, *Int. Arch. Occup. Environ. Health*, 71, 29–34, 1998.

Hadler, N.M., Coping with arm pain in the workplace, *Clinical Orthopaed. Relat. Res.*, 351, 57–62, 1998.

Hagberg, M., Morgenstern, H., and Kelsh, M., Impact of occupations and job tasks on the prevalence of carpal tunnel syndrome, *Scand. J. Work, Environ., Health*, 18, 337–345, 1992.

Hauck, W.W., Estimation of a common odds ratio, in *Biostatistics*, MacNeill, I. and Umphrey G.J., Eds., Vol. 38, University of Western Ontario series in philosophy of science, 1987, pp. 125–149.

Helander, M., *A Guide to the Ergonomics of Manufacturing*, Bristol, PA: Taylor & Francis, 1995.

Houtman, I.L.D., Bongers, P.M., Smulders, P.G.W., and Kompier, M.A., Psychosocial stressors at work and musculoskeletal problems, *Scand. J. Work, Environ., Health*, 20, 139–145, 1994.

Karwowski, W. and Marras, W., *The Occupational Ergonomics Handbook*, CRC Press, Salem, 1999.

Keller K., Corbett, J., and Nichols, D., Repetitive strain injury in computer keyboard users: patho-mechanics and treatment principles in individual and group intervention, *J. Hand Ther.*, 11, 9–26, 1998.

Kong, D., The Development of a Logistic Model for Carpal Tunnel Syndrome Risk Assessment Model, Ph.D. dissertation (unpublished), Pennsylvania State University, University Park, PA, 2002.

Konz, S. and Johnson, S., *Work Design*, 6th ed., Holcomb Hathaway, Publishers, Inc., Scottsdale, AZ, 2004.

Latko, W.A., Armstrong, T.J., Foulke, J.A., Herrin, G.D., Rabourn, R.A., and Ulin, S.S., Development of an observational method for assessing repetition in hand tasks, *Am. Industrial Hyg. Assoc. J.*, 58, 278–285, 1997.

Loisel, P., Gosselin, L., Durand, P., Lemaire, J., Poitras, S., and Abenhaim, L., Implementation of a participatory ergonomics program in the rehabilitation of workers suffering from subacute back pain, *Appl. Ergon.*, 32, 53–60, 2001.

Masear, V.R., Hayes, J.M., and Hyde, A.G., An industrial cause of carpal tunnel syndrome, *J. Hand Surg.*, 11A, 222–227, 1986.

Miller, R.F., Rapids, C., Lohman, W.H., Maldonado, G., and Mandel, J.S., An epidemiologic study of carpal tunnel syndrome and hand-arm vibration syndrome in relation to vibration exposure, *J. Hand Surg.*, 19A, 99–105, 1994.

Niebel, B. and Freivalds, A., *Methods, Standards, and Work Design*, 11th ed., McGraw-Hill, New York, NY, 2003.

Putz-Anderson, V., *Cumulative Trauma Disorders: A Manual for Musculoskeletal Diseases of the Upper Limbs*, Taylor & Francis, Bristol, PA, 1988.

Roquelaure, Y., Mechali, S., Dano, C., Fanello, S., Benetti, F., Bureau, D., Mariel, J., Martin, Y-H., Derriennic, F., and Penneau-Fontbonne, D., Occupational and personal risk factors for carpal tunnel syndrome in industrial workers, *Scand. J. Work, Environment, Health*, 23, 364–369, 1997.

Salvendy, G., *Handbook of Human Factors and Ergonomics*, 2nd ed., J. Wiley, New York, 1997.

Sanders, M.S. and McCormick, E.J., *Human Factors in Engineering and Design*, 7th ed., McGraw-Hill, Singapore, 1992.

Silverstein, B.A. and Hughes, R.E., Upper extremity musculoskeletal disorders at a pulp and paper mill, *Appl. Ergon.*, 27, 189–194, 1996.

Strom, G., When was the beginning of ergonomics and human factors? *Ergon. des.*, 11, 5–6, 2003.

Szabo, R.M., Carpal tunnel syndrome as a repetitive motion disorder, *Clin. Orthopaed. Relat. Res.*, 351, 78–89, 1998.

Viikari-Juntura, E., Risk factors for upper limb disorders, *Clin. Orthopaed. Relat. Res.*, 351, 39–43, 1998.

Wickens, C.D., Lee, J.D., Liu, Y., and Becker, S.E.G., *An Introduction to Human Factors Engineering*, 2nd ed., Prentice-Hall, Englewood Cliffs, NJ, 2004.

Woodson, W.E., *Human Factors Design Handbook*, McGraw-Hill, New York, 1981.

Yu, J.J., Prabhu, P.V., and Neale, W.C., A user-centered approach to designing a new top-level structure for a large and diverse corporate web site, *Proceedings of 4th Human Factors and the Web Conference*, Basking Ridge, New Jersey, 1998.

Zetterberg, C. and Öfverholm, T., Carpal tunnel syndrome and other wrist/hand symptoms and signs in male and female car assembly workers, *Int. J. Ind. Ergon.*, 23, 193–204, 1999.

3

Process Control for Quality Improvement

Wei Jiang
Stevens Institute of Technology

John V. Farr
Stevens Institute of Technology

3.1   Introduction

The goal of a manufacturing system is to produce multiple copies of the same product, each having attributes within specified tolerances. Variation reduction is one of the major techniques for achieving process stability and requires increasing amounts of process and equipment control at various levels of manufacturing systems. In particular, controlling complicated processes to produce smaller feature sizes is inherently difficult in semiconductor manufacturing. Moyne et al. (1993) attribute this difficulty to an insufficient number of sensors and actuators at each manufacturing process step for establishing a desired level of concurrent control over process parameters. Moreover, mathematical models incorporated into the control scheme rely on empirical data and are consequently imprecise. As pointed out by many researchers (Box and Kramer, 1992; Vander Wiel et al., 1992), it is highly desirable to investigate different control methods to detail the scope of their usage and limitations, and to address the complementary utilization of those methods in a control system.

Two categories of research and applications have been developed independently to achieve process control. Engineering process control (EPC) uses measurements to prescribe changes and *adjust* the process inputs with the intention of bringing the process outputs closer to targets. It employs feedback/feedforward controllers for process regulation and has gained a lot of popularity in continuous process industries. Statistical process control (SPC) uses measurements to *monitor* a process and look for major changes in order to eliminate the root causes of the changes. It has found widespread applications in discrete parts industries for process improvement, process parameter estimation, and process capability determination. Successful projects have also been developed in other industries such as hospital service, business marketing, and financial management for detecting important process changes to support decision making. Although both techniques aim at the same objective of reducing process variation, they have different origins and have used different implementation strategies for decades.

Practitioners of SPC argue that because of the complexity of manufacturing processes, EPC methods can very likely over control a process and increase process variability rather than decrease it, as demonstrated by Deming's (1986) funnel experiment. Moreover, important quality events may be masked by frequent adjustments and become difficult to detect and remove for ultimate quality improvement. On the other hand, practitioners of EPC criticize SPC methods for excluding the opportunities for reducing the variability in the process output. Owing to the stochastic nature of manufacturing processes, traditional SPC methods always generate too many false alarms and fail to discriminate quality deterioration from the in-control state defined by SPC rules. Recently, an integration of EPC and SPC methods has emerged in semiconductor manufacturing and has resulted in a tremendous improvement of industrial efficiency (Sachs et al., 1995).

The EPC/SPC integration employs an EPC control rule to regulate the system and superimposes SPC charts on the EPC system to detect process departures from the system model. Both academic research and industrial practice have shown the effectiveness of the EPC/SPC integration model when the process is subjected to both systematic variations and special-cause variations (Montgomery et al., 1994; Capilla et al., 1999; Jiang and Tsui, 2000). To avoid confusion, Box and Luceno (1997) refer to EPC activities as *process adjustment* and to SPC activities as *process monitoring*. While the two approaches have been applied independently in different areas for decades, the relationship between them has not yet been clearly explored.

This chapter reviews various cutting-edge models and techniques for industrial process control. A unified framework is developed to model the relationships among the well-known methods in EPC, SPC, and integrated EPC/SPC. An industrial quality control application for the chemical–mechanical planarization (CMP) process demonstrates the benefits of these methods.

3.2   Two Process-Control Approaches

3.2.1   Engineering Process Control

Engineering process control is a popular strategy for process optimization and improvement. It describes the manufacturing process as an input–output system where the input variables (recipes) can be manipulated (or adjusted) to coun teract the uncontrollable disturbances to maintain the process target. The output of the process can be measurements of the final product or critical in-process variables that need to be controlled. In general, without any control actions (adjustment of inputs), the output may shift or drift away from the desired quality target owing to disturbances (Box and Luceno, 1997). These disturbances often are not white noise but exhibit a dependence on past values. It is thus possible to anticipate the process behavior on the basis of past observations and to control the process by adjusting the input variables.

Engineering process control requires a process model. A simple but useful process model that describes a linear relationship between process inputs and outputs is (Vander Wiel et al., 1992),

$$e_t = gX_{t-1} + D_t \tag{3.1}$$

where e_t and X_t represent the process output and input (control) deviations from target, D_t the process disturbances that pass through part of the system and continue to affect the output, and g the process gain that measures the impact of input control to process outputs. To simplify our discussion, we assume that the process gain is unity, i.e., $g = -1$. When no process control is involved, the process output is simply the disturbance, and the variance of the output is obtained as σ_D^2. The objective of process control is to reduce process variations by adjusting inputs at the beginning of each run, i.e., $\sigma_e^2 < \sigma_D^2$, where σ_e^2 is the variance of the controlled output.

Feedforward control uses prediction of the disturbance to adjust the process, i.e., $X_{t-1} = \hat{D}_t$, where $\hat{D}_t$ is the prediction of disturbance at time t given process information up to $t-1$. It strongly relies on an accurate sensor system to capture the process disturbance. Another process control strategy widely adopted in industry is feedback control, which uses deviations of the output from the target (set-point) to indicate that a disturbance has upset the process and to calculate the amount of adjustment.

Figure 3.1 presents a typical process with a feedback control scheme. Since deviations or errors are used to compensate for the disturbance, the compensation scheme is essentially twofold. It is not perfect in maintaining the process on target, since any corrective action is taken only after the process deviates from its target first. On the other hand, as soon as the process output deviates from the target, corrective action is initiated regardless of the source and type of disturbance. It is important to note that the feedback scheme is beneficial only if there are autocorrelations among the outputs.

There is a rich body of research on feedback controllers (Astrom and Wittenmark, 1997). To minimize the variance of the output deviations from the quality target, two types of controllers are commonly used.

- *Minimum mean square error (MMSE) control.* From the time-series transfer function model that represents the relationship between the input X_t and output e_t, Box et al. (1994) develop the MMSE controller

$$X_t = -\frac{L_1(B)L_3(B)}{L_2(B)L_4(B)}e_t$$

  where B is backshift operator and $L_1(B)$, $L_2(B)$, $L_3(B)$ and $L_4(B)$ polynomials in B that are relevant to the process parameters. Theoretically, if the process can be accurately estimated, the output can be reduced to a white noise by the MMSE controller.
- *Proportional Integral Derivative (PID) Control.* This is a special class of the Autoregressive Integrated Moving Average (ARIMA) control model. The three-mode PID controller equation is formed by summing three methods of control, proportional (P), integral (I), and derivative (D). The discrete version of the PID controller is

$$X_t = k_0 + k_P e_t + k_I \sum_{i=0}^{\infty} e_{t-i} + k_D(e_t - e_{t-1}) \tag{3.2}$$

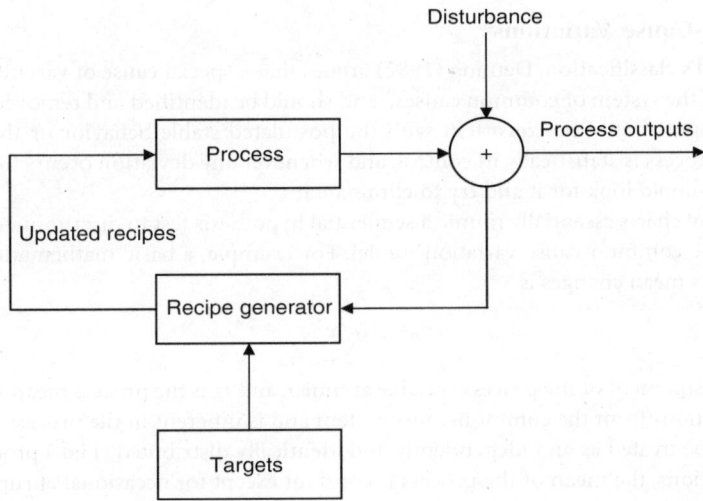

FIGURE 3.1 A feedback-controlled process.

where k_0 is always set to zero. The proportional control action is intuitive but is not able to eliminate steady-state errors, i.e., an offset will exist after a set-point change or a sustained load disturbance. The integral control action is often used because it can eliminate offset through continuously adjusting the controller output until the error reaches zero. The function of the derivative control action is to anticipate the future behavior of the error signal by considering its rate of change (Seborg et al., 1989). Tsung et al. (1998) discuss the design of PID controllers when the disturbance follows an ARIMA (1,1,1) model. Generally, other rules of thumb have to be used for designing a PID controller (Ziegler and Nichols, 1942; Astrom and Hagglund, 1988).

The MMSE control is optimal in terms of minimizing mean squared residual errors. However, this is only true for the idealized situation in which the model and model parameters are known exactly. If the process model is not known precisely, it has a serious robustness problem when the model is close to non-stationarity. As shown in Tsung et al. (1998) and Luceno (1998), the PID controller is very efficient and also robust against nonstationarity as that it can continuously adjust the process whenever there is an offset.

Theoretically, only predictable deviations can be reduced by EPC methods. Modeling errors due to process changes are generally hard to capture in real time and to compensate for with EPC schemes. Various adaptive EPC schemes that dynamically adjust control parameters have been investigated. Recently, an adaptive framework has been proposed in semiconductor manufacturing (by superimposing) an SPC scheme to monitor modeling errors and revise the process models (Sachs et al., 1995).

3.2.2 Statistical Process Control

The basic idea in SPC is a binary view of the state of a process, i.e., either it is running satisfactorily or not. As developed by Shewhart (1931), the two states are classified as having: (1) a common cause of variations and (2) an assignable/special cause of variations, respectively.

3.2.2.1 Common-Cause Variations

Common-cause variations are the basic assumption on which the SPC methods are based. This assumes that the sample comes from a *known* probability distribution, and the process is classified as "statistically" in control. In other words, "the future behavior can be predicted within probability limits determined by the common-cause system" (Box and Kramer, 1992). This kind of variation, from a management point of view, is inherent in the process and is difficult or impossible to eliminate.

3.2.2.2 Special-Cause Variations

Based on Shewhart's classification, Deming (1982) argues that a special cause of variations is "something special, not part of the system of common causes," and should be identified and removed at the root. That is, the process output should be consistent with the postulated stable behavior or the common-cause model when the process is statistically in control, and whenever any deviation occurs from the common-cause model, one should look for it and try to eliminate it.

Statistical control charts essentially mimic a sequential hypothesis test to discriminate special causes of variations from the common-cause variation model. For example, a basic mathematical model behind monitoring process mean changes is

$$e_t = \eta_t + X_t$$

where e_t is the measurement of the process variable at time t, and η_t is the process mean at that time. Here, X_t represents variations from the common-cause system and is inherent in the process. In some applications, X_t is or can be treated as an independently and identically distributed (i.i.d.) process.

With few exceptions, the mean of the process is constant except for occasional abrupt changes, i.e.,

$$\eta_t = \eta + \mu_t$$

where η is the mean target, and μ_t is zero for $t < t_0$ and has nonzero values for $t \geq t_0$. For example, if the special cause is a step-like change, μ_t is a constant μ after t_0. The goal of SPC charts is to detect the change point t_0 as quickly as possible so that corrective actions can be taken before quality deteriorates and defective units are produced. Among many others, the Shewhart chart, the exponentially weighted moving average (EWMA) chart, and the cumulative sum (CUSUM) chart are three important and widely used control charts.

- *Shewhart chart.* Process observations are tested against control limits $|e_t| > L \cdot \sigma_e$, where σ_e is the standard deviation of the chart statistic estimated by moving range and L is prespecified to maintain particular probability properties.
- *EWMA chart.* Roberts (1959) proposed monitoring the EWMA statistic of the process observations, $Z_t = \sum_{i=0}^{\infty} w_i e_{t-i}$, where $w_i = \lambda(1-\lambda)^i$ $(0 < \lambda \leq 1)$. The EWMA statistic utilizes previous information with the discount factor $(1-\lambda)$ and includes the Shewhart chart as a special case when $\lambda = 1$. It has a recursive form

$$Z_t = (1-\lambda)Z_{t-1} + \lambda e_t \tag{3.3}$$

where Z_0 is zero or the process mean. The stopping rule of the EWMA chart is $|Z_t| > L\sigma_z$.

CUSUM chart. Page (1954) introduced the CUSUM chart as a sequential probability test, which can also be obtained by letting λ approach 0 in Equation (3.3), i.e., the CUSUM algorithm assigns equal weights to previous observations. The tabular form of a CUSUM chart consists of two quantities,

$$Z_t^+ = \max[0, e_t + Z_{t-1}^+ - K] \text{ and } Z_t^- = \min[0, -e_t + Z_{t-1}^- - K]$$

where $Z_t^+ = Z_t^- = 0$. It has been shown that the CUSUM chart with $K = \mu/2$ is optimal for detecting a mean change of μ when the observations are i.i.d.

Although the purpose of these procedures is to detect process changes, it is well known that they may signal even when the process remains on target owing to the randomness of observations. The expected length of period between two successive false alarms is called *in-control* average run-length (ARL$_0$). When a special cause presents, the expected period before a signal is triggered is called *out-of-control* average run-length (ARL$_1$). A control chart is desired with a shorter ARL$_1$ but longer ARL$_0$. In practice, the Shewhart chart is sensitive in detecting large shifts while the EWMA and CUSUM charts are sensitive to small shifts (Lucas and Saccucci, 1990).

In typical applications of SPC charts, a fundamental assumption is that the common-cause variation is free of serial correlation. Unfortunately, the assumption of independence is often invalid in many manufacturing processes. For example, in discrete parts industries, the development of sensing and measurement technology has made it possible to measure critical dimensions on every unit produced, and in continuous process industry, the presence of inertial elements, such as tanks, reactors, and recycle streams, results in significant serial correlation in measurement variables. Serial correlations call for EPC techniques to reduce variations and present new challenges and opportunities to SPC for quality improvement.

3.3 Integration of Engineering Process Control/Statistical Process Control — Run-to-Run Control

Engineering process control and SPC are two complementary strategies developed in different industries for quality improvement. There is a corresponding relationship between them through prediction. Consider a pure-gain dynamic feedback-controlled process

$$e_t = X_{t-1} + D_t$$

Suppose $\hat{D}_{t+1}$ is an estimator (i.e., prediction) of D_{t+1} at time t, a realizable form of control could be obtained by setting $X_t = -\hat{D}_{t+1}$ and the output error at time $t+1$ becomes $e_{t+1} = D_{t+1} - \hat{D}_{t+1}$, which equals to the "prediction error". For example, when the process can be described by an ARIMA model, the MMSE control has a form identical to that of the MMSE predictor (Box et al., 1994). Similarly, as discussed in Section 3.5, a forecast-based special-cause chart (SCC) essentially monitors the MMSE prediction errors of an autocorrelated process.

As an alternative, an EWMA predictor, which corresponds to the integral (I) control, is one of the most frequently used prediction methods in business and industry, mainly because of its simplicity and efficiency. Box et al. (1994) and others have studied its optimality in terms of minimizing mean-squared prediction errors for integratedmoving average (IMA)(1) models; Cox (1961) shows that it is effective for AR(1) models when parameter ϕ is larger than 1/3. In SPC the EWMA statistic is also an effective control chart for detecting small and medium mean shifts for both i.i.d. and autocorrelated processes (Lucas and Saccucci, 1990; Montgomery and Mastrangelo, 1991).

The relationship between EPC and SPC through prediction has been recently explored in many industrial applications. To make an appropriate selection between the two approaches in practice, it is important to identify disturbance structures and strengths of the two control methods to influence the process. Figure 3.2 presents four categories of ongoing research and application of the two quality-control approaches.

- If a process is not correlated, there is no need to employ EPC schemes. Traditional SPC control charts should be used for identifying assignable cause variations.
- When data are correlated, the possibility of employing EPC techniques should be examined. SPC control charts are called for to monitor autocorrelated processes if no feasible EPC controller exists.
- If appropriate controllers are available, EPC control schemes can be employed to compensate for the autocorrelated disturbance. However, no single EPC controller system can compensate for all kinds of potential variations.

To identify and understand the cause of process changes, a unified control framework should be applied to regulate a process using feedback control while using the diagnostic capability of SPC to detect sudden shift disturbances to the process. The integration of EPC/SPC looks for the best opportunities of quality improvement by integrating and combining the strengths of EPC and SPC among the various levels of control that may be incorporated into a manufacturing system. Run-to-run (R2R) or sequential optimization and control is a typical realization of EPC/SPC integration in semiconductor manufacturing (Moyne et al., 1993; Rashap et al., 1995; Ruegsegger et al., 1999). The R2R controller is a model-based process control system in which the controller provides recipes (inputs) based on post-process measurements at the beginning of each run, updates the process model according to the measurements at the end of the run, and provides new recipes for the next run of the process. It generally does not modify recipes during a run because obtaining real-time information is usually very expensive in a semiconductor process and because frequent changes of inputs to the process may increase the variability of the process's outputs and possibly even make the process unstable. A block diagram of such an R2R controller is shown in Figure 3.3.

A good R2R controller should be able to compensate for various disturbances, such as process drifts, process shifts due to maintenance or other factors, model or sensor errors, etc. Moreover, it should be able to deal with the limitations, bounds, cost requirements, multiple targets, and time delays that are often encountered in real processes. The initial R2R process control model can be derived from former experiments using statistical methods such as the response surface model (RSM). When the controller is employed online, the model within the controller is updated according to the new measurements from run to run. A typical R2R system consists of three components: diagnosis module, gradual module, and rapid module (Sachs et al., 1995).

3.3.1 Diagnosis Module

This is a generalized SPC to distinguish between slow drifts and rapid shifts and decide if the process is running in accordance with the current process model. Since the inputs experience small changes, it is

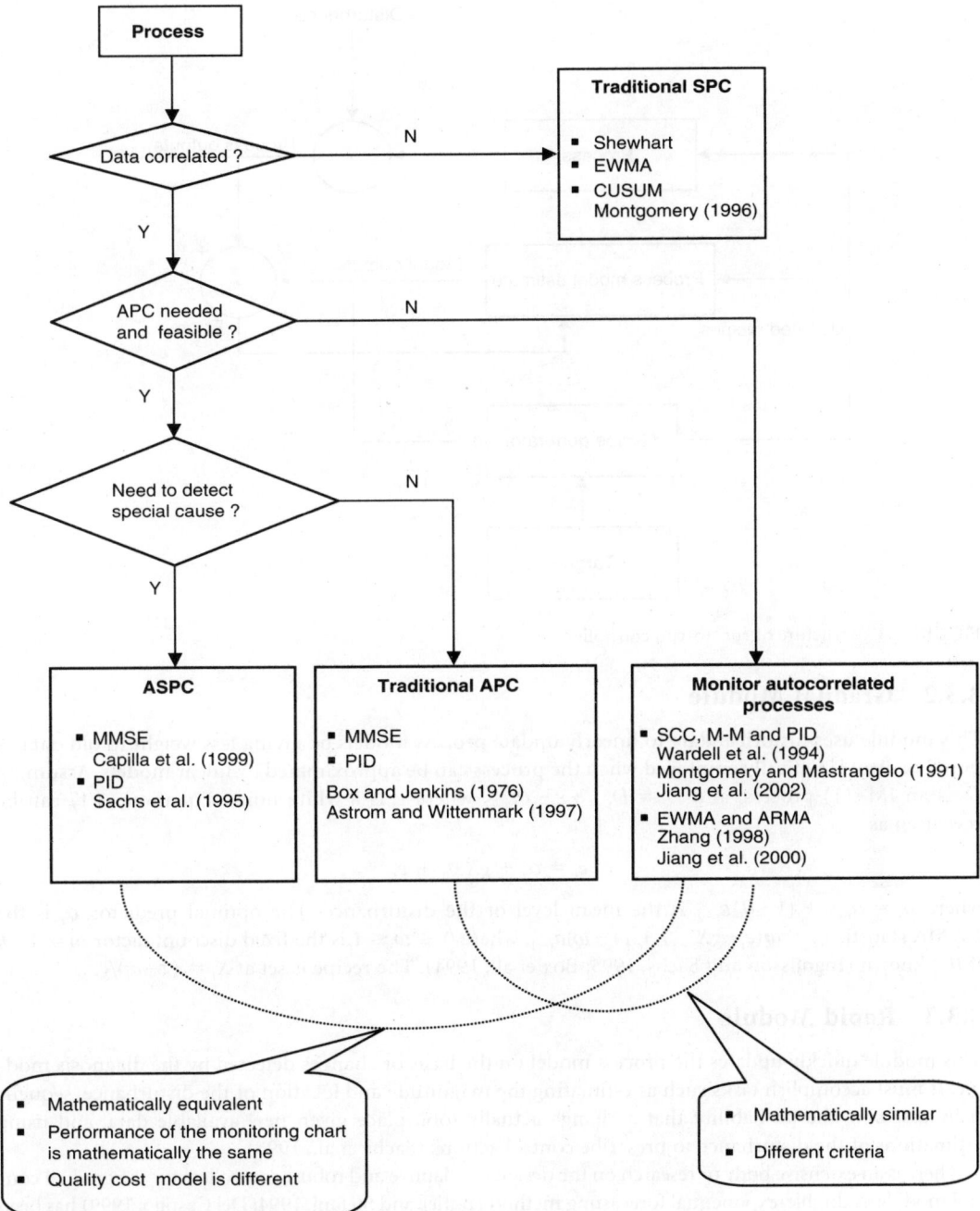

FIGURE 3.2　An overview of EPC and SPC methods.

generally impossible to apply standard control charts to monitor the outputs. Mandel (1969) suggests monitoring the prediction errors; and Zhang (1984) proposes cause-selecting control charts to determine which of the inputs or outputs is responsible for the out-of-control situation. This module determines which of the following gradual or rapid modes is engaged.

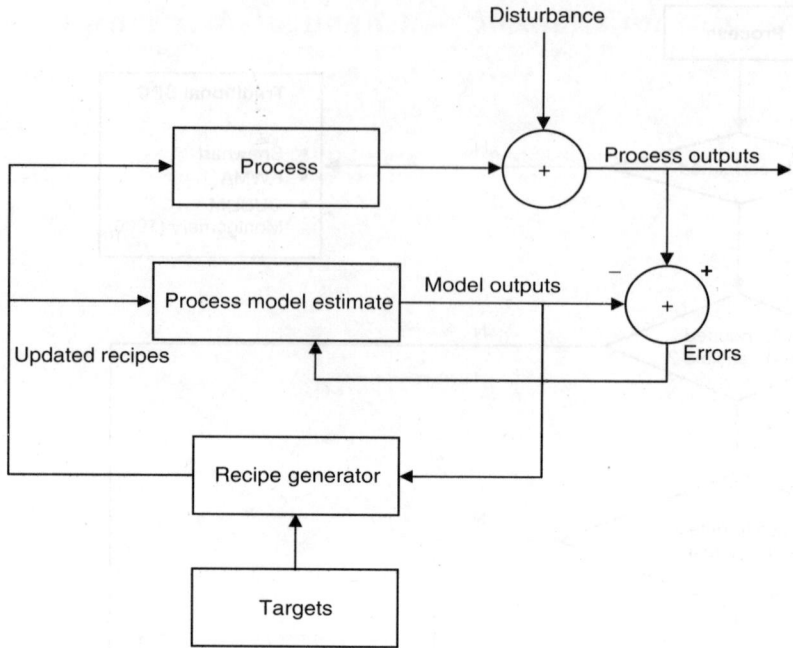

FIGURE 3.3   Structure of run-to-run controller.

3.3.2   Gradual Module

This module uses historical data to linearly update process models by giving less weight to old data. A pure I control is typically employed when the process can be approximated by linear models. Assuming D_t is an IMA(1) process, and $D_t = D_{t-1} + \varepsilon_t - \theta\varepsilon_{t-1}$ where ε_t is a white noise, Equation (3.1) can be rewritten as

$$e_t = \alpha_t + gX_{t-1} + \varepsilon_t$$

where $\alpha_t = \alpha_{t-1} + (1-\theta)\varepsilon_{t-1}$ is the mean level of the disturbance. The optimal predictor α_t is the EWMA statistic $a_t = \omega(e_t - gX_{t-1}) + (1-\omega)a_{t-1}$, where $0 \le \omega \le 1$ is the fixed discount factor $\omega = 1 - \theta$ if θ is known (Ingolfsson and Sachs, 1993; Box et al., 1994). The recipe is set at $X_t = (\tau - a_t)/g$.

3.3.3   Rapid Module

This module quickly updates the process model on the basis of changes detected by the diagnosis module. It must accomplish tasks such as estimating the magnitude and location of the disturbance, sequentially assessing the probability that a change actually took place given new available data, and using estimations of the disturbance to prescribe control actions (Sachs et al., 1995).

There is an extensive body of research on the design of adaptive and robust controllers for the gradual control module. A double-exponential forecasting method (Butler and Stefani, 1994; Del Castillo, 1999) has been proposed using a predictor corrector controller (PCC) to eliminate the impact of machine and process drift. Other control methods include optimized adaptive quality control (Del Castillo and Yeh, 1997), Kalman filter (Palmer et al., 1996), set-value methods (Baras and Patel, 1996), and machine learning methods such as artificial neural network (Smith and Boning, 1997). To facilitate the rapid module, Chen and Elsayed (2000) provide a Bayesian estimation method for detecting the shift size and estimating the time of the shift; Yashchin (1995) proposes an adaptive EWMA estimator of the process mean; and Pan and Del Castillo (2003) investigate using CUSUM charts in conjunction with sequential adjustments to improve the average squared deviations. The following section uses a CMP process to demonstrate the effectiveness of R2R control systems.

3.4 A Run-to-Run Example — Chemical–Mechanical Planarization

Chemical-mechanical planarization of dielectric films is basically a surface planarization method in which a wafer is affixed to a carrier and pressed facedown on a rotating platen holding a polishing pad as shown in Figure 3.4. Silica-based alkaline slurry is applied during polishing thus providing a chemical and mechanical component to the polishing process. The primary function of CMP is to smooth a nominally macroscopically flat wafer at the feature (or microlevel), i.e., to planarize its features. Therefore, to evenly planarize features across the whole wafer, it is crucial to have a uniform material removal rate across the wafer. This removal rate uniformity ensures that the entire wafer is uniformly reduced in height. The wafer is held on a rotating carrier facedown and is pressed against a polishing pad attached to a rotating disk.

A multilevel input–output control system of the CMP process is shown in Figure 3.5. The primary inputs to the CMP process are (1) rotational speeds of the pad and wafer (both constant), (2) load pressure

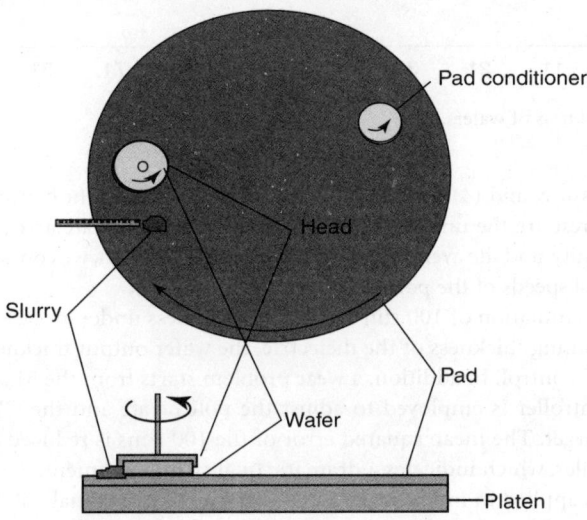

FIGURE 3.4 Schematic of a CMP system.

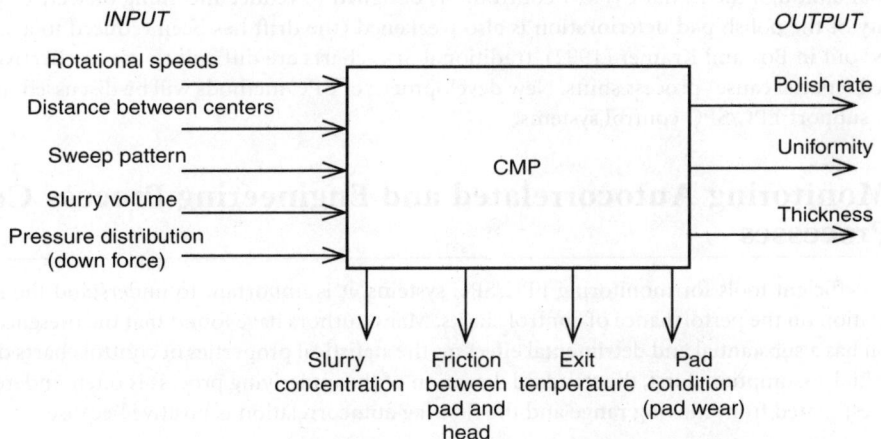

FIGURE 3.5 CMP process inputs and outputs.

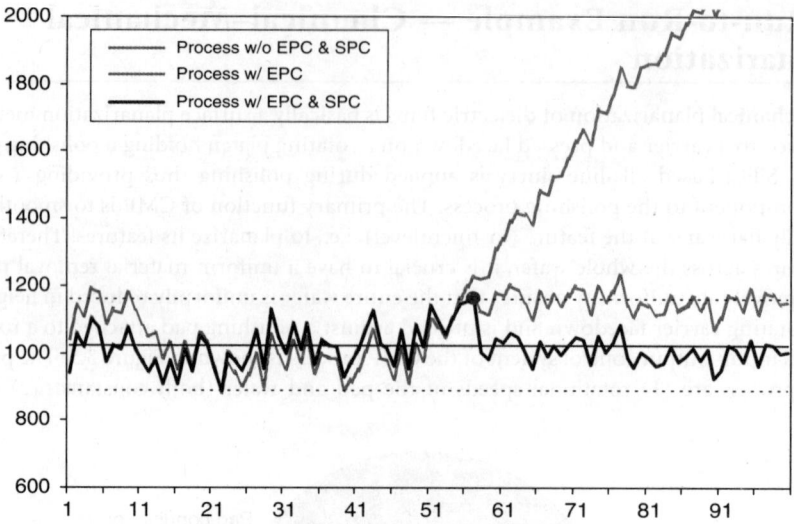

FIGURE 3.6   Output thickness of wafers.

magnitude, (3) ring pressure, and (4) pad conditioning (friction coefficient between pad and wafer). The primary outputs of interest are the uniformity of the material removal rate across the wafer as measured within wafer nonuniformity, and the average removal rate. For illustration, we consider control output wafer thickness using rotational speeds of the polish pad.

Figure 3.6 presents a simulation of 100 runs of output thickness under an R2R control system. Owing to variations of the incoming thickness of the dielectric, the wafer output thickness may drift away from target without EPC/SPC control. In addition, a wear problem starts from the 51st run on the polish pad. Now an EWMA (I) controller is employed to adjust the polish rate and the CMP output thickness is found much closer to target. The mean squared error of the 100 runs is reduced from 304,763 to 10,091 with the EWMA controller, which indicates a dramatic quality improvement.

If a Shewhart chart is applied to monitor the EPC–CMP process, a signal will be triggered at the 57th run and the polish rate model can be updated to take into consideration the polish pad deterioration. The mean squared error is then further reduced to 3,177, showing the effectiveness of SPC methods in improving product quality.

Note that although the initial EWMA controller is designed to reduce incoming dielectric variations, the severity of the polish pad deterioration is also weakened (the drift has been reduced to a step shift). As pointed out in Box and Kramer (1992), traditional SPC charts are difficult to use to effectively detect the masked (special cause) process shifts. New development of SPC methods will be discussed in the next section to support EPC/SPC control systems.

3.5 Monitoring Autocorrelated and Engineering Process Control Processes

To develop efficient tools for monitoring EPC/SPC systems, it is important to understand the impact of autocorrelation on the performance of control charts. Many authors have found that the presence of autocorrelation has a substantial and detrimental effect on the statistical properties of control charts developed under the i.i.d. assumption. First, the standard deviation of the underlying process is often underestimated when it is estimated from moving range and the first-lag autocorrelation is positive because

$$E(\hat{\sigma}_{MR}) = E(\overline{MR}/d_2) = \sigma\sqrt{1-\rho_1} \tag{3.4}$$

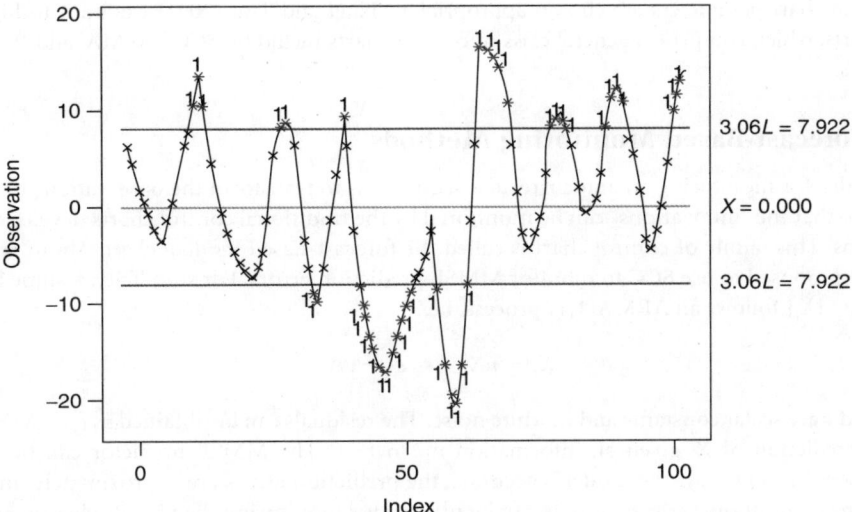

FIGURE 3.7 Mechanical vibratory displacement.

where ρ_1 is the first-lag correlation coefficient of the underlying process (Cryer and Ryan, 1990). For example, the process (see Figure 3.7) taken from Pandit and Wu (1983, p. 291) is highly autocorrelated, with $\rho_1 = 0.90$. The moving range estimate of the standard deviation is only 0.32σ, which results in a higher frequency of alarm signals than that expected from usual Shewhart charts.

Second, because of the systematic nonrandom patterns of the autocorrelated data, it becomes difficult either to recognize a state of statistical control or to identify departures from the in-control state. Alwan and Roberts (1988) point out that the individual X chart based on the assumption of i.i.d. observations can be misleading if they are actually autocorrelated. Maragah and Woodall (1992) quantify the effect of autocorrelation on the retrospective X chart with and without supplementary rules. Therefore, to accommodate autocorrelations among observations, development of new control charts has received considerable attention in the last decade.

3.5.1 Modifications of Traditional Methods

One common SPC strategy for monitoring autocorrelated processes is to modify the control limits of traditional charts and then to apply the modified charts to the original autocorrelated data. Vasilopoulos and Stamboulis (1978) provide an adjustment of control limits for Shewhart charts when monitoring autocorrelated processes. Johnson and Bagshaw (1974) and Bagshaw and Johnson (1975) provide the factor to adjust the critical boundary of CUSUM charts to correct the test procedure in the presence of correlation.

The out-of-control performance of these adjustments has been investigated recently. Yashchin (1993) shows that the CUSUM chart after adjustments can be seriously affected by mild autocorrelations. Zhang (1998) studies the EWMA chart for stationary processes. Jiang et al. (2000) extend the EWMA chart to a general class of control charts based on the autoregressive moving average transformation (ARMA), the ARMA charts. The monitoring statistic of an ARMA chart is defined to be the result of a *generalized* ARMA(1,1) process applied to the underlying process $\{X_t\}$, i.e.,

$$Z_t = \theta_0 X_t - \theta X_{t-1} + \phi Z_{t-1} = \sum_{i=0}^{t-1} w_i X_{t-i}$$

where $w_0 = \theta_0$, $w_i = \theta_0(\phi - \beta)\phi^{i-1}\theta_0(i \geq 1)$ and $\beta = \theta/\theta_0$. θ_0 is chosen so that the sum of all coefficients w_i is unity when $t \to \infty$, i.e., $\theta_0 = 1 + \theta - \phi$. The authors show that these charts can yield good performance when certain chart parameters are chosen appropriately. Jiang and Tsui (2001) extend it to higher order ARMA charts, which comprise a general class of control charts including SCC, EWMA, and PID charts as special cases.

3.5.2 Forecast-Based Monitoring Methods

A natural idea for monitoring an autocorrelated sequence is to transform the observations to an i.i.d. or near i.i.d. so that the "innovations" can be monitored by the traditional control charts developed for i.i.d. observations. This family of control chart is called the forecast-based *residual chart*. Alwan and Roberts (1988) first proposed to use SCC to monitor MMSE prediction errors. For simplicity, assume the underlying process $\{X_t\}$ follows an ARMA(1,1) process, i.e.,

$$X_t - uX_{t-1} = \varepsilon_t - v\varepsilon_{t-1}$$

where u and v are scalar constants and ε_t white noise. The residuals can be obtained as $e_t = X_t - \hat{X}_t$ where $\hat{X}_t$ is the prediction of X_t given all information up to $t-1$. The MMSE predictor can be written as $\hat{X}_t = v\hat{X}_{t-1} + (u-v)X_{t-1}$. If the model is accurate, the prediction errors are approximately uncorrelated and then any conventional control charts can be utilized for monitoring the i.i.d. prediction errors.

The SCC method has attracted considerable attention and has been further studied by many authors. Wardell et al. (1992, 1994) derive the run-length distribution of the Shewhart chart as applied to the residuals of an ARMA process; Vander Wiel (1996) studied the performance of SCCs for integrated moving average IMA(0,1,1) models. In general, monitoring the i.i.d. residuals gives SCC charts the advantage that the control limits can be easily determined by means of traditional control charts such as the Shewhart chart, the EWMA chart, and the CUSUM chart. Another advantage of the SCC chart is that its performance can be analytically approximated.

The EWMA predictor is another alternative proposed by Montgomery and Mastrangelo (1991, M–M chart). Jiang et al. (2002) further generalized the use of PID predictors with subsequent monitoring of the prediction errors, i.e.,

$$\hat{X}_{t+1} = \hat{X}_t + \lambda_1 e_t + \lambda_2 e_{t-1} + \lambda_3 e_{t-2} \tag{3.5}$$

where $e_t = X_t - \hat{X}_t$, $\lambda_1 = k_P + k_I + k_D$, $\lambda_2 = -(k_P + 2k_D)$, and $\lambda_3 = k_D$. The PID-based charts monitor e_t and include the SCC, EWMA, and M–M charts as special cases. Jiang et al. (2002) shows that the predictors of the EWMA chart and M–M chart may sometimes be inefficient and the SCC may be too sensitive to model deviation. On the other hand, the performance of the PID-based chart can be predicted via chart parameters through measures of two "capability indices." As a result, for any given underlying process, one can tune the parameters of the PID-based chart to optimize its performance.

3.5.3 Generalized Likelihood Ratio Test-Based Multivariate Methods

Forecast-based residual methods involve only a single testing statistic and often suffer from the problem of a narrow "window of opportunity" when the underlying process is positively correlated (Vander Wiel, 1996). For example, for monitoring an AR(1) process with $\rho_1 = 0.9$, a shift with size $\delta = 1$ will decrease to 0.1 from the second run after the shift occurrence owing to forecast recovery. If an SCC misses the detection in the first place, it will become very difficult to signal since the mean deviation shrinks to only 10% of the original size.

If the shift occurrence time is known, the "window of opportunity" problem is expected to be alleviated by including more historical observations/residuals in the statistical hypothesis test. For the above AR(1) example, if a mean shift is suspected to take place at $t-1$, then residuals at both time t and $t-1$ can be used to obtain a likelihood ratio (LR) test for the hypothesis instead of e_t only, i.e., the test statistic is $(0.1e_t + e_{t-1})/\sqrt{1.01}$. If the hypothesis is true, this LR test has a signal-to-noise (SN) ratio of $1.1/\sqrt{1.01} \approx 1.09$ and consequently is more powerful than e_t whose SN ratio is 1.

A generalized likelihood ratio test (GLRT) procedure can be obtained to test multiple shift locations (Vander Wiel, 1996; Apley and Shi, 1999). Assuming that the residual signature is $\{\delta_i\}$ $(t \geq 0)$ when a shift occurs, a GLRT based on residuals with window p is

$$\lambda_R = \max_{0 \leq k \leq p-1} \left| \sum_{i=0}^{k} \delta_i e_{t-k+i} \Big/ \sqrt{\sum_{i=0}^{k} \delta_i^2} \right|$$

This GLRT statistic, called *residual* GLRT, has been shown to be very effective for detecting mean shifts if p is sufficiently large. However, Apley and Shi (1999) indicate that it strongly depends on the accuracy of signature. If a shift is not detected in the window, the signature applied in λ_R might no longer be valid and the test statistic is no longer efficient. Consequently, this GLRT procedure is insensitive for detecting small shifts since they are very likely to be missed in the window.

Jiang (2005a) derives a GLRT based on the original observations for different change point locations. Consider a p-variate random vector transformed from the univariate autocorrelated observations, $Y_t = (X_{t-p+1}, X_{t-p+2}, \ldots, X_t)'$. A step shift that occurred at time $t - k + 1$ has a signature $d_k = (0,\ldots,0,\overset{k}{\overbrace{1,\ldots,1}})'$ $(1 \leq k \leq p)$ and $d_k = (1,1,\ldots,1)'$ $(k > p)$. The GLRT procedure (called *observational* GLRT) for testing these signatures is

$$\lambda_0 = \max_{1 \leq k \leq p} \left| d_k' \Sigma^{-1} Y_t \Big/ \sqrt{d_k' \Sigma^{-1} d_k} \right| \tag{3.6}$$

where Σ is the covariance matrix of Y_t. It is important to note that unlike the residual GLRT chart, one of d_k's always matches the true signature of Y_t regardless of the change point time. This grants a higher efficiency of the observational GLRT chart than the residual GLRT chart no matter how wide the window is. More importantly, the observational GLRT chart is essentially model free while the residual GLRT chart is model-based. When other shift patterns present, a multivariate T^2 chart can be developed on the basis of $T_t^2 = Y_t' \Sigma^{-1} Y_t$ (Apley and Tsung, 2002).

3.5.4 Batch Means Monitoring Methods

Other than the forecast-based residual methods, other methods, such as batch means techniques, can also be used in simulation research. Runger and Willemain (1995) propose to use a weighted batch mean (WBM) for monitoring autocorrelated observations, where the weights depend on the underlying process model. They show that the SCC chart is a special case of the WBM method and monitoring the batch means can be more efficient in detecting small process mean shifts than the forecast-based charts. They also consider classical/unweighted batch means where a batch size is determined based on a simple iterative procedure. Using the method of WBM requires knowledge of the underlying process model, whereas the classical method is "model-free." Their study suggested that the classical method yields in-control ARLs that are comparable to those of i.i.d. processes and is more sensitive than the spaced-batch method in detecting small-mean shifts.

As discussed in Alexopoulos et al. (2004), batch means methods have two different applications in SPC monitoring: estimation of the variance of the monitoring statistics of some commonly used SPC charts, and creation of new monitoring statistics based on batch means. To obtain accurate and precise estimators for the variance of batch means, they propose to use nonoverlapping batch means (NBM), overlapping batch means(OBM), and standardized time series (STS) methods well developed in simulation research.

3.5.5 Monitoring Integrated Engineering Process Control/Statistical Process Control Systems

Control charts developed for monitoring autocorrelated observations shed light on monitoring integrated EPC/SPC systems. As shown in Figure 3.2, the essential idea behind the forecast-based residual charts is mathematically similar to the pure EPC control strategy when the same forecasting scheme is used. In particular, monitoring the output of an MMSE controlled process has the same performance as the corresponding SCC charts. Similarly, the residual chart is equivalent to the associated monitoring component of the EPC/SPC system.

Similar to the forecast-based methods, assignable causes have an effect that is always contaminated by the EPC control action and result in a small "window of opportunity" for detection (Vander Wiel, 1996; Box et al., 1997). As an alternative, some authors suggest that monitoring the EPC action may improve the chance of detection (Box and Kramer, 1992; Tsung and Shi, 1998; Capilla et al., 1999). Kourti et al. (1995) propose a method of monitoring process outputs conditional on the inputs or other changing process parameters. Jiang and Tsui (2002) and Tsung and Tsui (2003) demonstrate that monitoring the control action may be more efficient than monitoring the output of the EPC/SPC system for some autocorrelated processes and vice versa for others.

To integrate the information provided by process inputs and outputs, Tsung et al. (1999) developed multivariate techniques based on Hotelling's T^2 chart and Bonferroni's approach. Denoting the multivariate vector by $Z_t = (e_t, X_{t-1})'$, which has covariance matrix Σ_Z, the T^2 chart monitor statistic

$$T_t^2 = Z_t \Sigma_Z^{-1} Z_t$$

When mean shift patterns are known, similar to the GLRT procedures for monitoring autocorrelated processes, more efficient monitoring statistics can be developed following the available signatures (Jiang, 2005b).

3.5.6   Design of Statistical Process Control Methods: Efficiency vs. Robustness

Although EPC and SPC techniques share the same objective of reducing process variations and many similarities in implementation, the criterion for selecting SPC monitoring charts is fundamentally different from corresponding EPC. For example, instead of minimizing the mean squared error/prediction error of a PID controller, maximization of the chance of detecting shifts is always desirable when designing a PID chart. Therefore, SN ratios developed in Jiang et al. (2000) have to be used and an ad hoc procedure is proposed for designing appropriate charts.

Taking the PID chart , shown in Figure 3.8 as an example, two SN ratios are crucial to the statistical performance of a PID chart. The standard deviation of charting statistic Z_t, is denoted by σ_Z and μ_T $(/\mu_S)$ denotes the shift levels of Z_t at the first step (long enough) after the shift happens. The transient-state ratio is defined by $C_T = \mu_T/\sigma_Z$, which measures the capability of the control chart to detect a shift in its first few steps. The steady-state ratio is defined by $C_S = \mu_S/\sigma_Z$, which measures the capability of the control chart to detect a shift in its steady state. By selecting control chart parameters, these two ratios can be manipulated in the desired way so that the chance of detection is maximized.

In general, if the transient ratio can be tuned to a value high enough (say 4 to 5) by choosing appropriate PID parameters, the corresponding PID chart will be able to detect the shift quickly. On the other hand, if this ratio is smaller than three, the shift will likely be missed at the transient state and needs to be detected in the later runs. In this case, the steady-state ratio becomes more important for detecting the shift efficiently at the steady state. Although a high steady-state ratio is helpful in detecting the shift in the steady state, it may result in an extremely small transient ratio and make the transition of the shifts from the transient state to the steady state very slow. To enable the chart to detect the shift efficiently in the steady state, a balance is needed, i.e., there must be a trade-off between the transient ratio and the steady-state ratio when choosing the charting parameters. Generally, Jiang et al. (2000) recommend the appropriate selection of chart parameter values to achieve C_S around three for balancing the values of C_T and C_S. This heuristic algorithm is also helpful in designing other types of SPC charts for autocorrelated or EPC processes, e.g., the EWMA and ARMA charts.

One of the obstacles that prohibit the usage of SPC methods in monitoring autocorrelated or EPC processes is the robustness of a control chart. It is defined by how its ARL changes when the process model is specified. Since residuals are no longer i.i.d., reliable estimates of process variations should be

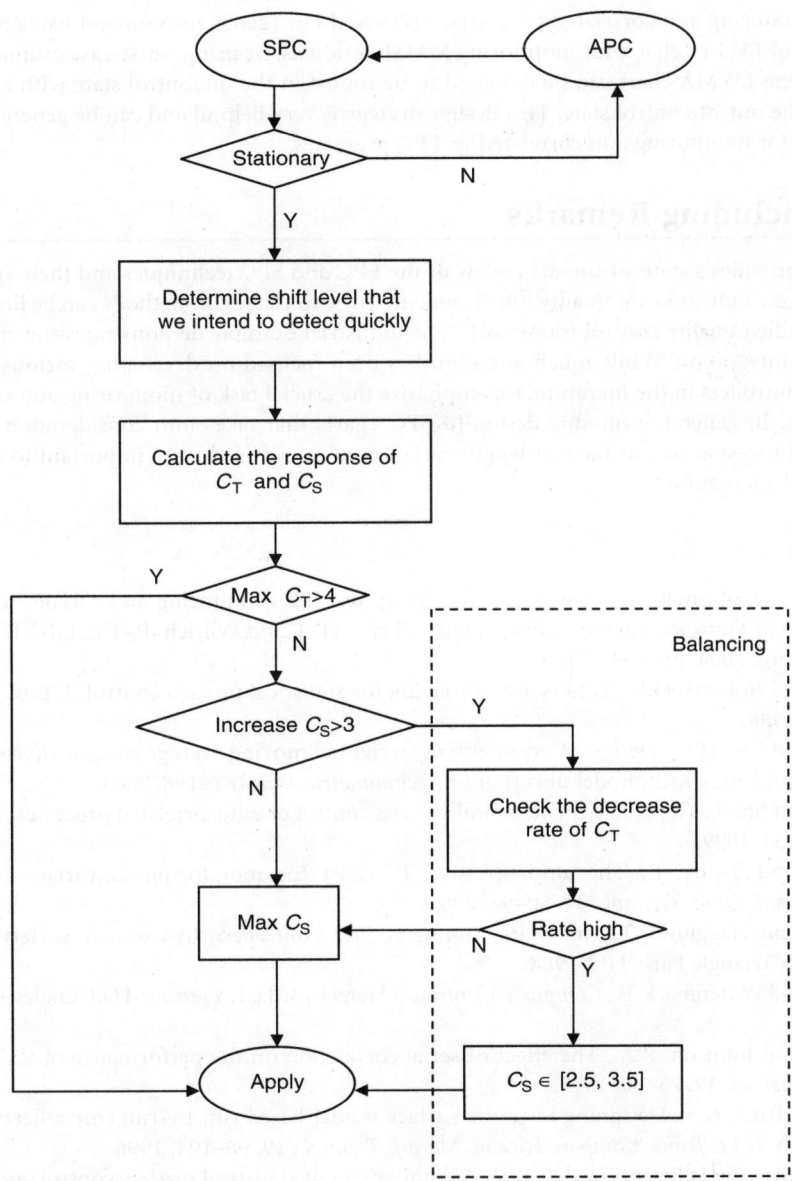

FIGURE 3.8 Design of PID charts.

used (Boyles, 2000; Alexopoulos et al., 2004). Moreover, even though a robust estimator of standard deviations can be obtained, a more sensitive control chart, such as a PID chart may still be less robust compared with less-sensitive control charts such as MMSE-based SCC charts. For example, Tsung et al. (1998) and Luceno (1998) conclude that PID controllers are generally more robust than MMSE controllers against model specification errors. However, Jiang et al. (2002) show that PID charts tend to have a shorter "in-control" ARL when the process model is misspecified, since model errors can be viewed as a kind of "shift/deviation" from the "true" process model.

The nonrobustness of sensitive control charts seems to discourage development of more efficient control charts, and a trade off between sensitivity and robustness becomes necessary when selecting control charts for monitoring autocorrelated processes. Apley and Lee (2003) recommend using a conservative control limit for EWMA charts for monitoring MMSE residuals. By using worst-case estimation of residual variance, the EWMA chart can be designed to be robust in the in-control state with a slight loss of efficiency in the out-of-control state. This design strategy is very helpful and can be generalized to other SPC methods for monitoring autocorrelated or EPC processes.

3.6 Concluding Remarks

This chapter provides a state-of-the-art review of the EPC and SPC techniques and their applications in parts and process industries for quality improvement. The two classes of methods can be linked and integrated in a unified quality control framework. The industrial example demonstrates the effectiveness of the EPC/SPC integration. While much attention has been focused on developing various efficient and robust EPC controllers in the literature, we emphasize the crucial task of monitoring autocorrelated and EPC processes. In general, economic design of SPC charts that takes into consideration the dynamic nature of the process as well as the run-length variation of a control chart is important to the success of EPC/SPC implementation.

References

Alexopoulos, C., Goldsman, D., Tsui, K.-L., and Jiang, W., SPC monitoring and variance estimation, in *Frontiers in Statistical Quality Control*, Vol. 7, Lenz, H.-J. and Wilrich, P.-Th., Eds., Physica-Verlag, Heidelberg, 2004, pp. 194–210. ·

Alwan, L.C. and Roberts, H.V., Time-series modeling for statistical process control, *J. Bus. Econ. Stat.*, 6, 87–95, 1988.

Apley, D.W. and Lee, H.C., Design of exponentially weighted moving average control charts for autocorrelated processes with model uncertainty, *Technometrics*, 45, 187–198, 2003.

Apley, D.W. and Shi, J., The GLRT for statistical process control of autocorrelated processes, *IIE Trans.*, 31, 1123–1134, 1999.

Apley, D.W. and Tsung, F., The autoregressive T^2 chart for monitoring univariate autocorrelated processes, *J. Qual. Technol.*, 34, 80–96, 2002.

Astrom, K.J. and Hagglund, T., *Automatic Tuning of PID Controllers*, Instrument Society of America, Research Triangle Park, NC, 1988.

Astrom, K. J. and Wittenmark, B., *Computer-Controlled Systems*, 3rd ed., Prentice-Hall, Englewood Cliffs, NJ, 1997.

Bagshaw, M. and Johnson, R.A., The effect of serial correlation on the performance of CUSUM Test II, *Technometrics*, 17, 73–80, 1975.

Baras, J.S. and Patel, N.S., Designing response surface model-based run-by-run controllers: a worst case approach, *IEEE Trans. Compon. Packag. Manuf. Technol.*, 19, 98–104, 1996.

Box, G.E.P., Coleman, D.E., and Baxley, R.V., A comparison of statistical process control and engineering process control, *J. Qual. Technol.*, 29, 128–130, 1997.

Box, G.E.P., Jenkins, G.M., and Reinsel, G.C., *Time Series Analysis Forecasting and Control*, 3rd ed., Prentice-Hall, Englewood Cliffs, NJ, 1994.

Box, G.E.P. and Kramer, T., Statistical process monitoring and feedback adjustment — A discussion. *Technometrics*, 34, 251–285, 1992.

Box, G. E. P. and Luceno, A., *Statistical Control by Monitoring and Feedback Adjustment*, Wiley, New York, 1997.

Boyles, R.A., Phase I analysis for autocorrelated processes,. *J. Qual. Technol.*, 32(4), 395–409, 2000.

Butler, S.W. and Stefani, J.A., Supervisory run-to-run control of polysilicon gate etch using in situ ellipsometry, *IEEE Trans. Semicond. Manuf.*, 7, 193–201, 1994.

Capilla, C., Ferrer, A., Romero, R., and Hualda, A., Integration of statistical and engineering process control in a continuous polymerization process, *Technometrics*, 41, 14–28, 1999.

Chen, A. and Elsayed, E.A., An alternative mean estimator for processes monitored by SPC charts, *Int. J. Prod. Res.*, 38(13), 3093–3109, 2000.

Cox, D.R., Prediction by exponentially weighted moving average and related methods, *J. R. Stat. Soc. Ser. B*, 23, 414–442, 1961.

Cryer, J.D. and Ryan, T.P., The estimation of sigma for an X chart: MR/d_2 or S/d_4?, *J. Qual. Technol.*, 22, 187–192, 1990.

Del Castillo, E., Long run and transient analysis of a double EWMA feedback controller, *IIE Trans.*, 31, 1157–1169, 1999.

Del Castillo, E. and Yeh, J.Y., An adaptive run-to-run optimizing controller for linear and nonlinear semiconductor processes, *IEEE Trans. Semicond. Manuf.*, 11, 285–295, 1998.

Deming, W.E., *Quality, Productivity and Competitive Position*, MIT Center for Advanced Engineering Study, Cambridge, MA, 1982.

Deming, W.E., *Out of Crisis*, MIT Center for Advanced Engineering Study, Cambridge, MA, 1986.

Ingolfsson, A. and Sachs, E., Stability and sensitivity of an EWMA controller, *J. Qual. Technol.*, 25, 271–287, 1993.

Jiang, W., Multivariate control charts for monitoring autocorrelated processes, *J. Qual. Technol.*, 36, 367–379, 2005a.

Jiang, W., A joint SPC monitoring scheme for APC-controlled processes, *IIE Trans. Qual. Reliab.*, 1201–1210, 2005b.

Jiang, W. and Tsui, K.-L., An economic model for integrated APC and SPC control charts, *IIE Trans. Qual. Reliab.*, 32, 505–513, 2000.

Jiang, W. and Tsui, K.-L., Some properties of ARMA charts for time series, *Nonlinear Anal. Theor. Method. Appl.*, 47(3), 2073–2088, 2001.

Jiang, W. and Tsui, K.-L., SPC monitoring of MMSE- and PI-controlled processes, *J. Qual. Technol.*, 34(4), 384–398, 2002.

Jiang, W., Tsui, K.-L., and Woodall, W.H., A new SPC monitoring method: the ARMA chart, *Technometrics*, 42, 399–410, 2000.

Jiang, W., Wu, H., Tsung, F., Nair, V., and Tsui, K.-L., Proportional integral derivative charts for process monitoring, *Technometrics*, 44, 205–214, 2002.

Johnson, R.A. and Bagshaw, M., The effect of serial correlation on the performance of CUSUM test, *Technometrics*, 16, 103–112, 1974.

Kourti, T., Nomikos, P., and MacGregor, J.F., Analysis, monitoring and fault diagnosis of batch processes using multiblock and multiway PLS, *J. Process Control*, 5(4), 277–284, 1995.

Lucas, J.M. and Saccucci, M.S., Exponentially weighted moving average control schemes: properties and enhancements, *Technometrics*, 32, 1–12, 1990.

Luceno, A., Performance of discrete feedback adjustment schemes with dead band, under stationary versus nonstationary stochastic disturbance, *Technometrics*, 27, 223–233, 1998.

Mandel, B.J., The regression control chart, *J. Qual. Technol.*, 1, 1–9, 1969.

Maragah, H.D. and Woodall, W.H., The effect of autocorrelation on the retrospective X-chart, *J. Stat. Comput. Simul.*, 40, 29–42, 1992.

Montgomery, D.C., Keats, J.B., Runger, G.C., and Messina, W.S., Integrating statistical process control and engineering process control, *J. Qual. Technol.*, 26, 79–87, 1994.

Montgomery, D.C. and Mastrangelo, C.M., Some statistical process control methods for autocorrelated data, *J. Qual. Technol.*, 23, 179–204, 1991.

Moyne, J., Etemad, H., and Elta, M., Run-to-run control framework for VLSI manufacturing, *Microelectronic Processing '93 Conference Proceedings*, 1993.

Page, E.S., Continuous inspection schemes, *Biometrika*, 41, 100–115, 1954.

Palmer, E., Ren, W., and Spanos, C.J., Control of photoresist properties: a Kalman filter based approach, *IEEE Trans. Semicond. Manuf.*, 9, 208–214, 1996.

Pandit, S.M. and Wu, S.M., *Times Series and System Analysis, with Applications*, Wiley, New York, 1983.

Pan, R. and Del Castillo, E., Integration of sequential process adjustment and process monitoring techniques, *Qual. Reliab. Eng. Int.*, 19(4), 371–386, 2003.

Rashap, B., Elta, M., Etemad, H., Freudenberg, J., Fournier, J., Giles, M., Grizzle, J., Kabamba, P., Khargonekar, P., Lafortune, S., Moyne, J., Teneketzis, D., and Terry, F. Jr., Control of semiconductor manufacturing equipment: real-time feedback control of a reactive ion etcher, *IEEE Trans. Semicond. Manuf.*, 8, 286–297, 1995.

Roberts, S.W., Control chart tests based on geometric moving averages, *Technometrics*, 1, 239–250, 1959.

Ruegsegger, S., Wagner, A., Freudenberg, J.S., and Grimard, D.S., Feedforward control for reduced run-to-run variation in microelectronics manufacturing, *IEEE Trans. Semicond. Manuf.*, 12, 493–502, 1999.

Runger, G.C. and Willemain, T.R., Model-based and model-free control of autocorrelated processes, *J. Qual. Technol.*, 27, 283–292, 1995.

Sachs, E., Hu, A., and Ingolfsson, A., Run by run process control: combining SPC and feedback control, *IEEE Trans. Semicond. Manuf.*, 8, 26–43, 1995.

Seborg, D.E., Edgar, T.F., and Mellichamp, D.A., *Process Dynamics and Control*, Wiley, New York, 1989.

Shewhart, W.A., *Economic Control of Quality of Manufactured Product*, Van Nostrand, New York, 1931.

Smith, T.H. and Boning, D.S., Artificial neural network exponentially weighted moving average controller for semiconductor processes, *J. Vac. Sci. Technol. Ser. A*, 15, 236–239, 1997.

Tsung, F. and Shi, J., Integrated design of run-to-run PID controller and SPC monitoring for process disturbance rejection, *IIE Trans.*, 31, 517–527, 1998.

Tsung, F., Shi, J., and Wu, C.F.J., Joint monitoring of PID controlled processes, *J. Qual. Technol.*, 31, 275–285, 1999.

Tsung, F. and Tsui, K.-L., A study on integration of SPC and APC for process monitoring, *IIE Trans.*, 35, 231–242, 2003.

Tsung, F., Wu, H., and Nair, V.N., On efficiency and robustness of discrete proportional-integral control schemes, *Technometrics*, 40, 214–222, 1998.

Vander Wiel, S.A., Monitoring processes that wander using integrated moving average models, *Technometrics*, 38, 139–151, 1996.

Vander Wiel, S.A., Tucker, W.T., Faltin, F.W., and Doganaksoy, N., Algorithmic statistical process control: concepts and application, *Technometrics*, 34, 278–281, 1992.

Vasilopoulos, A.V. and Stamboulis, A.P., Modification of control chart limits in the presence of data correlation, *J. Qual. Technol.*, 10, 20–30, 1978.

Wardell, D.G., Moskowitz, H., and Plante, R.D., Control charts in the presence of data correlation, *Manage. Sci.*, 38, 1084–1105, 1992.

Wardell, D.G., Moskowitz, H., and Plante, R.D., Run-length distributions of special-cause control charts for correlated observations, *Technometrics*, 36, 3–17, 1994.

Yashchin, E., Performance of CUSUM control schemes for serially correlated observations, *Technometrics*, 35, 37–52, 1993.

Yashchin, E., Estimating the current mean of a process subject to abrupt changes, *Technometrics*, 37, 311–323, 1995.

Zhang, G.X., A new type of control chart and a theory of diagnosis with control charts, *World Quality Congress Transactions*, American Society for Quality Control, London, 1984, pp. 175–185.

Zhang, N.F., A statistical control chart for stationary process data, *Technometrics*, 40, 24–38, 1998.

Ziegler, J.G. and Nichols, N.B., Optimum settings for automatic controllers, *ASME Trans.*, 64, 759–768, 1942.

4

Project Scheduling

Jonathan F. Bard
University of Texas

4.1 Introduction

A schedule is a statement of the tasks and activities to be performed over time. Project scheduling deals with the establishment of timetables and dates during which various resources, such as equipment and personnel, will be used to perform the activities required to complete the project. Schedules are the cornerstones of the planning and control system; and because of their importance, they are often written into the contract by the customer.

The act of scheduling integrates information on several aspects of a project, including an estimate of how long the activities will last, the technological precedence relationships among various activities, the constraints imposed by the budget and the availability of resources and, if applicable, deadlines. This information is processed into an acceptable schedule with the help of a decision support system that may include network models, a resource database, cost-estimating relationships, and options for accelerating performance. The aim is to answer the following questions:

1. If each activity goes according to plan, when will the project be completed?
2. Which tasks are most critical to ensure the timely completion of the project?
3. Which tasks can be delayed, if necessary, without delaying project completion, and by how much?
4. More specifically, at what times should each activity begin and end?
5. At any given time during the project, how much money should have been spent?
6. Is it worthwhile to incur extra costs to accelerate some of the activities? If so, which ones?

The first four questions relate to time, which is the chief concern of this chapter; the last two deal with the possibility of trading off time for money and are taken up in another chapter in this handbook chapters 5 and 6.

A common way to present the schedule is as a Gantt chart, which is essentially a bar chart that shows the relationships between activities and milestones over time. Different schedules can be prepared for the various participants in the project. A functional manager may be interested in a schedule of tasks performed by members of his or her group. The project manager may need a detailed schedule for each work breakdown structure (WBS) element and a master schedule for the entire project. The vice president of finance may need a combined schedule for all projects that are under way in the organization in order to plan cash flows and capital requirements. Each person involved in the project may need a schedule with all of the activities in which he or she is involved.

Schedules provide an essential communication and coordination link between the individuals and organizations participating in the project. They facilitate coordination among people coming from different organizations and working on different elements of the project in different locations at different times. By developing a schedule, the project manager is *planning* the project. By authorizing work to start on each task according to the schedule, the project manager triggers *execution* of the project; and by comparing the actual execution dates of tasks with the scheduled dates, he or she *monitors* the project. When actual performance deviates from the plan to such an extent that corrective action must be taken, the project manager is exercising *control*.

Although schedules come in many forms and levels of detail, they should all relate to the master schedule, which gives a time-phased picture of the principal activities and highlights the major milestones associated with the project. For large programs, a modular approach is recommended in order to reduce the danger of getting bogged down in excessive detail. To implement this approach, the schedule should be partitioned according to its functions or phases and then disaggregated to reflect the various work packages (WPs). For example, consider the WBS shown in Figure 4.1 for the development of a microcomputer. One possible modular array of project schedules is depicted in Figure 4.2. The details of each module would have to be worked out by the individual project leaders and then integrated by the project manager to gain the full perspective.

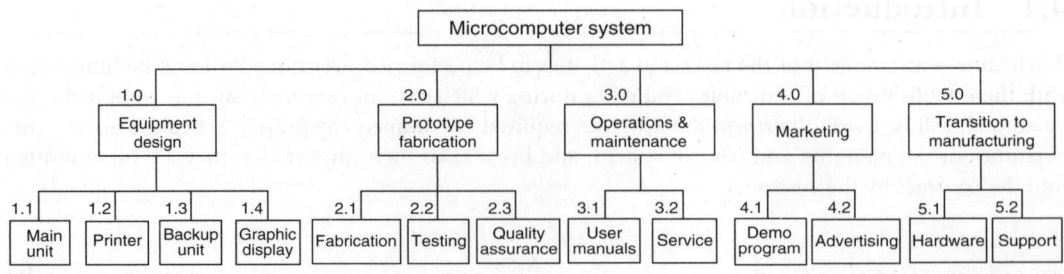

FIGURE 4.1   WBS for a microcomputer.

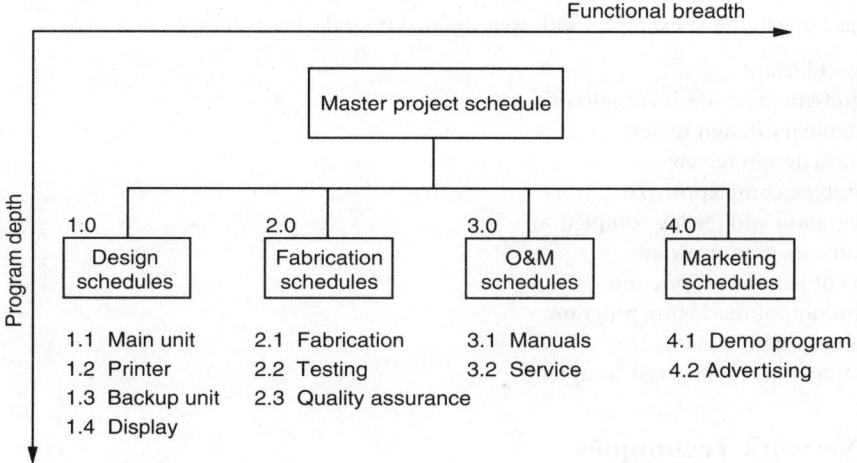

FIGURE 4.2 Modular array of project schedules.

Schedules are working tools for program planning, evaluation, and control. They are developed over many iterations with project team members and with continuous feedback from the client. The reality of changing circumstances requires that they remain dynamic throughout the project life cycle. When preparing the schedule, it is important that the dates and time allotments for the WPs be in precise agreement with those set forth in the master schedule. These times are control points for the project manager. It is his or her responsibility to insist on and maintain consistency, but the actual scheduling of tasks and WPs is usually done by those who are responsible for their accomplishment — after the project manager has approved the due dates. This procedure assures that the final schedule reflects the interdependencies among all of the tasks and participating units and that it is consistent with available resources and upper-management expectations.

It is worth noting that the most comprehensive schedule is not necessarily best in all situations. In fact, too much detail can impede communications and divert attention from critical activities. Nevertheless, the quality of a schedule has a major impact on the success of the project and frequently affects other projects that compete for the same resources.

4.1.1 Key Milestones

A place to begin the development of any schedule is to define the major milestones for the work to be accomplished. For ease of viewing, it is often convenient to array this information on a time line showing events and their due dates. Once agreed upon, the resulting milestone chart becomes the skeleton for the master schedule and its disaggregated components.

A key milestone is any important event in the project life cycle and may include, for instance, the fabrication of a prototype, the start of a new phase, a status review, a test, or the first shipment. Ideally, the completion of these milestones should be easily verifiable, but in reality, this may not be the case. Design, testing, and review tend to run together. There is always a desire to do a bit more work to correct superficial flaws or to get a marginal improvement in performance. This blurs the demarcation points and makes project control much more difficult.

Key milestones should be defined for all major phases of the project before start-up. Care must be taken to arrive at an appropriate level of detail. If the milestones are spread too far apart, continuity problems in tracking and control can arise. Conversely, too many milestones can result in unnecessary busy-work, overcontrol, confusion, and increased overhead costs. As a guideline for long-term projects, four key milestones per year seem to be sufficient for tracking without overburdening the system.

The project office, in close cooperation with the customer and the participating organizations, typically has the responsibility for defining key milestones. Selecting the right type and number is critical.

Every key milestone should represent a checkpoint for a collection of activities at the completion of a major project phase. Some examples with well-defined boundaries include

- Project kickoff
- Requirements analysis completion
- Preliminary design review
- Critical design review
- Prototype completion
- Integration and testing completion
- Quality assurance review
- Start of volume production
- Definition of marketing program
- First shipment
- Customer acceptance test completion

4.1.2 Network Techniques

The basic approach to all project scheduling is to form an actual or implied network that graphically portrays the relationships between the tasks and milestones in the project. Several techniques evolved in the late 1950s for organizing and representing this basic information. Best known today are the program evaluation and review technique (PERT) and the critical path method (CPM). PERT was developed by Booz, Allen, and Hamilton in conjunction with the U.S. Navy in 1958 as a tool for coordinating the activities of more than 11,000 contractors involved with the Polaris missile program. CPM was the result of a joint effort by DuPont and the UNIVAC division of Remington Rand to develop a procedure for scheduling maintenance shutdowns in chemical processing plants. The major difference between the two is that CPM assumes that activity times are deterministic whereas PERT views the time to complete an activity as a random variable that can be characterized by an optimistic, a pessimistic, and a most likely estimate of its durations. Over the years, a host of variants has arisen, mainly to address specific aspects of the tracking and control problem, such as budget fluctuations, complex intertask dependencies, and the multitude of uncertainties found in the research and development (R&D) environment.

The PERT/CPM is based on a diagram that represents the entire project as a network of arrows and nodes. The two most popular approaches are either to place the activities on the arrows (AOA) and have the nodes signify milestones, or to place activities on the nodes (AON) and let the arrows show precedence relations among activities. A precedence relation states, for example, that activity X must be completed before activity Y can begin, or that X and Y must end at the same time. It allows tasks that must precede or follow other tasks to be clearly identified, in time as well as in function. The resulting diagram can be used to identify potential scheduling difficulties to estimate the time needed to finish the entire project, and to improve coordination among the participants.

To apply PERT/CPM, a thorough understanding of the project's requirements and structure is needed. The effort spent in identifying activity relationships and constraints yields valuable insights. In particular, four questions must be answered to begin the modeling process:

1. What are the chief project activities?
2. What are the sequencing requirements or constraints for these activities?
3. Which activities can be conducted simultaneously?
4. What are the estimated time requirements for each activity?

The PERT/CPM networks are an integral component of project management and have been shown to provide the following benefits (Clark and Fujimoto, 1989; Meredith and Mantel, 1999):

1. They furnish a consistent framework for planning, scheduling, monitoring, and controlling projects.
2. They illustrate the interdependencies of all tasks, WPs, and work units.

3. They aid in setting up the proper communication channels between participating organizations and points of authority.
4. They can be used to estimate the expected project completion dates as well as the probability that the project will be completed by a specific date.
5. They identify so-called critical activities that, if delayed, will delay the completion of the entire project.
6. They also identify activities that have slack and so can be delayed for specific periods of time without penalty or from which resources may temporarily be borrowed without negative consequences.
7. They determine the dates on which tasks may be started or must be started if the project is to stay on schedule.
8. They illustrate which tasks must be coordinated to avoid resource or timing conflicts.
9. They also indicate which tasks may be run or must be run in parallel to achieve the predetermined completion date.

As we will see, PERT and CPM are easy to understand and use. Although computerized versions are available for both small and large projects, manual calculation is quite suitable for many everyday situations. Unfortunately, though, some managers have placed too much reliance on these techniques at the expense of good management practice. For example, when activities are scheduled for a designated time slot, there is a tendency to meet the schedule at all costs. This may divert resources from other activities and cause much more serious problems downstream, the effects of which may not be felt until a near catastrophe has set in. If tests are shortened or eliminated as a result of time pressure, design flaws may be discovered much later in the project. As a consequence, a project that seemed to be under control is suddenly several months behind schedule and substantially over budget. When this happens, it is convenient to blame PERT/CPM, even though the real cause is poor management.

In the remainder of this chapter, we discuss and illustrate the techniques used to estimate activity durations, to construct PERT/CPM networks, and to develop the project schedules. The focus is on the timing of activities. Issues related to resource and budget constraints as they affect the project's schedule are taken up in other chapters.

4.2 Estimating the Duration of Project Activities

A project is composed of a set of tasks. Each task is performed by one organizational unit and is part of a single WP. Most tasks can be broken down into activities, each of which is characterized by its technological specifications, drawings, lists of required materials, quality-control requirements, and so on. The technological processes selected for each activity affect the resources required, the materials needed, and the timetable. For example, to move a heavy piece of equipment from one point to another, resources such as a crane and a tractor-trailer might be called for as well as qualified operators. The time required to perform the activity may also be regarded as a resource. If the piece of equipment is mounted on a special fixture before moving, the required resources and the performance time may be affected. Thus, the schedule of the project as well as its cost and resource requirements are a function of the technological decisions.

Some activities cannot be performed unless other activities are completed beforehand. For example, if the piece of equipment to be moved is very large, then it might be necessary to disassemble it or at least remove a few of its parts before loading it onto the truck. Thus, the "moving" task has to be broken down into activities with precedence relations among them.

The processes of dividing a task into activities and dividing activities into subactivities should be performed carefully to strike a proper balance between size and duration. The following guidelines are recommended:

1. The length of each activity should be approximately in the range of 0.5 to 2% of the length of the project. Thus, if the project takes approximately 1 year, then each activity should be between a day and a week.

2. Critical activities that fall below this range should be included. For example, a critical design review that is scheduled to last 2 days on a 3-year project should be included in the activity list because of its pivotal importance.

3. If the number of activities is very large (e.g., above 250), then the project should be divided into subprojects, perhaps by functional area, and individual schedules should be developed for each. Schedules with too many activities quickly become unwieldy and are difficult to monitor and control.

Two approaches are used for estimating the length of an activity: the deterministic approach and the stochastic approach. The deterministic approach ignores uncertainty and thus results in a point estimate. The stochastic approach addresses the probabilistic elements in a project by estimating both the expected duration of each activity and its corresponding variance. Although tasks are subject to random forces and other uncertainties, the majority of project managers prefer the deterministic approach because of its simplicity and ease of understanding. A corollary benefit is that it yields satisfactory results in most instances.

4.2.1 Stochastic Approach

Only in rare circumstances is the exact duration of a planned activity known in advance. Therefore, to gain an understanding of how long it will take to perform the activity, it is logical to analyze past data and to construct a frequency distribution of related activity durations. An example of such a distribution is illustrated in Figure 4.3. From the plot, we observe that the activity under consideration was previously performed 40 times and required anywhere from 10 to 70 h. We also see that in 3 of the 40 observations, the actual duration was 45 h and that the most frequent duration was 35 h. That is, in 8 out of the 40 repetitions, the actual duration was 35 h.

The information in Figure 4.3 can be summarized by two measures: the first is associated with the center of the distribution (commonly used measures are the mean, the mode, and the median), and the second

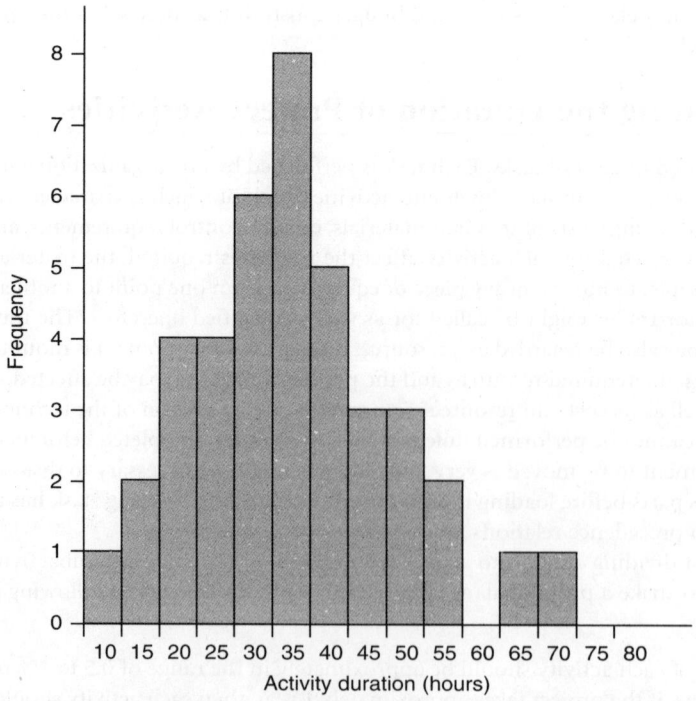

FIGURE 4.3 Frequency distribution of an activity duration.

is related to the spread of the distribution (commonly used measures are the variance, the standard deviation, and the interquartile range). The mean of the distribution in Figure 4.3 is 35.25, its mode is 35, and its median is also 35. The standard deviation is 13.3 and the variance is 176.8.

When working with empirical data, it is often desirable to fit the data with a continuous distribution that can be represented mathematically in closed form. This approach facilitates the analysis. Figure 4.4 shows the superposition of a normal distribution with the parameters $\mu = 35.25$ and $\sigma = 13.3$ on the original data.

Whereas the normal distribution is symmetrical and easy to work with, the distribution of activity durations is likely to be skewed. Furthermore, the normal distribution has a long left-hand tail, whereas actual performance time cannot be negative. A better model of the distribution of activity lengths has proved to be the beta distribution, which is illustrated in Figure 4.5.

A visual comparison between Figure 4.4 and Figure 4.5 reveals that the beta distribution provides a closer fit to the frequency data depicted in Figure 4.3. The left-hand tail of the beta distribution does not cross the zero duration point, and neither is it necessarily symmetric. Nevertheless, in practice, a statistical test (e.g., the χ^2 goodness-of-fit test or the Kolmogorov–Smirnov test; Banks et al., 2001) must be used to determine whether a theoretical distribution is a valid representation of the actual data.

In project scheduling, probabilistic considerations are incorporated by assuming that the time estimate for each activity can be derived from three different values:

a = optimistic time, which will be required if execution goes extremely well
m = most likely time, which will be required if execution is normal
b = pessimistic time, which will be required if everything goes badly

Statistically speaking, a and b are estimates of the lower and upper bounds of the frequency distribution, respectively. If the activity is repeated a large number of times, then only in $\sim$0.5% of the cases would the

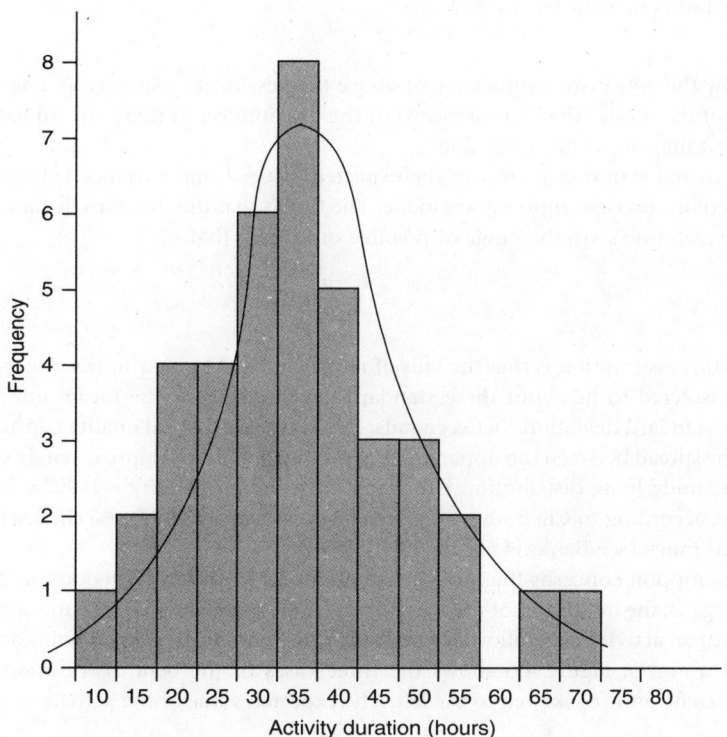

FIGURE 4.4   Normal distribution fitted to the data.

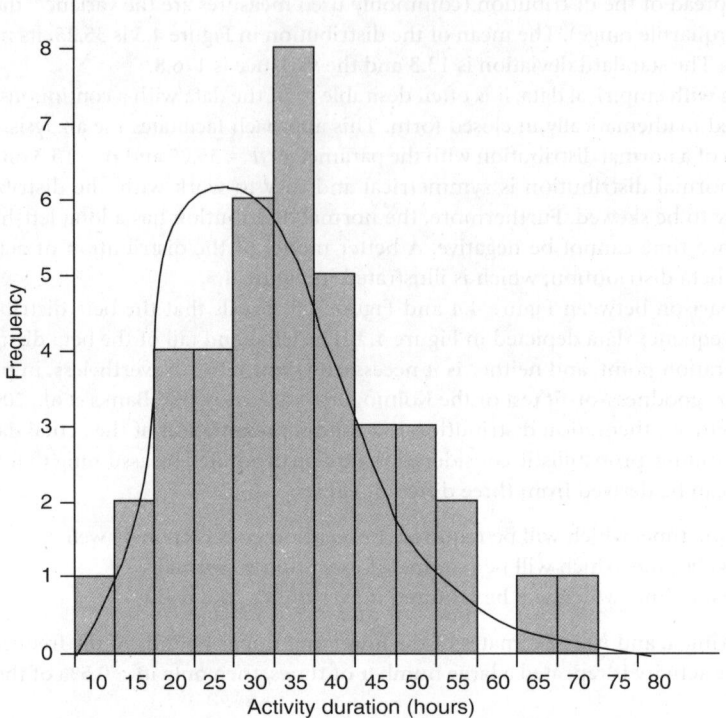

FIGURE 4.5　Beta distribution fitted to the data.

duration fall below the optimistic estimate, *a*, or above the pessimistic estimate, *b*. The most likely time, *m*, is an estimate of the mode (the highest point) of the distribution. It need not coincide with the midpoint $(a + b)/2$ but may occur on either side.

To convert *m*, *a*, and *b* into estimates of the expected value $\hat{d}$ and variance $(\hat{v})$ of the elapsed time required by the activity, two assumptions are made. The first is that the standard deviation $\hat{s}$ (square root of the variance) equals one sixth the range of possible outcomes, that is,

$$\hat{s} = \frac{b-a}{6} \qquad (4.1)$$

The rationale for this assumption is that the tails of many probability distributions (e.g., the normal distribution) are considered to lie about three standard deviations from the mean, implying a spread of approximately six standard deviations between tails. In industry, statistical quality control charts are constructed so that the spread between the upper and lower control limits is approximately six standard deviations (6σ). If the underlying distribution is normal, then the probability is 0.9973 that $\hat{d}$ falls within $b - a$. In any case, according to Chebyshev's inequality, there is at least an 89% chance that the duration will fall within this range (see Banks et al., 2001).

The second assumption concerns the form of the distribution and is needed to estimate the expected value, $\hat{d}$. In this regard, the definition of the three time estimates above provides an intuitive justification that the duration of an activity may follow a beta distribution with its unimodal point occurring at *m* and its end points at *a* and *b*. Figure 4.6 shows the three cases of the beta distribution: (a) symmetric, (b) skewed to the right, and (c) skewed to the left. The expected value of the activity duration is given by

$$\hat{d} = \frac{1}{3}\left[2m + \frac{1}{2}(a + b)\right] = \frac{a + 4m + b}{6} \qquad (4.2)$$

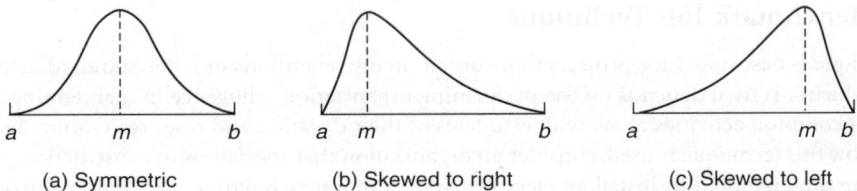

FIGURE 4.6   Three cases of the beta distribution: (a) symmetric; (b) skewed to the right; (c) skewed to the left.

Notice that $\hat{d}$ is a weighted average of the mode, m, and the midpoint $(a + b)/2$, where the former is given twice as much weight as the latter. Although the assumption of the beta distribution is an arbitrary one and its validity has been challenged from the start (Grubbs, 1962), it serves the purpose of locating $\hat{d}$ with respect to m, a, and b in what seems to be a reasonable way (Hillier and Lieberman, 2001).

The following calculations are based on the data in Figure 4.3, from which we observe that $a = 10$, $b = 70$, and $m = 35$:

$$\hat{d} = \frac{10 + (4)(35) + 70}{6} = 36.6 \quad \text{and} \quad \hat{s} = \frac{70 - 10}{6} = 10$$

Thus, assuming that the beta distribution is appropriate, the expected time to perform the activity is 36.6 h with an estimated standard deviation of 10 h.

4.2.2   Deterministic Approach

When past data for an activity similar to the one under consideration are available and the variability in performance time is negligible, the duration of the activity may be estimated by its mean; that is, the average time it took to perform the activity in the past. A problem arises when no past data exist. This problem is common in organizations that do not have an adequate information system to collect and store past data and in R&D projects in which an activity is performed for the first time. To deal with this situation, three techniques are available: the modular technique, the benchmark job technique, and the parametric technique. Each of these techniques is discussed below.

4.2.3   Modular Technique

This technique is based on decomposing each activity into subactivities (or modules), estimating the performance time of each module, and then totaling the results to get an approximate performance time for the activity. As an example, consider a project to install a new flexible manufacturing system (FMS). A training program for employees has to be developed as part of the project. The associated task can be broken down into the following activities:

1. Definition of goals for the training program
2. Study of the potential participants in the program and their qualifications
3. Detailed analysis of the FMS and its operation
4. Definition of required topics to be covered
5. Preparation of a syllabus for each topic
6. Preparation of handouts, transparencies, and so on
7. Evaluation of the proposed program (a pilot study)
8. Improvements and modifications

If possible, the time required to perform each activity is estimated directly. If not, the activity is broken into modules, and the time to perform each module is estimated on the basis of past experience. Although the new training task may not be wholly identical to previous tasks undertaken by the company, the modules themselves should be common to many training programs, so historical data may be available.

4.2.4   Benchmark Job Technique

This technique is best suited for projects that contain many repetitions of some standard activities. The extent to which it is used depends on the performing organization's diligence in maintaining a database of the most common activities along with estimates of their duration and resource requirements.

To see how this technique is used, consider an organization that specializes in construction projects. To estimate the time required to install an electrical system in a new building, the time required to install each component of the system would be multiplied by the number of components of that type in the new building. If, for example, the installation of an electrical outlet takes on average 10 min and there are 80 outlets in the new building, then a total of $80 \times 10 = 800$ min is required for this type of component. After performing similar calculations for each component type or job, the total time to install the electrical system would be determined by summing the resultant times.

The benchmark job technique is most appropriate when a project is composed of a set of basic elements whose execution time is additive. If the nature of the work does not support the additivity assumption, then another method — the parametric technique — should be used.

4.2.5   Parametric Technique

This technique is based on cause–effect analysis. The first step is to identify the independent variables. For example, in digging a tunnel, an independent variable might be the length of the tunnel. If it takes, on average, 20 h to dig 1 ft, then the time to dig a tunnel of length L can be estimated by $T(L) = 20L$, where time is considered the dependent variable and the length of the tunnel is considered the independent variable.

When the relationship between the dependent variable and the independent variable is known exactly, as it is in many physical systems, one can plot a response curve in two dimensions. Figure 4.7 depicts two examples of length vs time: line (a) represents a linear relationship between the independent and dependent variables, and line (b) a nonlinear one. In general, if the dependent variable, Y, is believed to be a linear function of the independent variable, X, then regression analysis can be used to estimate the parameters of the line

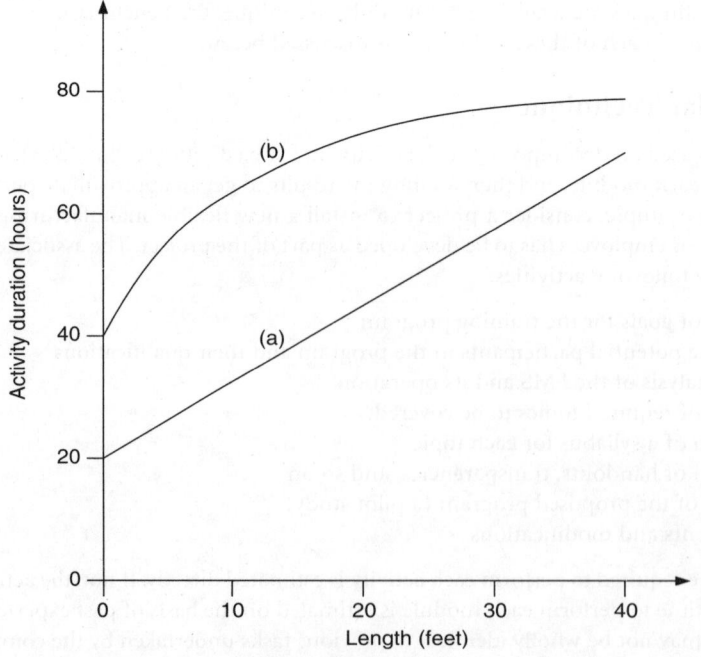

FIGURE 4.7   Two examples of activity duration as a function of length.

$Y = b_0 + b_1X$. Otherwise, either a transformation is performed on one or both of the variables to establish a linear relationship and then regression analysis is applied, or a nonlinear, curve-fitting technique is used.

In the simple case, we have n pairs of sample observations on X and Y, which can be represented on a scatter diagram as in Figure 4.8. Because the line $Y = b_0 + b_1X$ is unknown, we hypothesize that

$$Y_i = b_0 + b_1X_i + u_i, \quad i = 1, \ldots, n$$

$$E[u_i] = 0, \quad i = 1, \ldots, n$$

$$E[u_iu_j] = \begin{cases} 0 & \text{for } i \neq j; \, i, j = 1, \ldots, n \\ \sigma_u^2 & \text{for } i = j; \, i, j = 1, \ldots, n \end{cases}$$

where $E[\cdot]$ is the expected value operator and b_0, b_1, and σ_u^2 are the unknown parameters that must be estimated from the sample observations $X_1, \ldots, X_n$ and $Y_1, \ldots, Y_n$. It is usually assumed that $u_i \sim N(0, \sigma_u^2)$, i.e., u_i is normally distributed with mean 0 and variance σ_u^2.

To begin, denote the regression line by

$$\hat{Y} = \hat{b}_0 + \hat{b}_1X$$

where $\hat{b}_0$ and $\hat{b}_1$ are estimates of the unknown parameters b_0 and b_1, and $\hat{Y}$ the value of the dependent variable for any given value of X. To fit such a line, we must develop formulas for $\hat{b}_0$ and $\hat{b}_1$ in terms of the sample observations. This is done by the principle of least squares (Draper and Smith, 1998).

With some activities, more than one independent variable is required to estimate the performance time. For example, consider the activity of populating a printed circuit board. The use of three independent variables might be appropriate, the first being the number of components to be inserted, the second being the number of setups or tool changes required, and the third being the type of equipment used (here, a qualitative rather than a quantitative measure is called for).

In general, if we start with m independent variables, then the regression line is

$$Y = b_0 + b_1X_1 + b_2X_2 + \ldots + b_mX_m + u$$

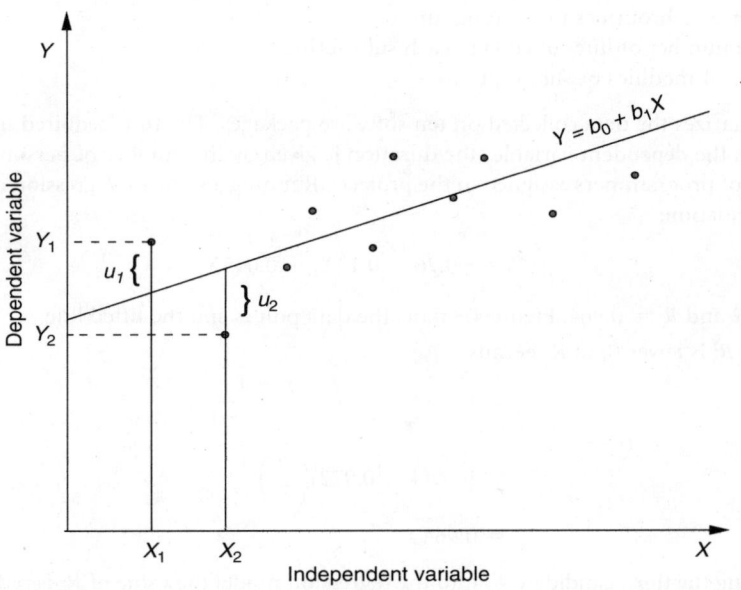

FIGURE 4.8 Typical scatter diagram.

The coefficients b_0, b_1, ..., b_m are also estimated by using the principle of least squares. Goodness-of-fit is measured by the R^2 value, which ranges from 0 (no correlation) to 1 (perfect correlation). However, some analysts prefer to use a normalized version of R^2 known as *adjusted* R^2 given by

$$R_a^2 = 1 - (1 - R^2)\left(\frac{n-1}{n-m-1}\right)$$

where n is the total number of observations and $m + 1$ the number of coefficients to be estimated. By working with the adjusted R^2 it is possible to compare regression models used to estimate the same dependent variable using different numbers of independent variables.

Guidelines for developing a regression equation include the following steps:

1. Identify the independent variables that affect activity duration.
2. Collect data on past performance time of the activity for different values of the independent variables.
3. Check the correlation between the variables. If necessary, use appropriate transformations and only then generate the regression equation.

In the case where several potential independent variables are considered, a technique called *stepwise regression analysis* can be used. This technique is designed to select the independent variables to be included in the model. At each step, at most one independent variable is added to the model. In the first step, a simple regression equation is developed with the independent variable that is the best predictor of the dependent variable (i.e., the one that yields the highest value of R^2). Next, a second variable is introduced. This process continues until no improvement in the regression equation is observed. The final form of the model includes only those independent variables that entered the regression equation during the stepwise iterations.

The quality of a regression model is assessed by analysis of residuals. These residuals ($e_i = Y_i - \hat{Y}_i$) are assumed to be normally distributed with a mean of zero. If this is not the case, or a trend in the value of the residuals as a function of any independent variable exists, then the dependent variable or some of the independent variables may require a transformation.

Example 1 An organization decides to use a regression equation to estimate the time required to develop a new software package. The candidate list of independent variables includes:

X_1 = number of subroutines in the program
X_2 = average number of lines of code in each subroutine
X_3 = number of modules or subprograms

Table 4.1 summarizes the data collected on ten software packages. The time required in person-months denoted by Y, is the dependent variable (the duration is given by the number of person-months divided by the number of programmers assigned to the project). Running a stepwise regression on the data yields the following equation:

$$Y = -0.76 + 0.13X_1 + 0.045X_2$$

with $R^2 = 0.972$ and $R_a^2 = 0.964$. Figure 4.9 plots the data points and the fitted line.

The value of R_a^2 is lower than R^2 because

$$R_a^2 = 1 - (1 - R^2)\left(\frac{n-1}{n-m-1}\right)$$

$$= 1 - (1 - 0.972)\left(\frac{9}{7}\right)$$

$$= 0.964$$

By introducing the third candidate X_3 into the regression model the value of R_a^2 is reduced to 0.963; consequently, it is best to use only the independent variables X_1 and X_2 as predictors, although the difference is minimal.

TABLE 4.1 Data for Regression Analysis

Package Number	Time Required, Y	X_1	X_2	X_3
1	7.9	50	100	4
2	6.8	30	60	2
3	16.9	90	120	7
4	26.1	110	280	9
5	14.4	65	140	8
6	17.5	70	170	7
7	7.8	40	60	2
8	19.3	80	195	7
9	21.3	100	180	6
10	14.3	75	120	3

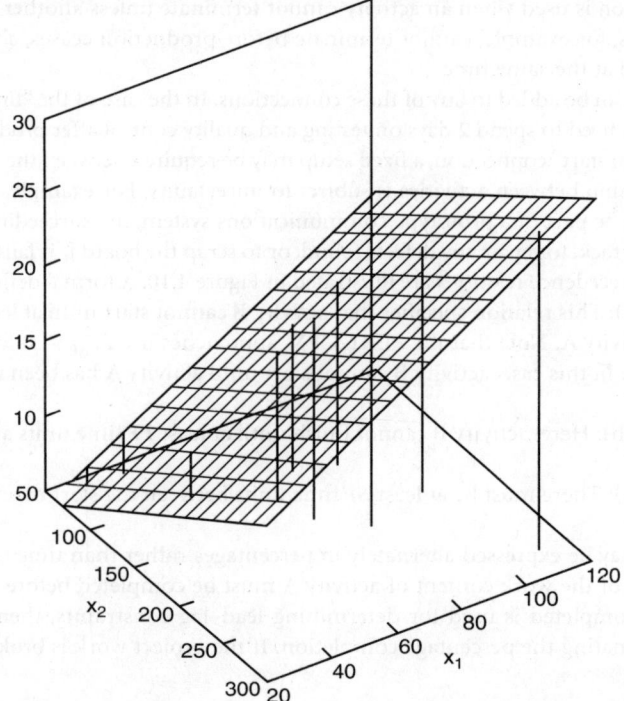

FIGURE 4.9 Data points and regression surface for example 1.

If a new software package similar to the previous ten is to be developed and it contains $X_1 = 45$ subroutines with an average of $X_2 = 170$ lines of code in each, then the estimated development time is

$$Y = -0.76 + (0.13)(45 + 0.045)(170) = 12.7 \text{ person-months}$$

In general, the following points should be taken into account when using and evaluating the results of a regression analysis:

- For the activity under investigation, only data collected on similar activities performed by the same work methods should be used in the calculations.
- When the value of R^2 or R_a^2 is low (below 0.5), the independent variables may not be appropriate.
- If the distribution of the residuals is not close to normal or there is a trend in the residuals as a function of any independent variable, then the regression model may not be appropriate.

4.3 Precedence Relations among Activities

The schedule of activities is constrained by the availability of resources required to perform each activity and by technological constraints known as *precedence relations*. Four general types of precedence relations exist among activities. The most common, termed "finish to start," requires that an activity can start only after its predecessor has been completed. For example, it is possible to lift a piece of equipment by a crane only after the equipment is secured to the hoist.

A "start to start" relationship exists when an activity can start only after a specified activity has already begun. For example, in projects in which concurrent engineering is applied, logistic support analysis starts as soon as the detailed design phase begins. The "start to finish" connection occurs when an activity cannot end until another activity has begun. This would be the case in a project of building a nuclear reactor and charging it with fuel, in which one industrial robot transfers radioactive material to another. The first robot can release the material only after the second robot achieves a tight enough grip. The "finish to finish" connection is used when an activity cannot terminate unless another activity is completed. Quality control efforts, for example, cannot terminate before production ceases, although the two activities can be performed at the same time.

A lag or time delay can be added to any of these connections. In the case of the "finish to finish" arrangement, there might be a need to spend 2 days on testing and quality control after production shuts down. In the case of the "finish to start" connection, a fixed setup may be required between the two activities. In some situations the relationship between activities is subject to uncertainty. For example, after testing a printed circuit board that is to be part of a prototype communications system, the succeeding activity might be to install the board on its rack, to repair any defects found, or to scrap the board if it fails the functionality test.

The four types of precedence relations are illustrated in Figure 4.10. A formal definition of each follows:

FS_{AB} (finish to start): This relation specifies that activity B cannot start until at least FS time units after the completion of activity A. Note that the PERT/CPM approaches use $FS_{AB} = 0$ for network analysis.

SS_{AB} (start to start): In this case, activity B cannot start until activity A has been in progress for at least SS time units.

FF_{AB} (finish to finish): Here, activity B cannot finish until at least FF time units after the completion of activity A.

SF_{AB} (start to finish): There must be at least SF time units between the start of activity A and the completion of activity B.

The leads or lags may be expressed alternately in percentages rather than time units. For example, we may specify that 20% of the work content of activity A must be completed before activity B can start. If percentage of work completed is used for determining lead–lag constraints, then a reliable procedure must be used for estimating the percentage completion. If the project work is broken up properly in the

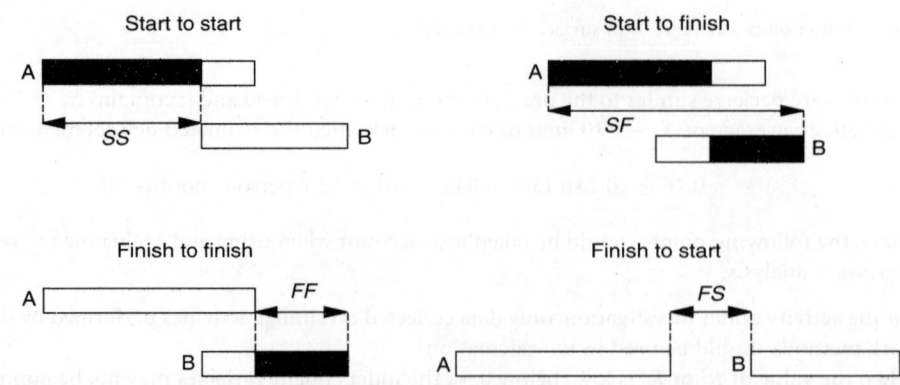

FIGURE 4.10 Lead–lag relationships in precedence diagramming (from Badiru and Pulat, 1995).

TABLE 4.2   Data for Example Project

Activity	Immediate Predecessors	Duration (weeks)
A	–	5
B	–	3
C	A	8
D	A,B	7
E	–	7
F	C,E,D	4
G	F	5

WBS, then it will be much easier to estimate percentage completion by evaluating the work completed at the elementary task levels. The lead–lag relationships may also be specified in terms of *at most* relationships instead of *at least* relationships. For example, we may have at most an *FF* lag requirement between the finish time of one activity and the finish time of another activity.

In the following sections, we concentrate on the analysis of "finish to start" connections, which are the most prevalent. Other types of connections are examined in Section 4.8, and the effect of uncertainty on precedence relations is discussed in Section 4.10. Uncertainty gives rise to probabilistic networks.

The large number of precedence relations among activities makes it difficult to rely on verbal descriptions alone to convey the effect of technological constraints on scheduling, so graphical representations are frequently used. In subsequent sections, a number of such representations are illustrated with the help of an example project. Table 4.2 contains the relevant activity data.

In this project, only "finish to start" precedence relations are considered. From Table 4.2, we see that activities A, B, and E do not have any predecessors, and thus can start at any time. Activity C, however, can start only after A finishes, whereas D can start after the completion of A and B. Further examination reveals that F can start only after C, E, and D are finished and that G must follow F. Because activity A precedes C, and C precedes F, A must also precede F by transitivity. Nevertheless, when using a network representation, it is necessary to list only immediate or direct precedence relations; implied relations are taken care of automatically.

The three models used to analyze precedence relations and their effects on the schedule are the Gantt chart, CPM, and PERT. As mentioned earlier, the last two are based on network techniques in which the activities are placed either on the nodes or on the arrows, depending on which is more intuitive for the analyst.

4.4   Gantt Chart

The most widely used management tool for project scheduling and control is a version of the bar chart developed during World War I by Henry L. Gantt. The Gantt chart, as it is called, enumerates the activities to be performed on the vertical axis and their corresponding duration on the horizontal axis. It is possible to schedule activities by either early-start or late-start logic. In the early-start approach, each activity is initiated as early as possible without violating the precedence relations. In the late-start approach, each activity is delayed as much as possible as long as the earliest finish time of the project is not compromised.

A range of schedules is generated on the Gantt chart when a combination of early and late starts is applied. The early-start schedule is performed first and yields the earliest finish time of the project. That time is then used as the required finish time for the late-start schedule. Figure 4.11 depicts the early-start Gantt chart schedule for the example above. The bars denote the activities; their location with respect to the time axis indicates the time over which the corresponding activity is performed. For example, activity D can start only after activities A and B finish, which happens at the end of week 5. A direct output of this schedule is the earliest finish time for the project (22 weeks for the example).

On the basis of the earliest finish time, the late-start schedule can be generated. This is done by shifting each activity to the right as much as possible while still starting the project at time zero and completing it

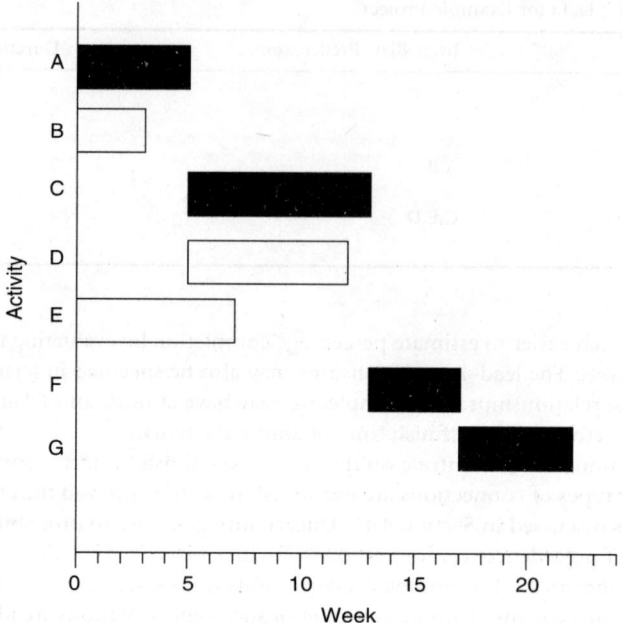

FIGURE 4.11   Gantt chart for an early-start schedule.

in 22 weeks. The resultant schedule is depicted in Figure 4.12. The difference between the start (or the finish) times of an activity on the two schedules is called the slack (or float) of the activity. Activities that do not have any slack are denoted by a shaded bar and are termed *critical*. The sequence of critical activities connecting the start and end points of the project is known as the *critical path*, which logically turns out to be the *longest path* in the network. A delay in any activity along the critical path delays the entire project. Put another way, the sum of durations for critical activities represents the *shortest* possible time to complete the project.

Gantt charts are simple to generate and interpret. In the construction, there should be a one-to-one correspondence between the listed tasks and the WBS and its numbering scheme. As shown in Figure 4.13, which depicts the Gantt chart for the microcomputer development project, a separate column can be added for this purpose. In fact, the schedule should not contain any tasks that do not appear in the WBS. Often, however, the Gantt chart includes milestones such as project kickoff and design review, which are listed along with the tasks.

In addition to showing the critical path, Gantt charts can be modified to indicate project and activity status. In Figure 4.13, a bold border is used to identify a critical activity, and a shaded area indicates the approximate completion status at the August review. Accordingly, we see that tasks 2, 5, and 8 are critical, falling on the longest path. Task 2 is 100% complete, task 4 65% complete, and task 7 55% complete; tasks 5, 6, and 8 have not yet been started.

Gantt charts can be modified further to show budget status by adding a column that lists planned and actual expenditures for each task. Many variations of the original bar graph have been developed to provide more detailed information for the project manager. One commonly used variation that replaces the bars with lines and adds triangles to indicate project status and revision points is shown in Figure 4.14. To explain the features, let us examine task 2, equipment design. According to the code given in the lower left-hand corner of the figure, this task was rescheduled three times, finally starting in February, and finishing at the end of June. Note the two rescheduled start milestones and the two rescheduled finish milestones.

The problem with adding features to the bar graph is that they detract from the clarity and simplicity of the basic form. Nevertheless, the additional information conveyed to the user may offset the additional effort

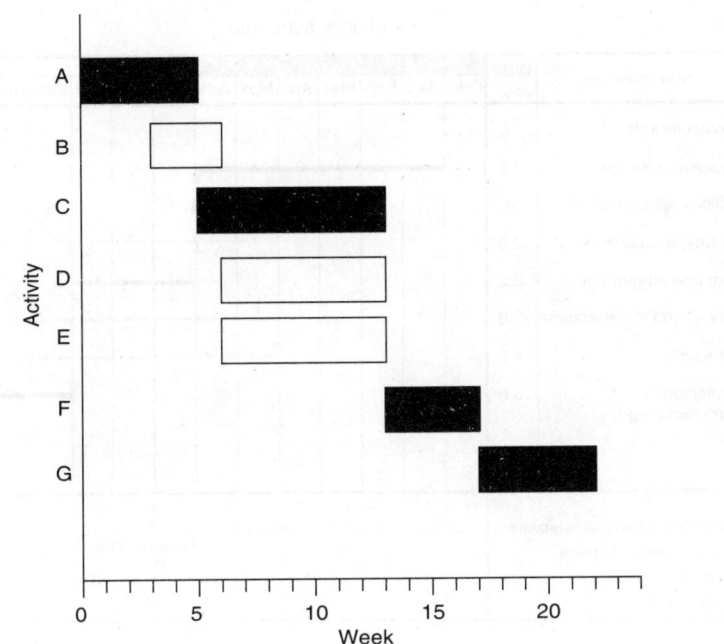

FIGURE 4.12 Gantt chart for a late-start schedule.

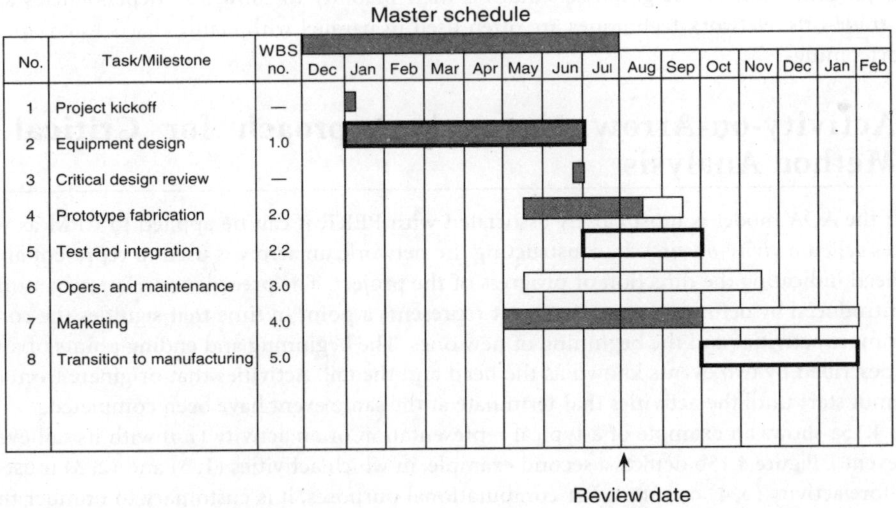

Master schedule

No.	Task/Milestone	WBS no.	Dec	Jan	Feb	Mar	Apr	May	Jun	Jul	Aug	Sep	Oct	Nov	Dec	Jan	Feb
1	Project kickoff	—															
2	Equipment design	1.0															
3	Critical design review	—															
4	Prototype fabrication	2.0															
5	Test and integration	2.2															
6	Opers. and maintenance	3.0															
7	Marketing	4.0															
8	Transition to manufacturing	5.0															

Review date

FIGURE 4.13 Gantt chart for the microcomputer development example.

required in generating and interpreting the data. A common modification of the analysis is the case when a milestone has a contractual due date. Consider, for example, activity 8 (WBS No. 5.0) in Figure 4.14. If management decides that the required due date for the termination of this activity is the end of February (instead of the end of January), then a slack of 1 month will be added to each activity in the project. If, however, the due date of activity 8 is the end of December, then the schedule in Figure 4.14 is no longer feasible because the sequence of activities 2, 5, and 8 (the critical sequence) cannot be completed by the end of December. Section 4.12 contains a discussion related to scheduling conflicts and their management.

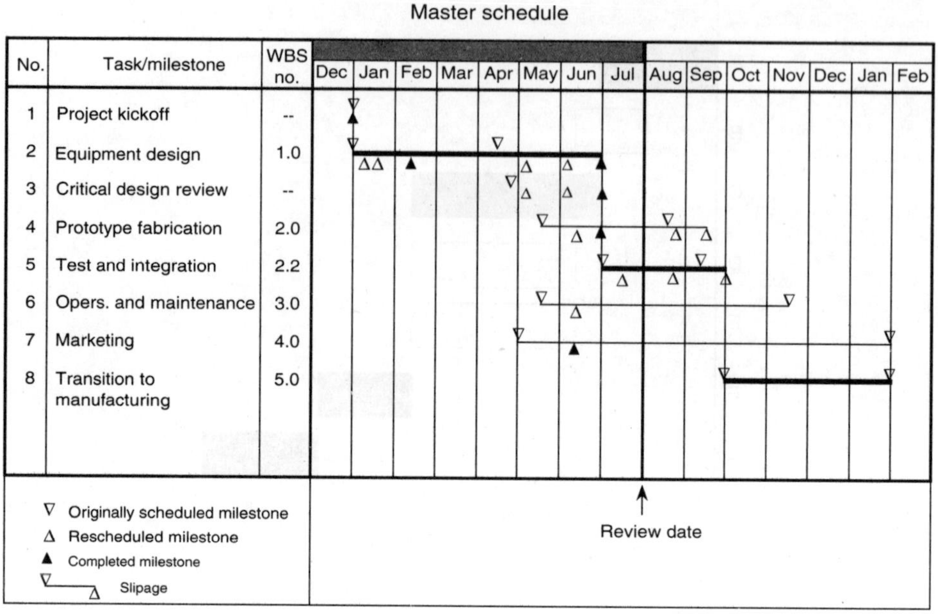

Master schedule

No.	Task/milestone	WBS no.	Dec	Jan	Feb	Mar	Apr	May	Jun	Jul	Aug	Sep	Oct	Nov	Dec	Jan	Feb
1	Project kickoff	--															
2	Equipment design	1.0															
3	Critical design review	--															
4	Prototype fabrication	2.0															
5	Test and integration	2.2															
6	Opers. and maintenance	3.0															
7	Marketing	4.0															
8	Transition to manufacturing	5.0															

∇　Originally scheduled milestone
△　Rescheduled milestone
▲　Completed milestone
∇—△　Slipage

Review date

FIGURE 4.14　Extended Gantt chart with task details.

The major limitation of bar-graph schedules is their inability to show task dependencies and time-resource trade-offs. Network techniques are often used in parallel with Gantt charts to compensate for these shortcomings.

4.5　Activity-on-Arrow Network Approach for Critical Path Method Analysis

Although the AOA model is most closely associated with PERT, it can be applied to CPM as well (it is sometimes called *activity-on-arc*). In constructing the network, an arrow is used to represent an activity, with its head indicating the direction of progress of the project. The precedence relations among activities are introduced by defining events. An event represents a point in time that signifies the completion of one or more activities and the beginning of new ones. The beginning and ending points of an activity are thus described by two events known as the head and the tail. Activities that originate from a certain event cannot start until the activities that terminate at the same event have been completed.

Figure 4.15a shows an example of a typical representation of an activity (i, j) with its tail event i and its head event j. Figure 4.15b depicts a second example, in which activities $(1, 3)$ and $(2, 3)$ must be completed before activity $(3, 4)$ can start. For computational purposes, it is customary to number the events in ascending order so that compared with the head event, a smaller number is always assigned to the tail event of an activity.

The rules for constructing a diagram are summarized below.

Rule 1. Each activity is represented by one and only one arrow in the network. No single activity can be represented twice in the network. This is to be differentiated from the case in which one activity is broken down into segments, wherein each segment may then be represented by separate arrows. For example, in designing a new computer architecture, the controller might be developed first followed by the arithmetic unit, the I/O processor, and so on.

Rule 2. No two activities can be identified by the same head and tail events. A situation such as this may arise when two or more activities can be performed in parallel. As an example, consider Figure 4.16a,

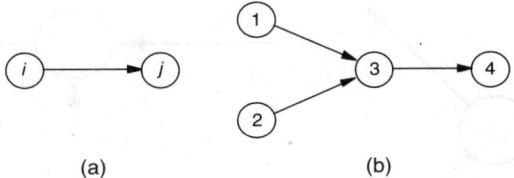

FIGURE 4.15   Network components.

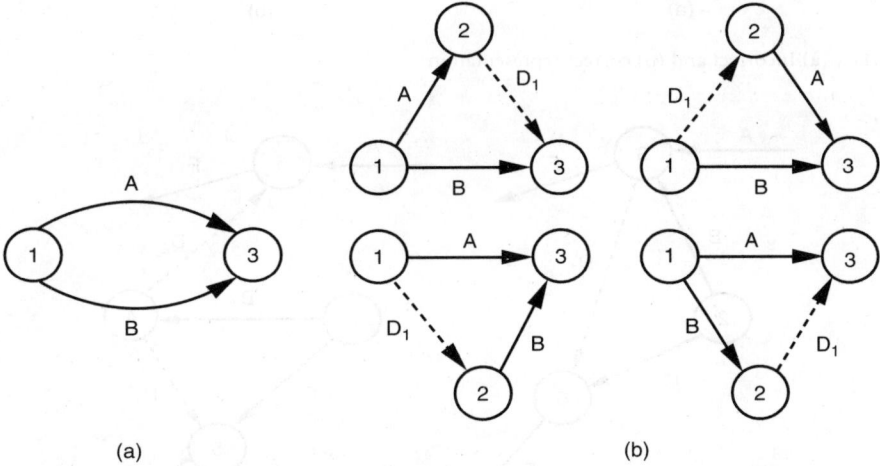

FIGURE 4.16   Use of a dummy arc between two nodes.

which shows activities A and B running in parallel. The procedure used to circumvent this difficulty is to introduce a dummy activity between either A or B. The four equivalent ways of doing this are shown in Figure 4.16b, where D_1 is the dummy activity. As a result of using D_1, activities A and B can now be identified by a unique set of events. It should be noted that dummy activities do not consume time or resources. Typically, they are represented by dashed lines in the network.

Dummy activities are also necessary in establishing logical relationships that cannot otherwise be represented correctly. Suppose that in a certain project, tasks A and B must precede C, whereas task E is preceded only by B. Figure 4.17a shows an incorrect but quite common way many beginners would draw this part of the network. The difficulty is that although the relationship among A, B, and C is correct, the diagram implies that E must be preceded by both A and B. The correct representation using dummy D_1 is depicted in Figure 4.17b.

Rule 3. To ensure the correct representation in the AOA diagram, the following questions must be answered as each activity is added to the network:

1. Which activities must be completed immediately before this activity can start?
2. Which activities must immediately follow this activity?
3. Which activities must occur concurrently with this activity?

This rule is self-explanatory. It provides guidance for checking and rechecking the precedence relations as the network is constructed.

The following examples further illustrate the use of dummy activities.

Example 2   Draw the AOA diagram so that the following precedence relations are satisfied:

1. E is preceded by B and C.
2. F is preceded by A and B.

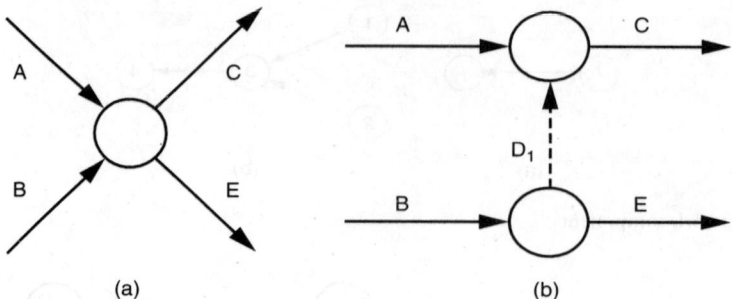

FIGURE 4.17   (a) Incorrect and (b) correct representation.

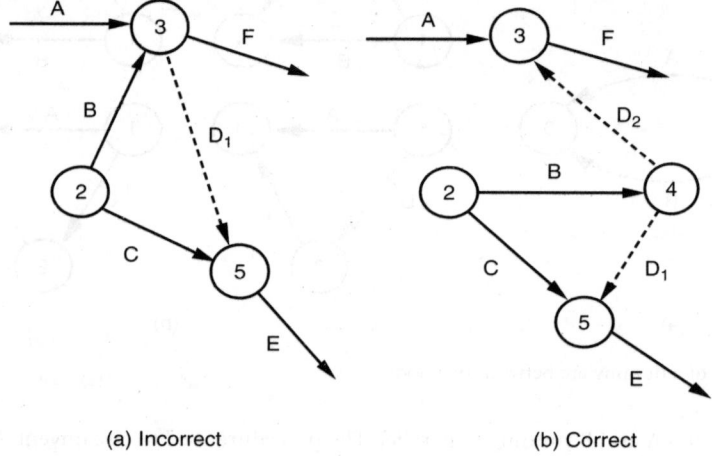

FIGURE 4.18   Subnetwork with two dummy arcs: (a) incorrect (b) correct.

Solution

Figure 4.18a shows an incorrect precedence relation for activity E. According to the requirements, B and C are to precede E, and A and B are to precede F. The dummy D_1 therefore is inserted to allow B to precede E. Doing so, however, implies that A must *also* precede E, which is incorrect. Figure 4.18b shows the correct relationships.

Example 3   Draw the precedence diagram for the following conditions:

1. G is preceded by A.
2. E is preceded by A and B.
3. F is preceded by B and C.

Solution

An incorrect and a correct representation are given in Figure 4.19. The diagram in part (a) of the figure is wrong because it implies that A precedes F.

It is a good practice to have a single start event common to all activities that have no predecessors and a single end event for all activities that have no successors. The actual mechanics of drawing the AOA network are illustrated using the data in Table 4.2.

The process begins by identifying all activities that have no predecessors and joining them to a unique start node. This is shown in Figure 4.20. Each activity terminates at a node. Only the first node in the

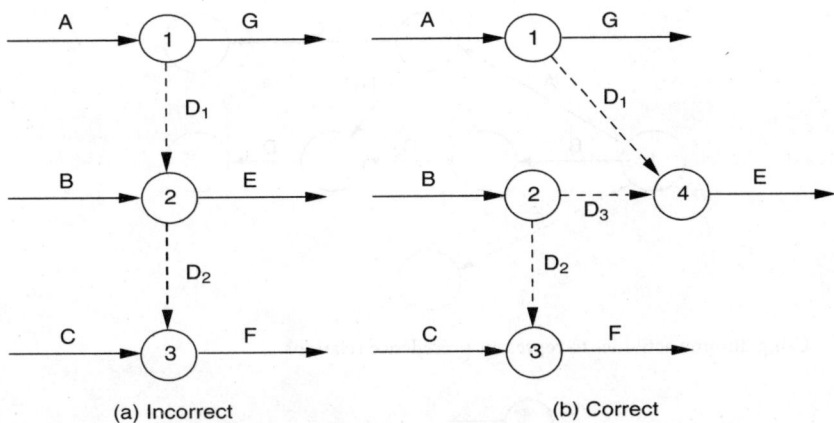

FIGURE 4.19 Subnetwork with complicated precedence relations: (a) incorrect (b) correct.

TABLE 4.2 Data for Example Project

Activity	Immediate Predecessors	Duration (weeks)
A	–	5
B	–	3
C	A	8
D	A,B	7
E	–	7
F	C,E,D	4
G	F	5

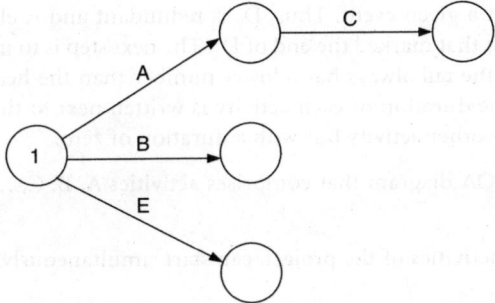

FIGURE 4.20 Partial plot of the example AOA network.

network is assigned a number (1); all other nodes are labeled only when network construction is completed, as explained presently. Because activity C has only one predecessor (A), it can be immediately added to the diagram (see Figure 4.20).

Activity D has both A and B as predecessors; thus, there is a need for an event that represents the completion of A and B. We begin by adding two dummy activities D_1 and D_2. The common end event of D_1 and D_2 is now the start event of D, as depicted in Figure 4.21. As we progress, it may happen that one or more dummy activities are added that are really not necessary. To correct this situation, a check will be made at completion and redundant dummies will be eliminated.

Before starting activity F, activities C, E, and D must be completed. Therefore, an event that represents the terminal point of these activities should be introduced. Notice that C, E, and D are not predecessors

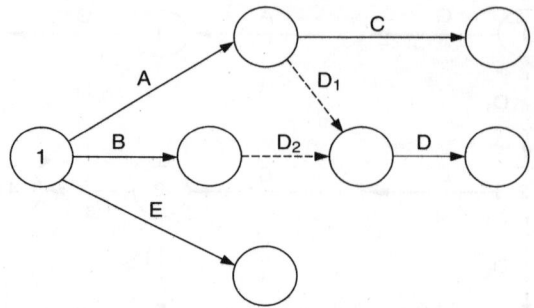

FIGURE 4.21   Using dummy activities to represent precedence relations.

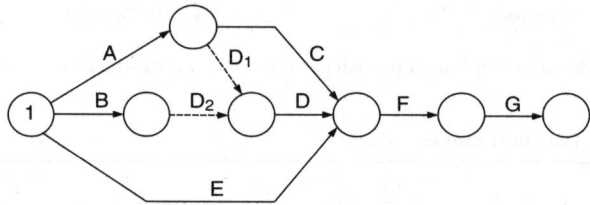

FIGURE 4.22   Network with activities F and G included.

of any other activity but F. This implies that the three arrows representing these activities can terminate at the same node (event) — the tail of F. Activity G, which has only F as a predecessor, can start from the head of F (see Figure 4.22).

Once all of the activities and their precedence relations have been included in the network diagram, it is possible to eliminate redundant dummy activities. A dummy activity is redundant when it is the only activity that starts or ends at a given event. Thus, D_2 is redundant and is eliminated by connecting the head of activity B to the event that marked the end of D_2. The next step is to number the events in ascending order, making sure that the tail always has a lower number than the head. The resulting network is illustrated in Figure 4.23. The duration of each activity is written next to the corresponding arrow. The dummy D_1 is shown like any other activity but with a duration of zero.

Example 4   Construct an AOA diagram that comprises activities A, B, C, ..., L such that the following relationships are satisfied:

1. A, B, and C, the first activities of the project, can start simultaneously.
2. A and B precede D.
3. B precedes E, F, and H.
4. F and C precede G.
5. E and H precede I and J.
6. C, D, F, and J precede K.
7. K precedes L.
8. I, G, and L are the terminal activities of the project.

Solution

The resulting diagram is shown in Figure 4.24. The dummy activities D_1 and D_2 are needed to establish correct precedence relations, and D_3 is introduced to ensure that the parallel activities E and H have unique finish events. Note that the events in the project are numbered in such a way that if there is a path connecting nodes i and j, then $i < j$. In fact, there is a basic result from graph theory that states that a directed graph is acyclic if and only if its nodes can be numbered, so that for all arcs (i, j), $i < j$.

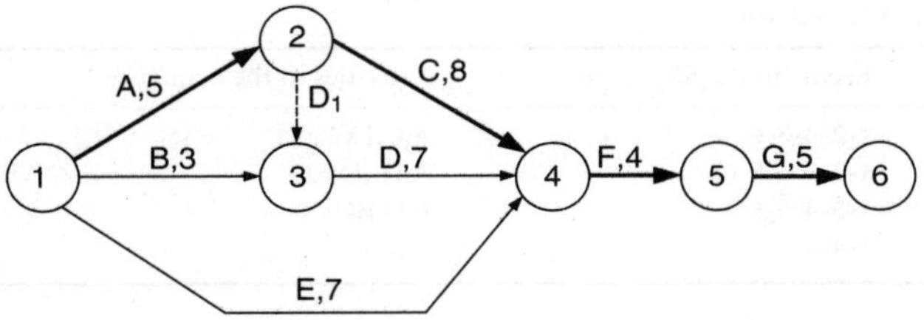

FIGURE 4.23 Complete AOA project network.

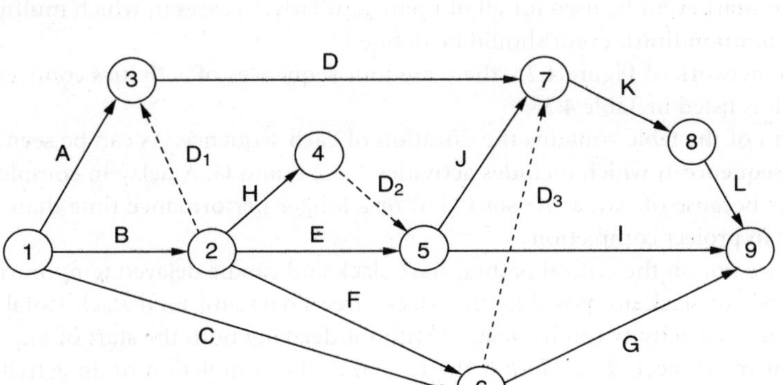

FIGURE 4.24 Network for Example 4.

Once the nodes are numbered, the network can be represented by a matrix whose respective rows and columns correspond to the start and finish events of a particular activity. The matrix for the example in Figure 4.23 is as follows:

Finishing Event

Starting Event	1	2	3	4	5	6
1		×	×	×		
2			×	×		
3				×		
4					×	
5						×
6						

where the entry "×" means that there is an activity connecting the two events (instead of an ×, it may be more efficient to use the activity number or its duration). For example, the × in row 3, column 4 indicates that an activity starts at event 3 and finishes at event 4, that is, activity D. The absence of an entry in the second row and fifth column means that no activity starts at event 2 and finishes at event 5.

Because the numbering scheme used ensures that if activity (i, j) exists, then $i < j$, it is sufficient to store only that portion of the matrix that is above the diagonal. Alternatively, the lower portion of the matrix can be used to store other information about an activity, such as resource requirements or budget. This conveniently represents computer input.

From the network diagram, it is easy to see the sequences of activities that connect the start of the project to its terminal node. As explained earlier, the longest sequence is called the *critical path.* The total time required to perform all of the activities on the critical path is the minimum duration of the project because these activities cannot be performed in parallel as a result of precedence relations among them.

TABLE 4.3 Sequences in the Network

Sequence Number	Events in the Sequence	Activities in the Sequence	Sum of Activity Times
1	1-2-4-5-6	A,C,F,G	22
2	1-2-3-4-5-6	A,D_1,D,F,G	21
3	1-3-4-5-6	B,D,F,G	19
4	1-4-5-6	E,F,G	16

To simplify the analysis, it is recommended that in the case of multiple activities that have no predecessors, a common start event be used for all of them. Similarly, in cases in which multiple activities have no successors, a common finish event should be defined.

In the example network of Figure 4.23, there are four sequences of activities connecting the start and finish nodes. Each is listed in Table 4.3.

The last column of the table contains the duration of each sequence. As can be seen, the longest path (critical path) is sequence 1, which includes activities A, C, F, and G. A delay in completing any of these (critical) activities because of, say, a late start (LS) or a longer performance time than initially expected will cause a delay in project completion.

Activities that are not on the critical path(s) have slack and can be delayed temporarily on an individual basis. Two types of slack are possible: free slack (free float) and total slack (total float). *Free slack* denotes the time that an activity can be delayed without delaying both the start of any succeeding activity and the end of the project. *Total slack* is the time that the completion of an activity can be delayed without delaying the end of the project. The delay of an activity that has total slack but no free slack reduces the slack of other activities in the project.

A simple rule can be used to identify the type of slack. A noncritical activity whose finish event is on the critical path has both total and free slack, and the two are equal. For example, noncritical activity E, whose event 4 is on the critical path, has total slack = free slack = 6, as we will see shortly. In contrast, the head of noncritical activity B is not on the critical path; its total slack = 3, and its free slack = 2. The head of activity B is the start event of activity D, which is also noncritical. The difference between the length of the critical sequence (A–C) and the noncritical sequence (B–D), which runs in parallel to (A–C), is the total slack of B and D and is equal to $(5 + 8) - (3 + 7) = 3$. Any delay in activity B will reduce the remaining slack for activity D. Therefore, the person responsible for performing D should be notified.

The roles of the total and free slacks in scheduling noncritical activities can be explained in terms of two general rules:

1. If the total slack *equals* the free slack, then the noncritical activity can be scheduled anywhere between its early-start (ES) and late-finish (LF) times.
2. If the free slack is *less than* the total slack, then the noncritical activity can be delayed relative to its ES time by no more than the amount of its free slack without affecting the schedule of those activities that immediately succeed it.

Further elaboration and an exact mathematical expression for calculating activity slacks are presented in the following subsections.

4.5.1 Calculating Event Times and Critical Path

Important scheduling information for the project manager is the earliest and latest times when each event can take place without causing a schedule overrun. This information is needed to compute the critical path. The *early time* of an event i is determined by the length of the longest sequence from the start node (event 1) to event i. Denote t_i as the early time of event i, and let $t_1 = 0$, implying that activities without precedence constraints begin as early as possible. If a starting date is given, then t_1 is adjusted accordingly.

To determine t_i for each event i, a *forward* pass is made through the network. Let L_{ij} be the duration or length of activity (i, j). The following formula is used for the calculations:

$$t_j = \max_i\{t_i + L_{ij}\} \quad \text{for all } (i, j) \text{ activities defined} \tag{4.3}$$

where $t_1 = 0$. Thus, to compute t_j for event j, t_i for the tail events of all incoming activities (i, j) must be computed first. In words, the early time of each event is the latest of the early times of its immediate predecessors plus the duration of the connecting activity.

The forward-pass calculations for the example network in Figure 4.23 will now be given. The early time for event 2 is simply

$$t_2 = t_1 + L_{12} = 0 + 5 = 5$$

where $L_{12} = 5$ is the duration of the activity connecting event 1 to event 2 (activity A).

Early-time calculations for event 3 are a bit more complicated because event 3 marks the completion of the two activities D_1 and B. By implication, there are two sequences connecting the start of the project to event 3. The first comprises activities A and D_1 and is of length 5; the second includes activity B only and has $L_{13} = 3$. Using Equation (4.3), we get

$$t_3 = \max \begin{Bmatrix} t_1 + L_{13} \\ t_1 + L_{23} \end{Bmatrix}$$

$$= \max \begin{Bmatrix} 0 + 3 \\ 5 + 0 \end{Bmatrix} = 5$$

so the early time of event 3 is $t_3 = 5$.

The remaining calculations are performed as follows:

$$t_4 = \max \begin{Bmatrix} t_1 + L_{14} \\ t_2 + L_{24} \\ t_3 + L_{34} \end{Bmatrix} = \max \begin{Bmatrix} 0 + 7 \\ 5 + 8 \\ 5 + 7 \end{Bmatrix} = 13$$

$$t_5 = t_4 + L_{45} = 13 + 4 = 17$$

$$t_6 = t_5 + L_{56} = 17 + 5 = 22$$

This confirms that the earliest that the project can finish is in 22 weeks.

The late time of each event is calculated next by making a *backward* pass through the network. Let T_i denote the late time of event i. If n is the finish event, then the calculations are generally initiated by setting $T_n = t_n$ and working backward toward the start event using the following formula:

$$T_i = \min_j[T_j - L_{ij}] \quad \text{for all } (i, j) \text{ activities defined} \tag{4.4}$$

If, however, a required project completion date given is later than the early time of event n, then it is possible to assign that time as the late time for the finish event. If the required date is earlier than the early time of the finish event, then no feasible schedule exists. This case is discussed later in the chapter.

In our example, $T_6 = t_6 = 22$. The late time for event 5 is calculated as follows:

$$T_5 = T_6 - L_{56} = 22 - 5 = 17$$

Similarly,

$$T_4 = T_5 - L_{45} = 17 - 4 = 13$$

$$T_3 = T_4 - L_{34} = 13 - 7 = 6$$

TABLE 4.4 Summary of Event Time Calculations

Event, i	Early Time, t_i	Late Time, T_i
1	0	0
2	5	5
3	5	6
4	13	13
5	17	17
6	22	22

Event 2 is connected by sequences of activities to both events 3 and 4. Thus, applying Equation (4.4), the late time of event 2 is the minimum among the late times dictated by the two sequences; that is,

$$T_2 = \min \left\{ \begin{array}{c} T_3 - L_{23} \\ T_4 - L_{24} \end{array} \right\} = \min \left\{ \begin{array}{c} 6 - 0 = 6 \\ 13 - 8 = 5 \end{array} \right\} = 5$$

The late time of event 1 is calculated in a similar manner:

$$T_1 = \min \left\{ \begin{array}{c} 6 - 3 = 3 \\ 5 - 5 = 0 \\ 13 - 7 = 6 \end{array} \right\} = 0$$

The results are summarized in Table 4.4.

The critical activities can now be identified by using the results of the forward and backward passes. An activity (i, j) lies on the critical path if it satisfies the following three conditions:

$$t_i = T_i$$

$$t_j = T_j$$

$$t_j - T_i = T_j - T_i = L_{ij}$$

These conditions actually indicate that there is no float or slack time between the earliest start (completion) and the latest start (completion) of the critical activities. In Figure 4.23, activities (0,2), (2,4), (4,5), and (5,6) define the critical path forming a chain that spans the network from node 1 (start) to node 6 (finish).

4.5.2 Calculating Activity Start and Finish Times

In addition to scheduling the events of a project, detailed scheduling of activities is performed by calculating the following four times (or dates) for each activity (i, j):

ES_{ij} = *early-start* time: the earliest time when activity (i, j) can start without violating any precedence relations.

EF_{ij} = *early-finish* time: the earliest time when activity (i, j) can finish without violating any precedence relations.

LS_{ij} = *late-start* time: the latest time when activity (i, j) can start without delaying the completion of the project.

LF_{ij} = *late-finish* time: the latest time when activity (i, j) can finish without delaying the completion of the project.

The calculations proceed as follows:

$$\text{ES}_{ij} = t_i \quad \text{for all } i$$

$$\text{EF}_{ij} = \text{ES}_{ij} + L_{ij} \quad \text{for all } (i, j) \text{ defined}$$

TABLE 4.5 Summary of Start and Finish Time Analysis

Activity	(i, j)	L_{ij}	$ES_{ij} = t_i$	$EF_{ij} = ES_{ij} + L_{ij}$	$LF_{ij} = T_j$	$LS_{ij} = LF_{ij} - L_{ij}$	$TS_{ij} = LS_{ij} - ES_{ij}$	$FS_{ij} = t_j - t_i - L_{ij}$
A	(1, 2)	5	0	5	5	0	0	0
B	(1, 3)	3	0	3	6	3	3	2
C	(2, 4)	8	5	13	13	5	0	0
D	(3, 4)	7	5	12	13	6	1	1
E	(1, 4)	7	0	7	13	6	6	6
F	(4, 5)	4	13	17	17	13	0	0
G	(5, 6)	5	17	22	22	17	0	0
D_1	(2, 3)	0	5	5	6	6	1	0

$$LF_{ij} = T_j \quad \text{for all } j$$

$$LS_{ij} = LF_{ij} - L_{ij} \quad \text{for all } (i, j) \text{ defined}$$

Thus, the earliest time an activity can begin is equal to the early time of its start event; the latest an activity can finish is equal to the LF of its finish event. For activity D in the example, which is denoted by arc (3,4) in the network, we have $ES_{34} = t_3 = 5$ and $LF_{34} = T_4 = 13$.

The earliest time an activity can finish is given by its ES plus its duration; the latest time when an activity can start is equal to its LF minus its duration. For activity D, this implies that $EF_{34} = ES_{34} + L_{34} = 5 + 7 = 12$, and $LS_{34} = LF_{34} - L_{34} = 13 - 7 = 6$. The full set of calculations is presented in Table 4.5.

4.5.3 Calculating Slacks

As mentioned earlier, there are two types of slack associated with an activity: total slack and free slack. Information about slack is important to the project manager, who may have to adjust budgets and resource allocations to stay on schedule. Knowing the amount of slack in an activity is essential if he or she is to do this without delaying the completion of the project. In a multiproject environment, slack in one project can be used temporarily to free up resources needed for other projects that are behind schedule or overly constrained.

Because of the importance of slack, project management is sometimes referred to as slack management, which will be elaborated on in the chapter that deals with resources and budgets. The total slack TS_{ij} (or total float TF_{ij}) of activity (i, j) is equal to the difference between its late start (LS_{ij}) and its early start (ES_{ij}) or the difference between its late finish (LF_{ij}) and its early finish (EF_{ij}), that is,

$$TS_{ij} = TF_{ij} = LS_{ij} - ES_{ij} = LF_{ij} - EF_{ij}$$

This is equivalent to the difference between the maximum time available to perform the activity ($T_j - t_i$) and its duration (L_{ij}). The total slack of activity D (3, 4) in the example is $TS_{34} = LS_{34} - ES_{34} = 6 - 5 = 1$.

The free slack (or free float) is defined by assuming that all activities start as early as possible. In this case, the free slack, FS_{ij}, for activity (i, j) is the difference between the early time of its finish event j and the sum of the early time of its start event i plus its length; that is,

$$FS_{ij} = t_j - (t_i + L_{ij}).$$

For the example, the free slack for activity D (3, 4) is $FS_{34} = t_4 - (t_3 + L_{34}) = 13 - (5 + 7) = 1$. Thus, it is possible to delay activity D by 1 week without affecting the start of any other activity. The times and slacks for the events and activities of the example are summarized in Table 4.5.

Activities with a total slack equal to zero are critical because any delay in these activities will lead to a delay in the completion of the project. The total slack is either equal to or larger than the free slack because the total slack of an activity is composed of its free slack plus the slack shared with other activities. For example, activity B denoted by (1, 3) has a free slack of 2 weeks. Thus, it can be delayed up to 2 weeks without affecting its successor D. If, however, B is delayed by 3 weeks, the project can still be finished on time provided that D starts immediately after B finishes. This follows because activities B and D share 1 week of total slack. Finally, notice that activity D_1 has a total slack of 1 and a free slack of 0, implying that noncritical activities may have zero free slack.

In an AOA network, the length of the arrows is not necessarily proportional to the duration of the activities. When developing a graphical representation of the problem, it is convenient to write the duration of each activity next to the corresponding arrow. Most software packages that are based on the AOA model follow this convention. In addition, they typically provide the user with the option of placing a subset of activity parameters above or below the arrows. We have intentionally omitted placing this information on our diagrams because of the clutter that it occasions. Nevertheless, it is good practice when manually performing the forward and backward calculations to write the ES and LS times above the corresponding nodes.

4.6 Activity-on-Node Network Approach for Critical Path Method Analysis

The AON model is an alternative approach of representing project activities and their interrelationships. It is most closely associated with CPM analysis and is the basis for most computer implementations. In the AON model, the arrows are used to denote the precedence relations among activities. Its basic advantage is that there is no need for dummy arrows and it is very easy to construct. In developing the network, it is convenient to add a single start node and a single finish node that uniquely identify these milestones. This is illustrated in Figure 4.25 for the example.

Some additional network construction rules include:

1. All nodes, with the exception of the terminal node, must have at least one successor.
2. All nodes, except the first, must have at least one predecessor.
3. There should be only one initial and one terminal node.
4. No arrows should be left dangling. Notwithstanding rules 1 and 2, every arrow must have a head and a tail.
5. An arrow specifies only precedence relations; its length has no significance with respect to the time duration accompanying either of the activities that it connects.
6. Cycles or closed-loop paths through the network are not permitted. They imply that an activity is a successor of another activity that depends on it.

As with the AOA model, the computational procedure involves forward and backward passes through the network. This is discussed next.

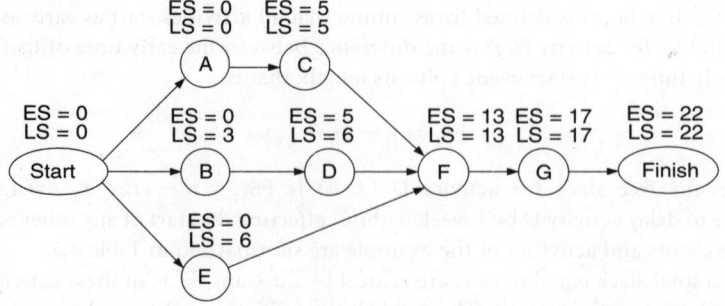

FIGURE 4.25   AON network for the example project.

4.6.1 Calculating Early-Start and Early-Finish Times of Activities

A forward pass is used to determine the earliest start time and the earliest finish time for each activity. During the forward pass, it is assumed that each activity begins as soon as possible; that is, as soon as the last of its predecessors is completed. Thus the ES time of an activity is equal to the maximum early-finish (EF) time of all of the activities immediately preceding it. The ES time of the initial activity is assumed to be zero, as is its EF. For all other activities, the EF time is equal to its ES time plus its duration.

Using slightly different notation to distinguish the AON calculations from those prescribed for the AOA model, we have

$$ES(K) = \max\{EF(J):J \text{ an immediate predecessor of } K\} \tag{4.5}$$

$$EF(K) = ES(K) + L(K) \tag{4.6}$$

where $L(K)$ denotes the duration of activity K.

Returning once again to the example, activities A, B, and E do not have predecessors (except the start node), and thus their ES times are zero; that is, $ES(A) = ES(B) = ES(E) = 0$. The EF time of these activities is equal to their ES time plus their duration, so $EF(A) = 0 + 5 = 5$, $EF(B) = 0 + 3 = 3$, and $EF(E) = 0 + 7 = 7$.

From Equation (4.5), the ES of any other activity is determined by the latest (the maximum) EF time of its predecessors. For activity D, the calculations are

$$ES(D) = \max\left\{\begin{array}{c} EF(A) \\ EF(B) \end{array}\right\} = \max\left\{\begin{array}{c} 5 \\ 3 \end{array}\right\} = 5$$

The ES and EF times of the remaining activities are computed in a similar manner. Table 4.6 summarizes the results.

4.6.2 Calculating Late Start and Finish Times of Activities

The calculation of late times on the AON network is performed in the reverse order of the calculation of early times. As with the AOA model, a backward pass is made beginning at the expected completion time and concluding at the earliest start time. To complete the project as soon as possible, the LF of the last activity is set equal to its EF time calculated in the forward pass. Alternatively, the latest allowable completion time may be fixed by a contractual deadline, if one exists, or some other rationale.

In general, the LF time of an activity with more than one successor is the earliest of the succeeding LS times. The LS time of an activity is its LF time minus its duration. Computational expressions for LF and LS are

$$LF(K) = \min\{LS(J):J \text{ is a successor of } K\} \tag{4.7}$$

$$LS(K) = LF(K) - L(K) \tag{4.8}$$

TABLE 4.6 Early Start and Early Finish of Project Activities

Activity	Early Start	Early Finish
A	0	5
B	0	3
C	5	13
D	5	12
E	0	7
F	13	17
G	17	22

TABLE 4.7   Late Finish and Late Start of Project Activities

Activity	Late Finish	Late Start
A	5	0
B	6	3
C	13	5
D	13	6
E	13	6
F	17	13
G	22	17

To begin the calculations for the example network in Figure 4.24, we set LF(G) = EF(G) = 22 and apply Equation (4.8) to get LS(G) = LF(G) − L(G) = 22 − 5 = 17. The LF of any other activity is equal to the earliest (or the minimum) among the LS time of its succeeding activities. Because activity F has only one successor (G), we get

$$\text{LF(F)} = \text{LS(G)} = 17 \quad \text{and} \quad \text{LS(F)} = 17 - 4 = 13$$

Continuing with activities C and D yield

$$\text{LF(C)} = \text{LS(F)} = 13 \quad \text{and} \quad \text{LS(C)} = 13 - 8 = 5,$$

$$\text{LF(D)} = \text{LS(F)} = 13 \quad \text{and} \quad \text{LS(D)} = 13 - 7 = 6$$

Because A has two successors, we get

$$\text{LF(A)} = \min\left\{ \begin{array}{c} \text{LS(C)} \\ \text{LS(D)} \end{array} \right\} = \min\left\{ \begin{array}{c} 5 \\ 6 \end{array} \right\} = 5$$

$$\text{and} \quad \text{LS(A)} = \text{LF(A)} - \text{L(A)} = 5 - 5 = 0$$

The LS and LF times of activities in the example project are summarized in Table 4.7. As expected, these results are identical to those of the AOA model.

The total slack of an activity is calculated as the difference between its LS (or finish) and its ES (or finish). The free slack of an activity is the difference between the earliest among the ES times of its successors and its EF time. That is, for each activity K,

$$\text{TS}(K) = \text{LS}(K) - \text{ES}(K)$$

$$\text{FS}(K) = \min\{\text{ES}(J) : J \text{ is successor of } K\} - \text{EF}(K)$$

Activities with zero total slack fall on the critical path. When performing the calculations manually, it is convenient to write the corresponding ES and LS times above each node to help identify the critical path.

4.7   Precedence Diagramming with Lead–Lag Relationships

When lead or lag constraints exist between the start and finish of activities or when precedence relations other than "finish to start" are present, it is often possible to split activities to simplify the analysis. Some of the factors that determine whether an activity can be split are technical or logical limitations, setup times required to restart split tasks, difficulty involved in managing resources for split tasks, loss of consistency of work, and management policy about splitting jobs.

Figure 4.26 presents a simple AON network that consists of three activities. The two top numbers on either side of the nodes correspond to ES and EF times, whereas the two bottom numbers correspond to LS and LF times. The activities are to be performed serially, and each has an expected duration of 10 days. The conventional CPM analysis indicates that the duration of the network is 30 days.

The Gantt chart for the example is shown in Figure 4.27. For comparison, Figure 4.28 displays the same network but with lead–lag constraints. For example, there is an *SS* constraint of 2 days and an *FF* constraint of 2 days between activities A and B. Thus, activity B can start as early as 2 days after activity A starts, but it cannot finish until 2 days after the completion of A. In other words, *at least* 2 days must separate the start times of A and B. Similarly, at least 2 days must separate the finish times of A and B. A similar precedence relation exists between activities B and C. The earliest and latest times obtained by considering the lag constraints are indicated in Figure 4.27.

The calculations show that if B is started just 2 days after A is started, then it can be completed as early as 12 days as opposed to the 20 days required in the case of conventional CPM. Similarly, activity C can finish in 14 days, which is considerably less than the 30 days calculated by conventional CPM. The lead–lag constraints allow us to compress or overlap activities. Depending on the nature of the tasks involved, an activity does not have to wait until its predecessor finishes before it can start. Figure 4.29 depicts the Gantt chart for the example incorporating the lead–lag constraints. As we can see, a portion of a succeeding activity can be performed simultaneously with a portion of a preceding activity.

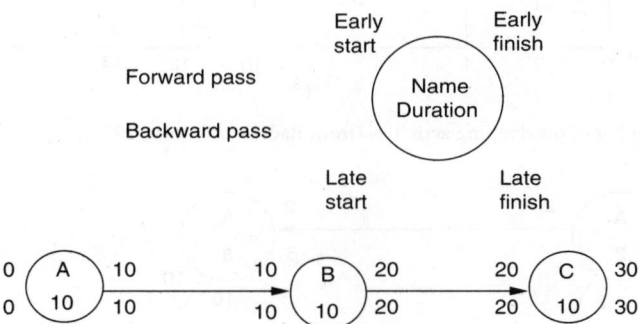

FIGURE 4.26 Serial activities in simple CPM network (from Badiru and Pulat, 1995).

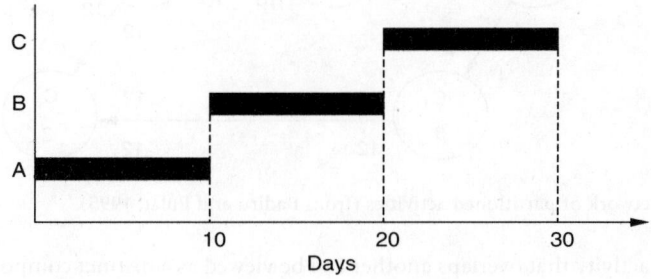

FIGURE 4.27 Gantt chart for serial network (from Badiru and Pulat, 1995).

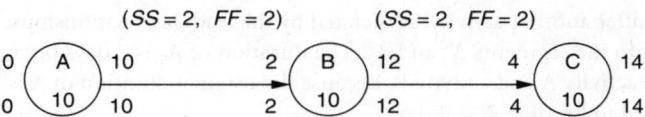

FIGURE 4.28 Serial network with lead and lag constraints (from Badiru and Pulat, 1995).

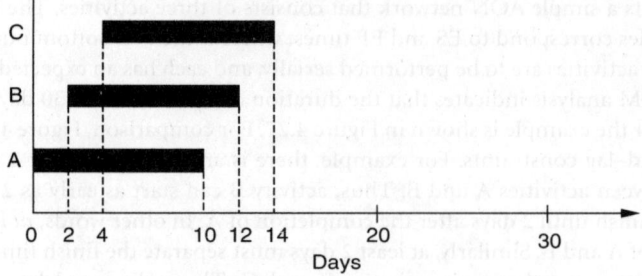

FIGURE 4.29 Gantt chart for network with lead and lag constraints (from Badiru and Pulat, 1995).

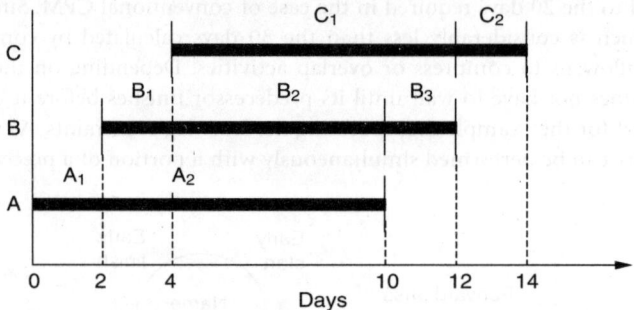

FIGURE 4.30 Partitioning of overlapping activities (from Badiru and Pulat, 1995).

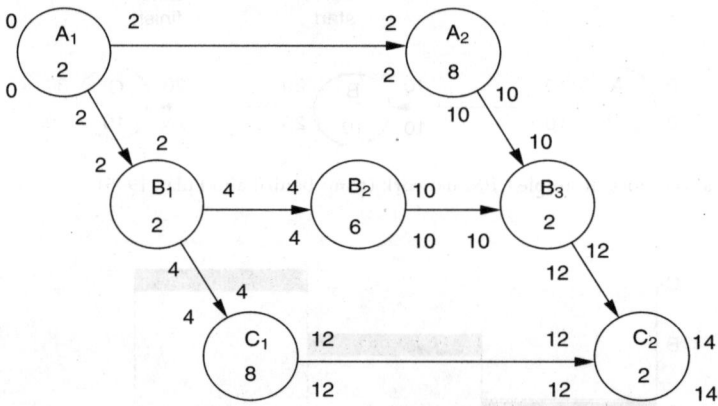

FIGURE 4.31 AON network of partitioned activities (from Badiru and Pulat, 1995).

The portion of an activity that overlaps another can be viewed as a distinct component of the required work. Thus, partial completion of an activity may be evaluated. Figure 4.30 shows how each of the three activities is partitioned into contiguous parts. Even though there is no physical break or termination of work in any activity, the distinct parts are determined on the basis of the amount of work that must be completed before or after another activity, as dictated by the lead–lag relationships. In Figure 4.30, activity A is partitioned into the segments A_1 and A_2. The duration of A_1 is 2 days because there is an $SS = 2$ relationship between activity A and activity B. Because the original duration of A is 10 days, the duration of A_2 is then calculated to be $10 - 2 = 8$ days.

Activity B is similarly partitioned into segments B_1, B_2, and B_3. The duration of B_1 is 2 days because there is an $SS = 2$ relationship between activity B and activity C. The duration of B_3 is also 2 days because there is an $FF = 2$ relationship between activities A and B. Because the original duration of B is 10 days, the duration of B_2 is calculated to be $10 - (2 + 2) = 6$ days. In a similar manner, activity C is partitioned into C_1 and C_2. The duration of C_2 is 2 days because there is an $FF = 2$ relationship between activity A and activity C. Given that the original duration of C is 10 days, the duration of C_1 is then calculated to be $10 - 2 = 8$ days. Figure 4.31 shows a conventional AON network drawn for the three activities after they are partitioned into distinct parts. The conventional forward and backward passes reveal that all of the activity parts are on the critical path. This makes sense, because the original three activities are performed serially and none of them has been physically split. Note that there are three critical paths in Figure 4.31, each with a length of 14 days. It should also be noted that the distinct segments of each activity are performed contiguously.

4.8 Linear Programming Approach for Critical Path Method Analysis

Many classical network problems can be formulated as linear programs and solved using standard algorithms. Finding the shortest and longest paths through a network are two such examples. Of course, the latter is exactly the problem that is solved in CPM analysis. To see its linear programming representation, we make use of the following notation, and assume an AOA model:

i, j = indices for nodes in the network; each node corresponds to an event; $i = 1$ is the unique project start node

N = set of nodes or events

n = number of events in the network; n is the unique node marking the end of the project

A = set of arcs in the network; each arc (i, j) corresponds to a project activity, where i denotes its start event and j its end event

L_{ij} = the length of the activity that starts at node i and terminates at node j

t_i = decision variable associated with the start time of event $i \in N$

The following linear program (LP) schedules all events and all activities in a feasible manner such that the project finishes as early as possible, assuming that work begins at time $t_1 = 0$:

$$\text{Minimize } t_n \tag{4.9a}$$

$$\text{subject to } t_j - t_i \geq L_{ij} \quad \text{for all activities } (i, j) \in A \tag{4.9b}$$

$$t_1 = 0, t_i \geq 0 \quad \text{for all } i \in N \tag{4.9c}$$

Note that the nonnegativity condition $t_i \geq 0$ is redundant, and that the last event t_n denotes the completion time of the project.

The slack associated with a nonbinding constraint in Equation (4.9b) represents the slack of the corresponding activity given the start times t_i found by the LP. These values may not coincide with the CPM calculations. To find the total slack of an activity it is necessary to perform sensitivity (ranging) analysis on the LP solution. The amount that each right-hand side (L_{ij}) can be increased without changing the optimal solution is equivalent to the total slack of activity (i, j).

The LP formulation for the example project is

Minimize t_6

subject to

$$\begin{array}{lll} t_2 - t_1 \geq 5 & \text{activity A} \\ t_3 - t_1 \geq 3 & \text{activity B} \\ t_4 - t_2 \geq 8 & \text{activity C} \\ t_4 - t_3 \geq 7 & \text{activity D} \\ t_4 - t_1 \geq 7 & \text{activity E} \\ t_5 - t_4 \geq 4 & \text{activity F} \\ t_6 - t_5 \geq 5 & \text{activity G} \end{array}$$

$$t_3 - t_2 \geq 0 \quad \text{dummy } D_1$$
$$t_1 = 0$$

Using the Excel add-in that comes with the book by Jensen and Bard (2003), we find the solution to be $t = (0, 5, 6, 13, 17, 22)$. The slack vector for the first eight rows is $(0, 3, 0, 0, 6, 0, 0, 1)$. Notice that these results differ slightly from those in Tables 4.4 and 4.5. To guarantee that the LP (4.9a)–(4.9c) finds the earliest time when each event can start, as was done in Section 4.6.1, the following penalty term must be added to the objective function (4.9a):

$$\varepsilon \sum_{i=2}^{n-1} t_i$$

where $\varepsilon > 0$ is an arbitrarily small constant. Conceptually, in the augmented formulation, the computations are made in two stages. First, t_n is found. Then, given this value, a search is conducted over the set of alternative optima to find the minimum values of t_i, $i = 2, \ldots, n - 1$. In reality, all the computations are made in one stage, not two.

4.9 Aggregating Activities in the Network

The detailed network model of a project is very useful in scheduling and monitoring progress at the operational (short-term) level. Management concerns at the tactical or strategic level, however, create a need for a focused presentation that eliminates unnecessary clutter. For projects that span a number of years and include hundreds of activities, it is likely that only a portion of those activities will be active or require close control at any point in time. To facilitate the management function, there is a need to condense information and aggregate tasks. The two common tools used for this purpose are hammock activities and milestones.

4.9.1 Hammock Activities

When a group of activities has a common start and a common end point, it is possible to replace the entire group with a single activity, called a hammock activity. For example, in the network depicted in Figure 4.32, it is possible to use a hammock activity between events 4 and 6. In so doing, activities F and G are collapsed into FG, whose duration is the sum of L_{45} and L_{56}.

In general, the duration of a hammock activity is equal to the duration of the longest sequence of activities that it replaces. If another hammock activity is used to represent A, B, C, D, and E, then its length would be

$$\max \begin{Bmatrix} L_{12} + L_{24} \\ L_{13} + L_{34} \\ L_{12} + L_{23} + L_{34} \\ L_{14} \end{Bmatrix} = \max \begin{Bmatrix} 5 + 8 \\ 3 + 7 \\ 5 + 0 + 7 \\ 7 \end{Bmatrix} = 13$$

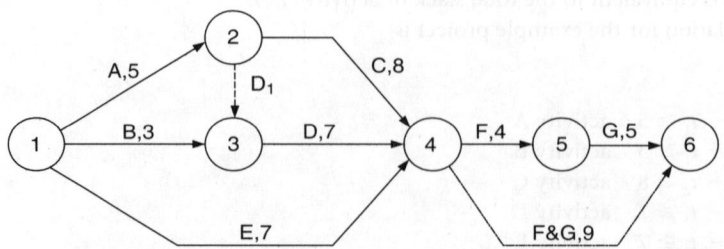

FIGURE 4.32 Example of hammock activity.

Hammock activities reduce the size of a network while preserving, in general, information on precedence relations and activity durations. By using hammock activities, an upper-level network that presents a synoptic view of the project can be created. Such networks are useful for medium (tactical) and long-range (strategic) planning. The common practice is to develop a hierarchy of networks in which the various levels correspond to the levels of either the WBS or the (organizational breakdown structure) OBS. Higher-level networks contain many hammock activities and provide upper management with a general picture of flows, milestones, and overall status. Lower-level networks consist of single activities and provide detailed schedule information for team leaders. Proper use of hammock activities can help in providing the right level of detail to each participant in the project.

4.9.2 Milestones

A higher level of aggregation is also possible by introducing milestones to mark the completion of significant activities. As explained in Section 4.1.1, milestones are commonly used to mark the delivery of goods and services, to denote points in time when payments are due, and to flag important events such as the successful completion of a critical design review. In the simplest case, a milestone can mark the completion of a single activity, as event 2 in our example marks the completion of activity A. It can also mark the completion of several activities, as exemplified by event 4, which denotes the completion of C, D, and E.

By using several levels of aggregation, that is, networks with various layers of hammock activities and milestones, it is possible to design the most appropriate decision support tool for each level of management. Such an exercise should take into account the WBS and the OBS. At the lowest levels of these structures, a detailed network is essential; at higher levels, aggregation by hammock activities and milestones is the norm.

4.10 Dealing with Uncertainty

The critical path method either assumes that the duration of an activity is known and deterministic or that a point estimate such as the mean or mode can be used in its place. It makes no allowance for activity variance. When fluctuations in performance time are low, this assumption is logically justified and has empirically been shown to produce accurate results. When high levels of uncertainty exist, however, CPM may not provide a very good estimate of the project completion time. In these situations, there is a need to account explicitly for the effects of uncertainty. Monte Carlo simulation and PERT are the two most common approaches that have been developed for this purpose.

4.10.1 Simulation Approach

This approach is based on simulating the project by randomly generating performance times for each activity from their perceived distributions. In most cases it is assumed that activity times follow a beta distribution, as discussed in Section 4.2.1. In each simulation run, a sample of the performance time of each activity is taken and a CPM analysis is conducted to determine the critical path and the project finish time for that realization. By repeating the process a large number of times, it is possible to construct a frequency distribution or histogram of the project completion time. This distribution then may be used to calculate the probability that the project finishes by a given date as well as the expected error of each such estimate.

A single simulation run would consist of the following steps:

1. Generate a random value for the duration of each activity from the appropriate distribution.
2. Determine the critical path and its duration using CPM.
3. Record the results.

The number of times that this procedure must be repeated depends on the error tolerances deemed acceptable. Standard statistical tests can be used to verify the accuracy of the estimates.

To understand the calculations, let us focus on the AOA network in Figure 4.23 for the example project and assume that each activity follows a beta distribution with parameter values given in Table 4.8. After performing ten simulation runs, the results listed in Table 4.9 for activity durations, critical path, and project completion time were obtained. Additional data collected but not presented include the earliest and latest start and completion times of each event and activity slacks.

Looking at the first run in Table 4.9, we see that the realized duration of activity A is 6.3, whereas the duration of activity B is 2.2. In the second run, the duration of A is 2.1, and so on. Note that the critical path differs from one replication to the next depending on the randomly generated durations of the activities. In the ten runs reported, the sequence A–D–F–G is the longest (critical) in two replications, whereas the sequence A–C–F–G is critical in the other eight. Activities A, F, and G are critical in 100% of the replications, whereas activity C is critical in 80% and activity D is critical in 20%.

A principal output of the simulation runs is a frequency distribution of the project length (the length of the critical path). Figure 4.33 plots the results of some 50 replications for the example. As can be seen, the project length varied from 17 to 29 weeks, with a mean of 22.5 weeks and a standard deviation of 2.9 weeks. Now let X be a random variable associated with project completion time. The probability of finishing the project within, say, τ weeks can be estimated from the following ratio:

$$P(X \le \tau) = \frac{\text{number of times project finished in} \le \tau \text{ weeks}}{\text{total number of replications}}$$

For the example, if $\tau = 20$ weeks, then the number of runs in which the length of the critical path was $\le$ 20 weeks is seen to be 13, so $P(X \le 20) = 13/50 = 26\%$.

In addition, it is possible to estimate the criticality of each activity. The *criticality index* (CI) of an activity is defined as the proportion of runs in which the activity was on the critical path (i.e., it had a zero slack). Dodin and Elmaghraby (1985) provided some theoretical background on this problem as well as extensive test results for large PERT networks.

TABLE 4.8   Statistics for Example Activities

Activity	Optimistic Time, a	Most Likely Time, m	Pessimistic Time, b	Expected Value, $\hat{d}$	Standard Deviation, $\hat{s}$
A	2	5	8	5	1
B	1	3	5	3	0.66
C	7	8	9	8	0.33
D	4	7	10	7	1
E	6	7	8	7	0.33
F	2	4	6	4	0.66
G	4	5	6	5	0.33

TABLE 4.9   Summary of Simulation Runs for Example Project

Run Number	Activity Duration							Critical Path	Completion Time
	A	B	C	D	E	F	G		
1	6.3	2.2	8.8	6.6	7.6	5.7	4.6	A-C-F-G	25.4
2	2.1	1.8	7.4	8.0	6.6	2.7	4.6	A-D-F-G	17.4
3	7.8	4.9	8.8	7.0	6.7	5.0	4.9	A-C-F-G	26.5
4	5.3	2.3	8.9	9.5	6.2	4.8	5.4	A-D-F-G	25.0
5	4.5	2.6	7.6	7.2	7.2	5.3	5.6	A-C-F-G	23.0
6	7.1	0.4	7.2	5.8	6.1	2.8	5.2	A-C-F-G	22.3
7	5.2	4.7	8.9	6.6	7.3	4.6	5.5	A-C-F-G	24.2
8	6.2	4.4	8.9	4.0	6.7	3.0	4.0	A-C-F-G	22.1
9	2.7	1.1	7.4	5.9	7.9	2.9	5.9	A-C-F-G	18.9
10	4.0	3.6	8.3	4.3	7.1	3.1	4.3	A-C-F-G	19.7

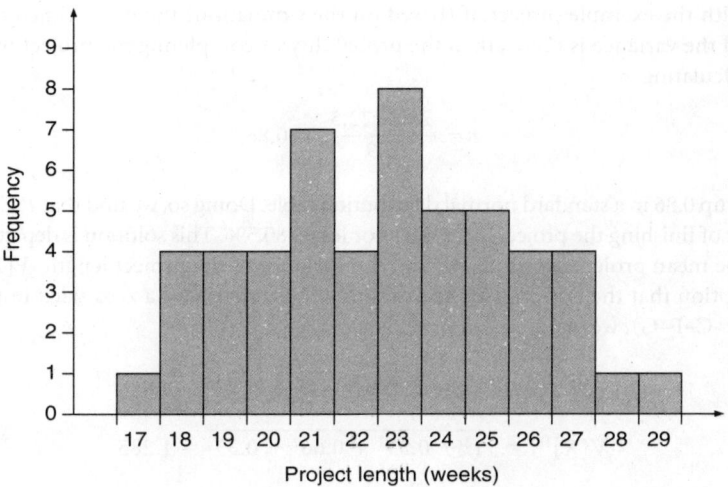

FIGURE 4.33 Frequency distribution of project length for simulation runs.

The simulation approach is easy to implement and has the advantage that it produces arbitrarily accurate results as the number of runs increases. However, for problems of realistic size, the computational burden may be significant for each run, so a balance must be reached between accuracy and effort.

4.10.2 Program Evaluation and Review Technique and Extensions

Two common analytical approaches are used to assess uncertainty in projects. Both are based on the *central limit theorem*, which states that the distribution of the sum of independent random variables is approximately normal when the number of terms in the sum is sufficiently large.

The first approach yields a rough estimate and assumes that the duration of each project activity is an independent random variable. Given probabilistic durations of activities along specific paths, it follows that elapsed times for achieving events along those paths are also probabilistic. Now, suppose that there are n activities in the project, k of which are critical. Denote the durations of the critical activities by the random variables d_i with mean $\bar{d}_i$ and variance s_i^2, $i = 1, \ldots, k$. Then the total project length is the random variable

$$X = d_1 + d_2 + \ldots + d_k$$

It follows that the mean project length, $E[X]$, and the variance of the project length, $V[X]$, are given by

$$E[X] = \bar{d}_1 + \bar{d}_2 + \ldots + \bar{d}_k$$

$$V[X] = s_1^2 + s_2^2 + \ldots + s_k^2$$

These formulas are based on elementary probability theory, which tells us that the expected value of the sum of any set of random variables is the sum of their expected values, and the variance of the sum of independent random variables is the sum of the variances.

Now, invoking the central limit theorem, we can use normal distribution theory to find the probability of completing the project in less than or equal to some given time τ as follows:

$$P(X \le \tau) = P\left(\frac{X - E[X]}{V[X]^{1/2}} \le \frac{\tau - E[X]}{V[X]^{1/2}}\right) = P\left(Z \le \frac{\tau - E[X]}{V[X]^{1/2}}\right) \qquad (4.10)$$

where Z is the standard normal deviate with mean 0 and variance 1. The desired probability in Equation (4.10) can be looked up in any statistics book.

Continuing with the example project, if (based on the simulation) the mean time of the critical path is 22.5 weeks and the variance is $(2.9)^2$, then the probability of completing the project within 25 weeks is found by first calculating

$$z = \frac{25 - 22.5}{2.9} = 0.86$$

and then looking up 0.86 in a standard normal distribution table. Doing so, we find that $P(Z \le 0.86) = 0.805$, so the probability of finishing the project in 25 weeks or less is 80.5%. This solution is depicted in Figure 4.34.

If, however, the mean project length, $E[X]$, and the variance of the project length, $V[X]$, are calculated using the assumption that the critical activities are only those that have a zero slack in the deterministic CPM analysis (A–C–F–G), we get

$$E[X] = 5 + 8 + 4 + 5 = 22$$

$$V[X]^{1\backslash 2} = \sqrt{1^2 + 0.33^2 + 0.66^2 + 0.33^2} = 1.285$$

On the basis of this assumption, the probability of completing the project within 25 weeks is

$$P\left(Z \le \frac{2522}{1.285}\right) = P(Z \le 2.33) = 0.99$$

This probability is higher than 0.805, which was computed using data from the simulation in which both sequences A–C–F–G and A–D–F–G were critical.

The procedure above is, in essence, PERT. Summarizing for an AON network:

1. For each activity i, assess its probability distribution or assume a beta distribution and obtain estimates of a_i, b_i and m_i. These values should be supplied by the project manager or experts who work in the field.
2. If a beta distribution is assumed for activity i, then use the estimates a, b, and m to compute the variance $\hat{s}_i^2$ and mean $\hat{d}_i$ from Equations (4.1) and (4.2). These values then are used in place of the true but unknown values of s_i^2 and $\bar{d}_i$, respectively, in the above formulas for $V[X]$ and $E[X]$.
3. Use CPM to determine the critical path given $\hat{d}_i$, $i = 1, \ldots, n$.
4. Once the critical activities are identified, sum their means and variances to find the mean and variance of the project length.
5. Use Equation (4.10) with the statistics computed in step 4 to evaluate the probability that the project finishes within some desired time.

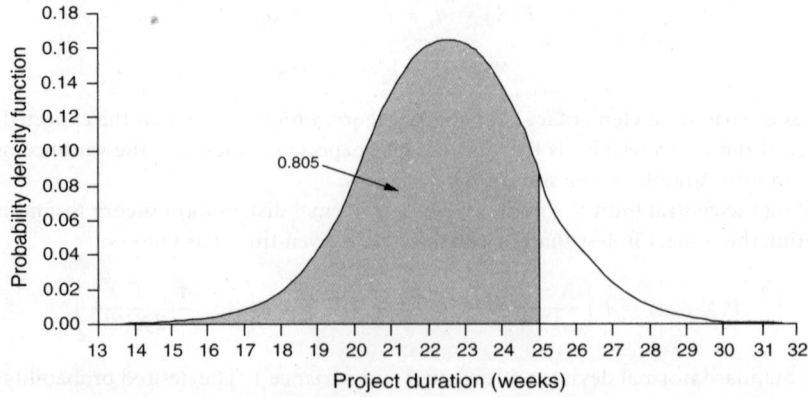

FIGURE 4.34   Example of probabilistic analysis with PERT.

Using PERT, it is possible to estimate completion time for a desired completion probability. For example, for a 95% probability the corresponding z value is $z_{95} = 1.64$. Solving for the time τ, for which the probability to complete the project is 95%, we get

$$z_{0.95} = \frac{\tau - 22.5}{2.9} = 1.64$$

or

$$\tau = (1.64)(2.5) + 22.5 = 27.256 \text{ weeks}$$

A shortcoming of the standard PERT calculations is that they ignore all activities that are not on the critical path. A more accurate analytical approach is to identify each sequence of activities that lead from the start node of the project to the finish event, and then to calculate separately the probability that the activities that compose each sequence will be completed by a given date. This step can be done as above by assuming that the central limit theorem holds for each sequence and then applying normal distribution theory to calculate the individual path probabilities. It is necessary, though, to make the additional assumption that the sequences themselves are statistically independent to proceed. This means that the time to traverse each path in the network is independent of what happens on the other paths. Although it is easy to see that this is rarely true because some activities are sure to be on more than one path, empirical evidence suggests that good results can be obtained if there is not too much overlap.

Once these calculations are performed, assuming that the various sequences are independent of each other, the probability of completing the project by a given date is set equal to the product of the individual probabilities that each sequence is finished by that date. That is, given n sequences with completion times $X_1, X_2, \ldots, X_n$, the probability that X is $\leq \tau$ is found from

$$P(X_1 \leq \tau) = P(X_1 \leq \tau)P(X_2 \leq \tau) \ldots P(X_n \leq \tau) \tag{4.11}$$

where now the random variable $X = \max\{X_1, X_2, \ldots, X_n\}$.

Example 5   Consider the simple project in Figure 4.35. If no uncertainty exists in activity durations, then the critical path is A-B and exactly 17 weeks are required to finish the project. Now if we assume that the durations of all four activities are normally distributed (the corresponding means and standard deviations are listed under the arrows in Figure 4.35), then the durations of the two sequences are also normally distributed (i.e., $N(\mu, \sigma)$), with the following parameters:

$$\text{length}(A-B) = X_1 \sim N(17, 3.61)$$
$$\text{length}(C-D) = X_2 \sim N(16, 3.35)$$

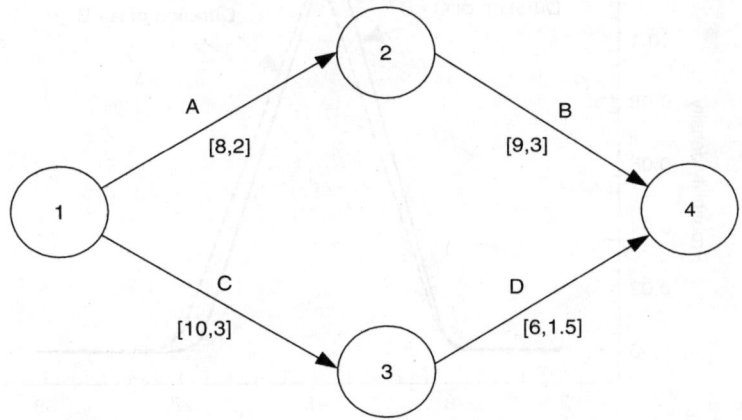

FIGURE 4.35   Stochastic network.

The accompanying probability density functions are plotted in Figure 4.36. It should be clear that the project can end in 17 weeks only if both A–B and C–D are completed within that time. The probability that A–B finishes within 17 weeks is

$$P(X_1 \le 17) = P\left(Z \le \frac{17 - 17}{3.61}\right) = P(Z \le 0) = 0.5$$

and similarly for C–D,

$$P(X_2 \le 17) = P\left(Z \le \frac{17 - 16}{3.35}\right) = P(Z \le 0.299) = 0.62$$

Using Equation (4.11), we can now determine the probability that both sequences finish within 17 weeks:

$$P(X \le 17) = P(X_1 \le 17)P(X_2 \le 17) = (0.5)(0.62) = 0.31$$

Thus, the probability that the project will finish by week 17 is ~31%. A similar analysis for 20 weeks yields $P(X \le 20) = 0.7 = 70\%$.

The approach that is based on calculating the probability of each sequence completing by a given due date is accurate only if the sequences are independent. This is not the case when one or more activities are members of two or more sequences. Consider, for example, the project in Figure 4.37. Here, activity E is a member of the two sequences that connect the start of the project (event 1) to its termination node (event 5). The expected lengths and standard deviations of these sequences are

Sequence	Expected Length	Standard Deviation
A–B–E	$8 + 9 + 3 = 20$	$\sqrt{2^2 + 3^2 + 4^2} = 5.39$
C–D–E	$10 + 6 + 3 = 19$	$\sqrt{3^2 + 1.5^2 + 4^2} = 5.22$

The probability that the sequence A–B–E will be completed in 17 days is calculated as follows:

$$z = \frac{17 - 20}{5.39} = -0.5565 \quad \text{implying that } P = 0.29$$

which is obtained from a standard normal distribution table by noting that

$$P(Z \le -z) = 1 - P(Z \le z)$$

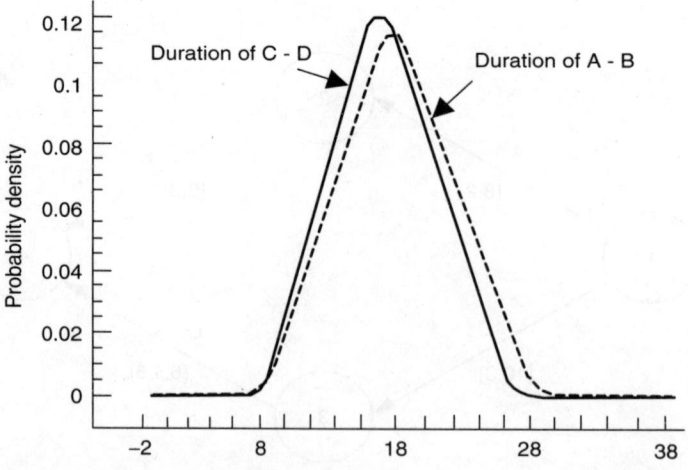

FIGURE 4.36 Performance time distribution for the two sequences.

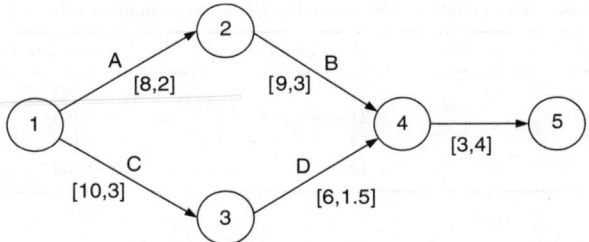

FIGURE 4.37 Stochastic network with dependent sequences.

Similarly, the probability that the sequence C–D–E will be completed in 17 days is calculated by determining $z = (17 - 19)/5.22 = -0.383$ and then using Table 4.9 (C-1) a standard normal distribution table to find $P = 0.35$.

Thus, the simple PERT estimate (based on the critical sequence A–B–E) indicates that the probability of completing the project in 17 days is 29%. If both sequences A–B–E and C–D–E are taken into account, then the probability of completing the project in 17 days is estimated as

$$P(X_{ABE} \leq 17)P(X_{CDE} \leq 17) = (0.29)(0.35) = 0.1 \text{ or } 10\%$$

assuming that the two sequences are independent. However, because activity E is common to both sequences, the true probability of completing the project in 17 days is somewhere between 10 and 29%.

The next question that naturally arises is what to do if only the parameters of the distribution are known but not its form (e.g., beta, normal), and the number of activities is too small to rely on the central limit theorem to give accurate results. In this case, Chebyshev's inequality can be used to calculate project duration probabilities (see Montgomery and Runger, 2003). The underlying theorem states that if X is a random variable with mean μ and variance σ^2, then for any $k > 0$,

$$P(|X - \mu| \geq k\sigma) \leq \frac{1}{k^2}$$

An alternative form is

$$P(|X - \mu| < k\sigma) \geq 1 - \frac{1}{k^2}$$

Based on the second inequality, the probability of a random variable being within ±3 standard deviations of its mean is at least 8/9, or 89%. Although this might not be a tight bound in all cases, it is surprising that such a bound can be found to hold for all possible discrete and continuous distributions.

To illustrate the effect of uncertainty, consider the example project. Four sequences connect the start node to the finish node. The mean length and the standard deviation of each sequence are summarized in Table 4.10.

The probability of completing each sequence in 22 weeks is computed next and summarized in Table 4.11.

Based on the simple PERT analysis, the probability of completing the project in 22 weeks is 0.5. If both sequences A–C–F–G and A–D–F–G are considered and assumed to be independent, the probability is reduced to $(0.5)(0.73) = 0.365$.

Because three activities (A, F, G) are common to both sequences, the actual probability of completing the project in 22 weeks is closer to 0.5 than to 0.365. Based on the data in Figure 4.33, we see that in 24 of 50 simulation runs, the project duration was 22 weeks or less. This implies that the probability of completing the project in 22 weeks is $24/50 = 0.48$, or 48%.

Continuing with this example, if the Chebyshev's inequality is used for the critical path ($\mu = 22$, $\sigma = 1.285$), then the probability of completing the project in, say, $22 + (2)(1.285) = 24.57$ weeks is approximately

$$1 - \left(\frac{1}{2}\right)^2 = \frac{3}{4} = 0.75$$

TABLE 4.10 Mean Length and Standard Deviation for Sequences in Example Project

Sequence	Mean Length	Standard Deviation
A–C–F–G	22	1.285
A–D–F–G	21	1.595
B–D–F–G	19	1.407
E–F–G	16	0.808

TABLE 4.11 Probability of Completing Each Sequence in 22 Weeks

Sequence	z-Value	Probability
A–C–F–G	$\dfrac{22 - 22}{1.285} = 0$	0.5
A–D–F–G	$\dfrac{22 - 21}{1.595} = 0.626$	0.73
B–D–F–G	$\dfrac{22 - 19}{1.407} = 2.13$	0.98
E–F–G	$\dfrac{22 - 16}{0.808} = 7.42$	1.0

By way of comparison, using the normal distribution assumption, the corresponding probability is

$$P\left(Z \le \frac{24.57 - 22}{1.285}\right) = P(Z \le 2) = 0.97$$

Of the two, the Chebyshev estimate is likely to be more reliable given that there are only a few activities on the critical path.

Because uncertainty is bound to be present in most activities, it is possible that after determining the critical path with CPM, a noncritical activity may become critical as certain tasks are completed. From a practical point of view, this suggests the basic advantage of ES schedules. Starting each activity as soon as possible reduces the chances of a noncritical activity becoming critical and delaying the project.

4.11 Critique of Program Evaluation and Review Technique and Critical Path Method Assumptions

Both PERT and CPM are models of projects and are hence open to a wide range of technical criticism including (1) the difficulty in accurately estimating durations, variances, and costs; (2) the validity of using the beta distribution in representing durations; (3) the validity of applying the central limit theorem; and (4) the heavy focus on the critical path for project control. In addition, PERT and CPM analysis is based on the precedence graph, which contains only two types of information: activity times and precedence constraints. The results may be highly sensitive to the data estimates and defining relationships.

Moreover, Schonberger (1981) showed that a PERT estimate that is based on the assumption that the variance of a sequence of activities is equal to the sum of the activity variances (i.e., that activities and sequences are independent) can lead to a consistent error in estimating the completion time of a project. A related problem, investigated by Britney (1976), concerns the cost of over- and underestimating activity duration times. He found that underestimates precipitate the reallocation of resources and, in many cases, engender costly project delays. Overestimates, conversely, result in inactivity and tend to misdirect management's attention to relatively unfruitful areas, causing planning losses. (Britney recommends a modification of PERT called BPERT, which uses concepts from Bayesian decision theory to consider these two categories of cost explicitly in deriving a project network plan.)

Another problem that sometimes arises, especially when PERT is used by subcontractors who work with the government, is the attempt to "beat" the network in order to get on or off the critical path. Many government contracts provide cost incentives for finishing a project early or are negotiated on a "cost-plus-fixed-fee" basis. The contractor who is on the critical path generally has more leverage in obtaining additional funds from these contracts because he or she has a major influence in determining the duration of the project. In contrast, some contractors deem it desirable to be less "visible" and therefore adjust their time estimates and activity descriptions in such a way as to ensure that they will not be on the critical path. This criticism, of course, reflects more on the use of the method than on the method itself, but PERT and CPM, by virtue of their focus on the critical path, enable such ploys to be used.

Finally, the cost of applying CPMs to a project is sometimes used as a basis for criticism. However, the cost of applying PERT or CPM rarely exceeds 2% of total project cost. Thus, this added cost is generally outweighed by the savings from improved scheduling and reduced project time.

As with any analytic technique, it is important, when using CPM and PERT, to fully understand the underlying assumptions and limitations that they impose. Management must be sure that the people who are charged with monitoring and controlling activity performance have a working knowledge of the statistical features of PERT as well as the general nature of critical path scheduling. Correct application of these techniques can provide a significant benefit in each phase of the project's life cycle as long as the above-mentioned pitfalls are avoided.

4.12 Critical Chain Process

Partially in response to these criticisms, Goldratt (1997) developed the critical chain buffer management (CCBM) process, which is an application of his theory of constraints to managing and scheduling projects. With CCBM, several alterations are made to traditional PERT. First, all individual activity slack, or "buffer," becomes project buffer. Each team member, responsible for his or her component of the activity network, creates a duration estimate free from any padding — one, say, that is based on a 50% probability of success. All activities on the critical chain (path) and feeder chains (noncritical chains in the network) are then linked with minimal time padding. The project buffer is aggregated and some proportion of the saved time (Goldratt uses a 50% rule of thumb) is added to the project.

Even adding 50% of the saved time significantly reduces the overall project schedule while requiring team members to be concerned less with activity padding and more with task completion. Even if they miss their delivery date (as they are likely to do) 50% of the time, the overall effect on the project's duration is minimized because of the downstream aggregated buffer.

The same approach can also be used for tasks that are not on the critical chain. Accordingly, all feeder path activities are reduced by the same order of magnitude and a feeder buffer is constructed for the overall noncritical chain of activities. Finally, CCBM distinguishes between its use of buffer and the traditional PERT use of project slack. With the PERT approach, project slack is a function of the overall completed activity network. In other words, slack is an outcome of the task dependencies, whereas CCBM's buffer is used as an *a priori* planning input that is based on a reasoned cut in each activity and the application of an aggregated project buffer at the end.

Proponents of CCBM argue that it is more than a new scheduling technique, representing instead a different paradigm by which project management should be viewed. The CCBM paradigm argues for truth in activity duration estimation, a "just in time" approach to scheduling noncritical activities, and greater discipline in project scheduling and control as a result of more open communication among internal project stakeholders.

The newness of CCBM is a point refuted by some who see the technique as either ill suited to many types of projects or simply a reconceptualization of well-understood scheduling methodologies (e.g., PERT). Nevertheless, a growing body of case studies and proponents is emerging to champion the CCBM process as it continues to diffuse throughout project organizations.

Even so, critical chain project management is not without its critics. Several arguments against the process include the following charges and perceived weaknesses in the methodology:

1. Lack of project milestones makes coordinated scheduling, particularly with external suppliers, highly problematic. Critics contend that the lack of in-process project milestones adversely affects the ability to coordinate schedule dates with suppliers who provide the external delivery of critical components.
2. Although it may be true that CCBM brings increased discipline to project scheduling, efficient methods for applying this technique to a firm's portfolio of projects are unclear; that is, CCBM seems to offer benefits on a project-by-project basis, but its usefulness at the program level has not been proved. Furthermore, because CCBM argues for dedicated resources in a multiproject environment where resources are shared, it is impossible to avoid multitasking, which severely limits its power.
3. Evidence of its success is still almost exclusively anecdotal and based on single-case studies. Debating the merits and pitfalls of CCBM has remained largely an intellectual exercise among academics and writers of project management theory. No large-scale empirical research exists to either confirm or refute its efficacy.
4. Critics also charge that Goldratt's evaluation of duration estimation is overly negative and critical, suggesting that his contention of huge levels of activity duration estimation "padding" is exaggerated.

Of course, it must be remembered that models, whether associated with CPM, PERT, or CCBM, are simplifications of reality designed to support analysis and decision making by focusing on the most important aspects of the problem. They should be judged not so much by their fidelity to the actual system but by the insight that they provide, by the certainty with which they show the correct consequences of the working assumptions, and by the ease with which the problem structure can be communicated.

4.13 Scheduling Conflicts

The discussion so far assumed that the only constraints on the schedule are precedence relations among activities. On the basis of these constraints, the early and late time of each event and the early and late start and finish of each activity are calculated.

In most projects, there are additional constraints that must be addressed, such as those associated with resource availability and the budget. In some cases, ready time and due-date constraints also exist. These constraints specify a time window in which an activity must be performed. In addition, there may be a target completion date for the project or a due date for a milestone. If these due dates are earlier than the corresponding dates derived from the CPM analysis, then the accompanying schedule will not be feasible.

There are several ways to handle these types of infeasibilities, such as

1. Reducing some activity durations by allocating more resources to them.
2. Eliminating some activities or reducing their lengths by using a more effective technology. For example, conventional painting, which requires the application of several layers of paint and a long drying time, may be replaced by anodizing — a faster but more expensive process.
3. Replacing some precedence relations of the "finish to start" type by other precedence relations, such as "start to start," without affecting quality, cost, or performance. When this is possible, a significant amount of time may be saved.

It is common to start the scheduling analysis with each activity being performed in the most economical way and assuming "finish to start" precedence relations. If infeasibility is detected, then one or more of the foregoing courses of action can be used to circumnavigate the cause of the problem.

References and Further Readings

Estimating the Duration of Project Activities

Banks, J., Carson J.S., Nelson, B.L., and Nicol, D.M., *Discrete-Event System Simulation*, 3rd ed., Prentice-Hall, Upper Saddle River, NJ, 2001.

Britney, R.R., Bayesian point estimation and the PERT scheduling of stochastic activities, *Manage. Sci.*, 22, 938–948, 1976.

Dodin, B., Bounding the project completion time distribution in PERT networks, *Oper. Res.*, 33, 862–881, 1985.

Grubbs, F., Attempts to validate certain PERT statistics or 'picking on PERT,' *Oper. Res.*, 10, 912–915, 1962.

Hershauer, J.C. and Nabielsky, G., Estimating activity times, *J. Syst. Manage.*, 23, 17–21, 1972.

Montgomery, D.C. and Runger, G.C., *Applied Statistics and Probability for Engineers*, 3rd ed., Wiley, New York, 2003.

Perry, C. and Greig, I.D., Estimating the mean and variance of subjective distributions in PERT and decision analysis, *Manage. Sci.*, 21, 1477–1480, 1975.

Project Scheduling

Clark, K.B. and Fujimoto, T., Overlapping problem solving in product development, in *Managing International Manufacturing*, Ferdows, K., Ed., North-Holland, New York, 1989.

Goldratt, E., *Critical Chain*, North River Press, Great Barrington, MA, 1997.

Hartley, K.O., The project schedule, in *Project Management: A Reference for Professionals*, Kimmon, R.L. and Lowree, J.H., Eds., Marcel Dekker, New York, 1989.

Hillier, F.S. and Lieberman, G.J., *Introduction to Operations Research*, 7th ed., McGraw-Hill, Boston, 2001.

Meredith, J.R. and Mantel, S.J., Jr., *Project Management: A Managerial Approach*, 4th ed., Wiley, New York, 1999.

Neumann, K., Schwindt, C., and Zimmermann, J., *Project Scheduling with Time Windows and Scarce Resources: Temporal and Resource Constrained Project Scheduling with Regular and Nonregular Objective Functions, Lecture Notes in Economics and Mathematical Systems*, Vol. 508, Springer, Amsterdam, 2002.

Steyn, H., An investigation into the fundamentals of critical chain project scheduling, *Int. J. Proj. Sched.*, 19, 363–369, 2000.

Vazsonyi, A., The history of the rise and fall of the PERT method, *Manage. Sci.*, 16, B449–B455, 1970.

Webster, F.M., *Survey of CPM Scheduling Packages and Related Project Control Programs*, Project Management Institute, Drexel Hill, PA, 1991.

CPM Approach

Badiru, A.B. and Pulat, P.S., *Comprehensive Project Management: Integrating Optimization Models, Management Principles, and Computers*, Prentice-Hall, Englewood Cliffs, NJ, 1995.

Cornell, D.G., Gotlieb, C.C., and Lee, Y.M., Minimal event-node network of project precedence relations, *Commun. ACM*, 16, 296–298, 1973.

Jewell, W.S., Divisible activities in critical path analysis, *Oper. Res.*, 13, 747–760, 1965.

Kelley, J.E., Jr. and Walker, M.R., Critical path planning and scheduling, *Proceedings of the Eastern Joint Computer Conference*, Boston, pp. 160–173, 1979.

PERT Approach

Burgher, P.H., PERT and the auditor, *Account. Rev.*, 39, 103–120, 1964.

Dodin, M.B., Determining the K most critical paths in PERT networks, *Oper. Res.*, 32, 859–877, 1984.

Dodin, M.B. and Elmaghraby, S.E., Approximating the criticality indices of the activities in PERT networks, *Manage. Sci.*, 31, 207–223, 1985.

Fazar, W., Program evaluation and review technique, *Am. Stat.*, 13, 10, 1959.

Fisher, D.L., Saisi, D., and Goldstein, W.M., Stochastic PERT networks: OP diagrams, critical paths and the project completion time, *Comp. Oper. Res.*, 12, 471–482, 1985.

PERT, *Program Evaluation Research Task, Phase I Summary Report,* Vol. 7, Special Projects Office, Bureau of Ordinance, Department of the Navy, Washington, DC, 1958, pp. 646–669.

Van Slyke, R.M., Monte Carlo methods and the PERT problem, *Oper. Res.*, 11, 839–860, 1963.

PERT and CPM Assumptions

Chase, R.B., Jacobs, F.R., and Aquilano, N.J., *Operations Management for Competitive Advantage,* 10th ed., McGraw-Hill, Boston, 2003.

Golenko-Ginzburg, D., On the distribution of activity time in PERT, *J. Oper. Res. Soc.*, 39, 767–771, 1988.

Littlefield, T.K. and Randolph, P.H., PERT duration times: mathematics or MBO, *Interfaces*, 21, 92–95, 1991.

Sasieni, M.W., A note on PERT times, *Manage. Sci.*, 16, 1652–1653, 1986.

Schonberger, R.J., Why projects are always late: a rationale based on manual simulation of a PERT/CPM network, *Interfaces*, 11, 66–70, 1981.

Wiest, J.D. and Levy, F.K., *A Management Guide to PERT/CPM,* 2nd ed., Prentice-Hall, Englewood Cliffs, NJ, 1977.

Computational Issues

Draper, N. and Smith, H., *Applied Regression Analysis,* 3rd ed., John Wiley & Sons, New York, 1998.

Hindelang, T.J. and Muth, J.F., A dynamic programming algorithm for decision CPM networks, *Oper. Res.*, 27, 225–241, 1979.

Jensen, P.A. and Bard, J.F., *Operations Research Models and Methods,* John Wiley & Sons, New York, 2003.

Kulkarni, V.G. and Provan, J.S., An improved implementation of conditional Monte Carlo estimation of path lengths in stochastic networks, *Oper. Res.*, 33, 1389–1393, 1985.

5

Cost Concepts and Estimation

Adedeji B. Badiru
The University of Tennessee

This chapter covers economic aspects of manufacturing projects. Basic cash flow analysis is presented. Other topics covered include cost concepts, cost estimation, cost monitoring, budgeting allocation, and inflation.

5.1 Cost Concepts and Definitions

Cost management in a project environment refers to the functions required to maintain effective financial control over the project throughout its life cycle. There are several cost concepts that influence the economic aspects of managing projects. Within a given scope of analysis, there may be a combination of different types of cost aspects to consider. These cost aspects include the ones discussed below:

Actual cost of work performed. This represents the cost actually incurred and recorded in accomplishing the work performed within a given time period.

Applied direct cost. This represents the amounts recognized in the time period associated with the consumption of labor, material, and other direct resources, without regard to the date of commitment or the date of payment. These amounts are to be charged to work-in-process (WIP) when resources are actually consumed, material resources are withdrawn from inventory for use, or material resources are received and scheduled for use within 60 days.

Budgeted cost for work performed.   This is the sum of the budgets for completed work plus the appropriate portion of the budgets for level of effort and apportioned effort. Apportioned effort is effort that by itself is not readily divisible into short-span work packages but is related in direct proportion to measured effort.

Budgeted cost for work scheduled.   This is the sum of budgets for all work packages and planning packages scheduled to be accomplished (including WIP) plus the level of effort and apportioned effort scheduled to be accomplished within a given period of time.

Direct cost.   This is a cost that is directly associated with the actual operations of a project. Typical sources of direct costs are direct material costs and direct labor costs. Direct costs are those that can be reasonably measured and allocated to a specific component of a project.

Economies of scale.   This refers to a reduction of the relative weight of the fixed cost in total cost by increasing output quantity. This helps to reduce the final unit cost of a product. Economies of scale are often simply referred to as the savings due to *mass production.*

Estimated cost at completion.   This is the sum of the actual direct cost, indirect costs that can be allocated to the contract, and the estimate of costs (direct and indirect) for authorized work remaining.

First cost.   This is the total initial investment required to initiate a project or the total initial cost of the equipment needed to start the project.

Fixed cost.   This is a cost incurred irrespective of the level of operation of a project. Fixed costs do not vary in proportion to the quantity of output. Examples of costs that make up the fixed cost of a project are administrative expenses, certain types of taxes, insurance cost, depreciation cost, and debt servicing cost. These costs usually do not vary in proportion to the quantity of output.

Incremental cost.   This refers to the additional cost of changing the production output from one level to another. Incremental costs are normally variable costs.

Indirect cost.   This is a cost that is indirectly associated with project operations. Indirect costs are those that are difficult to assign to specific components of a project. An example of an indirect cost is the cost of computer hardware and software needed to manage project operations. Indirect costs are usually calculated as a percentage of a component of direct costs. For example, the indirect costs in an organization may be computed as 10% of direct labor costs.

Life-cycle cost.   This is the sum of all costs, recurring and nonrecurring, associated with a project during its entire life cycle.

Maintenance cost.   This is a cost that occurs intermittently or periodically for the purpose of keeping project equipment in good operating condition.

Marginal cost.   This is the additional cost of increasing production output by one additional unit. The marginal cost is equal to the slope of the total cost curve or line at the current operating level.

Operating cost.   This is a recurring cost needed to keep a project in operation during its life cycle. Operating costs may include labor, material, and energy costs.

Opportunity cost.   This is the cost of forgoing the opportunity to invest in a venture that would have produced an economic advantage. Opportunity costs are usually incurred owing to limited resources that make it impossible to take advantage of all investment opportunities. They are often defined as the cost of the best rejected opportunity. Opportunity costs can be incurred because of a missed opportunity rather than due to an intentional rejection. In many cases, opportunity costs are hidden or implied because they typically relate to future events that cannot be accurately predicted.

Overhead cost.   This is a cost incurred for activities performed in support of the operations of a project. The activities that generate overhead costs support the project efforts rather than contribute directly to the project goal. The handling of overhead costs varies widely from company to company. Typical overhead items are electric power cost, insurance premiums, cost of security, and inventory carrying cost.

Standard cost.   This is a cost that represents the normal or expected cost of a unit of the output of an operation. Standard costs are established in advance. They are developed as a composite of several component costs such as direct labor cost per unit, material cost per unit, and allowable overhead charge per unit.

Sunk cost.   This is a cost that occurred in the past and cannot be recovered under the present analysis. Sunk costs should have no bearing on the prevailing economic analysis and project decisions. Ignoring sunk costs is always a difficult task for analysts. For example, if $950,000 was spent 4 years ago to buy a piece of equipment for a technology-based project, a decision on whether or not to replace the equipment now should not consider that initial cost. But uncompromising analysts might find it difficult to ignore that much money. Similarly, an individual making a decision on selling a personal automobile would typically try to relate the asking price to what was paid for the automobile when it was acquired. This is wrong under the strict concept of sunk costs.

Total cost. This is the sum of all the variable and fixed costs associated with a project.

Variable cost. This is a cost that varies in direct proportion to the level of operation or quantity of output. For example, the costs of material and labor required to make an item will be classified as variable costs because they vary with changes in level of output.

5.2 Cash Flow Analysis

The basic reason for performing economic analysis is to make a choice between mutually exclusive projects that are competing for limited resources. The cost performance of each project will depend on the timing and levels of its expenditures. The techniques of computing cash flow equivalence permit us to bring competing project cash flows to a common basis for comparison. The common basis depends on the prevailing interest rate. Two cash flows that are equivalent at a given interest rate will not be equivalent at a different interest rate. The basic techniques for converting cash flows from one point in time to another are presented in the next section.

5.2.1 Time Value of Money Calculations

Cash flow conversion involves the transfer of project funds from one point in time to another. The following notation is used for the variables involved in the conversion process:

i = interest rate per period
n = number of interest periods
P = a present sum of money
F = a future sum of money
A = a uniform end-of-period cash receipt or disbursement
G = a uniform arithmetic gradient increase in period-by-period payments or disbursements

In many cases, the interest rate used in performing economic analysis is set equal to the minimum attractive rate of return (MARR) of the decision maker. The MARR is also sometimes referred to as *hurdle rate*, *required internal rate of return* (IRR), *return on investment* (ROI), or *discount rate*. The value of MARR is chosen with the objective of maximizing the economic performance of a project.

Compound amount factor. The procedure for the single payment compound amount factor finds a future sum of money, F, that is equivalent to a present sum of money, P, at a specified interest rate, i, after n periods. This is calculated as

$$F = P(1 + i)^n$$

A graphic representation of the relationship between P and F is shown in Figure 5.1.

Example A sum of $5000 is deposited in a project account and left there to earn interest for 15 years. If the interest rate per year is 12%, the compound amount after 15 years can be calculated as follows:

$$F = \$5000(1 + 0.12)^{15}$$

$$= \$27,367.85$$

Present worth factor. The present worth factor computes P when F is given. The present worth factor is obtained by solving for P in the equation for the compound-amount factor, i.e.,

$$P = F(1 + i)^{-n}$$

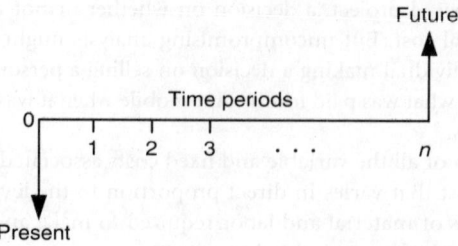

FIGURE 5.1   Single payment compound-amount cash flow.

Suppose it is estimated that $15,000 would be needed to complete the implementation of a project 5 years from now. How much should be deposited in a special project fund now so that the fund would accrue to the required $15,000 exactly 5 years from now? If the special project fund pays interest at 9.2% per year, the required deposit would be

$$P = \$15,000(1 + 0.092)^{-5}$$

$$= \$9660.03$$

Uniform series present worth factor.   The uniform series present worth factor is used to calculate the present worth equivalent, P, of a series of equal end-of-period amounts, A. Figure 5.2 shows the uniform series cash flow. The derivation of the formula uses the finite sum of the present worths of the individual amounts in the uniform series cash flow as shown below. Some formulae for series and summation operations are presented in the Appendix at the end of the book.

$$P = \sum_{t=1}^{n} A(1 + i)^{-1}$$

$$= A\left[\frac{(1 + i)^n - 1}{i(1 + i)^n}\right]$$

Example   Suppose the sum of $12,000 must be withdrawn from an account to meet the annual operating expenses of a multiyear project. The project account pays interest at 7.5% per year compounded on an annual basis. If the project is expected to last 10 years, how much must be deposited in the project account now so that the operating expenses of $12,000 can be withdrawn at the end of every year for 10 years? The project fund is expected to be depleted to zero by the end of the last year of the project. The first withdrawal will be made 1 year after the project account is opened, and no additional deposits will be made in the account during the project life cycle. The required deposit is calculated to be

$$P = \$12,000\left[\frac{(1 + 0.075)^{10} - 1}{0.075(1 + 0.075)^{10}}\right]$$

$$= \$82,368.92$$

Uniform series capital recovery factor.   The capital recovery formula is used to calculate the uniform series of equal end-of-period payments, A, that are equivalent to a given present amount, P. This is the converse of the uniform series present amount factor. The equation for the uniform series capital recovery factor is obtained by solving for A in the uniform series present-amount factor, i.e.,

$$A = P\left[\frac{i(1 + i)^n}{(1 + i)^n - 1}\right]$$

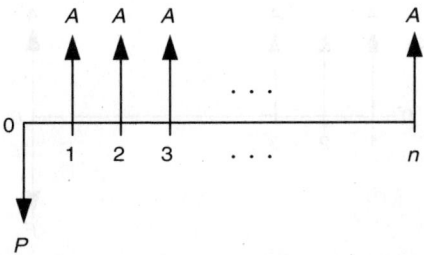

FIGURE 5.2 Uniform series cash flow.

Example Suppose a piece of equipment needed to launch a project must be purchased at a cost of $50,000. The entire cost is to be financed at 13.5% per year and repaid on a monthly installment schedule over 4 years. It is desired to calculate what the monthly loan payments will be. It is assumed that the first loan payment will be made exactly 1 month after the equipment is financed. If the interest rate of 13.5% per year is compounded monthly, then the interest rate per month will be 13.5%/12 = 1.125% per month. The number of interest periods over which the loan will be repaid is 4(12) = 48 months. Consequently, the monthly loan payments are calculated to be

$$A = \$50{,}000 \left[\frac{0.01125(1 + 0.01125)^{48}}{(1 + 0.01125)^{48} - 1} \right]$$

$$= \$1{,}353.82$$

Uniform series compound amount factor. The series compound amount factor is used to calculate a single future amount that is equivalent to a uniform series of equal end-of-period payments. The cash flow is shown in Figure 5.3. Note that the future amount occurs at the same point in time as the last amount in the uniform series of payments. The factor is derived as

$$F = \sum_{t=1}^{n} A(1 + i)^{n-1}$$

$$= A \left[\frac{(1 + i)^n - 1}{i} \right]$$

Example If equal end-of-year deposits of $5,000 are made to a project fund paying 8% per year for 10 years, how much can be expected to be available for withdrawal from the account for capital expenditure immediately after the last deposit is made?

$$F = \$5000 \left[\frac{(1 + 0.08)^{10} - 1}{0.08} \right]$$

$$= \$72{,}432.50$$

Uniform series sinking fund factor. The sinking fund factor is used to calculate the uniform series of equal end-of-period amounts, A, that are equivalent to a single future amount, F. This is the reverse of the uniform series compound amount factor. The formula for the sinking fund is obtained by solving for A in the formula for the uniform series compound amount factor, i.e.,

$$A = F \left[\frac{i}{(1 + i)^n - 1} \right]$$

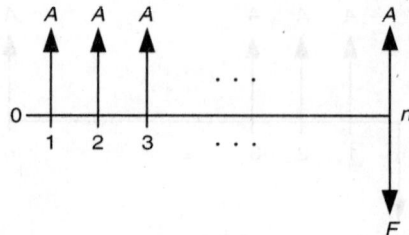

FIGURE 5.3   Uniform series compound-amount cash flow.

Example   How large are the end-of-year equal amounts that must be deposited into a project account so that a balance of $75,000 will be available for withdrawal immediately after the 12th annual deposit is made? The initial balance in the account is zero at the beginning of the first year. The account pays 10% interest per year. Using the formula for the sinking fund factor, the required annual deposits are

$$A = \$75,000 \left[\frac{0.10}{(1 + 0.10)^{12} - 1} \right]$$

$$= \$3,507.25$$

Capitalized cost formula.   *Capitalized* cost refers to the present value of a single amount that is equivalent to a perpetual series of equal end-of-period payments. This is an extension of the series present worth factor with an infinitely large number of periods. This is shown graphically in Figure 5.4.

Using the limit theorem from calculus as n approaches infinity, the series present worth factor reduces to the following formula for the capitalized cost:

$$P = \lim_{n \to \infty} A \left[\frac{(1 + i)^n - 1}{i(1 + i)^n} \right]$$

$$= A \left\{ \lim_{n \to \infty} \left[\frac{(1 + i)^n - 1}{i(1 + i)^n} \right] \right\}$$

$$= A \left(\frac{1}{i} \right)$$

Example   How much should be deposited in a general fund to service a recurring public service project to the tune of $6500 per year forever if the fund yields an annual interest rate of 11%? Using the capitalized cost formula, the required one-time deposit to the general fund is

$$P = \frac{\$6500}{0.11}$$

$$= \$59,090.91$$

The formulae presented above represent the basic cash flow conversion factors. The factors are widely tabulated, for convenience, in engineering economy books. Several variations and extensions of the factors are available. Such extensions include the arithmetic gradient series factor and the geometric series factor. Variations in the cash flow profiles include situations where payments are made at the beginning of each period rather than at the end and situations where a series of payments contains unequal amounts. Conversion formulae can be derived mathematically for those special cases by using the basic factors presented above.

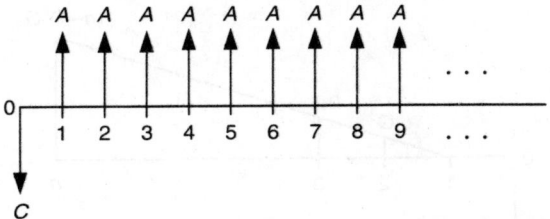

FIGURE 5.4 Capitalized cost cash flow.

Arithmetic gradient series. The gradient series cash flow involves an increase of a fixed amount in the cash flow at the end of each period. Thus, the amount at a given point in time is greater than the amount at the preceding period by a constant amount. This constant amount is denoted by G. Figure 5.5 shows the basic gradient series in which the base amount at the end of the first period is zero. The size of the cash flow in the gradient series at the end of period t is calculated as

$$A_t = (t - 1)G, \quad t = 1, 2, ..., n$$

The total present value of the gradient series is calculated by using the present-amount factor to convert each individual amount from time t to time 0 at an interest rate of $i\%$ per period and summing up the resulting present values. The finite summation reduces to a closed form as shown in the following equation:

$$P = \sum_{t=1}^{n} A_t(1 + i)^{-t}$$

$$= \sum_{t=1}^{n} (t - 1)G(1 + i)^{-t}$$

$$= G\sum_{t=1}^{n} (t - 1)(1 + i)^{-t}$$

$$= G\left[\frac{(1 + i)^n - (1 + ni)}{i^2(1 + i)^n}\right]$$

Example The cost of supplies for a 10-year period increases by $1500 every year, starting at the end of year 2. There is no supplies cost at the end of the first year. If the interest rate is 8% per year, determine the present amount that must be set aside at time zero to take care of all the future supplies expenditures. We have $G = 1,500$, $i = 0.08$, and $n = 10$. Using the arithmetic gradient formula, we obtain

$$P = 1500\left[\frac{1 - (1 + 10(0.08))(1 + 0.08)^{-10}}{(0.08)^2}\right]$$

$$= \$1,500(25.9768)$$

$$= \$38,965.20$$

In many cases, an arithmetic gradient starts with some base amount at the end of the first period and then increases by a constant amount thereafter. The nonzero base amount is denoted as A_1. Figure 5.6 shows this type of cash flow.

The calculation of the present amount for such cash flows requires breaking the cash flow into a uniform series cash flow of amount A_1 and an arithmetic gradient cash flow with zero base amount. The uniform

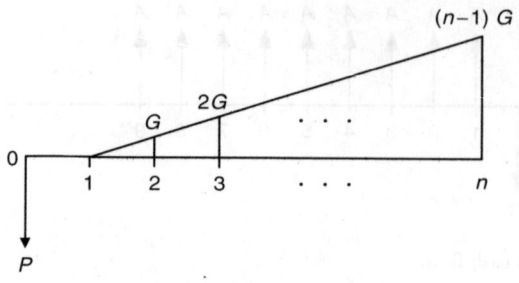

FIGURE 5.5 Arithmetic gradient cash flow with zero base amount.

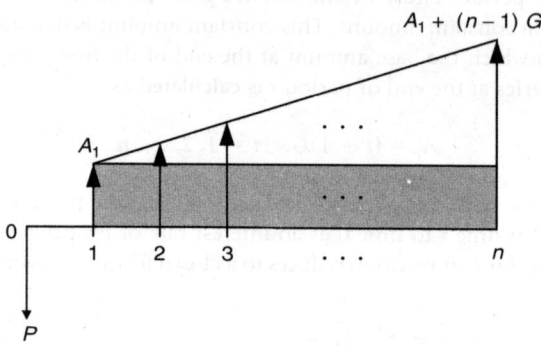

FIGURE 5.6 Arithmetic gradient cash flow with nonzero base amount.

series present worth formula is used to calculate the present worth of the uniform series portion, while the basic gradient series formula is used to calculate the gradient portion. The overall present worth is then calculated as

$$P = P_{\text{uniform series}} + P_{\text{gradient series}}$$

$$= A_1\left[\frac{(1 + i)^n - 1}{i(1 + i)^n}\right] + G\left[\frac{(1 + i)^n - (1 + ni)}{i^2(1 + i)^n}\right]$$

Increasing geometric series cash flow. In an increasing geometric series cash flow, the amounts in the cash flow increase by a constant percentage from period to period. There is a positive base amount, A_1, at the end of period 1. Figure 5.7 shows an increasing geometric series. The amount at time t is denoted as

$$A_t = A_{t-1}(1 + j) \quad t = 2, 3, \ldots, n$$

where j is the percentage increase in the cash flow from period to period. By doing a series of back substitutions, we can represent A_t in terms of A_1 instead of in terms of A_{t-1} as follows:

$$A_2 = A_1(1 + j)$$

$$A_3 = A_2(1 + j) = A_1(1 + j)(1 + j)$$

$$\vdots$$

$$A_t = A_1(1 + j)^{t-1}, \quad t = 1, 2, 3, \ldots, n$$

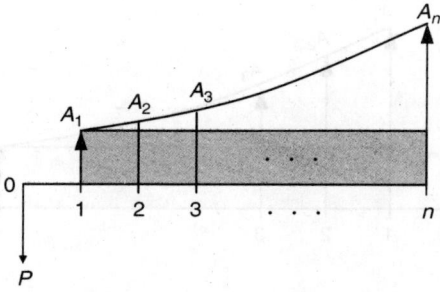

FIGURE 5.7 Increasing geometric series cash flow.

The formula for calculating the present worth of the increasing geometric series cash flow is derived by summing the present values of the individual cash flow amounts, i.e.,

$$P = \sum_{t=1}^{n} A_t (1 + i)^{-t}$$

$$= \sum_{t=1}^{n} [A_1 (1 + j)^{t-1}](1 + i)^{-t}$$

$$= \frac{A_1}{(1 + j)} \sum_{t=1}^{n} \left(\frac{1 + j}{1 + i} \right)^t$$

$$= A_1 \left[\frac{1 - (1 + j)^n (1 + i)^{-n}}{i - j} \right], \quad i \neq j$$

If $i = j$, the formula above reduces to the limit as $i \rightarrow j$, as shown below

$$P = \frac{n A_1}{1 + i}, \quad i = j$$

Example   Suppose funding for a 5-year project is to increase by 6% every year with an initial funding of $20,000 at the end of the first year. Determine how much must be deposited into a budget account at time zero in order to cover the anticipated funding levels if the budget account pays 10% interest per year. We have $j = 6\%$, $i = 10\%$, $n = 5$, $A_1 = \$20,000$. Therefore,

$$P = 20,000 \left[\frac{1 - (1 + 0.06)^5 (1 + 0.10)^{-5}}{0.10 - 0.06} \right]$$

$$= \$20,000(4.2267)$$

$$= \$84,533.60$$

Decreasing geometric series cash flow.   In a decreasing geometric series cash flow, the amounts in the cash flow decrease by a constant percentage from period to period. The cash flow starts at some positive base amount, A_1, at the end of period 1. Figure 5.8 shows a decreasing geometric series. The amount of time t is denoted as

$$A_t = A_{t-1}(1 - j), \quad t = 2, 3, \ldots, n$$

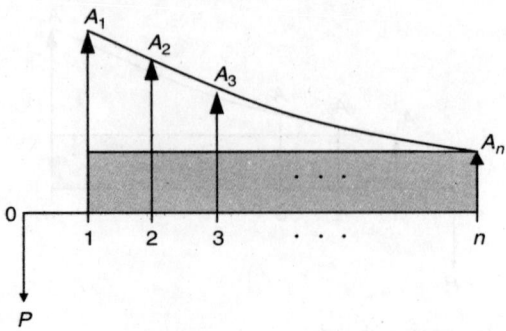

FIGURE 5.8   Decreasing geometric series cash flow.

where j is the percentage decrease in the cash flow from period to period. As in the case of the increasing geometric series, we can represent A_t in terms of A_1:

$$A_2 = A_1(1 - j)$$

$$A_3 = A_2(1 - j) = A_1(1 - j)(1 - j)$$

$$\dots$$

$$A_t = A_1(1 - j)^{t-1}, \quad t = 1, 2, 3, \dots, n$$

The formula for calculating the present worth of the decreasing geometric series cash flow is derived by finite summation as in the case of the increasing geometric series. The final formula is

$$P = A_1 \left[\frac{1 - (1 - j)^n (1 + i)^{-n}}{i + j} \right]$$

Example   The contract amount for a 3-year project is expected to decrease by 10% every year with an initial contract of $100,000 at the end of the first year. Determine how much must be available in a contract reservoir fund at time zero in order to cover the contract amounts. The fund pays 10% interest per year. Because $j = 10\%$, $i = 10\%$, $n = 3$, and $A_1 = \$100,000$, we should have

$$P = 100,000 \left[\frac{1 - (1 - 0.10)^3 (1 + 0.10)^{-3}}{0.10 + 0.10} \right]$$

$$= \$100,000(2.2615)$$

$$= \$226,150$$

Internal rate of return.   The IRR for a cash flow is defined as the interest rate that equates the future worth at time n or present worth at time 0 of the cash flow to zero. If we let i^* denote the IRR, then we have

$$FW_{t=n} = \sum_{t=0}^{n} (\pm A_t)(1 + i^*)^{n-t} = 0$$

$$PW_{t=0} = \sum_{t=0}^{n} (\pm A_t)(1 + i^*)^{-t} = 0$$

where "+" is used in the summation for positive cash flow amounts or receipts and "−" is used for negative cash flow amounts or disbursements. A_t denotes the cash flow amount at time t, which may be a receipt (+) or a disbursement (−). The value of i^* is referred to as *discounted cash flow rate of return, IRR*, or *true rate of return*. The procedure above essentially calculates the net future worth or the net present worth of the cash flow, i.e.,

Net future worth = future worth or the receipts − future worth of disbursements

NFW = FWreceipts − FWdisbursements

Net present worth = present worth or the receipts − present worth of disbursements

NPW = PWreceipts − PWdisbursements

Setting the NPW or NFW equal to zero and solving for the unknown variable i determines the IRR of the cash flow.

Benefit–cost ratio.   The benefit–cost ratio of a cash flow is the ratio of the present worth of benefits to the present worth of costs. This is defined as

$$B/C = \frac{\sum_{t=0}^{n} B_t (1 + i)^{-t}}{\sum_{t=0}^{n} C_t (1 + i)^{-t}}$$

$$= \frac{PW_{benefits}}{PW_{costs}}$$

where B_t is the benefit (receipt) at time t and C_t is the cost (disbursement) at time t. If the benefit–cost ratio is greater than 1, then the investment is acceptable. If the ratio is less than 1, the investment is not acceptable. A ratio of 1 indicates a break-even situation for the project.

Simple payback period.   Payback period refers to the length of time it will take to recover an initial investment. The approach does not consider the impact of the time value of money. Consequently, it is not an accurate method of evaluating the worth of an investment. However, it is a simple technique that is used widely to perform a "quick-and-dirty" assessment of investment performance. Also, the technique considers only the initial cost. Other costs that may occur after time zero are not included in the calculation. The payback period is defined as the smallest value of n (n_{min}) that satisfies the following expression:

$$\sum_{t=1}^{n_{min}} R_t \geq C_0$$

where R_t is the revenue at time t and C_0 the initial investment. The procedure calls for a simple addition of the revenues period by period until enough total has been accumulated to offset the initial investment.

Example   An organization is considering installing a new computer system that will generate significant savings in material and labor requirements for order processing. The system has an initial cost of $50,000. It is expected to save the organization $20,000 a year. The system has an anticipated useful life of 5 years with a salvage value of $5,000. Determine how long it would take for the system to pay for itself from the savings it is expected to generate. Because the annual savings are uniform, we can calculate the payback period by simply dividing the initial cost by the annual savings, i.e.,

$$n_{min} = \frac{\$50,000}{\$20,000}$$

$$= 2.5 \text{ years}$$

Note that the salvage value of $5,000 is not included in the above calculation because the amount is not realized until the end of the useful life of the asset (i.e., after 5 years). In some cases, it may be desired to consider the salvage value. In that case, the amount to be offset by the annual savings will be the net cost of the asset. In that case, we would have

$$n_{min} = \frac{\$50,000 - \$5000}{\$20,000}$$

$$= 2.25 \text{ years}$$

If there are tax liabilities associated with the annual savings, those liabilities must be deducted from the savings before calculating the payback period.

Discounted payback period. In this book, we introduce the *discounted payback period* approach in which the revenues are reinvested at a certain interest rate. The payback period is determined when enough money has been accumulated at the given interest rate to offset the initial cost as well as other interim costs. In this case, the calculation is made by the following expression:

$$\sum_{t=1}^{n_{min}} R_t(1 + i)^{n_{min}-1} \geq \sum_{t=0}^{n_{min}} C_t$$

Example A new solar cell unit is to be installed in an office complex at an initial cost of $150,000. It is expected that the system will generate annual cost savings of $22,500 on the electricity bill. The solar cell unit will need to be overhauled every 5 years at a cost of $5,000 per overhaul. If the annual interest rate is 10%, find the *discounted payback period* for the solar cell unit considering the time value of money. The costs of overhaul are to be considered in calculating the discounted payback period.

Solution Using the single payment compound-amount factor for one period iteratively, the following solution is obtained.

Time	Cumulative Savings
1	$22,500
2	$22,500 + $22,500(1.10)^1 = $47,250
3	$22,500 + $47,250(1.10)^1 = $74,475
4	$22,500 + $74,475(1.10)^1 = $104,422.50
5	$22,500 + $104,422.50(1.10)^1 - $5,000 = $132,364.75
6	$22,500 + $132,364.75(1.10)^1 = $168,101.23

The initial investment is $150,000. By the end of period 6, we have accumulated $168,101.23, more than the initial cost. Interpolating between periods 5 and 6, we obtain

$$n_{min} = 5 + \frac{150,000 - 132,364.75}{168,101.23 - 132,364.75}(6 - 5)$$

$$= 5.49$$

That is, it will take 5.49 years or 5 years and 6 months to recover the initial investment.

Investment life for multiple returns. The time it takes an amount to reach a certain multiple of its initial level is often of interest in many investment scenarios. The "Rule of 72" is one simple approach to calculating how long it will take an investment to double in value at a given interest rate per period. The Rule of 72 gives the following formula for estimating the doubling period:

$$n = \frac{72}{i}$$

where i is the interest rate expressed in percentage. Referring to the single payment compound-amount factor, we can set the future amount equal to twice the present amount and then solve for n, the number of periods. That is, $F = 2P$. Thus,

$$2P = P(1 + i)^n$$

Solving for n in the above equation yields an expression for calculating the exact number of periods required to double P:

$$n = \frac{\ln(2)}{\ln(1 + i)}$$

where i is the interest rate expressed in decimals. In the general case, for exact computation, the length of time it would take to accumulate m multiples of P is expressed as

$$n = \frac{\ln(m)}{\ln(1 + i)}$$

where m is the desired multiple. For example, at an interest rate of 5% per year, the time it would take an amount, P, to double in value ($m = 2$) is 14.21 years. This, of course, assumes that the interest rate will remain constant throughout the planning horizon. Table 5.1 presents a tabulation of the values calculated from both approaches. Figure 5.9 shows a graphical comparison of the Rule of 72 to the exact calculation.

5.2.2 Effects of Inflation

Inflation is a major player in financial and economic analyses of projects. Multiyear projects are particularly subject to the effects of inflation. Inflation can be defined as the decline in purchasing power of money.

Some of the most common causes of inflation are:

- Increase in amount of currency in circulation
- Shortage of consumer goods
- Escalation of the cost of production
- Arbitrary increase of prices by resellers

The general effects of inflation are felt in terms of an increase in the prices of goods and decrease in the worth of currency. In cash flow analysis, ROI for a project will be affected by time value of money as well as inflation. The *real interest rate* (d) is defined as the desired rate of return in the absence of inflation. When we talk of "today's dollars" or "constant dollars," we are referring to the use of real interest rate.

TABLE 5.1 Evaluation of the Rule of 72

$i\%$	n (Rule of 72)	N (Exact Value)
0.25	288.00	277.61
0.50	144.00	138.98
1.00	72.00	69.66
2.00	36.00	35.00
5.00	14.20	17.67
8.00	9.00	9.01
10.00	7.20	7.27
12.00	6.00	6.12
15.00	4.80	4.96
18.00	4.00	4.19
20.00	3.60	3.80
25.00	2.88	3.12
30.00	2.40	2.64

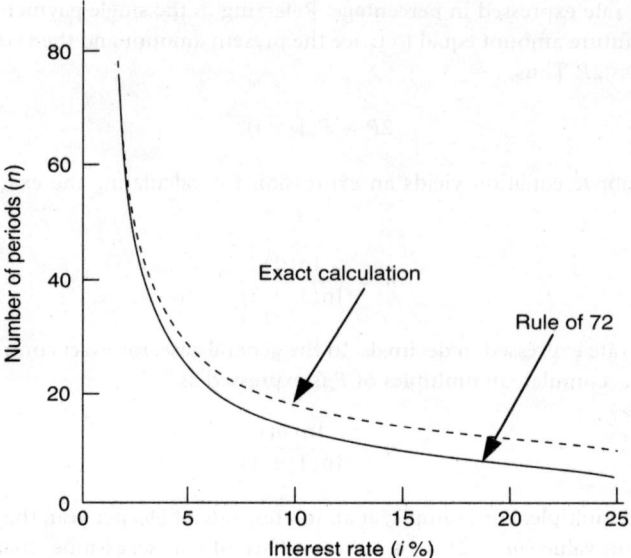

FIGURE 5.9 Evaluation of investment life for double return.

Combined interest rate (i) is the rate of return combining real interest rate and inflation rate. If we denote the *inflation rate* as j, then the relationship between the different rates can be expressed as

$$1 + i = (1 + d)(1 + j)$$

Thus, the combined interest rate can be expressed as

$$i = d + j + dj$$

Note that if $j = 0$ (i.e., no inflation), then $i = d$. We can also define *commodity escalation rate* (g) as the rate at which individual commodity prices escalate. This may be greater than or less than the overall inflation rate. In practice, several measures are used to convey inflationary effects. Some of these are *consumer price index, producer price index*, and *wholesale price index*. A "*market basket*" *rate* is defined as the estimate of inflation based on a weighted average of the annual rates of change in the costs of a wide range of representative commodities. A "then-current" cash flow is a cash flow that explicitly incorporates the impact of inflation. A "constant worth" cash flow is a cash flow that does not incorporate the effect of inflation. The real interest rate, d, is used for analyzing constant worth cash flows. Figure 5.10 shows constant worth and then-current cash flows.

The then-current cash flow in the figure is the equivalent cash flow considering the effect of inflation. C_k is what it would take to buy a certain "basket" of goods after k time periods if there was no inflation. T_k is what it would take to buy the same "basket" in k time period if inflation was taken into account. For the constant worth cash flow, we have

$$C_k = T_0, \quad k = 1, 2, \ldots, n$$

and for the then-current cash flow, we have

$$T_k = T_0(1 + j)k, \quad k = 1, 2, \ldots, n$$

where j is the inflation rate. If $C_k = T_0 = \$100$ under the constant worth cash flow, then we mean \$100 worth of buying power. If we are using the commodity escalation rate, g, then we will have

$$T_k = T_0(1 + g)k, \quad k = 1, 2, \ldots, n$$

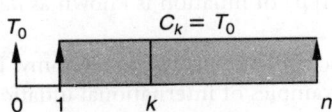
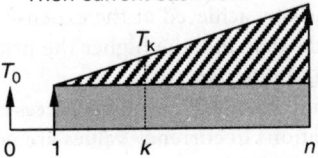

FIGURE 5.10 Cash flows for effects of inflation.

Thus, a then-current cash flow may increase based on both a regular inflation rate (j) and a commodity escalation rate (g). We can convert a then-current cash flow to a constant worth cash flow by using the following relationship:

$$C_k = T_k(1 + j)^{-k}, \quad k = 1, 2, ..., n$$

If we substitute T_k from the commodity escalation cash flow into the expression for C_k above, we get

$$C_k = T_k(1 + j)^{-k}$$

$$= T_0(1 + g)^k(1 + j)^{-k}$$

$$= T_0[(1 + g)/(1 + j)]^k, \quad k = 1, 2, ..., n$$

Note that if $g = 0$ and $j = 0$, the $C_k = T_0$. That is, there is no inflationary effect. We now define effective commodity escalation rate (v) as

$$v = [(1 + g)/(1 + j)] - 1$$

and we can express the commodity escalation rate (g) as

$$g = v + j + vj$$

Inflation can have a significant impact on the financial and economic aspects of a project. Inflation may be defined, in economic terms, as the increase in the amount of currency in circulation, resulting in a relatively high and sudden fall in its value. To a producer, inflation means a sudden increase in the cost of items that serve as inputs for the production process (equipment, labor, materials, etc.). To the retailer, inflation implies an imposed higher cost of finished products. To an ordinary citizen, inflation portends an unbearable escalation of prices of consumer goods. All these views are interrelated in a project management environment.

The amount of money supply, as a measure of a country's wealth, is controlled by the government. With no other choice, governments often feel impelled to create more money or credit to take care of old debts and pay for social programs. When money is generated at a faster rate than the growth of goods and services, it becomes a surplus commodity, and its value (purchasing power) will fall. This means that there will be too much money available to buy only a few goods and services. When the purchasing power of a currency falls, each individual in a product's life cycle has to dispense more of the currency in order to obtain the product. Some of the classic concepts of inflation are discussed below:

1. Increases in the producer's costs are passed on to consumers. At each stage of the product's journey from producer to consumer, prices are escalated disproportionately in order to make a good profit. The overall increase in the product's price is directly proportional to the number of intermediaries it encounters on its way to the consumer. This type of inflation is called *cost-driven (or cost-push) inflation*.

2. Excessive spending power of consumers forces an upward trend in prices. This high spending power is usually achieved at the expense of savings. The law of supply and demand dictates that the more the demand, the higher the price. This type of inflation is known as *demand-driven (or demand-pull) inflation.*

3. Impact of international economic forces can induce inflation in a local economy. Trade imbalances and fluctuations in currency values are notable examples of international inflationary factors.

4. Increasing base wages of workers generate more disposable income and, hence, higher demands for goods and services. The high demand, consequently, creates a pull on prices. Coupled with this, employers pass on the additional wage cost to consumers through higher prices. This type of inflation is, perhaps, the most difficult to solve because wages set by union contracts and prices set by producers almost never fall — at least not permanently. This type of inflation may be referred to as *wage-driven (or wage-push) inflation.*

5. Easy availability of credit leads consumers to "buy now and pay later" and, thereby, creates another loophole for inflation. This is a dangerous type of inflation because the credit not only pushes prices up, it also leaves consumers with less money later on to pay for the credit. Eventually, many credits become uncollectible debts, which may then drive the economy into recession.

6. Deficit spending results in an increase in money supply and, thereby, creates less room for each dollar to get around. The popular saying, "a dollar does not go far anymore," simply refers to inflation in layman's terms. The different levels of inflation may be categorized as discussed below.

Mild inflation.   When inflation is mild (2 to 4%), the economy actually prospers. Producers strive to produce at full capacity in order to take advantage of the high prices to the consumer. Private investments tend to be brisk and more jobs become available. However, the good fortune may only be temporary. Prompted by the prevailing success, employers are tempted to seek larger profits and workers begin to ask for higher wages. They cite their employer's prosperous business as a reason to bargain for bigger shares of the business profit. Thus, we end up with a vicious cycle where the producer asks for higher prices, the unions ask for higher wages, and inflation starts an upward trend.

Moderate inflation.   Moderate inflation occurs when prices increase at 5 to 9%. Consumers start purchasing more as an edge against inflation. They would rather spend their money now than watch it decline further in purchasing power. The increased market activity serves to fuel further inflation.

Severe inflation.   Severe inflation is indicated by price escalations of 10% or more. Double-digit inflation implies that prices rise much faster than wages do. Debtors tend to be the ones who benefit from this level of inflation because they repay debts with money that is less valuable then the money borrowed.

Hyperinflation.   When each price increase signals the increase in wages and costs, which again sends prices further up, the economy has reached a stage of malignant galloping inflation or hyperinflation. Rapid and uncontrollable inflation destroys the economy. The currency becomes economically useless as the government prints it excessively to pay for obligations.

Inflation can affect any project in terms of raw materials procurement, salaries and wages, and cost-tracking dilemmas. Some effects are immediate and easily observable. Other effects are subtle and pervasive. Whatever form it takes, inflation must be taken into account in long-term project planning and control. Large projects may be adversely affected by the effects of inflation in terms of cost overruns and poor resource utilization. The level of inflation will determine the severity of the impact on projects.

5.3 Break-Even Analysis

Break-even analysis refers to the determination of the balanced performance level where project income is equal to project expenditure. The total cost of an operation is expressed as the sum of the fixed and variable costs with respect to output quantity, i.e.,

$$TC(x) = FC + VC(x)$$

where x is the number of units produced, $\text{TC}(x)$ the total cost of producing x units, FC the total fixed cost, and $\text{VC}(x)$ the total variable cost associated with producing x units. The total revenue resulting from the sale of x units is defined as

$$\text{TR}(x) = px$$

where p is the price per unit. The profit due to the production and sale of x units of the product is calculated as

$$P(x) = \text{TR}(x) - \text{TC}(x)$$

The break-even point of an operation is defined as the value of a given parameter that will result in neither profit nor loss. The parameter of interest may be the number of units produced, the number of hours of operation, the number of units of a resource type allocated, or any other measure of interest. At the break-even point, we have the following relationship:

$$\text{TR}(x) = \text{TC}(x) \quad \text{or} \quad P(x) = 0$$

In some cases, there may be a known mathematical relationship between cost and the parameter of interest. For example, there may be a linear cost relationship between the total cost of a project and the number of units produced. The cost expressions facilitate straightforward break-even analysis. Figure 5.11 shows an example of a break-even point for a single project. Figure 5.12 shows examples of multiple break-even points that exist when multiple projects are compared. When two project alternatives are compared, the break-even point refers to the point of indifference between the two alternatives. In Figure 5.12, $x1$ represents the point where projects A and B are equally desirable, $x2$ represents where A and C are equally desirable, and $x3$ represents where B and C are equally desirable. The figure shows that if we are operating below a production level of $x2$ units, then project C is the preferred project among the three. If we are operating at a level more than $x2$ units, then project A is the best choice.

Example   Three project alternatives are being considered for producing a new product. The required analysis involves determining which alternative should be selected on the basis of how many units of the product are produced per year. Based on past records, there is a known relationship between the number of units produced per year, x, and the net annual profit, $P(x)$, from each alternative. The level of production is

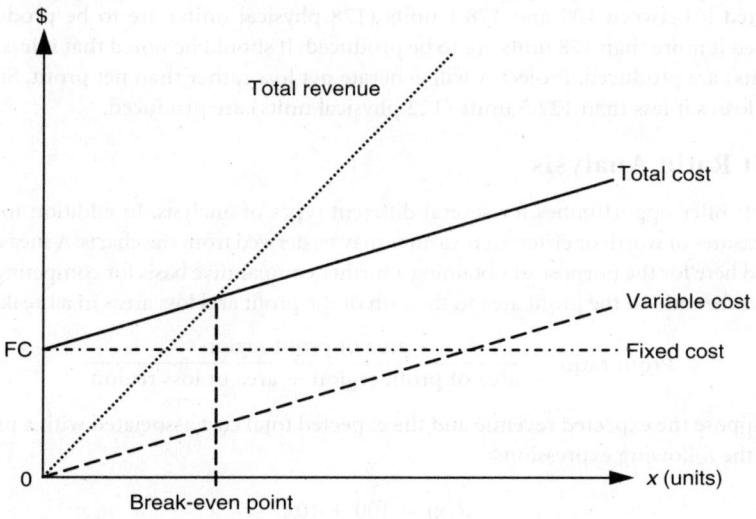

FIGURE 5.11   Break-even point for a single project.

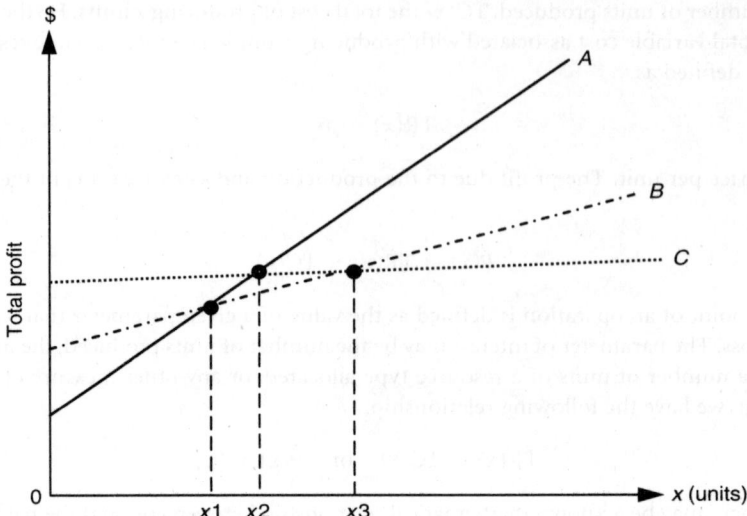

FIGURE 5.12 Break-even points for multiple projects.

expected to be between 0 and 250 units per year. The net annual profits (in thousands of dollars) are given below for each alternative:

$$\text{Project A:} \quad P(x) = 3x - 200$$

$$\text{Project B:} \quad P(x) = x$$

$$\text{Project C:} \quad P(x) = (1/50)x^2 - 300$$

This problem can be solved mathematically by finding the intersection points of the profit functions and evaluating the respective profits over the given range of product units. It can also be solved by a graphical approach. Figure 5.13 shows a plot of the profit functions. Such a plot is called a *break-even chart*. The plot shows that Project B should be selected if between 0 and 100 units are to be produced. Project A should be selected if between 100 and 178.1 units (178 physical units) are to be produced. Project C should be selected if more than 178 units are to be produced. It should be noted that if less than 66.7 units (66 physical units) are produced, Project A will generate net loss rather than net profit. Similarly, Project C will generate losses if less than 122.5 units (122 physical units) are produced.

5.3.1 Profit Ratio Analysis

Break-even charts offer opportunities for several different types of analysis. In addition to the break-even points, other measures of worth or criterion measures may be derived from the charts. A measure, called *profit ratio*, is presented here for the purpose of obtaining a further comparative basis for competing projects. Profit ratio is defined as the ratio of the profit area to the sum of the profit and loss areas in a break-even chart, i.e.,

$$\text{Profit ratio} = \frac{\text{area of profit region}}{\text{area of profit region} + \text{area of loss region}}$$

For example, suppose the expected revenue and the expected total cost associated with a project are given, respectively, by the following expressions:

$$R(x) = 100 + 10x$$

$$TC(x) = 2.5x + 250$$

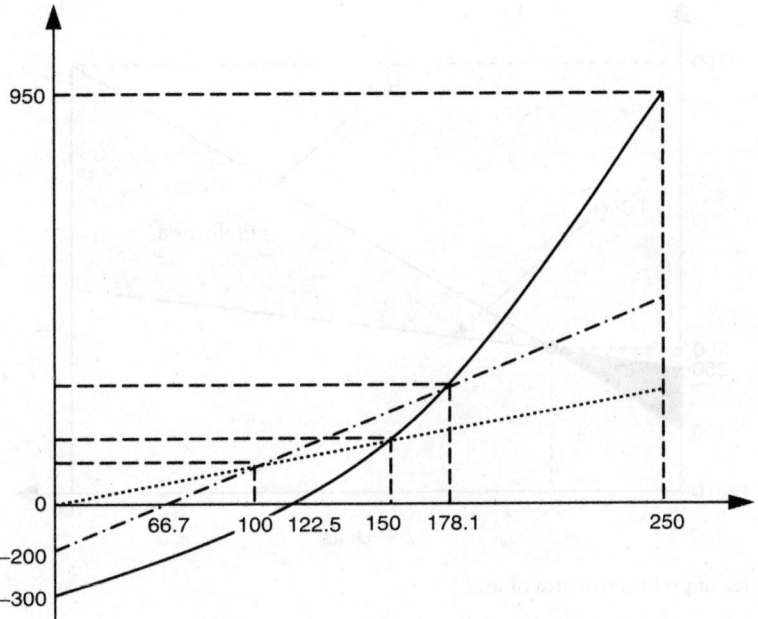

FIGURE 5.13   Plot of profit functions.

where x is the number of units produced and sold from the project. Figure 5.14 shows the break-even chart for the project. The break-even point is shown to be 20 units. Net profits are realized from the project if more than 20 units are produced, and net losses are realized if less than 20 units are produced. It should be noted that the revenue function in Figure 5.14 represents an unusual case where a revenue of $100 is realized when zero units are produced.

Suppose it is desired to calculate the profit ratio for this project if the number of units that can be produced is limited to between 0 and 100 units. From Figure 5.14, the surface area of the profit region and the area of the loss region can be calculated by using the standard formula for finding the area of a triangle: Area = (1/2)(Base)(Height). Using this formula, we have the following:

$$\text{Area of profit region} = 1/2(\text{Base})(\text{Height})$$

$$= 1/2(1100 - 500)(100 - 20)$$

$$= 24,000 \text{ square units}$$

$$\text{Area of loss region} = 1/2(\text{Base})(\text{Height})$$

$$= 1/2(250 - 100)(20)$$

$$= 1,500 \text{ square units}$$

Thus, the profit ratio is computed as

$$\text{Profit ratio} = \frac{24,000}{24,000 + 1,500}$$

$$= 0.9411$$

$$= 94.11\%$$

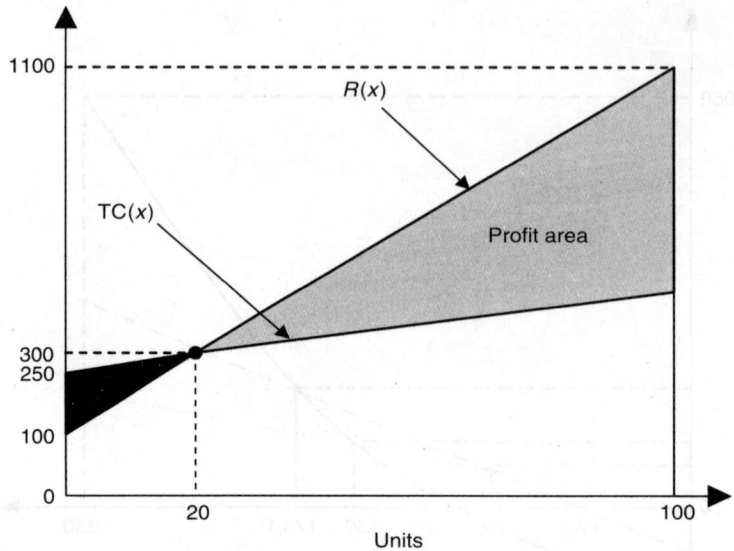

FIGURE 5.14   Area of profit versus area of loss.

The profit ratio may be used as a criterion for selecting among project alternatives. If this is done, the profit ratios for all the alternatives must be calculated over the same values of the independent variable. The project with the highest profit ratio will be selected as the desired project. For example, Figure 5.15 presents the break-even chart for an alternate project, say Project II. It is seen that both the revenue and cost functions for the project are nonlinear. The revenue and cost are defined as follows:

$$R(x) = 160x - x^2$$

$$TC(x) = 500 + x^2$$

If the cost and/or revenue functions for a project are not linear, the areas bounded by the functions may not be easily determined. For those cases, it may be necessary to use techniques such as definite integrals to find the areas. Figure 5.15 indicates that the project generates a loss if less than 3.3 units (3 actual units) or more than 76.8 (76 actual units) are produced. The respective profit and loss areas on the chart are calculated as follows:

$$\text{Area 1 (loss)} = \int_{0}^{3.3} [(500 + x^2) - (160x - x^2)]dx$$

$$= 802.8 \text{ unit-dollars}$$

$$\text{Area 2 (profit)} = \int_{3.3}^{76.8} [(160x - x^2) - (500 + x^2)]dx$$

$$= 132{,}272.08 \text{ unit-dollars}$$

$$\text{Area 3 (loss)} = \int_{76.8}^{100} [(500 + x^2) - (160x - x^2)]dx$$

$$= 48{,}135.98 \text{ unit-dollars}$$

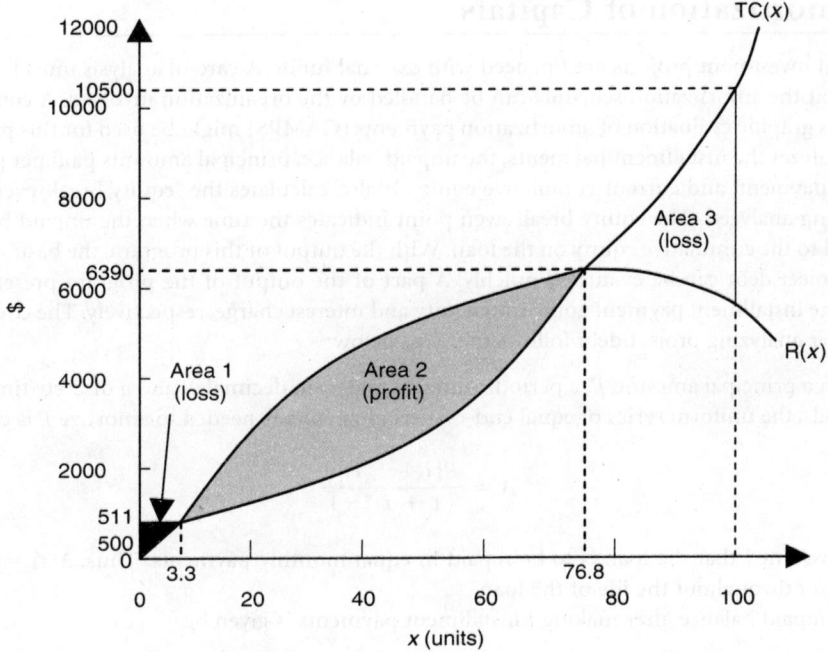

FIGURE 5.15 Break-even chart for revenue and cost functions.

Consequently, the profit ratio for Project II is computed as

$$\text{profit ratio} = \frac{\text{total area of profit region}}{\text{total area of profit region} + \text{total area of loss region}}$$

$$= \frac{132{,}272.08}{802.76 + 132{,}272.08 + 48{,}135.98}$$

$$= 0.7299$$

$$= 72.99\%$$

The profit ratio approach evaluates the performance of each alternative over a specified range of operating levels. Most of the existing evaluation methods use single-point analysis with the assumption that the operating condition is fixed at a given production level. The profit ratio measure allows an analyst to evaluate the net yield of an alternative given that the production level may shift from one level to another. An alternative, for example, may operate at a loss for most of its early life, while it may generate large incomes to offset the losses in its later stages. Conventional methods cannot easily capture this type of transition from one performance level to another. In addition to being used to compare alternate projects, the profit ratio may also be used for evaluating the economic feasibility of a single project. In such a case, a decision rule may be developed. An example of such a decision rule is

If profit ratio is greater than 75%, accept the project.
If profit ratio is less than or equal to 75%, reject the project.

5.4 Amortization of Capitals

Many capital investment projects are financed with external funds. A careful analysis must be conducted to ensure that the amortization schedule can be handled by the organization involved. A computer program such as graphic evaluation of amortization payments (GAMPS) might be used for this purpose. The program analyzes the installment payments, the unpaid balance, principal amounts paid per period, total installment payment, and current cumulative equity. It also calculates the "equity break-even point" for the debt being analyzed. The equity break-even point indicates the time when the unpaid balance on a loan is equal to the cumulative equity on the loan. With the output of this program, the basic cost of servicing the project debt can be evaluated quickly. A part of the output of the program presents the percentage of the installment payment going into equity and interest charge, respectively. The computational procedure for analyzing project debt follows the steps below:

1. Given a principal amount, P, a periodic interest rate, i (in decimals), and a discrete time span of n periods, the uniform series of equal end-of-period payments needed to amortize P is computed as

$$A = \frac{P[i(1 + i)^n]}{(1 + i)^n - 1}$$

It is assumed that the loan is to be repaid in equal monthly payments. Thus, $A(t) = A$, for each period t throughout the life of the loan.

2. The unpaid balance after making t installment payments is given by

$$U(t) = \frac{A[1 - (1 + i)^{t-n}]}{i}$$

3. The amount of equity or principal amount paid with installment payment number t is given by

$$E(t) = A(1 + i)^{t-n-1}$$

4. The amount of interest charge contained in installment payment number t is derived to be

$$I(t) = A[1 - (1 + i)^{t-n-1}]$$

where $A = E(t) + I(t)$.

5. The cumulative total payment made after t periods is denoted by

$$C(t) = \sum_{k=1}^{t} A(k)$$

$$= \sum_{k=1}^{t} A$$

$$= (A)(t)$$

6. The cumulative interest payment after t periods is given by

$$Q(t) = \sum_{x=1}^{t} I(x)$$

7. The cumulative principal payment after t periods is computed as

$$S(t) = \sum_{k=1}^{t} E(k)$$

$$= A \sum_{k=1}^{t} (1 + i)^{-(n-k+1)}$$

$$= A\left[\frac{(1 + i)^t - 1}{i(1 + i)^n}\right]$$

where

$$\sum_{n-1}^{t} x^n = \frac{x^{t+1} - x}{x - 1}$$

8. The percentage of interest charge contained in installment payment number t is

$$f(t) = \frac{I(t)}{A}(100\%)$$

9. The percentage of cumulative interest charge contained in the cumulative total payment up to and including payment number t is

$$F(t) = \frac{Q(t)}{C(t)}(100\%)$$

10. The percentage of cumulative principal payment contained in the cumulative total payment up to and including payment number t is

$$H(t) = \frac{S(t)}{C(t)}$$

$$= \frac{C(t) - Q(t)}{C(t)}$$

$$= 1 - \frac{Q(t)}{C(t)}$$

$$= 1 - F(t)$$

Example   Suppose that a manufacturing productivity improvement project is to be financed by borrowing $500,000 from an industrial development bank. The annual nominal interest rate for the loan is 10%. The loan is to be repaid in equal monthly installments over a period of 15 years. The first payment on the loan is to be made exactly one month after financing is approved. It is desired to perform a detailed analysis of the loan schedule. Table 5.2 presents a partial listing of the loan repayment schedule.

The tabulated result shows a monthly payment of $5373.04 on the loan. Considering time $t = 10$ months, one can see the following results:

$U(10) = \$487,475.13$ (unpaid balance)
$A(10) = \$5373.04$ (monthly payment)
$E(10) = \$1299.91$ (equity portion of the tenth payment)
$I(10) = \$4073.13$ (interest charge contained in the tenth payment)
$C(10) = \$53,730.40$ (total payment to date)
$S(10) = \$12,526.21$ (total equity to date)
$f(10) = 75.81\%$ (percentage of the tenth payment going into interest charge)
$F(10) = 76.69\%$ (percentage of the total payment going into interest charge)

Thus, over 76% of the sum of the first ten installment payments goes into interest charges. The analysis shows that by time $t = 180$, the unpaid balance has been reduced to zero, i.e., $U(180) = 0.0$. The total payment made on the loan is $967,148.40 and the total interest charge is $967,148.20 - \$500,000 = \$467,148.20$. Thus, 48.30% of the total payment goes into interest charges. The information about interest

TABLE 5.2   Amortization Schedule for Financed Project

t	$U(t)$	$A(t)$	$E(t)$	$I(t)$	$C(t)$	$S(t)$	$f(t)$	$F(t)$
1	498,794.98	5,373.04	1,206.36	4166.68	5,373.04	1,206.36	77.6	77.6
2	497,578.56	5,373.04	1,216.42	4156.62	10,746.08	2,422.78	77.4	77.5
3	496,352.01	5,373.04	1,226.55	4146.49	16,119.12	3,649.33	77.2	77.4
4	495,115.24	5,373.04	1,236.77	4136.27	21,492.16	4,886.10	76.9	77.3
5	493,868.16	5,373.04	1,247.08	4125.96	26,865.20	6,133.18	76.8	77.2
6	492,610.69	5,373.04	1,257.47	4115.57	32,238.24	7,390.65	76.6	77.1
7	491,342.74	5,373.04	1,267.95	4105.09	37,611.28	8,658.61	76.4	76.9
8	490,064.22	5,373.04	1,278.52	4094.52	42,984.32	9,937.12	76.2	76.9
9	488,775.05	5,373.04	1,289.17	4083.87	48,357.36	11,226.29	76.0	76.8
10	487,475.13	5,373.04	1,299.91	4073.13	53,730.40	12,526.21	75.8	76.7
.	.	.	.	.	.	.	.	.
.	.	.	.	.	.	.	.	.
.	.	.	.	.	.	.	.	.
170	51,347.67	5,373.04	4,904.27	468.77	913,416.80	448,656.40	8.7	50.9
171	46,402.53	5,373.04	4,945.14	427.90	918,789.84	453,601.54	7.9	50.6
172	41,416.18	5,373.04	4,986.35	386.69	924,162.88	458,587.89	7.2	50.4
173	36,388.27	5,373.04	5,027.91	345.13	929,535.92	463,615.80	6.4	50.1
174	31,318.47	5,373.04	5,069.80	303.24	934,908.96	468,685.60	5.6	49.9
175	26,206.42	5,373.04	5,112.05	260.99	940,282.00	473,797.66	4.9	49.6
176	21,051.76	5,373.04	5,154.65	218.39	945,655.04	478,952.31	4.1	49.4
177	15,854.15	5,373.04	5,197.61	175.43	951,028.08	484,149.92	3.3	49.1
178	10,613.23	5,373.04	5,240.92	132.12	956,401.12	489,390.84	2.5	48.8
179	5,328.63	5,373.04	5,284.60	88.44	961,774.16	494,675.44	1.7	48.6
180	0.00	5,373.04	5,328.63	44.41	967,147.20	500,004.07	0.8	48.3

charges might be very useful for tax purposes. The tabulated output shows that equity builds up slowly while unpaid balance decreases slowly. Note that very little equity is accumulated during the first 3 years of the loan schedule. This is shown graphically in Figure 5.16. The effects of inflation, depreciation, property appreciation, and other economic factors are not included in the analysis presented above. A project analyst should include such factors whenever they are relevant to the loan situation.

The point at which the curves intersect is referred to as the *equity break-even point*. It indicates when the unpaid balance is exactly equal to the accumulated equity or the cumulative principal payment. For the example, the equity break-even point is 120.9 months (over 10 years). The importance of the equity break-even point is that any equity accumulated after that point represents the amount of ownership or equity that the debtor is entitled to after the unpaid balance on the loan is settled with project collateral. The implication of this is very important, particularly in the case of mortgage loans. "Mortgage" is a word with French origin, meaning *death pledge* — perhaps a sarcastic reference to the burden of mortgage loans. The equity break-even point can be calculated directly from the formula derived below.

Let the equity break-even point, x, be defined as the point where $U(x) = S(x)$, i.e.,

$$A\left[\frac{1 - (1 + i)^{-(n-x)}}{i}\right] = A\left[\frac{(1 + i)^x - 1}{i(1 + i)^n}\right]$$

Multiplying both the numerator and denominator of the left-hand side of the above expression by $(1 + i)^n$ and simplifying yields

$$\frac{(1 + i)^n - (1 + i)^x}{i(1 + i)^n}$$

on the left-hand side. Consequently, we have

$$(1 + i)^n - (1 + i)^x = (1 + i)^x - 1$$

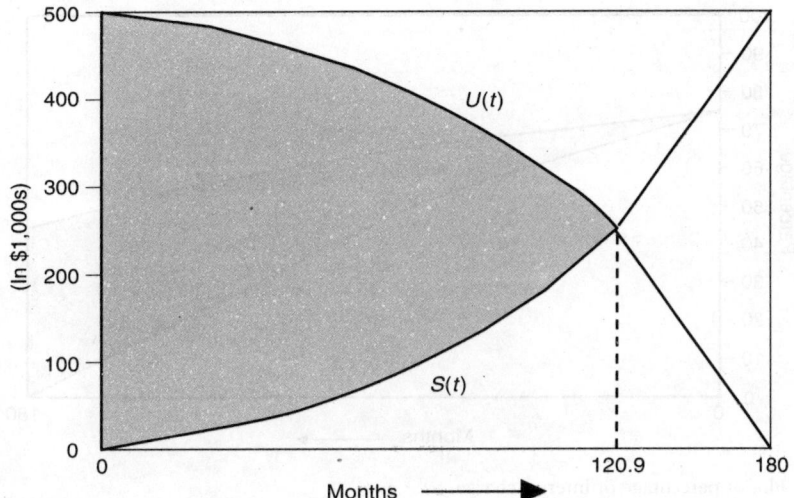

FIGURE 5.16 Plot of unpaid balance and cumulative equity.

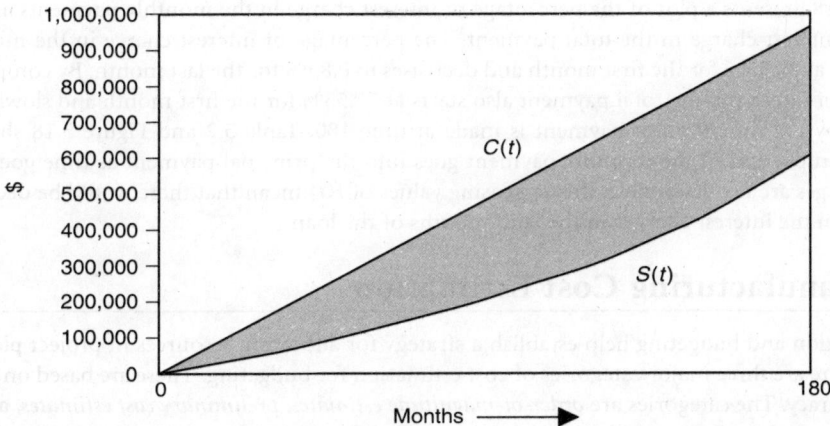

FIGURE 5.17 Plot of total loan payment and total equity.

$$(1 + i)^x = \frac{(1 + i)^n + 1}{2}$$

which yields the equity break-even expression

$$x = \frac{\ln[0.5(1 + i)^n + 0.5]}{\ln(1 + i)}$$

where ln is the natural log function, n the number of periods in the life of the loan and i the interest rate per period.

Figure 5.17 presents a plot of the total loan payment and the cumulative equity with respect to time. The total payment starts from $0.0 at time 0 and goes up to $967,147.20 by the end of the last month of the installment payments. Because only $500,000 was borrowed, the total interest payment on the loan is $967,147.20 − $500,00 = $467,147.20. The cumulative principal payment starts at $0.0 at time 0 and

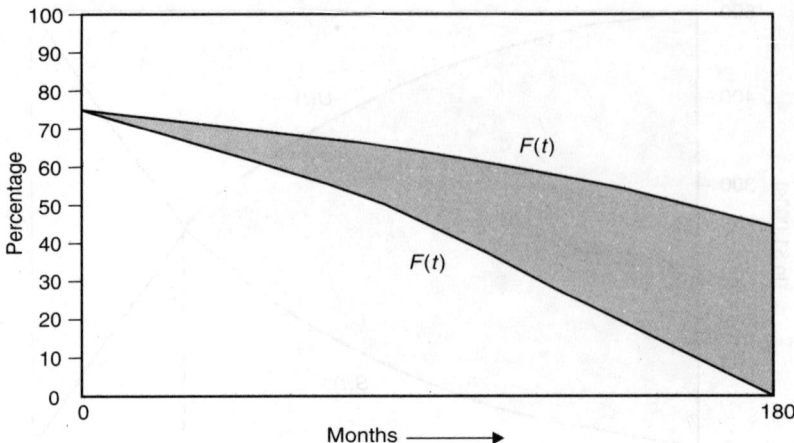

FIGURE 5.18　Plot of percentage of interest charge.

slowly builds up to $500,001.34, which is the original loan amount. The extra $1.34 is due to round-off error in the calculations.

Figure 5.18 presents a plot of the percentage of interest charge in the monthly payments and the percentage of interest charge in the total payment. The percentage of interest charge in the monthly payments starts at 77.55% for the first month and decreases to 0.83% for the last month. By comparison, the percentage of interest in the total payment also starts at 77.55% for the first month and slowly decreases to 48.30% by the time the last payment is made at time 180. Table 5.2 and Figure 5.18 show that an increasing proportion of the monthly payment goes into the principal payment as time goes on. If the interest charges are tax deductible, the decreasing values of $f(t)$ mean that there would be decreasing tax benefits from the interest charges in the later months of the loan.

5.5　Manufacturing Cost Estimation

Cost estimation and budgeting help establish a strategy for allocating resources in project planning and control. There are three major categories of cost estimation for budgeting. These are based on the desired level of accuracy. The categories are *order-of-magnitude estimates, preliminary cost estimates,* and *detailed cost estimates.* Order-of-magnitude cost estimates are usually gross estimates based on the experience and judgment of the estimator. They are sometimes called "ballpark" figures. These estimates are made without a formal evaluation of the details involved in the project. The level of accuracy associated with order-of-magnitude estimates can range from −50 to +50% of the actual cost. These estimates provide a quick way of getting cost information during the initial stages of a project.

$$50\% \text{ (Actual cost)} \leq \text{Order-of-magnitude estimate} \leq 150\% \text{ (Actual cost)}$$

Preliminary cost estimates are also gross estimates, but with a higher level of accuracy. In developing preliminary cost estimates, more attention is paid to some selected details of the project. An example of a preliminary cost estimate is the estimation of expected labor cost. Preliminary estimates are useful for evaluating project alternatives before final commitments are made. The level of accuracy associated with preliminary estimates can range from −20 to +20% of the actual cost.

$$80\% \text{ (Actual cost)} \leq \text{Preliminary estimate} \leq 120\% \text{ (Actual cost)}$$

Detailed cost estimates are developed after careful consideration is given to all the major details of a project. Considerable time is typically needed to obtain detailed cost estimates. Because of the amount of time and effort needed to develop detailed cost estimates, the estimates are usually developed after there is firm commitment that the project will take off. Detailed cost estimates are important for evaluating actual cost performance during the project. The level of accuracy associated with detailed estimates normally range from -5 to $+5\%$ of the actual cost.

$$95\% \text{ (Actual cost)} = \text{Detailed cost} = 105\% \text{ (Actual cost)}$$

There are two basic approaches to generating cost estimates. The first one is a variant approach, in which cost estimates are based on variations of previous cost records. The other approach is the generative cost estimation, in which cost estimates are developed from scratch without taking previous cost records into consideration.

Optimistic and pessimistic cost estimates. Using an adaptation of the PERT formula, we can combine optimistic and pessimistic cost estimates. Let O be the optimistic cost estimate, M the most likely cost estimate, and P the pessimistic cost estimate.

Then, the estimated cost can be estimated as

$$E[C] = \frac{O + 4M + P}{6}$$

and the cost variance can be estimated as

$$V[C] = \left[\frac{P - O}{6} \right]^2$$

5.6 Budgeting and Capital Allocation

Budgeting involves sharing limited resources between several project groups or functions in a project environment. Budget analysis can serve any of the following purposes:

- A plan for resources expenditure
- A project selection criterion
- A projection of project policy
- A basis for project control
- A performance measure
- A standardization of resource allocation
- An incentive for improvement

Top-down budgeting. Top-down budgeting involves collecting data from upper-level sources such as top and middle managers. The figures supplied by the managers may come from their personal judgment, past experience, or past data on similar project activities. The cost estimates are passed to lower-level managers, who then break the estimates down into specific work components within the project. These estimates may, in turn, be given to line managers, supervisors, and lead workers to continue the process until individual activity costs are obtained. Top management provides the global budget, while the functional level worker provides specific budget requirements for the project items.

Bottom-up budgeting. In this method, elemental activities their schedules, descriptions, and labor skill requirements are used to construct detailed budget requests. Line workers familiar with specific activities are requested to provide cost estimates. Estimates are made for each activity in terms of labor time, materials, and machine time. The estimates are then converted to an appropriate cost basis. The dollar estimates are combined into composite budgets at each successive level up the budgeting hierarchy. If estimate discrepancies develop, they can be resolved through the intervention of senior management, middle management, functional managers, project manager, accountants, or standard cost consultants. Figure 5.19 shows the breakdown of a project into phases and parts to facilitate bottom-up budgeting and improve both schedule and cost control.

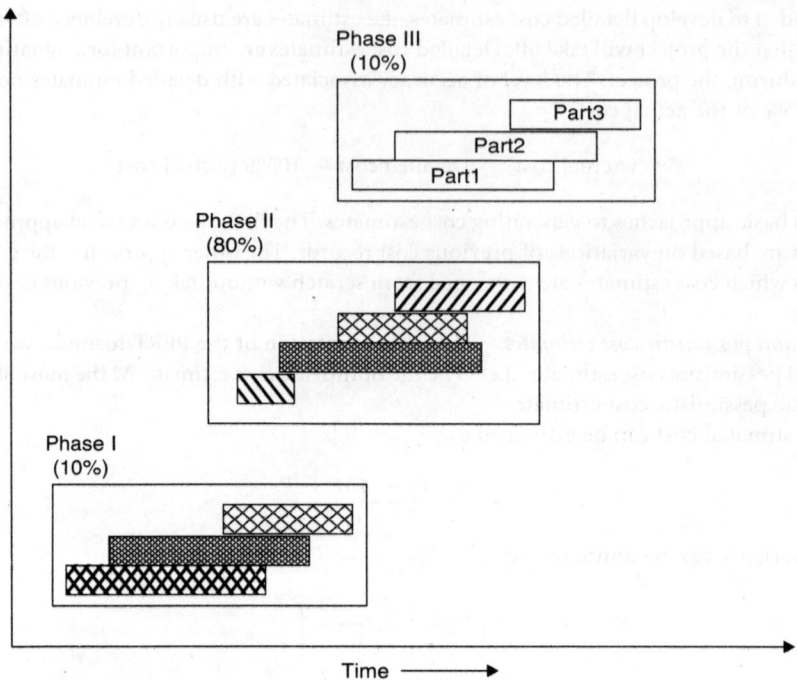

FIGURE 5.19 Budgeting by project phases.

Elemental budgets may be developed on the basis of the progress of each part of the project over time. When all the individual estimates are gathered, we obtain a composite budget estimate. Figure 5.20 shows an example of the various components that may be involved in an overall budget. The bar chart appended to a segment of the pie chart indicates the individual cost components making up that particular segment. Analytical tools such as learning curve analysis, work sampling, and statistical estimation may be employed in the cost estimation and budgeting processes.

Mathematical formulation of capital allocation.   Capital rationing involves selecting a combination of projects that will optimize the ROI. A mathematical formulation of the capital budgeting problem is presented below

$$\text{Maximize} \quad z = \sum_{i=1}^{n} v_i x_i$$

$$\text{Subject to} \quad \sum_{i=1}^{n} c_i x_i \le B$$

$$x_i = 0, 1; \quad i = 1, \dots, n$$

where n is number of projects, v_i the measure of performance for project i (e.g., present value), c_i the cost of project i, x_i the indicator variable for project i, and B the budget availability level.

A solution of the above model will indicate which projects should be selected in combination with which projects. The example that follows illustrates a capital rationing problem.

Example of a capital rationing problem.   Planning a portfolio of projects is essential in resource-limited projects, and a planning model for implementing a number of projects simultaneously exists in the literature on this subject. The capital rationing example presented here involves the determination of the optimal combination of project investments so as to maximize total ROI. Suppose that a project analyst

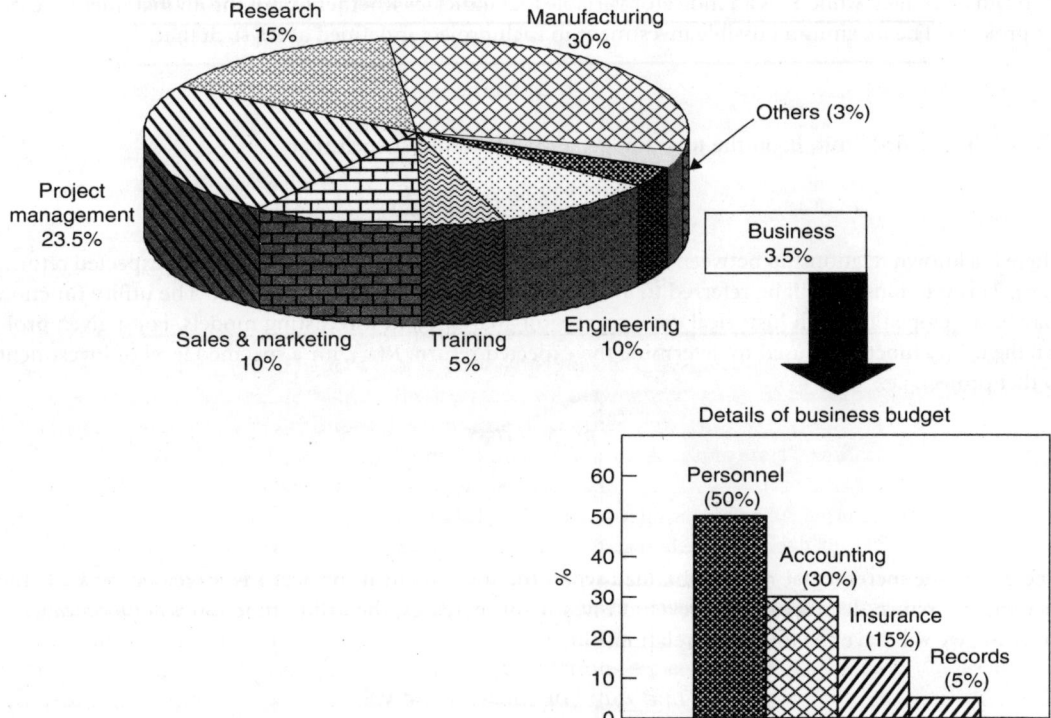

FIGURE 5.20 Budget breakdown and distribution.

is given N projects, $X_1, X_2, X_3, \ldots, X_N$, with the requirement to determine the level of investment in each project so that total investment return is maximized subject to a specified limit on available budget. The projects are not mutually exclusive.

The investment in each project starts at a base level b_i ($i = 1, 2, \ldots, N$) and increases by a variable increments k_{ij} ($j = 1, 2, 3, \ldots, K_i$), where K_i is the number of increments used for project i. Consequently, the level of investment in project X_i is defined as

$$x_i = b_i + \sum_{j=1}^{K_i} k_{ij}$$

where

$$x_i \geq 0 \quad \forall i$$

For most cases, the base investment will be zero. In those cases, we will have $b_i = 0$. In the modeling procedure used for this problem, we have

$$X_i = \begin{cases} 1 & \text{if the investment in project } i \text{ is greater than zero} \\ 0 & \text{otherwise} \end{cases}$$

and

$$Y_{ij} = \begin{cases} 1 & \text{if } j\text{th increment of alternative } i \text{ is used} \\ 0 & \text{otherwise} \end{cases}$$

The variable x_i is the actual level of investment in project i, while X_i is an indicator variable indicating whether or not project i is one of the projects selected for investment. Similarly, k_{ij} is the actual magnitude of the jth increment while Y_{ij} is an indicator variable that indicates whether or not the jth increment is used for project i. The maximum possible investment in each project is defined as M_i such that

$$b_i \le x_i \le M_i$$

There is a specified limit, B, on the total budget available to invest such that

$$\sum_i x_i \le B$$

There is a known relationship between the level of investment, x_i, in each project and the expected return, $R(x_i)$. This relationship will be referred to as the *utility function, $f(\cdot)$*, for the project. The utility function may be developed through historical data, regression analysis, and forecasting models. For a given project, the utility function is used to determine the expected return, $R(x_i)$, for a specified level of investment in that project, i.e.,

$$R(x_i) = f(x_i)$$

$$= \sum_{j=1}^{K_i} r_{ij} Y_{ij}$$

where r_{ij} is the incremental return obtained when the investment in project i is increased by k_{ij}. If the incremental return decreases as the level of investment increases, the utility function will be *concave*. In that case, we will have the following relationship:

$$r_{ij} \ge r_{ij+1} \quad \text{or} \quad r_{if} - r_{ij+1} \ge 0$$

Thus,

$$Y_{ij} \ge Y_{ij+1} \quad \text{or} \quad Y_{ij} - Y_{ij+1} \ge 0$$

so that only the first n increments ($j = 1, 2, \ldots, n$) that produce the highest returns are used for project i. Figure 5.21 shows an example of a concave investment utility function.

If the incremental returns do not define a concave function, $f(x_i)$, then one has to introduce the inequality constraints presented above into the optimization model. Otherwise, the inequality constraints may be left out of the model, since the first inequality, $Y_{ij} \ge Y_{ij+1}$, is always implicitly satisfied for concave functions. Our objective is to maximize the total return, i.e.,

$$\text{Maximize } Z = \sum_i \sum_j r_{ij} Y_{ij}$$

Subject to the following constraints:

$$x_i = b_i + \sum_j k_{ij} Y_{ij} \quad \forall i$$

$$b_i \le x_i \le M_i \quad \forall i$$

$$Y_{ij} \ge Y_{ij+1} \quad \forall i, j$$

$$\sum_i x_i \le B$$

$$x_i \ge 0 \quad \forall i$$

$$Y_{ij} = 0 \text{ or } 1 \quad \forall i, j$$

Now suppose we are given four projects (i.e., $N = 4$) and a budget limit of \$10 million. The respective investments and returns are shown in Tables 5.3–5.6.

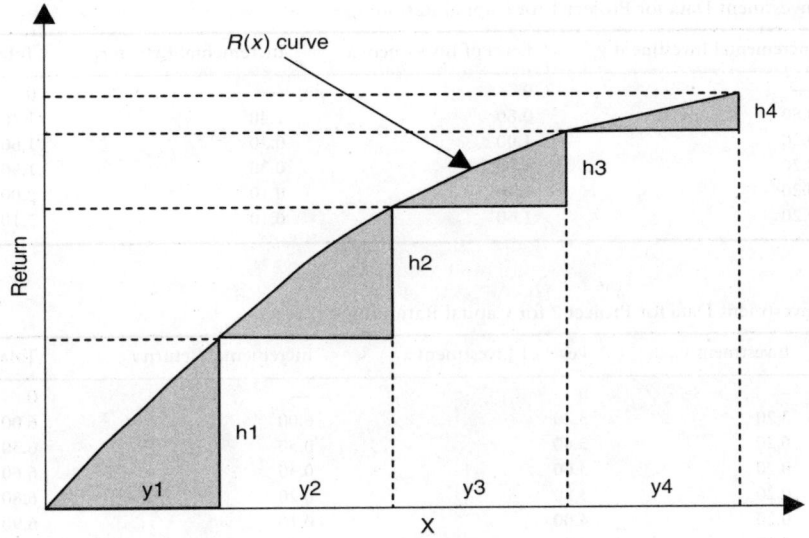

FIGURE 5.21 Utility curve for investment yield.

All the values are in millions of dollars. For example, in Table 5.3, if an incremental investment of $0.20 million from stage 2 to stage 3 is made in project 1, the expected incremental return from the project will be $0.30 million. Thus, a total investment of $1.20 million in project 1 will yield a total return of $1.90 million.

The question addressed by the optimization model is to determine how many investment increments should be used for each project, i.e., when should we stop increasing the investments in a given project? Obviously, for a single project, we would continue to invest as long as the incremental returns are larger than the incremental investments. However, for multiple projects, investment interactions complicate the decision so that investment in one project cannot be independent of the other projects. The LP model of the capital rationing example was solved with LINDO software. The solution indicates the following values for Y_{ij}

Project 1:

$$Y11 = 1, \quad Y12 = 1, \quad Y13 = 1, \quad Y14 = 0, \quad Y15 = 0$$

Thus, the investment in project 1 is X1 = $1.20 million. The corresponding return is $1.90 million.

Project 2:

$$Y21 = 1, \quad Y22 = 1, \quad Y23 = 1, \quad Y24 = 1, \quad Y25 = 0, \quad Y26 = 0, \quad Y27 = 0$$

Thus, the investment in project 2 is X2 = $3.80 million. The corresponding return is $6.80 million.

Project 3:

$$Y31 = 1, \quad Y32 = 1, \quad Y33 = 1, \quad Y34 = 1, \quad Y35 = 0, \quad Y36 = 0, \quad Y37 = 0$$

Thus, the investment in project 3 is X3 = $2.60 million. The corresponding return is $5.90 million.

Project 4:

$$Y41 = 1, \quad Y42 = 1, \quad Y43 = 1$$

Thus, the investment in project 4 is X4 = $2.35 million. The corresponding return is $3.70 million.

TABLE 5.3 Investment Data for Project 1 for Capital Rationing

Stage (j)	Incremental Investment y_{1j}	Level of Investment x_1	Incremental Return r_{1j}	Total Return $R(x_1)$
0	—	0	—	0
1	0.80	0.80	1.40	1.40
2	0.20	1.00	0.20	1.60
3	0.20	1.20	0.30	1.90
4	0.20	1.40	0.10	2.00
5	0.20	1.60	0.10	2.10

TABLE 5.4 Investment Data for Project 2 for Capital Rationing

Stage (j)	Investment y_{2j}	Level of Investment x_2	Incremental Return r_{2j}	Total Return $R(x_2)$
0	—	0	—	0
1	3.20	3.20	6.00	6.00
2	0.20	3.40	0.30	6.30
3	0.20	3.60	0.30	6.60
4	0.20	3.80	0.20	6.80
5	0.20	4.00	0.10	6.90
6	0.20	4.20	0.05	6.95
7	0.20	4.40	0.05	7.00

TABLE 5.5 Investment Data for Project 3 for Capital Rationing

Stage (j)	Incremental Investment Y_{3j}	Level of Investment x_3	Incremental Return R_{3j}	Total Return $R(x_3)$
0	—	0	—	0
1	2.00	2.00	4.90	4.90
2	0.20	2.20	0.30	5.20
3	0.20	2.40	0.40	5.60
4	0.20	2.60	0.30	5.90
5	0.20	2.80	0.20	6.10
6	0.20	3.00	0.10	6.20
7	0.20	3.20	0.10	6.30
8	0.20	3.40	0.10	6.40

TABLE 5.6 Investment Data for Project 4 for Capital Rationing

Stage (j)	Incremental Investment y_{4j}	Level of Investment x_4	Incremental Return R_{4j}	Total Return $R(x_4)$
0	—	0	—	0
1	1.95	1.95	3.00	3.00
2	0.20	2.15	0.50	3.50
3	0.20	2.35	0.20	3.70
4	0.20	2.55	0.10	3.80
5	0.20	2.75	0.05	3.85
6	0.20	2.95	0.15	4.00
7	0.20	3.15	0.00	4.00

The total investment in all four projects is $9,950,000. Thus, the optimal solution indicates that not all of the $10,000,000 available should be invested. The expected return from the total investment is $18,300,000. This translates into 83.92% ROI. Figure 5.22 presents histograms of the investments and the returns for the four projects. The individual returns on investment from the projects are shown graphically in Figure 5.23.

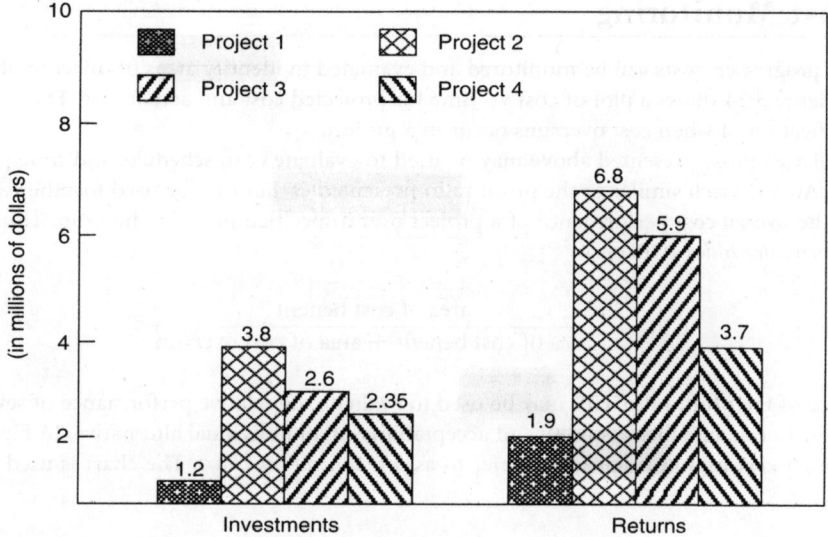

FIGURE 5.22 Histogram of capital rationing example.

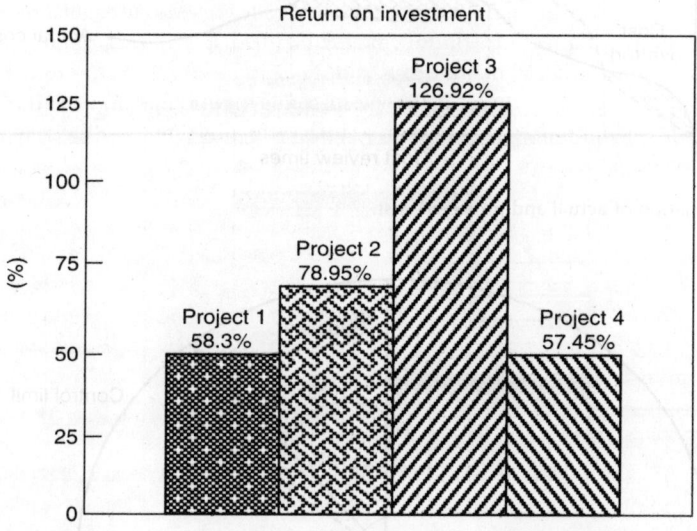

FIGURE 5.23 Histogram of returns on investments.

The optimal solution indicates an unusually large return on total investment. In a practical setting, expectations may need to be scaled down to fit the realities of the project environment. Not all optimization results will be directly applicable to real situations. Possible extensions of the above model of capital rationing include the incorporations of risk and time value of money into the solution procedure. Risk analysis would be relevant, particularly for cases where the levels of returns for the various levels of investment are not known with certainty. The incorporation of time value of money would be useful if the investment analysis is to be performed for a given planning horizon. For example, we might need to make investment decisions to cover the next 5 years rather than just the current time.

5.7 Cost Monitoring

As a project progresses, costs can be monitored and evaluated to identify areas of unacceptable cost performance. Figure 5.24 shows a plot of cost vs. time for projected cost and actual cost. The plot permits a quick identification of when cost overruns occur in a project.

Plots similar to those presented above may be used to evaluate cost, schedule, and time performance of a project. An approach similar to the profit ratio presented earlier may be used together with the plot to evaluate the overall cost performance of a project over a specified planning horizon. Thus, a formula for *cost performance index* (CPI) is

$$\text{CPI} = \frac{\text{area of cost benefit}}{\text{area of cost benefit} + \text{area of cost overrun}}$$

As in the case of the profit ratio, CPI may be used to evaluate the relative performance of several project alternatives or to evaluate the feasibility and acceptability of an individual alternative. In Figure 5.25, we present another cost monitoring tool we refer to as *cost control pie chart*. The chart is used to track the

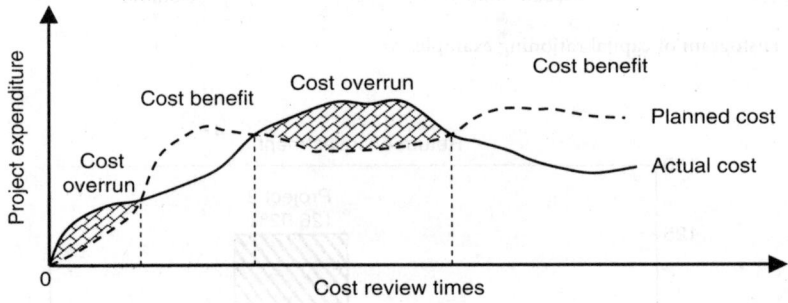

FIGURE 5.24 Evaluation of actual and projected cost.

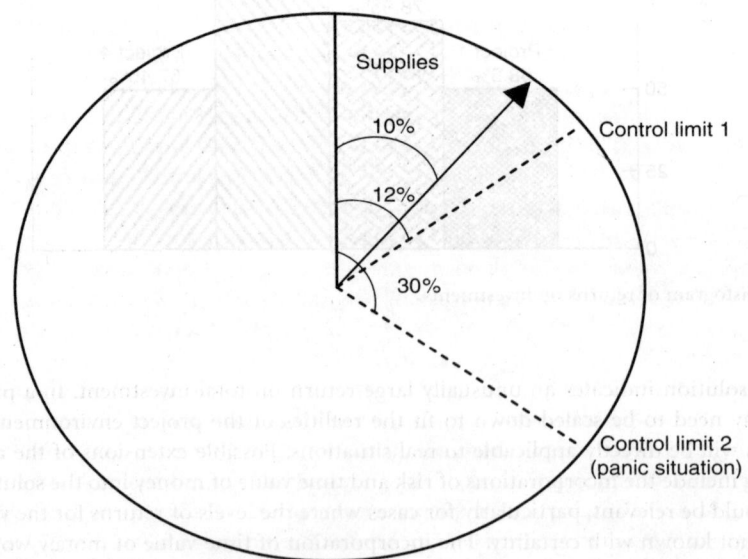

FIGURE 5.25 Cost control pie chart.

percentage of cost going into a specific component of a project. Control limits can be included in the pie chart to identify out-of-control cost situations. The example in Figure 5.25 shows that 10% of total cost is tied up in supplies. The control limit is located at 12% of total cost. Hence, the supplies expenditure is within control (so far, at least).

5.8 Project Balance Technique

One other approach to monitoring cost performance is the project balance technique. The technique helps in assessing the economic state of a project at a desired point in time in the life cycle of the project. It calculates the net cash flow of a project up to a given point in time. The project balance is calculated as

$$B(i)_t = S_t - P(1 + i)^t + \sum_{k=1}^{t} \text{PW}_{\text{income}}(i)_k$$

where $B(i)_t$ is the project balance at time t at an interest rate of i% per period, $\text{PW}_{\text{income}}(i)_t$ the present worth of net income from the project up to time t, P the initial cost of the project and S_t the salvage value at time t.

The project balanced at time t gives the net loss or net project associated with the project up to that time.

5.9 Cost and Schedule Control Systems Criteria

Contract management involves the process by which goods and services are acquired, utilized, monitored, and controlled in a project. Contract management addresses the contractual relationships from the initiation of a project to the completion of the project (i.e., completion of services and handover of deliverables). Some of the important aspects of contract management are

- Principles of contract law
- Bidding process and evaluation
- Contract and procurement strategies
- Selection of source and contractors
- Negotiation
- Worker safety considerations
- Product liability
- Uncertainty and risk management
- Conflict resolution

In 1967, the U.S. Department of Defense (DOD) introduced a set of 35 standards or criteria that contractors must comply with under cost or incentive contracts. The system of criteria is referred to as the *cost and schedule control systems criteria* (C/SCSC). Many government agencies now require compliance with C/SCSC for major contracts. The purpose is to manage the risk of cost overrun to the government. The system presents an integrated approach to cost and schedule management. It is now widely recognized and used in major project environments. It is intended to facilitate greater uniformity and provide advance warning about impending schedule or cost overruns.

The topics covered by C/SCSC include cost estimating and forecasting, budgeting, cost control, cost reporting, earned value analysis, resource allocation and management, and schedule adjustments. The important link between all of these is the dynamism of the relationship between performance, time, and cost. Such a relationship is represented in Figure 5.26. This is essentially a multiobjective problem. Because performance, time, and cost objectives cannot be satisfied equally well, concessions or compromises would need to be worked out in implementing C/SCSC.

Another dimension of the performance–time–cost relationship is the U.S. Air Force's R&M 2000 standard, which addresses *reliability* and *maintainability* of systems. R&M 2000 is intended to integrate reliability and maintainability into the performance, cost, and schedule management for government contracts. Cost and

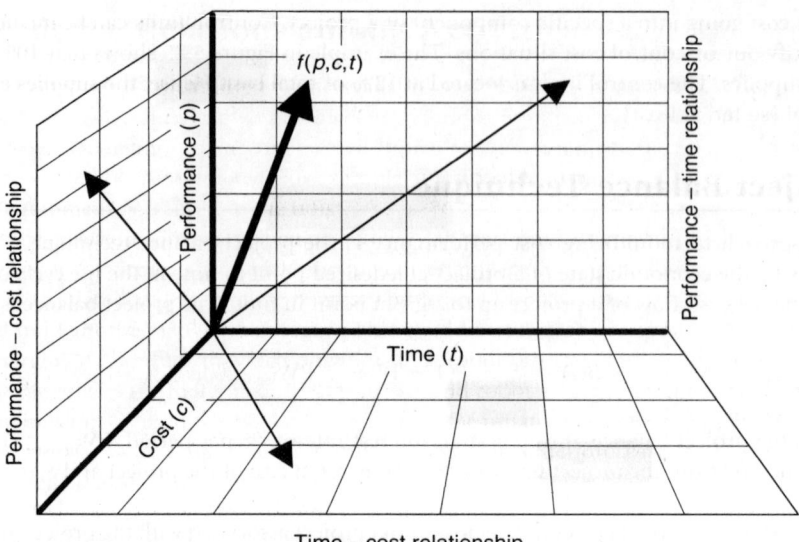

FIGURE 5.26   Performance–cost–time relationships for C/SCSC.

schedule control systems criteria and R&M 2000 constitute an effective guide for project design. Further details on cost and risk aspects of project management with respect to C/SCSC can be found elsewhere.

To comply with C/SCSC, contractors must use standardized planning and control methods based on *earned value*. Earned value refers to the actual dollar value of work performed at a given point in time compared with the planned cost for the work.

This is different from the conventional approach of measuring actual vs. planned costs, which is explicitly forbidden by C/SCSC. In the conventional approach, it is possible to misrepresent the actual content (or value) of the work accomplished. The work-rate analysis technique presented in section 35.3 can be useful in overcoming the deficiencies of the conventional approach. Cost and schedule control systems criteria is developed on a work content basis using the following factors:

- The actual cost of work performed (ACWP), which is determined on the basis of the data from cost accounting and information systems
- The budgeted cost of work scheduled (BCWS) or baseline cost determined by the costs of scheduled accomplishments
- The budgeted cost of work performed (BCWP) or earned value, the actual work of effort completed as of a specific point in time

The following equations can be used to calculate cost and schedule variances for work packages at any point in time.

$$\text{cost variance} = \text{BCWP} - \text{ACWP}$$

$$\text{percent cost variance} = (\text{Cost variance}/\text{BCWP}) \cdot 100$$

$$\text{schedule variance} = \text{BCWP} - \text{BCWS}$$

$$\text{percent schedule variance} = (\text{Schedule variance}/\text{BCWS}) \cdot 100$$

$$\text{ACWP and remaining funds} = \text{Target cost (TC)}$$

$$\text{ACWP} + \text{cost to complete} = \text{Estimated cost at completion (EAC)}$$

5.10 Sources of Capital

Financing a project means raising capital for the project. Capital is a resource consisting of funds available to execute a project. Capital includes not only privately owned production facilities, but also public investment. Public investments provide the infrastructure of the economy such as roads, bridges, and water supply. Other public capital that indirectly supports production and private enterprise includes schools, police stations, central financial institutions, and postal facilities.

If the physical infrastructure of the economy is lacking, the incentive for private entrepreneurs to invest in production facilities is likely to be lacking also. The government or community leaders can create the atmosphere for free enterprise by constructing better roads, providing better public safety facilities, and encouraging ventures that assure adequate support services.

As far as project investment is concerned, what can be achieved with project capital is very important. The avenues for raising capital funds include banks, government loans or grants, business partners, cash reserves, and other financial institutions. The key to the success of the free enterprise system is the availability of capital funds and the availability of sources to invest the funds in ventures that yield products needed by the society. Some specific ways that funds can be made available for business investments are discussed below:

Commercial loans. Commercial loans are the most common sources of project capital. Banks should be encouraged to loan money to entrepreneurs, particularly those just starting business. Government guarantees may be provided to make it easier for the enterprise to obtain the needed funds.

Bonds and stocks. Bonds and stocks are also common sources of capital. National policies regarding the issuance of bonds and stocks can be developed to target specific project types to encourage entrepreneurs.

Interpersonal loans. Interpersonal loans are unofficial means of raising capital. In some cases, there may be individuals with enough personal funds to provide personal loans to aspiring entrepreneurs. But presently, there is no official mechanism that handles the supervision of interpersonal business loans. If a supervisory body exists at a national level, wealthy citizens will be less apprehensive about loaning money to friends and relatives for business purposes. Thus, the wealthy citizens can become a strong source of business capital.

Foreign investment. Foreign investments can be attracted for local enterprises through government incentives. The incentives may be in terms of attractive zoning permits, foreign exchange permits, or tax breaks.

Investment banks. The operations of investment banks are often established to raise capital for specific projects. Investment banks buy securities from enterprises and resell them to other investors. Proceeds from these investments may serve as a source of business capital.

Mutual funds. Mutual funds represent collective funds from a group of individuals. The collective funds are often large enough to provide capital for business investments. Mutual funds may be established by individuals or under the sponsorship of a government agency. Encouragement and support should be provided for the group to spend the money for business investment purposes.

Supporting resources. A clearinghouse of potential goods and services that a new project can provide may be established by the government. New entrepreneurs interested in providing the goods and services should be encouraged to start relevant enterprises. They should be given access to technical, financial, and information resources to facilitate starting production operations. As an example, the state of Oklahoma, under the auspices of the Oklahoma Center for the Advancement of Science and Technology (OCAST), has established a resource database system. The system, named Technical Resource Access Center (TRAC), provides information about resources and services available to entrepreneurs in Oklahoma. The system is linked to the statewide economic development information system. This is a clearinghouse arrangement that will facilitate access to resources for project management.

The time value of money is an important factor in project planning and control. This is particularly crucial for long-term projects that are subject to changes in several cost parameters. Both the timing and quantity of cash flows are important for project management. The evaluation of a project alternative requires consideration of the initial investment, depreciation, taxes, inflation, economic life of the project, salvage value, and cash flows.

5.11 Activity-Based Costing

Activity-based costing (ABC) has emerged as an appealing costing technique in industry. The major motivation for ABC is that it offers an improved method to achieve enhancements in operational and strategic decisions. Activity-based costing offers a mechanism to allocate costs in direct proportion to the activities that are actually performed. This is an improvement over the traditional way of generically allocating costs to departments. It also improves the conventional approaches to allocating overhead costs.

The use of PERT/CPM, precedence diagramming, and the recently developed approach of critical resource diagramming can facilitate task decomposition to provide information for ABC. Some of the potential impacts of ABC on a production line are:

- Identification and removal of unnecessary costs
- Identification of the cost impact of adding specific attributes to a product
- Indication of the incremental cost of improved quality
- Identification of the value-added points in a production process
- Inclusion of specific inventory-carrying costs
- Provision of a basis for comparing production alternatives
- Ability to assess "what-if" scenarios for specific tasks

Activity-based costing is just one component of the overall activity-based management in an organization. Activity-based management involves a more global management approach to planning and control of organizational endeavors. This requires consideration for product planning, resource allocation, productivity management, quality control, training, line balancing, value analysis, and a host of other organizational responsibilities. Thus, while ABC is important, one must not lose sight of the universality of the environment in which it is expected to operate. And, frankly, there are some processes where functions are so intermingled that decomposition into specific activities may be difficult. Major considerations in the implementation of ABC are:

- Resources committed to developing activity-based information and cost
- Duration and level of effort needed to achieve ABC objectives
- Level of cost accuracy that can be achieved by ABC
- Ability to track activities based on ABC requirements
- Handling the volume of detailed information provided by ABC
- Sensitivity of the ABC system to changes in activity configuration

TABLE 5.7 Sample Income Statement

Statement of Income (In thousands, except per share amounts)		
Two years ended December 31	1996	1997
Net sales ($)	1,918,265	1,515,861
Costs and expenses ($)		
Cost of sales	1,057,849	878,571
Research and development	72,511	71,121
Marketing and distribution	470,573	392,851
General and administrative	110,062	81,825
	1,710,995	1,424,268
Operating income ($)	207,270	91,493
Consolidation of operations ($)	(36,981)	
Interest and other income, net ($)	9,771	17,722
Income before taxes ($)	180,060	109,215
Provision for income taxes ($)	58,807	45,115
Net income ($)	121,253	64,100
Equivalent shares ($)	61,880	60,872
Earnings per common share ($)	1.96	1.05

Income analysis can be enhanced by the ABC approach as shown in Table 5.7. Similarly, instead of allocating manufacturing overhead on the basis of direct labor costs, an ABC analysis could be conducted as illustrated in the example presented in Table 5.8. Table 5.9 shows a more comprehensive use of ABC to compare product lines. The specific ABC cost components shown in the table can be further broken down if needed. A spreadsheet analysis would indicate the impact on net profit as specific cost elements

TABLE 5.8 Activity-Based Cost Details for Manufacturing Project

	Unit Cost ($)	Cost Basis	Days Worked	Cost ($)
Labor				
Design engineer	200	Day	34	6,800
Carpenter	150	Day	27	4,050
Plumber	175	Day	2	350
Electrician	175	Day	82	14,350
IS engineer	200	Day	81	16,200
Labor subtotal				41,750
Contractor				
Air conditioning	10,000	Fixed	5	10,000
Access flooring	5,000	Fixed	5	5,000
Fire suppression	7,000	Fixed	5	7,000
AT&T	1,000	Fixed	50	1,000
DEC deinstall	4,000	Fixed	2	4,000
DEC install	8,000	Fixed	7	8,000
VAX mover	1,100	Fixed	7	1,100
Transformer mover	300	Fixed	7	300
Contractor subtotal			7	36,400
Materials				
Site preparation	2,500	Fixed	—	2,500
Hardware	31,900	Fixed	—	31,900
Software	42,290	Fixed	—	42,290
Other	10,860	Fixed	—	10,860
Materials subtotal				87,550
Grand total				165,700

TABLE 5.9 Activity-Based Costing Comparison of Product Lines

ABC Cost Components	Product A	Product B	Product C	Product D
Direct labor	27,000.00	37,000.00	12,500.00	16,000.00
Direct materials	37,250.00	52,600.00	31,000.00	35,000.00
Supplies	1,500.00	1,300.00	3,200.00	2,500.00
Engineering	7,200.00	8,100.00	18,500.00	17,250.00
Material handling	4,000.00	4,200.00	5,000.00	5,200.00
Quality assurance	5,200.00	6,000.00	9,800.00	8,300.00
Inventory cost	13,300.00	17,500.00	10,250.00	11,200.00
Marketing	3,000.00	2,700.00	4,000.00	4,300.00
Equipment depreciation	2,700.00	3,900.00	6,100.00	6,750.00
Utilities	950.00	700.00	2,300.00	2,800.00
Taxes and insurance	3,500.00	4,500.00	2,700.00	3,000.00
Total line cost	105,600.00	138,500.00	105,350.00	112,300.00
Annual production	13,000.00	18,000.00	7,500.00	8,500.00
Cost/unit	8.12	7.69	14.05	13.21
Price/unit	9.25	8.15	13.25	11.59
Net profit/unit	1.13	0.46	−0.80	−1.62
Total line revenue	120,250.00	146,700.00	99,375.00	98,515.00
Net line profit	14,650.00	8,200.00	−5,975.00	−13,785.00

are manipulated. On the basis of this analysis, it is seen that Product Line A is the most profitable. Product Line B comes in second even though it has the highest total line cost. Figure 5.27 presents a graphical comparison of the ABC cost elements for the product lines.

5.12 Cost, Time, and Productivity Formulae

This section presents a collection of common formulae useful for cost, time, and productivity analysis in manufacturing projects.

Average time to perform a task. On the basis of learning curve analysis, the average time required to perform a repetitive task is given by

$$t_n = an^{-b}$$

where t_n is the cumulative average time resulting from performing the task n times, t_1 the time required to perform the task the first time, and k the learning factor for the task (usually known or assumed).

The parameter k is a positive real constant whose magnitude is a function of the type of task being performed. A large value of k would cause the overall average time to drop quickly after just a few repetitions of the task. Thus, simple tasks tend to have large learning factors. Complex tasks tend to have smaller learning factors, thereby requiring several repetitions before significant reduction in time could be achieved.

Calculating the learning factor. If the learning factor is not known, it may be estimated from time observations by the following formula:

$$k = \frac{\log t_1 - \log t_n}{\log n}$$

Calculating total time. Total time, T_n, to complete a task n times, if the learning factor and the initial time are known, is obtained by multiplying the average time by the number of times. Thus,

$$T_n = t_1 n^{(1-k)}$$

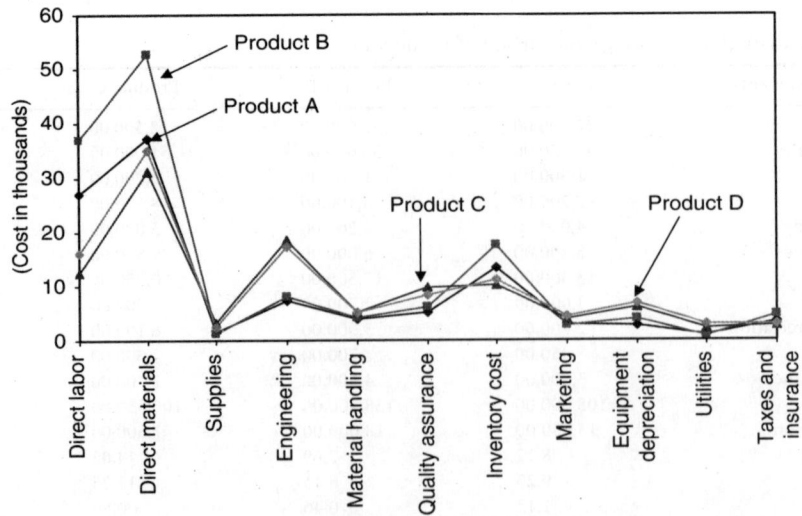

FIGURE 5.27   Activity-based comparison of product lines.

Determining time for nth performance of a task. The time required to perform a task the nth time is given by

$$x_n = t_1(1 - k)n^{-k}$$

Determining limit of learning effect. The limit of learning effect indicates the number of times of performance of a task at which no further improvement is achieved. This is often called the improvement ratio and is represented as

$$n \geq \frac{1}{1 - r^{1/k}}$$

Determining improvement target. It is sometimes desired to achieve a certain level of improvement after so many performances of a task, given a certain learning factor, k. Suppose that it takes so many trials, n_1, to achieve a certain average time performance, y_1, and it is desired to calculate how many trials, n_2, would be needed to achieve a given average time performance, y_2, the following formula would be used:

$$n_2 = n_1 y_1^{1/k} y_2^{-1/k} = n_1(y_2/y_1)^{-1/k} = n_1 r^{-1/k}$$

where the parameter r is referred to as the time improvement factor.

Calculation of the number of machines needed to meet output. The number of machines needed to achieve a specified total output is calculated from the following formula:

$$N = \frac{1.67t(O_T)}{uH}$$

where N is the number of machines, t the processing time per unit (in min), O_T the total output per shift, u the machine utilization ratio (in decimals), and H the hours worked per day (8 times number of shifts).

Calculation of machine utilization. Machine idle times adversely affect utilization. The fraction of the time a machine is productively engaged is referred to as the utilization ratio and is calculated as

$$u = \frac{h_a}{h_m}$$

where h_a is the actual hours worked and h_m the maximum hours the machine could work.

The *percent utilization* is obtained as 100% times u.

Calculation of output to allow for defects. To allow for a certain fraction of defects in total output, use the following formula to calculate starting output.

$$Y = \frac{X}{1 - f}$$

where Y is the starting output, X the target output, and f the fraction defective.

Calculation of machine availability. The percent of time that a machine is available for productive work is calculated as

$$A = \frac{o - u}{o}(100\%)$$

where o is the operator time and u the unplanned downtime.

6

Engineering Economic Evaluation and Cost Estimation

Olufemi A. Omitaomu
The University of Tennessee

6.1   Introduction

This chapter of the Handbook is about the application of engineering economic evaluation to decision-making. Decision-making is a part of our daily existence and it answers the questions "what investment alternative(s) should be selected, and what is the economic impact of such selection." People make decisions, but engineering economic analysis techniques help people to make better investment decisions. A variety of problems face decision makers: simple, intermediate, and complex problems. However, the problems most suitable for the application of engineering economic evaluation techniques must satisfy these requirements (Newnan et al., 2004):

1. The problems should be very important to justify the time spent using any of these techniques.
2. They must include some economic factors that can be used for the analysis.
3. They should require careful analysis and evaluation of the various consequences.
4. The problems should add some economic value to the decision makers.

Engineering economic evaluation is appropriate if these conditions are satisfied. Economic problems usually exist as alternatives, and these alternatives can be mutually exclusive, independent, or interdependent. Mutually exclusive alternatives involve economic problems in which only one of the proposed alternatives can be chosen; for example, a clear (mutually exclusive) choice is whether to build a 75-, 80-, or 85-story building. Therefore, these alternatives are usually analyzed using incremental methods. Economic alternatives are independent if more than one viable project may be selected. A lack of sufficient funds is usually the obstacle to implementing all selected independent projects: for example, a university must often choose between constructing a football stadium and a basketball arena. Investments are interdependent if an investment in one assumes an investment in the other, such as in the case of buying a car and buying auto insurance. However, an interdependent investment may also be mutually exclusive or independent.

Making final decisions about investment alternatives involves considering several factors, which can be classified into economic, noneconomic, and intangible factors (Blank and Tarquin, 2002). Economic factors include investment cost, interest rate, life of the investment, annual operation cost, annual revenue, effect of inflation, rate of return (ROR), exchange rates, and risk and uncertainty involved in the estimation of parameters. Noneconomic factors include personal preferences, social factors, environmental factors, professional requirements, legal implications, and political factors. Intangible factors include goodwill, friendship, and convenience. Both noneconomic and intangible factors play crucial roles in the final decision-making processes.

Economic analysis evaluation techniques are applicable for private as well as public investments. Private investments are for profit-making purposes. Therefore, private investments are not generally infrastructural in nature, that is, they are not concerned with investments in roads, education, or parks. Sources of capital for private investments include private investing and lending, and such investments are short-term, usually between 5 and 15 years. Public investments, on the other hand, are owned, financed, and used by the public; therefore, they are not for profit making. Their end result is to benefit members of the society through the provision of jobs, protection of lives and properties, provision of affordable health-care services, and major infrastructures such as roads, bridges, parks, and sewer systems. They are generally long-term, between 20 and 60 years, and are multipurpose in nature. Sources of capital include private lending and taxation. Conflicts of interest are very common in public investments because of their public nature. The most widely used engineering economic evaluation technique for evaluating public projects is benefit/cost (B/C) ratio analysis. This technique is described in Section 6.5.

The objective of this chapter is to provide a concise description of equations and techniques that arise in engineering economic analysis applications. Section 6.2 covers cost estimation techniques. Section 6.3 deals with nominal and effective interest rates. Section 6.4 describes cash-flow analysis. Techniques for evaluating investment alternatives are covered in Section 6.5. Section 6.6 is about replacement and retention analysis. Depreciation and depletion methods are delineated in Section 6.7, and the effects of inflation on economic analysis are discussed in Section 6.8. Section 6.9 covers after-tax economic analysis. Decision-making under conditions of risk, sensitivity, and uncertainty is the focus of Section 6.10. Some of the software tools used for engineering economic evaluations are presented in Appendix A, and a bibliography of some of the excellent textbooks on engineering economic analysis techniques is in the References.

6.2 Cost Estimation Techniques

Cost estimation is the process of forecasting the present and future cash-flow consequences of engineering designs and investments (Canada et al., 1996). The process is useful if it reduces the uncertainty surrounding a revenue or cost element. In doing this, a decision should result that creates increased value relative to the cost of making the estimate. Three groups of estimating have proven to be very useful in preparing estimates for economic analysis: time-series, subjective, and cost engineering techniques.

6.2.1 Time-Series Techniques

Time-series data are revenue and cost elements that are functions of time, e.g., unit sales per month and annual operating cost.

6.2.1.1 Correlation and Regression Analysis

Regression is a statistical analysis of fitting a line through data to minimize squared error. With linear regression, approximated model coefficients can be used to obtain an estimate of a revenue/cost element.

The relationship between x and y used to fit n data points ($1 \leq i \leq n$) is

$$y = a + bx \tag{6.1}$$

where
x = independent variable
y = dependent variable
$\bar{x}$ = average of independent variable
$\bar{y}$ = average of dependent variable
The mathematical expressions used to estimate a and b in Equation (6.1) are

$$b = \frac{\sum_{i=1}^{n} x_i y_i - \bar{x} \sum_{i=1}^{n} y_i}{\sum_{i=1}^{n} x_i^2 - \bar{x} \sum_{i=1}^{n} x_i} \tag{6.2}$$

$$a = \bar{y} - b\bar{x}$$

The correlation coefficient is a measure of the strength of the relationship between two variables only if the variables are linearly related.

Let

$$r = \frac{S_{xy}}{S_{xx} S_{yy}} \quad (-1 \leq r \leq 1) \tag{6.3}$$

where

$$S_{xy} = \sum_{i=1}^{n} x_i y_i - \frac{\left(\sum_{i=1}^{n} x_i\right)\left(\sum_{i=1}^{n} y_i\right)}{n}$$

$$S_{xx} = \sum_{i=1}^{n} x_i^2 - \frac{\left(\sum_{i=1}^{n} x_i\right)^2}{n} \tag{6.4}$$

$$S_{yy} = \sum_{i=1}^{n} y_i^2 - \frac{\left(\sum_{i=1}^{n} y_i\right)^2}{n}$$

r = the correlation coefficient and measures the degree of strength

r^2 = the coefficient of determination that measures the proportion of the total variation that is explained by the regression line

A positive value of r indicates that the independent and the dependent variables increase at the same rate. When r is negative, one variable decreases as the other increases. If there is no relationship between these variables, r will be zero.

6.2.1.2 Exponential Smoothing

An advantage of the exponential smoothing method compared with the simple linear regression for time-series estimates is that it permits the estimator to place relatively more weight on current data rather than treating all prior data points with equal importance. In addition, exponential smoothing is more sensitive to changes than linear regression. However, the basic assumption that trends and patterns of the past will continue into the future is a disadvantage. Hence, expert judgment should be used in interpreting results.

Let

$$S_t = \alpha' x_t + (1 - \alpha')S_{t-1} \quad (0 \le \alpha' \le 1) \tag{6.5}$$

where

S_t = forecast for period $t+1$, made in period t

α = smoothing constant

x_t = the actual data point in period t

S_{t-1} = forecast for period t, made in period $t - 1$

Intermediate choices for α' between 0 and 1 provide forecasts that have more or less emphasis on long-run average outcomes vs current outcomes.

6.2.2 Subjective Techniques

These techniques are subjective in nature; however, they are widely used in cases where the current investment events are not well enough understood to apply cost engineering or time-series techniques.

6.2.2.1 The Delphi Method

This method is a progressive practice to develop consensus from different experts. The experts are usually selected on the basis of their experience and visibility in the organization. The Delphi method is usually both complex and poorly understood. In it, experts are asked to make forecasts anonymously and through an intermediary. The process involved can be summarized as follows (Canada et al., 1996):

- Each invited participant is given an unambiguous description of the forecasting problem and the necessary background information.
- The participants are asked to provide their estimates based on the presented problem scenarios.
- An interquartile range of the opinions is computed and presented to the participants at the beginning of the second round.
- The participants are asked in the second round to review their responses in the first round in relation to the interquartile range from that round.
- The participants can, at this stage, request additional information. They may maintain or change their responses.
- If there is a significant deviation in opinion in the second round, a third-round questionnaire may be given to the participants. During this round, participants receive a summary of the second-round responses and a request to reconsider and explain their decisions in view of the second-round responses.

6.2.2.2 Technological Forecasting

This method can be used to estimate the growth and direction of a technology. It uses historical information of a selected technological parameter to extrapolate future trends. This method assumes that factors that affect historical data will remain constant into the future. Some of the commonly predicted parameters are speed, horsepower, and weight. This method cannot predict accurately when there are unforeseen changes in technology interactions.

6.2.3 Cost Engineering Techniques

Cost engineering techniques are usually used for estimating investment and working capital parameters. They can be easily applied because they make use of various cost/revenue indexes.

6.2.3.1 Unit Technique

This is the most popular of the cost engineering techniques. It uses an assumed or estimated "per unit" factor such as, for example, maintenance cost per month. This factor is multiplied by the appropriate number of units to provide the total estimate. It is usually used for preliminary estimating.

6.2.3.2 Ratio Technique

This technique is used for updating costs through the use of a cost index over a period of time.
Let

$$C_t = C_0 \left(\frac{I_t}{I_0} \right)$$ (6.6)

where
  C_t = estimated cost at present time t
  C_0 = cost at previous time t_0
  I_t = index value at time t
  I_0 = index value at time t_0

6.2.3.3 Factor Technique

This is an extension of the unit technique in which one sums the product of one or more quantities involving unit factors and adds these to any components estimated directly.
Let

$$C = \sum C_d + \sum f_i U_i$$ (6.7)

where
  C = value estimated
  C_d = cost of selected components estimated directly
  f_i = cost per unit of component i
  U_i = number of units of component i

6.2.3.4 Exponential Costing

This is also called "cost-capacity equation." It is good for estimating costs from design variables for equipment, materials, and construction. It recognizes that cost varies as some power of the change in capacity of size.
Let

$$C_A = C_B \left(\frac{S_A}{S_B} \right)^X = C_B \left(\frac{S_A}{S_B} \right)^X \left(\frac{I_t}{I_0} \right)$$ (6.8)

where
  C_A = cost of plant A
  C_B = cost of plant B
  S_A = size of plant A
  S_B = size of plant B

I_t = index value at time t

I_0 = index value at time t_0

X = cost-exponent factor, usually between 0.5 and 0.8

The accuracy of the exponential costing method depends largely on the similarity between the two projects and the accuracy of the cost-exponent factor. Generally, error ranges from ±10 to ±30% of the actual cost.

6.2.3.5 Learning Curves

In repetitive operations involving direct labor, the average time to produce an item or provide a service is typically found to decrease over time as workers learn their tasks better. As a result, cumulative average and unit times required to complete a task will drop considerably as output increases.

Let

$$Y_i = Y_1 i^b$$

$$b = \frac{\log(p)}{\log(2)} \tag{6.9}$$

where

Y_1 = direct labor hours (or cost) for the first unit

Y_i = direct labor hours (or cost) for the ith production unit

i = cumulative count of units of output

b = learning curve exponent

p = learning rate percentage

6.2.3.6 A Range of Estimates

To reduce the uncertainties surrounding estimating future values, a range of possible values rather than a single value is usually more realistic. A range could include an optimistic estimate (O), the most likely estimate (M), and a pessimistic estimate (P). Hence, the estimated mean cost or revenue value can be estimated as (Badiru, 1996; Canada et al., 1996)

$$\text{mean value} = \frac{O + 4M + P}{6} \tag{6.10}$$

6.3 Interest Rates in Economic Analysis

Interest rates are used to express the time value of money. Interest *paid* is the cost on borrowed money and interest *earned* is the benefit on saved or invested money. Interest rates can be quoted as simple interest or compound interest. *Simple interest* is interest paid only on the principal, while *compound interest* is interest paid on both the principal and the accrued interest. The expressions for calculating future amounts based on simple interest and compound interest are

$$F_n = P(1 + ni) \quad \text{For simple interest}$$

$$F_n = P(1 + i)^n \quad \text{For compound interest} \tag{6.11}$$

where

F_n = future value after n periods

P = initial investment amount

i = interest rate

n = investment or loan periods

At $n = 0$ and 1, the future values for both simple and compound interests are equal. Simple interest rate is not used in economic analysis and will not be discussed any further.

6.3.1 Nominal and Effective Interest Rates

Compound interest rate (or interest rate for short) is used in economic analysis to account for the time value of money. Interest rates are usually expressed as a percentage, and the interest period (the time unit of the rate) is usually a year. However, interest rates can also be computed more than once a year. Compound interest rates can be quoted as *nominal interest rates* or as *effective interest rates*.

Let

r = nominal interest rate per year

m = number of compounding periods per year

i = effective interest rate per compounding period

i_a = effective interest rate per year

Nominal interest rate is the interest rate without considering the effect of any compounding. It is not the real interest rate used for economic analysis; however, it is usually the quoted interest rate. It is equivalent to annual percentage rate (APR), which is usually quoted for loan and credit-card purposes. The expression for calculating nominal interest rate is

$$r = (\text{interest rate per period})(\text{number of periods}) \tag{6.12}$$

The format for expressing r is "$r\%$ per time period t."

Effective interest rate can be expressed per year or per compounding period. Effective interest rate per year is used in engineering economic analysis calculations. It is the annual interest rate taking into account the effect of any compounding during the year. It accounts for both the nominal rate and the compounding frequency. Effective interest rate *per year* is given by

$$i_a = (1 + i)^m - 1$$
$$i_a = \left(1 + \frac{r}{m}\right)^m - 1 \tag{6.13}$$
$$i_a = (F/P, r/m\%, m) - 1$$

Effective interest rate *per compounding period* is given by

$$i = (1 + i_a)^{1/m} - 1 = \frac{r}{m} \tag{6.14}$$

When compounding occurs more frequently, the compounding period becomes shorter and, therefore, becomes a continuous compounding phenomenon. This situation can be seen in the stock market and international trade. Effective interest rate per year *for continuous compounding* is given by

$$i_a = e^r - 1 \tag{6.15}$$

Note that the time period for i_a and r in Equations (6.13)–(6.15) must be the same.

6.3.2 Fixed and Variable Interest Rates

Interest rates may be fixed over the useful life of an investment or they may vary from year to year. This situation also applies to the interest rates corporations use to evaluate alternatives. The reasons for varying interest rates include fluctuations in national and international economies, effects of inflation, and changes in market share (Blank and Tarquin, 2002). Loan rates, such as mortgage loans, may be adjusted from year to year on the basis of the inflation index of the consumer price index (CPI). If the variations in the interest rates are not large, cash-flow calculations usually neglect the effect of interest rate changes. However, the equivalent values will vary considerably if the variations in interest rates are large. In such cases, the varying interest rates should be accounted for in economic analysis.

6.3.3   Rule of 72 and Rule of 100

Knowing the compound interest rate or simple interest rate, it is possible to estimate the number of years it will take an investment to double without resulting in complex computation. The *Rule of 72* is used when the interest rate is quoted as compound interest rate and it is given as

$$\text{Estimated number of years} = \frac{72}{i} \qquad (6.16)$$

where i is in percentage.

When the interest rate is quoted as simple interest, replace "72" in Equation (6.16) with "100," hence *Rule of 100* (Blank and Tarquin, 2002). The equation is also useful if the number of years is known but i is not known.

6.4   Cash-Flow Analysis

A *cash-flow diagram* (CFD), as used in engineering economic analysis, is a graphical representation of when all cash flows occur. Cash flows can be positive or negative. Positive cash flows (cash inflows) increase the funds available to the company; therefore, they include both receipts and revenue. Negative cash flows (cash outflows) are deductions from the company's funds; hence, they include first cost, annual expenses, and other cash disbursements. The difference between several receipts and disbursements that occur within a given interest period is called the net cash flow. Cash-flow diagrams can be drawn on the basis of cash inflows, cash outflows, and net cash flows. In addition, they can be drawn from either the lender's perspective or the borrower's perspective on the basis of actual or estimated information. Whichever perspective is chosen, CFDs are based on the following assumptions (Eschenbach, 2003; Sullivan et al., 2003):

- Interest rate is computed once in a period.
- All cash flows occur at the end of the period.
- All periods are of the same length.
- The interest rate and the number of periods are of the same length.
- Negative cash flows are drawn downward from the time line.
- Positive cash flows are drawn upward from the time line.

Engineering economic analysis utilizes the following terms and symbols for CFDs:

- P = cash-flow value at a time designated as the present. This is usually at time 0. It may also be called the present value (PV) or the present worth (PW) dollars.
- F = cash-flow value at some time in the future. It is also called the future value (FV) or the future worth (FW) dollars.
- A = a series of equal, consecutive, end-of-period amounts of money. This is also called the annual worth (AW) or the equivalent uniform annual worth (EUAW) dollars per period.
- G = a uniform arithmetic gradient increase in period-by-period payments or disbursements.
- n = number of interest periods (days, weeks, months, or years).
- i = interest rate per time period expressed as a percentage.
- t = time, stated in periods (years, months, days).

Cash flows occur in many configurations; however, this chapter develops factors for computing cash-flow equivalence of the general cash-flow profiles. Two or more cash flows are equivalent if they are equal in terms of their economic values and if this equality only occurs at a particular interest rate.

6.4.1   Compound Amount Factor

This factor determines an FV of money, F, that is equivalent to a PV of money, P, after n periods at interest rate i. The FV, F, is given as

$$F = P(1 + i)^n = P(F/P, i, n) \qquad (6.17)$$

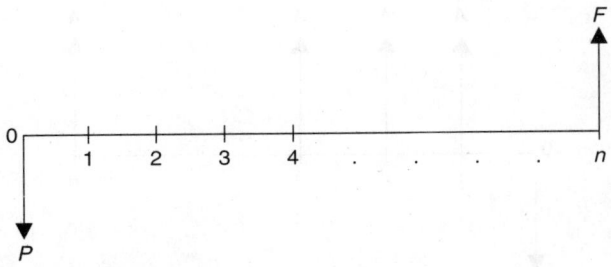

FIGURE 6.1 Single payment cash flow.

The expression $(1 + i)^n$ is called the *single-payment compound amount factor* and is usually referred to as the *F/P factor* in engineering economic analysis. The cash-flow representation of this factor is shown in Figure 6.1.

6.4.2 Present Worth Factor

The PW factor is the opposite of the compound amount factor. This factor computes the PV, P, of a given FV, F, after n years at interest rate i. The mathematical expression for the unknown PV is

$$P = F\left[\frac{1}{(1 + i)^n}\right] = F(P/F, i, n)$$ (6.18)

The expression in the brackets is the *single-payment present worth factor* or the *P/F factor*.

6.4.3 Uniform-Series Present Worth Factor

This factor is used to determine the equivalent PW, P, of a series of equal end-of-period amounts, A. The expression for this equivalence is

$$P = \sum_{t=1}^{n} A(1 + i)^{-t}$$

$$P = A\left[\frac{(1 + i)^n - 1}{i(1 + i)^n}\right] = A(P/A, i, n), \quad i \neq 0$$ (6.19)

The expression in the bracket is the *uniform-series present worth factor* or the *P/A factor*. It should be noted that to use this formula, the PW factor, P, must be one period prior to the first end-of-period amount, as shown in Figure 6.2.

6.4.4 Capitalized Cost Formula

Capitalized cost is a special case of uniform-series PW factor when the number of periods is infinitely long. Therefore, it refers to the PW equivalent of a perpetual series of equal end-of-period amounts. From Equation (6.19), as n approaches infinity, the capitalized cost is given as

$$C = \lim_{n \to \infty} A\left[\frac{(1 + i)^n - 1}{i(1 + i)^n}\right]$$

$$= A\left\{\lim_{n \to \infty}\left[\frac{(1 + i)^n - 1}{i(1 + i)^n}\right]\right\}$$

$$= A\left(\frac{1}{i}\right)$$ (6.20)

The cash flow for this formula is shown in Figure 6.3.

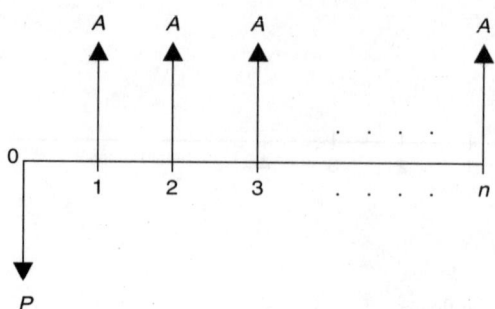

FIGURE 6.2   Uniform-series cash flow.

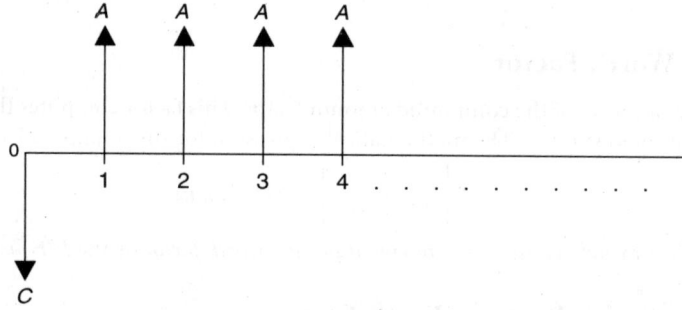

FIGURE 6.3   Capitalized cost cash flow.

6.4.5   Uniform-Series Capital Recovery Factor

This factor is the converse of the uniform-series PW factor; it calculates the uniform series of equal end-of-period amounts, A, given a present amount, P. Solving Equation (6.19) for A gives

$$A = P \left[\frac{i(1 + i)^n}{(1 + i)^n - 1} \right] = P(A/P, i, n) \tag{6.21}$$

The expression in the bracket is the *uniform-series PW factor* or the *A/P factor*.

6.4.6   Uniform-Series Sinking Fund Factor

This factor calculates the uniform series of equal end-of-period amounts, A, that are equivalent to a single FV, F. The uniform series starts at the end of period 1 and ends at the period of the given F value as shown in Figure 6.4. Substituting P from Equation (6.18) into Equation (6.21) gives this factor as follows:

$$A = F \left[\frac{1}{(1 + i)^n} \right] \left[\frac{i(1 + i)^n}{(1 + i)^n - 1} \right]$$

$$A = F \left[\frac{i}{(1 + i)^n - 1} \right] = F(A/F, i, n) \tag{6.22}$$

The term in brackets is called the *uniform-series sinking fund factor or A/F factor*.

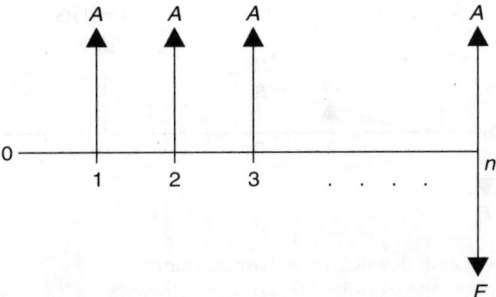

FIGURE 6.4   Uniform-series cash flow.

6.4.7   Uniform-Series Compound Amount Factor

This factor computes the FV, F, of a uniform series of equal end-of-period amounts, A. The FV, in this case, occurs at the same period as the last end-of-period amount. This factor is the reverse of the uniform-series sinking fund factor. Solving for F from Equation (6.22) gives

$$F = A\left[\frac{(1+i)^n - 1}{i}\right] = A(F/A, i, n) \tag{6.23}$$

The expression in brackets is the *uniform-series compound amount factor or F/A factor*.

6.4.8   Arithmetic Gradient Series Factor

An *arithmetic gradient series* is a cash-flow series that either increases or decreases by a fixed amount at the end of each period. Thus, the cash flow changes by a constant amount each period. This constant amount is called the *gradient* and is denoted by G. The basic arithmetic gradient series with a base amount of 0 is shown in Figure 6.5.

The closed form of the PW of an arithmetic gradient series with a zero base amount is

$$P = G\left[\frac{(1+i)^n - (1+ni)}{i^2(1+i)^n}\right] = G(P/G, i, n) \tag{6.24}$$

When the base amount is not 0, as shown in Figure 6.6, the following closed form can be used for the PW:

$$P = A_1\left[\frac{(1+i)^n - 1}{i(1+i)^n}\right] + G\left[\frac{(1+i)^n - (1+ni)}{i^2(1+i)^n}\right] \tag{6.25}$$

$$P = A_1(P/A, i, n) + G(P/G, i, n)$$

where A_1 is the base amount at the end of the first period. When the base amount at the end of period 1 is 0, Equation (6.25) reduces to Equation (6.24). Both Equations (6.24) and (6.25) can be used for both increasing and decreasing gradient amounts. In the case of increasing gradient, G is treated as a positive amount, while for a decreasing gradient, G is treated as a negative amount.

The AW equivalent of an arithmetic gradient series can also be found by using the following equation:

$$A = A_1 + G\left[\frac{1}{i} - \frac{n}{(1+i)^n - 1}\right] \tag{6.26}$$

$$A = A_1 + G(A/G, i, n)$$

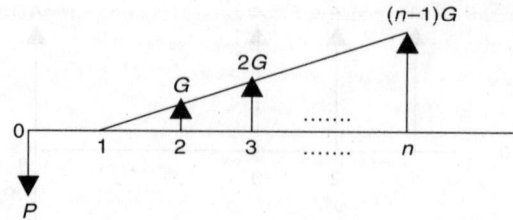

FIGURE 6.5   Arithmetic gradient cash flow with zero base amount.

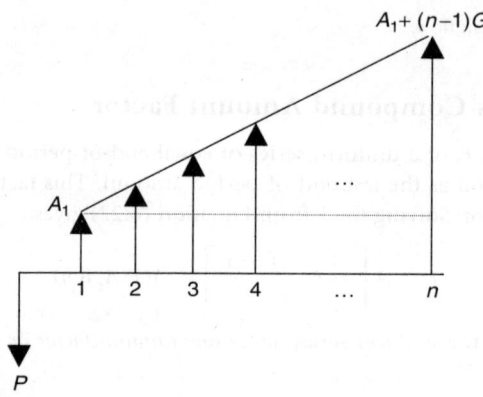

FIGURE 6.6   Arithmetic gradient cash flow with nonzero base amount.

where A_1 is the base amount at the end of the first period. It must be noted that using Equation (6.24), the PW must be two periods prior to the first nonzero amount; however, using Equations (6.25) and (6.26), the PW is one period prior to the first nonzero amount, as shown in Figures 6.5 and 6.6, respectively.

6.4.9   The Tent Cash-Flow Equation

Several real-world applications of the arithmetic gradient factor go beyond the one-sided cash-flow profile shown in Figure 6.6; they usually tend toward a combination of several one-sided cash flows (Badiru and Omitaomu, 2003), as shown in Figure 6.7. Several other profiles of the cash flow are also possible.

To facilitate a less convoluted use of the arithmetic gradient series factor, Badiru and Omitaomu (2003) developed a general tent equation (GTE) for tent cash flows based on the basic tent (BT) cash-flow profile shown in Figure 6.8.

The GTE is given as

$$P_0 = A_0 + A_1 \left[\frac{(1+i)^T - 1}{i(1+i)^T} \right] + G_1 \left[\frac{(1+i)^T - (1+Ti)}{i^2(1+i)^T} \right]$$

$$= A_0 + A_1 \, (P/A, i, T) + G_1 \, (P/G, i, T)$$

$$P_T = A_{T+1} \left[\frac{(1+i)^x - 1}{i(1+i)^x} \right] + G_2 \left[\frac{(1+i)^x - (1+xi)}{i^2(1+i)^x} \right]$$

$$= A_{T+1} \, (P/A, i, x) + G_2 \, (P/G, i, x)$$

(6.27)

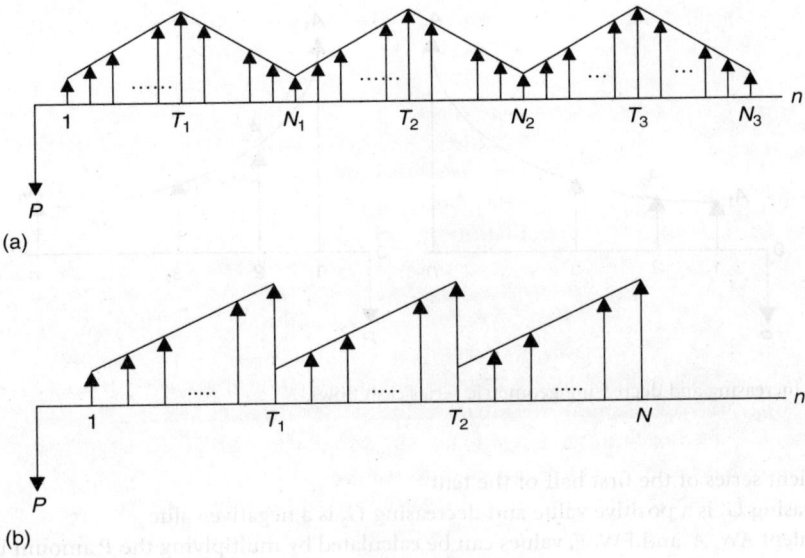

(a)

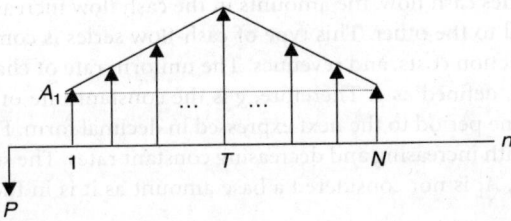

(b)

FIGURE 6.7 (a) Executive tent (ET) cash-flow profile (Adopted from Badiru and Omitaomu, 2003). (b) Saw-tooth tent (STT) cash-flow profile (Adopted from Badiru and Omitaomu, 2003).

FIGURE 6.8 Basic tent (BT) cash-flow profile (Adopted from Badiru and Omitaomu, 2003).

The equivalent PW of the tent cash flow is given by

$$P = \begin{cases} P_0 + P_T(1 + i)^{-T} & \text{when } G_2 < 0 \\ P_0 - P_T(1 + i)^{-T} & \text{when } G_2 > 0 \end{cases} \tag{6.28}$$

where

P = total PV at time $t = 0$

P_0 = PV for the first half of the tent at time $t = 0$

P_T = PV for the second half of the tent at time $t = T$

i = interest rate in fraction

n = number of periods

T = the center time value of the tent

$x = (n - T)$

A_0 = amount of $n = 0$

A_1 = amount of $n = 1$

A_{T+1} = amount of $n = T + 1$

G_1 = gradient series of the first half of the tent

$\Rightarrow$ increasing G_1 is a positive value and decreasing G_1 is a negative value

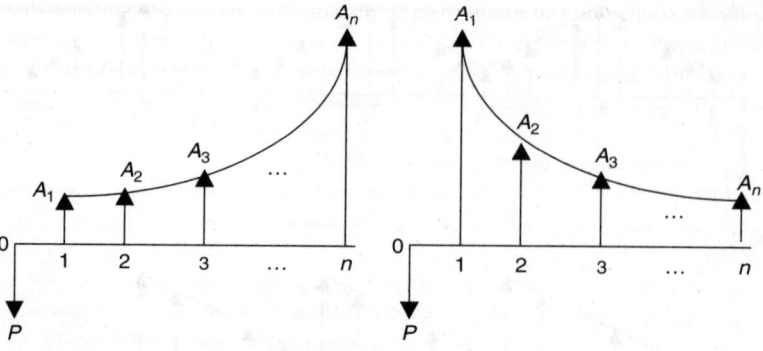

FIGURE 6.9   Increasing and decreasing geometric series cash flows.

G_2 = gradient series of the first half of the tent
   ⇒ increasing G_2 is a positive value and decreasing G_2 is a negative value
   The equivalent AW, A, and FW, F, values can be calculated by multiplying the P amount by the A/P or F/P factor, respectively.

6.4.10   Geometric Gradient Series Factor

In the case of geometric series cash flow, the amounts in the cash flow increase or decrease by a constant percentage from one period to the other. This type of cash-flow series is common in operations involving operating costs, construction costs, and revenues. The uniform rate of change is the geometric gradient series of the cash flows, defined as g. Therefore, g is the constant rate of change by which amounts increase or decrease from one period to the next expressed in decimal form. Figure 6.9 presents CFDs for geometric gradient series with increasing and decreasing constant rates. The series starts in year 1 with an initial amount A_1; however, A_1 is not considered a base amount as it is in the arithmetic gradient series cash-flow equation.

The formula for calculating the PW of the *increasing* geometric series cash flow is

$$P = A_1 \left[\frac{1 - (1 + g)^n (1 + i)^{-n}}{(i - g)} \right] \qquad i \neq g \text{ and } g \text{ is increasing}$$

$$(6.29)$$

$$P = \frac{nA_1}{(1 + i)} \qquad i = g \text{ and } g \text{ is increasing}$$

However, the formula for calculating the PW of the *decreasing* geometric series cash flow is (Badiru, 1996)

$$P = A_1 \left[\frac{1 - (1 - g)^n (1 + i)^{-n}}{(i + g)} \right], \text{ when } g \text{ is decreasing} \qquad (6.30)$$

The equivalent AW, A, and FW, F, values can also be derived; however, it is usually easier to determine the P amount and multiply it by the A/P or F/P factor, respectively.

6.5   Economic Methods of Comparing Alternatives

The objective of performing an economic analysis is to make a choice or choices between investments that are competing for limited financial resources. Investment opportunities are either mutually exclusive or independent. For mutually exclusive investment opportunities, only one viable project can be selected; therefore, each project is an alternative and competes with others. On the other hand, more than one project may be selected if the investment opportunities are independent; hence, the alternatives do not compete with one another in the evaluation.

There are several methods for comparing investment alternatives: PW analysis, AW analysis, ROR analysis, and B/C ratio analysis. In using these methods, the *do-nothing* (DN) option is a viable alternative that must be considered unless otherwise stated. The selection of DN as the accepted alternative means that no new investment will be initiated, that is, the current state of affairs continues. Three different analysis periods are used for evaluating alternatives: equal service life for all alternatives, different service life for all alternatives, and infinite service life for all alternatives. The kind of analysis period may influence the method of evaluation used.

6.5.1 Present Worth Analysis

The *present worth* analysis is the application of some of the engineering economic analysis factors in which the present amount is unknown. It is used for projects with equal service life and can be used for evaluating one alternative, two or more mutually exclusive opportunities, and independent alternatives. Present worth analysis evaluates projects by converting all future cash flows into present amount. The guidelines for using PW analysis projects are:

- *For one alternative:* Calculate PW at the minimum attractive rate of return (MARR). If PW ≥ 0, the requested MARR is met or exceeded and the alternative is economically viable.
- *For two or more alternatives:* Calculate the PW of each alternative at the MARR. Select the alternative with the *numerically largest* PW value. Numerically largest indicates a lower PW of cost cash flows (less negative) or larger PW of net cash flows (more positive).
- *For independent projects:* Calculate the PW of each alternative. Select all projects with PW ≥ 0 at the given MARR.

The general equation for PW analysis is

$$PW = A_0 + A(P/A, i, n) + F(P/F, i, n) + [A_1(P/A, i, n) + G(P/G, i, n)] + A_1(P/A, g, i, n) \quad (6.31)$$

This equation reduces to a manageable size depending on the cash-flow profiles of the alternatives. For example, for bonds, the PW analysis equation becomes

$$PW = A(P/A, i, n) + F(P/F, i, n) = rZ(P/A, i, n) + C(P/F, i, n) \quad (6.32)$$

where

Z = face, or par, value of the bond
C = redemption or disposal price (usually equal to Z)
r = bond rate (nominal interest rate) per interest period
n = number of periods before redemption
i = bond yield rate per period

Two extensions of the PW analysis are capitalized cost and discounted payback period (Badiru, 1996).

6.5.2 Capitalized Cost

Capitalized cost is a special case of uniform-series PW factor when the number of periods is infinitely long. Therefore, it refers to the PW equivalent of a perpetual series of equal end-of-period amounts. The PW in the case of an infinitely long period is given as

$$C = PW = A\left[\frac{1}{i}\right] \quad (6.33)$$

where

C = present worth of the asset
A = equal end-of-period amounts
i = investment interest rate

6.5.3 Discounted Payback Period Analysis

The discounted payback period, n_p, is the estimated number of years it will take for the estimated revenues and other economic benefits to recover the initial investment at a stated ROR (Badiru, 1996). This method is used as an initial screening technique for providing additional insight into the various investment opportunities. It is not used as a primary measure of worth to select an alternative because it does not take cash flows after the payback period into consideration in the evaluation process (Blank and Tarquin, 2002). The payback period can be calculated at $i \geq 0\%$. Calculations at $i > 0\%$ take the time value of money into account and are more economically correct. This calculation is called discounted payback period and its general expression is given as

$$0 = -P + \sum_{t=1}^{n_p} \text{NCF}_t (P/F, i, t) + \text{NCF}_A (P/A, i, n_p) \quad \text{when } i > 0\% \tag{6.34}$$

where
- P = initial investment or first cost
- NCF_t = estimated net cash flow for each year t
- NCF_A = estimated equal amount net cash flow for each year
- n_p = discounted payback period

When $i = 0\%$ in Equation (6.34), the expression becomes

$$n_p = \frac{P}{\text{NCF}} \quad \text{when } i = 0\% \tag{6.35}$$

Equation (6.35) is the simple (no-return) payback period. The simple payback period does not take the time value of money into consideration; however, it is a readily understood concept by people not familiar with economic analysis.

To facilitate the computation of payback and discounted payback periods using Equations (6.34) and (6.35), the following tabular headings can be used (Sullivan et al., 2003):

End of Year j	Net Cash Flow	Cumulative PW at $i = 0\%$/yr through Year j	PW of Cash Flow at $i > 0\%$/yr	Cumulative PW at $i > 0\%$/yr through Year j
	(A)	(B)	(C)	(D)

Column A: Net cash flow for the alternative.

Column B: Cumulative of the net cash flow in column A. The jth year at which the cumulative balance in this column turns positive is the payback period for the alternative.

Column C: PW at the given interest rate of the respective net cash flow in column A.

Column D: Cumulative of the PW in column C. The jth year at which the cumulative balance in this column turns positive is the discounted payback period for the alternative.

6.5.4 Annual Worth Analysis

The AW method of evaluating investment opportunities is the most readily used of all the measures because people easily understand what it means. This method is mostly used for projects with unequal service life since it requires the computation of the equivalent amount of the initial investment and the future amounts for only one project life cycle. Annual worth analysis converts all future and present cash flows into equal end-of-period amounts. For mutually exclusive alternatives, AW can be calculated at the MARR and viable alternative(s) selected on the basis of the following guidelines:

- *One alternative*: Select alternative with AW ≥ 0 since MARR is met or exceeded.
- *Two or more alternatives*: Choose alternative with the lowest cost or the highest revenue AW value.

The AW amount for an alternative consists of two components: capital recovery for the initial investment P at a stated interest rate (usually at the MARR) and the equivalent annual amount A. Therefore, the general equation for AW analysis is

$$AW = -CR - A$$
$$= -[P(A/P, i, n) - S(A/F, i, n)] - A \quad (6.36)$$

where
CR = capital recovery component
A = annual amount component of other cash flows
P = initial investment (first cost) of all assets
S = estimated salvage value of the assets at the end of their useful life

Annual worth analysis is specifically useful in areas such as asset replacement and retention, breakeven studies, make-or-buy decisions, and all studies relating to profit measure. It should be noted that expenditures of money increase the AW, while receipts of money such as selling an asset for its salvage value decrease AW. The assumptions of the AW method are that (Blank and Tarquin, 2002) the following:

1. The service provided will be needed forever since it computes the annual value per cycle.
2. The alternatives will be repeated exactly the same in succeeding life cycles. This is especially important when the service life is several years into the future.
3. All cash flows will change by the same amount as the inflation or deflation rate.

The validity of these assumptions is based on the accuracy of the cash-flow estimates. If the cash-flow estimates are very accurate, then these assumptions will be valid and will minimize the degree of uncertainty surrounding the final decisions based on this method.

6.5.5 Permanent Investments

This measure is the reverse of capitalized cost. It is the AW of an alternative that has an infinitely long period. Public projects such as bridges, dams, irrigation systems, and railroads fall into this category. In addition, permanent and charitable organization endowments are evaluated using this approach. The AW in the case of permanent investments is given by

$$AW = A = Pi \quad (6.37)$$

where
A = capital recovery amount
P = initial investment of the asset
i = investment interest rate

6.5.6 Internal Rate of Return Analysis

Internal rate of return (IRR) is the third and most widely used method of measure in the industry. It is also referred to as simply ROR or return on investment (ROI). It is defined as the interest rate that equates the equivalent worth of investment cash inflows (receipts and savings) to the equivalent worth of cash outflows (expenditures); that is, the interest rate at which the benefits are equivalent to the costs. If $i*$ denotes the IRR, then the unknown interest rate can be solved by using any of the following expressions:

$$PW(\text{Benefits}) - PW(\text{Costs}) = 0$$
$$EUAB - EUAC = 0 \quad (6.38)$$

where
PW = present worth
EUAB = equivalent uniform annual benefits

EUAC = equivalent uniform annual costs
The procedure for selecting the viable alternative(s) is:

- If $i* \geq$ MARR, accept the alternative as an economically viable project.
- If $i* <$ MARR, the alternative is not economically viable.

When applied correctly, IRR analysis will always result in the same decision as with PW or AW analysis. However, there are some difficulties with IRR analysis: multiple $i*$, reinvestment at $i*$, and computational difficulty. Multiple $i*$ usually occurs whenever there is more than one sign change in the cash-flow profile; hence, there is no unique $i*$ value. In addition, there may be no real value of $i*$ that will solve Equation (6.38), but only real values of $i*$ are valid in economic analysis. Moreover, IRR analysis usually assumes that the selected project can be reinvested at the calculated $i*$, but this assumption is not valid in economic analysis. These difficulties have given rise to an extension of IRR analysis called external rate of return (ERR) analysis (Sullivan et al., 2003).

6.5.7 External Rate of Return Analysis

The difference between ERR and IRR is that ERR takes into account the interest rate external to the project at which the net cash flow generated or required by the project over its useful life can be reinvested or borrowed. Therefore, this method requires the knowledge of an external MARR for a similar project under evaluation. The expression for calculating ERR is given by

$$F = P(1 + i')^n \tag{6.39}$$

where
P = the present value of all cash outflows at the MARR
F = the future value of all cash inflows at the MARR
i' = the unknown ERR
n = the useful life or evaluation project of the project
    Using this method, a project is acceptable when the calculated i' is greater than the MARR. However, if i' is equal to the MARR (breakeven situation), noneconomic factors may be used to justify the final decision. The ERR method has two advantages over the IRR method: it does not result in trial and error in determining the unknown ROR and it is not subject to the possibility of multiple rates of return even when there are several sign changes in the cash-flow profile.

6.5.8 Benefit/Cost Ratio Analysis

The three methods of analysis described above are mostly used for private projects since the objective of most private projects is to maximize profits. Public projects, on the other hand, are executed to provide services to the citizenry at no profit; therefore, they require a special method of analysis. The B/C ratio analysis is normally used for evaluating public projects. It has its roots in The Flood Act of 1936, which requires that for a federally financed project to be justified, its benefits must exceed its costs (Blank and Tarquin, 2002). The B/C ratio analysis is the systematic method of calculating the ratio of project benefits to project costs at a discounted rate. For over 60 years, the B/C ratio method has been the accepted procedure for making go/no-go decisions on independent and mutually exclusive projects in the public sector.
    The B/C ratio is defined as

$$\text{B/C} = \frac{\text{PW(Benefits)}}{\text{PW(Costs)}} = \frac{\sum_{t=0}^{n} B_t(1 + i)^{-t}}{\sum_{t=0}^{n} C_t(1 + i)^{-t}} \tag{6.40}$$

where
B_t = is benefit (revenue) at time t
C_t = is cost at time t

If the B/C ratio is $>$ 1, then the investment is viable; if the ratio is $<$ 1, the project is not acceptable. A ratio of 1 indicates a breakeven situation for the project and noneconomic factors may be considered to validate the final decision about the project.

6.5.9 Incremental Analysis

Under some circumstances, IRR analysis does not provide the same ranking of alternatives as do PW and AW analyses for multiple alternatives. Hence, there is a need for a better approach for analyzing multiple alternatives using the IRR method. Incremental analysis can be defined as the evaluation of the differences between alternatives. The procedure essentially decides whether or not differential costs are justified by differential benefits. Incremental analysis is mandatory for economic analysis involving the use of IRR and B/C ratio analyses that evaluate three or more mutually exclusive alternatives. It is not used for independent projects since more than one project can be selected. The steps involved in using incremental analysis are:

1. If IRR (B/C ratio) for each alternative is given, reject all alternatives with IRR$<$MARR (B/C$<$ 1.0).
2. Arrange other alternatives in increasing order of initial cost (total costs).
3. Compute incremental cash flow pairwise starting with the first two alternatives.
4. Compute incremental measures of worth using the appropriate equations.
5. Use the following criteria for selecting the alternatives that will advance to the next stage of comparisons:
   (i) If ΔIRR$>$MARR, select higher-cost alternative.
   (ii) If ΔB/C$>$1.0, select higher-cost alternative.
6. Eliminate the defeated alternative and repeat steps 3–5 for the remaining alternatives.
7. Continue until only one alternative remains. This last alternative is the most economically viable alternative.

6.6 Replacement or Retention Analysis

Replacement or retention analysis is one of the commonly performed economic analyses in the industry. This is an application of the AW analysis. A replacement occurs when an asset is retired for specific reasons and another asset is acquired in its place to continue providing the required service. This is also called a "like-for-like exchange;" therefore, no gain or loss is realized and no tax (credit) is paid (received) on the exchange. A retirement occurs when an asset is salvaged and the service rendered by the asset is discontinued. This is also called a "disposal." In this case, a gain or loss is realized, defining a tax liability or credit.

Several factors are responsible for evaluating the replacement of an asset, including (Sullivan et al., 2003):

- *Deteriorating factor*: Changes that occur in the physical condition of an asset as a result of aging, unexpected accidents, and other factors that affect the physical condition of the asset.
- *Requirements factor*: Changes in production plans that affect the economics of use of the asset.
- *Technological factor*: The impact of changes in technology can also be a factor.
- *Financial factor*: The lease of an asset may become more attractive than ownership, for example.

Whatever the reason for a replacement evaluation, it is usually designed to answer the following fundamental question (Newnan et al., 2004):

Should we replace the current asset (defender) now because it is no longer economical to keep it, or should we keep it for one or more additional year(s) before replacing it with a more economical alternative?

Therefore, the question is not *if* the asset should be replaced (because it would be replaced eventually) but *when* it should be replaced. This evaluation study is different from the other studies considered in Section 6.5, where all the alternatives are new. In this case, the defender is competing against a challenger that has been selected from a mutually exclusive set of challengers. Replacement analysis can be based on before-tax cash flows (BTCF) or after-tax cash flows (ATCF); however, it is better to use ATCF to account for the effect of taxes on decisions. Replacement analysis involves several terms, including (Blank and Tarquin, 2002):

- *Defender*: This is the currently installed asset being considered for replacement.
- *Challenger*: This is the potential replacement.
- *Defender first cost*: The current market value (MV) of the defender is the correct estimate for this term in the replacement study. However, if the defender must be upgraded to make it equivalent to the challenger, the cost of the upgrade is added to the MV to obtain the correct estimate for this term.
- *Challenger first cost*: This is the amount that must be recovered when replacing a defender with a challenger. This may be equal to the first cost of the challenger. However, if trade-in is involved, this will be the first cost minus the difference between the trade-in value and the MV of the defender.
- *First cost*: This is the total cost of preparing the asset for economic use. It includes the purchase price, delivery cost, and installation cost.
- *Sunk cost*: This is the difference between an asset's book value (BV) and its MV at a particular period. Sunk costs have no relevance to the replacement decisions and must be neglected.
- *Outsider viewpoint*: This is the perspective that would have been taken by an impartial third party to establish the fair MV of a used asset. This perspective forces the analyst to focus on the present and future cash flows in a replacement study, hence avoiding the temptation to dwell on past (sunk) costs.
- *Asset life*: The life of an asset can be divided into three categories: ownership life, useful life, and economic life. The ownership life of an asset is the period between when an owner acquired it and when he disposed of it. The useful life, on the other hand, is the period an asset is kept in productive service. In addition, the economic service life (ESL) of an asset is the number of periods that results in the EUAC of owning and operating the asset. The economic life is often shorter than the useful life and it is usually 1 year for the defender.
- *Marginal costs*: These are the year-by-year costs associated with keeping an asset. In replacement problems, the total marginal cost for any year may include the capital recovery, operating and maintenance costs, yearly taxes and insurance, and other expenses that occur during that particular year.

6.6.1 Economic Service Life

Of all the various forms of asset life, the ESL is the most important for a replacement analysis. This life is also called the "minimum cost life." This is the number of remaining periods that result in the minimum equivalent annual cost of owning and operating an asset. This value is not usually known; therefore, it must be determined in order to perform a replacement analysis. Economic service life is determined by calculating the total AW of costs for the years the asset is in useful service.

6.6.2 Replacement Analysis

Replacement studies can be performed under three different circumstances (Newnan et al., 2004):

1. Defender marginal costs can be computed and are increasing.
2. Defender marginal costs can be computed but are not increasing.
3. Defender marginal costs cannot be computed.

6.6.2.1 When Defender Marginal Costs can be Computed and are Increasing

In this case, the defender is retained for as long as the marginal cost of keeping it for one more year is less than the minimum EUAC of the challenger. The steps involved are:

1. Compute the marginal costs of the defender.
2. Compute the EUAC for the challenger and determine the minimum EUAC for the challenger.

3. Make a plot of cost against both the minimum EUAC of the challenger and the marginal costs of the defender.
4. Keep the defender for at least the number of years that its marginal cost is less than the minimum EUAC of the challenger.

6.6.2.2 When Defender Marginal Costs can be Computed but are not Increasing

Since the marginal cost of the defender is not increasing, the minimum EUAC of the defender is compared directly against the minimum EUAC of the challenger. The steps involved are:

1. Compute the EUAC for the challenger and determine the ESL for the challenger.
2. Compute the EUAC for the defender and determine the ESL for the defender.
3. If the minimum EUAC of the defender is greater than the minimum EUAC of the challenger, replace the defender *now*; otherwise, keep the defender for at least the length of its ESL if the data used for the analysis are very accurate. If the data are not reliable or a lot of changes are envisaged in the estimation in the near future, keep the defender for at least for one more year and repeat the analysis.

6.6.2.3 When Defender Marginal Costs cannot be Computed

Since the marginal costs of the defender cannot be computed, the EUAC of the defender over its useful life is compared against the minimum EUAC of the challenger. The steps are:

1. Compute the EUAC of the defender over its remaining useful life.
2. Compute the EUAC of the challenger and determine the minimum EUAC.
3. Compare these two values directly.
4. Choose the lesser of the two values.

6.7 Depreciation and Depletion Methods

Depreciation is important in economic analysis because it is a tax-allowed deduction included in tax calculations. Depreciation is used in relation to tangible assets such as equipment, computers, machinery, buildings, and vehicles. Depletion, on the other hand, is used in relation to investments in natural resources such as minerals, ores, and timber. Almost everything depreciates as time proceeds; however, land is considered a nondepreciable asset. Depreciation can be defined as (Newnan et al., 2004):

- A decline in the MV of an asset (deterioration).
- A decline in the value of an asset to its owner (obsolescence).
- Allocation of the cost of an asset over its depreciable or useful life. Accountants usually use this definition, and it is employed in economic analysis for income-tax computation purposes. Therefore, depreciation is a way to claim over time an already paid expense for a depreciable asset.

For an asset to be depreciated, it must satisfy these three requirements (Blank and Tarquin, 2002):

1. The asset must be used for business purposes to generate income.
2. The asset must have a useful life that can be determined and is longer than 1 year.
3. The asset must be one that decays, gets used up, wears out, becomes obsolete, or loses value to the owner over time as a result of natural causes.

Depreciation Notations

Let

n = recovery period in years
B = first cost, unadjusted basis, or basis
S = estimated salvage value
D_t = annual depreciable charge
MV = market value
BV_t = book value after period, t

d = depreciation rate = $1/n$

t = year ($t = 1,2,3,\ldots,n$).

Depreciation Terminology

- *Depreciation*: The annual depreciation amount, D_t, is the decreasing value of the asset to the owner. It does not represent an actual cash flow or actual usage pattern.
- *Book depreciation*: This is an internal description of depreciation. It is the reduction in the asset investment due to its usage pattern and expected useful life.
- *Tax depreciation*: This is used for after-tax economic analysis. In the United States and many other countries, the annual tax depreciation is tax deductible using the approved method of computation.
- *First cost or unadjusted basis*: This is the cost of preparing the asset for economic use and is also called the "basis." This term is used when an asset is new. Adjusted basis is used after some depreciation has been charged.
- *Book value*: This represents the remaining undepreciated capital investment after the total amount of depreciation charges to date have been subtracted from the basis. It is usually calculated at the end of each year.
- *Salvage value*: Estimated trade-in or MV at the end of the asset's useful life. It may be positive, negative, or zero. It can be expressed as a dollar amount or as a percentage of the first cost.
- *Market value*: This is the estimated amount realizable if the asset were sold in an open market. This amount may be different from the BV.
- *Recovery period*: This is the depreciable life of an asset in years. There are often different n values for book and tax depreciations. Both values may be different from the asset's estimated productive life.
- *Depreciation or recovery rate*: This is the fraction of the first cost removed by depreciation each year. Depending on the method of depreciation, this rate may be different for each recovery period.
- *Half-year convention*: This is used with the modified accelerated cost recovery system (MACRS) depreciation method. It assumes that assets are placed in service or disposed of midyear, regardless of when these placements actually occur during the year. There are also midquarter and midmonth conventions.

6.7.1 Depreciation Methods

There are five principal depreciation methods:

- Classical (historical) depreciation methods
  - Sraight line (SL)
  - Declining balance (DB)
  - Sum-of-years'-digits (SOYD).
- MACRS.

6.7.1.1 Straight-Line Method

This is the simplest and the best-known method of depreciation. It assumes that a constant amount is depreciated each year over the depreciable (useful) life of the asset; hence, the BV decreases linearly with time. The SL method is considered the standard against which other depreciation models are compared. It offers an excellent representation of an asset used regularly over an estimated period, especially for book depreciation purposes. The annual depreciation charge is given as

$$D_t = \frac{B - S}{n} = (B - S)d \tag{6.41}$$

The BV after t year(s) is given as

$$\text{BV}_t = B - \frac{t}{n}(B - S) = B - tD_t \tag{6.42}$$

6.7.1.2 Declining Balance Method

This method is commonly applied as the book depreciation method in the industry because it accelerates the write-off of asset value. It is also called fixed (uniform) percentage method; therefore, a constant depreciation rate is applied to the BV of the asset. According to the Tax Reform Act of 1986, two rates are applied to the SL rate; these are 150 and 200%. If 150% is used, it is called the DB method, and if 200% is used, it is called the double declining balance (DDB) method. The DB annual depreciation charge is

$$D_t = \frac{1.5B}{n}\left(1 - \frac{1.5B}{n}\right)^{t-1} \tag{6.43}$$

Total DB depreciation at the end of t years is

$$B\left[1 - \left(1 - \frac{1.5}{n}\right)^t\right] = B[1 - (1 - d)^t] \tag{6.44}$$

Book value at the end of t years is

$$B\left(1 - \frac{1.5}{n}\right)^t = B(1 - d)^t \tag{6.45}$$

For DDB (200% depreciation) method, substitute 2.0 for 1.5 in Equations (6.43)–(6.45).

It should be noted that salvage value is not used in equations for DB and DDB methods; therefore, these methods are independent of the salvage value of the asset. The implication of this is that the depreciation schedule may be below an implied salvage value, above an implied salvage value, or just at the level of the implied salvage value. Any of these three situations is possible in the real world. However, the U.S. Internal Revenue Service (IRS) does not permit the deduction of depreciation charges below the salvage value, while companies will not like to deduct depreciation charges that would keep the BV above the salvage value. The solution to this problem is to use a composite depreciation method. The IRS provides that a taxpayer may change from DB or DDB to SL at any time during the life of an asset. However, the question is when to switch. The criterion used to answer this question is to maximize the PW of the total depreciation.

6.7.1.3 Sum-Of-Years'-Digits Method

This method results in larger depreciation charges during the early years of an asset (than SL) and smaller charges at the latter part of the estimated useful life; however, write-off is not as rapid as for DDB or MACRS. Similar to the SL method, this method uses the salvage value in computing the annual depreciation charge. The annual depreciation charge is

$$D_t = \frac{n - t + 1}{\text{SUM}}(B - S) = d_t(B - S)$$
$$\text{SUM} = \frac{n(n + 1)}{2} \tag{6.46}$$

The BV at the end of t years is

$$\text{BV}_t = B - \frac{t(n - (t/2) + 0.5)}{\text{SUM}}(B - S) \tag{6.47}$$

6.7.1.4 Modified Accelerated Cost Recovery System Method

This is the only approved tax depreciation method in the United States. It is a composite method that automatically switches from DB or DDB to SL depreciation. The switch usually takes place whenever the SL depreciation results in larger depreciation charges, that is, a more rapid reduction in the BV of the asset. One advantage of the MACRS method is that it assumes that the salvage value is 0; therefore, it always depreciates to 0. Another outstanding advantage of this method is that it uses property classes, which specify the recovery periods, n. The method adopts the half-year convention, which makes the actual recovery period to be 1 year longer than the specified period. The half-year convention means that the IRS assumes that the assets are placed in service halfway through the year, no matter when the assets were actually placed in service. This convention is also applicable when the asset is disposed of before the end of the depreciation period. The MACRS method consists of two systems for computing depreciation deductions: general depreciation systems (GDS) and alternative depreciation systems (ADS). Alternative depreciation systems are used for properties placed in any tax-exempt use as well as properties used predominantly outside the United States. The system provides a longer recovery period and uses only SL method of depreciation. Therefore, this system is generally not considered an option for economic analysis. However, any property that qualifies for GDS can be depreciated under ADS, if preferred.

The following information is required to depreciate an asset using the MACRS method:

- The cost basis
- The date the property was placed in service
- The property class and recovery period
- The MACRS depreciation system to be used (GDS or ADS)
- The time convention that applies (e.g., half or quarter-year convention).

The steps involved in using the MACRS depreciation method are:

1. Determine the property class of the asset being depreciated using published tables. Any asset not in any of the stated classes is automatically assigned a 7-year recovery period under the GDS system.
2. After the property class is known, read off the appropriate published depreciation rates.
3. The last step is to multiply the asset's cost basis by the depreciation rate for each year to get the annual depreciation charge.

The MACRS annual depreciation amount is

$$D_t = \text{(first cost)(tabulated depreciation rate)} = d_t B \qquad (6.48)$$

The annual BV is

$$BV_t = \text{first cost} - \text{sum of accumulated depreciation}$$

$$= B - \sum_{j=1}^{t} D_j \qquad (6.49)$$

6.7.2 Depletion Methods

Depletion is applicable to natural resources from places such as mines, wells, and forests. It recovers investment in natural resources; therefore, it is the exhaustion of natural resources as a result of their removal. There are two methods of calculating depletion: cost (factor) depletion and percentage depletion (Park, 2001; Newnan et al., 2004). In the United States, except for standing timber and most oil and gas wells, depletion is calculated by both methods and the larger value is taken as the depletion for the year.

6.7.2.1 Cost Depletion

This is based on the level of activities; however, the total cost of depletion cannot exceed the first cost of the resource. The annual depletion charge is

$$\frac{\text{first cost}}{\text{resource capacity}} \text{ (year's usage or volume)} \qquad (6.50)$$

6.7.2.2 Percentage Depletion

This is a special consideration for natural resources. It is an annual allowance of a percentage of the gross income from the property. Since it is based on income rather than the cost of the property, the total depletion on a property may exceed the cost of the property. The percentage depletion allowance in any year is limited to not more than 50% of the taxable income from the property. The percentage depletion amount is

$$\text{percentage} \times \text{gross income from property} \tag{6.51}$$

The percentages are usually published and they change from time to time.

6.8 Effects of Inflation on Economic Analysis

Inflation plays a crucial role in economic analysis of projects and investments, especially multiyear projects and investments. All cash-flow estimates must account for the effects of inflation for economic analysis results to be valid. Estimates without consideration of the effects of inflation will result in wrong economic decisions. Inflation is characterized by rising prices for goods and services, that is, a decline in the purchasing power of money. An inflationary trend gives today's dollars greater purchasing power than the future dollars; hence, inflation is an increase in the amount of money necessary to obtain the same amount of goods or service in the future. This phenomenon helps long-term borrowers of money because they may repay a loan with dollars of reduced buying power. Deflation is the opposite of inflation; it is a situation in which money borrowed is repaid with dollars of greater purchase power. This effect is to the advantage of the lenders. Some of the causes of inflation are (Badiru, 1996):

- Increase in the amount of currency in circulation
- Shortage of consumer goods
- Arbitrary increase of prices
- Escalation of the cost of production

The effects of inflation on economic analysis will depend on the level of the inflationary trend. The common levels of inflation are (Badiru, 1996):

- *Mild inflation*: When the inflation rate is between 2 and 4%.
- *Moderate inflation*: The inflation rate is between 5 and 9%.
- *Severe inflation*: When the inflation rate is in the double digits (10% or more).
- *Hyperinflation*: The inflation rate is in the higher double digits to three digits.

Whatever the level of inflation, inflation must be accounted for in long-term economic analyses. Some of the effects of the neglect of inflation include cost overruns, selection of the wrong alternative, decrease in estimated ROR, and poor resource utilization. To account for the effects of inflation, some definitions are necessary:

- *Real dollars (R$)*: Dollars expressed in terms of the same purchasing power relative to a particular time. These are also called "today's dollars" or "constant-value dollars."
- *Actual dollars (A$)*: The number of dollars associated with a cash flow as of the time it occurs. These are also known as the "future dollars" or the "then-current dollars."

The relationship between the actual dollars and the future dollars is

$$\text{future dollars} = \text{today's dollars } (1 + f)^n \tag{6.52}$$

where
f = inflation rate per period (year).
n = the number of periods between time t and the base time period. The base time period is usually the present time (time $t = 0$).

Three different rates are usually considered in addition to these definitions:

- *Real or inflation-free interest rate, i*: The desired ROR that does not account for the anticipated effect of inflation; therefore, it presents an actual gain in purchasing power:

$$i = \frac{(i_f - f)}{(1 + f)} \tag{6.53}$$

- *Inflation rate, f:* This is a measure of the average rate of change in the value of the currency during a specified period of time.

- *Inflation-adjusted or combined interest rate, i_f:* This is the desired ROR that accounts for the anticipated effect of inflation. It combines the real interest rate and the inflation rate and is also called "inflated interest rate." The MARR adjusted for inflation is called "inflation-adjusted MARR."

$$i_f = i + f + i(f) \tag{6.54}$$

Measures of price changes, such as the CPI and the producer price index (PPI), are used to convey inflationary trends and are estimates of general price inflation or deflation. The consumer price index is a composite price index that measures changes in the prices paid for food, shelter, medical care, transportation, and other selected goods and services used by individuals and families. It measures price change from the purchaser's perspective. Producer price index, on the other hand, is a measure of average changes in the selling prices of items used in the production of goods and services and measures price changes from the seller's perspective. Both measures are monthly measures based on survey information and are published by the Bureau of Labor Statistics in the U.S. Department of Labor. They are based on current and historical information and may be used, as appropriate, to represent future economic conditions or for short-term forecasting purposes. Annual change in price indexes are calculated using the following relations:

$$\text{(CPI or PPI annual change rate, \%)}_k = \frac{(\text{Index})_k - (\text{Index})_{k-1}}{(\text{Index})_{k-1}} \times 100\% \tag{6.55}$$

where k is the based year.

The effects of inflation on economic quantities can be significant. Some of the responsive economic quantities are labor, material, and equipment costs. However, the unresponsive quantities include depreciation, lease, bond, fixed annuity, and interest rates based on existing loans or contract agreements. Either i or i_f can be introduced into economic analysis factors, depending upon whether the cash flow is expressed in today's dollars or future dollars. If the series is expressed in today's dollars, then the discounted cash flow uses the real interest rate. However, the combined interest rate is used if the series is expressed in future dollars. This is applicable to both the PW and the AW methods. It is especially important that AW analysis account for the effects of inflation since current dollars must be recovered with future inflated dollars.

6.8.1 Foreign Exchange Rates

The idea of accounting for the effects of inflation on local investments can be extended to account for the effects of devalued currency on foreign investments. When local businesses invest in a foreign country, several factors come into consideration, such as when the initial investment is going to be made, when the benefits are going to be accrued to the local business, and at what rate the benefits are going to be accrued. Therefore, the ROR of a local business (say, in the United States) with respect to a foreign country can be given by (Sullivan et al., 2003)

$$i_{us} = \frac{i_{fc} - f_e}{1 + f_e}$$

$$i_{fc} = i_{us} + f_e + f_e(i_{us}) \tag{6.56}$$

where

i_{us} = ROR in terms of a market interest rate relative to US\$
i_{fc} = ROR in terms of a market interest rate relative to the currency of a foreign country
f_e = annual devaluation rate between the currency of a foreign country and the US\$. A positive value means that the foreign currency is being devalued relative to the US\$. A negative value means that the US\$ is being devalued relative to the foreign currency

6.9 After-Tax Economic Analysis

For a complete and accurate economic analysis result, both the effects of inflation and taxes must be taken into consideration when evaluating alternatives. Taxes are an inevitable burden, and their effects must be accounted for in economic analysis. There are several types of taxes:

- *Income taxes*: These are taxes assessed as a function of gross revenue less allowable deductions and are levied by the federal, most state, and municipal governments.
- *Property taxes*: They are assessed as a function of the value of property owned, such as land, buildings, and equipment, and are mostly levied by municipal, county, or state governments.
- *Sales taxes*: These are assessed on purchases of goods and services; hence, they are independent of gross income or profits. They are normally levied by state, municipal, or county governments. Sales taxes are relevant in economic analysis only to the extent that they add to the cost of items purchased.
- *Excise taxes*: These are federal taxes assessed as a function of the sale of certain goods or services often considered nonnecessities. They are usually charged to the manufacturer of the goods and services, but a portion of the cost is passed on to the purchaser.

Income taxes are the most significant type of tax encountered in economic analysis; therefore, the effects of income taxes can be accounted for using these relations:

$$TI = \text{gross income} - \text{expenses} - \text{depreciation (depletion) deductions}$$
$$T = TI \times \text{applicable tax rate} \tag{6.57}$$
$$NPAT = TI(1 - T)$$

where

TI = taxable income (amount upon which taxes are based)

T = tax rate (percentage of taxable income owed in taxes)

NPAT = net profit after taxes (taxable income less income taxes each year. This amount is returned to the company)

The tax rate used in economic analysis is usually the effective tax rate, and it is computed using this relation:

$$\text{effective tax rates } (Te) = \text{state rate} + (1 - \text{state rate})(\text{federal rate}) \tag{6.58}$$

therefore,

$$T = (\text{taxable income}) \, (Te)$$

6.9.1 Before-Tax and After-Tax Cash Flow

The only difference between a BTCF and an ATCF is that ATCF includes expenses (or savings) due to income taxes and uses after-tax MARR to calculate equivalent worth. Hence, ATCF is the BTCF less taxes. The after-tax MARR is usually smaller than the before-tax MARR, and they are related by the following equation:

$$\text{after-tax MARR} \cong (\text{before-tax MARR})(1 - Te) \tag{6.59}$$

6.9.2 Effects of Taxes on Capital Gain

Capital gain is the amount incurred when the selling price of a property exceeds its first cost. Since future capital gains are difficult to estimate, they are not detailed in after-tax study. However in actual tax law, there is no difference between short-term and long-term gain.

Capital loss is the loss incurred when a depreciable asset is disposed of for less than its current BV. An economic analysis does not usually account for capital loss because it is not easily estimated for alternatives. However, after-tax replacement analysis should account for any capital loss. For economic analysis, this loss provides a tax savings in the year of replacement.

Depreciation recapture occurs when a depreciable asset is sold for more than its current BV. Therefore, depreciation recapture is the selling price less the BV. This is often present in after-tax analysis. When the MACRS depreciation method is used, the estimated salvage value of an asset can be anticipated as the depreciation recapture because MACRS assumes zero salvage value.

Therefore, the TI equation can be rewritten as

$$TI = \text{gross income} - \text{expenses} - \text{depreciation (depletion)deductions} \\ + \text{depreciation recapture} + \text{capital gain} - ca \tag{6.60}$$

6.9.3 After-Tax Economic Analysis

The ATCF estimates are used to compute the PW, AW, or FW at the after-tax MARR. The same logic as for before-tax evaluation methods discussed in Section 6.5 also applies; however, the calculations required for after-tax computations are certainly more involved than those for before-tax analysis. The major elements in an after-tax economic analysis are:

- Before-tax cash flow
- Depreciation
- Taxable income
- Income taxes
- After-tax cash flow

Therefore, to facilitate the computation of after-tax economic evaluation using Equations (6.57)–(6.60), the following tabular headings can be used (Sullivan et al., 2003):

Year	Before-Tax Cash Flow	Depreciation	Taxable Income	Income Taxes	After-Tax Cash Flow
	(A)	(B)	(C)	(D)	(E)
			(A)−(B)	−T(C)	(A)+(D)

Column A: Same information used for before-tax analyses.
Column B: Depreciation that can be claimed for tax purposes.
Column C: Amount subject to income taxes. It is the difference between columns A and B.
Column D: Income taxes paid (negative amount) or income taxes saved (positive amount).
Column E: After-tax cash flow to be used directly in economic evaluation techniques or after-tax analyses. It is the addition of columns A and D.

6.10 Decision-Making Under Risk, Sensitivity, and Uncertainty

Economic analysis is usually concerned with the present and future consequences of investment alternatives. However, estimating the consequence of future costs and benefits is not always easy, but it must be done for decision purposes. The accuracy of such estimates is an important element of the results of economic analyses. When the degree of confidence on data, information, and techniques used in estimating future cash flows is 100%, then this kind of analysis is called *decisions under certainty*. However, there is hardly any situation in which the confidence level is 100%; there always exist some elements of uncertainty attached to every decision. The uncertainty surrounding cash-flow data for project evaluation is multidimensional in nature, and the vagueness of one factor interplays with the vagueness of the other factors to create an even more complicated decision-making scenario. There are several techniques for handling uncertainty in economic analysis and they can be classified into probabilistic, nonprobabilistic, and fuzzy techniques.

Both risk and uncertainty in cash-flow estimates are caused by a lack of accurate data and information regarding the future conditions of the investments under consideration. Such future conditions may include changes in technology, the relationship between independent projects, the impacts of international trades and businesses, and several other peculiar conditions. The terms "risk" and "uncertainty" are used interchangeably in economic analysis. However, *decision under risk* involves situations in which the future estimates can be estimated in terms of probability of occurrences. *Decision under uncertainty*, on the other hand, involves situations in which such future estimates cannot be estimated in terms of probability of occurrences. Therefore, it is usually helpful to determine the impact of a change in a cash-flow estimate on the overall capital investment decision; that is, to determine how *sensitive* an investment is to changes in a given cash-flow estimate that is not known with certainty.

6.10.1 Sources of Uncertainty in Project Cash Flows

Some of the aspects of investment cash flows that contain imprecision include the economic life of a project, cash-flow estimates, estimating the MARR, estimating the effects of inflation, and timing of cash flows. The economic life of a project may not be known with certainty because the life of the technology used in developing the project may not be known with certainty either. In most cases, the technology may be new and there may not be enough information to make a probabilistic approximation. In addition, some of the technology may be shorter lived than initially anticipated. Cash-flow estimates for each project phase are a difficult problem. Costs and benefits streams for projects are usually indefinite. In addition, the benefits may also be delayed as a result of project extension and unexpected additional project costs. The MARR for projects is usually project-dependent; however, the method of determining project MARR is better modeled with uncertainty because there may be a delay in accruing project benefits, the project costs may exceed budgeted values, the benefits may not be as huge as initially thought, or the technology may be shorter lived than initially anticipated. Inflation becomes a major concern in projects that take several years to complete. Therefore, estimating the impact of inflation is not possible with certainty. Shifted cash flows are another issue when projects are not completed on schedule and may have some elements of probability attached to it. A reasonable probabilistic assumption of how long the project will take may help in accounting for shifted cash flows.

6.10.2 Non-Probabilistic Models

There are several nonprobabilistic techniques, such as breakeven analysis, sensitivity graph (spiderplot), and the use of a combination of these two factors (Sullivan et al., 2003). However, breakeven analysis remains the most popular.

Breakeven analysis is used when the decision about a project is very sensitive to a single factor and this factor is not known with certainty. Therefore, the objective of this technique is to determine the breakeven point (Q_{BE}) for this decision variable. The approach can be used for a single project or for two projects. The breakeven technique usually assumes a linear revenue relation, but a nonlinear relation is often more realistic. When one of the parameters of an evaluating technique, such as P, F, A, I, or n for a single project, is not known or not estimated with certainty, a breakeven technique can be used by setting the equivalent relation for the PW, AW, ROR, or B/C equal to 0 in order to determine the breakeven point for the unknown parameter. The project may also be modeled in terms of its total revenue and total cost (fixed cost plus variable cost). Therefore, at some unit of product quantity, the revenue and the total cost relations intersect to identify the breakeven quantity. This identified quantity is an excellent starting target for planning purposes. Product quantity less than the breakeven quantity indicates a loss; while product quantity greater than the breakeven quantity indicates a profit. Figure 6.10 shows linear and nonlinear breakeven graphs for a single project.

The breakeven technique can also be used to determine the common economic parameters between two competing projects. Some of the parameters that may be involved are the interest rate, the first cost, the annual operating cost, the useful life, the salvage value, and the ROR, among others. The steps used in this case can be summarized as follows:

1. Define the parameter of interest and its dimension.
2. Compute the PW or AW equation for each alternative as a function of the parameter of interest.

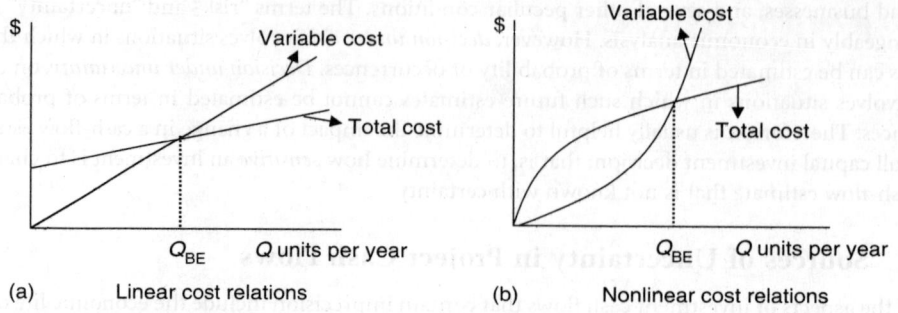

FIGURE 6.10   Linear and nonlinear breakeven graphs.

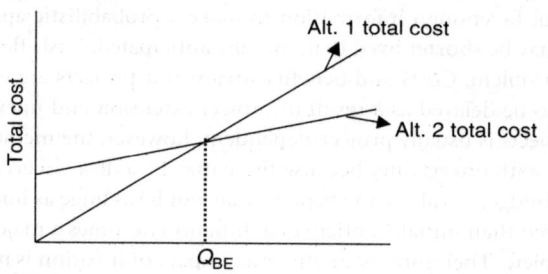

FIGURE 6.11   Breakeven between two alternatives with linear relations.

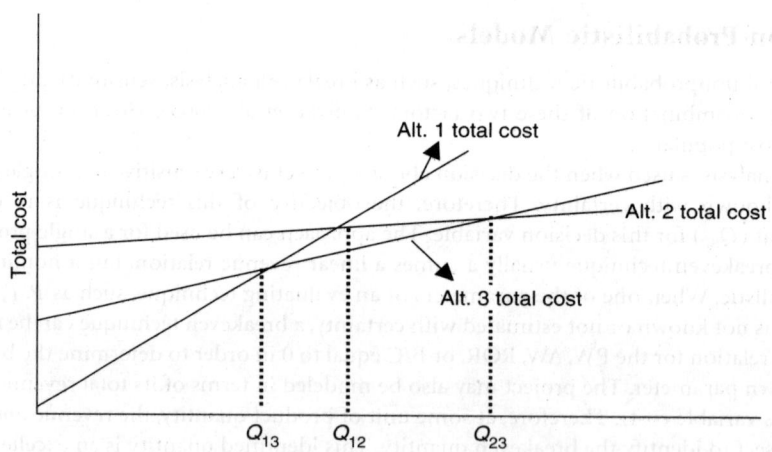

FIGURE 6.12   Breakeven between three alternatives with linear relations.

3. Equate the two equations and solve for the breakeven value of the parameter of interest.
4. If the anticipated value is above this calculated breakeven value, select the alternative with the lower parameter cost (smaller slope). If otherwise, select the alternative with the higher parameter cost (larger slope). Figure 6.11 shows a graphical example.

This approach can also be used for three or more alternatives by comparing the alternatives in pairs to find their respective breakeven points. Figure 6.12 shows a graphical example for three alternatives.

6.10.3 Probabilistic Models

A probabilistic model is decision making under risk and involves the use of statistics and probability. The most popular is Monte Carlo sampling and simulation analysis.

The simulation approach to engineering economic analysis is summarized as follows (Blank and Tarquin, 2002):

1. Formulate alternative(s) and select the measure of worth to be used.
2. Select the parameters in each alternative to be treated as random variables and estimate values for other definite parameters.
3. Determine whether each variable is discrete or continuous and describe a probability distribution for each variable in each alternative.
4. Develop random samples.
5. Compute *n* values of the selected measure of worth from the relation(s) in step 1 using the definite estimates made and *n* sample values for the varying parameters.
6. Construct the probability distribution of the measure computed in step 5 using between 10 and 20 cells of data and calculate measures such as the mean, the root-mean-square deviation, and other relevant probabilities.
7. Draw conclusions about each alternative and decide which is to be selected.

The results of this approach can be compared with decision-making when parameter estimates are made with certainty.

Appendix A: Software Tools for Economic Analysis

This appendix presents some of the software tools used in engineering economic evaluation. The use and importance of software tools in engineering economic evaluation, especially spreadsheet functions, has been emphasized in the literature (Alloway, 1994; Lavelle, 1996). Spreadsheets provide rapid solutions, and the results of the analyses can be saved for easy presentation and reference. Microsoft is the most widely used computer software, and only Microsoft Excel functions are considered in this appendix.

The most commonly used Excel functions in economic evaluation are (Blank and Tarquin, 2002):

DB (Declining Balance)

- DB (cost, salvage, life, period, month).
- Calculates the depreciation amount for an asset using the DB method:
    - Cost: First cost or basis of the asset.
    - Salvage: Salvage value.
    - Life: Recovery period.
    - Period: The year for which the depreciation is to be calculated.
    - Month: (optional) A full year is assumed for the omitted first year.

DDB (Double Declining Balance)

- DDB (cost, salvage, life, period, factor).
- Calculates the depreciation amount for an asset using the DDB method:
    - Cost: First cost or basis of the asset.
    - Salvage: Salvage value.
    - Life: Recovery period.
    - Period: The year for which the depreciation is to be calculated.
    - Factor: (optional) Enter 1.5 for 150% DB and so on. The function will use 2.0 for 200% DB if omitted.

FV (Future Value)

- FV (rate, nper, pmt, pv, type).
- Calculates the FW for a periodic payment at a specific interest rate:
  - rate: Interest rate per compounding period.
  - nper: Number of compounding periods.
  - pmt: Constant payment amount.
  - pv: The PV amount. The function will assume that pv is zero if omitted.
  - type: (optional) Either 0 or 1. A 0 represents end-of-the-period payment, and 1 represents beginning-of-the-period payment. If omitted, 0 is assumed.

IPMT (Interest Payment)

- IPMT (rate, per, nper, pv, fv, type).
- Calculates the interest accrued for a given period on the basis of constant periodic payment and interest rate:
  - rate: Interest rate per compounding period.
  - per: Period for which interest is to be calculated.
  - nper: Number of compounding periods.
  - pv: The PV amount. The function will assume that pv is zero if omitted.
  - fv: The FV (a cash balance after the last payment is made). If omitted, the function will assume it to be 0.
  - type: (optional) Either 0 or 1. A 0 represents end-of-the-period payment, and 1 represents beginning-of-the-period payment. If omitted, 0 is assumed.

IRR (Internal Rate of Return)

- IRR (values, guess).
- Calculates the IRR between 100% and infinity for a series of cash flows at regular periods:
  - values: A set of numbers in a spreadsheet row or column for which the rate of return will be calculated. There must be at least one positive (cash inflow) and one negative (cash outflow) number.
  - guess: (optional) Guess a rate of return to reduce the number of iterations. Change the guess if #NUM! error appears.

MIRR (Modified Internal Rate of Return)

- MIRR (values, finance rate, reinvest rate).
- Calculates the MIRR for a series of cash flows and reinvestment of income and interest at a stated rate:
  - values: A set of numbers in a spreadsheet row or column for which the ROR will be calculated. There must be at least one positive (cash inflow) and one negative (cash outflow) number.
  - finance rate: Interest rate of money used in the cash flows.
  - reinvest rate: Interest rate for reinvestment on positive cash flows.

NPER (Number of Periods)

- NPER (rate, pmt, pv, fv, type).
- Calculates the number of periods for the PW of an investment to equal the FV specified:
  - rate: Interest rate per compounding period.
  - pmt: Amount paid during each compounding period.
  - pv: Present values.
  - fv: (optional) The future value (a cash balance after the last payment is made). If omitted, the function will assume it to be 0.
  - type: (optional) Either 0 or 1. A 0 represents end-of-the-period payment, and 1 represents beginning-of-the-period payment. If omitted, 0 is assumed.

NPV (Net Present Value)

- NPV (rate, series).
- Calculates the net present worth (NPW) of a series of future cash flows at a particular interest rate:
  - rate: Interest rate per compounding period.
  - series: Series of inflow and outflow setup in a range of cells in the spreadsheet.

PMT (Payments)

- PMT (rate, nper, pv, fv, type).
- Calculates equivalent periodic amounts based on PW or FW at a stated interest rate:
  - rate: Interest rate per compounding period.
  - nper: Number of compounding periods.
  - pv: The PV amount. The function will assume that pv is 0 if omitted.
  - fv: The future value (a cash balance after the last payment is made). If omitted, the function will assume it to be 0.
  - type: (optional) Either 0 or 1. A 0 represents end-of-the-period payment, and 1 represents beginning-of-the-period payment. If omitted, 0 is assumed.

PPMT (Principal Payment)

- PPMT (rate, per, nper, pv, fv, type).
- Calculates the PMT on the principal based on uniform payments at a stated interest rate:
  - rate: Interest rate per compounding period.
  - per: Period for which interest is to be calculated.
  - nper: Number of compounding periods.
  - pv: The PV amount. The function will assume that pv is 0 if omitted.
  - fv: The FV (a cash balance after the last payment is made). If omitted, the function will assume it to be 0.
  - type: (optional) Either 0 or 1. A 0 represents end-of-the-period payment, and 1 represents beginning-of-the-period payment. If omitted, 0 is assumed.

PV (Present Value)

- PV (rate, nper, pmt, fv, type).
- Calculates the PW of a future series of equal cash flows and a single lump sum in the last period at a stated interest rate:
  - rate: Interest rate per compounding period.
  - nper: Number of compounding periods.
  - pmt: Cash flow at regular intervals. Inflows are positive and outflows are negative.
  - fv: The FV (a cash balance after the last payment is made). If omitted, the function will assume it to be 0.
  - type: (optional) Either 0 or 1. A 0 represents end-of-the-period payment, and 1 represents beginning-of-the-period payment. If omitted, 0 is assumed.

RATE (Interest Rate)

- RATE (nper, pmt, pv, fv, type, guess).
- Calculates the interest rate per compounding period for a series of payments or incomes:
  - nper: Number of compounding periods.
  - pmt: Cash flow at regular intervals. Inflows are positive and outflows are negative.
  - pv: The PV amount. The function will assume that pv is 0 if omitted.
  - fv: The FV (a cash balance after the last payment is made). If omitted, the function will assume it to be 0.
  - type: (optional) Either 0 or 1. A 0 represents end-of-the-period payment, and 1 represents beginning-of-the-period payment. If omitted, 0 is assumed.
  - guess: (optional) Guess a ROR to reduce the number of iterations. Change the guess is #NUM! error appears.

SLN (Straight-Line Depreciation)

- SLN (cost, salvage, life).
- Calculates the straight-line depreciation of an asset for a given year:
  - cost: First cost or basis of the asset.
  - salvage: Salvage value.
  - life: Recovery period.

SYD (Sum-Of-Year-Digits Depreciation)

- SYD (cost, salvage, life, period).
- Calculates the SOYD depreciation of an asset for a given year:
  - cost: First cost or basis of the asset.
  - salvage: Salvage value.
  - life: Recovery period.
  - period: The year for which the depreciation is to be calculated.

VDB (Variable Declining Balance)

- VDB (cost, salvage, life, start-period, end-period, factor, no-switch).
- Calculates the depreciation schedule using the DB method with a switch to SLN in the year in which straight line has a larger depreciation amount. This function can be used for MACRS depreciation schedule computations.
  - cost: First cost or basis of the asset.
  - salvage: Salvage value.
  - life: Recovery period.
  - start-period: First period for depreciation to be calculated.
  - end-period: Last period for depreciation to be calculated.
  - factor: (optional) Enter 1.5 for 150% DB and so on. The function will use 2.0 for 200% DB if omitted.
  - no-switch: (optional) If omitted or entered as FALSE, the function will switch from DB or DDB to SLN depreciation when the latter is greater than DB depreciation. If entered as TRUE, the function will not switch to SLN depreciation at any time during the depreciation life.

The Other Software Tool

The other software tool discussed in this appendix is the *ENGINeering Economic Analysis* (*ENGINEA*) software for economic evaluation (Omitaomu et al., 2005). The software is developed to help users solve problems in engineering economic analysis and financial management. The software provides users the ability to perform cash-flow analysis, depreciation analysis, replacement analysis, and interest factor calculations. It can be used for both undergraduate and graduate engineering economy and financial management courses. This software is unique because of its many capabilities; it is a one-stop shop of techniques for solving engineering economic analysis problems. The software requires an IBM-compatible personal computer running Windows 95 or better. For a detailed description of the software, see Omitaomu et al. (2005).

To obtain a free copy of the ENGINEA software, please write or send email to the author of the software at the following addresses: Olufemi A. Omitaomu, Department of Industrial and Information Engineering, The University of Tennessee, 416 East Stadium Hall, Knoxville, TN 37996-0700, USA. E-mail: omitaomu@utk.edu

References

Alloway, J.A., Jr., Spreadsheets: enhancing learning and application of engineering economy techniques, *Eng. Econ.*, 3, 263–274, 1994.

Badiru, A.B., *Project Management in Manufacturing and High Technology Operations*, 2nd ed., Wiley, New York, 1996.

Badiru, A.B. and Omitaomu, H.O., Design and analysis of tent cash flow models for engineering economy lectures, *Eng. Econ.*, 48, 363–374, 2003.

Blank, L.T. and Tarquin, A., *Engineering Economy*, 5th ed., McGraw-Hill, New York, 2002.

Canada, J.R., Sullivan, W.G., and White, J.A., *Capital Investment Analysis for Engineering and Management*, 2nd ed., Prentice-Hall, Englewood Cliffs, NJ, 1996.

Eschenbach, T.G., *Engineering Economy: Applying Theory to Practice*, 2nd ed., Oxford University Press, New York, 2003.

Lavelle, J.P., Reader's forum: enhancing engineering economy concepts with computer spreadsheets, *Eng. Econ.*, 4, 381–386, 1996.

Newnan, D.G., Eschebach, T.G., and Lavelle, J.P., *Engineering Economic Analysis*, 9th ed., Oxford University Press, New York, 2004.

Omitaomu, O.A., Smith, L.D., and Badiru, A.B., The ENGINeering Economic Analysis (ENGINEA) software: enhancing teaching and application of economic analysis techniques. *Comput. Educ. J.*, in press.

Park, C.S., *Contemporary Engineering Economics*, 3rd ed., Prentice-Hall, Englewood Cliffs, NJ, 2001.

Sullivan, W.G., Wicks, M.E., and Luxhoj, T.J., *Engineering Economy*, 12th ed., Prentice-Hall, New Jersey, 2003.

7

Work Sampling

Paul S. Ray
The University of Alabama

7.1 Introduction

One very useful technique of work measurement is work sampling, which can be applied in a wide variety of work situations. In its simplest form, it is used by the shop supervisor or foreman to estimate idle times of machines, e.g., if he notices a machine idle for two out of his ten trips, he estimates that the machine is idle for 20% of the time.

The technique was first applied in the British textile industry by L.H.C. Tippett (Tippett, 1953) under the name of "ratio delay." R.L. Morrow (Morrow, 1941) introduced the technique in the United States in 1941. The name "work sampling" was introduced by C.L. Brisley (Brisley, 1952) and H.L. Waddell (Waddell, 1952) in 1952, Ralph M. Barnes reproduced the name work sampling with permission of Tippett in his book on work sampling (Barnes, 1980). Several other names, e.g., "activity sampling," "ratio delay," "snap readings," or "occurrence sampling," (Konz and Johnson, 2000) are occasionally used synonymously for "work sampling." The technique is particularly useful in measuring indirect work, in service activities, and in cases where a stopwatch method is not acceptable. Since the 1950s, the technique has become a standard tool used by industrial engineers for measuring indirect and service jobs. Performance sampling may be applied to fine tune the ratios more accurately in work sampling (Meyers, 1999).

7.2 Basic Concepts of Work Sampling

Work sampling is based on the laws of probability and is used to determine the proportions of the total time devoted to the various components of a task. The probability of an event occurring or not occurring is obtained from the statistical binomial distribution. When the number of observations is large, the binomial distribution can be approximated to normal distribution. The binomial probability of x occurrences is calculated as follows:

$$b(x) = {}^nC_x\, p^x q^{(n-x)} \tag{7.1}$$

where p = probability of x occurrences
$\quad q = 1-p$ = probability of no x occurrence
$\quad n$ = number of observations

Normal distribution is used instead of binomial distribution in work sampling for convenience. The normal distribution of a proportion p has an average value of ratio $= p$ and standard deviation

$$\sigma_p = \sqrt{p(1-p)/n} \tag{7.2}$$

where p = proportion of occurrence of an event
$\quad n$ = number of observations

$$A = sp = z\sigma_p$$

where A = desired absolute accuracy
$\quad s$ = relative accuracy for a propfortion p

7.3 Accuracy

Absolute accuracy indicates the range within which the value of p is expected to lie. It is the closeness of the ratio p to the true ratio p. If $p = 30\%$ and relative accuracy $= 10\%$, then $A = (0.3)(0.1) = 0.03$ or 3%.

Example 1   For a study where the true p value $= 30\%$, a $\pm 3\%$ absolute accuracy level indicates that the calculated value of p will be between $(30 \pm (30 \times 0.03)) = 30 \pm 0.75$ or between 29.25 and 30.75.

7.4 Confidence Interval

"Confidence interval" denotes the long-term probability that the ratio p will be within the accuracy limits. The concept relies on the relative frequency interpretation of probability. Thus, 95% confidence means that if a large number of samples is taken, 95 out of 100 of these will contain the true value of p. The probability value is given by the proportion of the area under the normal curve included for a specified value of standard deviation (z). The usual confidence levels and the corresponding "z" values are given in Table 7.1.

7.5 Sample Size

The number of observations or sample size can be determined using Equation (7.1) as follows:

$$A = z\sigma_p = z\sqrt{p(1-p)/n} \text{ or } n = (z^2/A^2)(p(1-p)) \tag{7.3}$$

TABLE 7.1   Confidence Levels and Corresponding "z" Values

Confidence Level (%)	Standard Deviations (z)
68	± 1.00
90	± 1.64
95	± 1.96
99.73	± 3.00

where A is usually 0.05 or 5% for industrial work

p = percentage of total work time a component occurs

z = number of standard deviations depending on the confidence level desired

Example 2   Determine the idle percentage of a milling machine.

Relative accuracy desired = 0.05%.

Confidence level desired = 95%.

Preliminary study indicated p = 30%.

For this, accuracy $A = (0.05)(0.3) = 0.015$, and $z = 1.96$

$n = (z^2/A^2)\,(p(1-p)) = (1.96^2/0.015^2)\,(0.30 \times 0.70)$

$n = 3585.49 \cong 3586$

7.6   Random Observation Times

To be statistically acceptable, it is essential that the work sampling procedure give each individual moment during observation an equal chance of being observed. The observations have to be random, unbiased, and independent so that the assumption of the binomial theory of constant probability of event occurrence is attained. Hence, it is essential that observations be taken at random times when conducting a work sampling study. To ensure the randomness of observations, a convenient manual way is to use the random number table published in many handbooks. Random numbers can also be obtained by random number generators programmed in many handheld engineering calculators. Another way of obtaining random numbers is to write and place a large number of valid times mixed up in a hat, and pick up slips of paper at random. The required number of times of observations may be determined from a random number table as follows: for a picked number 859, the time may be taken as 8.59 A.M. if the start time of shift is 8.00 A.M. Another way of getting trip times is to multiply two-digit random numbers by ten (Niebel and Freivalds, 2003) to get time values in minutes after the start of shift. Only the time values falling within the work time are accepted for planning trips.

Example 3   For two-digit random numbers 04, 31, and 17, the observation times will be $4 \times 10 = 40$ min, $31 \times 10 = 310$ min, and $17 \times 10 = 170$ min after start of the shift, typically 8:00 A.M. The observation times are then 8.40 A.M., 1.10 P.M., and 10.50 A.M., respectively. The process is repeated until the required numbers of valid observation times within the shift work time are obtained.

7.7   Control Charts

Control charts are extensively used in quality-control work to identify when the system has gone out of control. The same principle is used to control the quality of the work sampling study. The $3\,\sigma$ limit is normally used in work sampling to set the upper and lower limits of control. First, the value of p is plotted in the chart as the centerline of the p-chart. The variability of p is then found for the control limits.

Example 4   For $p = 0.3$ and sample $n = 500$, $\sigma = \sqrt{p(1-p)/n} = 0.0205$ and $3\sigma = 0.0615$. The limits are then 0.3 ± 0.0615 or 0.2385 and 0.3615 as shown in Figure 7.1.

On the second Friday, the calculated value of p, which was based on observations, fell beyond limits, indicating the need for investigation and initiation of corrective action.

7.8   Plan of a Typical Work Sampling Study

A typical work sampling plan has the following steps:

- Determine the objective.
- Identify the elements of the study.
- Conduct a preliminary study to estimate the ratio percentages.
- Determine the desired accuracy and confidence levels.

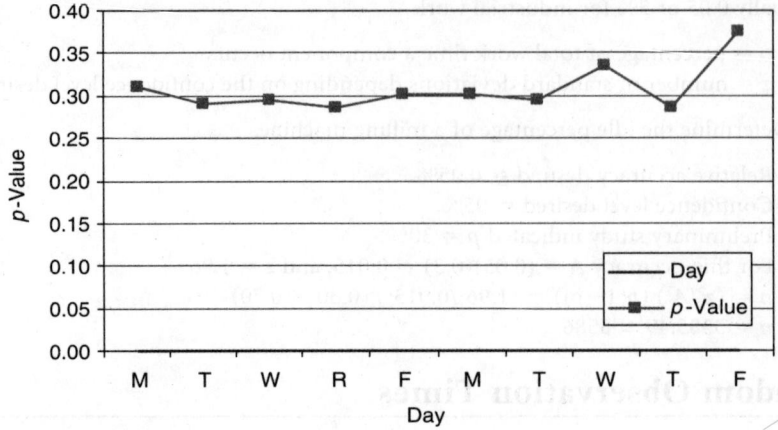

FIGURE 7.1   Control chart on daily basis.

- Determine the required number of observations.
- Schedule the random trip times and routes.
- Design the observation form.
- Collect sampling data.
- Summarize the study.

7.9   Applications of Work Sampling

Work sampling is most suitable for determining (1) machine utilization, (2) allowances for unavoidable delays, and (3) work standards for direct and indirect work. The technique is particularly suitable for determining standards for indirect and service work. Each type of application is illustrated below with an example.

7.9.1   Machine Utilization

Example 5   ABC Company was concerned about the utilization of forklift trucks and wanted to determine the average idle time of the forklift trucks in their plant. A work sampling study was conducted for 5 weeks. The data collected for 5 weeks are given in Table 7.2. The desired confidence level and relative accuracy were 95 and 5.0%, respectively.

Forklift truck utilization = (1500/2000) × 100% = 75%.
Idle percentage = (500/2000) × 100% = 25%.

Assuming that the desired confidence level is 95%, the accuracy of idle percentage is determined as follows:

$$A = s* \, 0.25 = 1.96 \sqrt{p(1-p)/n} = 1.96 \sqrt{0.25 \times 0.75/2000} = 0.0190$$

relative accuracy (s) = (0.0190/0.25) × 100% = 7.60%.

TABLE 7.2   Work Sampling Summary Sheet

Elements	Number of Observations	Total Numbers
Working	11111 11111 11111 11111	1500
Idle	11111 11111 11111	500
		2000

The desired accuracy required was 5%. The number of observations required to achieve the accuracy was found as follows:

$$0.05* 0.25 = 1.96\sqrt{0.25 \times 0.75/n} \text{ or } n = 1.96/(0.05\times0.25)^2 (0.05\times0.25) = 3074$$

The number of observations per day during the preliminary study was 2000/25 = 80.

The same sampling group was assigned to collect additional sampling data at 80 observations per day.

Hence, the additional number of days required for study to achieve the desired accuracy was (3074−2000)/80 = 13.4 d ≅ 14 d.

7.9.2 Allowances for Personal and Unavoidable Delays

Example 6 presents a method for determining allowances for personal and unavoidable delays.

Objective The industrial engineering department of the ABC Company wanted to determine the allowance for personal reasons and unavoidable delays for their machine shop.

Identify the Elements of the Study The elements were (1) working, (2) personal delay, and (3) unavoidable delay.

Design the Observation Form The observation form is task specific and has to be designed specifically for each job. For the ABC study in Example 5, a sampling form was developed as shown in Table 7.3 for the elements required to be observed: (1) working, (2) personal activities, e.g., drinking water, and (3) unavoidable delay, e.g., stopped for answering to foreman. Each tally mark indicated one observation of the corresponding element. At times, more elements may be included in the work sampling form to meet some future requirements of an organization.

Preliminary Study to Estimate the Ratio Percentages The preliminary work sampling study was conducted to estimate the approximate percentage values for the elements as given in Table 7.3.

Confidence Level and Accuracy The desired levels of confidence and accuracy were 95 and 0.05%, respectively.

Sample Size The preliminary estimates were used to determine the sample size required. The smallest value of the percentage occurrence was used for computation to ensure the desired level or higher accuracy for all elements.

$$A = 0.05\times0.05 = 0.0025, \text{ and } z = 1.96$$
$$n = (z/A)^2(p)(1-p) = (1.96/0.0025)^2(0.05)(0.95)$$
$$= 29,197$$

There were ten workers working on similar machines. Hence the number of observation per trip was ten. The number of trips required was 29,197/10 = 2,920. One work sampler could make 5 trips/h or 40 trips/day of 8 h shift. Three persons were available for sampling. Together, they could make 120 trips/day.

Hence, the duration of study required was 2920/120 = 24.33 d ≅ 5 weeks.

Schedule of Trips A random number table was determined to schedule 40 random trips per day for each observer. The procedure has been explained in Section 7.6. In addition, the routes of the observers were changed randomly each day.

TABLE 7.3 Pilot Work Sampling Summary (ABC Company)

Work Elements	Observations	Total Number of Observations	Percentage Occurrence (%)
1. Working	1111 1111...............	102	85
2. Personal activities	1111.......................	6	5
3. Unavoidable delays	1111 1111...............	12	10
Total		120	100

TABLE 7.4 Study Summary

Work Elements	Total Number of Observations	Percentage Occurrence
Working	24,730	24,730/29,200 = 0.8469 = 84.69%≅84.7%
Personal delay	1,470	1,470/29,200 = 0.503 = 5.03% ≅5.0%
Unavoidable delay	3,000	3,000/29,200 = 0.1027 = 10.27% ≅10.3%
Total	29,200	

Collecting Sampling Data   The observers were trained to be objective and not anticipate any expected observation. Each trip had a randomly selected route in addition to the random times. Video cameras may be used to minimize bias in collecting data, as the camera records any ongoing activity accurately.

7.9.3 Summarizing the Sampling Data

The ABC study is summarized in Table 7.4.

Certain unavoidable delay should be based on the work element data alone (24,730), if the unavoidable delay under study happens to be highly dependent on work time, as in case of fatigue.

7.9.4 Determining Work Standards

For the ABC Company, work standards for the above machine shop were developed as follows:

Given that during the study period, the machine shop produced 100,000 pieces of fan motor shafts, a fatigue allowance of 8% is allowed as per company policy in the machine shop. The section had five workers working on similar machines. The performance rating was found to be 110%, determined by estimating the pace of work periodically during the sampling study, and thought to be reasonable.

$$\text{Total work time} = 5 \text{ weeks} \times 5 \text{ days/week} \times 40 \text{ h/day} \times 5 \text{ operators} = 5000 \text{ m h}$$
$$\text{Observed time per piece} = 5,000 \times 60/100,000 = 3.0 \text{ min/pc}$$
$$\text{Normal time} = 1.10 \times 3.00 = 3.30 \text{ min}$$
$$\text{Total allowance} = 5 + 10.3 + 8.0 = 23.30\%$$
$$\text{Standard time per pc} = 3.30 \times (1.233) = 4.07 \text{ min/piece of motor shaft}$$

7.10   Computerized Work Sampling

A number of software packages with a variety of features are available for work sampling. Application of these packages saves the clerical time associated with recording and summarization of sampling data. The well-known ones among them are WorkSamp, by the Royal J. Dossett Corp.; CAWS/E, by C-Four; and PalmCAWS, by C-Four. These software packages reduce time for the clerical routines of work sampling and allow faster processing with greater accuracy. Use of computers may save about 35% of total work sampling study cost (Niebel and Freivalds, 2003) by eliminating the clerical work time, which is comparatively high relative to actual observation time.

7.11   Conclusion

Work sampling is a valuable technique for determining equipment utilization and the allowances that should be assigned for unavoidable delays in production operations. Determining standards for service and indirect work is another area in which work sampling has been found to be practical and economical. The trend of continued increase in service jobs, e.g., plant maintenance activities, custodial tasks, and nonindustrial jobs, where work sampling is the only practical and economical tool for establishing standards, is enhancing the value of work sampling as an industrial engineering tool in the 21st century.

References

Barnes, R.M., *Work Sampling*, 7th ed., Wiley, New York, 1980.

Brisley, C.L., How you can put work sampling to work, *Factory Manage. Maint.*, 110, 83–89, 1952.

Konz, S. and Johnson, S., *Work Design Industrial Ergonomics*, 5th ed., Holcomb Hathaway, Arizona, 2000.

Morrow, R.L., Ratio delay study, *Mech. Eng.*, 63, 302–303, 1941.

Meyers, F.E., *Motion and Time Study for Lean Manufacturing*, 2nd ed., Prentice-Hall, Englewood Cliffs, NJ, 1999.

Niebel, B. and Freivalds A., *Methods. Standards, and Work Design*, 11th ed., McGraw-Hill, New York, 2003.

Tippett, L.H.C., The ratio delay technique, *Time and Motion Study*, May 1953, pp. 10–19.

Waddell, H.L., Work sampling — a new tool to help cut costs, boost productivity, make decisions, *Factory Manage. Maint.*, 110, 1952.

PART III

FUNDAMENTALS OF SYSTEMS ENGINEERING

8

An Overview of Industrial and Systems Engineering

S. A. Oke
University of Lagos

Summary

There is currently a great need for a published work that presents a holistic overview of industrial and systems engineering. Such a publication should address the role and importance of industrial and systems engineers in today's society. The usefulness of such a publication would be enhanced if it contained information on the challenges that the industrial and systems engineers face, their ways of solving problems, and the impacts of their proffered solutions on business improvement. Such documentation should be captured in an integrated, concise, and elegantly distilled way. The current chapter aims at bridging these important gaps in current approaches to industrial and systems engineering. The chapter is developed in a creative and innovative way, with pointers on what lies ahead for an industrial and systems engineer. The opportunities that the industrial and systems engineering graduates have on a worldwide scale are also discussed.

8.1 Introduction

Industrial and systems engineers (ISEs) are perhaps the most preferred engineering professionals because of their ability to manage complex organizations. They are trained to design, develop and install optimal methods for coordinating people, materials, equipment, energy, and information. The integration of these resources is needed in order to create products and services in a business world that is becoming increasingly complex and globalized (see Figure 8.1). Industrial and systems engineers oversee management goals and operational performance. Their aims are the effective management of people, coordinating techniques in business organization, and adapting technological innovations toward achieving increased performance. They also stimulate awareness of the legal, environmental, and socioeconomic factors that have a

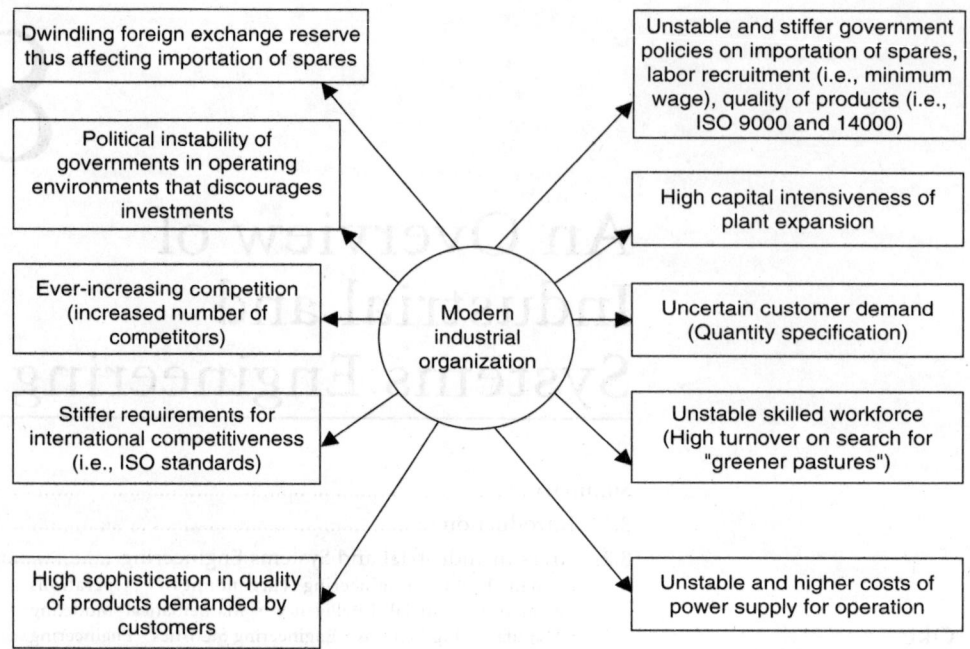

FIGURE 8.1   The complex nature of today's industrial organizational environment.

significant impact on engineering systems. Industrial and systems engineers can apply creative values in solving complex and unstructured problems in order to synthesize and design potential solutions and organize, coordinate, lead, facilitate, and participate in teamwork. They possess good mathematical skills, a strong desire for organizational performance, and a sustained drive for organizational improvement.

In deriving efficient solutions to manufacturing, organizational, and associated problems, ISEs analyze products and their requirements. They utilize mathematical techniques such as operations research (OR) to meet those requirements, and to plan production and information systems. They implement activities to achieve product quality, reliability, and safety by developing effective management control systems to meet financial and production planning needs. Systems design and development for the continual distribution of the product or service is also carried out by ISEs to enhance an organization's ability to satisfy their customers. Industrial and systems engineers focus on optimal integration of raw materials available, transportation options, and costs in deciding plant location. They coordinate various activities and devices on the assembly lines through simulations and other applications.

The organization's wage and salary administration systems and job evaluation programs can also be developed by them, leading to their eventual absorption into management positions. They share similar goals with health and safety engineers in promoting product safety and health in the whole production process through the application of knowledge of industrial processes and such areas as mechanical, chemical, and psychological principles. They are well grounded in the application of health and safety regulations while anticipating, recognizing, and evaluating hazardous conditions and developing hazard-control techniques.

Industrial and systems engineers can assist in developing efficient and profitable business practice by improving customer services and the quality of products. This would improve the competitiveness and resource utilization in organizations. From another perspective, ISEs are engaged in setting traditional labor or time standards and in the redesign of organizational structure in order to eliminate or reduce some forms of frustration or wastes in manufacturing. This is essential for the long-term survivability and the health of the business.

Another aspect of the business that the ISEs could be useful in is making work safer, easier, more rewarding, and faster through better designs that reduce production cost and allow the introduction of new technologies. This improves the lifestyle of the populace by making it possible for them to afford and use technological advanced goods and services. In addition, they offer ways of improving the working environment, thereby improving efficiencies and increasing cycle time and throughput, and helping manufacturing organizations to obtain their products more quickly. Also, ISEs have provided methods by which businesses can analyze their processes and try to make improvements upon them. They focus on optimization – doing more with less – and help to reduce waste in the society.[1] The ISEs give assistance in guiding the society and business to care more for their workforce while improving the bottom line.

Since this handbook deals with two associated fields – industrial and systems engineering – there is a strong need to define these two professions in order to have a clear perspective about them and to appreciate their interrelationships. Throughout this chapter, these two fields are used together and the discussions that follow are applicable to either. Perhaps the first classic and widely accepted definition of Industrial Engineering (IE) was offered by the then American Institute of Industrial Engineering (AIIE) in 1948.[2] Others have extended the definition. "Industrial Engineering is uniquely concerned with the analysis, design, installation, control, evaluation, and improvement of sociotechnical systems in a manner that protects the integrity and health of human, social, and natural ecologies. A sociotechnical system can be viewed as any organization in which people, materials, information, equipment, procedures, and energy interact in an integrated fashion throughout the life cycles of its associated products, services, or programs (see foot note 2). Through a global system's perspective of such organizations, industrial engineering draws upon specialized knowledge and skills in the mathematical, physical, and social sciences, together with the principles and methods of engineering analysis and design, to specify the product and evaluate the results obtained from such systems, thereby assuring such objectives as performance, reliability, maintainability, schedule adherence, and cost control (Figure 8.2).

As shown in Figure 8.2, there are five general areas of industrial and systems engineering. Each of these areas specifically makes out some positive contributions to the growth of industrial and systems engineering. The first area shown in the diagram is twofold, and comprises sociology and economics. The combination of the knowledge from these two areas helps in the area of supply chain. The second area is, mathematics, which is a powerful tool of ISEs. Operations research is an important part of this area. The third area is psychology, which is a strong pillar for ergonomics. Accounting and economics both constitute the fourth area. These are useful subjects in the area of engineering economics. The fifth area is computer. Computers are helpful in CAD/CAM, which is an important area of industrial and systems engineering.

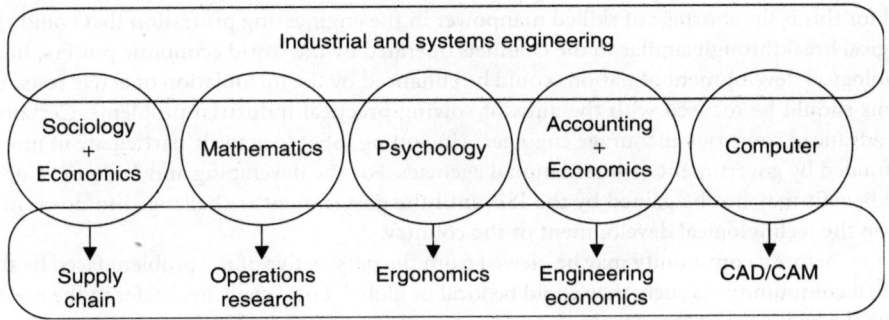

FIGURE 8.2   Some areas of industrial and systems engineering and related disciplines.

---

[1] http://www.orie.cornell.edu/~IIE
[2] http://www.iienet.org

According to the International Council on Systems Engineering (INCOSE),[3,4] systems engineering is an interdisciplinary approach and means to enable the realization of successful systems. Such systems can be diverse, encompassing people and organizations, software and data, equipment and hardware, facilities and materials, and services and techniques. The system's components are interrelated and employ organized interaction toward a common purpose. From the viewpoint of INCOSE (see foot note 3), systems engineering focuses on defining customer needs and required functionality early in the development cycle, documenting requirements, and then proceeding with design synthesis and systems validation while considering the complete problem. The philosophy of systems engineering teaches that attention should be focused on what the entities do before determining what the entities are. A good example to illustrate this point may be drawn from the transportation system. In solving a problem in this area, instead of beginning the problem-solving process by thinking of a bridge and how it will be designed, the systems engineer is trained to conceptualize the need to cross a body of water with certain cargo in a certain way.

The systems engineer then looks at bridge design from the point of view of the type of bridge to be built (see foot note 4). For example, is it going to have a suspension or superstructure design? From this stage he would work down to the design detail level where systems engineer gets involved, considering foundation soil mechanics and the placement of structures. The contemporary business is characterized by several challenges. This requires the ISEs to have skills, knowledge, and technical know-how in the collection, analysis, and interpretation of data relevant to problems that arise in the workplace. This places the organization well above the competition.

The radical growth in global competition, constantly and rapidly evolving corporate needs, and the dynamic changes in technology are some of the important forces shaping the world of business. Thus, stakeholders in the economy are expected to operate within a complex but ever-changing business environment. Against this backdrop, the dire need for professionals who are reliable, current, and relevant becomes obvious. Industrial and systems engineers are certainly needed in the economy for bringing about radical change, value creation, and significant improvement in productive activities.

The ISE must be focused and have the ability to think broadly in order to make a unique contribution to the society. To complement this effort, the organization itself must be able to develop effective marketing strategies (aided by a powerful tool, the Internet) as a competitive advantage so that the organization could position itself as the best in the industry.

The challenges facing the ISE may be divided into two categories: those faced by ISEs in developing and underdeveloped countries, and those faced by engineers in the developed countries. In the developed countries, there is a high level of technological sophistication that promotes and enhances the professional skills of the ISE. Unfortunately, the reverse is the case in some developing and underdeveloped countries. Engineers in underdeveloped countries, for instance, rarely practice technological development, possibly owing to the high level of poverty in such environments. Another reason that could be advanced for this is the shortage of skilled manpower in the engineering profession that could champion technological breakthrough similar to the channels operated by the world economic powers. In addition, the technological development of nations could be enhanced by the formulation of active research teams. Such teams should be focused with the aims of solving practical industrial problems. Certain governments in advanced countries encourage engineers (including ISEs) to actively participate in international projects funded by government or international agencies. For the developing and underdeveloped countries, this benefit may not be gained by the ISE until the government is challenged to do so in order to improve on the technological development of the country.

Challenges before a community may be viewed from the perspective of the problem faced by the inhabitants of that community. As such, they could be local or global. Local challenges refer to the need must be satisfied by the engineers in that community. These needs may not be relevant to other communities, for example, the ISE may be in a position to advise the local government chairman of a community on the

---

[3]http:// www.hra-incose.org
[4]http:// www.incose.org

disbursement of funds on roads within the powers of the local government. Decision-science models could be used to prioritize certain criteria, such as the number of users, the economic indexes of the various towns and villages, the level of business activities, the number of active industries, the length of the road, and the topography or the shape of the road.

Soon after graduation, an ISE is expected to tackle a myriad of social, political, and economic problems. This presents a great challenge to the professionals who live in a society where these problems exist. Consider the social problems of electricity generation, water provision, flood control, etc. The ISE in a society where these problems exist is expected to work together with other engineers in order to solve these problems. They are expected to design, improve on existing designs, and install integrated systems of men, materials, and equipment so as to optimize the use of resources. For electricity distribution, the ISE should be able to develop scientific tools for the distribution of power generation as well as for the proper scheduling of the maintenance tasks to which the facilities must be subjected.

The distribution network should minimize the cost. Loss prevention should be a key factor to consider. As such, the quality of the materials purchased for maintenance should be controlled, and a minimum acceptable standard should be established. In solving water problems, for instance, the primary distribution route should be a major concern. The ISE may need to develop reliability models that could be applied to predict the life of components used in the system. The scope of activities of the ISE should be wide enough for them to work with other scientists in the health sector on modeling and control of diseases caused by water-distribution problems. The ISE should be able to solve problems under uncertain conditions and limited budgets.

The ISE can work in a wide range of industries, such as the manufacturing, logistics, service, and defense industries. In manufacturing, the ISE must ensure that the equipment, manpower, and other resources in the process are integrated in such a manner that efficient operation is maintained and continuous improvement is ensured. The ISE functions in the logistics industry through the management of supply-chain systems (e.g., manufacturing facilities, transportation carriers, distribution hubs, retailers) to fulfill customer orders in the most cost-effective way (see foot note 1). In the service industry, the ISE provides consultancies in areas related to organizational effectiveness, service quality, information systems, project management, banking, service strategy, etc. In the defense industry, the ISE provides tools to support the management of military assets and military operations in an effective and efficient manner. The ISE works with a variety of job titles. The typical job titles of an ISE graduate include industrial engineer, manufacturing engineer, logistics engineer, supply-chain engineer, quality engineer, systems engineer, operations analyst, management engineer, and management consultant (Figure 8.3).

Experiences in the United States and other countries show that a large proportion of ISE graduates work in consultancy firms or as independent consultants, helping companies to engineer processes and systems to improve productivity, effect efficient operation of complex systems, and manage and optimize these processes and systems.

After completing their university education, ISEs acquire skills from practical exposure in an industry. Depending on the organization that an industrial or systems engineer works for, the experience may differ in depth or coverage. The trend of professional development in industrial and systems engineering is rapidly changing in recent times. This is enhanced by the ever-increasing development in the Information, Communication and Technology (ICT) sector of the economy.

Industrial and systems engineering is methodology-based and is one of the fastest growing areas of engineering. It provides a framework that can be focused on any area of interest, and incorporates inputs from a variety of disciplines, while maintaining the engineer's familiarity and grasp of physical processes. The honor of discovering industrial engineering belongs to a large number of individuals. The eminent scholars in industrial engineering are Henry Gantt (the inventor of the Gantt chart) and Lillian Gilbreth (a coinventor of time and motion studies). Some other scientists have also contributed immensely to its growth over the years. The original application of industrial engineering at the turn of the century was in manufacturing a technology-based orientation, which gradually changed with the development of OR, cybernetics, modern control theory, and computing power.

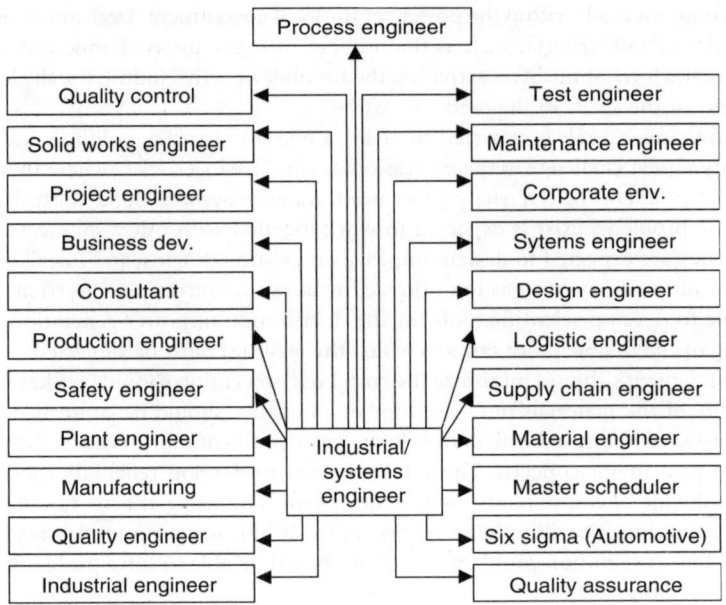

FIGURE 8.3 Job titles of industrial and systems engineers.

Computers and information systems have changed the way industrial engineers do business. The unique competencies of an ISE can be enhanced by the powers of the computer. Today, the fields of application have widened dramatically, ranging from the traditional areas of production engineering, facilities planning, and material handling to the design and optimization of more broadly defined systems. An ISE is a versatile professional who uses scientific tools in problem solving through a holistic and integrated approach. The main objective of an ISE is to optimize performance through the design, improvement, and installation of integrated of human, machine, and equipment systems. The uniqueness of industrial and systems engineering among engineering disciplines lies in the fact that it is not restricted to technological or industrial problems alone. It also covers nontechnological or non-industry-oriented problems also. The training of ISEs positions them to look at the total picture of what makes a system work best. They question themselves about the right combination of human and natural resources, technology and equipment, and information and finance. The ISEs make the system function well. They design and implement innovative processes and systems that improve quality and productivity, eliminate waste in organizations, and help them to save money or increase profitability.

Industrial and systems engineers are the bridges between management and engineering in situations where scientific methods are used heavily in making managerial decisions. The industrial and systems engineering field provides the theoretical and intellectual framework for translating designs into economic products and services, rather than the fundamental mechanics of design. Industrial and systems engineering is vital in solving today's critical and complex problems in manufacturing, distribution of goods and services, health care, utilities, transportation, entertainment, and the environment. The ISEs design and refine processes and systems to improve quality, safety, and productivity. The field provides a perfect blend of technical skills and people orientation. An industrial engineer addresses the overall system performance and productivity, responsiveness to customer needs, and the quality of the products or services produced by an enterprise. Also, they are the specialists who ensure that people can safely perform their required tasks in the workplace environment. Basically, the field deals with analyzing complex systems, formulating abstract models of these systems, and solving them with the intention of improving system performance.

8.2   Areas of Industrial and Systems Engineering

The discussions under this section mainly consist of some explanations of the areas that exist for industrial and systems engineering programs in major higher institutions the world over.

8.2.1   Human Factors Engineering (Ergonomics)

Human Factors engineering is a practical discipline dealing with the design and improvement of productivity and safety in the workplace. It is concerns the relationship of manufacturing and service technologies interacting with humans. Its focus is not restricted to manufacturing alone — it extends to service systems as well. The main methodology of ergonomics involves the mutual adaptation of the components of human-machine-environment systems by means of human-centered design of machines in production systems. Ergonomics studies human perceptions, motions, workstations, machines, products, and work environments (see Figure 8.4).

Today's ever-increasing concerns about humans in the technological domain make this field very appropriate. People in their everyday lives or in carrying out their work activities create many of the man-made products and environments for use. In many instances, the nature of these products and environments directly influences the extent to which they serve their intended human use. The discipline of human factors deals with the problems and processes that are involved in man's efforts to design these products and environments so that they optimally serve their intended use by humans. This general area of human endeavor (and its various facets) has come to be known as human factors engineering or, simply, Human Factors, biomechanics, engineering psychology, or ergonomics.

8.2.2   Operations Research

Operations research specifically provides the mathematical tools required by ISEs in order to carry out their task efficiently. Its aims are to optimize system performance and predict system behavior using rational decision making, and to analyze and evaluate complex conditions and systems (see Table 8.1 and Figure 8.5).

This area of industrial and systems engineering deals with the application of scientific methods in decision making, especially in the allocation of scarce human resources, money, materials, equipment, or facilities. It

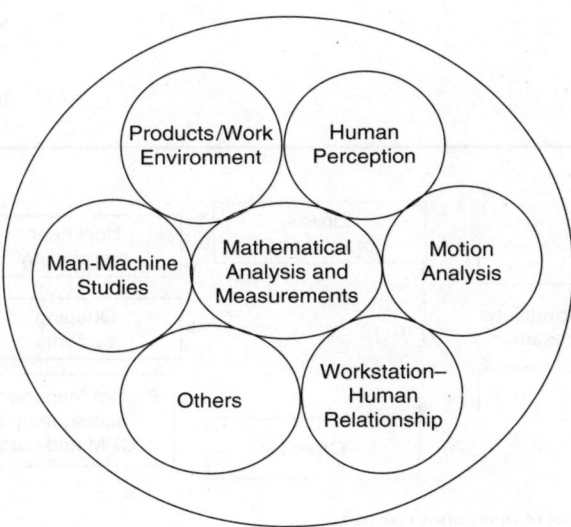

FIGURE 8.4   Major areas of ergonomics.

covers such areas as mathematical and computer modeling and information technology. It could be applied to managerial decision making in the areas of staff and machine scheduling, vehicle routing, warehouse location, product distribution, quality control, traffic-light phasing, and police patrolling. Preventive maintenance scheduling, economic forecasting, experiment design, power plant fuel allocation, stock portfolio optimization, cost-effective environmental protection, inventory control and university course scheduling are some of the other problems that could be addressed by employing OR.

Subjects such as mathematics and computer modeling can forecast the implications of various choices and identify the best alternatives. The OR methodology is applied to a wide range of problems in both public and private sectors. These problems often involve designing systems to operate in the most effective way. Operations research is interdisciplinary and draws heavily on mathematics. It exposes graduates in the field of industrial and systems engineering to a wide variety of opportunities in areas such as pharmaceuticals, ICT, financial consultancy services, manufacturing, research, logistics and supply-chain management, and health. These graduates are employed as technical analysts with prospects for managerial positions. Operations research adopts courses from computer science, engineering management, and other engineering programs to train students to become highly skilled in quantitative and qualitative modeling and the analysis of a wide range of systems-level decision problems. It focuses on to productivity, efficiency and quality.[5] It also affects

TABLE 8.1　Operations Research in Industrial and Systems Engineering

Purpose	Aims	Resources Analyzed	Major Aspects	Applications
To carry out efficient tasks	Optimize system performance	Human resources	Mathematical and computer modeling	Staff and machine scheduling
	Prediction of system behavior	Money	Information technology	Vehicle routing
		Materials		Ware house location
		Equipment or facilities		Product distribution
		Others		Quality control
				Traffic light phasing
				Police patrolling
				Preventive maintenance scheduling
				Economic forecasting
				Design of experiments
				Power plant/fuel allocation
				Stock portfolio optimization
				Cost-effective environmental protection
				Inventory control
				University course scheduling

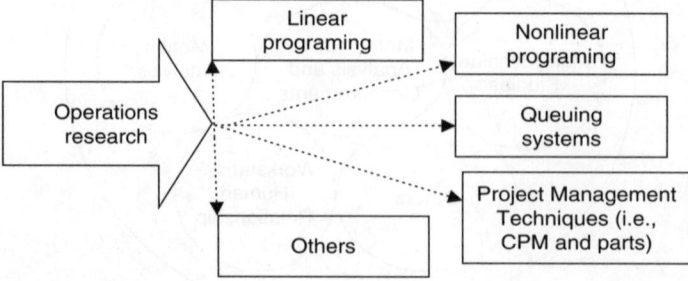

FIGURE 8.5　The main areas of operations research.

[5]http://stats.bls.gov/oco/ocos032.htm

the creative utilization of analytical and computational skills in problem solving, while increasing the knowledge necessary to become truly competent in today's highly competitive business environment. Operations research has had a tremendous impact on almost every facet of modern life, including marketing, the oil and gas industry, the judiciary, defense, computer operations, inventory planning, the airline system, and international banking. It is a subject of beauty whose applications seem endless.

8.2.3 Artificial Intelligence

The aim of studying artificial intelligence (AI) is to understand how the human mind works, thereby fostering leading to an appreciation of the nature of intelligence, and to engineer systems that exhibit intelligence. Some of the basic keys to understanding intelligence are vision, robotics, and language. Other aspects related to AI include reasoning, knowledge representation, natural language generation (NLG), genetic algorithms, and expert systems. Studies on reasoning have evolved from the following dimensions: case-based, nonmonotonic, model, qualitative, automated, spatial, temporal, and common sense. For knowledge representation, knowledge bases are used to model application domains and to facilitate access to stored information. Knowledge representation originally concentrated around protocols that were typically tuned to deal with relatively small knowledge bases, but that provided powerful and highly expressive reasoning services. Natural language generation systems are computer software systems that produce texts in English and other human languages, often from nonlinguistic input data. Natural language generation systems, like most AI systems, need substantial amounts of knowledge that is difficult to acquire. In general terms, these problems were due to the complexity, novelty, and poorly understood nature of the tasks our systems attempted, and were worsened by the fact that people write so differently. A genetic algorithm is a search algorithm based on the mechanics of natural selection and natural genetics. It is an iterative procedure that maintains a population of structures that are candidate solutions to specific domain challenges. During each generation the structures in the current population are rated for their effectiveness as solutions, and on the basis of these evaluations, a new population of candidate structures is formed by using specific genetic operators such as reproduction, cross over, and mutation. An expert system is a computer software that can solve a narrowly defined set of problems using information and reasoning techniques normally associated with a human expert. It could also be viewed as a computer system that performs at or near the level of a human expert in a particular field of endeavor.

8.2.4 Mathematical Modeling

A model is a simplified representation of a real system or phenomenon. Models are abstractions revealing only the features that are relevant to the real system behavior under study. In industrial and systems engineering, virtually all areas of the disciplines have concepts that can be modeled in one form or the other. In particular, mathematical models are elements, concepts, and attributes of a real system represented by using mathematical symbols, e.g., $v = u + at$, $A = \pi r^2$, etc. Models are powerful tools for predicting the behavior of a real system by changing some items in the models to detect the reaction of changes in the behavior of other variations. They provide frames of reference by which the performance of the real system can be measured. They articulate abstractions, thereby enabling us to distinguish between relevant and irrelevant features of the real system. Models are prone to manipulations more easily in a way that the real systems are often not.

8.2.5 Manufacturing Systems

In order to survive in the competitive environment, significant changes should be made in the ways of preparing organizations' design, and manufacturing, selling, and servicing their goods and commodities. Manufacturers are committed to continuous improvement in product design, defect levels, and costs (see foot note 1). This is achieved by fusing the designing, manufacturing, and marketing into a complete whole. Manufacturing system consists two parts: its science and automation. Manufacturing science refers to investigations on the processes involved in the transformation of raw materials into finished products. This involves the traditional aspects. Traditionally, manufacturing science may refer to the techniques of work-study, inventory systems, material-requirement planning, etc. On the other hand, the automation aspect of manufacturing covers issues like e-manufacturing, Toyota system, the use of computer-assisted manufacturing systems (NC, CNC, and DNC), automated material handling systems, group technology, flexible manufacturing systems, process planning and control, etc.

Industrial and systems engineering students conduct research in the areas of manufacturing in combination with courses in finance, manufacturing processes, and personnel management. They also do research in manufacturing-design projects. This exposes the students to a manufacturing environment with activities in the design or improvement of manufacturing systems, product design, and quality.

8.2.6 Engineering Statistics

Recent years have experienced increasing use of statistics in the industrial and systems engineering field. Industrial and systems engineers need to understand the basic statistical tools to function in a world that is becoming increasingly dependent on quantitative information. This clearly shows that the interpretation of practical and research results in industrial and systems engineering depends to a large extent on statistical methods. Statistics is used used in almost every area relevant to these fields. It is utilized as a tool for evaluating economic data in "financial engineering". For this reason, ISEs are exposed to statistical reasoning early in their careers. Industrial and systems engineers also employ statistical techniques to establish quality control techniques. This involves detecting an abnormal increase in defects, which reflects equipment malfunction. The question of what, how, and when do we apply statistical techniques in practical situations and how to interpret the results are answered in the topics related to statistics.

8.2.7 Engineering Computing

The impact industrial and systems of computers on engineering is complex and many-sided. The practitioners of data analysis in industrial and systems engineering rely on a computer as it is an important and powerful tool for collecting, recording, retrieving, analyzing simple and complex problems, as well as distributing huge information in industrial and systems engineering. It saves countless years of tedious work by the ISEs. The computer removes the necessity for men to monitor and control tedious and repetitive processes. Despite the importance of computers, its potential is so little explored that its full impact is yet to be realized. There are several powerful computer programs that can reduce the complexity of solving engineering problems.

9

System Safety Engineering

Paul S. Ray
The University of Alabama at Tuscaloosa

9.1   Introduction

The concern for safety has existed since the beginning of human history, the first recorded evidence being the Code of Hammurabi, which was written down in the ancient kingdom of Babylon around 1750 B.C. The code contained clauses pertaining to injuries and monetary damages against those who were responsible for the injuries. The safety movement has continued through the ages. But in the United States, since the start of the Industrial Revolution in the early 1800s, in the U.S., people have been exposed to more and increasingly complex hazards. Industry attempted to control the obvious hazards by a trial-and-error approach, which was based on an after-the-fact philosophy of ensuring safety from lessons learned. The cost of mishaps was manageable until the introduction of manned spaceflight programs in 1950s and early 1960s.

   At that time, the cost of even a single accident became prohibitively high, and lessons learned from accidents were no longer acceptable. There needed to be a new approach to safety, a proactive rather than the traditional reactive approach. Industrial safety has developed on the basis of mishap records and has been classified by hazard types. System safety, on the other hand, has developed to identify the hazards that may be inherent in a product or operation before any mishap happens, and as such is based on analytical procedures. The goal of system safety is to make a product "first-time safe," i.e., to identify and mitigate any hazard that may result in mishap later on.

Early safety requirements were generated by the development of the military nuclear weapons systems, aerospace industry, and space programs. As these systems became more complex, the consequence of accidents became unacceptable. All of these systems were supported by the Department of Defense (DOD); hence, the first effort to develop system safety as a separate discipline came from one of the military departments, namely, the U.S. Air Force (USAF). In 1962, the USAF published "System Safety Engineering for the Development of Air Force Ballistic Missiles." The Minuteman Intercontinental Ballistic Missile (ICBM) program was one of the first programs to have a formal, disciplined, system safety program associated with it (Roland and Moriarty, 1990).

The word "system" is defined as a set of interrelated components working together toward some common objective (Blanchard and Fabrycky, 1981). System safety may be defined as a discipline that uses a systematic engineering approach to make a system, product, or process safe through its life cycle. The heart of the system safety approach is the hazard analysis. A number of analytical procedures have been developed that are helpful in analyzing and identifying hazards present in a system design. These specialized safety analysis techniques, along with the safety disciplines, principles, and procedures, have evolved into the new discipline of System Safety Engineering.

9.2 Historical Background

The transformation of trial-and-error or "fly-fix-fly" (Roland and Moriarty, 1990; Stephenson, 1991) into the "first-time safe" approach started with the introduction of increased complexity in aircrafts and military weapon design in 1940s. The first article on system safety, "Engineering for Safety," was presented to the Institute of Aeronautical Scienes in September 1947 (Roland and Moriarty, 1990). However, only in the early 1960s did the concept of system safety begin to evolve as a separate discipline. In 1962, the USAF published "System Safety Engineering for the Development of Air Force Ballistic Missiles," and in 1963, published the MIL-S-38130, which covered the general requirements for engineering systems. The DOD adopted this standard as MIL-S-38130A in 1966, and in 1969, it was revised and adopted as MIL-STD-882B with the title, "System Safety Program Requirements." The first recorded formal application of System Safety as a discipline was by the USAF for the Minuteman ICBM Program in 1987. The early applications were also in the areas of aerospace and space programs.

The System Safety Society, the professional organization for the discipline, emerged in the early 1960s and was founded in 1962 in Los Angeles by Roger Lockwood. The society was chartered in California as the Aerospace System Safety Society in 1964 and renamed the System Safety Society in 1967. As the international membership grew, the society was incorporated as an international organization to represent its international nature.

In the late 1960s, the Atomic Energy Commission (AEC) made an effort to improve the approach to system safety and the effort resulted in a revision of the Management Oversight and Risk Tree (MORT) Manual (Johnson, 1973) in 1973. In 1970, the National Aeronautics and Space Administration (NASA) published the System Safety Manual, NHB 1700.1 (V3). In the 1980s, the U.S. Army Corps of Engineers developed MIL-STD-882A and B (1987). The Naval Facilities Command sponsored system safety courses in 1984. The American Institute of Chemical Engineers (AICHE) developed guidelines for hazard evaluation procedures in 1985. In the 1990s, organizations became aware of the software hazard and its impact on future systems. The System Safety Society extended its memberships in overseas countries and became an international organization.

The updated MIL-STD-882 issued in 2000 under the name "Standard Practice for System Safety" allows for flexibility in implementation while maintaining the basic requirements of system safety (Stephans, 2004).

9.3 Industrial Safety vs System Safety

- Industrial safety has traditionally been reactive in nature. It is said on the shop floor that "the safety rules and regulations are written in blood," indicating that safety rules and regulations are developed on the basis of the injury experiences of the workers. This reactive approach was acceptable to industry when the cost of accident and loss of the product/ system was not prohibitive. With the development of space vehicles, nuclear plants, and complicated military equipment, the cost of loss of vehicles and systems has become prohibitive, and the reactive approach of correcting deficiency of design on the basis of failure experience is no longer acceptable. In addition, safety programs in industry aimed at complying with the safety regulations have been based on the minimum acceptable level of performance. By contrast, the system safety effort aims to achieve the highest level of safety feasible for a project.

9.4   Definition of Basic Terminology

Some basic definitions in the system safety field include:

- Hazard: Some aspects of a system that have potential for causing mishap.
- Safety: Degree of freedom from hazard.
- System: A system is a set of people, hardware, software, and procedures that work together to achieve a common goal.
- System safety: The discipline that applies engineering and management techniques to make systems safe throughout their life cycle.

9.5   System Safety Precedence

The basic steps of system safety precedence as prescribed by the MIL-STD-882B are given below:

1. Design for minimum risk.
2. Incorporate safety devices.
3. Provide warning devices.
4. Develop procedures and training.
5. Accept remaining risks if the hazard level is now acceptable.

In order to ensure a minimum level of residual hazard in a system, the priority of the system safety effort has been to design the system for minimum risk at the very beginning. However, if an identified hazard cannot be reduced to an acceptable level, effort should be directed to reduce hazard by providing safety devices; if the hazard level still remains unacceptable, warning devices should be provided to alert people of the hazard. If the level of the remaining hazard is still not acceptable, administrative controls such as procedures and training should be provided. If the remaining level reaches an acceptable level, risk acceptance steps should be initiated. In case, the hazard level still remains unacceptable, the system has to be discarded, or major modifications of the design are to be arranged to bring it to an acceptable level.

9.6   System Safety Criteria

Hazard severity and hazard probability are the two main criteria for measuring risk. MIL-STD 882B (1984) guidelines are generally followed for these criteria, as given in Table 9.1.

Five qualitative hazard-probability levels have been specified in MIL-STD-882B on the basis of the likelihood of occurrence of each category due to unmitigated hazard, and are given in Table 9.2.

The hazard assessment matrix uses the elements incorporated in Table 9.1 and Table 9.2 to estimate the risk of a hazard. Even a low-likelihood event with very high severity can be comparable to a medium-frequency, medium-severity hazard or a high-frequency but low-severity hazard. A method of establishing the hazard risk indices is illustrated in the Hazard Risk Assessment Matrix in Table 9.3.

Criteria for the different hazard risk indices are as follows:

Red	Unacceptable
Teal	Undesirable
Yellow	Acceptable with management review
Green	Acceptable without management review.

TABLE 9.1   Hazard Severity

Category	Name	Characteristics
I	Catastrophic	Death or system loss
II	Critical	Severe injury or morbidity
III	Marginal	Major injury or damage
IV	Negligible	Less than minor injury or damage to system

TABLE 9.2   Hazard Probability Levels

Description of Hazard	Level	Mishap Description
Frequent	A	Likely to occur frequently
Probable	B	Will occur several times
Occasional	C	Occurs sometimes during the life of the system
Remote	D	Unlikely but may occur reasonably
Improbable	E	Unlikely to occur, but possible

TABLE 9.3   Hazard Risk Assessment Matrix

Frequency of Occurrence	Hazard I Catastrophic	Hazard II Critical	Hazard III Marginal	Hazard IV Negligible
A–Frequent	Red	Red	Red	Yellow
B–Probable	Red	Red	Teal	Yellow
C–Occasional	Red	Teal	Teal	Green
D–Remote	Yellow	Teal	Yellow	Green
E–Improbable	Yellow	Yellow	Yellow	Green

9.7   System Life Cycle and Type of Analysis

System development may be divided into five phases: concept, definition, development, production, and deployment (MIL-STD-882B, 1987; Roland et al., 1990). The types of analysis suitable for each phase include the following:

9.7.1   Conceptual Stage

This is the step when the basic design of the system or product is established. Preliminary Hazard Analysis (PHA) is the most common analysis conducted at this stage. At times, for large projects, a Preliminary Hazard List (PHL) may be prepared and an Energy Trace Barrier Analysis (ETBA) may also be conducted.

A PHL provides the initial identification of hazards and is a feeder document to PHA. This is the first step of systematic analysis of potential hazards based on the experience and expertise of the analyst, and is used to prioritize the hazards according to their individual risks. A review of the conceptual design is conducted at the end of this stage to approve the basic design of a product. The generic format of the PHL and PHA is given in Figure 9.1 and Figure 9.2.

9.7.2   Definition Stage

Preliminary design of a specific system or product that meets the criteria developed during the conceptual stage is established during this stage. The definitions of the subsystems and assemblies of the system are also produced at this stage. The PHA continues, and as the specific design data become available, subsystem hazard analysis (SSHA) commences. At the end of the stage, a Preliminary Design Review (PDR) is conducted

Preliminary hazard list				
Project _____ Prepared by_____ Date_____				
Technique used: Checklist, Brainstorming, ETBA, Other Page____of____				
Hazard	Cause	Effect	Risk code	Comments

FIGURE 9.1   Preliminary hazard list.

Preliminary hazard analysis				
Project: Garden tractor Prepared by_____ Date_____				
Technique used: Checklist, _____ Page 1 of 4				
Hazard	Cause	System effect	Category	Preventive measures
Overturn	Improper steering on an incline	Loss of control, overturning and injury to head	III- IV	Exercise proper steering, keep speed within safe limit
Collision	Brake failure	Injury and damage to the vehicle	IV	Check brake prior to starting mowing
		Continued		

FIGURE 9.2   Preliminary hazard analysis.

to ensure the safety provisions. A generic format of subsystem/system hazard analysis (SSHA / SHA) is illustrated in Figure 9.3.

9.7.3   Development Stage

The specific design for a product is developed during this stage. Integration of all safety features, engineering specifications, and operational requirements are finalized at this stage. The PHA is completed. The SSHA started during the definition stage also has to be completed during this stage while SHA should be completed prior to system production. Operational hazard analysis (OHA) starts late at this stage to identify the hazards associated with operation and maintenance of the system, particularly the human–machine interfaces. Critical design review (CDR) is conducted at the end of this stage to ensure that all hazards have been controlled and the product is approved for production.

9.7.4   Production Stage

During this stage, monitoring the tests that involves safety and check on proper functioning of safety devices are to be conducted by system-safety group. The final acceptance review (FAR) is conducted at the end of this stage to release the product for production.

Subsystem / system hazard analysis							
Subsystem_____				Prepared by_____			
Project _____				Date _____			
Method used: FMEA___, FTA___, Other___				Page____of_____			
Item	Hazard	Cause	Effects	Risk code	Control steps	Controlled risk code	Signature
			Continued				

FIGURE 9.3   Subsystem/system hazard analysis.

9.7.5 Deployment Stage

 The system becomes operational at this stage. The follow-up work relating to operational and maintenance hazards comprises the major safety activities during this stage. Reviews of the corrective design
changes are conducted by the system-safety group. The operating hazard analysis (OHA) is completed
before the end of production. Maintenance hazard analysis (MHA) and OHA are conducted periodically
throughout the life cycle of a product. The final stage of termination of a product should be monitored
by a system safety person to ensure that approved safety procedures are followed for termination.

9.8   Common Analytical Tools

There are several analytical tools in the field, but only a few common tools appear to be currently applied
on most occasions. These are energy trace and barrier analysis (ETBA), failure modes and effects analysis (FMEA), and fault tree analysis (FTA).

9.8.1   Energy Trace and Barrier Analysis (ETBA)

According to this technique, an accident is defined as an incident due to an unwanted flow of energy that
overcomes inadequate barriers to strike people and property, causing adverse consequences. Energy trace
and barrier analysis is suitable for conducting several types of analyses, particularly for the PHL and PHA.
The amount of data needed depends on the type of analysis. The PHL may need only sketches and preliminary drawings, while PHA will need more details. The approach is to find the types of energy in a
project, their origins and paths, the barriers and their adequacy, and the control steps needed to improve
safety of a system. A generic worksheet is illustrated in Figure 9.4.

9.8.2   Failure Modes and Effects Analysis

The FMEA is a well-known technique, originally used by reliability engineers, that is now widely used for
system-safety analysis by NASA, DOE, DOD, and private industries. The approach is to break down the
system into subsystems and then into the individual components for the hardware FMEA. Analysis is then
continued to identify possible failure modes of each individual component and their impact on the overall system safety. Hardware FMEA is helpful to identify critical components and single-point failures. A
top-level analysis covering up to the level of subassemblies or assemblies can be done at the early stage of
design from configuration information and is called Functional FMEA. The limitations of FMEA include
the exclusion of operational interfaces, multiple failures, human factors, and environmental factors. A
generic worksheet of FMEA is illustrated in Figure 9.5.

Energy trace and barrier analysis							
Project _____ Prepared by_____							
Energy type_____ Date____Page____of____							
Source Documents_____							
Energy data	Barriers	Targets	RAC	Barrier evaluation	Control steps	Improved RAC	Standards

FIGURE 9.4 Energy trace and barrier analysis worksheet. Adapted from Stephans (2004).

Failure modes and effects analysis					
Project _____ Prepared by_____					
System _____ Date _____					
Subsystem _____ Page _____ of _____					
Component _____					
Drawing no. _____					
Part no.	Part description	Failure mode and causes	Failure effects on system	Risk code	Comments

FIGURE 9.5 Failure modes and effects analysis.

9.8.3 Fault Tree Analysis

This technique was developed by the Bell Telephone Laboratories in 1961–1962 for the ICBM launch control system, and has been a primary tool for system safety analysis. In FTA, the undesirable event is listed in the top block and deductive logic is used to determine all the subsequent events that can lead to the top event. The flow of logic is from top downwards through the events placed in branches of the tree-like information-flow network. The six basic symbols used in a fault tree are given in Figure 9.6.

A fault event is shown by a rectangle. It may be either a top or an intermediate event. A basic event is presented by a circle, which also indicates an end event. An undeveloped event is shown by a diamond, which may be chosen by an analyst. An AND gate requires all inputs to generate an output. An OR gate requires only one or any combination of inputs to generate an output. The procedure is illustrated in Figure 9.7.

9.8.4 Qualitative or Quantitative Fault Tree

A qualitative fault tree does not include any probability values. But the logic used to develop a fault tree provides a good insight into the project and identifies the priority and an idea of event severity even though numerical probability figures are not available.

A quantitative fault tree requires the probability values of the basic events. The probability values of events in a fault tree are very low and the approximation due to the assumption of independence of the events is not significant. Usually the events in the fault tree are assumed to be independent of each other

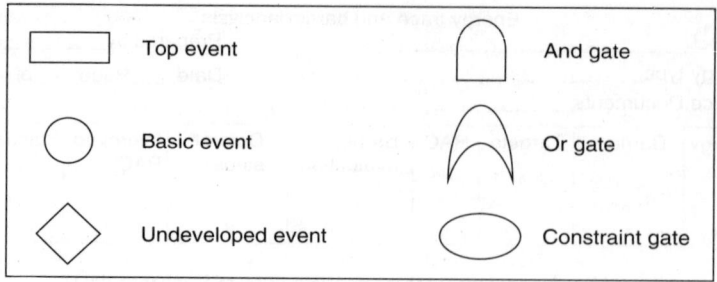

FIGURE 9.6   Basic fault tree symbols.

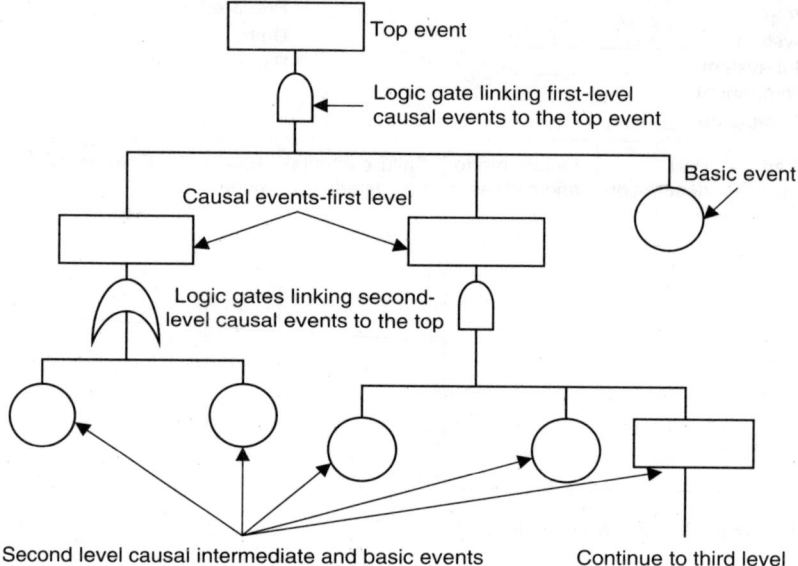

FIGURE 9.7   Fault tree procedural steps.

and the computation in this case is done by the multiplication rule for AND gates and addition rule for the OR gates. The computation is done using the following rules:

$$P \text{ (output from AND gate)} = P \text{ (input 1)} \times P \text{ (input 2)} \times \ldots P \text{ (input n)} \tag{9.1}$$

and

$$P \text{ (output from OR gate)} = P \text{ (input 1)} + \text{(input 2)} + \ldots P \text{ (input n)} \tag{9.2}$$

The computation of probability starts at the basic events and moves up progressively until the top event is reached.

9.8.5   Single-Point Failures and Cut Sets

The fault tree is useful in identification of single-point failures. A single-point failure is a basic event or events that can lead to the top event and cause system failure. Cut set analysis is conducted to identify all single-point failures. A cut set is a group of events that can lead to system failure. By eliminating the redundant basic events, minimum cut sets can be determined. Basic events that contribute the most to the unwanted top event are identified by analyzing the minimum cut sets. The most effective ways are determined to reduce the probability of system failure.

9.9 Other Emerging Analytical Tools

9.9.1 Management Oversight and Risk Tree

MORT, both a program and an assessment tool, was developed by W. G. Johnson (Johnson, 1973) for the Department of Energy (DOE) in 1970. MORT uses a logic tree approach to identify risks of an operation, and management oversights require action to ensure safety. The method consists of a program as it aids to control oversights as well as a logic tree, to identify the risks associated with an operation. The MORT diagram consists of three levels of relationships. The top level identifies 98 generic undesirable events, the second level contains 1500 possible specific causal factors to an accident or management oversight events, and the third level contains thousands of criteria to evaluate if the safety program is doing well. The major disadvantage of MORT is that it is very tedious and time-consuming to learn and apply. [A simpler version of the method, called Mini-Mort, has been developed by R. A. Stephans in 1988 to simplify MORT by reducing the number of events for evaluation from 1500 to 150.] This has been accomplished by eliminating the bottom tier of the diagram. MORT is primarily suitable for major accident investigation and has been included in the DOE's MORT program.

9.9.2 Software Hazard Analysis

Most complex systems are nowadays operated and controlled by some software. This tendency is expected to increase and enhance the importance of SHA in the future. Until now, the state of the art in SHA has been inadequate. The main difference between a piece of hardware and software is that software does not fail owing to wear out resulting from use, but fails owing to a fault embedded in it. The analysis should start at the requirement phase of software development with the help of a checklist, should continue in the definition phase while logic structure and interfaces are being developed, and then proceed to the development phase while codes are formulated. Fault trees can be developed for software modules to examine the fault paths. A fault tree presenting software events facilitates examinations of interfaces throughout the system. During the integration phase, software modules should be tested separately and in the system environment. MIL-STD-882B, Notice 1 specifies the sequence of analysis linked to each phase of system development. Although this standard was developed for DOD systems, it may be used for identifying software hazards. A mathematical model of a system called Petri net can model hardware, software, and human components. Petri net can be used to examine the synchronization in system modules.

9.9.3 Sneak Circuit Analysis

A sneak circuit is a latent functional path that can initiate an unintended and undesired action. Sneak circuit analysis is the process used to identify sneak circuits. The technique was standardized by Boeing in 1967 while working for the Minuteman missile project. The methodology consists of analyzing the topological patterns from detailed engineering schematics and drawings. It is believed that all circuitry can be resolved into five basic patterns. These are single line, ground dome, power dome, combination dome, and H dome (Roland et al., 1990). Of these the H dome is the most vulnerable and should be avoided in design (NASA, 1987).

9.10 Conclusions

System safety began to evolve in the 1960s and has now firmly established itself as a separate discipline for safety analysis of complex systems prevalent in space programs, nuclear weapons, and complex defense weapon systems. Numerous techniques have been added to the repertoire of a system safety analyst. For reasons of brevity, only the most commonly applied analysis techniques have been included in this chapter.

References

Blanchard, B.S. and Fabrycky, W.J., *Systems Engineering and Analysis*, Prentice-Hall, Englewood Cliffs, NJ, 1981.

Johnson, W.G., *The Management Oversight and Risk Tree*, Atomic Energy Commission, Washington, DC, 1973.

National Aeronautics and Space Administration, NHB 1700.1 (V3), System Safety, Safety Office, NASA, Washington, DC, 1970.

National Aeronautics and Space Administration, *Methodology for Conduct of NSTS Hazard Analyses. NSTS 22254*, NSTS Program Office, Johnson Space Center, NHB 1700.1 (V3), System Safety, Houston, 1987.

Roland, H.E. and Moriarty, B., *System Safety Engineering and Management*, Wiley, New York,, 1990.

Stephans A.R., *System Safety for the 21st Century*, Wiley, New York, 2004.

U.S. Air Force, *SDP 127-1:* System Safety Handbook for the Acquisition Manager, HQ Space Division/ SE, Los Angeles, 1987.

U.S. Department of Defense, *MIL-STD- 882B:* System Safety Program Requirements, U.S. Department of Defense, Washington, DC, 1984 (updated by Notice 1, 1987).

10

Metaheuristics: A Solution Methodology for Optimization Problems

Reinaldo J. Moraga
Northern Illinois University

Gail W. DePuy
University of Louisville

Gary E. Whitehouse
University of Central Florida

10.1 Introduction

Many practical problems in the field of industrial engineering can be modeled using operations research techniques. Such problems can be structured as the optimization of some decision variables that are restricted by a set of constraints. When the decision variables are discrete in nature, the problem of finding optimal solutions is known as combinatorial optimization. Examples of real-world combinatorial optimization problems include resource-constrained project scheduling problems, assembly-line balancing problems, vehicle routing and scheduling problems, facility location problems, facility layout design problems, job sequencing and machine scheduling problems, manpower planning problems, production planning and distribution, etc. Most of the combinatorial optimization problems found in practice are NP-complete, meaning that they cannot be solved to optimality in polynomial time. Therefore, heuristic solution approaches are commonly used to find solutions to many combinatorial optimization problems found in industrial engineering.

While the goal of combinatorial optimization research is to find an algorithm that guarantees an optimal solution in polynomial time with respect to the problem size, the main interest in practice is to find a nearly optimal or at least good-quality solution in a reasonable amount of time. Numerous approaches to solve combinatorial optimization problems have been proposed, varying from brute-force enumeration through highly esoteric optimization methods. The majority of these methods can be broadly classified as either "exact" algorithms or "heuristic" algorithms. Exact algorithms are those that yield an optimal solution. The branch-and-bound method, for example, is a quasi-enumerative approach that has been used in a wide variety of combinatorial problems. It is reasonably efficient for problems of modest size and, as a general methodology, constitutes an important part of the set of exact solution methods for the general class of integer linear programming problems (Ingnizio and Cavalier, 1994). The size of practical problems frequently precludes the application of exact methods, and thus, heuristic algorithms are often used for real-time solution of combinatorial optimization problems.

With heuristic algorithms, there is, in theory, a chance to find an optimal solution. But the chance of finding the optimal solution can become remote, because heuristics often get stuck in a local optimal solution. Metaheuristics or "modern heuristics" deal with this problem by introducing systematic rules to move out of a local minimum or avoid a local minimum. The common characteristic of these metaheuristics is the use of some mechanisms to avoid local optima. Metaheuristics such as genetic algorithms (GA), simulated annealing (SA), tabu search (TS), artificial neural networks (ANN), ant colony optimization (ACO), particle swarm optimization (PSO), and randomized search heuristics succeed in leaving the local optimum by temporarily accepting moves that cause a worsening of the objective function value.

Glover and Laguna (1997) state that the term "metaheuristic" refers to

> a master strategy that guides and modifies other heuristics to produce solutions beyond those that are normally generated in a quest for local optimality. The heuristics guided by such meta-strategy may be high level procedures or may embody nothing more than a description of available moves for transforming one solution into another, together with associated evaluation rule. (Glover and Laguna, 1997)

They also propose a classification method for metaheuristics in terms of three basic design choices: (1) the use of adaptive memory, (2) the kind of neighborhood exploration used, and (3) the number of current solutions carried from one iteration to the next. Therefore, the metaheuristic classification scheme is of the form $\alpha|\beta|\gamma$ was proposed by (Glover and Laguna, 1997) where choices for α are A (if the metaheuristic has adaptive memory) and M (if the method is memoryless). Choices for β are N (for a method used in a somehow systematic neighborhood search) and S (for a method using random sampling). Finally, γ may be 1 (if the method makes one move at each iteration) or P (for a population-based approach with population size of P).

This chapter summarizes several common metaheuristics including SA, TS, GA, ANN, ACO, PSO, and randomized search heuristics.

10.2   Metaheuristic Search Methods

10.2.1   Simulated Annealing

Simulated annealing is an approach based on the early research of Metropolis et al. (1953) for solving combinatorial optimization problems. Essentially, Metropolis's algorithm simulates the change in energy of a system when it is subjected to a cooling process until it converges to a steady "frozen" state. The name of the approach points to a direct analogy with the way that liquids freeze and crystallize or that metals cool and anneal. Kirkpatrick et al. (1983) and Cerny (1985) took these ideas and proposed this approach for combinatorial problems. Since then, SA has been extensively studied and also applied to numerous other areas. Some papers presenting good overviews as well as annotated bibliographies on the topic of SA are Collins et al. (1988), Tovey (1988), and Eglese (1990).

SA searches the set of all possible solutions, reducing the chance of getting stuck in a local optimum by allowing moves to inferior solutions under the control of a randomized scheme. Assuming a minimization problem with solution space S, objective function f, and neighborhood structure N, the general SA algorithm can be stated as follows (Dowsland, 1993):

Select an initial solution s_0;
Select an initial temperature $t_0 > 0$;
Set a temperature change function α;
Repeat
   Repeat
      Randomly select $s \in N(s_0)$;
      $\delta = f(s) - f(s_0)$;
      If $\delta < 0$ then $s_0 = s$

else generate random x uniformly in the range $(0,1)$;

if $x < \exp(-\delta/T)$ then $s = s_0$;
Until *iteration_count* = *nrep*;
Set $t = \alpha(t)$;
Until stopping criterion = true.
s_0 is the approximation to the optimal solution.

The comparison between the random number generated and the probability given by the Boltzmann distribution (Metropolis et al., 1953) is the SA's mechanism to avoid local optimum. According to the global classification scheme that Glover and Laguna (1997) have proposed, SA can be classified as either an M|S|1 or M|N|1 metaheuristic. In other words, SA is a memoryless approach that might employ either neighborhood search or random sampling as it moves from one solution to the next.

Reinelt (1994) states that SA can give very good-quality solutions, but that it may take a considerable amount of time, as the temperature has to drop off very slowly and many replications at each temperature step are necessary. In addition, an issue that should not be underestimated is the proper choice of the annealing scheme, which is highly problem dependent and subject to numerous experiments. An extensive number of SA applications in the combinatorial problem field can be found in literature. Some very good references on common applications are traveling salesman problem (Cerny, 1985; Bonomi and Lutton, 1984), vehicle routing problem (Osman, 1993; Thangiah et al., 1994; Chiang and Russell, 1996), quadratic assignment (Wilhelm and Ward, 1987), graph coloring (Chams et al., 1987), and machine scheduling (Perusch, 1987; Wright, 1988; Ogbu and Smith, 1990).

10.2.2   Tabu Search

The tabu search approach was introduced by Glover (1986). Seminal ideas regarding the method have also been developed by Hansen (1986). The philosophy behind TS search derives from and exploits a collection of principles of intelligent problem-solving. An essential building block underlying TS is the use of flexible memory. From the standpoint of TS, flexible memory consists of the dual processes of creating and exploiting structures for taking advantage of history.

Assuming a minimization problem, the general TS algorithm can be outlined as follows:
Select an initial solution s_0;
Initialize memory structures;
Repeat.
Generate a set A of non-tabu solutions $\in N(s_0)$;
s = best solution of A;
Update memory structures
If $f(s) < f(s_0)$ then $s_0 = s$;
Until stopping criterion = true.
s_0 is the approximation to the optimal solution.

As a metaheuristic, TS can be classified either as A|N|1 or A|N|P (Glover and Laguna, 1997), which means that it has adaptive memory using neighborhood search, and that it moves from one current solution to the next after every iteration or after a population-based approach. According to Laporte et al. (2000), TS-based methods have been the most successful metaheuristics for a wide variety of problems, especially for solving the vehicle routing problem. In general, Reinelt (1994) states that basic difficulties include the design of a reasonable tabu list, the efficient management of the list, and the selection of the most appropriate move that is not forbidden.

Tabu Search is the most widely recognized of the modern heuristics and has been extensively applied to a number of combinatorial problems such as the vehicle routing problem (Rochat and Taillard, 1995; Barbarosoglu and Osgur, 1999; Xu and Kelly, 1996; Willard, 1989; Toth and Vigo, 1998; Rego and Roucairol, 1996), the traveling salesman problem (Malek et al., 1989; Fiechter, 1994), the quadratic assignment problem (Skorin-Kapov, 1990, 1994), and the Multidimensional knapsack problem

(Dammeyer and Voss, 1993; Glover and Kochenberger, 1996). In their book, Glover and Laguna (1997) present a very good discussion of Tabu Search applications on scheduling, the fixed charge transportation problem, manufacturing, optimization on graphs, and others.

10.2.3 Genetic Algorithms

Genetic Algorithms were introduced by Holland (1975) to imitate some of the processes of natural evolution and selection. The GA approach resembles the way in which species adapt to a complicated and changing environment in order to maximize the probability of their survival. The physical attributes that species acquire are encoded in their chromosomes and transmitted to the next generation of species, which will have greater probabilities of survival. This approach could be classified as A|N|P.

The elements of the GA approach consist of the following main components:

- Chromosomal representation. Each chromosome represents a legal solution to the problem and is composed of a string of genes. Normally, the binary system {0,1} is used to represent genes, but sometimes, depending on the application, integers or real numbers are also used.
- Initial population. Once the most suitable representation has been decided for the chromosomes, it is necessary to create an initial population to serve as the starting point for the genetic algorithm. Often, this population is taken randomly or by using specialized, problem-specific information.
- Fitness evaluation. This involves defining an objective or fitness function against which each chromosome is tested for suitability for the environment under consideration. As the algorithm proceeds, the individual fitness of the best chromosome increases, as well as the total fitness of the entire population.
- Selection. The selection procedure consists of picking out two parent chromosomes, on the basis of their fitness value. They are then used by the crossover as mutation operators to produce two offspring for the new population.
- Crossover and mutation. Once a pair of chromosomes has been selected, crossover can take place to produce offspring. A crossover probability of 1 indicates that all the selected chromosomes are used in reproduction. However, it is recommended to use a crossover probability of between 0.65 and 0.85. One-point crossover involves taking the two selected parents and crossing them at a randomly chosen point. A problem may arise if just one crossover operator is used: if the initial population of chromosome has a value in some particular position, applying crossover to all the offspring will have the same value in the same position. To alter this undesirable situation, a mutation operator is used. This attempts to introduce some random alteration of the gene.

The general GA approach can be applied by using the steps shown below:

Choose an initial population of chromosomes;
    Repeat
        Repeat
        If crossover condition is satisfied, then
            Select parent chromosomes;
            Choose crossover parameters;
            Perform crossover;
        If mutation condition is satisfied, then
            Select chromosomes for mutation;
            Choose mutation points;
            Perform mutation;
        Evaluate fitness of offspring;
      Until sufficient offspring created;
      Select new population;

Until stopping criterion = true.

Report approximation population;

Schaffer and Eshelman (1996) state that it is recognized that GA has difficulty solving combinatorial problems because representations that induce good schemata are hard to find (Radcliffe and Surry, 1995). This shortcoming is magnified by the fact that there is not yet a clear theory of schemata (Radcliffe, 1990) (schemata are similarity templates that describe a subset of strings with similarities at certain string positions [Holland, 1975]). The possibility of finding very good, near-optimal, or optimal solutions depends on the size and quality of the initial population. In addition, Chatterjee et al. (1996) state that the crossover operation loses too much information by accident. Also, Reinelt (1994) states that finding optimal solutions for some TSP instances requires considerable CPU time. A classic book on GA was written by Goldberg (1989).

Some application of GA to the traveling salesman problem can be found in Chatterjee et al. (1996), Tamaki et al. (1994), and Whitley et al. (1991). Sequencing and scheduling problems have also been solved using this technique (Reeves, 1995; Lee et al., 1997). Other combinatorial problems that have been solved by using genetic algorithms include the multidimensional knapsack problem (Chu and Beasley, 1998), graph coloring (Davis, 1991), bin packing (Prosser, 1988), and the vehicle routing problem (Thangiah et al., 1994; Potvin and Bengio, 1996; Blanton and Wainwright, 1993; Van Breedam, 2001).

10.2.4 Artificial Neural Networks

It is well understood that all biological functions, including memory, are stored in neurons and in the connections between them. In addition, learning can be seen as the establishment of new connections between neurons or the modifications of existing connections. Biological neurons have been modeled using ANN, a novel contribution in the area of artificial intelligence. A neural network is basically a set of layers containing a number of nodes. Each node of a particular layer receives inputs from nodes belonging to the preceding layer and sends outputs to all nodes belonging to the following layer. A fully interconnected network results.

Although the application of ANN has focused on diagnosis, prediction, classification, pattern recognition, control, etc., there have been some applications to combinatorial problems. In the area of scheduling problems, Gulati and Iyengar (1987) and Arizona et al. (1992) have developed an ANN to solve the single machine problem. Both contributions are extensions of the Hopfield neural network, which is used to solve optimization problems (Hopfield and Tank, 1985). For the due-date tardiness problem, Sabuncuoglu and Hommertzheim (1992) use back propagation ANN. Hayes and Sayegh (1992) adopt a similar approach to solve the two-machine flow shop problem with the total flow time criterion. Kim and Lee (1993) combined a heuristic rule, the apparent tardiness cost (ATC) rule, with an ANN to solve single machine scheduling problems. Bernard et al. (1988) propose self-organizing feature maps to solve the TSP. They based their approach on the work of Kohonen (1984) on self-organizing feature maps. Potvin (1993) provides a very good review of the three types of ANN that can be used for solving the TSP: the Hopfield-Tank neural network, the elastic net, and self-organizing maps. Reinelt (1994) presents a summarized and basic ANN procedure for solving the TSP as follows:

Procedure

Initialize a cycle of m neurons (e.g., $m = 3n$) in the area of the points defining the problem instance.

As long as the stopping criterion is not satisfied, perform the following steps.

- Choose a TSP point i at random.
- According to certain rules, move the neuron whose position is closest to i and some of its neighbors in the neuron cycle towards i.

Construct a tour from the final configuration of neurons.

End of procedure

For the previous algorithm, an ANN can be classified as A|N|1. Reinelt (1994) argues that ANN results are not yet convincing, because the tours generated by his procedure are similar to the nearest insertion tour with respect to structure and quality. See Potvin and Smith (2003) for a further review and discussion of the neural networks used in optimization.

10.2.5   Ant Colony Optimization

The (ACO) metaheuristic defines a particular class of ant algorithms that have been inspired by the behavior of real ant colonies, to solve discrete optimization problems. This novel approach was introduced by Dorigo (1992) and was originally called ant system. If an experiment is conducted while a colony of ants searches for a food source on a ground connected to the colony's nest by a bridge with two branches of different length, after a temporary phase of few minutes, most ants start using the shortest branch. Initially, the ants search their local neighborhood randomly. While going from the nest to the food source and vice versa, ants deposit a certain pheromone on the ground. They make a probabilistic choice biased by the amount of pheromone they smell on the two branches. The pheromone is deposited on the shorter path at a higher rate than on the longer one, making the shorter path more likely to be selected until all ants end up using it (Dorigo and Di Caro, 1999). In general, this approach could be classified as A|N|P.

The pseudocode for the general ACO metaheuristic is given as follows:
Procedure ACO_Metaheuristic
parameter_initialization;
Repeat
    schedule_activities
    ants_generation_and_activity();
    pheromone_evaporation();
    daemon_actions(); //optional
    end schedule_activities
Until stopping criterion = true;
end Procedure.

The procedure ants_generation_and_activity refers to the process of constructing solutions to the optimization problem. The pheromone_evaporation is related to the process of updating the amount of pheromone on the ground. The procedure daemon_actions () of the above algorithm is optional and refers to centralized actions that cannot be executed by single ants (i.e., activation of a local procedure).

The ACO metaheuristic has been successfully applied to a number of classical problems, for example, the traveling salesman problem (Dorigo, 1992; Dorigo and Gambardella, 1997), multidimensional knapsack problems (Leguizamon and Michalewicz, 1999), job shop (Colorni et al., 1994), and the quadratic assignment problem (Maniezzo and Colorni, 1999; Stützle and Hoos, 2000). Dorigo and Stützle's book (Dorigo and Stützle, 2004) is recommended for a full overview of successful implementations.

10.2.6   Particle Swarm Optimization

Particle swarm optimization is an algorithm proposed by Kennedy and Eberhart (1995), motivated by such social behaviors of organisms, as the flocking together of birds or the schooling together of fish. Particle swarm optimization as an optimization tool provides a population-based search procedure in which individuals (here called "particles") change their positions (states) with time. In a PSO system, particles fly around in a multidimensional search space. During flight, each particle adjusts its position according to its own experience and the experience of a neighboring particle, making use of the best position encountered by itself and its neighbor. Thus a PSO system combines local search methods with global search methods, attempting to balance exploration and exploitation.

Particle swarm Optimization is an iterative algorithm that may be classified as A|N|P. Each particle in the swarm occupies a position and has a velocity or rate of change in position. Each particle knows its current position and velocity as well as the position and velocity of the other particles. Also, each particle knows its own previous best position (in term of objective value) as well as the best overall position of any particle (i.e., global best). Knowledge of the global best position, a particle's current position, and a particle's best position will determine a particle's rate of change in position, i.e., the particle's velocity, which in turn will determine the particle's new position. In its simplest form the algorithm can be written as follows:

$$V_{id} = W^*V_{id} + c1^*Rand_1(P_{id} - X_{id}) + c2^*Rand_2(P_{gd} - X_{id})$$

$$X_{id} = X_{id} + V_{id}$$

where V_{id} is the velocity of particle i at iteration d, P_{id} the best previous position of the particle, and P_{gd} the best position found so far or global best. W is the weight, representing how much a particle trusts its own velocity. The terms $c1$ and $c2$ are social cognitive coefficients, where the first is a measure of how much a particle trusts its own best position and the second is a measure of how a particle trusts its global best position. $Rand_{1/2}$ are two independent (0 … 1) random variables. At each iteration of the PSO algorithm, V_{id} is calculated and X_{id}, P_{id}, and P_{gd} are updated. The algorithms stop when a stopping criterion has been reached, or when the maximum number of iterations has been exhausted. The pseudocode for the general PSO metaheuristic is given as follows:

Initialize particle position and velocity for the swarm
Determine best position and assign it to global best
$i = 1$
Do until i = numiterations OR convergence criteria = TRUE
      Calculate new velocities for swarm
      Calculate new positions for swarm
      Calculate new objective function values for swarm
      Determine best previous position for each particle
      Determine global best
      $i = i+1$
      Check if convergence criteria = TRUE
Loop
Show results

Particle swarm optimization has been applied to the traveling salesman problem (Wang et al., 2003; Onwubolu and Clerc, 2004), machine scheduling (Tasgetiren et al., 2004; Xia et al., 2004), train scheduling (Chang and Kwan, 2004), and vehicle routing (Zhao et al., 2004).

10.2.7 Randomized Search Metaheuristics

Metaheuristic for randomized priority search (Meta-RaPS) (DePuy and Whitehouse, 2000) and Greedy randomized adaptive search procedure (GRASP) (Feo and Resende, 1989, 1995) are two generic, high-level search procedures that introduce randomness to a construction heuristic as a device to avoid getting trapped at a local optimal solution. Both approaches may be classified as M|S|1. Because Meta-RaPS is thought to be a more general form of GRASP and because Meta-RaPS was developed by the authors of this chapter, it will be the main focus of this section.

Meta-RaPS, which evolved from a computer heuristic COMSOAL (computer method of sequencing operations for assembly lines) designed by Arcus (1966), is an iterative procedure that constructs a feasible solution through the utilization of a construction heuristic infused with randomness and an improvement heuristic. The best solution is then reported after a number of iterations.

In general, a construction heuristic builds a solution by repeatedly adding feasible components or activities to the current solution until a stopping criterion is met. The order in which the activities are added to the solution is based on their priority values as determined by the construction heuristic's priority rule. The priority value of each feasible activity is calculated and the activity with the best or most desirable priority value is then added to the current solution. For example, a TSP construction heuristic builds a solution by repeatedly adding unvisited cities (i.e., feasible activities) to the route (i.e., solution) until all the cities are included in the route (i.e., stopping criteria). In most TSP construction heuristics, the order in which the cities are added to the route is based on minimizing the total route distance (i.e., priority rule) by calculating the increase in route distance (i.e., priority value) for each feasible city and inserting the city with the minimum increase in route distance (i.e., best priority value). Solutions generated using a construction heuristic are usually good, but rarely optimal, as the construction heuristic often generates a local optimum solution. The use of Meta-RaPS to include randomness in a construction heuristic frequently allows the global optimum solution to be found.

Meta-RaPS modifies the way a general construction heuristic chooses the next activity to add to the solution by occasionally choosing an activity that does not have the best priority value. Sometimes, the next activity is chosen randomly from those feasible activities with good, but not the best, priority values. Meta-RaPS uses two user-defined parameters, %priority and %restriction, to include randomness in a construction heuristic. The %priority parameter is used to determine how often the next activity added to the solution has the best priority value. The remaining time (100%–%priority), the next activity added to the solution, is randomly chosen from an "available" list of those feasible activities with good priority values. Good priority values are determined using the %restriction parameter. Those feasible activities whose priority values are within %restriction of the best priority value are included on the "available" list. Again, an activity is then randomly selected from this list to be added to the solution next. Therefore, each application or iteration of the heuristic will most likely produce a different solution because the same activity will not be randomly selected from the "available" list each time the heuristic is used.

The pseudocode for one iteration of the basic Meta-RaPS procedure is given as follows:

    Do until feasible solution generated
    Find priority value for each feasible activity
    Find best priority value
    P = RND(1,100)
    If $P \le$ %priority Then
            Add activity with best priority value to solution
    Else
    Form "available" list of all feasible activities whose priority values are within
            % restriction of best priority value
            Randomly choose activity from available list and add to solution
    End If
    End Until
    Calculate and Print solution

In addition, a solution improvement algorithm can be included in Meta-RaPS using a %improvement parameter. If an iteration's solution value is within %improvement of the best unimproved solution value found so far, then an improvement heuristic (neighborhood search) is performed.

Meta-RaPS is a general solution approach that has been successfully applied to a variety of combinatorial optimization problems including the traveling salesman problem (DePuy et al., 2005), the multidimensional knapsack problem (Moraga et al., 2005), the resource-constrained project scheduling problem (Whitehouse et al., 2002), and the set covering problem (Lan et al., 2004). As mentioned previously, Meta-RaPS and GRASP are similar in nature and GRASP has been applied to variety of combinatorial problems as well, including machine scheduling (Rios-Mercado and Bard, 1998; Feo et al., 1996), quadratic assignment problem (Mavridou et al, 1998), aircraft routing (Arguello, et al., 1997) and fleet scheduling (Sosnowska, 2000).

10.3 Metaheuristics as a Field

The field of metaheuristics has become relevant for the research community, to such an extent that it counts on its own biannual permanent conference since 1995, the Metaheuristics International Conference (MIC). A special volume is normally published that collects selected papers (Ibaraki et al., to appear; Resende M and Pinho de Sousa, 2003; Ribeiro and Hansen, 2001; Voss et al., 1998; Osman and Kelly, 1996). In more specific areas of metaheuristics, there are also important conferences such as the International Conference on Artificial Immune Systems (ICARIS), which has met annually since 2002 (www.artificial-immune-systems.org); the ANTS Workshop Series, which has met biannually since 1998 (http://iridia.ulb.ac.be); and the Genetic and Evolutionary Computation Conference (GECCO; http://isgec.org). In many IE-related conferences, there are also application-centered tracks focused on the use of metaheuristics, for example, the Industrial Engineering Research Conference (IERC), the International Conference on Computers and Industrial Engineering (ICC&IE), and the Winter Simulation Conference (WSC).

Glover and Kochenberger (2003) edited the *Handbook of Metaheuristics* in which they review a vast number of techniques and applications. Also, a book edited by Resende and Pinho de Sousa (2004) entitled *Metaheuristics: Computer Decision-Making* presents a very good review of various approaches and their applications. These sources are strongly recommended for a good background on the topic of metaheuristics.

10.4 Comments and Conclusions on Metaheuristics

Metaheuristics have gained a lot of attention from the research community and practitioners because they are not difficult to implement, because most of them can be seen as general purpose approaches, and because if they do not succeed in finding optimal solutions, they are often within a small deviation from the optimum. For example, in the case of the multidimensional knapsack problem, many of the revised approaches are within a deviation range of 0 to 1.7% with respect to the optimal solutions. For larger size and harder problems (Chu and Beasley, 1998), some of them fluctuate within a range of 0.05 to 12% (Moraga, et al., 2005). Therefore, metaheuristics constitute an important solution approach for many practical optimization problems.

References

Arcus A., COMSOAL: A computer method of sequencing operations for assembly lines, I the problem in simple form, in *Readings in Production and Operations Management*, Buffa, E., Ed., Wiley, New York, 1966.

Arguello, M., Bard, J., and Yu, G., A GRASP for aircraft routing in response to groundings and delays, *J. Combinatorial Optimization*, 1, 211–228, 1997.

Arizona, I., Yamamato, A., and Ohto, H., Scheduling for minimizing total actual flow time by neural networks, *Int. J. Prod. Res.* 30, 3, 1992.

Barbarosoglu, G. and Ozgur, D., A tabu search algorithm for the vehicle routing problem, *Comp. and Oper. Res.*, 26, 255–270, 1999.

Bernard, A., De La Croix, G., and Le Texier, J., Self–organizing feature maps and the travelling salesman problem, *Neural Networks*, 1, 289–293, 1988.

Blanton, J. and Wainwright, R., Multiple vehicle routing problems with times and capacity constraints using genetic algorithms, in *Proceedings of the 5th International Conference on Genetic Algorithms*, Forrest, S., Ed., Morgan Kaufmann, San Mateo, CA, 1993, pp. 452–459.

Bonomi, E. and Lutton, J., The N-city traveling salesman problem, statistical mechanics and the metropolis algorithm, *SIAM Rev.*, 26, 551–568, 1984.

Cerny, V., Thermodynamical approach to the traveling salesman problem: an efficient simulation algorithm, *J. Optim. Theory Appl.*, 45, 41–51, 1985.

Chams, M., Hertz, A., and Werra, D., Some experiments with simulated annealing for colouring graphs, *Eur. J. Oper. Res.*, 32, 260–266, 1987.

Chang, C.S. and Kwan, C.M., Evaluation of evolutionary algorithms for multi-objective train schedule optimization, *Lecture Notes Artificial Intelligence*, 3339, 803–815, 2004.

Chatterjee, S., Carrera, C., and Lynch, L., Genetic algorithms and traveling salesman problems, *Eur. J. Oper. Res.*, 93, 490–510, 1996.

Chiang, W.-C. and Russell, R., Simulated annealing metaheuristics for the vehicle routing problem with time windows, *Ann. Oper. Res.*, 63, 3–27, 1996.

Chu, P.C. and Beasley, J.E., A genetic algorithm for the multidimensional knapsack problem. *J. Heuristics*, 4, 63–86, 1998.

Collins, N., Eglese, R., and Golden, B., Simulated annealing — an annotated bibliography, *Am. J. Math. Manage. Sci.*, 8, 209–307, 1988.

Colorni, A., Dorigo, M., Maniezzo, V., and Trubian, M., Ant system for job-shop scheduling, *JORBEL — Belgian J. Oper. Res., Stat. Comp. Sci.*, 34, 39–53, 1994.

Dammeyer, F. and Voss, S., Dynamic tabu list management using the reverse elimination method, *Ann. Oper. Res.*, 41, 31–46, 1993.

Davis, L., *Handbook of Genetic Algorithms*, Van Nostrand Reinhold, New York, 1991.

DePuy, G.W. and Whitehouse, G.E., Applying the COMSOAL computer heuristic to the constrained resource allocation problem, *Comput. Ind. Eng.*, 38, 3, 413–422, 2000.

DePuy, G.W., Moraga, R.J., and Whitehouse, G.E., Meta-RaPS: A Simple And Effective Approach For Solving The Traveling salesman problem, *Transp. Res. Part E: Logistics Transp. Rev.*, 41, 2, 115 –130, 2005.

Dorigo, M. and Di Caro, G., The Ant Colony Optimization Metaheuristic, in *New Ideas in Optimization*, Corne, D., Dorigo, M., and Glover, F., Eds., McGraw-Hill, New York, 1999, pp. 11–32.

Dorigo, M. and Gambardella, L.M., Ant colonies for the traveling salesman problem, *BioSystems*, 43, 73–81, 1997.

Dorigo, M. and Stützle, T., *Ant Colony Optimization*, MIT Press, London, 2004.

Dorigo, M., Optimization, Learning and Natural Algorithms, Ph.D.thesis, Politecnico di Milano, Italy, 1992, (in Italian).

Dowsland, K.A., Simulated annealing, in *Modern Heuristic Techniques for Combinatorial Problems*, Reeves, C., Ed., Wiley, New York, 1993.

Eglese, R., Simulated annealing: a tool for operational research, *Eur. J. Oper. Res.*, 46, 271–281, 1990.

Feo, T. and Resende, M., A probabilistic heuristic for a computationally difficult set covering problem, *OR Lett.*, 8, 67–71, 1989.

Feo, T. and Resende, M., Greedy randomized adaptive search procedures, *J. Global Optimization*, 6, 109–133, 1995.

Feo, T., Sarathy, K., McGahan, J., A GRASP for single machine scheduling with sequence dependent setup costs and linear delay penalties, *Comput. Oper. Res.*, 23, 881–895, 1996.

Fiechter C.-N., A parallel tabu search algorithm for large traveling salesman problems, *Discrete Appl. Math.*, 51, 243–267, 1994.

Glover, F. and Kochenberger, G.A., *Handbook of Metaheuristics. International Series in Operations Research & Management Science*, vol. 57, Kluwer Academic Publishers, Boston, 2003.

Glover, F., and Kochenberger, G., Critical event tabu search for multidimentional knapsack problems, in *Metaheuristics: Theory and Applications*, Osman, I.H., and Kelly, J.P., Eds., Kluwer Academic Publishers, Boston, 1996, pp. 407–427.

Glover, F. and Laguna, M., *Tabu Search*, Kluwer Academic Publishers, Dordrecht, 1997.

Glover, F. Future paths for integer programming and links to artificial intelligence, *Comp. and Oper. Res.*, 5, 533–549, 1986.

Goldberg, D.E., *Genetic Algorithms in Search, Optimization, and Machine Learning*, Addison-Wesley, Reading, MA., 1989.

Gulati, D. and Iyengar, S., Nonlinear networks for deterministic scheduling, *Proceedings of the ICNN*, San Diego, CA, 1987, pp. 745–752.

Hansen, P., The Steepest Ascent Mildest Descent Heuristic for Combinatorial Programming, Talk presented at the Congress on Numerical Methods in Combinatorial Optimization, Capri, 1986.

Hayes, P. and Sayegh, S., A supervised neural network approach to optimization as applied to the n-job, m-machine job sequencing problem, *Proceedings of the 1992 Artificial Neural Networks in Engineering, ANNIE'92*, St. Louis, MO, 1992.

Holland, J., *Adaptation in Natural and Artificial Systems*, University of Michigan Press, Ann Arbor, MI, 1975.

Hopfield, J. and Tank, D., Neural computation of decisions in optimization problems, *Biol. Cybern.*, 52, 141–152, 1985.

Ibaraki, T., Nonobe, K., and Yagiura, M., Eds., *Metaheuristics: Progress as Real Problem Solvers*, Springer, Kluwer Academic Publishers, Boston, 2005.

Ingnizio, J. and Cavalier, T., *Linear Programming*, Prentice-Hall, Englemod Cliffs, NJ, 1994.

Kennedy, J. and Eberhart, R., Particle Swarm Optimization, *IEEE Int. Conf. Proc.*, 4, 942–1948, 1995.

Kim, S. and Lee, Y., Enhancement of a job sequencing rule using an artificial neural network, *Proceedings of the 2nd Industrial Engineering Research Conference*, Norcross, GA, 1993.

Kirkpatrick, S., Gelatt, C.,Jr., and Vecchi, M., Optimization by simulated annealing, *Science*, 222, 671–680, 1983.

Kohonen, T., Self-organized formation of feature maps, in Caianiello, E.R., and Musso, G., Eds., Cybernetic Systems: Recognition, Learning, Self-Organiszation, Research Studies Press, Letchworth, UK, 1984, pp. 3–12.

Lan, G., DePuy, G.W., and Whitehouse, G.E., An Effective and Simple Heuristic for the Set Covering Problem, in review by *Eur. J. Oper. Res.*, 2004.

Laporte, G., Gendreau, M., Potvin, J., and Semet, F., Classical and modern heuristics for the vehicle routing problem, *Int. Trans. Oper. Res.*, 7, 285–300, 2000.

Lee, C.-Y., Piramuthu, S., and Tsai, Y.-K., Job shop scheduling with a genetic algorithm and machine learning, *Int. J. Prod. Res.*, 35, 1171–1191, 1997.

Leguizamon, G. and Michalewicz, Z., A new version of ant system for subset problem, in *Proceedings of the Congress on Evolutionary Computation (CEC'99)*, Angeline, P.J., Michalewicz, Z., Schoenauer, M., Yao, X., and Zalzala, A., Eds., IEEE Press, Piscataway, NJ, 1999, pp. 1459–1464.

Malek, M., Guruswamy, M., Pandya, M., and Owens, H., Serial and parallel simulated annealing and tabu search algorithms for the traveling salesman problem, *Ann. Oper. Res.*, 21, 59–84, 1989.

Maniezzo, V. and Colorni, A. The ant system applied to the quadratic assignment problem, *IEEE Trans. Knowledge Data Eng.*, 11, 769–778, 1999.

Mavridou, T., Pardalos, P., Pitsoulis, L., Resende, M., A GRASP for the bi-quadratic assignment problem, *Eur. J. Oper. Res.*, 105, 613–621, 1998.

Metropolis, N., Rosenbluth, A., Rosenbluth, M., Teller, A., and Teller, E., Equation of state calculation by fast computing machines, *J. Chem. Phys.*, 21, 1087–1092, 1953.

Moraga, R.J., DePuy, G.W., and Whitehouse, G.W., Meta-RaPS approach for the 0-1 multidimensional knapsack problem, *Comput. Ind. Eng.*, 48(2), 83–96, 2005.

Ogbu, F., Smith, D., The application of the simulated annealing algorithm to the solutions of the n/m/Cmax flowshop problem, *Comp. and Oper. Res.*, 17, 243–253, 1990.

Onwubolu, G.C. and Clerc, M., Optimal path for automated drilling operations by a new heuristic approach using particle swarm optimization, *Int. J. Prod. Res.*, 42(3), 473–491, 2004.

Osman, I. and Kelly, J., Eds., *Metaheuristics: The Theory and Applications*, Kluwer Academic Publishers, Boston, 1996.

Osman, I., Metaestrategy simulated annealing and tabu search algorithms for the vehicle routing problem, *Ann. Oper. Res.*, 41, 421–451, 1993.

Perusch, M., Simulated annealing applied to a single machine scheduling problems with sequence-dependent setup times and due dates, *Belgian J. Oper. Res. Stat. Comp. Sci.*, 27, 1987.

Potvin, J.M. and Bengio, S., The vehicle routing problem with time windows — Part II: genetic search, *INFORMS J. Computing*, 8, 165–172, 1996.

Potvin, J.-Y. and Smith, K.A., Artificial neural networks for combinatorial optimization, in *Handbook of Metaheuristics*, Glover, F., and Kochenberger, G.A., Eds., International Series in Operations Research and Management Science, Kluwer Academic Publishers, Boston, 2003, p. 57.

Potvin, J.-Y., The traveling salesman problem: a neural network perspective, *ORSA J. Computing*, 5, 328–348, 1993.

Prosser, P., A hybrid genetic algorithm for pallet loading, *Proceedings of the 8th European Conference on Artificial Intelligence*, Pitman, London, 1988.

Radcliffe, N. and Surry, P., Fitness variance of formae and performance prediction, in *Foundations of Genetic Algorithms*, Whitley, D. and Vose, M., Eds., 3, Morgan Kaufmann, San Mateo, CA, 1995, pp. 51–72.

Radcliffe, N., Genetic Neural Networks on MIMD Computers, Ph.D. thesis, Department of Theoretical Physics, University of Edinburgh, Edinburgh, Scotland, 1990.

Reeves, C.R., A genetic algorithm for flowshop sequencing, *Comp. Oper. Res.*, 22, 5–13, 1995.

Rego, C. and Roucairol, C., A parallel tabu search algorithm using ejection chains for the vehicle routing problem, in *Metaheuristics: Theory and Applications*, Osman, I.H. and Kelly, J.P., Eds., Kluwer, Boston, 1996.

Reinelt, G., *The Traveling Salesman: Computational Solutions for TSP Applications*, Lectures Notes in Computer Science, Springer, Berlin, 1994.

Resende, M. and Pinho de Sousa, J., Eds., *Metaheuristics: Computer Decision-Making*, Applied Optimization, vol. 86, Kluwer Academic Publishers, Boston, USA, 2003.

Resende, M.G. and Pinho de Sousa, J., *Metaheuristics: Computer Decision-Making. Series: Applied Optimization*, vol. 86, Kluwer Academic Publishers, Boston, 2004.

Ribeiro, C. and Hansen, P., Eds., *Essays and Surveys in Metaheuristics*, Operations Research/Computer Science Interfaces, vol. 15, Kluwer Academic Publishers, Boston, 2001.

Rios-Mercado, R. and Bard, J., Heuristics for the flow line problem with setup costs, *Eur. J. Oper. Res.*, 110 (1), 76–98, 1998.

Rochat, Y. and Taillard, E., Probabilistic diversification and intensification in local search for vehicle routing, *J. Heuristics*, 1(1), 147–167, 1995.

Sabuncuoglu, I. and Hommertzheim, D., Artificial neural networks: Investigations and developments of neural network for scheduling problems, paper presented at 1992 TIMS/ORSA Joint National Meeting, Orlando, 1992.

Schaffer, J. and Eshelman, L., Combinatorial optimazation by genetic algorithms: the value of the genotype/phenotype distinction, in *Modern Search Methods*, Rayward-Smith, V.J., Osman, I.H., Reeves, C.R., and Smith, G.D., Eds., Wiley, New York, 1996.

Skorin-Kapov, J., Extensions of tabu search adaptation to the quadratic assignment problem, *Comp. Oper. Res.*, 21, 855–865, 1994.

Skorin-Kapov, J., Tabu search applied to the quadratic assignment problem, *ORSA J. Computing*, 2, 33–45, 1990.

Sosnowska, D., Optimization of a simplified fleet assignment problem with metaheuristics: simulated annealing and GRASP, in *Approximation and Complexity in Numerical Optimization*, Pardalos, P., Ed., Kluwer, Academic Publishers, Boston, 2000.

Stützle, T. and Hoos, H., Max-min ant system, *Future Generation Comp. Sys.*, 16, 8, 889–914, 2000.

Tamaki, H., Kita, H., Shimizu, N., Maekawa, K., Nishikawa, Y., Comparison study of genetic codings for the Traveling salesman problem, *IEEE Conference on Evolutionary Computation*, Piscataway, NJ, 1994, pp. 1–6.

Tasgetiren, M.F., Sevkli, M., Liang, Y.C., Gencyilmaz, G., Particle swarm optimization algorithm for permutation flowshop sequencing problem, in *Ant Colony Optimization and Swarm Intelligence*, Lecture Notes in Computer Science, vol. 3172, Springer, Berlin, 2004, pp. 382–389.

Thangiah, S., Osman, I., and Sun, T., Hybrid Genetic Algorithm, Simulated Annealing and Tabu Search Methods for Vehicle Routing Problem with Time Windows, Technical Report 27, Computer Science Department, Slippery Rock University, 1994.

Toth, P. and Vigo, D., Granular Tabu Search, Working Paper, DEIS, University of Bologna, 1998.

Tovey, C., Simulated annealing, *Am. J. Math. Manage. Sci.*, 8, 389–407, 1988.

Van Breedam, A., Comparing descent heuristics and metaheuristics for the vehicle routing problem, *Comp. Oper. Res.*, 28, 289–315, 2001.

Voss, S., Martello, S., Osman, I., and Roucairol, C., Eds., *Metaheuristics: Advances and Trends in Local Search Paradigms for Optimization*, Kluwer Academic Publishers, Boston, 1998.

Wang, K.-P., Huang, L., Zhou, C.-G., and Pang, W., Particle swarm optimization for traveling salesman problem, *International Conference on Machine Learning and Cybernetics*, vol. 3, 2003, pp. 1583–1585.

Whitehouse, G.E., DePuy, G.W., and Moraga, R.J., Meta-RaPS approach for solving the resource allocation problem, *Proceedings of the 2002 World Automation Congress*, June 9–13, Orlando, FL, 2002.

Whitley, D., Starkweather, T., and Shaner, D., The traveling salesman problem and sequence scheduling: quality solutions using genetic edge recombination, in *Handbook of Genetic Algorithms*, Davis, L., Ed., Van Nostrand Reinhold, New York, 1991.

Wilhelm, M. and Ward, T., Solving quadratic assignment problems by simulated annealing, *IIE Trans. Ind. Eng. Res. Dev.*, 19, 107–119, 1987.

Willard, J., Vehicle rRouting Using r-Optimal Tabu Search, M.Sc. thesis, The Management School, Imperial College, London, 1989.

Wright, M., Applying stochastic algorithms to a locomotive scheduling problem, *J. Oper. Res. Soc.*, 40, 101–106, 1988.

Xia, W., Wu, Z., Zhang, W., and Yang, G., Applying particle swarm optimization to job-shop scheduling problem, *Chinese J. Mech. Eng. (English Edition)*, 17(3), 437–441, 2004.

Xu, J. and Kelly, J.P., A network flow-based tabu search heuristic for the vehicle routing problem, *Transp. Sci.*, 30, 379–393, 1996.

Zhao, Y.W., Wu, B., Wang, W.L., Ma, Y.L., Wang, W.A., and Sun, H., Particle swarm optimization for vehicle routing problem with time windows, *Adv. Mat. Manuf. Sci. Tech. Mat. Sci. Forum*, 471–472, 801–805, 2004.

Multidisciplinary Systems Teams

Craig M. Harvey
Louisiana State University

Taren Daigle
Louisiana State University

Ashok Darisipudi
Louisiana State University

Ling Rothrock
Pennsylvania State University

Larry Nabatilan
Louisiana State University

11.1 Introduction

What makes a team successful? This is the million-dollar question, and it remains unanswered today. However, there are many factors believed to contribute to team success. Let us begin this chapter, by exploring three case examples where teams have been unsuccessful and the factors that led to their failure.

11.1.1 Case Study 1: Team Communication

Communication among teams and team members is crucial to team performance. One recent case within the medical domain illustrates this further.

Hospital Had Data Before Transplant Error
Associated Press
Thursday, February 20, 2003

DURHAM, N.C., Feb. 19 — Duke University Hospital surgeons declined a heart and lungs offered for transplant in two cases before one doctor requested them — but for a teenager with the wrong blood type, two organ-procurement agencies said.

The girl's surgeons may also have committed themselves to the transplant too early, removing her own damaged organs before the replacements arrived and the mismatch could be discovered.

Correct information about the blood type was given to a Duke surgical team that flew to Boston to extract the donor organs Feb. 7, said the New England Organ Bank, the Newton, Mass., organization that offered the organs. Duke hospital officials had no comment today on why doctors sought the type-A organs for Jesica Santillan, a type-O-positive patient who is now near death as her body rejects the transplants......(quote taken from Associated Press, 2003)

Source: Associated Press Report on Transplant Error

What can we draw from this case? First and foremost, it appears that there was miscommunication as to the blood type of the patient. While the article states that the surgical team that picked up the transplant had the correct information, a type-A blood donor's organs were harvested. Although organ transplants are conducted every day, they are still very risky procedures; however, their risks pale when compared with the alternative, which in this case was doing nothing and inviting certain death for the patient. Patients who have transplants are prone to rejection of the organ(s) and infections related to the surgery, as well as other medical complications that may be exacerbated by the surgery. Using organs of the wrong blood type simply adds to the already mountains of potential complications.

So why was there a miscommunication? Why was it not clear to all parties involved that these organs could not be used? Duke University later reported that the lead surgeon failed to confirm that the organs were of the right blood type (Stein, 2003). While the surgeon was ultimately the person responsible, there were many team members in the chain that handled this case who could have made this information available to the appropriate person. The surgeon was not the only person in the operating room and not the only person to come in contact with the information about these organs.

Second, some may question why the surgeons removed Jesica's organs prior to confirmation that the donated organs had been received and that they could be used. Once again, was this the result of poor communication, hospital procedures, or a combination of both?

11.1.2   Case Study 2: Shared Expectations

In team dynamics, teams consisting of group members with the same beliefs and expectations tend to perform better. The following case illustrates the impact of teams when they do not share the same beliefs and expectations.

On Dec. 11, 1998, atop a Delta II launch vehicle from the Cape Canaveral Air Force Station, Florida, the Mars Climate Orbiter was launched. After nine and a half months, on Sept. 23, 1999 the $125 million Mars Climate Orbiter mission was lost when it entered the Martian atmosphere on a lower than expected trajectory. Why this had happened was a mystery to NASA at first. A subsequent investigation of the Mars Climate Orbiter (NASA, 1999a) found that the root cause of the MCO spacecraft was "failure to use metric units in the coding of the round software file, 'Small Forces,' used in trajectory models."

Because the two virtual design teams working on this project did not have a shared perspective, and because one team used English units and one team used metric units, the Mars Climate Orbiter was destroyed during its mission to Mars. This event illustrates that teams have to have a shared understanding in order to perform successfully. This shared understanding can be explicit (e.g., through communication) or implicit (e.g., shared because the teams have had the same history).

11.1.3   Case Study 3: Team Conflict

Conflicts are a natural process within any team. In fact, conflicts can help teams question their decisions and ultimately assure that teams do not develop problems such as groupthink, or the unconscious molding of a team into the adoption of one "acceptable" opinion. It is vital that the management of teams listen to their teams' concerns so that they can be addressed instead of brushing them away without any consideration. The following tragedy illustrates what can happen when management is unable to listen to team concerns.

On January 28, 1986, the Space Shuttle *Challenger* exploded and seven astronauts died because two rubber O-rings leaked. These rings had lost their resiliency because the shuttle was launched on a very cold day. Ambient temperatures were in the low 30's and the O-rings themselves were much colder, less than 20°F.

Concerned that the rings would not seal at such a cold temperature, the engineers who designed the rocket opposed launching the *Challenger* the next day. Their misgivings derived from several sources: a history of O-ring damage during previous cool weather launches of the shuttle, the physics of resiliency (which declines exponentially with cooling), and experimental data. Presented in 13 charts this evidence was faxed to NASA where a high level official responded that he was "appalled" by the recommendation not to launch and indicated that the rocket maker Morton Thiokol should reconsider even though this was Thiokol's only no-launch recommendation in 12 years. Reassessing the situation after skeptical responses, the Thiokol managers changed their minds and decided that they now favored launching the next day. That morning the *Challenger* blew up 73 seconds after its rockets were ignited (adapted from Tufte's "Visual Explanations", 1997).

This accident, according to sociologists, is a symptom of structural history, bureaucracy, and conformity to organizational norms. Taken in small doses, the assorted interpretation of the launch decisions is plausible and rarely mutually exclusive. Breakdown of communications, different views of risk, and external and internal pressures to launch all contributed to the tragic event.

This case also illustrates the effect of conflicts and arguments, prevalent in teams, which may sometimes persist unresolved and lead to disastrous consequences. These conflicts placed two domains of authority with disparate spheres of influence, the engineers and the managers, in the position of arguing (over whether the *Challenger* could safely be launched in cold weather). Despite the engineers' knowledge of the possibility of failure, the managers, who were comfortable with their subjective methods of risk assessment, held misconceptions concerning the importance of the O-ring problem. Thiokol also felt the need to eliminate further delays with the *Challenger*. Managers were in the business of administration and not in engineering, as is illustrated by the comments of Jerald Mason, a manager at Thiokol, who said to a subordinate, "Take off your engineering hat and put on your management hat" (Adler, 1988, p. 29). This comment suggested that management should disregard the concerns of engineers in order to keep the program in motion and to prevent further delay of the launch of the *Challenger*. That decision had long-term repercussions in the management structure at NASA.

This chapter will examine the elements that impact team performance. We will use the input–process–output (IPO) model as a means of discussing the different elements that affect teams in the workplace. Paris et al. (2000, p. 1052) help set the agenda for this chapter on teams:

"Transforming teams of experts into expert teams necessarily begins with an understanding of what characteristics uniquely define the team and set it apart from small groups".

11.2 Historical Perspective

Organizations have used teams for centuries and longer. The great pyramids of Egypt could not have been constructed without teams of designers, engineers, masons, and laborers (Shenhar, 1999). However, the study of teams and their design structure is a much newer concept, and it was not until the 1900s that interest in this subject in any widespread sense began to grow. In the 1940s, group dynamics research became a popular field. The research focused mostly on psychological and emotional aspects of training groups, self-study groups, and therapy groups (Gersick, 1988). Group tasks were mostly studied in terms of personal gains such as relationships and interpersonal skills (Mills, 1979). A group's ability to handle conflict, maintain control, and develop a sense of intimacy was the main measure of group performance over the life span of the group (Bennis and Shepard, 1956; Bion, 1961; Mann et al., 1967).

During the 1950s and 1960s, a movement to further understand team performance based on several other process variables began (Paris et al., 2000; McGrath, 1990). Often military teams were chosen, in part owing to the high stress, extreme time/pressure constraints, and severe consequences for actions that accompanied this particular type of team (Ilgen, 1999; Annett and Stanton, 2000). Military teams continue to be a popular test group for team communication, training, coordination, leadership, and overall performance research (Bowers et al., 1994; Achille et al., 1995; Leedom and Simon, 1995; Salas et al., 1995; Ehrlich et al., 1997; Cannon-Bowers et al., 1998). In Salas et al.'s (1995) *Military Team Research: 10 years of progress,* presents a review of military team literature from the 1980s and early 1990s and notes some of the important research contributions to overall team knowledge such as theoretical advancements, a deeper understanding of process variables, including implicit coordination and communication, and individual and team training techniques.

Other areas of team research include understanding and generalizing team development and evaluating the team's overall effectiveness. Tuckman (1965) introduced a model of group development as a sequence of phases. This model outlined four phases of group progress, namely, forming, storming, norming, and performing. Later, the model was updated and a final stage was added: adjourning (Tuckman and Jensen, 1977). Tuckman indicates that the model offers a snapshot of groups in their ever-changing life spans; it cannot, however, encompass the actual transition periods between the stages, or predict how long the group will remain in each stage. Some researchers have criticized Tuckman's sequential model, arguing that groups evolve in iterative cycles rather than through a linear order (Scheidel and Crowell, 1964; Fisher, 1970; Gersick, 1988). Gersick (1988) also points out that Poole (1981, 1983a, 1983b) raised the most serious challenge by suggesting that there are several possible sequences that groups use in their development, not just one.

Another model developed to illustrate the varying factors in team design and performance (outcomes) is the IPO paradigm (McGrath, 1964). Inputs to the model include influences from external factors, such as individual team member characteristics, team design, and task design (McGrath, 1964). The middle phase consists of process variables that incorporate actions and interactions, such as cooperation, communication, a shared mental model, and problem-solving skills, into the team (Kinlaw, 1987; Lajoie and Sterling, 1999; Driskell and Salas, 1992). The output sector focuses on end results of the group, specifically member satisfaction, productivity, and overall team commitment, among others.

As researchers continue to expand existing models and develop new and different ways to classify, identify, and evaluate teams, we can expect that the areas of team research will also modify and adjust to meet the ever-changing demands and interests.

11.3   Defining Teams

The words "team" and "group" are both equally prevalent in team literature. But most of the popular management literature uses the term "team" (e.g., team effectiveness, marketing teams) where as the academic literature uses the word "group" (e.g., group cohesion, group dynamics). According to some researchers, groups vary in their degree of "groupness" (how much the team/group members are dependent on each other), with some groups being more interdependent and integrated than others. According to Brannick and Prince (1997), teams can be distinguished from small groups, as teams have unique requirements for coordination and task interdependency. Some authors use "team" for groups that have a high degree of "groupness" (Katzenbach and Smith, 1993). In other words, groups become teams when they develop a sense of shared commitment and strive for synergy among the members (Guzzo and Dickson, 1996). Before going into the details about teams, let us define the term "team." Multiple definitions exist, as is illustrated in Table 11.1.

Arrow et al.'s (2000) definition of team is a very comprehensive one. Their definition is based on the synthesis of a vast literature on teams and small groups. They take into account the complex, adaptive, and dynamic nature of teams along with coordination and relationships among team members to define teams. The most significant aspect of their approach is that they consider relationships among team

TABLE 11.1   Team Definitions

Team Definition	Reference
Teams consist of two or more individuals, who have specific role assignments, perform specific tasks and who must interact or coordinate to achieve a common goal or outcome.	Baker and Salas (1997)
Teams consist of two or more individuals who make decisions.	Orasanu and Salas (1993)
Teams consist of two or more individuals who have specialized knowledge and skills.	Cannon-Bowers et al. (1995)
A team is a bounded system composed of a set of interdependent individuals organized to perform specific tasks that affect others.	Guzzo and Dickson (1996)
A team is a collection of individuals who are interdependent in their tasks, who share responsibility for outcomes, who see themselves and who are seen by others as an intact social entity embedded in one or more larger social systems (e.g., business unit or the corporation), and who manage their relationships across organizational boundaries.	Cohen and Bailey (1997)
A team is a complex, adaptive, dynamic, coordinated, and bounded set of patterned relations among team members, tasks, and tools.	Arrow et al. (2000)

members, tasks, and tools in their definition. They define "teams" as not merely a group of people who work together on a common objective and share the work responsibilities, but instead consider the tasks of the team and tools available to the team.

11.3.1   Typology of Teams

According to Cohen and Bailey (1997), four types of teams can be identified in organizations today: (1) work teams; (2) parallel teams; (3) project teams; and (4) management teams. Other researchers, such as Sundstrom et al. (1990), used integration and differentiation as the taxonomy to differentiate the four types of groups. They define teams as (1) advice and involvement groups; (2) production and service teams; (3) project and development teams; and (4) action and negotiation teams. Although Cohen and Bailey (1997) and Sunderstrom et al. (1990) offer different typologies in identifying different teams, their categories overlap with each other. For example, work teams correspond to production and service teams, parallel teams correspond to advice and involvement teams, and project teams correspond to project and development teams. Similarly, management teams correspond to action and negotiation teams. Thus, while the names may differ, their definitions are very similar. Table 11.2 provides a brief explanation of Cohen and Bailey's (1997) four types of teams.

Teams come in many forms. Problem-solving, special-purpose, and self-managing teams are a few examples of how teams can be used to increase employee involvement in company decision making and improve quality, efficiency, and work environment (Lawler, 1986; Lawler, 1995; Hoerr, 1989). Handovers in the medical field, shift changes at a nuclear power plant, and new product design teams are some of the numerous uses for teams in various industries (Matthews et al., 2002).

Teams can be permanent or temporary, dependent or autonomous, small or large, homogeneous or heterogeneous, and real or virtual. Team members ultimately assume responsibility for their contributions to the overall team goal and share responsibilities for success or failure of the organization to which they belong (Dingus, 1990; Annett and Stanton, 2000). However, in order for teams to be successful, the organization itself must first encourage teamwork.

11.3.2   Changing Work Structures

The basic structure of the organization sets the foundation for the way that problems are addressed and for the solutions developed within that company (Compton, 1997; Lajoie and Sterling, 1999). Numerous layers of management, a hierarchical reporting structure, and rigidly defined employee roles and responsibilities typically characterize traditional work structures and a lack of shared vision for the organization's future

TABLE 11.2 Typology of Teams

Type of Team	Definition
Work	Work teams are work units responsible for producing goods or providing services where their membership is stable and well defined. Work teams are directed by supervisors who make most of the definitions about what is done, how is it done, and who does it. Self-managing or semi-autonomous or empowered work teams are special alternative form of work teams where employees involve in making decisions without the need of supervisors and managers. Examples for work teams include teams found in manufacturing and mining crews, etc.
Parallel	Parallel teams pull together people from different work units or jobs to perform functions that the regular organization is not equipped to perform well. In other words, they literally exist in parallel with the formal organization structure and used mostly for problem-solving and improvement-oriented activities. Examples include quality improvement teams and task forces, etc.
Project	Project teams are time-limited teams. They always produce one-time outputs like new product or service marketing or developing a new information system or setting up a new plant, etc. They are nonrepetitive in nature and require considerable application of knowledge, judgment and expertise. As they always work on new products and applications, they draw their members from different departments of the organization. Thus they can also be termed as cross-functional teams.
Management	Management teams coordinate and provide direction to the subunits under their authority and control, laterally integrating interdependent subunits across key business processes. The management team is responsible for the overall performance of a business unit in an organization. Most of the time they are composed of managers responsible for each subunit. Examples include strategic development teams of any organization that gives a competitive edge over its competitors.

(Dingus, 1990). However, as many U.S. companies came to realize during the 1970s and 1980s, this is not necessarily the formula for long-term success and productivity in ever more competitive world markets. Many manufacturers realized that in order to stay competitive they needed to implement changes to some or all aspects of their design structure.

The traditional work structures, although divisionalized and departmentalized into sectors of smaller and seemingly more manageable pieces, facilitate endless layers of management (Mills, 1998). The many layers of bureaucracy often "slow innovation, stifle creativity, and impair improvement" and are costly to the company's bottom line (Dingus, 1990). A top-heavy payroll means that the company must demand increasingly more efficient and cost-effective manufacturing and marketing techniques from the bottom in order to survive. Even the basic manner in which decisions are made within a traditional work structure is often hierarchy based: mostly from the top-down, following a unidirectional path that does not easily sanction communication between departments or the consideration of changes at lower levels (Prasad, 1996). These factors combined with unfocused and uncoordinated goals, interdivisional competition and decreased communication, as well as a lack of a homogeneous and consistent company vision were found to be costly and crippling to U.S. manufacturers during the 1980s (Dingus, 1990; Prasad, 1996).

In the 1970s, Japanese manufacturers understood the problems with traditional work structures and developed an approach entirely different from the one described above (Compton, 1997). In Japan, industries designed and manufactured products as a company-wide effort. They launched groups of engineers, marketing personnel, and manufacturers all collaborating and working together to advance the company and its products (Compton, 1997). The use of collaborative product development by Japan meant that laborers and management worked together in manufacturing the product. The result was a quicker design-to-production rate and often at a lower cost and higher quality than their traditionally structured U.S. competitors. U.S. manufacturers soon found that they had lost large segments of their markets (e.g., automobiles) or the entire market itself (e.g., consumer electronics) to Japanese competitors (Compton, 1997). In an effort to compete with Asian rivals, U.S. manufacturers adopted Japan's quality circles and employee involvement ideas and began to implement problem-solving teams, special-purpose teams, and self-managing teams in the late 1970s to mid-1980s (Hoerr, 1989).

The new problem-solving teams usually consisted of 5 to 12 volunteer members from different areas of the company who met once a week to discuss ways to improve efficiency, quality, and overall work environment (Hoerr, 1989). Results from this type of team included reduced manufacturing costs and improved product quality. But the teams usually lacked the power to implement ideas, and thus, management still maintained sole authority in approving and executing the team's recommendations. If management did not support and put into action the team's suggested solutions, participants began to lose interest and energy in furthering the team and may decide to halt meetings altogether (Hoerr, 1989).

Another type of team that was introduced in the 1980s was the special-purpose team. This team generally included union representatives, management, and laborers all collaborating to make operational decisions (Hauck, 1988; Hoerr, 1989). The function of special-purpose teams usually included designing and introducing new technologies, improving the quality of work life, and increasing product quality (Hoerr, 1989). Special-purpose teams are also noted for creating a foundation for self-managing work teams in the 1980s (Hoerr, 1989). Self-managing, or autonomous, work teams had the largest effect on the traditional work system; they changed the basic way workload and responsibilities were distributed within the organization. Self-managing teams were also based on Japan's quality circles process, but extended beyond the scope of circles and created flatter organizational work systems by allowing participants to self-govern (Donovan, 1986). Self-managing teams eliminated many layers of management by allowing teams to take over duties formerly regarded as managerial tasks (e.g., ordering materials, scheduling, performance evaluation, and discipline procedures) (Donovan, 1986; Hoerr, 1989). The work teams were staffed with all the technical, managerial, and interpersonal skills needed to perform all necessary tasks to complete their job (Donovan, 1986). The members were cross-trained to perform all tasks involved in completing the job, and may have rotated from task to task as decided upon (Hackman, 1976; Hoerr, 1989). Participants in autonomous work teams reported an increase in feelings of self-esteem, improved workmanship, more satisfying work life, and an overall increase in feelings of job security (Donovan, 1986). Organizations that used self-managing work teams reported increased worker flexibility, leaner staffing, improved productivity, and a lower employee turnover rate (Donovan, 1986).

Industry's movement to use teams as a method of reducing the hierarchical structure in organizations and increasing employee involvement, which began in the 1980s, continues today (Hoerr, 1989). However, it should be noted that many organizations face immense difficulty in making the transition from traditional work structures to participatory ones, and changes have occurred at a slow pace (e.g., Ford Motor Co., General Electric's Salisbury, N.C. plant, John Deere Horicon Works) (Shyne, 1987; Hoerr, 1989; Compton, 1997). If team-oriented methods such as quality circles, autonomous work teams, and others including total quality management and six-sigma, have been proven to increase productivity and quality, why is teamwork in U.S. corporations spreading so slowly?

Many companies fail to adequately implement team-based methodologies when redesigning their organization. Some companies meddle in the methods, picking and choosing the parts they want to integrate instead of emerging themselves in the ideals that these methods represent. Other companies use a cookbook approach in their attempt to change the existing organization's work system (Ginnodo, 1986). Still others fail because of limited support, inadequate funding, or poor planning by management (Ginnodo, 1986; Compton, 1997). In order for team-based methodologies to work, they must be custom-tailored for the company and nurtured from design to implementation to evaluation (Compton, 1997). Too often, companies do not fully understand the methods that they are trying to follow and the results can be disastrous. An organization and all of its employees must be fully committed to change if it is to expect any long-term success (Compton, 1997).

Each company/organization is unique and has its own culture, rules, and regulations; thus, organizational structure redesigning should be approached with caution. Many organizations have even found it beneficial to use a combination of several team-based methods in order to facilitate teamwork within their organization. Regardless of the specific team-based approach or method an organization uses to facilitate a teamwork-friendly environment, several basic design variables should be considered before the teams are implemented. The design of work teams will be the focus of the subsequent sections, with design recommendations outlined using the IPO model (Figure 11.1) as a guide.

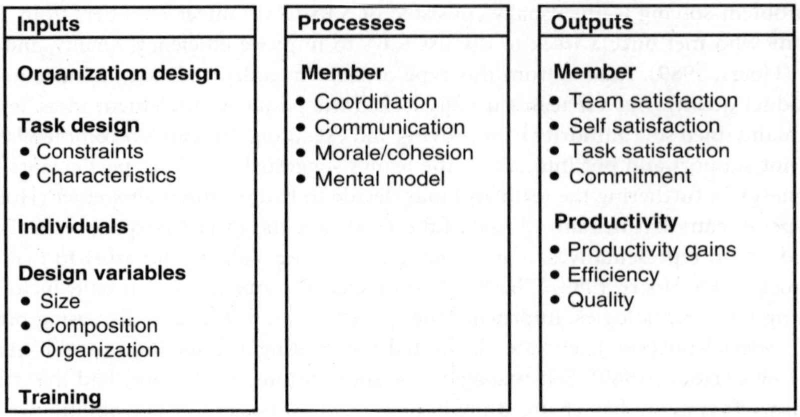

Inputs	Processes	Outputs
Organization design **Task design** • Constraints • Characteristics **Individuals** **Design variables** • Size • Composition • Organization **Training**	**Member** • Coordination • Communication • Morale/cohesion • Mental model	**Member** • Team satisfaction • Self satisfaction • Task satisfaction • Commitment **Productivity** • Productivity gains • Efficiency • Quality

FIGURE 11.1 IPO model.

11.4 Inputs

Team design inputs include variables that are decided upon or chosen before the team is assembled. These inputs include the organization/company's design structure, task design, individual participants, team design variables, and training initiatives.

11.4.1 Organization/Company Design Structure

The first step in designing an effective team is to ensure that the organization's work structure and environment facilitate teamwork (Johnson, 1986). The work structure and environment of a company dictate the manner in which problems are addressed and solutions developed; thus, they should be designed with teams in mind (Compton, 1997). The company's managers should be accessible for team questions and supportive of the team's recommendations for improvement (Paris et al., 2000). Top and middle-level management support is deemed important for team success. In order for teams to be effective, managers must relieve some of the burden of members' routine duties so they may participate in team meetings and projects. Management should also have an "open-door" policy with employees to encourage communication and participation from all levels of employment. In addition to the role of management in supporting teamwork, the organization must provide the necessary and adequate resources to enable teams to work efficiently and effectively (Coates, 1989; Paris et al., 2000).

11.4.2 Task Design

Beyond the company's work structure and environment, the task or job is another key element that ultimately affects team success. The team's responsibilities or job tasks should be designed to be both challenging and rewarding for its members. The team's task should be a whole and meaningful unit of work that allows the team to be responsible for an entire process or product (Donovan, 1986). The outcome of the team's work should have significant impact or consequences on others (e.g., the customer or end user of the product/service) (Hackman, 1987). To accomplish some complex tasks, teams face two issues: how to divide up the labor, and how to coordinate their efforts. In any organization this division of labor and its coordination is attributed mainly to its organizational structure. Many theories of organizational structures are proposed and used in different organizations. Some examples are matrix, project, and hierarchical organizational forms.

There are two aspects of division of labor (Mintzberg, 1992). First, there are technical aspects of the task which determine in what way and to what extent you can break up the task into subtasks that can be performed by a single person. This often determines what jobs or positions may exist in the organization. There is some discretion here, but in general there is not a lot that an organization can do to change how this is done short of adopting a different technology altogether. Second, there is the allocation of people to jobs. People have different competencies, and are better placed in certain jobs rather than others. They also have different interests, and so have different levels of motivation for different jobs. Placing people in the right jobs is a crucial strategic issue.

As organizations enter the 21st century, the source of competitive advantage is increasingly human resources. This may sound strange in a technological age where machines do more and more of the work, but this is because technology is knowledge-driven. It is all about understanding how things work and being able to exploit that knowledge to solve client problems. The most important resource most organizations have is human smarts. Given that the key problem in division of labor is the assignment of people with certain competencies and interests to tasks, part and parcel of the division of labor is the notion of specialization (Mintzberg, 1992).

Teams should be assigned tasks that are autonomous and interdependent in nature, require use of high levels of knowledge, skills, and abilities, and provide/promote communication and cooperation among the members (Campion et al., 1993). Task autonomy ensures that the job allows independence or authority in methods, procedures, scheduling, staffing, and other job related parameters (Donovan, 1986; Hackman, 1987; Campion et al., 1993). The assigning of interdependent tasks helps the team to work as a group effort; members can cover for one another and be cross-trained on all aspects of the final output. Depending on the nature of the task, members may choose to divide it into subtasks with each member working on a smaller piece of the whole project. Based on individual skills or preferences individuals may volunteer to perform certain subtasks. Other teams may rotate members between different subtasks to provide task variety or increased flexibility in the workplace (Medsker and Campion, 2001). In addition to the task design variables mentioned above, member workload, time constraints, governmental regulations, and company policies should also be considered (Paris et al., 2000).

Workload and time constraints are especially important variables of task design. If one team member is overloaded or overworked, the entire team's progress can be limited by this constraint. Time constraints guide the pace of the team. Without a clearly defined timeline or with severe limitations on time limits, the reliability of team decision-making decreases (Adelman et al., 1986).

11.4.3 Individuals

It is important to remember that many people have never experienced a work team environment, and thus it is not unusual for employees to be reluctant to participate in teams; they may even resist the changes being made to the organization's structure (Compton, 1997). With the implementation of teams, employee roles change dramatically.

Middle and lower levels of management are usually the most resistant to implementing teams (Bonvallet, 1990). Often they must relinquish their traditional roles of order giver, reviewer, and approver and instead assume responsibility as coordinator, communicator, objective setter, and resource generator (Bonvallet, 1990; Compton, 1997). Middle management's fears (associated with the implementation of self-managing work teams in particular) may stem from more than just decreased power and status. "A flatter organizational chart means fewer supervisors and managers. Many middle managers feel that 'participative' management styles mean that they will participate themselves right out of their job" (Yankelovich and Immerwahr, 1983, p. 37).

Many employees are also reluctant to accept the idea of work teams due to the changes in their traditional roles/responsibilities from simple order taker and passive worker, to active planner, developer, and implementer of ideas and resources (Compton, 1997). Still other employees simply prefer to work alone. Some individuals may experience feelings of reduced personal achievement and autonomy when working in a team setting (Medsker and Campion, 2001). And although management should encourage participation by all employees, they should not require or force it upon those not wishing to participate

(Coates, 1989). Instead, they should allocate these individuals separate jobs/tasks that are better suited to a single individual's efforts.

Motivation also plays a key role in an individual's willingness to participate in teamwork settings. Motivation is defined as the "sum of a person's aspirations, values, self-esteem, and sensibilities" (Latham, 1988, p. 207). Employee motivators were once thought to be dominated by monetary and status rewards, but are now believed to encompass much more (Latham, 1988). An individual's motivation to perform a task or participate in a team can derive from both extrinsic and intrinsic motivators such as pay, status, self-satisfaction/worth, goal achievement, and education advancements to name a few. Thus, when trying to motivate employees, it may be beneficial for management to design jobs appealing from a task-oriented, a social, and a human resource perspective (Medsker and Campion, 2001).

11.4.4 Design Variables

When designing teams, variables such as size, composition, and organization/structure play an important role in overall performance.

The physical size of a team has been shown to affect its performance. Thus, teams should only contain the minimum number of participants as required by the assigned task (Steiner, 1966, 1972; Campion et al., 1993). Paris et al. (2000, p. 1059) noted that the selection of team size "…becomes problematic when the task has not been performed before, or when it is artificially constrained by such factors as leader preferences, available resources, or the number of people free to participate." Slater (1958) concluded that teams of five members were the most effective in decision making when the gathering or sharing of information was involved. Other researchers have suggested that too large a team can result in more pooled resources, but decreased actual productivity (Williams and Sternberg, 1988). Others note that larger teams may improve the team's overall effectiveness, but at the same time decrease member involvement and coordination (Morgan and Lassiter, 1992; Campion et al., 1993). Also, with larger teams it becomes increasingly difficult to arrange meetings and allow equal participation time at these team meetings (Compton, 1997).

Besides team size, team composition should also be taken into consideration during the design process. The composition of a team includes the degree of member heterogeneity, combined knowledge, skills, and abilities of its individual members, training level, proximity, and many other variables. Member heterogeneity includes age, race, gender, status level in company, experience, attitudes, etc. (Paris et al., 2000). The degree of homogeneity/heterogeneity within the team has been shown to affect team performance. Hoffman et al. (1961) noted that increasing a group's heterogeneity might increase the potential problem-solving ability of the group. Hackman (1987) noted that a group's heterogeneity may mean individual skills and perceptions are too diverse to work effectively as a team. Hackman also notes that members of excessively homogeneous groups may lack the necessary diversity of skills to adequately perform the job. The right mix of the particular knowledge, skills, and abilities that will produce an effective team is difficult to establish (Paris et al., 2000). Groups should have enough diversity to effectively perform the job, but be similar enough in perceptions to agree on problem-solving decisions (Hackman, 1987).

Team member proximity has shown to affect member communication, cohesion, and overall team performance. Proximity involves the physical distances between members. With the surge of new technology and globalization of companies, virtual teams are becoming increasingly popular (Carletta et al., 2000). Virtual teams are those who use technology as a means of communication because proximity issues and other scheduling conflicts make physical meetings impractical (Annett and Stanton, 2000). Technology can be anything that is not considered face-to-face oriented, whether it is computers, audio equipment, telephones, video, or even paper-mediated forms of communication (Paris et al., 2000). Annett and Stanton (2000, p. 1049) note that technology may assist team performance, particularly where geographical factors make face-to-face meetings difficult. Other researchers cite that although advanced forms of communication technology may be available to the team, members should use the least technical form possible in order to build and maintain group solidarity (Carletta et al., 2000).

Another decision variable to be considered when designing work teams is the organization structure of the team itself. Organization structure includes elements such as whom to select, what to educate/teach, how to train, and how to facilitate pay and status ranks within the team. Other aspects of team organization may include who shall lead the team, how to replace/discipline members, and whom the team should report to. Entin et al. (1994) found that team training was an effective method for improving teamwork, communication, and overall performance, and thus should be considered in team design. But with that in mind, it should be noted that training presents an especially challenging obstacle in team building, because it is often difficult to decide what and how to train team members. Although team training may be difficult to organize, it is an excellent employee motivation tool. Training signifies that an organization is investing time, money, and resources in its employees to keep them from becoming obsolete. Employees in turn have increased loyalty to the organization and feelings of improved job security (Latham, 1988).

11.5 Processes

The general goal of teamwork theories is to be able to improve team performance (i.e., output) (Annett et al., 2000). The intermediate variables that influence team output are conventionally known as team processes (Brannick et al., 1995; Annett et al., 2000). Team process variables can be categorized into three sections: (1) behavioral; (2) cognitive; and (3) affective spheres (Annett and Stanton, 2000; Annett et al., 2000). The behavioral process variables include communication and coordination. The cognitive sphere entails a shared mental model and the affective sphere involves morale and group cohesiveness. Over the years, many researchers have attempted to classify individual process variables and understand how each affects the team's performance. Annett and Stanton (2000) note that one of the most important questions that researchers can ask is how process variables affect the overall product. Process variables include coordination, communication, skill usage, morale/cohesion, and shared mental model, to name a few.

11.5.1 Team Communication and Coordination

Communication effects on overall team performance were the focus of experiments led by Macy et al. (1953). The researchers gave experimental groups a set of marbles of various colors and streaking, and the control group was given solid colored marbles. Then the groups were instructed to find the color marble that each member had in common. The experiment found that only after the experimental group established a shared vocabulary were they able to succeed as well as the control group. Thus Macy et al. (1953) concluded that communication and a shared knowledge (i.e., vocabulary) are instrumental for group performance. Harvey and Koubek (2000) also note that a vocabulary schema, which includes the development of a common language, has important effects on team collaboration. In *Entrepreneurship Reconsidered:The Team as Hero* (1987, p. 81), Reich comments on the importance of organization-wide communication, stating " … [A] company's ability to adapt to new opportunities and capitalize on them depends on its capacity to share information …. " Communication within a team is the ability of members to send, receive, and discuss information (Annett et al., 2000). Mills (1998) notes that effective communication entails active information exchange between members and only then can the members progress on to more pursued/purposeful interaction.

Brannick et al. (1995, p. 641) define team coordination as " … the moment-to-moment behaviors, by which interdependent team members achieve important goals." Coordination means that interdependent members must share the workload in a balanced manner in order to meet predetermined deadlines in a timely fashion (Annett et al., 2000). Depending on the team's organizational workload distribution plan, members may work in parallel or sequential order. In sequential order, one or more member's actions are interrelated to another member, meaning that one member's incoming information needed to perform his or her own task is based on the findings/outcome of another member. In sequentially ordered member tasks, coordination among members is essential for effective team performance. Another important aspect of team processes is a shared mental model.

11.5.2   Team Mental Models

Team research in the 1990s introduced the concept of shared mental models. Mental models are mental simulations that " ... humans use to organize new information, to describe, explain and predict events" (Rouse and Morris, 1986; Paris et al., 2000, p. 1055). Team-shared mental models were developed by Cannon-Bowers and Salas (1990) as an extension of the individual based mental model. Shared mental models allow team members to recognize needs of teammates, facilitate information processing, share task information, and provide support (Kaplan, 1990; Paris et al., 2000; Hinsz, 2004). Cannon-Bowers et al. (1993) note that training may be able to build shared mental models. A study by Entin and Serfaty (1999) tested the effects of training on team performance, and hypothesized that highly developed shared mental models would aid effective teams in operating under high-stress conditions. They concluded that team training did improve team processes and outcomes. Other researchers have suggested that feedback and team structure are ways of improving shared mental models (Annett and Stanton, 2000; Rasker et al., 2000; Stanton and Ashleigh, 2000). Although much research has been devoted to understanding mental models in recent years, more investigations are needed to understand how they and other process variables affect team performance.

11.5.3   Team Cohesiveness and Group Bond

Why some groups succeed while others fail is still somewhat a mystery, yet one concept is thought to impact groups and their interaction. This concept is *group cohesiveness*. Cohesiveness is a complex concept to define; yet, most people can recognize whether it exists in groups in which they are participating.

Two distinct views of cohesiveness exist. One group views cohesion as a single construct. For example, Festinger (1950, p. 274) defined cohesiveness as "the resultant of all forces acting on the members to remain in the group." Festinger's definition, views cohesion as existing solely because of the socio-emotional aspects of cohesion based on the attraction among group members.

Another group views cohesion as a multidimensional construct (Hackman, 1976; Zaccaro and Lowe, 1988; Zaccaro, 1991). Here, cohesion is not only viewed as consisting of interpersonal or socio-emotional cohesiveness (Festinger, 1950; Festinger et al., 1950), but also task cohesiveness as well. Task cohesiveness is a result of groups attaining important goals together (Festinger et al., 1950) or when there is a "shared commitment to the task of the group" (Hackman, 1976, p. 1517).

There are four major consequences of group cohesiveness: interaction, group productivity, satisfaction, and social influence. Cohesiveness has been related to quality and quantity of interaction (Barker et al., 1987). Members of highly cohesive groups generally communicate with one another more often, whereas, groups with low cohesion interact less frequently and behave independently (Barker et al., 1987; Zaccaro and Lowe, 1988).

In terms of task performance, the results seem to be inconclusive due to the two views of cohesiveness mentioned earlier. If the group's focus is interpersonal cohesion, it might be normal to expect that their productivity would be less compared to a group where task cohesion is their focus.

Zaccaro and his colleagues (Zaccaro and Lowe, 1988; Zaccaro and McCoy, 1988; Zaccaro, 1991) have looked at productivity where interpersonal and task cohesion were manipulated independently. The effects of cohesion are expected to also vary based on the particular task. Therefore, Zaccaro and his colleagues (Zaccaro and Lowe, 1988; Zaccaro and McCoy, 1988) conducted studies for groups performing additive and disjunctive tasks.

Additive tasks use individual performance measures summed together to obtain a group score. For these tasks, success is obtained through maximum individual effort and by minimizing interactions that distract individuals from the task. High-task cohesion groups performed better. High interpersonal attraction groups, which were found to be associated with performance decrements, conversed more than groups low in interpersonal attraction. Zaccaro and Lowe (1988) also found that increased task commitment resulted in increased member attraction. Therefore, members interacted more; however, this interaction inhibited performance.

Disjunctive tasks require groups to adopt a single solution and therefore, the group process must allow the emergence of this solution. Success requires that the group must maintain a member(s) with the ability to solve the problem. The member must be able to defend the solution to the group and the group must be able to come to a consensus agreement of the solution. Zaccaro and McCoy (1988) found that groups with both high task and interpersonal cohesion performed better. High task cohesion allowed the high ability member to participate, while interpersonal cohesion facilitated group interaction. This suggests that the effect of cohesion depends on the task characteristics.

In a later study, Zaccaro (1991) once again looked at cohesion as a multidimensional construct. This time Zaccaro studied student military groups and examined the effect of cohesion on: group performance processes, role uncertainty, absenteeism, and individual performance. Zaccaro found that task cohesion had a stronger relationship with role uncertainty and absenteeism than interpersonal cohesion. Task cohesion was also significantly associated with individual performance, yet interpersonal cohesion had no effect.

The importance of task cohesion to role uncertainty was an important finding. As Zaccaro mentions, previous studies have stated that cohesion lowers uncertainty by enhancing the social support as a coping mechanism (Zaccaro, 1991). However, the results here imply that task cohesion is more important and as Hackman (1976) and Hackman and Morris (1975) have stated, groups held together by a strong task cohesion establish norms and strategies to deal with ambiguous role requirements.

In terms of satisfaction, the research has been clear that members of cohesive groups are better satisfied than noncohesive groups (Barker et al., 1987). The findings indicate individuals in highly attracted groups take on more responsibilities, participate more, attend more meetings, and work harder toward difficult goals. In addition, groups that are successful result in members who have an increased attraction. Therefore it seems important to ensure group members stay satisfied with goal performance.

Social influence also appears to be greater in highly cohesive groups. If members are highly attracted to one another, then individuals are more likely to listen to one another. However, this can have a negative influence as well as resulting in groupthink.

Interpersonal attraction and task cohesion may be only two factors that affect a group's cohesiveness. Szilagyi and Wallace (1980) state that within cohesive groups, there appears to be an atmosphere of closeness, common attitudes, behavior and performance that does not appear in other groups. They indicate that several factors (see Table 11.3) affect the level of cohesiveness both internal and external to the group. These include interpersonal attraction and task cohesion, but others as well.

As stated earlier, cohesiveness is not easily defined. Some suggest that the atmosphere of the group is driven by more than just cohesion. In their work with learning groups, Piper et al. (1983) suggest that there are three elements that form the group atmosphere: the bond between the participant and another participant, the bond between the participant and the leader, and the bond between the participant and his/her conception of the group as a whole. Although their study had some limitations based on the sample size, Piper et al. believe the group property, cohesion, emerges from the set of bonds that exist within the group.

Mullen and Cooper (1994) suggest that there are three elements as well: interpersonal attraction, task cohesion, and group pride based on belonging to a successful group. Mullen and Cooper's meta-analysis of 66 studies looked at two paradigms, experimental and correlational, used to determine the impact of cohesiveness on performance. The experimental paradigm introduces levels of cohesiveness into ad hoc groups. The correlational paradigm relates the perceived cohesiveness of the groups' members to their performance, usually in "real" groups. Mullen and Cooper found in the correlational studies that performance decreased as a result of interpersonal attraction and group pride and increased as a function of commitment to the task. In the experimental groups, performance increased as a result of all three factors. Thus, task cohesion emerged as the critical component of cohesiveness.

It should be noted that the intermediate results of process variables may aid in overall team performance. Process "paybacks" include better communication between team members, shared understanding of project goals/tasks, easier problem-solving decision process (consensus), and improved employee morale and responsibility (Mills, 1998).

TABLE 11.3　Factors Affecting Cohesiveness (Szilagyi and Wallace 1980)

Factors Increasing Cohesiveness	Factors Decreasing Cohesiveness
Agreement of group goals	Disagreement of group goals
Frequency of interaction	Group size
Personal attractiveness	Unpleasant experiences within the group
Intergroup competition	Intragroup competition
Favorable evaluation	Domination by one individual

11.6　Outputs

Outputs are the end-results of the team's efforts and overall performance. Teams must enhance its members and the organization itself, in addition to completing the assigned task/job, to truly be effective. Members of effective teams often report improved satisfaction in themselves, the team, and the assigned task as well as increased commitment to the team (Campion et al., 1993).

Organizations that utilize teams can expect, among other benefits, productivity gains, improved efficiency, and increased quality of product/service. In addition, considering the elements that affect team performance prior to forming the team may allow organizations to prevent problems such as those discussed in the introductory cases.

11.7　Conclusion

Organizations that use teams report increased employee flexibility between tasks, leaner staffing, increased productivity, improved product/service quality, and a better overall work environment (Donovan, 1986). Mills (1998) also indicates that organizations can expect shortened development cycles and faster time-to-market, improved product innovation, increased product quality, increased product value, lowered development costs, and lowered production costs.

Organizations as a whole are not alone in realizing the numerous advantages of teamwork. Team members themselves are also affected by the experience. Members often experience both extrinsic and intrinsic rewards as a result of team participation. Extrinsic rewards include those allocated by the organization, such as pay and job promotion (Sainfort et al., 2001). Self-satisfaction, motivation, job security, and job loyalty are examples of intrinsic rewards for team participation (Donovan, 1986; Medsker and Campion, 2001). Another advantage of employee participation in a team is the facilitation of effective communication between the layers of the company (e.g., management, floor level, supervisors, and board members) (Prasad, 1996). Johnson (1986, p. 48) notes the following:

> Improve the way team members interact, and you improve their ability to solve problems. Better problem-solving means better efficiency in general. Increased efficiency tends to boost morale and productivity. It also helps to decrease stress, turnover and operating costs. And all of these improvements bolster the organization's public image. Once established, an effective team becomes self-perpetuating.

If teams are so advantageous, why are employees and managers alike so reluctant to implement them in all aspects of their corporation? First, teams are time-consuming to design, implement, and evaluate (Newman, 1984). And in today's markets, time is money. Team participation also takes time away from the participant's normal job activities, and many times, skills training must be provided in order to complete the team's set task (Donovan, 1986; Newman, 1984). Employees themselves may be reluctant to participate in teams because of added responsibility, personal biases toward each other and other departments, fear of retaliation or apathy by management, and past team/group experiences (Baloff and Doherty, 1989).

Medsker and Campion (2001) note that advantages in team design can cause disadvantages in the organization itself. The unique interpersonal relationships developed within a team can increase

TABLE 11.4 Summary of Team Advantages and Disadvantages

Advantages of using teams	Disadvantages of using teams
Greater workforce flexibility	Groupthink
Improved work environment	Social loafing
Decreased turnover rate	Wasted resources
Leaner staffing	Increased competition between groups
Improved workmanship	Wasted time (due to socializing)
Increased productivity, quality of product, decreased costs	Increased time for decision-making (need to reach consensus)
Enhanced employee development	
Increased job security	
Increased employee self-esteem	

communication and coordination, but the increase in team loyalty is so much as to cause competition and a decrease in communication/cooperation between teams (Medsker and Campion, 2001). Other disadvantages can include "groupthink," "social loafing," and a waste of members' time and energy (Hackman, 1987; Annett and Stanton, 2000). Often in groups, a consensus must be reached at the expense of compromising the opinions of several members. Compton (1997) indicates that the risk involved with developing a consensus is that the decision will be generated from the " ... lowest common denominator of all possible options." Compton also points out that teams also face the risk of being dominated by the most vocal member (i.e., peer pressure) to conform to the "correct" position. Another disadvantage mentioned above was social loafing. Loafing, or free-riding, occurs when members do not see a direct correlation between their individual effort and the team's outcomes, or they feel the team is evaluated only as whole and not based on their individual performance/contributions.

Whether a team's advantages outweigh the disadvantages depends largely on the design, implementation, and evaluation techniques used. With so many variables to be considered when designing, implementing, and evaluating teams, it is easy to see the continued need for more research in order to understand the dynamic processes that teamwork entails. In summary, when implementing teams, one must carefully consider the team design elements as discussed in the IPO model along with the potential advantages and disadvantages as listed in Table 11.4.

References

Arrow, H., McGrath, J.E., and Berdahl, J.L., *Small Group as Complex Systems: Formation, Coordination, Development, and Adaptation*, Sage Publications, Inc, Thousand Oaks, CA, 2000.

Achille, L.B., Schulze, K.G., and Schmidt-Neilsen, A., An analysis of communication and the use of military teams in navy team training, *Mil. Psychol.*, 7, 95–107, 1995.

Adelman, L., Zirk, D.A., Lehner, P.E., Moffett, R.J., and Hall, R., Distributed tactical decision-making: Conceptual framework and empirical results, *IEEE Trans. Syst. Manuf. Cybern.*, SMC-16, 794–805, 1986.

Adler, J., After the *Challenger*: How NASA Struggled to Put Itself Back Together, *Newsweek*, 112, 28–36, 1988.

Annett, J., Cunningham, D., and Mathias-Jones, P., A method for measuring team skills, *Ergonomics*, 43, 1076–1094, 2000.

Annett, J. and Stanton, N.A., Editorial: team work-a problem for ergonomics? *Ergonomics*, 43, 1045–1051, 2000.

Baker, D.P. and Salas, E., Principles of measuring teamwork: A summary and look toward the future, in *Team Performance Assessment and Measurement*, Brannick, M.T., Salas, E., and Prince, C., Eds., Earlbaum, Mahwah, NJ, 1997, pp. 331–355.

Baloff, N. and Doherty, E.M., Potential pitfalls in employee participation, *Organizational Dynamics*, 17(3), 51–62, 1989.

Barker, L.L., Wahlers, K.J., Watson, K.W., and Kibler, R.J., *Groups in Process: An Introduction to Small Group Communication*, Prentice-Hall, Englewood Cliffs, NJ, 1987.

Bennis, W. and Shepard, H., A theory of group development, *Hum. Relat.*, 9, 415–437, 1956.

Bion, W.R., *Experiences in Groups*, Basic Books, New York, 1961.

Bonvallet, W.A., Learnings from implementing self-managing teams, in *Achieving High Commitment Work Systems: A practitioner's Guide to Sociotechnical System Implementation*, Hauck, W.C.D. and Victor, R., Eds., Institute of Industrial Engineers, Norcross, GA, 1990, pp. 307–310.

Bowers, C.A., Baker, D.P., and Salas, E., Measuring the importance of teamwork: the reliability and validity of job/task analysis indices for team-training design, *Mil. Psychol.*, 6, 205–214, 1994.

Brannick, M.T. and Prince, A., An overview of team performance measurement, in *Team Performance Assessment and Measurement*, Brannick, M.T., Salas, E., and Prince, C., Eds., Earlbaum, Mahwah, NJ, 1997.

Brannick, M.T., Prince, A., Prince, C., and Salas, E., The measurement of team process, *Hum. Factors*, 37, 641–651, 1995.

Campion, M.A., Medsker, G.J., and Higgs, A.C., Relations between work group characteristics and effectiveness: implementations for designing effective work groups, *Pers. Psychol.*, 46, 823–850, 1993

Cannon-Bowers, J.A. and Salas, E., Cognitive psychology and team training: shared mental models in complex systems, *Fifth Annual Conference of the Society for Industrial Organizational Psychology*, Miami, FL, 1990.

Cannon-Bowers, J.A., Salas, E., and Converse, S.A., Shared mental models in expert team decision making, in *Current Issues in Individual and Group Decision-Making*, Castellan, N.J., Jr., Ed., Erlbaum, Hillsdale, NJ, 1993, pp. 221–246.

Cannon-Bowers, J.A., Salas, E., Blickensderfer, E., and Bowers, C.A., The impact of cross-training and workload on team functioning: a replication and extension of initial findings, *Hum. Factors*, 40, 92–101, 1998.

Cannon-Bowers, J.A., Tannenbaum, S.I., Salas, E., and Volpe, C.E., Defining competencies and establishing team-training requirements, in *Team Effectiveness And Decision Making In Organizations*, Guzzo, R.A. and Salas, E., Eds., Jossey-Bass, San Francisco, 1995, pp. 333–380.

Carletta, J., Anderson, A.H., and McEwan R., The effects of multimedia communication technology on non-collocated teams: a case study, *Ergonomics*, 43, 1237–1251, 2000.

Coates, E.J., Employee participation: a basic link in the productivity chain, *Industrial Management*, 31(3), 2–4, 1989.

Cohen, S.G. and Bailey, D.E., What makes teams work: group effectiveness research from the shop floor to the executive suite, *J. Manage.*, 23, 239–290, 1997.

Compton, W.D., *Engineering Management: Creating and Managing World-Class Operations*, Prentice-Hall Inc, Upper Saddle River, NJ, 1997.

Dingus, V., Implementing work redesigns in established business, in *Achieving High Commitment Work Systems: A Practitioner's Guide to Sociotechnical System Implementation*, Hauck, W.C. and Victor, R.D., Eds., Industrial Engineering and Management Press, Norcross, GA, 1990, pp. 247–272.

Donovan, M.J., Self-managing work teams: extending the quality circle concept, *International Association of Quality Circles 1986 Conference Proceedings*, 1986.

Driskell, J.E. and Salas, E., Collective behavior and team performance, *Hum. Factors*, 34, 277–288, 1992.

Ehrlich, J.A., Knerr, B.W., Lampton, D.R., and McDonald, D.P., Team Situational Awareness Training in Virtual Environments: Potential Capabilities and Research Issues (Technical Report 1069), U.S. Army Research Institute for the Behavioral and Social Sciences, Alexandria, VA, 1997.

Entin, E. and Serfaty, D., Adaptive team coordination, *Hum. Factors*, 41, 312–325, 1999.

Entin, E., Serfaty, D., and Deckert, J.C., Team Adaptation and Coordination Training (Technical Report 648-1), Alpha Tech, Inc, Burlington, MA, 1994.

Festinger, L., Informal social communication, *Psychol. Rev.*, 57, 271–282, 1950.

Festinger, L., Schachter, S., and Back, K., *Social Pressures in Informal Groups*, Stanford University Press, Stanford, CA, 1950.

Fisher, B.A., Decision emergence: phases in group decision-making, *Speech Monogr.*, 37, 53–66, 1970.

Gersick, C.G., Time and transition in work teams: toward a new model of group development, *Acad. Manage. J.*, 31, 9–41, 1988.

Ginnodo, B., Getting started and keeping up the momentum (at Armco, Inc.), *Commitment Plus*, Pride Publications, Limited, Peterborough, UK, 1986.

Guzzo R.A. and Dickson M.W., Teams in organizations: recent research on performance and effectiveness, *Annu. Rev. Psychol.*, 47, 307–338, 1996.

Hackman, J.R., *The Design of Self-Managing Work Groups*, School of Organization and Management, Yale University, New Haven, CT, 1976.

Hackman, J.R., The design of work teams, in *Handbook of Organizational Behavior*, Lorsch, J., Ed., Prentice-Hall, Englewood Cliffs, NJ, 1987, pp. 315–342.

Hackman, J.R. and Morris, C.G., Group tasks, group interaction process, and group performance effectiveness: a review and proposed integration, in *Advances in Experimental Social Psychology*, Vol. 8, Berkowitz, L., Ed., 1975, pp. 45–99.

Harvey, C.M. and Koubek, R.J., Cognitive, social, and environmental attributes of distributed engineering collaboration: a review and proposed model of collaboration, *Hum. Factors Ergonomics Manuf.*, 10, 369–393, 2000.

Hauck, W.C., Employee involvement: two by two, *1988 IIE Integrated Systems Conference Proceedings*, 1988.

Hinsz, V.B., Metacognition and mental models in groups: an illustration with metamemory of group recognition memory, in *Team Cognition*, Salas, E. and Fiore, S., Ed., American Psychological Association, Washington, DC, 2004.

Hoerr, J., The payoff from teamwork: the gains in quality are substantial — so why isn't it spreading faster? *Business Week*, McGraw-Hill, Inc, New York, 1989.

Hoffman, L.R. and Maier, N.R.F., Quality and acceptance of problem solutions by members of homogeneous and heterogenous groups, *J. Abnormal Social Psychol.*, 63, 401–407, 1961.

Ilgen, D.R., Teams embedded in organizations: some implications, *Am. Psychol.*, 54, 129–139, 1999.

Johnson, C.R., An outline for team building, *Train. Dev. J.*, 48–52, 1986.

Kaplan, R., Collaboration from a cognitive perspective: sharing models across expertise, *EDRA*, 21, 45–51, 1990.

Katzenbach J. R. and Smith, D.K., *The Wisdom of Teams: Creating the High Performance Organization*, Harvard Business School Press, Boston, MA, 1993.

Kinlaw, D.C., Teaming up for management training: want to increase the payoffs in management training? Try integrating objectives in a team setting, *Train. Dev. J.*, 41(11), 44–46, 1987.

LaJoie, A.S. and Sterling, B.S., *A Review and Annotated Bibliography of the Literature Pertaining to Team and Small Group Performance (1989 to 1999)*, U.S. Army Research Institute for the Behavioral and Social Sciences, Arlington, VA, 1999.

Latham, G.P., Employee motivation: yesterday, today, and tomorrow, in *Futures of Organizations*, Hage, J., Ed., Lexington Books, Lexington, 1988, pp. 205–226.

Lawler, E.E., III, *High Involvement Management*, Jossey-Bass, San Francisco, CA, 1986.

Lawler, E.E., III and Morhman, S.A., Quality circles: after the honeymoon, *Organ. Dyn.*, 15, 42–54, 1987.

Leedom, D.K. and Simon, R., Improving team coordination: a case for behavior-based training, *Mil. Psychol.*, 7, 109–122, 1995.

Macy, J., Christie, L.S., and Luce, R.D., Coding noise in a task-oriented group, *J. Abnorm. Soc. Psychol.*, 48, 401–409, 1953.

Mann, R., Gibbard, G., and Hartman, J., *Interpersonal Styles and Group Development*, Wiley, New York, 1967.

Matthews, A.L., Harvey, C.M., Schuster, R.J., and Durso, F.T., Emergency physician to admitting physician handovers: an exploratory study. *Proceedings for the Human Factors and Ergonomics Society 46th Annual Meeting*, 2002.

McGrath, J.E., *Social Psychology: A Brief Introduction*, Holt, Reinhart, & Winston, New York, 1964.

McGrath, J.E., Time matters in groups. in *Intellectual Teamwork*, Galegher, J., Kraut, R., and Egido, C., Eds., Lawrence Erlbaum, Hillsdale, NJ, 1990, pp. 23–61.

Medsker, G.J. and Campion, M.A., Job and team design, in *Handbook of Industrial Engineering: Technology and Operations Management*, Salvendy, G., Ed., Wiley, New York, 2001, pp. 868–898.

Mills, A., *Collaborative Engineering and the Internet: Linking Product Development Partners via the Web*, Society of Manufacturing Engineers, Dearborn, MI, 1998.

Mills, T., Changing paradigms for studying human groups, *J. Appl. Behav. Sci.*, 15, 407–423, 1979.

Mintzberg, H., Structure in fives: designing effective organizations, Prentice-Hall, New York, 1992.

Morgan, B.B. and Lassister, D.L., Team composition and staffing, in *Teams: Their Training and Performance*, Swezwy, R.W. and Salas, E., Eds., Ablex, Norwood, NJ, 1992, pp. 75–100.

Mullen, B. and Cooper, C., The relation between group cohesiveness and performance: an integration, *Psychol. Bull.*, 115, 210–227, 1994.

Newman, B., Expediency as benefactor: how team building saves time and gets the job done, *Train. Dev. J.*, 38, 26–30, 1984.

Orasanu, J. and Salas, E., Team decision making in complex environments, in *Decision Making in Action: Models and Methods*, Klein, G., Orasanu, J., and Calderwood, R., Eds., Ablex Publishing, Westport, Connecticut, 1993, pp. 327–345.

Paris, C.R., Salas, E., and Cannon-Bowers, J.A., Teamwork in multi-person systems: a review and analysis, *Ergonomics*, 43, 1052–1075, 2000.

Piper, W.E., Marrache, M., Lacroix, R., Richardson, A.M., and Jones, B.D., Cohesion as a basic bond in groups, *Hum. Relat.*, 36, 93–108, 1983.

Poole, M.S., Decision development in small groups. I. A comparison of two models, *Commun. Monogr.*, 48, 1–24, 1981.

Poole, M.S., Decision development in small groups. II. A study of multiple sequences of decision making, *Commun. Monogr.*, 50, 206–232, 1983a.

Poole, M.S., Decision development in small groups. III. A multiple sequence model of group decision development, *Commun. Monogr.*, 50, 321–341, 1983b.

Prasad, B., *Concurrent Engineering Fundamentals: Integrated Product and Process Organization*, Prentice-Hall, Inc, Upper Saddle River, NJ, 1996.

Rasker, P.C., Post, W.M., and Schraagen, J.M.C., Effects of two types of intra-team feedback on developing a shared mental model in command and control systems, *Ergonomics*, 43, 1167–1189, 2000.

Reich, R.B., Entrepreneurship reconsidered: the team as hero, *Harvard Bus. Rev.*, 77–83, 1987.

Rouse, W.B. and Morris, N.M., On looking into the black box: prospects and limits in the search for mental models, *Psychol. Bull.*, 100, 349–363, 1986.

Sainfort, F., Taveira, A.D., Arora, N.K., and Smith, M.J., Teams and team management and leadership, in *Handbook of Industrial Engineering: Technology and Operations Management*, Salvendy, G., Ed., Wiley, New York, 2001, pp. 975–994.

Salas, E., Bowers, C.A., and Cannon-Bowers, J.A., Military team research: 10 years of progress, *Mil. Psychol.*, 7, 55–75, 1995.

Scheidel, T. and Crowell, L., Idea development in small discussion groups, *Q. J. Speech*, 50, 140–145, 1964.

Shenhar, A.J., Systems engineering management: the multidisciplinary discipline, in *Handbook of Systems Engineering and Management*, Sage, A.P. and Rouse, W.B., Eds., Wiley, New York, 1999, pp. 113–136.

Shyne, K.C., Participative management at John Deere, *Commitment Plus*, Pride Publications, Limited, Peterborough, UK, 1987.

Slater, P.E., Contrasting correlates of group size, *Sociometry*, 25, 129–139, 1958.

Stanton, N.A. and Ashleigh, M.J., A field study of team working in a new human supervisory control system, *Ergonomics*, 43, 1190–1209, 2000.

Steiner, I.D., Models for inferring relationships between group size and potential group productivity, *Behav. Sci.*, 11, 273–283, 1966.

Steiner, I.D., *Group Process and Productivity*, Academic, New York, 1972.

Stein, R., Girl has Second Transplant After Error, *The Washington Post*, February, 2003.

Szilagyi, A.D. and Wallace, M.J., *Organizational Behavior and Performance,* Goodyear Publishing Company, Inc., Santa Monica, CA, 1980.

Sundstrom, E., De Meuse, K.P., and Futrell, D., Work teams: applications and effectiveness, *Am. Psychol.,* 45, 120–133, 1990.

Tuckman, B., Developmental sequence in small groups, *Psychol. Bull.,* 63, 384–399, 1965.

Tuckman, B. and Jensen, M., Stages of small-group development, *Group Org. Stud.,* 2, 419–427, 1977.

Tufte, E., *Visual Explanations: Images and Quantities, Evidence and Narrative,* Graphics Press, Connecticut, 1997.

Williams, W.M. and Sternberg, R.J., Group intelligence: why some groups are better than others, *Intelligence,* 12, 351–377, 1988.

Yankelowich, D. and Immerwahr, J., Putting the work ethic to work, New York: Public Agenda Foundation, 1983.

Zaccaro, S.J., Nonequivalent associations between forms of cohesiveness and group-related outcomes: evidence for multidimensionality, *J. Soc. Psychol.,* 131, 387–399, 1991.

Zaccaro, S.J. and Lowe, C.A., Cohesiveness and performance on an additive task: evidence for multidimensionality, *J. Soc. Psychol.,* 128, 547–558, 1988.

Zaccaro, S.J. and McCoy, M.C., The effects of task and interpersonal cohesiveness on performance of a disjunctive group task, *J Appl. Soc. Psychol.,* 18, 837–851, 1988.

12

Strategic Performance Measurement

Garry D. Coleman
Transformation Systems, Inc.

12.1 What is Strategic Performance Measurement?

The focus of this chapter is *strategic performance measurement*, a key management system for performing the study (or check) function of Shewhart's plan-do-study-act cycle. Strategic performance measurement applies to a higher level system of interest (unit of analysis) and a longer term horizon than *operational performance measurement*. While the dividing line between these two types of performance measurement is not crystal clear, the following distinctions can be made:

- Strategic performance measurement applies to the organizational level, whether of a corporation, a business unit, a plant, or a department. Operational performance measurement applies to small groups or individuals, such as a work group, an assembly line, or a single employee.
- Strategic performance measurement is primarily concerned with performance that has medium- to long-term consequences; thus performance is measured and reported on a weekly, monthly, quarterly, or annual basis. More frequent, even daily, measurement and reporting may also be included, but only for the most important performance measures. Data may also be collected, daily or perhaps continually, but should be aggregated and reported weekly or monthly. Operational performance measurement focuses on immediate performance, with reporting on a continual, hourly, shift, or daily basis. Strategic performance measurement tends to measure performance on a periodic basis, while operational performance measurement tends to measure on a continual or even continuous basis.

- Strategic performance measurement is concerned with measuring the mission- or strategy-critical activities and results of an organization. These activities and results are key to the organization's success, and their measurements are referred to as strategic performance measures, key performance indicators, or mission-driven metrics. These measurements can be classified into a few key performance dimensions, such as Drucker's (1954) nine key results areas, the Balanced Scorecard's four performance perspectives (Kaplan and Norton, 1996), the Baldrige criteria's six business results items (Baldrige National Quality Program, 2004), or Sink's (1985) seven performance criteria.
- Strategic performance measurement tends to measure aspects of performance impacting the entire organization, while operational performance may be focused on a single product or service (out of many). In an organization with only one product, strategic and operational measurement may be similar. In an organization with multiple products or services, strategic performance measurement is likely to aggregate performance data from multiple operational sources.
- Strategic performance measurement is a popular topic in the management, accounting, industrial engineering, human-resources management, information technology, statistics and industrial and organizational psychology literature. Authors such as Neely(1999), Kaplan and Norton(1992, 1996); Thor(1995, 1998); Sink(1985); Wheeler(1993); and Brown(1996, 2000) have documented the need for and the challenges facing strategic performance measurement beyond traditional financial and accounting measures. Operational performance measurement has long been associated with pioneers such as Frederick Taylor, Frank and Lillian Gilbreth, Marvin Mundel, and others. Careful reading of their work often shows an appreciation for and some application to strategic performance measurement, yet they are remembered for their contributions to operational measurement.
- For the remainder of this chapter, strategic performance measurement will be referred to as performance measurement. The term "measurement" will be used to apply to both strategic and operational performance measurement.

Why is performance measurement important enough to warrant a chapter of its own? Andrew Neely (1999, p. 210) summarized the reasons for the current interest in performance measurement very well. His first reason is perhaps the most important for the industrial engineer: the "changing nature of work." As industrialized nations have seen their workforces shift to predominantly knowledge and service work, concerns have arisen about how to measure performance in these enterprises with less tangible products. Fierce competition and a history of measuring performance have facilitated steady productivity and quality improvement in the manufacturing sector in recent years. Productivity and quality improvement in the service sector has generally lagged that of the manufacturing sector. The shift to a knowledge and service dominated economy has led to increased interest in finding better ways to measure and then improve performance in these sectors. Other reasons for increased interest in performance measurement cited by Neely include increasing competition, specific improvement initiatives that require a strong measurement component (such as Six Sigma or business process reengineering), national and international awards (with their emphasis on results, information, and analysis), changing organizational roles (e.g., the introduction of the chief information officer or, more recently, the chief knowledge officer), changing external demands (by regulators and shareholders), and the power of information technology (enabling us to measure what was too expensive to measure or analyze in the past).

12.2 The Measurement and Evaluation Process

In an organizational setting, measurement is the codifying of observations into data that can be analyzed, portrayed as information, and evaluated to support the decision maker. The term "observation" is used broadly here and may include direct observation by a human, sensing by a machine, or document review. Document review may involve secondary measurement, relying on the recorded observations of another human or machine; or it may involve the direct measurement of some output or artifact contained in the documents. The act of measurement produces data ("evidence"), often but not always in quantified form. Quantitative data are often based on counts of observations (e.g., units, defects, person-hours) or scaling

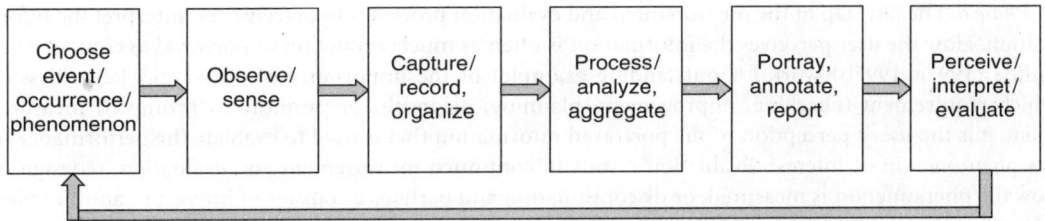

FIGURE 12.1   The measurement and evaluation process. (Adapted from Coleman, G.D. and Clark, L.A., *Proceedings of the Industrial Engineering Research Conference*, 2001.)

of attributes (e.g., volume, weight, speed). Qualitative data are often based on categorization of observations (e.g., poor/fair/good) or the confirmation (or not) of the presence of desired characteristics (e.g., yes/no, pass/fail). Such qualitative data are easily quantified by calculating the percentages in each category.

Measuring performance is a process (see Figure 12.1) that produces a codified representation of organizational performance. Assuming it was measured properly, this codified representation is simply a fact. This fact may exist in the form of a number, chart, picture, or text, and is descriptive of the phenomena being observed (i.e., organizational performance) and the process used to produce the fact prior to evaluation. Evaluation is the interpretation and judgment of the output of the measurement process (i.e., the number, chart, picture, or text). Evaluation results in a determination of the desirability of the level or trend of performance observed, typically on the basis of a comparison or expectation. Too often, those who are developing new or enhanced performance indicators jump to evaluation before fully completing the measurement step. They base the suitability of an indicator not on how well it represents the phenomena of interest, but on how it will be evaluated by those receiving reports of this indicator. As industrial engineers, we must know when to separate measurement from evaluation. Figure 12.1 illustrates the measurement and evaluation process as having six phases, described in the following excerpt from Coleman and Clark (2001). The model in Figure 12.1 is an expansion of the work of Kurstedt (1986).

Phase 1. The process begins by asking what should be measured. Management or other stakeholders are interested in some event, occurrence, or phenomenon. This interest may be driven by a need to check conformity, track improvement, develop expectations for planning, diagnose problems, or promote accomplishments. This phenomenon of interest is often described in terms of key performance areas or criteria, which represent the priorities associated with this phenomenon.

Phase 2. The phenomenon of interest is observed or sensed to measure each key performance area (KPA). One or more indicators may be measured to represent the KPA. Each indicator requires an operational definition (a defined procedure for how the observation will be converted into data). While the KPAs are "glittering generalities," the indicators are specific and reliable.

Phase 3. The output of the measurement procedure is data, which are then captured or recorded for further use. Capturing represents entering the data into the "system," whether a paper or an electronic system. This step includes ensuring that all the data generated are captured in a timely, consistent, and accurate manner. This often includes organizing or sorting the data (by time, place, person, product, etc.) to feed the analysis procedures.

Phase 4. Raw data are analyzed or processed to produce information. Manual calculations, spreadsheets, statistical software packages, and other tools are used to summarize and add value to the data. Summarizing often includes aggregating data across time or units. That is, individual values are captured and processed; then, totals or means are calculated for reporting.

Phase 5. The output of analyzing the data is information, portrayed in the format preferred by the user (manager). That is, when the values of the indicators representing KPAs for a particular phenomenon are measured, the portrayal should provide context that helps the user understand the information (Wheeler, 1993). Too often, the analyst chooses a portrayal reflecting his or her own preference rather than the user's preference.

Phase 6. The last step of the measurement and evaluation process is to perceive and interpret the information. How the user perceives the information is often as much a function of portrayal as content. (See Tufte's (1997a, 1997b) work for outstanding examples of the importance of portrayal.) Regardless of which requirement (checking, improvement, planning, diagnosis, or promotion) prompted measurement, it is the user's perception of the portrayed information that is used to evaluate the performance of the phenomenon of interest. Evaluation results in continued measurement and evaluation, redesign of how the phenomenon is measured, or discontinuation and perhaps a transfer of interest to another phenomenon (Coleman and Clark, 2001).

12.3 Purposes of Strategic Performance Measurement

The effectiveness of performance measurement is often dependent on its purpose. That is, when one is evaluating whether a particular indicator is a "good" performance measure, one must first ask what the intended purpose of the indicator is. An indicator that is good for one purpose may not be as effective for another. Alternatively, an indicator that is potentially good for two or more purposes may best be used for only one purpose at a time. The use of the same indicator for potentially competing purposes, even though it could meet either purpose under ideal conditions, may lead to distortion (tampering), reluctance to report performance, or unexpected consequences, such as a lack of cooperation among the units being measured. In organizations, performance is typically measured for one or more of the following purposes:

- Control
- Improvement
- Planning
- Diagnosis
- Promotion

12.3.1 Control

Measuring performance for control may be viewed as measuring to check that what is expected has in fact occurred. Typically, a manager uses control indicators to evaluate the performance of some part of the organization for which the manager is responsible, such as a plant or department. A higher level manager may have multiple units to control and require separate indicators from each unit. A lower level manager may use indicators to control the performance of the individuals who work directly for that manager. In either case, the individual or unit whose performance is being monitored and controlled reports performance "upline" to the manager. If another part of the organization has the measurement responsibility (e.g., accounting and finance, quality control, or internal audit), it reports the most recent value of the indicators to the manager. The manager then reviews the level of performance on these indicators to check if the expectations are being met. Depending on the results of the comparison of current performance to expectations, and the manager's personal preferences, the manager takes action (or not) to intervene with the unit for the purpose of changing future levels of performance. Too often, managers only provide substantial feedback to the unit being evaluated when performance does not meet expectations. Control can be better maintained and performance improved when managers also reinforce good performance by providing feedback on expectations that are being met.

Care should be taken to distinguish between using an indicator to *control* the performance of an organizational unit and using the same indicator to judge the performance of the individuals managing or working in that unit. Measures of performance needed by managers may include elements of performance not completely within the control of those managing and working in that unit. For example, an indicator of total revenue generated by a plant may reflect the effectiveness of ongoing sales efforts, unit pricing pressure in the market, or a temporary downturn in the economy. While taking action in response to any of these factors may be appropriate for the senior-level manager who checks this plant's performance, judging the performance of local managers at the plant level by total revenue could lead to an emphasis on "making the numbers" over focusing on the factors that the local managers do control. "Making the numbers" in this situation could lead to such potentially undesirable consequences as building to inventory or spending for overtime to meet increased production targets generated by lower sales prices. A good rule of thumb is to measure performance one level above the level of control over results to encourage strategic action and to avoid suboptimization. At the same time, judgment of the performance of individual managers should focus on the causes and effects they control within the context of overall organizational performance. It is leadership's job to assist these managers in dealing with the factors beyond their control that affect their unit's overall performance.

12.3.2   Improvement

Measuring performance for improvement is more internally focused than measuring for control. Measuring for improvement focuses on measuring the performance of the unit one is responsible for and obtaining information to establish current performance levels and trends. The emphasis here is less on evaluating something or someone's performance, and more on understanding current performance levels, how performance is changing over time, the impact of managerial actions, and identifying opportunities for improving performance. Managers often measure a number of things for use by themselves and their subordinates. An astute manager will identify drivers of end-result performance (e.g., sales, profits, customer warranty claims) and develop indicators that lead or predict eventual changes in these end results. Such leading indicators might include employee attitudes, customer satisfaction with service, compliance with quality-management systems, and percent product reworked. Sears found that changes in store-level financial results could be predicted by measuring improvements in employee attitudes toward their job and toward the company. This predicted employee behavior, which, in turn, influenced improvements in customer behavior (customer retention and referral to other customers), leading, finally, to increases in revenue and operating margin (Rucci, et al., 1998).

Employees, supervisors, and managers should be encouraged to establish and maintain indicators that they can use as yardsticks to understand and improve the performance of their units, regardless of whether these indicators are needed for reporting upline. Simply measuring a key performance indicator and making it promptly visible for those who deliver this performance can lead to improvement with little additional action from management. This assumes that those who deliver this performance know the desired direction for improvement on this indicator and have the resources and discretion to take actions for improvement. It is leadership's job to make sure the people in the organization have the knowledge, resources, discretion, and direction to use performance information to make improvements.

12.3.3   Planning

Measuring for the purpose of planning has at least two functions: (1) increasing understanding of current capabilities and the setting of realistic targets (i.e., goals) for future performance; and (2) monitoring progress toward meeting existing plans. One could argue that these simply represent planning-centric versions of measuring for improvement and then measuring for control. The role of measuring performance as part of a planning effort is important enough to warrant a separate discussion.

Nearly all strategic management or strategic planning efforts begin with understanding the organization and its environment. This effort is referred to as internal and external strategic analysis (Thompson and Strickland, 2003), organizational systems analysis (Sink and Tuttle, 1989), or, in plain words, "preparing to plan." A key part of internal analysis is understanding current performance levels, including the current value of key performance indicators and their recent trends. This provides the baseline for future performance evaluations of the effectiveness of the planned strategy and its deployment. Also, the choice of key performance indicators tells the organization what is important and is a specific form of direction often more carefully followed than narrative statements of goals and vision. Understanding current performance and its relation to current processes and resources provides managers with a realistic view of what is possible without having to make substantial changes to the system. Thus, setting intelligent targets for future performance requires an understanding of how implementation of the plan will change processes and resources to enable achievement of these targets. A key part of the external analysis is obtaining relevant comparisons so that the competitiveness of current performance levels and future performance targets can be evaluated. To answer the question of how good a particular performance level is, one must ask "compared to what?" Current competitor performance provides an answer to this question, but it must be assumed that competitors are also planning for improved performance. Setting future performance targets must take this moving competitive benchmark into account. Even the projected performance of your best current competitor may be inadequate as a future performance target to beat. The strategic management literature is full of examples of corporations that did not see their new competition coming and were blindsided by new competitors playing by different rules with substitutes for these corporations' bread-and-butter products (see Hamel and Prahalad, 1996; Hamel, 2002). As Drucker (1998) has pointed out, some of the most important information managers need comes from outside their organizations and even outside their industries. A challenge for performance measurement is to provide not only internal, but is also external performance information that provides competitive intelligence for making strategic decisions.

Most strategic management or strategic planning processes include a last or next to last step that serves to measure, evaluate, and take corrective action. Often, this step is expected to be occurring throughout the process, with the formal execution of the explicit step occurring after goals have been set, action plans deployed, and strategy implementation is underway. That is, periodic review of progress toward meeting goals is a regular part of a strategic management effort, and performance indicators can provide evidence of that progress. When the goal-setting process includes the identification of key performance indicators and future performance targets for each indicator, the decision of which indicators to review has largely been made. In cases where goals are perhaps more qualitative or include simple quantitative targets without an operationally defined performance indicator, the planning team must choose or develop a set of progress indicators for these periodic (e.g., monthly or quarterly) reviews. A rule of thumb for these cases, based on the work of Sink and Tuttle (1989), is to develop indicators that provide evidence of the effectiveness, efficiency, quality, and impact of progress on each goal. Each of these terms is defined in Section 12.4. Even when key performance indicators have been predetermined at the time of goal setting, additional "drill-down" indicators may be required to explain performance trends and illustrate perceived cause-and-effect relationships among managerial actions, environmental and competitor actions, and observed levels of performance on end-result indicators.

Once the indicators have been chosen or developed, the periodic reviews are much more than collecting data, reporting current performance levels, and comparing to plan. How these reviews are conducted has a major impact on the organization's approach and even success with strategic management. If the reviews overemphasize checking or making sure that the people responsible for each goal are making their numbers, then reviews run the risk of taking on a confrontational style and may lead to gaming, distortion, and hoarding of information. On the other hand, reviews that focus on what can be learned from the performance information and sharing lessons, and even resources when needed, can lead to better goal setting, improved action plans for implementing strategies, and increased sharing of performance information that may indicate future trends, good or bad. The type of review chosen is likely to reflect the organization's culture and the leadership's preferences. While either style may be used to drive performance, the two styles differ in the types of initiatives and actions leadership must take outside of and between periodic reviews to support performance improvement.

12.3.4   Diagnosis

Measuring performance for diagnosis or screening (Thor, 1998) is similar to the drill-down described for illustrating cause-and-effect relationships among controllable and noncontrollable factors and their impact on end results. When an undesired (or desirable but unexplainable) result on a key indicator is observed, exploring the recent history of related indicators may provide insight into the possible causes. Tools such as the cause-and-effect (fishbone) diagram (Goal/QPC, 1985; Ishikawa, 1985) or quality function deployment

(Akao, 1990) are useful in identifying drill-down metrics, likely to be at the cause of the observed effect. Unlike the previous methods, which are used for continual measurement of performance, measuring for diagnosis may be a one-time measurement activity with a start and an end. Thus, the resources devoted to systematizing or institutionalizing the new indicators required should be balanced against the likelihood that these indicators will be needed again in the near future.

12.3.5   Promotion

Measuring for promotion (an idea contributed by Joanne Alberto) is using performance indicators and historical data to illustrate the capabilities of an organization. The intent is to go beyond simple sales-pitch claims of cutting costs by X% or producing product twice as fast as the leading competitor. Here, the manager is using verifiable performance information to show the quantity and quality of product or service the organization is capable of delivering. Not only does this performance information show what is currently possible, it also provides a potential client with evidence that the organization measures (and improves) its performance as part of its management process. Thus, the customer can worry less about having to continually check this provider's performance and can rely upon the provider to manage its day-to-day performance. A caveat here is that it is important to balance the organization's need to protect proprietary performance information with the customer's need for evidence of competitive product and service delivery. Care should also be taken in supporting the validity of promotional performance information so that the claims of less scrupulous competitors, who may boast of better levels of performance but present poorly substantiated evidence, are discounted appropriately.

Once the manager or engineer has clarified why performance is being measured, the question of what to measure should be addressed. Organizational performance is multidimensional, and a single indicator rarely meets all the needs of the intended purpose.

12.4   Dimensions of Performance

While it has long been recognized that organizational performance is multidimensional, the practice of measuring multiple performance dimensions was popularized by the introduction of Kaplan and Norton's (1992) Balanced Scorecard. At its core, the Balanced Scorecard recognizes that organizations cannot be effectively managed with financial measures alone. While necessary for survival, financial measures tend to be lagging indicators of results and are frequently difficult to link to managerial actions aimed at improving medium- to long-term performance. Compounding this shortcoming, financial measurement systems are typically designed to meet reporting requirements for publicly traded companies or auditor's requirements for government agencies and privately held companies (i.e., financial accounting). Providing information to support managing the organization (i.e., managerial accounting) is an afterthought. This creates a situation where indicators developed for one purpose (fiscal control) are reused for another purpose (management and improvement), creating predictable problems.

This section describes a number of frameworks for organizing the multiple dimensions of organizational performance. Each framework is a useful tool for auditing an organization's collective set of indicators to identify potential gaps. The intent here is neither to advocate the adoption of a specific framework as the measurement categories for a given organization, nor to advocate that an organization has at least one indicator for every dimension of these frameworks. The astute management team must recognize that organizational performance is multidimensional and make sure their measurement system provides performance information on the dimensions key to the success of their organization (see Section 12.4.1).

The Balanced Scorecard views organizational performance from four perspectives, with the financial perspective being one of those four. The other three perspectives are the customer perspective, the internal process perspective, and the learning and growth perspective. Kaplan and Norton (1996) later suggested a general causal structure among the four perspectives. Thus, managerial actions to improve learning and growth, both at the individual and organizational level, should result in improved performance on indicators of internal process performance, assuming the learning and growth initiatives and indicators are aligned with the internal process objectives. Improved performance on internal process indicators should result in improved results of the customer perspective indicators, if the process indicators reflect performance that is ultimately important to customers. And finally, if the customer perspective indicators reflect customer behaviors likely to impact the organization, then it is reasonable to expect improved performance on these customer indicators to lead improved financial performance. For example, an initiative aimed at improving the quality assurance skills of quality technicians and quality management skills of production line supervisors might be indicated by increased numbers of Certified Quality Technicians and Certified Quality Managers (learning and growth indicators). Assuming this initiative was aimed at closing a relevant gap in skills, the application of these skills could be expected to improve levels of internal process indicators such as percent scrap and shift the discovery of defects further upline in the value stream (potentially reducing average cycle time for good product produced). Improvements in results on these internal process indicators could lead to fewer customer warranty returns, translating into direct financial savings. Improved performance on other customer-perspective indicators such as customer perceptions of quality and their likelihood to recommend the product to others, although less directly linked, may also be predictors of improved financial results such as increased sales.

While popular, the Balanced Scorecard has received some valid criticism. Nørreklit (2003) argues that the Balanced Scorecard has generated attention on the basis of persuasive rhetoric rather than on convincing theory. Theoretical shortcomings include suggested cause-and-effect relationships based on logic rather than empirical evidence and use of a strategic management system without addressing key contextual elements of strategic management (e.g., monitoring key aspects of the dynamic external environment or employing a top-down control model for implementation that appears to ignore organizational realities). Pfeffer and Sutton (2000, p. 148) point out that the Balanced Scorecard is "great in theory," but identify a number of problems in its implementation and use:

> "the system is too complex, with too many measures; the system is often highly subjective in its actual implementation; and precise metrics often miss important elements of performance that are more difficult to quantify but that may be critical to organizational success over the long term."

The industrial engineer's challenge is to sort through these shortcomings and address them with a well-designed measurement system that aligns with other management systems and balances practical managerial needs with theoretical purity. Practical issues related to designing and implementing a measurement system are described in Section 12.5.

The industrial engineer should recognize that the Balanced Scorecard is only one framework for organizing the dimensions of organizational performance, and should be familiar with various alternatives and develop or adapt a framework that fits the organization's needs. For example, Richard Barrett (1999a) proposed enhancing the Balanced Scorecard by expanding the customer perspective to include suppliers' perspectives and adding three additional perspectives: corporate culture, community contribution, and society contribution. Certainly the importance of supply chain management and partnering with suppliers warrants the inclusion of a supplier perspective in an organizational scorecard. Corporate culture has long been recognized as important to organizational success (Deal and Kennedy, 1982; Peters and Waterman, 1982) and appears as a key factor in the popular press accounts of great organizations. However, much work remains regarding how best to measure corporate culture and to use this information to manage the organization better. Management scholar Ralph Kilmann (1989; Kilmann and Saxton, 1983) and industrial engineer Larry Mallak (Mallak, et al., 1997; Mallak and Kurstedt, 1996) offer approaches to measuring corporate culture. Barrett's recommendation to measure community and societal contributions are similar dimensions measured at different levels. Community contribution includes not only the cities, counties, and states where the organization and its employees reside and sell their products, but also the industries and professions in which the organization operates. Societal contribution expands beyond local impact and measures the organization's immediate and longer term global impact.

A widely accepted performance dimensions framework that is updated annually is the Business Results category of the Baldrige Criteria for Performance Excellence (Baldrige National Quality Program, 2004).

This category consists of six items that may be thought of as performance dimensions: product and service outcomes; customer-focused results; financial and market results; human-resource results; organizational effectiveness results; and leadership and social responsibility results. When identifying indicators for each dimension, the Baldrige criteria stress choosing indicators that are linked to organizational priorities such as strategic goals and key customer requirements. The criteria also emphasize reporting results by product, service, customer, or market segment as appropriate and providing comparative data to facilitate the evaluation of levels and trends. The Baldrige criteria also include relative weights for each of these dimensions, although the 2005 version weights the last five dimensions equally at 70 out of 450 total points, and the product and service outcomes dimension is weighted slightly higher with 100 out of the total 450 points.

Indicators of product and service outcomes provide evidence of the performance of products and services that are important to customers. In the food service industry where customers want healthy eating alternatives, this might include providing comparisons of nutritional information of your leading products to those of key competitors. Indicators of customer-focused results provide evidence of the attitudes and behaviors of customers toward a company's products and services. This requires not only indicators of customer satisfaction and dissatisfaction, but also indicators of customer loyalty and their likelihood to recommend the company's products to others. Financial and market results include traditional financial indicators such as return on investment and profitability and market indicators such as market share (by segment) and growth. Human resource results are particular relevant to industrial engineers because they include indicators of the effectiveness of work systems. The first area to address under work systems is the organization and management of work, including how work and jobs are organized and managed "to promote cooperation, initiative, empowerment, innovation, and organizational culture" and "to achieve the agility to keep current with business needs and to achieve... action plans" (Baldrige National Quality Program, 2004, p. 21). Measuring the effectiveness of work systems should be at least one indicator of the performance of the industrial engineering function. Other items to be reported under human-resource results include indicators of employee learning and development and employee well-being, satisfaction and dissatisfaction. Such indicators are not just the domain of the human-resource manager, but include indicators that reflect the effectiveness of the work systems and supporting aids developed by the industrial engineers, such as indicators of safety, absenteeism, workers' compensation claims, cross-training rates, and success rates in adopting new work processes. The organizational effectiveness results performance dimension addresses the unique aspects of each organization by first focusing on indicators of the performance of key value-adding processes and then focusing on similar indicators for other key processes. Many of the indicators falling under this dimension drive performance in the other dimensions; examples include indicators of process cycle time, productivity, supplier performance, and efficiency. The final dimension in the Baldrige results framework is leadership and social responsibility results. This dimension starts with indicators of the accomplishment of strategy and action plans (earlier described as indicators of progress against plans) to show the effectiveness of leadership and leadership's direction. Evidence of appropriate leadership action and deployment of direction is then requested in the form of indicators of ethical behavior, stakeholder trust in leaders, and effective governance. These are further supported with indicators of fiscal accountability, and regulatory and legal compliance. Finally, this dimension asks for indicators of organizational citizenship via support for key communities. This dimension is relatively new, having been introduced after some of the corporate scandals of the late 20th century.

D. Scott Sink provides an industrial engineer's view of performance with his seven performance criteria (Sink, 1985; Sink and Tuttle, 1989). He suggests that organizational performance can be described in terms of seven interrelated criteria:

- *Effectiveness*: indicators of doing the correct things; a comparison of actual to planned outputs
- *Efficiency*: a resource-oriented criterion; a comparison of planned to actual resources used
- *Quality*: defined by one or more of David Garvin's (1984) five definitions of quality (transcendent, product-based, manufacturing-based, user-based, or value-based) and measured at up to five (or six) points throughout the value stream (more on quality later in this section)

- *Productivity*: an indicator based on a ratio of outputs to the inputs required to produce those outputs (more on productivity later)
- *Innovation*: indicators of organizational learning and growth as applied to the organization's current or future product and service offerings
- *Quality of work-life*: indicators of employee-centered results; preferably those predictive of higher levels of employee work performance
- *Profitability/budgetability*: indicators of the relationship of revenues to expenses; whether the goal is to make a net profit or to stay within budget (while delivering expected levels of service)

12.4.1 The Concept of Key Performance Areas

Regardless of which framework an organization chooses, the performance indicators should reflect the key performance areas of that organization. Key performance areas are the vital few categories or dimensions of performance for a specific organization. KPAs may or may not reflect a comprehensive view of performance, but they do represent those dimensions most critical to that organization's success. While the indicators used to report the performance of each KPA might change as strategy or the competitive environment changes, the KPAs are relatively constant. A personal example is that of the snow skier, whose skiing mission might be to have fun, build capability, and get some exercise. A skier's KPAs might include speed, fatigue, and capability. While speed might have once been indicated by success in racing against competitors (a relative rather than absolute indicator of speed), the older and wiser skier may prefer an indicator that reflects the fastest speed at which one can ski fully under control (e.g., perhaps time down a favorite slope). Rather than simply adopting one of the performance dimensions frameworks described previously, an organization's managers should familiarize themselves with the alternative frameworks and customize the dimensions of their organizational scoreboard to reflect their organization's KPAs. What is most important is that the measurement system provide the managers with the information necessary to evaluate the organization's performance in all key areas (i.e., KPAs) as opposed to conforming to someone else's definition of balance.

12.4.2 Productivity

Productivity is a particularly important concept for industrial engineers and warrants further discussion here. Productivity indicators reflect the ratio of an organization's or individual's outputs to the inputs required to produce those outputs. The challenge is determining which outputs and inputs to include and how to consolidate them into a single numerator and denominator. Outputs include all the products and services an organization produces and may even include by-products. Inputs include labor, capital, materials, energy, and information.

Many commonly used productivity indicators are actually partial measures of productivity. That is, only part of the total inputs used to produce the outputs are included in the denominator. The most common are measures of labor productivity, where the indicator is a ratio of outputs produced to the labor inputs used to produce them (e.g., tons of coal per man day, pieces of mail handled per hour). While relatively simple and seemingly useful, care should be taken in interpreting and evaluating the results of partial productivity indicators. The concept of input substitution such as increasing the use of capital (e.g., new equipment) or materials (e.g., buying finished components rather than raw materials) may cause labor productivity values to increase dramatically, owing to reasons other than more productive labor. A more recent shortcoming of measuring labor productivity is that direct labor has been steadily decreasing as a percent of total costs of many manufactured, mined, or grown products. In some cases, direct labor productivity today is at levels almost unimaginable 20 or 30 years ago. One might argue that the decades-long emphasis on measuring and managing labor productivity has succeeded, and that industrial engineers in these industries need to turn their attention to improving the productivity of materials and energy, and perhaps indirect labor. For more information, Sumanth (1998) provides a thoughtful summary of the limitations of partial productivity measures.

Total or multifactor productivity measurement approaches strive to address the limitations of partial productivity measures. Differing outputs are combined using a common scale such as constant value dollars to produce a single numerator, and a similar approach is used to combine inputs to produce a single denominator. Total factor approaches include all identifiable inputs, while multifactor approaches include two or more inputs, typically the inputs that make up the vast majority of total costs. The resulting ratio is compared with a baseline value to determine the percent change in productivity. Miller (1984) provides a relatively simple example using data available from most accounting systems to calculate the changes in profits due to any changes in productivity, as well as to separate out profit changes due to price recovery (i.e., net changes in selling prices of outputs relative to the changes in purchasing costs of inputs). Sink (1985) and Pineda (1996) describe multifactor models with additional analytical capabilities, useful for setting productivity targets based on budget targets and determining the relative contributions of specific inputs to any changes in overall productivity.

12.4.3 Quality

Quality, like productivity, deserves additional attention in an industrial engineer's view of measuring performance. Quality is ultimately determined by the end-user of the product or service. And often, there are many intermediate customers who will judge and perhaps influence the quality of the product before it reaches the end-user. As there are numerous definitions of quality, it is important to know which definition your customers are using. While your first customer downstream (e.g., an original equipment manufacturer or a distributor) might use a manufacturing-based (i.e., conformance to requirements) indicator such as measuring physical dimensions to confirm they fall within a specified range, the end-user may use a user-based (i.e., fitness-for-use) indicator such as reliability (e.g., measuring mean time between failures [MTBF]) to evaluate quality. A full discussion of the five common definitions of quality and the eight dimensions of quality (performance, features, reliability, conformance, durability, serviceability, aesthetics, and perceived quality) is found in Garvin (1984). While seemingly adding confusion to the definition of quality within a larger performance construct, Garvin's eight dimensions of quality can be thought of as differing perspectives from which quality is viewed. Without multiple perspectives, one may get an incomplete view of a product's quality. As Garvin points out, "a product can be ranked high on one dimension while being low on another" (p. 30).

Once one or more definitions of quality have been chosen, the industrial engineer must decide where to measure quality prior to finalizing the indicators to be used. Sink and Tuttle (1989) describe quality as being measured and managed at five (later six) checkpoints. The sixth, sometimes omitted, checkpoint is measuring the overall quality management or quality assurance process of the organization. Today we can easily relate this sixth checkpoint to the registration of an organization's quality management systems, as evidenced by receiving an ISO 9001 certificate. The five checkpoints correspond to key milestones in the value stream, with checkpoints two and four representing traditional incoming quality measurement (prior to or just as inputs enter the organization) and outgoing quality measurement (just before outputs leave the organization), respectively. Quality checkpoint three is in-process quality measurement, a near-discipline in its own right with statistical process control methods, gage calibration, certified quality technicians, certified quality engineers. At checkpoint three we are measuring the key variables and attributes of products and services that predict or directly lead to the desired characteristics at outgoing quality measurement (quality checkpoint four) as well as those that contribute to success on the quality dimensions that are important further downstream (see checkpoint five). Tracking such variables and attributes lends itself to statistical analysis. See other chapters for a discussion of statistical process control. For example, chapters 3 and 36. For an excellent introduction to applying statistical thinking and basic methods to management data, see Donald Wheeler's *Understanding Variation* (1993). The novice industrial engineer can benefit by taking heed of the late W. Edwards Deming's often stated admonition to begin by "plotting points" and utilizing the "most under-used tools" in management, a pencil and piece of grid paper. Quality checkpoint one is proactive management of suppliers and includes the indicators used to manage the supply chain. What might be incoming, in-process, outgoing, or overall quality management system indicators from the supplier's perspectives are quality checkpoint one indicators from the receiving organization's perspective. Quality checkpoint five is the measurement of product and service quality after it has left the organization's direct control and is in the hands of the customers. Quality checkpoint five might include indicators from the Baldrige items of product and service outcomes and customer-oriented results. Quality checkpoint five indicators provide evidence that products or services are achieving the outcomes desired by customers and the customer's reactions to those outcomes.

12.4.4　Human Capital

Industrial engineers have long been involved in the measurement and evaluation of the performance of individuals and groups. As the knowledge content of work has increased, the cost and value of knowledge workers has increased. Organizations spend substantial energy and resources to hire, grow, and retain skilled and knowledgeable employees. Although these expenditures are likely to appear in the income statement as operating costs, they are arguably investments that generate human capital. While an organization does not own human capital, the collective knowledge, skills, and abilities of its employees do represent an organizational asset — one that should be maintained or it can quickly lose value. Organizations need better measurement approaches and performance indicators to judge the relative value of alternative investments that can be made in human capital. They need to know which are the most effective options for hiring, growing, and keeping talent. The following paragraphs provide the industrial engineer with context and examples to help tailor their performance measurement toolkit to the unique challenges associated with measuring the return on investments in human capital.

Traditional human resource approaches to measuring human capital have focused on operational indicators of the performance of the human-resources function. In particular, these indicators have emphasized the input or cost side of developing human capital. Such indicators might include average cost to hire, number of days to fill an empty position, or cost of particular employee benefits programs. Current state-of-the-art approaches (Becker, et al., 2001) focus on business results first, and then link indicators of how well human capital is being managed to those results.

Assuming the organization has developed a multidimensional performance measurement system as described earlier in this chapter, the next step is to identify human capital-related drivers of the leading organizational performance indicators (e.g., product and service results, customer-oriented results as opposed to lagging performance results such as financial results or social responsibility results). Such drivers are likely to be related to employee attitudes and behaviors. Drivers of customer-oriented results might include employee attitudes towards their jobs or supervisors, or behaviors such as use of standard protocols and knowing when to escalate an issue to a customer-service manager. Drivers of product and service results might include behaviors such as use of prescribed quality assurance procedures, completing customer orientation upon delivery, or perhaps an organizational effectiveness indicator such as cycle time (that is, where cycle time is heavily dependent upon employee performance). Indicators of the health of an organization's human capital are likely to predict or at least lead performance on these human capital drivers of organizational performance. Indicators of the health of human capital reflect the value of the human asset. Examples of such indicators include average years of education among knowledge workers (assumes a relatively large pool of employees), a depth chart for key competencies (i.e., how many employees are fully qualified to fulfill each mission), attrition rates, or more sophisticated turnover curves that plot turnover rates in key positions by years of seniority. Finally, traditional cost-oriented measures of human resource programs can be evaluated in terms of their impact on the health of human capital and human capital drivers of organizational performance.

Human capital indicators should help answer questions such as: Does the new benefit program reduce turnover among engineers with 10 to 20 years of experience? Does the latest training initiative expand our depth chart in areas that were previously thin, thus reducing our risk of not being able to meet product and service commitments? Do changes to our performance management system improve employee attitudes among key customer interface employees? Do our initiatives aimed at improving employee attitudes and behaviors translate into better products and services as well as customers who increase the percentage of their business they give to our organization? Measuring human capital and the return on investments in human capital are new frontiers in measurement for industrial engineers, with the potential to make substantial contributions to organizational competitiveness.

Those interested in a philosophical discussion of performance dimensions and how to choose the appropriate unit of analysis should read Kizilos' (1984) *Kratylus automates his urnworks*. This thought-provoking article sometimes frustrates engineers who are looking for a single "correct" answer to the question of what dimensions of performance should be measured. The article is written in the form of a play with only four characters and makes an excellent group exercise.

12.5   Implementing a Measurement System

Once you are clear about why you are measuring performance and have a general idea about what dimensions of performance to measure, the question becomes how to make this a functioning measurement system. The measurement system includes not only the specific indicators, but also the plan and procedures for data gathering, data entry, data storage, data analysis, and information portrayal, reporting, and reviewing. A key recommendation is that those whose performance is being measured should have some involvement in developing the measurement system. The approaches that can be used to develop the measurement system include: (1) have internal or external experts develop it in consultation with those who will use the system; (2) have the management develop it for themselves and delegate implementation; (3) have the units being measured develop their own measurement systems and seek management's approval; or (4) use a collaborative approach involving the managers, the unit being measured, and subject matter expert assistance. This last approach can be accomplished by forming a small team, the measurement system design team.

12.5.1   The Measurement System Design Team

A "design team" is a team whose task is to design and perhaps develop the measurement system; however, day-to-day operation of the measurement system should be assigned to a function or individual whose regular duties include measurement and reporting (i.e., it should be an obvious fit with their job and be seen as job enlargement rather than an add-on duty unrelated to their regular work). When ongoing performance measurement is assigned as an extra duty, it tends to lose focus and energy over time and falls into a state of neglect. Depending on how work responsibility is broken down in an organization, it may make sense to assign responsibility for measurement system operation to industrial engineering, accounting and finance, the chief information officer, quality management/assurance, or even human resources. The design team should include the manager who "owns" the measurement system, a measurement expert (e.g., the industrial engineer), two or more employees representing the unit whose performance is being measured, and representatives from supporting functions such as accounting and information systems.

The team's first task is to clarify the purpose of the measurement system. While the purpose may have been stated when the team was formed, it is critical that everyone on the team understand why this system is being created and what is expected from it. Design questions that arise during measurement system development can often be answered by referring back to the purpose of the system. Next, the team should identify all the users of the system. If the system is being created for control purposes, then the manager or management team exerting control is the primary user. If the system is being created to support improvement, then most or all of the unit being measured may be users. The team's task is to ask the users how they will use this measurement system. Specifically, what kinds of decisions do they intend to make based upon the information they receive from this measurement system? What information (available now or not) do they feel they need to support those decisions? This may require developing trial or example indicators and showing them to the users for reaction and feedback. Before finalizing the list of indicators to be included, the team should develop examples of what the results reported for these indicators will look like, on the basis of clear operational definitions including formulae and identification of data sources.

12.5.2 Input/Output Analysis Using the SIPOC Model

A useful tool for helping users identify information needs at the organizational level is the input/output analysis or SIPOC (suppliers, inputs, processes, outputs, and customers) model. The intent is to get the users to describe their organization as an open system, recognizing that in reality there are many feedback loops within this system that make it at least a partially closed loop system. The SIPOC model is particularly useful for the design team approach to developing a measurement system. The model helps the team members gain a common understanding of the organization and provides a framework for discussing the role and appropriateness of candidate indicators.

The first step of the SIPOC model is to identify the organization's primary customers, where a customer is anyone who receives a product or service (including information) from the organization. Next it identifies the outputs, or specific products and services, provided to these customers: for an organization with a limited number of products and services, these can be identified on a customer-by-customer basis; for an organization with many products and services, it is more efficient to identify the products and services as a single comprehensive list and then audit this list customer by customer to make sure all relevant products and services are included.

The next step is not typically seen in the SIPOC model, but it is a critical part of any input/output analysis. It starts with the identification of the customers' desired outcomes, i.e., the results they want as a consequence of receiving the organization's products and services. A customer who purchases a car may want years of reliable transportation, a high resale value, and styling that endures changes in vogue. A customer who purchases support services may want low-cost operations, seamless interfaces with its end users, and a positive impact on its local community. While the organization may not have full control in helping its customers achieve these desired outcomes, it should consider (i.e., measure) how its performance contributes or influences the achievement of these outcomes. The identification of desired outcomes also includes identifying the desired outcomes of the organization, such as financial performance (e.g., target return on investment, market share), employee retention and growth, repeat customers, and social responsibility. Measuring and comparing customer's desired outcomes to the organization's desired outcomes often highlights key management challenges, such as balancing the customer's desire for low prices with the organization's financial return targets. Measuring outcomes helps the organization understand customer needs beyond simply ensuring that outputs meet explicit specifications.

At the heart of the SIPOC model is the identification of processes, particularly the processes that produce the products and services. A second list of support processes, those that provide internal services necessary to the functioning of the organization but are not directly involved in producing products or services for external consumption, should also be identified. Processes lend themselves to further analysis through common industrial engineering tools such as process flow charts, which are often used in value-stream mapping and other lean initiatives. Process flow charts are useful for identifying key measurement points in the flow of information and materials and thus the source of many operational performance indicators. Strategic performance measurement may include a few key process indicators, particularly those that predict the successful delivery of products and services. Once processes are identified, the inputs required for those processes are identified. As with outputs, it may be more efficient to identify inputs as a single list and then compare them to the processes to make sure all key inputs have been identified. The five generic categories of inputs that may be used to organize the list are labor, materials, capital, energy, and information. In order to be useful for identifying performance indicators, the inputs must be more specific than the five categories. For example, labor might include direct hourly labor, engineering labor, contracted labor, management, and indirect labor. These can be classified further if there is a need to measure and manage labor at a finer level, although this seems more operational than strategic. Relevant labor indicators might include cost, benefits, hours, percent of total cost, and absenteeism (in terms of percentage). Materials might include purchased components, raw materials, supplies, software, and purchased services such as shipping or travel. For some organizations, it may make sense to break purchased services out as a separate category. Capital includes working capital, depreciated equipment, facilities, and property. Energy includes electricity, natural gas, and fuel. Information may include decisions (e.g., from headquarters), regulations, specifications, and requests for quotations.

The last component of the SIPOC model is the identification of suppliers. While this component has always been important, the advent of overt improvement approaches such as supply-chain management and the increased reliance on outsourcing have made the selection and management of suppliers a key success factor for many organizations. Suppliers can also be viewed as a set of upstream processes that can be flow charted and measured like the organization's own processes. The design team may wish to work with key suppliers to identify indicators of supplier performance that predict the success of (i.e., assure) the products and services being provided as inputs in meeting the needs of the organization's processes and subsequent products and services.

12.5.3   Identifying Potential Indicators

Once the design team has developed its viewpoint on the measurement system users' information needs and a collective understanding of the organization's business, it is time to identify the first level indicators for the measurement system. These indicators have likely been emerging throughout the team's work to date. An inventory of existing performance indicators should be completed early in the team's project. User needs should be translated into candidate metrics. Process flow charts and the SIPOC model should be used to identify actions, products, services, and results that need to be measured and evaluated. At this point, the team should not allow itself to be limited by the feasibility of a particular indicator or the perception that an indicator represents performance beyond the organization's control. If an indicator is later deemed impractical to measure, a suitable surrogate can be developed. Indicators that reflect performance only partially controllable or influenced by the organization are often those most important to customers and end-users. When the organization has only partial control of a performance indicator of importance to customers, the organization needs to understand its contribution to that performance and how it interacts with factors beyond its control. The team should be able at this point to produce a list of candidate indicators for further consideration.

Before identifying additional candidate indicators, the team should decide whether to proceed in a bottom-up or top-down fashion. A bottom-up approach would have the team continue to identify candidate indicators, perhaps using a group technique such as brainstorming or the nominal group technique (Delbecq, et al., 1975). Once the team feels it has a relatively comprehensive list of candidate indicators, it should then consolidate the list using a technique such as affinity diagrams (Brassard, 1989) or prioritize the list with the nominal group technique or analytical hierarchy process. The aim here is to shorten the candidate list to a more manageable size by clustering the indicators into categories that form the foundation for the dimensions of the organization's scoreboard or a prioritized list from which the "vital few" indicators can be extracted and then categorized by one or more of the performance dimensions frameworks (see Section 12.4) to identify gaps. A top-down approach would begin by choosing one of the performance dimension frameworks (e.g., the six Baldrige Results Items) as the dimensions of the organization's scorecard, then identifying and categorizing candidate metrics by dimension. Dimensions with few or no candidate indicators will require additional analysis and synthesis to produce candidates for that dimension. The resulting lists for each dimension should then be prioritized to produce the vital few indicators for each dimension. The team is now ready to try the indicators out with users and obtain fitness-for-use feedback.

12.5.4   Portraying Information — Indicator Format

An important part of piloting and later institutionalizing the vital few indicators is to develop appropriate portrayal formats for each indicator. What is appropriate depends upon the user's preferences, the indicator's purpose, and how results on the indicator will be evaluated. User preferences may include charts versus tables, use of color (some users are partially or fully color-blind), and the ability to drill down and easily obtain additional detail. An indicator intended for control purposes must be easily transmissible in a report format and should not be dependent upon color (the chart maker often loses control of the chart once it is submitted, and color charts are often reproduced on black-and-white copiers), nor should it be dependent upon verbal explanation. Such an indicator should also support the application of statistical thinking so that common causes of variation are not treated as assignable causes, with the accompanying request for action. An indicator intended for feedback and improvement of the entire organization or a large group will need to be easily understood by a diverse audience, large enough to be seen from a distance, and easily dispersed widely and quickly. A detailed discussion of portrayal is beyond the scope of this chapter. Design teams should support themselves with materials such as Wheeler's *Understanding Variation* (1993) and Edward Tufte's booklet, *Visual and Statistical Thinking: Displays of Evidence for Decision Making* (1997a), a quick and entertaining read on the implications of proper portrayal.

Rules of thumb for portraying performance information include the following considerations:

- A picture is often worth a thousand words, so charts, sketches, and photographs should be used when they meet user needs.
- Start by developing the chart on paper (by hand), before moving to computer-generated graphics. Starting with computer-generated charts often leads to a portrayal based on what the tool can do rather than what the user desires.
- ALL CAPS IS HARDER TO READ AND IMPLIES SHOUTING; so use upper- and lowercase text.
- An accompanying table of the data used to produce the chart is desirable whenever possible.
- Longitudinal data is always preferable. If a change in process or product results in a capability that is no longer comparable, annotate this change in capability and continue to show historical performance until the new capability is well established.
- For high-level indicators that aggregate performance or only indicate end results, driver indicators that provide an explanation of changes observed in the high-level indicator should be provided as supporting material (to support cause-and-effect thinking).
- Indicators should help the user understand the current level of performance, the trend in performance, and provide appropriate comparisons for evaluation. Comparisons with the performance of competitors, customer expectations, or targets set by the organization provide context for judging the desirability of results.
- When using labels to note acceptable ranges of variability, clearly distinguish limits based on the capability of the process from limits established by customers (i.e., specifications) and limits established by management (i.e., targets).
- The date produced or revised and the owner (producer) of the indicator should be clearly labeled.
- Supporting information such as formulae used, data sources, and tools used to process the data should be available as a footnote or hyperlink, or in supporting information such as an appendix.
- To the extent possible, keep portrayal formats consistent from reporting period to reporting period. Continuous improvement is laudable, but users spend more time interpreting results and making decisions when they are familiar with the format of the indicator.
- Annotate charts with the initiation and completion of improvement interventions intended to change the level, trend, or variability of results.
- Acknowledge possible omissions or errors in the data as part of the portrayal.

It may not be possible to apply all of these guidelines to every chart. These guidelines should be used as a checklist and regularly refer back to each indicator's purpose when evaluating charts. Thanks to Dr. Altyn Clark for helping develop these guidelines.

12.6 Integrity Audits

Performance measurements should be scrutinized, just like other functions and processes. Financial indicators and the financial control and accounting system they are part of typically receive an annual audit by an external (third-party) firm. Nonfinancial strategic performance indicators do not consistently receive the same scrutiny. So how do managers know that these nonfinancial indicators are providing them with valid, accurate, and reliable information? Valid information here refers to face or content validity: Does the indicator measure what it purports to measure? Reliable information means consistency in producing the same measurement output (i.e., indicator value) when identical performance conditions are repeatedly measured. Accuracy refers to how close the measurement output values are to the true performance values. By assuming that the indicators are providing valid, accurate, and reliable information, what assurance do managers have that their measurement systems are clearly understood, useful, and add value to the organization? A certain amount of financial measurement is a necessary part of doing business, for quarterly and annual SEC filings, reports to shareholders, or mandated by legislation in order to continue receiving government funding. The nonfinancial components of the measurement system are not typically mandated by legislation with the exception of compliance statistics reported to Occupational Safety and Health Administration (OSHA) and environmental agencies. Organizations compelled to participate in supplier certification programs or achieve quality or environmental management systems certification may feel coerced to develop a rudimentary nonfinancial measurement system. However, they need to realize that the return from developing a strategic performance measurement system is not compliance, but is the provision of useful information that adds value to the organization through better and faster decision-making. After investing the time and resources to develop a strategic performance measurement system, organizations should periodically audit that system for validity, reliability, and accuracy and assess the system for continued relevance and value added.

It is beyond the scope of this chapter to describe the audit and assessment process in detail. The interested reader should refer to Coleman and Clark (2001). Figure 12.2 provides an overview of where the techniques suggested by Coleman and Clark can be applied to audit and assess the measurement process. The "Approach" in the figure includes deciding on the extent of the audit and assessment, balancing previous efforts with current needs, and choosing among the variety of techniques available. The techniques in the figure are shown at the phases of the measurement and evaluation process where they are most applicable. Table 12.1 provides brief descriptions of these techniques and sources for additional information.

Organizations concerned with the resource requirements to develop, operate, and maintain a measurement system may balk at the additional tasking of conducting a comprehensive audit and assessment. Such organizations should, at a minimum, subject their measurement system to a critical review, perhaps using a technique as simple as "start, stop, or continue." During or immediately following a periodic review of performance (where the current levels of performance on each key indicator are reviewed and

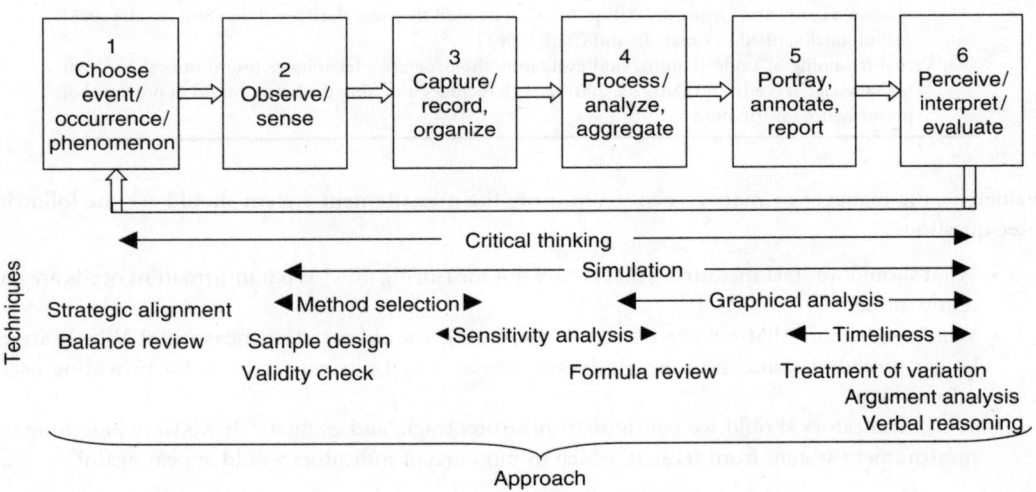

FIGURE 12.2 Auditing and assessing the measurement and evaluation process.(From Coleman, G.D. and Clark, L.A., *Proceedings of the Industrial Engineering Research Conference.*)

TABLE 12.1 Techniques Available for Auditing and Assessing Strategic Performance Measurement
Systems

1. Strategic alignment — audit against the organization's priorities, implicit and explicit
2. Balance review — assessment against the elements of one or more "balance" frameworks (e.g., Kaplan and Norton's Balanced Scorecard, Barrett's Balanced Needs Scorecard, Sink's Seven Criteria)
3. Critical thinking — scrutinizing for "faulty assumptions, questionable logic, weaknesses in methodology, inappropriate statistical analysis, and unwarranted conclusions" (Leedy, 2001, p. 36). Includes assessing the logic of the hierarchy of measures and the aggregation schemes. Assess value and usefulness by using Brown's (1996) or Sink's guidelines for the number of indicators used at one level in the organization
4. Sample design — assessing sample design and the appropriateness of the generalizations made from these samples (i.e., external validity). This is more than an issue of sample size. "The procedure of stratification, the choice of sampling unit, the formulas prescribed for the estimations, are more important than size in the determination of precision" (Deming 1960, p. 28)
5. Validity check — auditing for evidence of validity. What types of validity have been established for these measures: face, content, construct, or criterion validity?
6. Method selection — assessment of the appropriateness of the method(s) chosen for the data being used. Includes choice of quantitative and qualitative methods. Might include assessment of the reli ability of the methods. Internal validity might be addressed here
7. Simulation — observing or entering data of known properties (often repeatedly), then comparing the output (distribution) of the measurement process against expectations
8. Sensitivity analysis — varying input variables over predetermined ranges (typically plus and minus a fixed percent from a mean or median value) and evaluating the response (output) in terms of per centage change from the mean or median output value
9. Formula review — comparison of the mathematical formulae to the operational and conceptual def initions of the measure. Also includes auditing of replications of the formulae to ensure consistent application
10. Graphical analysis — at its simplest, plotting results and intermediate outputs to identify underlying pat terns. In more advanced forms, may include application of statistical techniques such as individual and moving range charts (Wheeler, 1993). Assess any underlying patterns for possible impact on validity
11. Timeliness — an assessment of the value of the information provided based on how quickly the measured results reach someone who can directly use the results to control and improve perform ance. One simple technique is to track the lag time between occurrence and reporting of per formance, then apply a user-based judgment of the acceptability of this lag
12. Treatment of variation — graphical analysis is one technique for addressing variation. More impor tantly, how do the users of the measurement information perceive or react to variation in results? Assess available evidence of statistical thinking and the likelihood of interpreting noise as a signal or failing to detect a signal when present
13. Argument analysis – "discriminating between reasons that do and do not support a particular con clusion" (Leedy and Ormrod, 2001, p. 36). Can be used to assess clarity with the Sink, et al. (1995) technique described in Coleman and Clark (2001)
14. Verbal reasoning – "understanding and evaluating the persuasive techniques found in oral and writ ten language" (Leedy and Ormrod, 2001, p. 36). Includes assessing the biases found in portrayal of performance information

evaluated), the manager or management team using the measurement system should ask the following three questions:

- What should we start measuring that we are not measuring now? What information needs are cur rently unmet?
- Which indicators that we are currently measuring should we stop measuring? Which are no longer providing value, no longer relevant, or never met our expectations for providing useful information?
- Which indicators should we continue to measure, track, and evaluate? If we were designing our measurement system from scratch, which of our current indicators would appear again?

Another less resource-intensive approach is to address the auditing and assessing of the measurement system as part of a periodic organizational assessment (see Section 12.7).

12.7 Organizational Assessments: Strategic Snapshots of Performance

Organizational assessments are a periodic snapshot form of strategic performance measurement. They are periodic in that they do not measure performance frequently: once a year to once every 5 or 10 years is common. They are snapshots because they reflect the organization's performance at a particular time and may not be fully evaluated until several weeks or months later. They are relatively comprehensive in scope, often measuring and evaluating all or most of the enterprise's activities and results, including the organization's measurement and evaluation system (as implied in Section 12.6). Preparing for an organizational assessment may require a review of the organization's measurement system, and the assessment process will provide both direct and indirect feedback on the usefulness and value of the measurement system. Organizational assessments are used for conformity, to ensure the organization meets some standard (e.g., accreditation, certification), or for improvement and recognition where the organization is compared with a standard and provided feedback for improvement. Those exhibiting the highest levels of performance against the standard are recognized with an organizational award (e.g., Baldrige Award, State or Corporate Awards for Excellence, European Quality Award).

Organizational assessment typically begins with a self-study comparing the organization and its goals against an established standard (i.e., criteria or guidelines). The completed self-study is then submitted to a third party (i.e., the accreditation, registration, or award body) for review and evaluation. This third-party review begins with an evaluation of the self-study and is often, but not always, followed by a visit to the organization. The purpose of the visit is to validate and clarify what was reported in the self-study. The third party then renders a judgment and provides feedback to the organization. Depending on the specific application, the third-party judgment may result in substantial consequences for the organization (e.g., winning an award, receiving accreditation, or failure to do so). Ideally, the feedback from the third party is fed into the organization's improvement cycle, implemented, measured, and reflected in future plans and results.

Organizations that operate an ongoing improvement cycle and feed the results of the assessment into that cycle are likely to receive the greatest return on the investment from the resources required to complete the self-study and assessment. Particularly in situations where the organizational assessments occur several years apart, having an ongoing improvement process maintains the momentum and focus on what is important and should make preparing for future assessments easier. The improvement process translates assessment findings into plans, actions, and targets, applies resources, and then follows up with regular review of results and then new or updated plans, actions, and targets. While the overall improvement process should be management-led, industrial engineers are often tasked as analysts and project managers to convert assessment findings into plans, actions, and results.

Organizations wishing to gain much of the benefit of a comprehensive assessment but concerned about the resource requirements should simply complete a five-page organizational profile, the preface of a Baldrige Award application (self-study) (Baldrige National Quality Program, 2004, pp. 10–12). The organizational profile asks the organization to document its organizational environment including products, services, culture, employees, facilities, equipment, and regulatory environment; its organizational relationships including organization structure, customers, suppliers, customer and supplier requirements, and how the organization partners with its customers and suppliers; its competitive environment including current market position(s), key success factors, and sources of comparative data for evaluating current performance; its key business, operational, and human resource challenges; and a description of its performance improvement system. For many organizations, particularly, smaller organizations and departments or functions within larger organizations, developing and collectively reviewing the organizational profile may provide more than 50% of the value of a complete organizational assessment. Too few management teams have developed consensus answers to the questions posed by the organizational profile. Developing the organizational profile as a team and keeping it current provides a key tool for providing organizational direction and furnishes an important input into the development and maintenance of the performance measurement system. Even organizations not interested in the Baldrige or other business excellence awards can use the profile as a resource for the development of management systems or the preparation of accreditation self-studies.

Organizational assessments, like other forms of performance measurement, should be subject to periodic audit and assessment (see Section 12.6). As a relatively new form of measurement, the reliability and validity of the results of organizational assessments are not well known. Few, if any, of the organizations that offer or manage these assessments provide statistics showing they periodically evaluate the efficacy of their assessment processes. Researchers (Coleman et al., 2001, 2002; Coleman and Koelling, 1998; Keinath and Gorski, 1999; van der Wiele et al., 1995) have begun to estimate some of the properties associated with the scores and feedback received from organizational assessments. Their findings suggest that training the assessors (a.k.a. evaluators, examiners) reduces scoring leniency; however, their findings are less conclusive regarding the effect of training on interrater reliability and accuracy. Those interested in interpreting the variability observed among results from organizational assessments should consult the above-cited sources.

References and Further Reading

Akao, Y., Ed., *QFD: Quality Function Deployment*, Productivity Press, Cambridge, MA, 1990.

Baldrige National Quality Program, *2005 Baldrige Criteria for Performance Excellence*, National Institute of Standards and Technology, Gaithersburg, MD, 2004.

Barrett, R., *Liberating the Corporate Soul*. Presentation to The Performance Center, Arlington, VA, March 4, 1999a.

Barrett, R., *Liberating the Corporate Soul*, Fulfilling Books, Alexandria, VA, 1999b.

Becker, B.E., Huselid, M.A., and Ulrich, D., *The HR Scorecard: Linking People, Strategy and Performance*, Harvard Business School, Boston, 2001.

Bitici, U.S., Turner, T., and Begemann, C., Dynamics of performance measurement systems, *Int. J. Oper. Prod. Manage.*, 20, 692–704, 2002.

Brassard, M., *The Memory Jogger Plus + Featuring the Seven Manage and Planning Tools*, GOAL/QPC, Methuen, MA, 1989.

Brown, M.G., *Winning Score: How to Design and Implement Organizational Scorecards*, Productivity Press, Portland, OR, 2000.

Brown, M.G., *Keeping Score: Using the Right Metrics to Drive World-Class Performance*, AMACOM Books, New York, 1996.

Boudreau, J.W. and Ramstad, P.M., Measuring intellectual capital: learning from financial history. *Human Resource Manage.*, 36, 343–356, 1997.

Bourne, M., Neely, A., Platts, K., and Mills, J., The success and failure of performance measurement initiatives: perceptions of participating managers, *Int. J. Oper. Prod. Manage.*, 22, 1288–1310, 2002.

Chen, G.J., Indicator development matrix: translating strategy into performance improvement actions, *Proceedings of the Industrial Engineering Research 2000 Conference*, 2000 [CD-ROM].

Coleman, G.D. and Clark, L.A., A framework for auditing and assessing non-financial performance measurement systems, *Proceedings of the Industrial Engineering Research Conference*, 2001, [CD-ROM].

Coleman, G.D., Costa, J., and Stetar, W., The measures of performance: Managing human capital is a natural fit for industrial engineers, *IE: Ind. Eng.*, 36, 40–44, 2004.

Coleman, G. D., Koelling, C.P., and Geller, E.S., Training and scoring accuracy of organisational self-assessments, *Int. J. Qual. Reliability Manage.*, 18, 512–527, 2001.

Coleman, G.D. and Koelling, C.P., Estimating the consistency of third-party evaluator scoring of organizational self-assessments, *Qual. Manage. J.*, 5, 31–53, 1998.

Coleman, G.D., van Aken, E.M., and Shen J., Estimating interrater reliability of examiner scoring for a state quality award, *Qual. Manage. J.*, 9, 39–58, 2002

Crandall, R.E., Keys to better performance measurement, *Ind. Manage.*, January/February, 19–24, 2002.

Deal, T.E. and Kennedy, A.A., *Corporate Cultures*, Addison-Wesley, Reading, MA, 1982.

Delbecq, A.L., van de Ven, A.H., and Gustafson, D.H., *Group Techniques for Program Planning: A Guide to Nominal Group and Delphi Processes*, Green Briar, Middleton, WI, 1975, 1986.

Deming, W.E., *The New Economics for Industry, Government, and Education*, MIT Center for Advanced Engineering Study, Cambridge, MA, 1993.

Deming, W.E., *Out of the Crisis*, MIT Center for Advanced Engineering Study, Cambridge, MA, 1986.

Deming, W.E., *Sample Design in Business Research*, Wiley, New York, 1960.

Dillman, D.A., *Mail and Telephone Surveys: The Total Design Method*, Wiley, Canada, 1978.

Drucker, P.F., The next information revolution, *Forbes ASAP*, 47–58, 1998.

Drucker, P.F., *The practice of management*, Harper & Row, New York, 1954.

Franco, M. and Bourne, M., Factors that play a role in "managing through measures," Manage. Dec., 41, 698–710, 2003.

Garvin, D.A., What does "product quality" really mean? *Sloan Manage. Rev.*, 26, 25–43, 1984.

GOAL/QPC, *The Memory Jogger*, Methuen, MA, 1985.

Hamel, G., *Leading the Revolution*, Harvard Business School Press, 2002, (Revised).

Hamel, G. and Prahalad, P.K., *Competing for the Future*, Harvard Business School Press, Cambridge, MA, 1996.

Ishikawa, K., *What is total quality control? The Japanese way* (trans. Lu, D.J.), Prentice-Hall, Englewood Cliffs, NJ, 1985.

Kaplan, R.S. and Norton, D.P., Using the balanced scorecard as a strategic management system, *Harv. Bus. Rev.*, 74, 75–85, 1996.

Kaplan, R.S. and Norton, D.P., The balanced scorecard: Measures that drive performance, *Harv. Bus. Rev.*, 70, 71–79, 1992.

Kaplan, R.S., Yesterday's accounting undermines production, *Harv. Bus. Rev.*, 62(4), 1984.

Keinath, B.J. and Gorski, B.A., An empirical study of the Minnesota quality award evaluation process, *Qual. Manage. J.*, 6, 29–38, 1999.

Kurstedt, H.A., The industrial engineer's systematic approach to management, *MSM Working Draft Articles and Responsive Systems Article*, Management Systems Laboratory, Virginia Tech, Blacksburg, VA, 1986.

Kennerley, M. and Neely., A., A framework of the factors affecting the evolution of performance measurement systems, *Int. J. Oper. Prod. Manage.*, 22, 1222–1245, 2002.

Kilmann, R., *Managing Beyond the Quick Fix*, Jossey-Bass, San Francisco, 1989.

Kilmann, R. and Saxton, M. J., *Kilmann-Saxton Culture Gap Survey*, XICOM, Tuxedo, NY, 1983.

Kizilos, T., Kratylus automates his urnworks, *Harv. Bus. Rev.*, 62 (May/June), 136–144, 1984.

Lawton, R., Balance your balanced scorecard, *Qual. Prog.*, 35, 66–71, 2002.

Lawton, R., Using measures to connect strategy with customers, *J. Qual. Participation*, 54–58, 2000.

Leedy, P.D. and Ormrod, J.E., *Practical Research: Planning and Design*, Merrill Prentice Hall, Upper Saddle River, NJ, 2001.

Mallak, L.A., Bringelson, L.S., and Lyth, D.M., A cultural study of ISO 9000 certification, *Int. J. Qual. Reliability Manage.*, 14, 328–348, 1997.

Mallak, L.A. and Kurstedt, H.A.,Jr., Using culture gap analysis to manage organizational change, *Eng. Manage. J.*, 8, 35–41, 1996.

Marr, B. and Andy, N., Automating the balanced scorecard - selection criteria to identify appropriate software applications. *Meas. Bus. Excellence Incorporating Qual. Focus*, 7, 29–36, 2003.

Miller, D., Profitability = productivity + price recovery, *Harv. Bus. Rev.*, 62, 145–153, 1984.

Muckler, F.A. and Seven, S.A., 1992.Selecting performance measures: "Objective" versus "subjective" measurement, *Hum. Factors*, 34, 441–455, 1984.

Neely, A., Bourne, M., and Adams, C., Better budgeting or beyond budgeting? *Meas. Bus. Excellence Incorporating Qual. Focus*, 7, 22–28, 2003.

Neely, A., The performance measurement revolution: Why now and what next? *Int. J. Oper. Prod. Manage.*, 19, 205–228, 1999.

Nørreklit, H., The Balanced Scorecard: what is the score? A rhetorical analysis of the Balanced Scorecard. *Acc. Organ. Soc.*, 28, 591–619, 2003.

Peters, T.J. and Waterman, R.H.,Jr., *In Search of Excellence: Lessons from America's Best-Run Companies*, Warner Books, New York, 1982.

Pfeffer, J. and Sutton, R.I., *The Knowing-Doing Gap: How Smart Companies Turn Knowledge into Action*, Harvard Business School Press, Boston, 2000.

Pineda, A. J., Productivity measurement and analysis (Module 4), in *Productivity and Quality Management: A Modular Programme*, Prokopenko, J. and North, K., Eds., International Labor Organization, Geneva, Switzerland, 1996.

Pyzdek, T., *Pyzdek's Guide to SPC, Volume One: Fundamentals*, Quality Publishing, Tucson, 1990.

Redman, T.C., Confronting data demons. *Six Sigma Forum Mag.*, 3 (3), 13–22, 2004.

Riggs, J.L., Monitoring with a matrix that motivates as it measures performance, *Ind. Eng.*, 34–43, 1986.

Rucci, A.J., Kirn, S.P., and Quinn, R.T., The employee-customer-profit chain at Sears, *Harv. Bus. Rev.*, 76, 82–97, 1998.

Sink, D.S., *Productivity Management: Planning, Measurement and Evaluation, Control and Improvement*, Wiley , New York, 1985.

Sink, D.S. and Tuttle, T.C., *Planning and Measurement in Your Organization of the Future*, IIE Press, Norcross, GA, 1989.

Sumanth, D.J., *Total Productivity Management: A Systematic and Quantitative Approach to Compete in Quality, Price, and Time*, St. Lucie Press, Boca Raton, FL, 1998.

Thompson, A.A.,Jr. and Strickland, A.J.,III., *Strategic Management: Concepts and Cases*, 13th ed., Irwin/McGraw-Hill, Boston, 2003.

Thor, C.G., *Designing Feedback*, Crisp Publications, Menlo Park, CA, 1998.

Thor, C.G., Using a family of measures to assess organizational performance, *Nati. Prod. Rev.*, 111–131, 1995.

Tufte, E. R., *Visual and Statistical Thinking: Displays of Evidence for Decision Making*, Graphics Press, Chesire, Connecticut, 1997a.

Tufte, E.R., *Visual Explanations: Images and Quantities, Evidence and Narrative*, Graphics Press, Chesire, Connecticut, 1997b.

van Aken, E.M. and Coleman, G.D., Building better measurement, *Ind. Manage.*, 28–33, 2002.

van der Wiele, T., Williams, R., Kolb, F., and Dale., B., Assessor training for the European quality award: An examination, *Qual. World Tech. Suppl.*, 53–62, 1995.

Wheeler, D.J., *Understanding Variation: The Key to Managing Chaos*, SPC Press, Knoxville, 1993.

13

Fundamentals of Project Management

Adedeji B. Badiru
University of Tennessee

13.1 Introduction

Project management represents an excellent basis for integrating various management techniques such as operations research, operations management, forecasting, quality control, and simulation. Traditional approaches to project management use these techniques in a disjointed fashion, thus ignoring the potential interplay among the techniques. The need for integrated project management worldwide is evidenced by a 1993 report by the World Bank. In the report, the bank, which has loaned more than $300 billion to developing countries over the last half century, acknowledges that there has been a dramatic rise in the number of failed projects around the world. Lack of an integrated approach to managing the projects was cited as one of the major causes of failure.

In modern project management, it is essential that related techniques be employed in an integrated fashion so as to maximize the total project output. Project management has been defined as:

The process of managing, allocating, and timing resources to achieve a specific goal in an efficient and expedient manner.

Alternatively, we can define project management as:

The systematic integration of technical, human, and financial resources to achieve goals and objectives.

This comprehensive definition requires an integrated approach to project management. This chapter presents such an integrated approach. To accomplish the goal of project management, an integrated use of managerial, mathematical, and computer tools must be developed. The first step in the project management process is to *set goals*.

Project management continues to grow as an effective means of managing functions in any organization. Project management should be an enterprise-wide endeavor. Enterprise-wide project management has been defined as the application of project management techniques and practices across the full scope of the enterprise. This concept is also referred to as management by project (MBP). Management by project is a recent concept that employs project management techniques in various functions within an organization. Management by project recommends pursuing endeavors as project-oriented activities. It is an effective way to conduct any business activity. It represents a disciplined approach that defines any work assignment as a project. Under MBP, every undertaking is viewed as a project that must be managed just like a traditional project. The characteristics required of each project so defined are:

1. An identified scope and a goal
2. A desired completion time
3. Availability of resources
4. A defined performance measure
5. A measurement scale for review of work

An MBP approach to operations helps in identifying unique entities within functional requirements. This identification helps determine where functions overlap and how they are interrelated, thus paving the way for better planning, scheduling, and control. Enterprise-wide project management facilitates a unified view of organizational goals and provides a way for project teams to use information generated by other departments to carry out their functions.

The use of project management continues to grow rapidly. The need to develop effective management tools increases with the increasing complexity of new technologies and processes. The life cycle of a new product to be introduced into a competitive market is a good example of a complex process that must be managed with integrative project management approaches. The product will encounter management functions as it goes from one stage to the next. Project management will be needed in developing, marketing, transportation, and delivery strategies for the product. When the product finally gets to the customer, project management will be needed to integrate its use with those of other products within the customer's organization.

The need for a project management approach is established by the fact that a project will always tend to increase in size even if its scope narrows. The following three literary laws are applicable to any project environment:

Parkinson's law: Work expands to fill the available time or space.
Peter's principle: People rise to their level of incompetence.
Murphy's law: Whatever can go wrong will.

An integrated project management approach can help diminish the impact of these laws through good project planning, organizing, scheduling, and control.

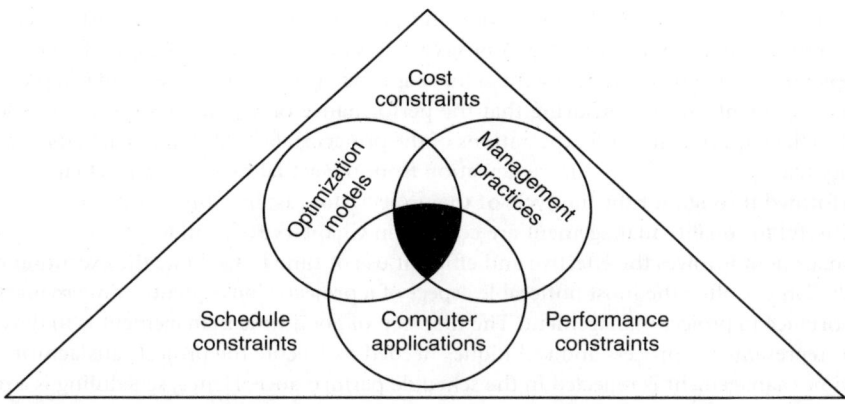

FIGURE 13.1 Integration of project management tools.

13.2 The Integrated Approach

Project management tools can be classified into three major categories:

1. *Qualitative tools.* These are the managerial tools that aid in the interpersonal and organizational processes required for project management.
2. *Quantitative tools.* These are analytical techniques that aid in the computational aspects of project management.
3. *Computer tools.* These are software and hardware tools that simplify the process of planning, organizing, scheduling, and controlling a project. Software tools can help in both the qualitative and quantitative analyses needed for project management.

Although individual books dealing with management principles, optimization models, and computer tools are available, there are few guidelines for the integration of the three areas for project management purposes. In this chapter, we integrate these three areas for a comprehensive guide to project management. The chapter introduces the Triad Approach to improve the effectiveness of project management with respect to schedule, cost, and performance constraints. Figure 13.1 illustrates this emphasis. The approach considers not only the management of the project itself, but also the management of all the functions that support the project.

It is one thing to have a quantitative model, but it is a different thing to be able to apply the model to real-world problems in a practical form. The systems approach presented in this chapter illustrates how to make the transitions from model to practice.

A systems approach helps increase the intersection of the three categories of project management tools and, hence, improve overall management effectiveness. Crisis should not be the instigator for the use of project management techniques. Project management approaches should be used upfront to prevent avoidable problems rather than to fight them when they develop. What is worth doing is worth doing well, right from the beginning.

13.3 Project Management Body of Knowledge

The Product Management Institute (PMI) defines the project management body of knowledge (PMBOK) as those topics, subject areas, and processes that are used in conjunction with sound project management principles to collectively execute a project. Eight major functional areas are identified in the PMBOK: *scope, quality, time, cost, risk, human resources, contract/procurement,* and *communications.*

Scope management refers to the process of directing and controlling the entire scope of the project with respect to a specific goal. The establishment and clear definition of project goals and objectives form the foundation of scope management. The scope and plans form the baseline against which changes or deviations can be monitored and controlled. A project that is out of scope may be out of luck as far as satisfactory completion is concerned. Topics essential for scope management are covered in Chapters 1–3 and 9.

Quality management involves ensuring that the performance of a project conforms to specifications with respect to the requirements and expectations of the project stakeholders and participants. The objective of quality management is to minimize deviation from the actual project plans. Quality management must be performed throughout the life cycle of the project, not just by a final inspection of the product. Techniques useful for quality management are covered in Chapters 1, 5, and 6.

Time management involves the effective and efficient use of time to facilitate the execution of a project expeditiously. Time is often the most noticeable aspect of a project. Consequently, time management is of utmost importance in project management. The first step of good time management is to develop a project plan that represents the process and techniques needed to execute the project satisfactorily. The effectiveness of time management is reflected in the schedule performance. Hence, scheduling is a major focus in project management. Chapters 4 and 7 present techniques needed for effective time management.

Cost management is a primary function in project management. Cost is a vital criterion for assessing project performance. Cost management involves having effective control over project costs through the use of reliable techniques of estimation, forecasting, budgeting, and reporting. Cost estimation requires collecting relevant data needed to estimate elemental costs during the life cycle of a project. Cost planning involves developing an adequate budget for the planned work. Cost control involves the continual process of monitoring, collecting, analyzing, and reporting cost data. Chapter 8 discusses cost management concepts and techniques.

Risk management is the process of identifying, analyzing, and recognizing the various risks and uncertainties that might affect a project. Change can be expected in any project environment. Change portends risk and uncertainty. Risk analysis outlines possible future events and the likelihood of their occurrence. With the information from risk analysis, the project team can be better prepared for change with good planning and control actions. By identifying the various project alternatives and their associated risks, the project team can select the most appropriate courses of action. Techniques relevant for risk management are presented in Chapters 4–6, 8, and 9.

Human resources management recognizes the fact that people make things happen. Even in highly automated environments, human resources are still a key element in accomplishing goals and objectives. Human resources management involves the function of directing human resources throughout a project's life cycle. This requires the art and science of behavioral knowledge to achieve project objectives. Employee involvement and empowerment are crucial elements for achieving the quality objectives of a project. The project manager is the key player in human resources management. Good leadership qualities and interpersonal skills are essential for dealing with both internal and external human resources associated with a project. The legal and safety aspects of employee welfare are important factors in human resources management. Chapters 1–6 and 10 present topics relevant to human resources management.

Contract/procurement management involves the process of acquiring the necessary equipment, tools, goods, services, and resources needed to successfully accomplish project goals. The buy, lease, or make options available to the project must be evaluated with respect to time, cost, and technical performance requirements. Contractual agreements in written or oral form constitute the legal document that defines the work obligation of each participant in a project. Procurement refers to the actual process of obtaining the needed services and resources. Concepts and techniques useful for contract/procurement management are presented in Chapters 2, 5, and 8.

Communications management refers to the functional interface among individuals and groups within the project environment. This involves proper organization, routing, and control of information needed to facilitate work. Good communication is in effect when there is a common understanding of information between the communicator and the target. Communications management facilitates unity of purpose in the project environment. The success of a project is directly related to the effectiveness of project communication. From the author's experience, most project problems can be traced to a lack of proper communication. Guidelines for improving project communication are presented in Chapters 1–3 and 6. Chapter 11 presents a case study that illustrates how the various elements in the PMBOK can be integrated.

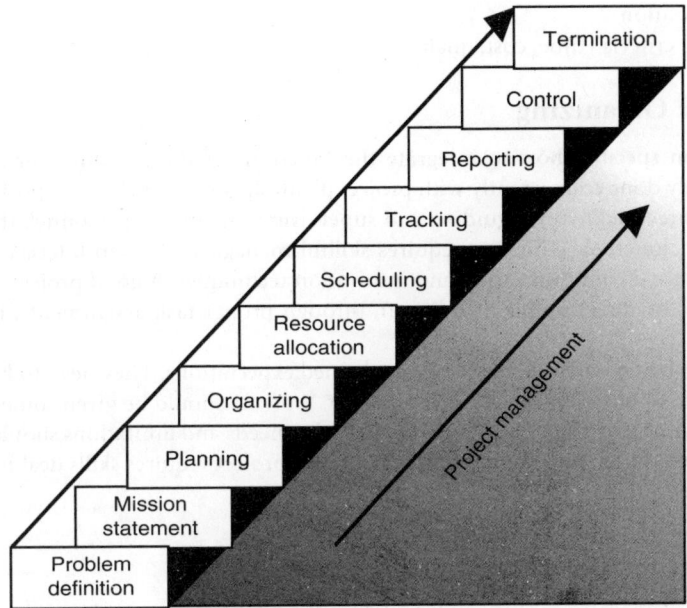

FIGURE 13.2   Project management steps.

13.4   Project Management Process

Organize, *prioritize*, and *optimize* the project. The project management process consists of several steps, starting from problem definition and going through project termination. Figure 13.2 presents the major steps. A brief overview of the steps is presented in this section. Some of the major steps are discussed in subsequent chapters.

13.4.1   Problem Identification

Problem identification is the stage where a need for a proposed project is identified, defined, and justified. A project may be concerned with the development of new products, implementation of new processes, or improvement of existing facilities.

13.4.2   Project Definition

Project definition is the phase at which the purpose of the project is clarified. A mission statement is the major output of this stage. For example, a prevailing low level of productivity may indicate a need for a new manufacturing technology. In general, the definition should specify how project management may be used to avoid missed deadlines, poor scheduling, inadequate resource allocation, lack of coordination, poor quality, and conflicting priorities.

13.4.3   Project Planning

A plan represents the outline of the series of actions needed to accomplish a goal. Project planning determines how to initiate a project and execute its objectives. It may be a simple statement of a project goal or it may be a detailed account of procedures to be followed during the project. Project planning is discussed in detail in Chapter 2. Planning can be summarized as

• Objectives
• Project definition

- Team organization
- Performance criteria (time, cost, quality)

13.4.4 Project Organizing

Project organization specifies how to integrate the functions of the personnel involved in a project. Organizing is usually done concurrently with project planning. Directing is an important aspect of project organization. Directing involves guiding and supervising the project personnel. It is a crucial aspect of the management function. Directing requires skillful managers who can interact with subordinates effectively through good communication and motivation techniques. A good project manager will facilitate project success by directing his or her staff, through proper task assignments, toward the project goal.

Workers perform better when there are clearly defined expectations. They need to know how their job functions contribute to the overall goals of the project. Workers should be given some flexibility for self-direction in performing their functions. Individual worker needs and limitations should be recognized by the manager when directing project functions. Directing a project requires skills dealing with motivating, supervising, and delegating.

13.4.5 Resource Allocation

Project goals and objectives are accomplished by allocating resources to functional requirements. Resources can consist of money, people, equipment, tools, facilities, information, skills, and so on. These are usually in short supply. The people needed for a particular task may be committed to other ongoing projects. A crucial piece of equipment may be under the control of another team. Chapter 5 addresses resource allocation in detail.

13.4.6 Project Scheduling

Timeliness is the essence of project management. Scheduling is often the major focus in project management. The main purpose of scheduling is to allocate resources so that the overall project objectives are achieved within a reasonable time span. Project objectives are generally conflicting in nature. For example, minimization of the project completion time and minimization of the project cost are conflicting objectives. That is, one objective is improved at the expense of worsening the other objective. Therefore, project scheduling is a multiple-objective decision-making problem.

In general, scheduling involves the assignment of time periods to specific tasks within the work schedule. Resource availability, time limitations, urgency level, required performance level, precedence requirements, work priorities, technical constraints, and other factors complicate the scheduling process. Thus, the assignment of a time slot to a task does not necessarily ensure that the task will be performed satisfactorily in accordance with the schedule. Consequently, careful control must be developed and maintained throughout the project scheduling process. Chapter 4 covers project scheduling in detail. Project scheduling involves:

- Resource availability (human, material, money)
- Scheduling techniques (critical path method (CPM), precedence diagramming method (PERT), Gantt charts)

13.4.7 Project Tracking and Reporting

This phase involves checking whether or not project results conform to project plans and performance specifications. Tracking and reporting are prerequisites for project control. A properly organized report of the project status will help identify any deficiencies in the progress of the project and help pinpoint corrective actions.

13.4.8 Project Control

Project control requires that appropriate actions be taken to correct unacceptable deviations from expected performance. Control is actuated through measurement, evaluation, and corrective action. Measurement is the process of measuring the relationship between planned performance and actual performance with respect to project objectives. The variables to be measured, the measurement scales, and the measuring approaches should be clearly specified during the planning stage. Corrective actions may involve rescheduling, reallocation of resources, or expedition of task performance. Project control is discussed in detail in Chapter 6. Control involves:

- Tracking and reporting
- Measurement and evaluation
- Corrective action (plan revision, rescheduling, updating)

13.4.9 Project Termination

Termination is the last stage of a project. The phaseout of a project is as important as its initiation. The termination of a project should be implemented expeditiously. A project should not be allowed to drag on after the expected completion time. A terminal activity should be defined for a project during the planning phase. An example of a terminal activity may be the submission of a final report, the power-on of new equipment, or the signing of a release order. The conclusion of such an activity should be viewed as the completion of the project. Arrangements may be made for follow-up activities that may improve or extend the outcome of the project. These follow-up or spin-off projects should be managed as new projects but with proper input–output relationships within the sequence of projects.

13.5 Project Management Outline

An outline of the functions to be carried out during a project should be made during the planning stage of the project. A model for such an outline is presented below. It may be necessary to rearrange the contents of the outline to fit the specific needs of a project.

1. *Planning*
    I. Specify project background
        A. Define current situation and process
            1. Understand the process
            2. Identify important variables
            3. Quantify variables
        B. Identify areas for improvement
            1. List and discuss the areas
            2. Study potential strategy for solution
    II. Define unique terminologies relevant to the project
        A. Industry-specific terminologies
        B. Company-specific terminologies
        C. Project-specific terminologies
    III. Define project goal and objectives
        A. Write mission statement
        B. Solicit inputs and ideas from personnel
    IV. Establish performance standards
        A. Schedule
        B. Performance
        C. Cost
    V. Conduct formal project feasibility study

  A. Determine impact on cost
  B. Determine impact on organization
  C. Determine impact on deliverables
VI. Secure management support

2. *Organizing*
   I. Identifying project management team
      A. Specify project organization structure
         1. Matrix structure
         2. Formal and informal structures
         3. Justify structure
      B. Specify departments involved and key personnel
         1. Purchasing
         2. Materials management
         3. Engineering, design, manufacturing, and so on
      C. Define project management responsibilities
         1. Select project manager
         2. Write project charter
         3. Establish project policies and procedures
  II. Implement Triple C Model
      A. Communication
         1. Determine communication interfaces
         2. Develop communication matrix
      B. Cooperation
         1. Outline cooperation requirements
      C. Coordination
         1. Develop work breakdown structure
         2. Assign task responsibilities
         3. Develop responsibility chart

3. *Scheduling and Resource Allocation*
   I. Develop master schedule
      A. Estimate task duration
      B. Identify task precedence requirements
         1. Technical precedence
         2. Resource-imposed precedence
         3. Procedural precedence
      C. Use analytical models
         1. CPM
         2. PERT
         3. Gantt Chart
         4. Optimization models

4. *Tracking, Reporting, and Control*
   I. Establish guidelines for tracking, reporting, and control
      A. Define data requirements
         1. Data categories
         2. Data characterization
         3. Measurement scales
      B. Develop data documentation
         1. Data update requirements

      2. Data quality control
      3. Establish data security measures
  II. Categorize control points
    A. Schedule audit
      1. Activity network and Gantt charts
      2. Milestones
      3. Delivery schedule
    B. Performance audit
      1. Employee performance
      2. Product quality
    C. Cost audit
      1. Cost containment measures
      2. Percent completion vs budget depletion
  III. Identify implementation process
    A. Comparison with targeted schedules
    B. Corrective course of action
      1. Rescheduling
      2. Reallocation of resources
  IV. Terminate the project
    A. Performance review
    B. Strategy for follow-up projects
    C. Personnel retention
  V. Document project and submit final report

13.6 Selecting the Project Manager

The role of a manager is to use available resources (manpower and tools) to accomplish goals and objectives. A project manager has the primary responsibility of ensuring that a project is implemented according to the project plan. The project manager has a wide span of interaction within and outside the project environment. He or she must be versatile, assertive, and effective in handling problems that develop during the execution phase of the project. Selecting a project manager requires careful consideration because the selection of the project manager is one of the most crucial project functions. The project manager should be someone who can get the job done promptly and satisfactorily; possess both technical and administrative credibility; be perceived as having the technical knowledge to direct the project; be current with the technologies pertinent to the project requirements; and be conversant with the industry's terminologies. The project manager must also be a good record keeper. Since the project manager is the vital link between the project and upper management, he or she must be able to convey information at various levels of detail. The project manager should have good leadership qualities, although leadership is an after-the-fact attribute. Therefore, caution should be exercised in extrapolating prior observations to future performance when evaluating candidates for the post of project manager.

The selection process should be as formal as a regular recruiting process. A pool of candidates may be developed through nominations, applications, eligibility records, short-listed groups, or administrative appointment. The candidates should be aware of the nature of the project and what they would be expected to do. Formal interviews may be required in some cases, particularly those involving large projects. In a few cases, the selection may have to be made by default if there are no suitably qualified candidates. Default appointment of a project manager implies that no formal evaluation process has been carried out. Political considerations and quota requirements often lead to default selection of project managers. As soon as a selection is made, an announcement should be made to inform the project team of the selection. The desirable attributes a project manager should possess are:

- Inquisitiveness
- Good labor relations

- Good motivational skills
- Availability and accessibility
- Versatility with company operations
- Good rapport with senior executives
- Good analytical and technical background
- Technical and administrative credibility
- Perseverance toward project goals
- Excellent communication skills
- Receptiveness to suggestions
- Good leadership qualities
- Good diplomatic skills
- Congenial personality

13.7 Selling the Project Plan

The project plan must be sold throughout the organization. Different levels of detail will be needed when presenting the project to various groups in the organization. The higher the level of management, the lower the level of detail. Top management will be more interested in the global aspects of the project. For example, when presenting the project to management, it is necessary to specify how the overall organization will be affected by the project. When presenting the project to the supervisory level staff, the most important aspect of the project may be the operational level of detail. At the worker or operator level, the individual will be more concerned about how he or she fits into the project. The project manager or analyst must be able to accommodate these various levels of detail when presenting the plan to both participants in and customers of the project. Regardless of the group being addressed, the project presentation should cover the following elements with appropriate levels of details:

- Executive summary
- Introduction
- Project description
  - Goals and objectives
  - Expected outcomes
- Performance measures
- Conclusion

The use of charts, figures, and tables is necessary for better communication with management. A presentation to middle-level managers may follow a more detailed outline that might include the following:

- Objectives
- Methodologies
- What has been done
- What is currently being done
- What remains to be done
- Problems encountered to date
- Results obtained to date
- Future work plan
- Conclusions and recommendations

13.8 Staffing the Project

Once the project manager has been selected and formally installed, one of his first tasks is the selection of the personnel for the project. In some cases, the project manager simply inherits a project team that was formed before he was selected as the project manager. In that case, the project manager's initial responsibility will be to ensure that a good project team has been formed. The project team should be chosen on the basis of skills relevant to the project requirements and team congeniality. The personnel required may be obtained either from within the organization or from outside sources. If outside sources are used, a clear statement should be made about the duration of the project assignment. If opportunities for permanent absorption into the organization exist, the project manager may use that fact as an incentive both in recruiting for the project and in running the project. An incentive for internal personnel may be the opportunity for advancement within the organization.

Job descriptions should be prepared in unambiguous terms. Formal employment announcements may be issued or direct contacts through functional departments may be utilized. The objective is to avoid having a pool of applicants that is either too large or too small. If job descriptions are too broad, many unqualified people will apply. If the descriptions are too restrictive, very few of those qualified will apply. Some skill tolerance or allowance should be established. Since it is nearly impossible to obtain the perfect candidate for each position, some preparation should be made for in-house specialized skill development to satisfy project objectives. Typical job classifications in a project environment include the following:

- Project administrator
- Project director
- Project coordinator
- Program manager
- Project manager
- Project engineer
- Project assistant
- Project specialist
- Task manager
- Project auditor

Staff selection criteria should be based on project requirements and the availability of a staff pool. Factors to consider in staff selection include:

- Recommendation letters and references
- Salary requirements
- Geographical preference
- Education and experience
- Past project performance
- Time frame of availability
- Frequency of previous job changes
- Versatility for project requirements
- Completeness and directness of responses
- Special project requirements (quotas, politics, etc.)

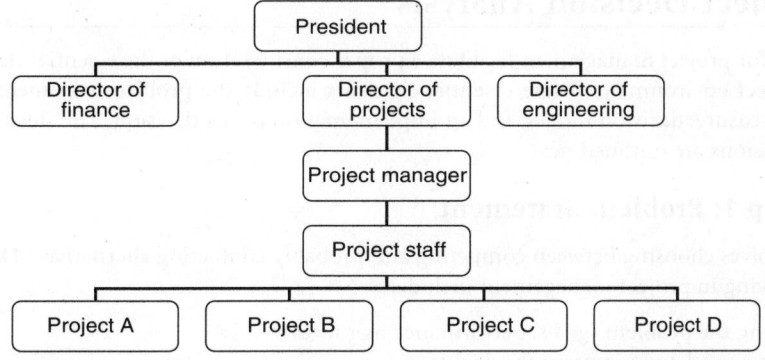

FIGURE 13.3   Organization of the project office.

- Overqualification (overqualified workers tend to be unhappy at lower job levels)
- Organizational skills

An initial screening of the applicants on the basis of the above factors may help reduce the applicant pool to a manageable level. If company policy permits, direct contact over the telephone or in person may then be used to further prune the pool of applicants. A direct conversation usually brings out more accurate information about applicants. In many cases, people fill out applications by writing what they feel the employer wants to read rather than what they want to say. Direct contact can help determine if applicants are really interested in the job, whether they will be available when needed, and whether they possess vital communication skills.

Confidentiality of applicants should be maintained, particularly for applicants who do not want a disclosure to their current employers. References should be checked out and the information obtained should be used with the utmost discretion. Interviews should then be arranged for the leading candidates. Final selection should be based on the merits of the applicants rather than mere personality appeal. Both the successful and the unsuccessful candidates should be informed of the outcome as soon as administrative policies permit.

In many technical fields, personnel shortage is a serious problem. The problem of recruiting in such circumstances becomes that of expanding the pool of applicants rather than pruning the pool. It is a big battle among employers to entice highly qualified technical personnel from one another. Some recruiters have even been known to resort to unethical means in the attempt to lure prospective employees. Project staffing involving skilled manpower can be enhanced by the following:

- Employee exchange programs
- Transfer from other projects
- In-house training for new employees
- Use of temporary project consultants
- Diversification of in-house job skills
- Cooperative arrangements among employers
- Continuing education for present employees

Committees may be set up to guide the project effort from the recruitment stage to the final implementation stage. Figure 13.3 shows a generic organizational chart for the project office and the role of a project committee. The primary role of a committee should be to provide supporting consultations to the project manager. Such a committee might use the steering committee model, which is formed by including representatives from different functional areas. The steering committee should serve as an advisory board for the project. A committee may be set up under one of the following two structures:

1. *Ad hoc committee.* This is set up for a more immediate and specific purpose (e.g., project feasibility study).
2. *Standing committee.* This is set up on a more permanent basis to oversee ongoing project activities.

13.9 Project Decision Analysis

Decision steps for project management facilitate a proper consideration of the essential elements of decisions in a project environment. These essential elements include the problem statement, information, performance measure, decision model, and an implementation of the decision. The steps recommended for project decisions are outlined next.

13.9.1 Step 1: Problem Statement

A problem involves choosing between competing and probably conflicting alternatives. The components of problem solving in project management include:

- Describing the problem (goals, performance measures)
- Defining a model to represent the problem
- Solving the model

- Testing the solution
- Implementing and maintaining the solution

Problem definition is very crucial. In many cases, *symptoms* of a problem are more readily recognized than its *cause* and *location*. Even after the problem is accurately identified and defined, a benefit/cost analysis may be needed to determine if the cost of solving the problem is justified.

13.9.2 Step 2: Data and Information Requirements

Information is the driving force for the project decision process. Information clarifies the relative states of past, present, and future events. The collection, storage, retrieval, organization, and processing of raw data are important components for generating information. Without data, there can be no information. Without good information, there cannot be a valid decision. The essential requirements for generating information are

- Ensuring that an effective data collection procedure is followed
- Determining the type and the appropriate amount of data to collect
- Evaluating the data collected with respect to information potential
- Evaluating the cost of collecting the required data

For example, suppose a manager is presented with a recorded fact that says, *"Sales for the last quarter are 10,000 units."* This constitutes ordinary data. There are many ways of using the above data to make a decision depending on the manager's value system. An analyst, however, can ensure the proper use of the data by transforming it into information, such as, *"Sales of 10,000 units for last quarter are within x percent of the targeted value."* This type of information is more useful to the manager for decision-making.

13.9.3 Step 3: Performance Measure

A performance measure for the competing alternatives should be specified. The decision maker assigns a perceived worth or value to the available alternatives. Setting measures of performance is crucial to the process of defining and selecting alternatives. Some performance measures commonly used in project management are project cost, completion time, resource usage, and stability in the workforce.

13.9.4 Step 4: Decision Model

A decision model provides the basis for the analysis and synthesis of information and is the mechanism by which competing alternatives are compared. To be effective, a decision model must be based on a systematic and logical framework for guiding project decisions. A decision model can be a verbal, graphical, or mathematical representation of the ideas in the decision-making process. A project decision model should have the following characteristics:

- Simplified representation of the actual situation
- Explanation and prediction of the actual situation
- Validity and appropriateness
- Applicability to similar problems

The formulation of a decision model involves three essential components:

- *Abstraction*: Determining the relevant factors
- *Construction*: Combining the factors into a logical model
- *Validation*: Assuring that the model adequately represents the problem

The basic types of decision models for project management are described next.

13.9.4.1 Descriptive Models

These models are directed at describing a decision scenario and identifying the associated problem. For example, a project analyst might use a CPM network model to identify bottleneck tasks in a project.

13.9.4.2 Prescriptive Models

These models furnish procedural guidelines for implementing actions. The Triple C approach, for example, is a model that prescribes the procedures for achieving communication, cooperation, and coordination in a project environment.

13.9.4.3 Predictive Models

These models are used to predict future events in a problem environment. They are typically based on historical data about the problem situation. For example, a regression model based on past data may be used to predict future productivity gains associated with expected levels of resource allocation. Simulation models can be used when uncertainties exist in the task durations or resource requirements.

13.9.4.4 Satisficing Models

These are models that provide trade-off strategies for achieving a satisfactory solution to a problem within given constraints. Goal programming and other multicriteria techniques provide good satisficing solutions. For example, these models are helpful in cases where time limitations, resource shortages, and performance requirements constrain the implementation of a project.

13.9.4.5 Optimization Models

These models are designed to find the best-available solution to a problem subject to a certain set of constraints. For example, a linear programming model can be used to determine the optimal product mix in a production environment.

In many situations, two or more of the above models may be involved in the solution of a problem. For example, a descriptive model might provide insights into the nature of the problem; an optimization model might provide the optimal set of actions to take in solving the problem; a satisficing model might temper the optimal solution with reality; and a predictive model might predict the expected outcome of implementing the solution.

13.9.5 Step 5: Making the Decision

Using the available data, information, and the decision model, the decision maker will determine the real-world actions that are needed to solve the stated problem. A sensitivity analysis may be useful for determining what changes in parameter values might cause a change in the decision.

13.9.6 Step 6: Implementing the Decision

A decision represents the selection of an alternative that satisfies the objective stated in the problem statement. A good decision is useless until it is implemented. An important aspect of a decision is to specify how it is to be implemented. Selling the decision and the project to management requires a well-organized persuasive presentation. The way a decision is presented can directly influence whether or not it is adopted. The presentation of a decision should include at least the following: an executive summary, technical aspects of the decision, managerial aspects of the decision, resources required to implement the decision, cost of the decision, the time frame for implementing the decision, and the risks associated with the decision.

13.10 Conducting Project Meetings

Meetings are one avenue for information flow in project decision-making. Effective management of meetings is an important skill for any managerial staff. Employees often feel that meetings waste time and obstruct productivity. This is because most meetings are poorly organized, improperly managed, called at the wrong time, or even unnecessary. In some organizations, meetings are conducted as routing requirements rather than from necessity. Meetings are essential for communication and decision-making. Unfortunately, many meetings accomplish nothing and waste everyone's time. A meeting of 30 people lasting only 30 min in effect wastes 15 full hours of employee time. That much time, in a corporate setting, may amount to thousands of dollars in lost time. It does not make sense to use a 1-h meeting to discuss a task that will take only 5 min to perform. That is like hiring someone at a $50,000 annual salary to manage an annual budget of $20,000. One humorous satire says:

"Management meetings are rapidly becoming this country's biggest growth industry. As nearly as I can determine, the working day of a typical middle manager consists of seven hours of meetings, plus lunch. Half a dozen years ago at my newspaper, we hired a new middle management editor with an impressive reputation. Unfortunately, I haven't met her yet. On her first day at work, she went into a meeting and has never come out."

The satire concludes with:

"I'm expected to attend the next meeting. I'm not sure when it's scheduled exactly. I think they're having a meeting this afternoon about that."

In the past, when an employee had a request, he went to the boss, who would say yes or no right away. The whole process might have taken less than 1 min of the employee's day. Nowadays several hierarchies of meetings may need to be held to review the request. Thus, we may have a departmental meeting, a middle-management staff meeting, upper-management meeting, executive meeting, steering committee meeting, ad hoc committee meeting, and a meeting with outside consultants all for the purpose of reviewing that simple request. The following observations have been made about project meetings:

1. Most of the information passed out at meetings can be more effectively disseminated through an ordinary memo. The proliferation of desktop computers and electronic mail should be fully exploited to replace most meetings.
2. The point of diminishing return for any meeting is equal to the number of people who are actually needed for the meeting. The more people at a meeting, the lower the meeting's productivity. The extra attendees serve only to generate unconstructive and conflicting ideas that only impede the meeting.
3. Not being invited to a meeting could be viewed as an indication of the high value placed on an individual's time within the organization.
4. Regularly scheduled meetings with specific time slots often become a forum for social assemblies.
5. The optimal adjournment time of a meeting is equal to the scheduled start time plus five times the number of agenda items minus the start-up time. Mathematically, this is expressed as

$$L = (T + 5N) - S$$

where
L = optimal length in minutes
T = scheduled time
N = number of agenda items
S = meeting start-up time (i.e., time taken to actually call the meeting to order)

Since it is difficult to do away with meetings (the necessary and the unnecessary), we must attempt to maximize their output. Some guidelines for running meetings more effectively are presented next.

1. Do premeeting homework.
   - List topics to be discussed (agenda).
   - Establish the desired outcome for each topic.
   - Determine how the outcome will be verified.
   - Determine who really needs to attend the meeting.
   - Evaluate the suitability of meeting time and venue.

         – Categorize meeting topics (e.g., announcements, important, urgent).
         – Assign a time duration to each topic.
         – Verify that the meeting is really needed.
         – Consider alternatives to the meeting (e.g., memo, telephone, and electronic mail).

2. Circulate a written agenda prior to the meeting.
3. Start meeting on time.
4. Review agenda at the beginning.
5. Get everyone involved; if necessary, employ direct questions and eye contact.
6. Keep to the agenda; do not add new items unless absolutely essential.
7. Be a facilitator for meeting discussions.
8. Quickly terminate conflicts that develop from routine discussions.
9. Redirect irrelevant discussions back to the topic of the meeting.
10. Retain leadership and control of the meeting.
11. Recap the accomplishments of each topic before going to the next. Let those who have made commitments (e.g., promise to look into certain issues) know what is expected of them.
12. End meeting on time.
13. Prepare and distribute minutes. Emphasize the outcome and success of the meeting.

The economic impact of poorly managed meetings has been analyzed and guidelines for project managers to improve meetings have been provided. It has been suggested that managers evaluate meetings by asking the following postmeeting questions:

1. What did we do well in this meeting?
2. What can we improve next time?

Despite the shortcomings of poorly managed meetings, meetings offer a suitable avenue for group decision-making.

13.11 Group Decision-Making

Many decision situations are complex and poorly understood. No one person has all the information to make all decisions accurately. As a result, crucial decisions are made by a group of people. Some organizations use outside consultants with appropriate expertise to make recommendations for important decisions. Other organizations set up their own internal consulting groups without having to go outside the organization. Decisions can be made through linear responsibility, in which case one person makes the final decision based on inputs from other people. Decisions can also be made through shared responsibility, in which case a group of people share the responsibility for making joint decisions. The major advantages of group decision-making are:

1. *Ability to share experience, knowledge, and resources.* Many heads are better than one. A group will possess greater collective ability to solve a given decision problem.
2. *Increased credibility.* Decisions made by a group of people often carry more weight in an organization.
3. *Improved morale.* Personnel morale can be positively influenced because many people have the opportunity to participate in the decision-making process.
4. *Better rationalization.* The opportunity to observe other people's views can lead to an improvement in an individual's reasoning process.

Some disadvantages of group decision-making are:

1. Difficulty in arriving at a decision. Individuals may have conflicting objectives.
2. Reluctance of some individuals in implementing the decisions.
3. Potential for conflicts among members of the decision group.
4. Loss of productive employee time.

13.11.1 Brainstorming

Brainstorming is a way of generating many new ideas. In brainstorming, the decision group comes together to discuss alternate ways of solving a problem. The members of the brainstorming group may be from different departments, may have different backgrounds and training, and may not even know one another. The diversity of the participants helps create a stimulating environment for generating different ideas from different viewpoints. The technique encourages free outward expression of new ideas no matter how far-fetched the ideas might appear. No criticism of any new idea is permitted during the brainstorming session. A major concern in brainstorming is that extroverts may take control of the discussions. For this reason, an experienced and respected individual should manage the brainstorming discussions. The group leader establishes the procedure for proposing ideas, keeps the discussions in line with the group's mission, discourages disruptive statements, and encourages the participation of all members.

After the group runs out of ideas, open discussions are held to weed out the unsuitable ones. It is expected that even the rejected ideas may stimulate the generation of other ideas, which may eventually lead to other favored ideas. Guidelines for improving brainstorming sessions are presented as follows:

- Focus on a specific problem.
- Keep ideas relevant to the intended decision.
- Be receptive to all new ideas.
- Evaluate the ideas on a relative basis after exhausting new ideas.
- Maintain an atmosphere conducive to cooperative discussions.
- Maintain a record of the ideas generated.

13.11.2 Delphi Method

The traditional approach to group decision-making is to obtain the opinion of experienced participants through open discussions. An attempt is made to reach a consensus among the participants. However, open group discussions are often biased because of the influence or subtle intimidation from dominant individuals. Even when the threat of a dominant individual is not present, opinions may still be swayed by group pressure. This is called the "bandwagon effect" of group decision-making.

The Delphi method, developed in 1964, attempts to overcome these difficulties by requiring individuals to present their opinions anonymously through an intermediary. The method differs from the other interactive group methods because it eliminates face-to-face confrontations. It was originally developed for forecasting applications, but it has been modified in various ways for application to different types of decision-making. The method can be quite useful for project management decisions. It is particularly effective when decisions must be based on a broad set of factors. The Delphi method is normally implemented as follows:

1. *Problem definition.* A decision problem that is considered significant is identified and clearly described.
2. *Group selection.* An appropriate group of experts or experienced individuals is formed to address the particular decision problem. Both internal and external experts may be involved in the Delphi process. A leading individual is appointed to serve as the administrator of the decision process. The group may operate through the mail or gather together in a room. In either case, all opinions are expressed anonymously on paper. If the group meets in the same room, care should be taken to provide enough room so that each member does not have the feeling that someone may accidentally or deliberately observe their responses.
3. *Initial opinion poll.* The technique is initiated by describing the problem to be addressed in unambiguous terms. The group members are requested to submit a list of major areas of concern in their specialty areas as they relate to the decision problem.
4. *Questionnaire design and distribution.* Questionnaires are prepared to address the areas of concern related to the decision problem. The written responses to the questionnaires are collected and organized by the administrator. The administrator aggregates the responses in a statistical format.

For example, the average, mode, and median of the responses may be computed. This analysis is distributed to the decision group. Each member can then see how his or her responses compare with the anonymous views of the other members.

5. *Iterative balloting.* Additional questionnaires based on the previous responses are passed to the members. The members submit their responses again. They may choose to alter or not to alter their previous responses.

6. *Silent discussions and consensus.* The iterative balloting may involve anonymous written discussions of why some responses are correct or incorrect. The process is continued until a consensus is reached. A consensus may be declared after five or six iterations of the balloting or when a specified percentage (e.g., 80%) of the group agrees on the questionnaires. If a consensus cannot be declared on a particular point, it may be displayed to the whole group with a note that it does not represent a consensus.

In addition to its use in technological forecasting, the Delphi method has been widely used in other general decision-making processes. Its major characteristics of anonymity of responses, statistical summary of responses, and controlled procedure make it a reliable mechanism for obtaining numeric data from subjective opinion. The major limitations of the Delphi method are:

1. Its effectiveness may be limited in cultures where strict hierarchy, seniority, and age influence decision-making processes.
2. Some experts may not readily accept the contribution of nonexperts to the group decision-making process.
3. Since opinions are expressed anonymously, some members may take the liberty of making ludicrous statements. However, if the group composition is carefully reviewed, this problem may be avoided.

13.11.3 Nominal Group Technique

The nominal group technique is a silent version of brainstorming. It is a method of reaching consensus. Rather than asking people to state their ideas aloud, the team leader asks each member to jot down a minimum number of ideas, for example, five or six. A single list of ideas is then written on a chalkboard for the whole group to see. The group then discusses the ideas and weeds out some iteratively until a final decision is made. The nominal group technique is easier to control. Unlike brainstorming where members may get into shouting matches, the nominal group technique permits members to silently present their views. In addition, it allows introverted members to contribute to the decision without the pressure of having to speak out too often.

In all of the group decision-making techniques, an important aspect that can enhance and expedite the decision-making process is to require that members review all pertinent data before coming to the group meeting. This will ensure that the decision process is not impeded by trivial preliminary discussions. Some disadvantages of group decision-making are

1. Peer pressure in a group situation may influence a member's opinion or discussions.
2. In a large group, some members may not get to participate effectively in the discussions.
3. A member's relative reputation in the group may influence how well his or her opinion is rated
4. A member with a dominant personality may overwhelm the other members in the discussions.
5. The limited time available to the group may create a time pressure that forces some members to present their opinions without fully evaluating the ramifications of the available data.
6. It is often difficult to get all members of a decision group together at the same time.

Despite the noted disadvantages, group decision-making definitely has many advantages that may nullify shortcomings. The advantages as presented earlier will have varying levels of effect from one organization to another. The Triple C principle presented in Chapter 2 may also be used to improve the success of decision teams. Teamwork can be enhanced in group decision-making by adhering to the following guidelines:

1. Get a willing group of people together.
2. Set an achievable goal for the group.
3. Determine the limitations of the group.
4. Develop a set of guiding rules for the group.
5. Create an atmosphere conducive to group synergism.
6. Identify the questions to be addressed in advance.
7. Plan to address only one topic per meeting.

For major decisions and long-term group activities, arrange for team training, which allows the group to learn the decision rules and responsibilities together. The steps for the nominal group technique are:

1. Silently generate ideas, in writing.
2. Record ideas without discussion.
3. Conduct group discussion for clarification of meaning, not argument.
4. Vote to establish the priority or rank of each item.
5. Discuss vote.
6. Cast final vote.

13.11.4 Interviews, Surveys, and Questionnaires

Interviews, surveys, and questionnaires are important information-gathering techniques. They also foster cooperative working relationships. They encourage direct participation and inputs into project decision-making processes. They provide an opportunity for employees at the lower levels of an organization to contribute ideas and inputs for decision-making. The greater the number of people involved in the interviews, surveys, and questionnaires, the more valid the final decision. The following guidelines are useful for conducting interviews, surveys, and questionnaires to collect data and information for project decisions:

1. Collect and organize background information and supporting documents on the items to be covered by the interview, survey, or questionnaire.
2. Outline the items to be covered and list the major questions to be asked.
3. Use a suitable medium of interaction and communication: telephone, fax, electronic mail, face-to-face, observation, meeting venue, poster, or memo.
4. Tell the respondent the purpose of the interview, survey, or questionnaire, and indicate how long it will take.
5. Use open-ended questions that stimulate ideas from the respondents.
6. Minimize the use of yes or no types of questions.
7. Encourage expressive statements that indicate the respondent's views.
8. Use the who, what, where, when, why, and how approach to elicit specific information.
9. Thank the respondents for their participation.
10. Let the respondents know the outcome of the exercise.

13.11.5 Multivote

Multivoting is a series of votes used to arrive at a group decision. It can be used to assign priorities to a list of items. It can be used at team meetings after a brainstorming session has generated a long list of items. Multivoting helps reduce such long lists to a few items, usually three to five. The steps for multivoting are

1. Take a first vote. Each person votes as many times as desired, but only once per item.
2. Circle the items receiving a relatively higher number of votes (i.e., majority vote) than the other items.
3. Take a second vote. Each person votes for a number of items equal to one half the total number of items circled in step 2. Only one vote per item is permitted.

4. Repeat steps 2 and 3 until the list is reduced to three to five items, depending on the needs of the group. It is not recommended to multivote down to only one item.
5. Perform further analysis of the items selected in step 4, if needed.

13.12 Project Leadership

Some leaders lead by setting a good example. Others attempt to lead by dictating. People learn and act best when good examples are available to emulate. Examples learned in childhood can last a lifetime. A leader should have a spirit of performance, which stimulates his or her subordinates to perform at their own best. Rather than dictating what needs to be done, a good leader would show what needs to be done. Showing in this case does not necessarily imply an actual physical demonstration of what is to be done. Rather, it implies projecting a commitment to the function at hand and a readiness to participate as appropriate. Traditional managers manage workers to work. So, there is no point of convergence or active participation. Modern managers team up with workers to get the job done. Figure 13.4 presents a leadership model for project management. The model suggests starting by listening and asking questions, specifying objectives, developing clear directions, removing obstacles, encouraging individual initiatives, learning from past experiences, and repeating the loop by listening some more and asking more questions.

Good leadership is an essential component of project management. Project leadership involves dealing with managers and supporting personnel across the functional lines of the project. It is a misconception to think that a leader leads only his or her own subordinates. Leadership responsibilities can cover vertically up or down functions. A good project leader can lead not only his or her subordinates, but also the entire project organization including the highest superiors. A 3D leadership model that consists of self-leadership, team leadership, and leadership-oriented teamwork has been suggested. Leadership involves recognizing an opportunity to make an improvement in a project and taking the initiative to implement the improvement. In addition to inherent personal qualities, leadership style can be influenced by training, experience, and dedication. Some pitfalls to avoid in project leadership are

Politics and Egotism

- Forget personal ego.
- Do not glamorize personality.

FIGURE 13.4 Project leadership loop.

- Focus on the big picture of project goals.
- Build up credibility with successful leadership actions.
- Cut out politics and develop a spirit of cooperation.
- Back up words with action.
- Adopt a "do as I do" attitude.
- Avoid a "do as I say" attitude.
- Participate in joint problem solving.
- Develop and implement workable ideas.

13.13   Personnel Management

Positive personnel management and interactions are essential for project success. Effective personnel management can enhance team building and coordination. The following guidelines are offered for personnel management in a project environment:

1. Leadership style
   - Lead the team rather than manage the team.
   - Display self-confidence.
   - Establish self-concept of your job functions.
   - Engage in professional networking without being pushy.
   - Be discrete with personal discussions.
   - Perform a self-assessment of professional strengths.
   - Dress professionally without being flashy.
   - Be assertive without being autocratic.
   - Keep up with the developments in the technical field.
   - Work hard without overexerting.
   - Take positive initiative where others procrastinate.
2. Supervision
   - Delegate when appropriate.
   - Motivate subordinates with vigor and an objective approach.
   - Set goals and prioritize them.
   - Develop objective performance-appraisal mechanisms.
   - Discipline promptly, as required.
   - Do not overmanage.
   - Do not shy away from mentoring or being mentored.
   - Establish credibility and decisiveness.
   - Do not be intimidated by difficult employees.
   - Use empathy in decision-making processes.
3. Communication
   - Be professional in communication approaches.
   - Do homework about the communication needs.
   - Contribute constructively to meaningful discussions.
   - Exhibit knowledge without being patronizing.
   - Convey ideas effectively to gain respect.
   - Cultivate good listening habits.
   - Incorporate charisma into communication approaches.
4. Handling Conflicts
   - Learn the politics and policies of the organization.
   - Align project goals with organizational goals.
   - Overcome fear of confrontation.
   - Form a mediating liaison among peers, subordinates, and superiors.
   - Control emotions in tense situations.

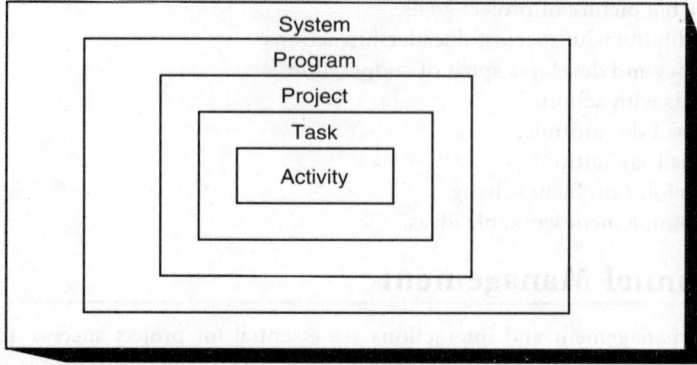

FIGURE 13.5   Systems structure of a project.

– Do not take office conflicts home and do not take home conflicts to work.
– Avoid a power struggle but claim functional rights.
– Handle mistakes honestly without being condescending.

13.14   Integrated Systems Approach

The traditional concepts of systems analysis are applicable to the project process. The definitions of a project system and its components are presented next.

System. A project system consists of interrelated elements organized for the purpose of achieving a common goal. The elements are organized to work synergistically to generate a unified output that is greater than the sum of the individual outputs of the components.

Program. A program is a very large and prolonged undertaking. Such endeavors often span several years. Programs are usually associated with particular systems. For example, we may have a space exploration program within a national defense system.

Project. A project is a time-phased effort of much smaller scope and duration than a program. Programs are sometimes viewed as consisting of a set of projects. Government projects are often called *programs* because of their broad and comprehensive nature. Industry tends to use the term *project* because of the short-term and focused nature of most industrial efforts.

Task. A task is a functional element of a project. A project is composed of a sequence of tasks that all contribute to the overall project goal.

Activity. An activity can be defined as a single element of a project. Activities are generally smaller in scope than tasks. In a detailed analysis of a project, an activity may be viewed as the smallest, practically indivisible work element of the project. For example, we can regard a manufacturing plant as a system. A plant-wide endeavor to improve productivity can be viewed as a program. The installation of a flexible manufacturing system is a project within the productivity improvement program. The process of identifying and selecting equipment vendors is a task, and the actual process of placing an order with a preferred vendor is an activity. The systems structure of a project is illustrated in Figure 13.5.

The emergence of systems development has had an extensive effect on project management in recent years. A system can be defined as a collection of interrelated elements brought together to achieve a specified objective. In a management context, the purposes of a system are to develop and manage operational procedures and to facilitate an effective decision-making process. Some of the common characteristics of a system include:

1. Interaction with the environment
2. Objective
3. Self-regulation
4. Self-adjustment

Representative components of a project system are the organizational subsystem, planning subsystem, scheduling subsystem, information management subsystem, control subsystem, and project delivery subsystem. The primary responsibilities of project analysts involve ensuring the proper flow of information throughout the project system. The classical approach to the decision process follows rigid lines of organizational charts. By contrast, the systems approach considers all the interactions necessary among the various elements of an organization in the decision process.

The various elements (or subsystems) of the organization act simultaneously in a separate but interrelated fashion to achieve a common goal. This synergism helps to expedite the decision process and to enhance the effectiveness of decisions. The supporting commitments from other subsystems of the organization serve to counter-balance the weaknesses of a given subsystem. Thus, the overall effectiveness of the system is greater than the sum of the individual results from the subsystems.

The increasing complexity of organizations and projects makes the systems approach essential in today's management environment. As the number of complex projects increases, there will be an increasing need for project management professionals who can function as systems integrators. Project management techniques can be applied to the various stages of implementing a system as shown in the following guidelines:

1. *Stage definition.* Define the system and associated problems using keywords that signify the importance of the problem to the overall organization. Locate experts in this area who are willing to contribute to the effort. Prepare and announce the development plan.
2. *Personnel assignment.* The project group and the respective tasks should be announced, a qualified project manager should be appointed, and a solid line of command should be established and enforced.
3. *Project initiation.* Arrange an organizational meeting during which a general approach to the problem should be discussed. Prepare a specific development plan and arrange for the installation of needed hardware and tools.
4. *System prototype.* Develop a prototype system, test it, and learn more about the problem from the test results.
5. *Full system development.* Expand the prototype to a full system, evaluate the user interface structure, and incorporate user-training facilities and documentation.
6. *System verification.* Get experts and potential users involved, ensure that the system performs as designed, and debug the system as needed.
7. *System validation.* Ensure that the system yields expected outputs. Validate the system by evaluating performance level, such as percentage of success in so many trials, measuring the level of deviation from expected outputs, and measuring the effectiveness of the system output in solving the problem.
8. *System integration.* Implement the full system as planned, ensure that the system can coexist with systems already in operation, and arrange for technology transfer to other projects.
9. *System maintenance.* Arrange for continuing maintenance of the system. Update solution procedures as new pieces of information become available. Retain responsibility for system performance of delegate to well-trained and authorized personnel.
10. *Documentation.* Prepare full documentation of the system, prepare a user's guide, and appoint a user consultant.

Systems integration permits sharing of resources. Physical equipment, concepts, information, and skills may be shared as resources. Systems integration is now a major concern of many organizations. Even some of the organizations that traditionally compete and typically shun cooperative efforts are beginning to appreciate the value of integrating their operations. For these reasons, systems integration has emerged as a major interest in business. Systems integration may involve the physical integration of management processes or a combination of any of these.

Systems integration involves the linking of components to form subsystems and the linking of subsystems to form composite systems either within a single department or across departments. It facilitates the coordination of technical and managerial efforts to enhance organizational functions, reduce cost, save energy, improve productivity, and increase the utilization of resources. Systems integration emphasizes the identification and coordination of the interface requirements among the components in an integrated system. The components and subsystems operate synergistically to optimize the performance of the total system. Systems integration ensures that all performance goals are satisfied with a minimum expenditure of time and resources. Integration can be achieved in several forms, including the following:

1. *Dual-use integration*: This involves the use of a single component by separate subsystems to reduce both the initial cost and the operating cost during the project life cycle.
2. *Dynamic resource integration*: This involves integrating the resource flows of two normally separate subsystems so that the resource flow from one to or through the other minimizes the total resource requirements in a project.
3. *Restructuring of functions*: This involves the restructuring of functions and reintegration of subsystems to optimize costs when a new subsystem is introduced into the project environment.

Systems integration is particularly important when introducing new technology into an existing system. It involves coordinating new operations to coexist with existing operations. It may require the

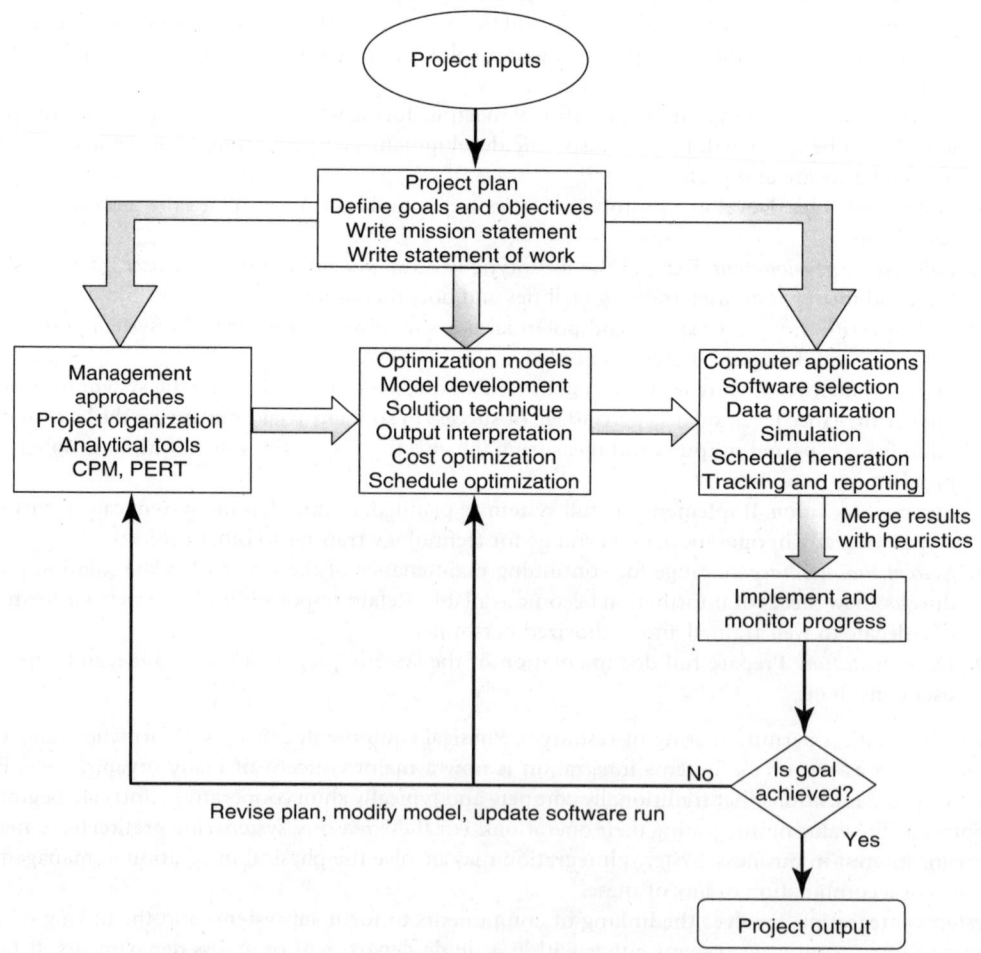

FIGURE 13.6 Flowchart of integrated project management.

adjustment of functions to permit the sharing of resources, development of new policies to accommodate product integration, or realignment of managerial responsibilities. It can affect both hardware and software components of an organization. Presented below are guidelines and important questions relevant for systems integration.

- What are the unique characteristics of each component in the integrated system?
- What are the characteristics of each component in the integrated system?
- What physical interfaces exist among the components?
- What data/information interfaces exist among the components?
- What ideological differences exist among the components?
- What are the data flow requirements for the components?
- Are there similar integrated systems operating elsewhere?
- What are the reporting requirements in the integrated system?
- Are there any hierarchical restrictions on the operations of the components of the integrated system?
- What internal and external factors are expected to influence the integrated system?
- How can the performance of the integrated system be measured?
- What benefit/cost documentations are required for the integrated system?
- What is the cost of designing and implementing the integrated system?
- What are the relative priorities assigned to each component of the integrated system?
- What are the strengths of the integrated system?
- What are the weaknesses of the integrated system?
- What resources are needed to keep the integrated system operating satisfactorily?
- Which section of the organization will have primary responsibility for the operation of the integrated system?
- What are the quality specifications and requirements for the integrated system?

The integrated approach to project management recommended in this chapter is represented by the flowchart in Figure 13.6. The process starts with a managerial analysis of the project effort. Goals and objectives are defined, a mission statement is written, and the statement of work is developed. After these, traditional project management approaches, such as the selection of an organization structure, are employed. Conventional analytical tools including the CPM and the PERT are then mobilized. The use of optimization models is then called upon as appropriate. Some of the parameters to be optimized are cost, resource allocation, and schedule length. It should be understood that not all project parameters will be amenable to optimization. The use of commercial project management software should start only after the managerial functions have been completed. Some project management software programs have built-in capabilities for planning and optimization needs.

A frequent mistake in project management is the rush to use a project management software without first completing the planning and analytical studies required by the project. Project management software should be used as a management tool, the same way a word processor is used as a writing tool. It will not be effective to start using the word processor without first organizing the thoughts about what is to be written. Project management is much more than just the project management software. If project management is carried out in accordance with the integration approach presented in Figure 13.6, the odds of success will be increased. Of course, the structure of the flowchart should not be rigid. Flows and interfaces among the blocks in the flowchart may need to be altered or modified depending on specific project needs.

14

Modeling, Identification/ Estimation in Stochastic Systems

O. Ibidapo-Obe
University of Lagos

14.1   Introduction to Model Types

Proper modeling of a system is the first step towards formulating an optimization strategy for the system. There are different types of models. The sections below present the most common types of models.

14.1.1   Material (Iconic) Models

The material or iconic models simulate the actual system as a prototype in the physical space. It could be a scaled model of an empirical system or a direct physical analog. The study of its behavior under various conditions possible is undertaken. Examples include wind-tunnel laboratories, the linear accelerator laboratory, fatigue-testing equipment, etc.

14.1.2   Mathematical Models

Mathematical modeling involves the application of mathematical/empirical knowledge to the problems of the environment, industry, biosystems, etc. Stimulators to this approach include the advent of high-speed electronic computers and developments in computer technology, progress in applied mathematics (numerical methods), and progress in empirical knowledge (engineering). A mathematical model consists of a set of mathematical formulae giving the validity of certain fundamental "natural laws" and various hypotheses relating to physical processes. Engineering problems are classified into direct and inverse problems (Sage and Melsa, 1971).

14.1.2.1 Direct Engineering Problems

The direct engineering problem is to find the output of a system given the input (see Figure 14.1).

14.1.2.2 Inverse Direct Engineering Problems

The inverse problems are of three main types – design/synthesis, control/instrumentation, and modeling/identification (Lee, 1964).

- *Design/synthesis*: given an input and output, find a system description that fits a physically realizable relationship optimally.
- *Control/instrumentation*: given a system description and a response, find the input that is responsible for the response (output).
- *Modeling/identification*: given a set of inputs and corresponding outputs from a system, find a mathematical description (model) of the system (see Figure 14.2).

14.1.2.3 Objective/Cost Function

The criteria for objective/cost function selection would be to minimize the errors between the model and actual system (Liebelt, 1967). The "goodness of fit" of the criteria can be evaluated when both the model and system are forced by sample inputs (see Figure 14.3).

14.1.2.4 General Problem Formulation

Let

$$\frac{dx(t)}{dt} = f(x(t),\, u(t),\, w(t),\, p(t),\, t) \tag{14.1}$$

be the system equation where

$x(t)$ is the system state vector, $u(t)$ the input signal/control, $w(t)$ the input disturbance/noise, and $p(t)$ the unknown parameter.

Assume that the observation is of the form

$$z(t) = h(x(t),\, u(t),\, w(t),\, p(t),\, v(t)t) \tag{14.2}$$

where $v(t)$ is the observation noise.

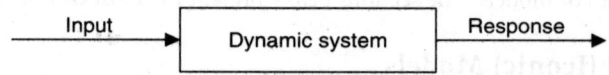

FIGURE 14.1 Input–response relationship in system modeling.

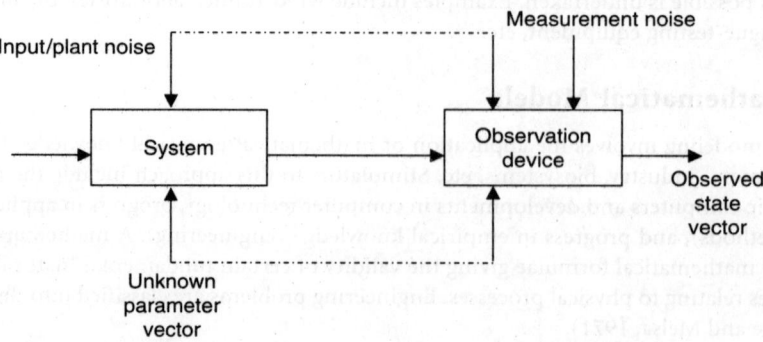

FIGURE 14.2 General system configuration.

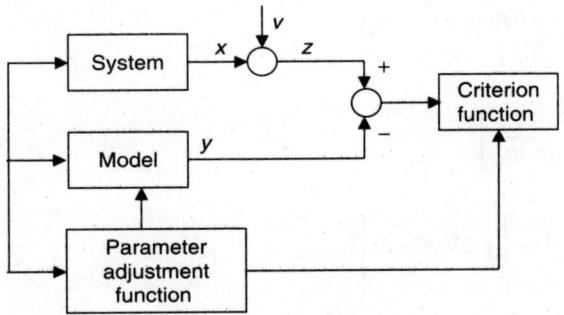

FIGURE 14.3 Parameter models.

The identification/estimation problem is to determine $p(t)$ (and perhaps $x(t)$ as well as the mean and variance coefficients of system noise $w(t)$ and observation noise $v(t)$):

system: $\dfrac{dx(t)}{dt} = f(x(t), u(t), w(t), p(t), t)$

observation: $z(t) = Dy + Eu, \quad D, E$ are matrices

model: $\dfrac{dy(t)}{dt} = g(y(t), u(t), w(t), p', t)$

criterion function: $J(T, p') = \displaystyle\int_0^T \|x(t) - y(t)\|_w dt$

W is an appropriate weighing matrix

Problem:
Seek an optimum set of parameters p^* that minimizes J, i.e.,

$$J(T, p^*) = \min J(T, p')$$

Analytical expressions for p^* are possible:

$$\frac{dJ}{dp'} = 0 \text{ provided } \frac{d^2J}{dp'^2} > 0 \text{ in special cases.}$$

Search techniques are useful when the number of parameters is small. The technique consists of:

(i) random selection of pre-selected grid pattern for parameters $p'_1, p'_2, \ldots$ and corresponding $J_1, J_2, \ldots$ and

(ii) simple comparison test for the determination of minimum J.

Gradient methods are based on finding the values of p' for which the gradient vector equals zero:

$$\nabla_0 J = \left[\frac{\partial J}{\partial p_1}, \ \frac{\partial J}{\partial p_2}, \ \ldots, \ \frac{\partial J}{\partial p_k} \right] = 0$$

and

$$p^{(i+1)} = p^{(i)} - K\nabla_0 J(p^{(i)})$$

where for
steepest descent, $K = kJ$, where k is a constant.

Newton–Raphson, $K = \dfrac{J(p)}{\|\nabla J(p)\|^2}$

Newton, $K = H^{-1} = \left[\dfrac{\partial^2 J}{\partial p_i \partial p_k}\right]^{-1}$

Gauss–Newton, $K = \delta^{-1} = \left[\displaystyle\int_0^T 2\nabla y \nabla y'\, dt\right]^{-1}$

It is desirable to have online or recursive identification so as to make optimum adaptation to the system goal possible in the face of uncertainty and change in the environmental conditions.

14.1.3 Systems Identification

Identification problems can be categorized into two broad areas, such as the total ignorance/"black-box" identification and the gray-box identification. In the gray-box identification, the system equations may be known or deductible from the basic physics or chemistry of the process up to the coefficients or parameters of the equation. The methods of solution consist of classical and modern techniques (Bekey, 1970).

14.1.3.1 Classical Methods

Deconvolution methods (see Figure 14.4):
Given $u(t)$ and $y(t)$ for $0 \le t \le T$, determine $h(t)$.

  (i) Observe input and output at N periodical sampled time intervals, say Δ sec apart in $[0,T]$ such that
      $N\Delta = T$
 (ii) It is known that

$$y(t) = \int_0^t h(t - \tau)u(\tau)d\tau \quad \text{(convolution integral)} \tag{14.3}$$

(iii) Assume that

$$u(t) = u(n\Delta) \text{ or } u(t) \approx \tfrac{1}{2}\{u(n\Delta) + u(n+1)\Delta\} \tag{14.4}$$

for $n\Delta < t < (n+1)\Delta$

$$h(t) \approx h\left(\frac{2n+1}{2}\Delta\right), \; n\Delta <= t < (n+1)\Delta \tag{14.5}$$

(iv) Let

$$y(n\Delta) = \Delta \sum_{i=0}^{n-1} h\left(\frac{2n-1}{2}\Delta - i\Delta\right)u(i\Delta) \tag{14.6}$$

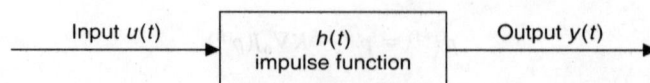

FIGURE 14.4 Input–output relationship for impulse function.

(v) If

$$
y(T) = \begin{bmatrix} y(\Delta) \\ y(2\Delta) \\ \cdot \\ y(N\Delta) \end{bmatrix}, \quad
h(T) = \begin{bmatrix} h\left(\dfrac{\Delta}{2}\right) \\ h\left(\dfrac{3\Delta}{2}\right) \\ \cdot \\ h\left(\dfrac{(2N-1)\Delta}{2}\right) \end{bmatrix}
\tag{14.8}
$$

then $y(T) = \Delta U h(T)$

(14.8)

where

$$
U = \begin{bmatrix}
u(0) & 0 & 0 & 0 & 0 \\
u(\Delta) & u(0) & 0 & 0 & 0 \\
u(2\Delta) & u(\Delta) & u(0) & \cdot & 0 \\
\cdot & \cdot & \cdot & \cdot & 0 \\
u((N-1)\Delta) & u((n-2)\Delta) & \cdot & u(0) & (0)
\end{bmatrix}
\tag{14.9}
$$

(vi) From Equation (14.8),
$h(T) = = U^{-1}\Delta^{-1}y(T)$ so that

$$
h_n \cong h\left(\frac{2n-1}{2}\Delta\right), \quad h_1 = \frac{y(\Delta)}{\Delta u(0)}
$$

$$
= \frac{1}{u(0)}\left\{ \frac{y(n\Delta)}{\Delta} - \sum_{i=1}^{n-1} h_{n-i}u(i\Delta) \right\}
$$

Advantages:

(i) Simple.
(ii) Quite effective for many identification problems.
(iii) Fast Fourier transform (FFT) may be used to reduce the computational requirements.
(iv) Any input may be used (no need for special test inputs).

Disadvantages:

(i) Sequential/online use of algorithm impossible unless time interval of interest is short.
(ii) Numerical round-off errors make the technique inaccurate for $m \rightarrow \infty$.

Correlation techniques (see Figure 14.5): Correlation techniques use white noise test signal, hence it is necessary to have wide bandwidth to detect high-frequency components of $h(t)$. For zero error, $u(t)$ must be proper "white" (infinite bandwidth) (see Figure 14.6):

(i) It is assumed that steady state is reached.
(ii) Noise $u(t)$ and $v(t)$ ergodic and Gaussian with zero mean:

$$
x_0(t) = \frac{1}{t}\int_0^t x(\lambda)d\lambda
\tag{14.10}
$$

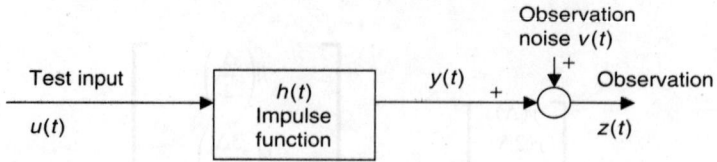

FIGURE 14.5   Configuration for signal identification.

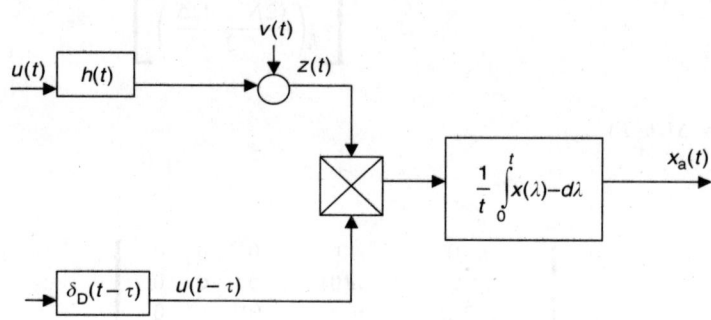

FIGURE 14.6   Identification correlator.

$$x(t) = z(t)u(t - \tau) \tag{14.11}$$

$$z(t) = y(t) + v(t) \tag{14.12}$$

$$y(t) = \int_0^t h(\eta)u(t - \eta)\mathrm{d}\eta \tag{14.13}$$

Now,

$$E\{x_a(t)\} = \frac{1}{t}\int_0^t E\{x(\lambda)\}\mathrm{d}\lambda$$

$$= E\{x\}$$

$$= E\{z(t)u(t - \tau)\}$$

$$= R_{uz}(\tau)$$

$$R_u(\tau) = E\{u(t)z(t + \tau)\}$$

From Equations (14.12) and (14.13),

$$E\{x_a(t)\} = R_{uz}(\tau) = \int_0^\infty h(\eta)R_u(t - \eta)\mathrm{d}\eta$$

Taking Fourier transforms, $R_{uz}(s) = h(s)R_u(s)$. If the assumption on bandwidth holds, $R_{uz}(s) = kh(s)$, $R_{uz}(\tau) = kh(\tau)$; and if $u(t)$ is white δ_D is a Dirac delta, $R_u(\tau) = R_u\delta(1)$, and $R_u(s) = R_w$; and then, for $R_u = 1$, $R_{uz}(\tau) = E\{x_a(t)\} = h(\tau)$.

Complete system identification is subsequently obtained by using N correlators in parallel, such that the quantities

$$R_{uz}(\tau_i) = h(\tau_i), \quad i = 1, 2, \ldots, N$$

are measured.

Advantages:

(i) Not critically dependent on normal operating record.
(ii) By correlating over a sufficiently long period of time, the amplitude of the test signal can be set very low such that the plant is essentially undisturbed by the white-noise test signal.
(iii) No *a priori* knowledge of the system to be tested is required.

14.1.4 System Estimation

Kalman (1960) initiated a new formulation of the Wiener (classical) theory expressing the results of the estimation in the time domain rather than in the frequency domain. The modern theory is more fundamental, requires minimum mathematical background, is perfect for digital computation, and provides a general estimator, whereas the classical method can only deal with restricted dimensions, is rigorous, and has limited applicability to nonlinear systems.

14.1.4.1 Problem Formulation

Let $\Theta = \Theta(x)$, where Θ is a vector of m observations and x a vector whose variables are to be estimated. The estimation problem is continuous if Θ is a continuous function of time; otherwise, it is a discrete estimation problem. Estimating the past is known as smoothing, estimating the present as filtering, while estimating the future is prediction/forecasting.

Nomenclature:

(i) An estimate $\hat{x}$ of x is unbiased if $E(\hat{x}) = x$.
(ii) Let $e = \hat{x} - x$ and $C_e = E[(\hat{x} - x)(\hat{x} - x)']$.

Maximum likelihood:

Let $\Theta = Bx + v$, where v is the noise. The maximum likelihood method takes $\hat{x}$ as the value that maximizes the probability of measurements that actually occurred taking into account the known statistical properties of v. The conditional probability density function for Θ, given x, is the density of v centered around Bx. If v is zero mean Gaussian-distributed with covariant matrix C_v then

$$p(\Theta|x) = \frac{1}{(2\pi)^{1/2}|C_v|^{1/2}} e\left[\frac{-1}{2}(\Theta - Bx)C_v^{-1}(\Theta - Bx)\right]$$

so that

$$\text{Maximum } p(\Theta|x) = \text{Max}\left[\frac{-1}{2}(\Theta - Bx)C_v^{-1}(\Theta - Bx)\right]$$

$$\text{Hence, } x = (B'C_v^{-1}B)^{-1}B'C_v^{-1}\Theta.$$

14.1.4.2 Least-Squares/Weighted Least Squares

The least squares choose $\hat{x}$ as that value that minimizes the sum of squares of the deviations $\theta_i - \hat{\theta}_p$, i.e.,
Minimize $J = (\Theta - Bx)'(\Theta - Bx)$.
Thus, setting $\partial J \partial x = 0$ yields $\hat{x} = (B'B)^{-1}B'\Theta$.
For weighted least squares,
Minimize $J = (\Theta - Bx)'W^{-1}(\Theta - Bx)$ or $J = \|\Theta - Bx\|_W^{-1}$
yielding $\hat{x} = (B'W^{-1}B)^{-1}B'W^{-1}\Theta$.

Bayes estimators:
For Bayes estimators, statistical models for both x and Θ are assumed available. The *a posteri* conditional density function $p(x|\Theta)$

$$p(x|\Theta) = \frac{p(\Theta|x)p(x)}{p(\Theta)} \quad \text{(Bayes rule)}$$

since it contains all the statistical information of interest. $\hat{x}$ is computed from $p(x|\Theta)$.

Minimum variance:

$$\text{Minimize } J = \int_{-\infty}^{\infty}\int_{-\infty}^{\infty} \cdots \int_{-\infty}^{\infty} (\hat{x} - x)' S(\hat{x} - x)p(x|\Theta)dx_1 \cdots dx_n$$

where S is an arbitrary, positive, semidefinite matrix.
   Set $\partial J/\partial x = 0$ to yield

$$\hat{x} = \int_{-\infty}^{\infty}\int_{-\infty}^{\infty} \cdots \int_{-\infty}^{\infty} xp(x|\Theta)dx_1 \cdots dx_n$$

Therefore, $\hat{x} = E[x|\Theta]$.

For linear minimum variance unbiased (Gauss–Markov), let $\hat{x} = A\theta$, where A is an unknown parameter and $C_x = E(xx')$, $C_{x\theta} = E(x\theta')$, and $C_\theta = E(\theta\theta')$.

   Now,

$$
\begin{aligned}
C_e &= E(ee') \\
    &= E[(\hat{x} - x)(\hat{x} - x)'] \\
    &= E[(A\theta - x)(A\theta - x)'] \\
    &= E[(A\theta\theta'A' - x\theta'A' - A\theta x' + xx')] \\
    &= E[A\theta\theta'A'] - E[x\theta'A'] - E[A\theta x'] + E[xx'] \\
    &= AC_\theta A' - C_{x\theta}A' - AC_\theta x + C_x \\
    &= (A - C_{x\theta}C_\theta^{-1})C_\theta(A - C_{x\theta}C_\theta^{-1}) - C_{x\theta}C_\theta^{-1}C'_{x\theta} + C_x
\end{aligned}
$$

The minimum is obtained when $A - C_{x\theta}C_\theta^{-1} = 0$, i.e., $A = C_{x\theta}C_\theta^{-1}$. Therefore $x = C_{x\theta}C_\theta^{-1}\theta$, $C_e = C_x - C_{x\theta}C_\theta^{-1}C'_{x\theta}$.

   (i) If $\theta = Bx + v$, then
$$
\begin{aligned}
C_x\theta &= E[x\theta'] \\
          &= E[x(Bx + v)'] \\
          &= E[xx'B'] + E[xv'] \\
C_{x\theta} &= C_xB' + C_{xv} \\
C_\theta &= E[\theta\theta'] \\
         &= [(Bx + v)(Bx + v)'] \\
         &= BC_xB' + C'_{xv}B' + BC_{xv} + C_v \\
         &= BC_xB' + C'_{xv}B' + BC_{xv} + C_v \\
\hat{x} &= [C_xB' + C_{xv}][BC_xB' + (BC_{xv})' + BC_{xv} + C_v]^{-1}\theta \\
C_e &= C_x - (C_xB' + C_{xv})(BC_xB' + (BC_{xv})' + BC_{xv} + C_v)^{-1}(C_xB' + C_{xv})'
\end{aligned}
$$

   (ii) If $\theta = Bx + v$, $C_{xv} = 0$ then
$$
\begin{aligned}
\hat{x} &= (C_xB')[BC_xB' + C_v]^{-1}\theta \\
C_e &= C_x - (C_xB')(BC_xB' + C_v)^{-1}(C_xB')'
\end{aligned}
$$
   or
$$
\begin{aligned}
\hat{x} &= (B'C_v^{-1}B)^{-1}B'C_v^{-1}\theta \\
\hat{x} &= (C_x^{-1} + B'C_vB)^{-1}B'C_v^{-1}\theta \\
C_e &= (C_x^{-1} + B'C_v^{-1}B)^{-1}
\end{aligned}
$$

   (iii) If in (ii), $C_x \to \infty$ (no information on state) then
$$
\begin{aligned}
\hat{x} &= (B'C_v^{-1}B)^{-1}B'C_v^{-1}\theta \\
C_e &= (B'C_v^{-1}B)^{-1}
\end{aligned}
$$

14.2 Partitioned Data Sets

Let θ^r be a set of measurements of dimension r and $\hat{x}^r$ the estimate obtained using Θ^r. Let $\Theta^r = B^r x + v^r$ such that $(C_v^r)^{-1}$ exists; $C_{xv}^r = C_{vx}^r = 0$; and $C_x^{-1} \rightarrow 0$. Then

$$\hat{x}^r = (B'^r C_v^r B^r)^{-1} B'^r (C_v^r)^{-1} \theta^r$$
$$C_e = (B'^r (C_v^r)^{-1} B^r)^{-1}$$

Suppose now that an additional set of data Θ^s is taken:

$\theta^s = B^s x + v^s$ provided $C_{xv}^s = 0$, $C_{vv}^{rs} = 0$, so that

$$\begin{bmatrix} \theta^r \\ \theta^s \end{bmatrix} = \begin{bmatrix} B^r \\ B^s \end{bmatrix} x + \begin{bmatrix} v^r \\ v^s \end{bmatrix}$$

Let $r + s = m$, then $\theta^m = (\theta^r, \theta^s)'$, $B^m = (B^r, B^s)'$, and $v^m = (v^r, v^s)'$:

$$C_v^m = E(v^m, v^{m\prime})$$

$$= E\left[\begin{bmatrix} v^r \\ v^r \end{bmatrix} [v^{r\prime} \; v^{s\prime}] \right]$$

$$= E\begin{bmatrix} v^r v^{r\prime} & v^s v^{r\prime} \\ v^s v^{r\prime} & v^s v^{s\prime} \end{bmatrix}$$

$$= \begin{bmatrix} C_v^r & C_{vv}^{rs} \\ C_{vv}^{sr} & C_v^s \end{bmatrix}$$

$$= \begin{bmatrix} C_v^r & 0 \\ 0 & C_v^s \end{bmatrix}$$

$$\hat{x}^m = (B'^m (C_v^m)^{-1} B^m)^{-1} B'^m (C_v^m)^{-1} \theta^m$$
$$C_e^m = (B'^m (C_v^m)^{-1} B^m)^{-1}$$

$$\hat{x}^m = \left[(B^{r\prime} B^{s\prime}) \begin{bmatrix} C_v^r & 0 \\ 0 & C_v^s \end{bmatrix}^{-1} \begin{bmatrix} B^r \\ B^s \end{bmatrix} \right]^{-1} (B^{r\prime} B^{s\prime}) \begin{bmatrix} C_v^r & 0 \\ 0 & C_v^s \end{bmatrix}$$

thus yielding

$$\hat{x}^m = (B'^r (C_v^r)^{-1} B^r + B^{s\prime} (C_v^s)^{-1} B^s)^{-1} (B^{r\prime} (C_v^r)^{-1} \Theta^r + B^{s\prime} C_v^s)$$
$$C_e^m = [(C_e^r)^{-1} + B^{s\prime} (C_v^s)^{-1} B^s]^{-1}$$

Kalman form:

Given an estimate x^r, the old error matrix C_e^r, the new data $\theta^s = B^s x + v^s$, the new estimate $\hat{x}^m$ based on all the data is found by the sequence

$$k = C_e^r B^{s\prime} (C_v^s + B^s C_e^r B^{s\prime})^{-1}$$
$$\hat{x}^m = \hat{x}^r + k(\theta^s - B^s \hat{x}^r)$$
$$C_e^m = C_e^r - kB^s C_e^r$$

14.2.1 Discrete Dynamic Linear System Estimation

Dynamic system:

$$x_{i+1} = s(i+1, i)x_i + w_i, \quad s(i+1, i) \text{ is the transformation matrix}$$
$$E(w_i) = 0, \; \forall i$$
$$E(w_i w_j^T) = 0, \; \forall i \neq j$$
$$E(w_i w_j^T) = 0, \; \forall i \leq j$$
$$W_i = E(w_i w_j^T)$$

Observation vector $\theta = A_i x_i + q_i$, A_i is transformation matrix

$$E(q_i q_j^T) = 0, \forall i \neq j$$
$$E(q_i w_j^T) = 0, \forall i, j$$
$$E(q_i x_j^T) = 0, \forall i, j$$
$$Q_i = E(q_i q_i^T)$$

Prediction:

$$\hat{X}_p^m = S(p, n)\hat{X}_n^m, \ p \geq n, \ S(.,.) \text{ is the transformation matrix } \overrightarrow{AB}$$

$$C_p^m = S(p, n)C_n^m S^T(p, n) + \sum_{k=n}^{p-1} S(p, k+1)W_k S^T(p, k+1), \ p > n \geq m$$

Filtering:

$$\hat{X}_{m+1}^m = S(m+1, m)\hat{X}_m^m$$

$$C_{m+1}^m = S(m+1, m)C_m^m S^T S(m+1, m) + W_{m+1}$$

$$K = C_{m+1}^m \Delta_{m+1}^T (Q_{m+1} + \Delta_{m+1} C_{m+1}^m \Delta_{m+1}^T)^{-1}$$

$$\hat{X}_{m+1}^{m+1} = \hat{X}_m^m + K(\theta_{m+1} - \Delta_{m+1}\hat{X}_{m+1}^m)$$

$$C_{m+1}^{m+1} = C_{m+1}^m - K\Delta_{m+1}C_{m+1}^m$$

Smoothing:

$$\hat{X}_r^m = \hat{X}_r^m + J[\hat{X}_{r+1}^m - S(r+1, r)\hat{X}_r^r]$$

$$C_r^m = C_r^r + J[C_{r+1}^m - C_{r+1}^r]J^T$$

$$J = C_r^r S^T(r+1, r)(C_{r+1}^r)^{-1}$$

$$C_{r+1}^r = S(r+1, r)C_r^r S^T(r+1, r) + W_r$$

14.2.2 Continuous Dynamic Linear System

Let

$$\dot{x}(t) = Ax(t) + Bu(t) \tag{14.14}$$

be the system equation and

$$y(t) = Cx(t) + Dv(t) \tag{14.15}$$

be the observation equation. The estimate $\hat{x}(t)$ is restricted as a linear function of $y(\tau), 0 \leq \tau \leq t$; thus

$$\hat{x}(t) = \int_0^t \alpha(\tau)y(\tau)d\tau \tag{14.16}$$

The solution to Equation (14.14) is

$$x(t) = \Phi(t)x(0) + \int_0^t \Phi(t)\Phi^{-1}(s)Bu(s)ds \tag{14.17}$$

where $\Phi(\cdot)$ is the transition matrix. From Equation (14.16)

$$\hat{x}(t+\delta) = \int_0^{t+\delta} \alpha(\tau)y(\tau)d\tau$$

and from Equation (14.17),

$$\hat{x}(t+\delta) = \Phi(t+\delta)\left\{x(0) + \int_0^{t+\delta} \Phi^{-1}(\sigma)B(\sigma)u(\sigma)d\sigma\right\}$$

$$= \Phi(t+\delta)\Phi^{-1}(t)\Phi(t)\left\{x(0) + \int_0^t \Phi^{-1}(\sigma)B(\sigma)u(\sigma)d\sigma + \int_t^{t+\delta} \Phi^{-1}(\sigma)B(\sigma)u(\sigma)d\sigma\right\}$$

$$x(t + \delta) = \Phi(t + \delta)\Phi^{-1}(t)\left\{x(t) + \int_t^{t+\delta} \Phi(t)\Phi^{-1}(\sigma)B(\sigma)u(\sigma)d\sigma\right\}$$

Using the orthogonality principle,

$$E\{[x(t + \delta) - \hat{x}(t + \delta)y'(\tau)]\} = 0 \quad \text{for } 0 \leq \tau \leq t$$

and recalling that

$$E\{u(\sigma)y'(\tau)\} = 0, \sigma > t$$

Hence,

$$E\{[\Phi(t + \delta)\Phi^{-1}(t)x(t) - \hat{x}(t + \delta)]y'(\tau)\} = 0, 0 \leq \tau \leq t$$

thus

$$\hat{x}(t + \delta) = \Phi(t + \delta)\Phi^{-1}(t)\hat{x}(t),$$

If

$$u(t) \approx N(0, Q(t)),$$

$$v(t) \approx N(0, R(t))$$

Given that

$$E\{x(0)\} = \hat{x}_0$$

$$E\{[x(0) - \hat{x}_0][x(0) - \hat{x}_0]'\} = P_0$$

and $R^{-1}(t)$ exists. The Kalman filter consists of

Estimate: $\dot{\hat{x}}(t) = A\hat{x}(t) + K(t)[y(t) - C\hat{x}(t)], \hat{x}(0) = \hat{x}_0$

Error Covariance: $\dot{\hat{P}}(t) = AP(t) + P(t)A' + BQB' - KRK'$

Propagation: $P(0) = P_0$. For steady-state $\dot{\hat{P}}(t) = 0$

Kalman gain matrix:

$\dot{K}(t) = P(T)C'R^{-1}(t)$ when $E[u(t)'v(\tau)] = 0$ and $\dot{K}(t) = [P(T)G' + BG]R^{-1}(t)$
when $E[u(t)v'(\tau)] = G(t)\delta(t - \tau)$.

The fixed-time smoothing algorithm $\hat{x}_{t|T}$ is as follows:

$P(t|T) = (A + BQB'P^{-1})' - BQB'$
with $\hat{x}(T|T) = \hat{x}(t = T)$ and $P(T|T) = P(t = T)$ as initial conditions.

14.2.3 Continuous Nonlinear Estimation

The analysis of stochastic dynamic systems often leads to differential equations of the form $\dot{x}(t) = f(t, x)$ $+ G(t, x)u(t), x_{t_0} = c, 0 \leq t \leq T \leq \infty$ or in integral form

$$x(t) = c + \int_0^t f(s, x)ds + \int_0^t G(s, x)dw(s), 0 \leq t \leq T \leq \infty$$

where $dw(t)/dt = u(t)$, $w(t)$ the Wiener process, $x(t)$ and $f(t, s)$ n-dimensional while $G(t, x)$ is $n \times m$ matrix function and $u(t)$ is m-dimensional.

Ito rule:

$$\int_0^T G(t, x)dw(t) = \lim_{\Delta \to 0, i=0} \sum^N G(t_i, x(t_i))(w(t_{i+1}) - w(t_i))$$

For the partition $t_0 < t_1 < t_i \ldots < t_i < t_{i+1}\ldots t_N = T$ and $\Delta = \max_i(t_{i+1} - t_i)$.

Stratonovich rule:

$$\int_0^T G(t, x)dw(t) = \lim_{\Delta \to 0, i=0} \sum^{N-1} G\left(t_i, \frac{x(t_{i+1}) - x(t_i)}{2}\right)(w(t_{i+1}) - w(t_i))$$

See Figure 14.7 for functional configuration.

Let the observation be of the form $y(t) = z(t) + v(t)$ where $z(t) = \phi(x(s), s \leq t)$ and $v(t)$ are p-dimensional vectors.

It is further assumed that $E[z(t)z'(t)] < \infty$ and $E[z(t)v'(t)] = 0$ for all t.

Doob (1949, 1953) obtained the estimator $\hat{x}(t|T) = E[x(t)|y(s), t_0 \leq s \leq \tau]$, where for $t = \tau$ (filtering), $t < \tau$ (smoothing) and $t > \tau$ (prediction):

$$\hat{x}(t) = E[x(t)|y(s), t_0 \leq s \leq \tau]$$
$$= \int_{-\infty}^{\infty} x(t)P_r(x(t)|y(t))dx(t) \tag{14.18}$$

Let

$$P_t = E[(x(t) - \hat{x}(t))(x(t) - \hat{x}(t))']$$
$$= \int_{-\infty}^{\infty} x(t) - \hat{x}(t))(x(t) - \hat{x}(t))' P_r(x(t)|y(t))dx(t) \tag{14.19}$$

Assume that

$$E\{u(t) = E\{v(t) = E\{u(t)v'(t) = 0$$
$$E\{u(t)u'(t)\} = Q_t, \; E\{v(t)v(t)'\} = R_t$$

The Folker–Plank stochastic differential equations for the probability density function P_r:

$$\frac{\partial P_r}{\partial t} = -\text{trace}\left(\frac{\partial}{\partial x}\{f(t, x)P_r\}\right) + \frac{1}{2}\text{trace}\left(\frac{\partial}{\partial x}\left(\left[\frac{\partial}{\partial x}\right]'\{G(t, x)QG'(t, x)P_r\}\right)\right)$$
$$+ P_r(y - \Phi(t, x))R_t^{-1}(\Phi(t, x) - \Phi(t, \hat{x})) \tag{14.20}$$

Using Equation (14.20) in Equations (14.18) and (14.19),
$$d\hat{x}(t) = \hat{f}(t, x)dt + E(x - \hat{x})\Phi'(t, x)|y(t)R_t^{-1}(y(t) - \Phi(t, x))dt$$
$$dP_t + d\hat{x}d\hat{x} = E\{f(t, x)(x - \hat{x})|y(t)\}dt$$
$$\quad + E\{(x - \hat{x})f'(t, x)|y(t)\}dt$$
$$\quad + E\{G(t, x)Q_tG'(t, x)|y(t)\}dt$$
$$\quad + E\{(x - \hat{x})(x - \hat{x})'[\Phi(t, x) - \Phi(t, \hat{x})]R_t^{-1}(y(t) - \hat{\Phi}(t, x))|y(t)\}dt$$

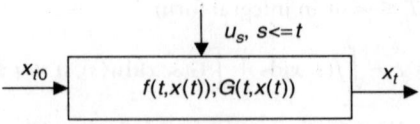

FIGURE 14.7   Functional configuration.

14.3   Extended Kalman Filter

The extended Kalman filter results from application of the linear Kalman–Bucy filter to a linearized non-linear system, where the nonlinear system is relinearized after each observation.

Let

$$f(t, x) = f(t, \hat{x}) - \frac{\partial f}{\partial \hat{x}}(x - \hat{x})$$

$$\Phi(t, x) = \Phi(t, \hat{x}) - \frac{\partial \Phi}{\partial \hat{x}}(x - \hat{x})$$

and

$$G(t, x)Q_tG'(t, x) = G(t, \hat{x})Q_tG'(t, \hat{x}) + (x - \hat{x})\left\{\left(\frac{\partial}{\partial \hat{x}}\right)G'(t, \hat{x})Q_tG'(t, \hat{x})\right\}$$

Substituting the above equation into the previous equation we obtain

$$\frac{d\hat{x}}{dt} = f(t, \hat{x}) + P_t\frac{\partial \Phi}{\partial \hat{x}}R_t^{-1}(y(t) - \Phi(t, x))$$

$$\frac{dP_t}{dt} = \frac{\partial f}{\partial \hat{x}}P_t + P_t\frac{\partial f'}{\partial \hat{x}}G(t, \hat{x})Q_tG'(t, \hat{x}) - P_t\frac{\partial \Phi'}{\partial \hat{x}}R_t^{-1}\frac{\partial \Phi'}{\partial \hat{x}}P_t$$

which can now be solved with appropriate initial conditions

$$\hat{x}\big|_{t_0} = \hat{x}(t_0) \quad \text{and} \quad P_t\big|_{t_0} = P(t_0)$$

This is the extended Kalman filter for nonlinear systems.

14.4   Partitional Estimation

Lainiotis (1974) proposed the Partition Theorem, general continuous-data Bayes rule for the posterior probability density.

Let

$$P_r(x(\tau)|t, t_0) = \frac{\Lambda(t, t_0|x(\tau))P_r(x(\tau))}{\int \Lambda(t, t_0|x(\tau))P_r(x(\tau))d\tau}$$

where
(i) $\Lambda(t, t_0|x(t)) = \exp\left\{\int_{t_0}^t \hat{h}'(\sigma|\sigma, t_0; x(t))R_\sigma^{-1}y(\sigma)d\sigma - \frac{1}{2}\int_{t_0}^t \|\hat{h}(\sigma|\sigma, t_0; x(t))\|^2R_\sigma^{-1}d\sigma\right\}$

(ii) $\hat{h}(\sigma|\sigma, t_0; x(\tau)) = E\{h(\sigma, x(\sigma))|y(\sigma); x(\tau).$

(iii) $P_r(x(\tau))$ is the *a priori* density of $x(\tau)$.

The partitioned algorithm for filtering is given by

$$\hat{x}(t) = \int \hat{x}(\tau)P_r(x(\tau)|t, t_0)dx(\tau)$$

and

$$P_t = \int \{P_\tau + [\hat{x}(\tau) - \hat{x}(t)][\hat{x}(\tau) - \hat{x}(t)]'P_r(x(\tau)|t, t_0)dx(\tau)\}$$

where $P_r(x(\tau)|t, t_0)$ is given previously and both $\hat{x}(\tau)$ and P_τ are the "anchored" or conditional mean-square error estimate and error-covariance matrices, respectively.

The partitioned algorithm takes its name from the fact that if the observation interval is partitioned into several small subintervals, repeated use of the filtering equations for each subinterval leads to an effective and computationally efficient algorithm for the general estimation problem.

14.5 Invariant Imbedding

The invariant imbedding approach provides a sequential estimation scheme, which does not depend on *a priori* noise statistical assumptions. The concept in invariant imbedding is to find the estimate $\hat{x}(\tau)$ of $x(t)$ such that the cost function

$$J = \frac{1}{2} \int_0^T \left\{ \|y(t) - \Phi(t, \hat{x}(t))\|_{W_1}^2 + \|\dot{x}(t) - f(t, \hat{x}(t))\|_{W_2}^2 \right\} dt$$

is minimized where W_1 and W_2 are weighing matrices that afford the opportunity to place more emphasis on the most reliable measurements. The Hamiltonian

$$H = \frac{1}{2} W_1 (y(t) - \Phi(t, \hat{x}(t)))^2 + \frac{1}{2} W_2 G^2(t, x(t)) u^2(t) + \lambda^2(t)(f(t, \hat{x}(t)) + G(t, \hat{x}(t)) u(t))$$

The necessary conditions for a minimum are

$$\dot{x}(t) = \frac{\partial H}{\partial \lambda}$$

$$\dot{\lambda}(t) = -\frac{\partial H}{\partial \hat{x}}$$

$$\frac{\partial H}{\partial u} = 0$$

which yield the filtering equations

$$\frac{d\hat{x}}{dT} = f(T, \hat{x}(T)) + P\frac{\partial \Phi'}{\partial \hat{x}}(y(T) - \Phi(T, \hat{x}(T)))$$

$$\frac{dP}{dT} = \frac{\partial f'}{\partial \hat{x}} P + P\frac{\partial f'}{\partial \hat{x}} + P\left(\left[\frac{\partial^2 \Phi'}{\partial \hat{x}^2} \right](y(T) - \Phi(T, \hat{x}(T))) - \frac{\partial \Phi'}{\partial \hat{x}}\frac{\partial \Phi}{\partial \hat{x}} \right)P + \frac{1}{W}$$

14.6 Stochastic Approximations/Innovations Concept

Stochastic approximation is a scheme for successive approximation of a sought quantity when the observation and the system dynamics involve random errors (Albert and Gardner, 1967). It is applicable to the statistical problem of

   (i) finding the value of a parameter that causes an unknown noisy-regression function to take on some preassigned value,
  (ii) finding the value of a parameter that minimizes an unknown noisy-regression function.

Stochastic approximation has wide applications to system modeling, data filtering, and data prediction. It is known that a procedure that is optimal in the decision theoretic sense can be nonoptimal. Sometimes, the algorithm is too complex to implement, for example, in situations where the nonlinear effects cannot be accurately approximated by linearization or the noise processes are strictly non-Gaussian. A theoretical solution is obtained by using the concepts of innovations and martingales. Subsequently, a numerically feasible solution is achieved through stochastic approximation. The innovations approach separates the task of obtaining a more tractable expression for the equation

$$\hat{x}(t|T) = E\{x(t)|y(s), 0 \leq s \leq \tau\}$$

into two parts:

  (i) The data process $\{y(t), 0 \leq t \leq T\}$ is transformed through a causal and causally invertible filter $v(t) = y(t) - \Phi(\hat{x}(s), s \leq t)$ (the innovations process) with the same intensity as the observation process.
 (ii) The optimal estimator is determined as a functional of the innovations process.

The algorithm given below has been used for several problems (Gelb, 1974):

  (i) Pick an α_t^i gain-matrix function, such that for each element $(\alpha_t^i)_{kl}$,

$$\int_0^\infty (\alpha_t^i)_{kl} dt = \infty, \quad i = 1 \quad \text{and} \quad \int_0^\infty (\alpha_t^i)_{kl}^2 \, dt < \infty$$

 (ii) Solve the suboptimal problem

$$\frac{d\hat{x}}{dt} = f(t, \hat{x}) + \alpha_t^i G((t, \hat{x})(y(t) - \Phi(t, \hat{x}))$$

where it is assumed, without any loss of generality, with entries $(\alpha_1^i, \alpha_2^i, ..., \alpha_n^i)$. The 1st component of point (i) is

$$\frac{d\hat{x}_l}{dt} = f_l(t, \hat{x}) + \sum_{k=1}^m \alpha_l^i g_{lk}(t, \hat{x})(y_k(t) - \Phi_k(t, \hat{x}))$$

 (iii) Compute the innovations process
       $v^i(t) = y(t) - \Phi(t, \hat{x}^i)$ and check for its whiteness (within a prescribed tolerance level) by computing the autocorrelation function as well as the power spectrum.
 (iv) If the result of the test conducted in step (iii) is positive, stop.
       Else, iterate on α_t^i. Thus

$$\alpha^{i+1}(t) = \alpha^i(t) + \gamma^i(t)\Psi(v^i(t))$$

where

$$\alpha^1(t) = \gamma^i(t)$$

$$= \left\{ -\frac{a}{t} \text{ or } -\frac{a}{b+t} \text{ or } -\frac{a+t}{b+t^2} \right\}$$

and

$$\Psi(v^i(t)) = v^i(t) - E\{v^i(t)\}$$

  (v) Go to step (ii).

The optimal trajectories constitute a martingale process, and the convergence of the approximate algorithm depends on the assumption that the innovations of the observations are a martingale process. According to the martingale convergence theorem, if $\{x^n\}_k$ is a submartingale, and if l.u.b. $E\{x^n\}_k < \infty$, then there is a random discretized systems holds.

References

Albert A.E. and Gardner, L.A., Jr., *Stochastic Approximation and Nonlinear Regression*, 1967.
Bekey, G.A., *System Identification – An Introduction and A Survey, Simulation*, 1970.
Doob, J.L., Heuristic approach to the Kolmogorov–Smirnov theorems, *Ann. Math. Stat.*, 20, 393–403, 1949.
Doob, J.L., *Stochastic Processes, Mathematical Reviews*, Wiley, 1953.

Gelb, A., *Applied Optimal Estimation*, MIT Press, Cambridge, MA, 1974.

Kalman, R.E., A new approach to linear filtering and prediction problems, transaction of the ASME, *J. Basic Eng.*, 82, 35–45, 1960.

Lainiotis, D.G., Partioned estimation algorithm, 11: linear estimation, *Inform. Sci.*, 7, 317–340, 1974.

Lee, R.C.K., *Optimal Estimation, Identification and Control*, MIT Press, Cambridge, MA, 1964.

Liebelt, P.B., *An Introduction to Optimal Estimation*, Addison-Wiley, New York, 1967.

Sage, A.P. and Melsa, J.L., *System Identification*, Academic Press, New York, 1971.

PART IV

MANUFACTURING AND PRODUCTION SYSTEMS

15

Manufacturing Technology

Shivakumar Raman
University of Oklahoma

Aashish Wadke
University of Oklahoma

15.1   Manufacturing

The term "manufacturing" originates from the Latin word *manufactus,* which means "made by hand." Manufacturing has seen several advances during the last three centuries: mechanization, automation, and, most recently, computerization. Processes that used to be predominantly done by hand and hand tools have evolved into sophisticated processes making use of cutting-edge technology and machinery. Quality has improved steadily, with today's specifications even those for simple toys, exceeding those achievable just a few years ago. Mass production, a concept developed by Henry Ford, has advanced so much that it is now a complex, highly agile, and highly automated manufacturing enterprise.

From a technological standpoint, manufacturing involves the making of products from raw materials through the use of human labor and resources such as machines, tools, and facilities. It could be more generally regarded as the conversion of an unusable state into a usable state by adding value along the way. For instance, a log of wood serves as the raw material for making wood planks, which in turn provide the raw material for producing chairs. The value added is usually represented in terms of cost and time.

From a broad or systemic point of view, manufacturing encompasses design, processes, controls, quality assurance, and all aspects related to the life cycle of the product from concept to product to recycling. The systemic view has been realized and improved over 75 years, and the context and concept of integrating the various functions have become even more paramount today. Thus, designs for manufacturability, assembly, quality, and environmentability are gathered under the general umbrella of "Design for X." This is important, considering that the majority of expenditures toward the development of a product occur during its initial stages and during product design. Concurrent Engineering and Simultaneous Engineering are two other life-cycle-based concepts that also promote integration rather than fragmentation. One reason for the systemic point of view stems from the demands placed on products today. It is not uncommon to find stringent tolerance and finish requirements on a small piece of lever that goes into a toy assembly that costs only a few dollars. Aesthetics, ergonomics, safety, and creativity have each experienced significant growth in the development of products.

15.2 Interchangeability

Interchangeable manufacture relies largely on standardization of products and processes. In the systemic sense, interchangeability facilitates "pluggability" or modularity and de-emphasizes design by product. Interchangeability must facilitate easy replacement of parts within assemblies at reasonable costs. The key words for interchangeability are "tolerances" and "allowances." Tolerance is the permissible variation on manufactured parts with the understanding that perfect parts are impossible to make with currently available machines. Even if they could be manufactured perfectly, current measurement tools cannot suitably verify that. Instruments are limited by a least count or resolution, the best achievable with modern-day instruments being a nanometer. Allowances quantify the degree of looseness or tightness of a fit or assembly. Depending on the allowances, the fits are classified into "clearance," "interference," and "transition" fits. Since most commercial products, including automobiles and airplanes, are assemblies, tolerances and allowances must be suitably specified to promote interchangeable manufacture. Stated otherwise, if an assembly is completely composed of standard parts that could each be procured "off the shelf," the labor involved in designing dimensions of the piece parts can be minimized quite significantly. The final product will also have good longevity, as part replacement will become simpler.

15.3 Manufacturing Processes and Process Planning

Currently, several alternatives are available for processing; thus, there is usually more than one way to manufacture any given part. The processes can be classified as:

1. Casting and foundry processes
2. Forming and shaping processes
3. Machining processes
4. Joining processes
5. Finishing processes
6. Nontraditional manufacturing processes

Every candidate process is suited to promote a particular functionality characteristic and is capable of generating a certain level of quality. Quality is a subjective metric, and, in the context of manufacturing, often implies surface finish and tolerances. The economics of any process is also very important and can be conveniently decomposed with the analysis of manufacturing operations. (Operations are subsets of processes and tasks are subsets of operations.) A manufactured part can be decomposed into several features, the features into operations, and operations into tasks. Since several candidate operations may be selected, and many sequences of operations exist, several viable process plans can be made.

Process planning is the coordinated selection and sequencing of manufacturing operations to manufacture a part. It is expected to provide detailed documentation of the human power and resources while rendering a product. Since the number of alternatives to manufacture a product may be very large, an exhaustive enumeration of all possible or feasible plans could be prohibitive; hence, a subset of the total number of viable sequences is derived from the grand list utilizing manufacturing precedence information. Objectives and constraints are identified and formulated suitably and weights drawn to evaluate and compare alternative plans. Considering the enormous effort involved, Computer aided process planning systems have become very attractive in order to generate feasible sequences and to minimize the lead-time and nonvalue-added costs.

15.4 Casting and Foundry Processes

Casting processes (Figure 15.1 and Figure 15.2) can be further classified into permanent and expendable mold-type processes. The basic idea is to superheat a metal or alloy well beyond its melting point (often 3–5 times the melting point to increase the fluidity of the metal), pour it into a mold, and allow it to solidify within the mold. Upon solidification, the part is retrieved from the mold and finished suitably. The expendable mold processes destroy the mold after solidification; they include, for example, sand casting, plaster molding, and investment casting. The last mentioned can result in better finishes and tighter tolerances than the former two. High- and low-pressure die casting, and centrifugal casting processes also result in good finishes, but are permanent mold processes. In these processes, the mold preservation is of principal concern since molds are reused for each part cast within them.

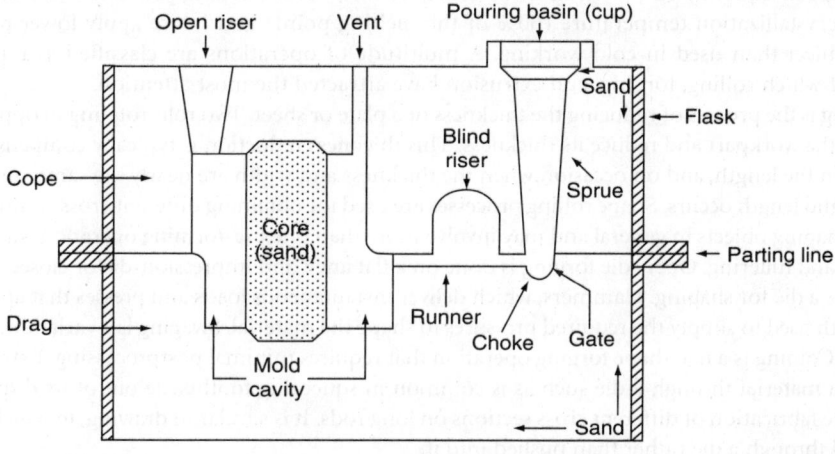

FIGURE 15.1 Sand casting.

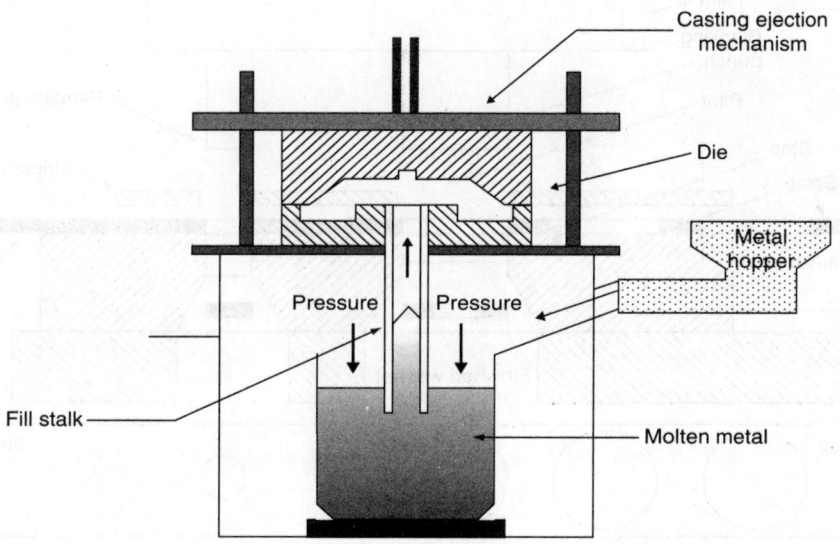

FIGURE 15.2 Die casting.

Common materials that are cast include metals such as aluminum, copper and their alloys, low-melting-point alloys, cast iron, and steel. The molds have a complex gating system with gates and runners that allow for the delivery of liquid metal to the cavity, risers to compensate for shrinkage porosity, and a pouring cup. The mold design, metal fluidity, and solidification patterns are all very important to obtain high-quality castings devoid of defects. Suitable provisions are made through allowances to compensate for shrinkage and finishing.

15.4.1 Forming and Working Processes

Forming processes (Figures 15.3–15.6) include bulk-metal forming as well as sheet-metal operations. Forming is largely applicable to metals that are either workable or malleable. Thus, brittle materials are not suitable for forming. Forming is the combined application of temperature and pressure to shape an object to specifications in solid state. Cold forming includes operations performed close to room temperature and consequently employs higher pressures to form. Hot working processes heat the workpart above its recrystallization temperature (60% of the melting point) and hence apply lower pressures to shape the object than used in cold working. A multitude of operations are classified as bulk forming processes, of which rolling, forging, and extrusion have attracted the most attention.

Flat rolling is the process of reducing the thickness of a plate or sheet. Two rolls rotating in opposite directions scoop the workpart and reduce its thickness. This thickness reduction is typically compensated for by an increase in the length, and on occasion when the thickness and width are nearly the same, an increase in both width and length occurs. Shape rolling processes are used for obtaining different cross sections. Forging is used for shaping objects in general and may involve more than one pre-forming operation, such as blocking, edging, and fullering. Open-die forging is done on a flat anvil and impression-die or closed-die forging processes use a die for shaping. Hammers, which deliver instantaneous loads and presses that apply gradual loads are both used to supply the required pressures to shape the material. Swaging is a variation of the forging process. Coining is a net-shape forging operation that requires minimal postprocessing. Extrusion is the pushing of a material through a die such as is common in squeezing toothpaste out of its dispenser. This allows for the fabrication of different cross sections on long rods. It is similar to drawing, in which the material is pulled through a die rather than pushed into it.

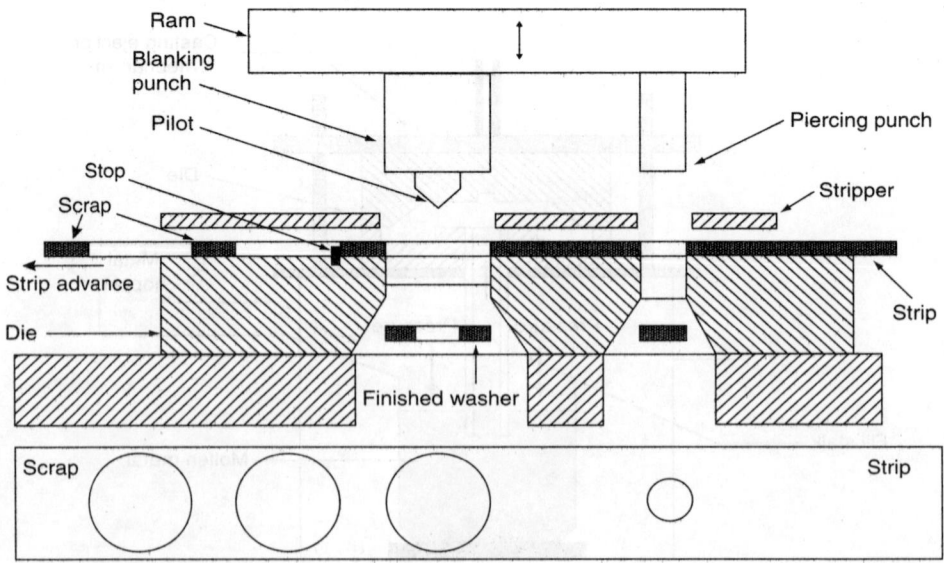

FIGURE 15.3 Shearing die.

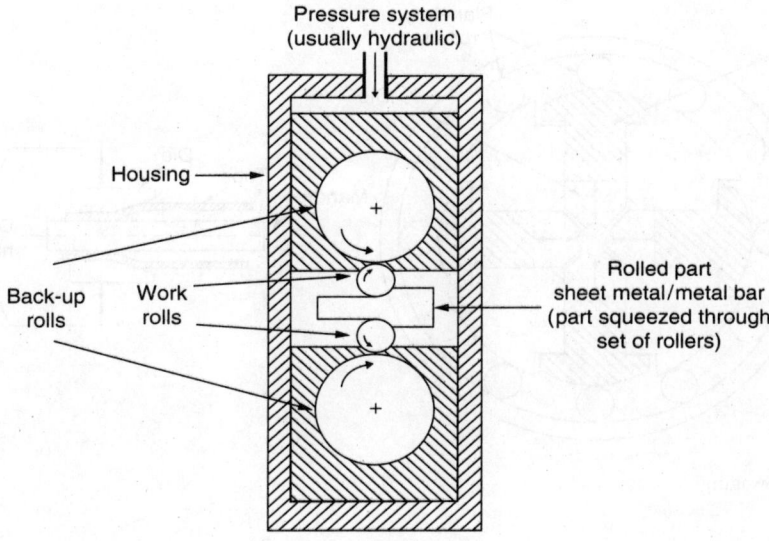

FIGURE 15.4 Rolling mill (4-high).

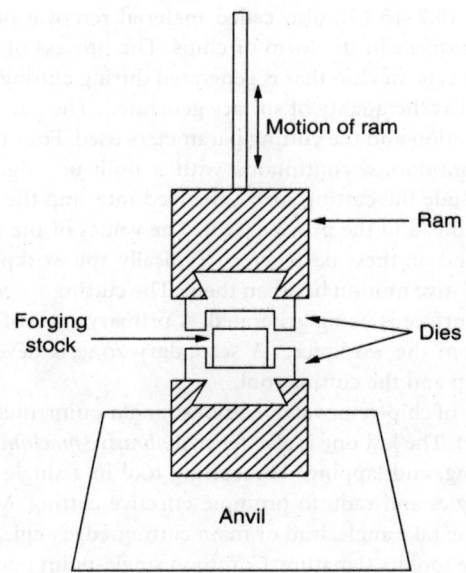

FIGURE 15.5 Forging.

Sheet-metal operations include shearing, bending, stretch forming, spinning, and explosive forming. Shearing is usually the first operation performed in sheet-metal fabrication and can be categorized into blanking and piercing (punching). In the former operation, the slug cutout is important, whereas in the latter operation, the sheet from which the slug is cut is important. Sheet-metal forming utilizes progressive dies or compound dies such that multiple operations can be combined in producing a single component.

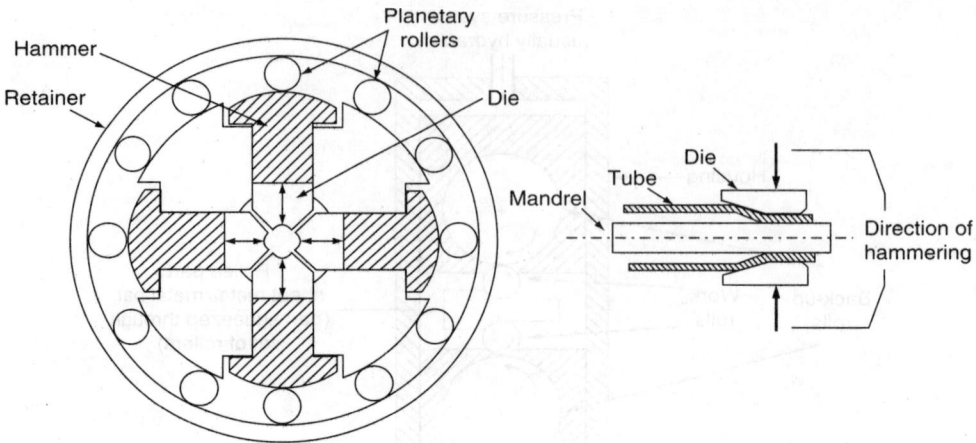

FIGURE 15.6   Swaging.

15.5   Machining Processes

Machining processes (Figures 15.7–15.12), also called material removal processes, use a sharp tool to remove material from the workpiece in the form of chips. The process of cutting involves both plastic deformation and fracture. The type of chip that is generated during cutting significantly affects both the removal of the material as well as the quality of surface generated. The size and the type of the chip vary depending on the type of operation and the cutting parameters used. Four types of chips are common in machining: continuous, discontinuous, continuous with a built-up edge, and serrated. The cutting parameters of importance include the cutting speed, the feed rate, and the depth of cut. These parameters affect the workpiece, the tool, and the process itself. The values of the forces, stresses, and temperatures on the cutting tool depend on these parameters. Typically, the workpiece and tool are rotated and translated such that there is relative motion between them. The cutting speed is the circumferential speed or the speed at which a new surface is being generated. A primary zone of deformation causes shear of material separating a chip from the workpiece. A secondary zone is developed owing to the friction between the newly formed chip and the cutting tool.

There are usually three types of chip-removal operations: single-point, multipoint (fixed geometry), and multipoint (random geometry). The last one is also termed *abrasive machining process*, and includes operations such as grinding, honing, and lapping. The cutting tool in a single-point operation resembles a wedge and is given several angles and radii to promote effective cutting. Most notably, the cutting-tool geometry is characterized by the rake angle, lead or main cutting-edge angle, nose radius, and edge radius which when combined give the tool its signature. Common single-point operations include turning, boring, and facing. Turning is performed to make round parts, facing makes flat features, and boring turns nonstandard diameter, internal cylindrical surfaces. Multipoint (fixed geometry) operations include milling and drilling. Milling operations can be further categorized into face milling, peripheral or slab milling, and end milling. The face milling uses the face of the tool, while slab milling uses the periphery of the cutter to effect the cutting action. These are typically applied to make flat features at a rate of material removal significantly higher than single-point operations such as shaping and planing. End milling cuts along the face as well as the periphery and is used for making slots and extensive contours. Drilling is used to make standard-sized holes with a cutter that has more than one active cutting edge. Reaming is a hole-finishing operation that follows drilling.

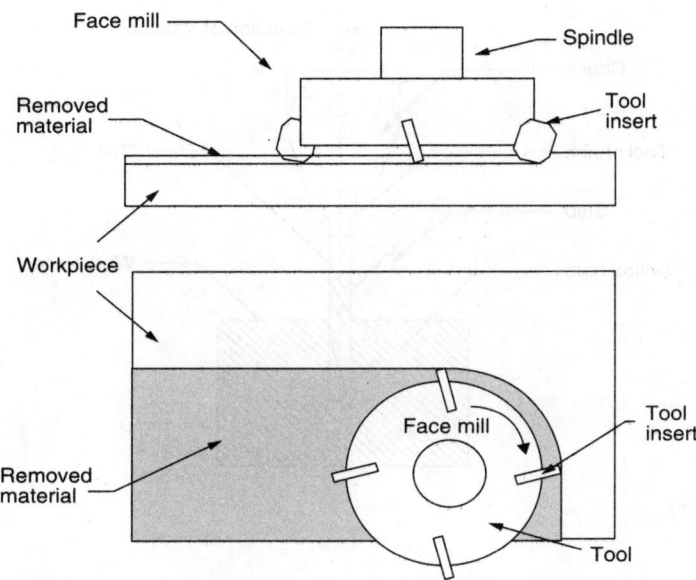

FIGURE 15.7   Face milling.

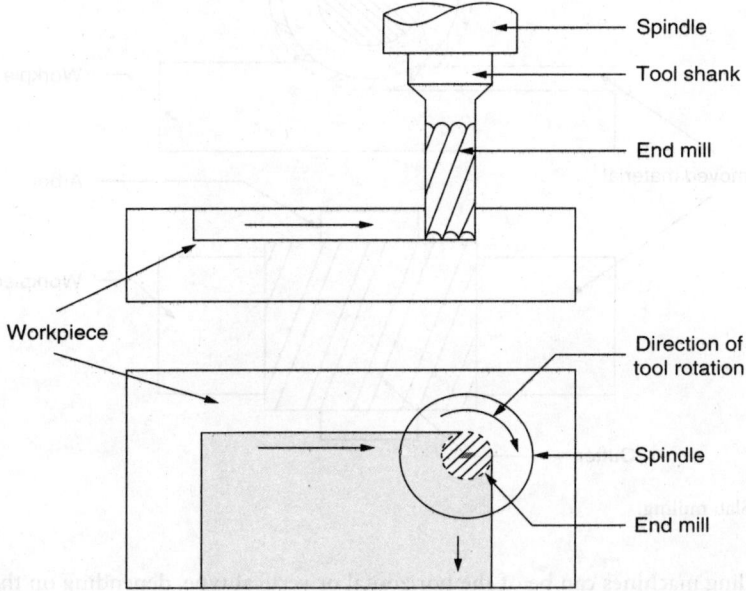

FIGURE 15.8   End milling.

15.6   Machines and Computer Integration

Machine tools are used to carry out machining operations and are selected on the basis of on their ability to achieve a certain level of accuracy in size and geometry. Engine lathes are the most versatile machines and can perform several operations, including turning, facing, drilling, boring, threading, and

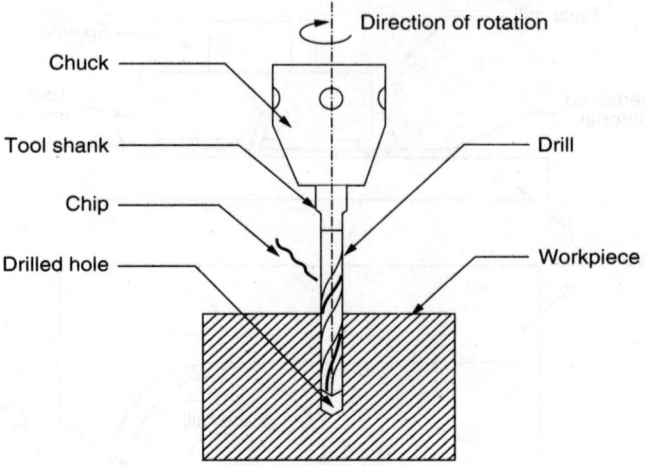

FIGURE 15.9 Drilling.

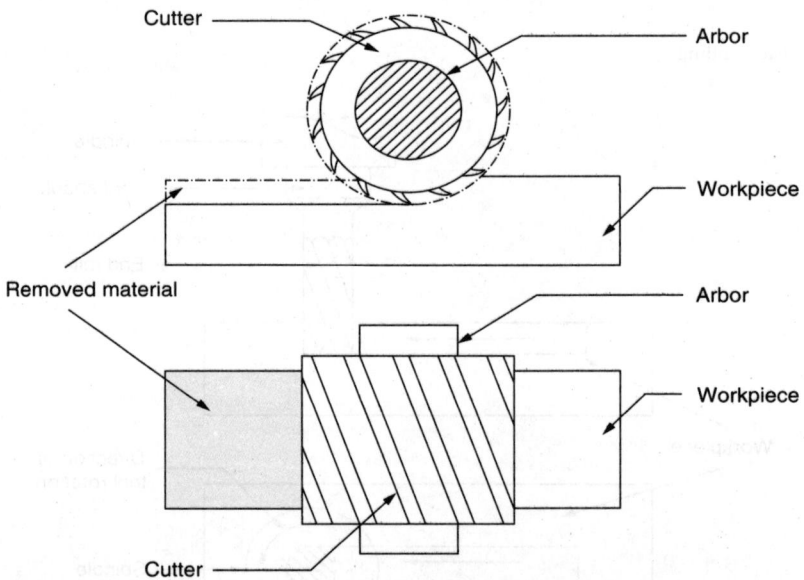

FIGURE 15.10 Slab milling.

chamfering. Milling machines can be of the horizontal or vertical type, depending on the configuration of the cutter relative to the workpiece.

Numerical control (NC) machines add programmable flexibility to standard lathes and mills. The tape media used in traditional NC machines have been replaced by inexpensive computers in the newer computer numerical control (CNC) machines of today. Machining centers with multitool magazines and mill-turn arrangements have advanced the state of the art further. Industrial robots are primarily material handlers, but are also used for painting, welding, drilling, etc. The industrial robots lend programmable flexibility similar to NC machines. Programming allows for the minimization of routine actions, such as tool repositioning to beginning of cut in NCs, and standard pick-and-place motions in industrial robots.

Flexible machine cells and flexible manufacturing systems configure multiple CNC machines and automated material handlers that include robots, conveyors, and automated guided vehicle systems

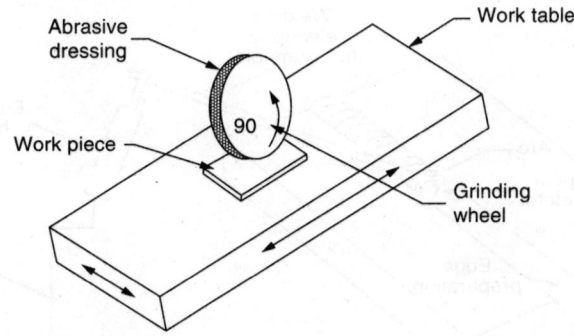

FIGURE 15.11 Grinding.

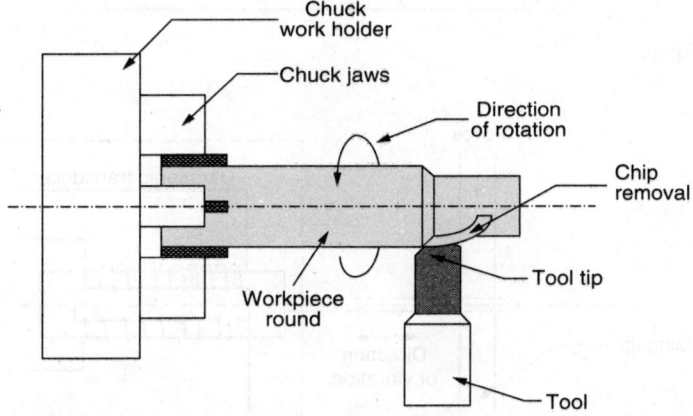

FIGURE 15.12 Turning.

within a single system to extend their flexibility while producing goods. Computer aided design/computer aided manufacturing is replaced by computer integrated manufacturing (CIM) to improve integration at a systemic level. The flexibility of CIM systems combined with rapid prototyping have led to agile manufacturing enterprises. These systems employ flexible fixturing and automated inspection. The most popular automated inspection systems employed in industry are the coordinate measuring machines (CMMs), machine vision systems, and optical projectors. The CMMs are also used in the Reverse Engineering enterprise to reconstruct the design from the product. Parts are reverse engineered whenever the original drawings are no longer available or when the original equipment manufacturer does not support the product anymore.

15.6.1 Joining Processes

Joining processes (Figures 15.13–15.15) are employed in the manufacture of multipiece parts and assemblies. Joining processes include mechanical fastening (bolting and riveting), adhesive bonding, and welding processes. Welding processes use different sources of heat to cause localized melting of the metal to be joined or the melting of a filler to develop a joint between two metallic parts. Clean faying surfaces are joined together through a butt weld or a lap weld, although other configurations are also possible. Two other joining processes are brazing and soldering, which differ from each other in the temperature applied.

The source of heating differs in different welding operations such as arc welding, gas welding, and resistance and solid-state welding. Arc welding strikes an arc between two electrodes to generate the required heat. One electrode is typically the plate to be joined. The other electrode could be consumable

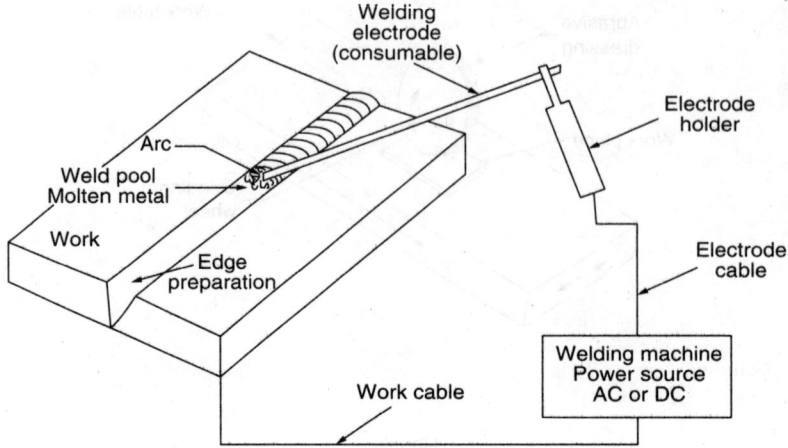

FIGURE 15.13 Welding.

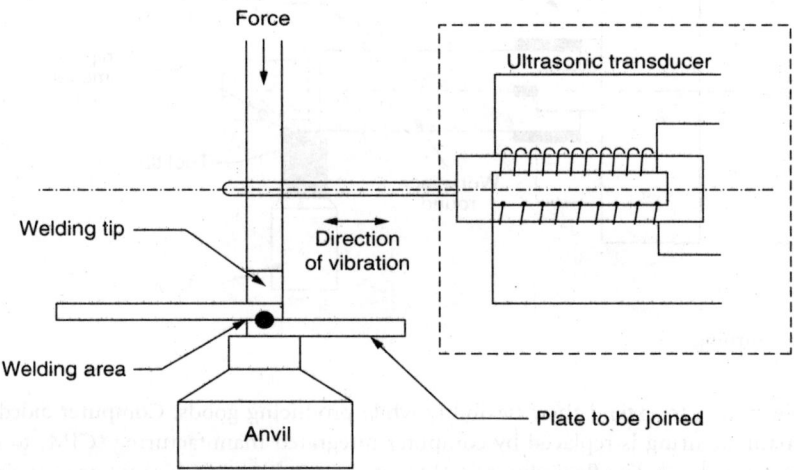

FIGURE 15.14 Ultrasonic welding (solid state).

or nonconsumable. Stick welding is most common, and is also called shielded metal arc welding (SMAW). Metal inert gas or gas metal arc welding, like SMAW, uses a consumable electrode. The electrode provides the filler and the inert gas provides an atmosphere such that contamination of the weld pool is prevented. A steady rate of flow of electrode is often made through automated means to maintain the arc gap, thereby controlling the temperature of the arc. Gas tungsten arc welding or tungsten inert gas welding uses a nonconsumable electrode, and a separate filler must be supplied for welding.

In resistance welding, the resistance offered by the air gap between the faying surfaces to the flow of electric current between two fixed electrodes is used to generate the heat required for welding. The focused heat can be used to make spot or seam welds. Gas welding typically employs acetylene and oxygen in various proportions to develop different temperatures to heat workpieces or fillers for welding, brazing, and soldering. If the acetylene is in excess, a reducing flame is obtained, whereas if oxygen is in excess, then an oxidizing flame is generated; if equal proportions of the two are used, a neutral flame results. The oxidizing flames generate the highest temperatures. Other solid-state processes include thermit welding, ultrasonic welding, and friction welding.

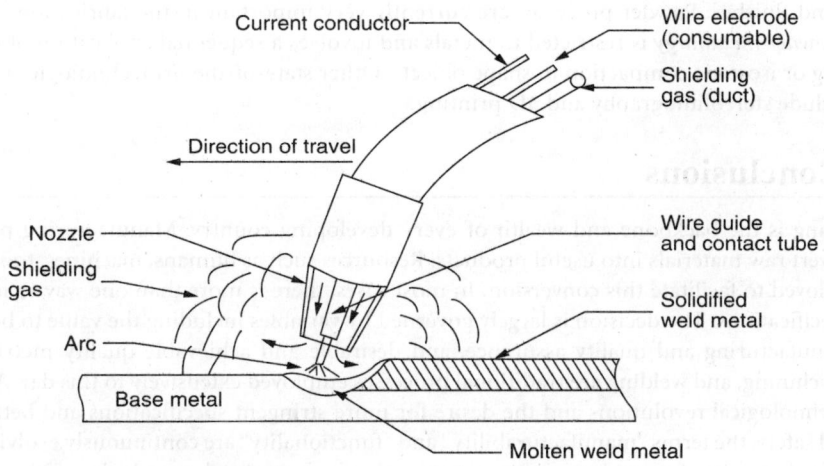

FIGURE 15.15   Gas metal arc welding (GMAW).

15.7   Finishing Processes

Finishing processes include polishing, shot-peening, sand blasting, cladding and electroplating, and coating and painting. Polishing may involve very minor material removal and is hence on occasion classified under machining operations. Shot and sand blasting are typically used to improve cleanliness and surface properties. Coatings are applied through Chemical Vapor Deposition or Physical Vapor Deposition to improve surface properties. For instance, in some cases, hard coatings are applied to softer and tougher substrates to improve wear resistance while retaining fracture resistance. The coatings are less than 10 μm thick in many cases. Cladding is done as in aluminum cladding on stainless steel to improve its heat conductivity.

15.8   Nontraditional Manufacturing Processes

Nontraditional processes include electro-discharge machining (EDM), electrochemical machining, chemical grinding, abrasive water-jet machining, and laser machining. The EDM, termed earlier as spark erosion machining, uses a dielectric to control the machining rate. Softer electrodes may be employed to obtain complex-shaped (nonround) holes using EDM. Wire EDM is a variation of this process. *Abrasive jet machining* uses an abrasive slurry in a forced jet of water to machine hard and otherwise nonmachinable materials.

Even newer cutting-edge technologies have emerged, most notably, *nanotechnology or molecular manufacturing*. A nanometer is a billionth of a meter. The ability to modify and construct products at a molecular level by moving atoms and molecules to desired positions makes nanotechnology very attractive. Single-wall nanotubes are one of the biggest innovations that have been envisioned for building future transistors and sensors. Variants of the nano theme include nanometric fabrication, such as ultra-high precision machining, where very small depths of cut are consistently taken to fabricate components. Biotechnology, another emerging technology, also benefits significantly from nanotechnology and nanoscience.

Near-net-shape manufacturing is common in many applications, such as in the production of coins like dimes and nickels. The idea is to complete much of the processing in a single step without requiring significant finishing. Injection molding, used for the fabrication of plastics; investment and impression-die casting; and precision forging are all considered to be near-net fabrication processes. Machining, once considered wasteful and expensive, has once again proven its immense worth in producing very tight tolerances and finishes. Powder processes are currently very important in the fabrication of very hard materials. *Powder metallurgy* is restricted to metals and involves a sequential application of compaction and sintering or isostatic compaction to shape objects. Other state-of-the-art technologies in rapid prototyping include stereolithography and 3D printing.

15.9 Conclusions

Manufacturing is the backbone and wealth of every developing country. Manufacturing processes are used to convert raw materials into useful products. Resources such as humans, machines, tools, and tooling are employed to facilitate this conversion. In most cases, there is more than one way to manufacture a part to specifications. The decision is largely governed by variables including the value to be added, the costs of manufacturing and quality assurance, and desirable and achievable quality metrics. Casting, forming, machining, and welding are traditional processes employed extensively to this day. All the same, owing to technological revolutions and the desire for more stringent specifications and better comfort, security, and safety, the terms "manufacturability" and "functionality" are continuously evolving. This has led to the development of several newer processes, such as nanotechnology, which could not have even been imagined 50 years ago. As our expectations continue to evolve, this trend of growth in processes and machines is expected to continue.

16

Cross-Training in Production Systems with Human Learning and Forgetting

David A. Nembhard
Pennsylvania State University

Bryan A. Norman
University of Pittsburgh

16.1 Introduction

Organizations have been making use of workforce cross-training, also known as job rotation, job flexibility, and worker multifunctionality, for a long time. A number of related concepts have been widely recognized by both researchers and practitioners in a variety of theoretical models and organizational designs. In the context of these systems, in which workers are trained on alternative tasks in addition to their primary tasks, much of the workforce flexibility literature has attempted to address the fundamental question of how much cross-training should be given to each worker, i.e., how many different tasks each worker should master under certain assumptions (e.g., Russell et al., 1991; Bokhorst and Slomp, 2000; Slomp and Molleman, 2000; Cesani and Steudel, 2000a, 2000b). A key difficulty in determining the appropriate number of tasks is the fact that there is significant heterogeneity within many organizations with respect to human performance and capabilities (see, for example, Buzacott, 2002). Specifically, levels of human learning and forgetting play an important role in shaping productivity outcomes (Shafer et al., 2001).

Manufacturing and service industries face increasingly intense competition and must respond to a broad spectrum of market challenges. These forces are moving businesses toward greater agility and flexibility, thereby increasing product and service diversity and accelerating product and service innovation

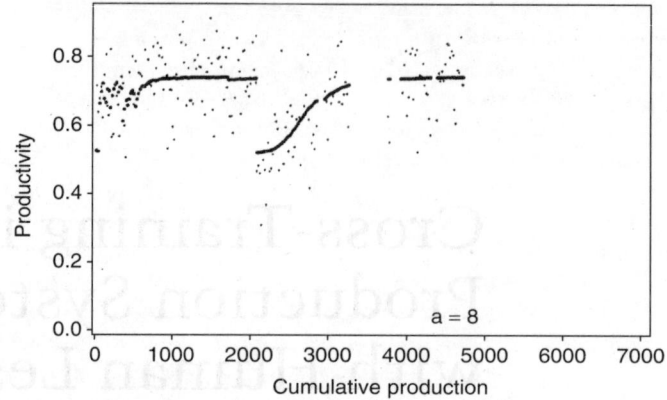

FIGURE 16.1 Sample learning/forgetting episode with a 2-month break after 2000 units.

(Pine, 1993). As a result, many organizations are focusing more attention on cross-training as an approach to meeting these challenges (McCune, 1994). It has been evident for some time that pure specialization has its limitations; furthermore, there is growing evidence that its modern replacement, job rotation, has its limitations as well. Whereas managers formerly relied on specialization based on the principles of Taylorism, today they hope that the benefits of job rotation will compensate for the loss of long periods of intensive training. However, at present, few guidelines exist to aid managers in decisions on job rotation and job sequencing.

Figure 16.1 illustrates a period of learning, forgetting, and subsequent relearning for one worker. Note that there is a large relative drop in performance after 2000 units of work, corresponding to the worker's 2-month absence from the task. Human responses such as these can have both significant and complex effects on productivity as they interact with numerous operational characteristics and decisions.

16.2 Overview of the Literature

Dar-El (2000) describes a model for optimizing training schedules for a single task. Although the quantitative study of cross-training in cellular manufacturing (CM) settings incorporating individual learning and forgetting behaviors has been relatively sparse, numerous human factors studies have identified key knowledge and models about human performance. Engineering studies that integrate previously developed models have generally been based on arbitrary experimental conditions of learning, forgetting, and performance. Thus, while many studies have provided some general direction, there continues to be a need for empirically based experiments and simulations, as a number of researchers have pointed out (e.g., Kher, 2000; Kher et al., 1999; McCreery and Krajewski, 1999; Malhotra et al., 1993)

16.2.1 Dual-Resource-Constrained Problems

The topic of creating a multiskilled workforce has been extensively investigated in a traditional manufacturing job shop system. Several researchers have illustrated the effect of cross-training when learning and forgetting effects are present. In a dual-resource-constrained (DRC) shop setting, Malhotra et al. (1993) investigated the relationship between learning losses and the benefits of a cross-trained workforce through a series of simulations. McCreery and Krajewski (1999) conducted a simulation study and found that when learning and forgetting effects exist, the degree of cross-training and the worker deployment policy should depend primarily on the degree of task complexity and the degree of product variety. Kher (2000) conducted a full factorial designed experiment to investigate the training schemes obtained by cross-trained workers in a DRC shop under learning, relearning, and attrition conditions. Kher (2000) and Kher et al. (1999) concluded that the effectiveness of cross-training depended significantly on the existing forgetting rate of the workers. In addition, Kher et al. (1999) remarked on the significant relationship between batch size and worker flexibility. The paper also suggested conditions under which worker flexibility would be suitable that were based on production batch size and learning behaviors.

In a DRC job shop with operator absenteeism and assumed steady-state productivity, Molleman and Slomp (1999) suggested that the distribution of skills within teams and the degree of workforce cross-training had a significant impact on system performance. Their findings indicated that a uniform workforce skill distribution resulted in better system performance. In other words, each worker should, optimally, master the same number and types of tasks. Slomp and Molleman (2000) compared four cross-training policies based on the workload of the "bottleneck" worker in both static and dynamic circumstances. The results confirmed the intuition that the higher the level of cross-training, the better the team performance (i.e., the lower the workload for the bottleneck worker). On the other hand, consistent with other cross-training investigations, Slomp and Mollman also indicated that as the degree of cross-training increases, the improvement of the system performance decreases. Brusco and Johns (1997) developed a mixed integer programming (MIP) model to examine the cross-training configurations in a maintenance operation in a paper-production system. The mathematical model was developed based on different levels of worker productivity in order to minimize the training cost. Under the assumption of a heterogeneous workforce in which each worker has a varying productivity level with respect to the task, each worker is assumed to have full capability (i.e., 100% production) in one of four available tasks and less capability in the secondary tasks. On the basis of the overall cost savings, they found that cross-training employees in a secondary task at a 50% production level was adequate when compared with cross-training at a 100% production level. Brusco et al. (1998) investigated a cross-training policy to minimize workforce requirements based on two-skilled workforce utilization. The result agreed with Brusco and Johns (1997) that the worker should not necessarily be cross-trained at the 100% production level. Moreover, they also found that demand variation has a significant impact on the number of workers needed to meet the system requirements.

In the health care application domain, Campbell (1999) constructed a nonlinear programming model to explore and quantify the value of cross-training 20 workers in four departments. The model was constructed on the basis of departmental requirements and varying workforce capability. In a manner similar to Brusco and Johns (1997), each worker was assigned to be less proficient in secondary departments than in the primary one. A full factorial experimental design was conducted to investigate the effect of numerous factors on system performance. Among those factors were level of cross-training, capability level, skill homogeneity, and demand variation. The experimental results concluded that the degree of cross-training depends significantly on the demand variation. Under high-demand conditions, the improvement of system performance due to an increase in the cross-training level was greater than in the low-demand scenario. Moreover, in many circumstances, small amounts of cross-training were more beneficial. The result was consistent with previous research (Brusco and Johns, 1997; Brusco et al., 1998) that indicated that each worker need not necessarily be fully cross-trained in all operations in order to achieve substantial benefits from cross-training.

16.2.2 Worker-Task Assignment

Wisner and Pearson (1993) indicated the necessity of incorporating worker learning and relearning behavior in worker-task assignment studies. Their study showed that worker-relearning behavior has a significant impact on labor assignment policy, especially when a worker is required to transfer between tasks regularly. However, the assignment of individual workers to tasks that are based on learning and forgetting behavioral characteristics has received relatively little attention in the literature. This is partially due to the difficulty in collecting detailed performance data at the individual level. Without a complete understanding of the distribution of individual learning characteristics, worker-task assignment had been based on things that did not necessarily relate to productivity, let alone productivity under conditions of continuous change. As a result, research in the area of worker-task assignment has generally assumed steady-state productivity (i.e., no learning or forgetting). For example, Mazzola and Neebe (1988) modeled the assignment of jobs to agents under agent capacity constraints. More recently, Davis and Moore (1997) constructed an optimization model for assigning workers that was based on minimizing the number of task reassignments. Brusco and Johns (1997) developed an MIP assignment model in order to minimize total staffing cost. Molleman and Slomp (1999) presented a linear goal-programming method to assign workers to specific tasks in order to minimize shortages, makespan, and production time.

Many optimization-based assignment studies have been based on the classic assignment problem, in which people are assigned to perform tasks in order to minimize or maximize a given system performance measure. In addition to the common constraints of worker and task requirements, recent studies have introduced "side-constraints" to make the problem more practical (Caron et al., 1999). For example, Molleman and Slomp (1999) developed a linear goal-programming model subject to labor and skill requirements, (e.g., the shortage of capacity and the load of the busiest team member). Human factors, such as learning ability, motivational issues, and worker attitude, should also be considered to make the study more applicable. Caron et al. (1999) developed an assignment model in order to maximize the number of jobs assigned, assuming two side constraints — seniority and job priority.

Other worker-task assignment studies involved worker heterogeneity, such as varying worker-task efficiency, to represent the productivity level among workers performing a specific task. Several studies defined worker efficiency as the fraction of standard production time (e.g., Nelson, 1970; Park, 1991; Bobrowski and Park, 1993; Malhotra and Kher, 1994; Molleman and Slomp, 1999; Bokhorst and Slomp, 2000). In addition, Campbell (1999) developed a nonlinear model for assigning cross-trained workers in a multidepartment service industry based on worker capability. Nevertheless, these solution methods often assume constant worker-task efficiency.

16.2.3 Cellular Task Assignment

Both qualitative and quantitative studies have shown that labor-related issues have a significant effect on achieving optimal system performance in a CM shop. Owing to the substantial amount of interaction between labor skill and machining technology in a CM system, some researchers have taken into account workforce skill-related issues in the worker-task assignment problem. Warner et al. (1997) proposed a procedure for assigning workers to cells on the basis of their human and technological skills. "Technological skills" were described as mechanical, mathematical, and measurement ability, while "human skills" referred to communication skills, leadership, teamwork, and decision-making ability.

There has been increasing interest in the subject of labor-task assignment for manufacturing cells. For example, Askin and Huang (1997) and Norman et al. (2002) proposed solution procedures based on MIP models for assigning workers to cells and selecting a specific training program for each worker. Bhaskar and Srinivasan (1997) presented mathematical models for static and dynamic worker-assignment problems in order to balance workload and to minimize makespan. Through a series of simulation experiments, Russell et al. (1991) examined labor-assignment strategies based on a hypothetical group technology (GT) shop of three cells. The results favored a completely flexible, homogeneous workforce. In other words, each worker should be completely and equally cross-trained and allowed to be assigned to any machine in any cell. Cesani and Steudel (2000b) conducted an empirical investigation in a two-operator cell and found that when the degree of task sharing among the operators increases (i.e., more than one operator is responsible for a machine), the system performance improves significantly. Nevertheless, their notable finding, which coincides with the results in the DRC shop, indicates that as the cross-training level increases, the improvement of system performance decreases (Slomp and Molleman, 2000; Cesani and Steudel, 2000b). This raises the question, what level of cross-training and worker flexibility is required in order to achieve optimal cell performance?

There are important limitations in past CM assignment research. For instance, past studies have had the objective of balancing workload (an important objective), but not directly improving cell productivity. Also, many have assumed identical productivity levels for each operator over time. The impacts of individual learning and forgetting characteristics have been relatively absent in the investigation of cellular worker-task assignment problems.

A number of studies have suggested that the need for workforce flexibility was not only driven by customer demand fluctuation or low staffing levels but also by the variation in the workforce supply due to operator absenteeism (Bokhorst and Slomp, 2000; Molleman and Slomp, 1999; Van den Beukel and Molleman, 1998). Van den Beukel and Molleman (1998) conducted a survey study in ten different organizations, and results indicated that the need for a flexible, multiskilled workforce in response to the problem of operator absenteeism is more significant when rapid response to customer demand is crucial, but relatively less important when external resources, such as temporary workers, or other equipment are available. Although the negative impact of operator absenteeism in work cells has been recognized in the literature, few quantitative studies have been conducted to investigate specifically how cross-training policies could effectively respond to the problem (e.g., Bokhorst and Slomp, 2000; Molleman and Slomp, 1999).

16.2.4 Models of Individual Learning and Forgetting

Although worker-to-task assignment has been considered in the literature, researchers typically assume that workers operate at one productivity level or at a discrete set of productivity levels. However, in reality, workers learn and forget tasks depending on how often they perform a task and how much time has elapsed since they last did the task. Thus, it is important to consider the issue of worker learning and forgetting when determining the best worker-assignment strategy.

Patterns of learning and forgetting have been studied from a variety of perspectives. For example, psychologists have studied and modeled individual learning processes (Mazur and Hastie, 1978), while engineers have modeled learning as it relates to manufacturing costs (Yelle, 1979), process times (Adler and Nanda, 1974a, 1974b; Axsater and Elmaghraby, 1981; Sule, 1981; Pratsini et al., 1993; Smunt, 1987), setup time learning (Karwan et al., 1988; Pratsini et al., 1994), and line balancing (Dar-El and Rubinovitz, 1991).

At the organizational level, research has centered on the overall implications of learning across large organizational units separated by functional, hierarchical, or geographical boundaries. These organizational learning curves have allowed managers to examine experiential productivity improvements, transfer of knowledge between parallel units (Epple et al., 1991), and the ability of an organization to retain knowledge over time (Argote et al., 1990; Epple et al., 1991). Organizational learning curves are particularly appropriate for measuring improvements in plant-wide productivity over relatively long periods of time. Learning in organizations, characterized by make-to-order operations, has led to the development of more general models of organizational learning (Womer, 1979, Dorroh et al., 1994). A comprehensive overview of the organizational learning literature can be found in Argote (1999).

Another stream of research has examined learning at the micro-level in order to investigate the mechanisms by which learning occurs in individuals (e.g., Anderson, 1982; Hancock, 1967; Mazur and Hastie, 1978). Much of this research has examined the determination of a most appropriate mathematical form for individual learning. While the majority of work has centered on the log-linear model (e.g., Badiru, 1992; Buck and Cheng, 1993; Glover, 1966; Hancock, 1967), there have been continuing efforts to identify alternative formulations. Indeed, some alternative forms were found to represent observed behavior better than the log-linear model in their respective industrial settings (e.g., Asher, 1956; Carr, 1946; Ebbinghaus, 1885; Knecht, 1974; Levy, 1965; Pegels, 1969).

Research addressing the retention of learning (forgetting) suggests that the longer a person studies, the longer the retention (Ebbinghaus, 1885). The forgetting process has been shown to be describable by the traditional log-linear performance function (Anderson, 1985). On the basis of individuals' ability to recall nonsense syllables, Farr (1987) found that as the meaningfulness of the task increased, retention also increased. The retention of learning at an organizational level has also been found to decline rapidly during extended production interruption (Argote, 1999). Several research efforts have attempted to explain the impact that forgetting has on production scheduling (Smunt, 1987; Fisk and Ballou, 1982; Sule, 1983; Khoshnevis and Wolfe, 1983a, 1983b). Globerson (1989) developed a model of learning and forgetting for a data-correction task. The model indicates that forgetting behaves much like the mirror image of the learning process. Bailey (1989) found that the forgetting of a procedural task could be expressed as a linear function of the product of the amount learned and the log of the elapsed time.

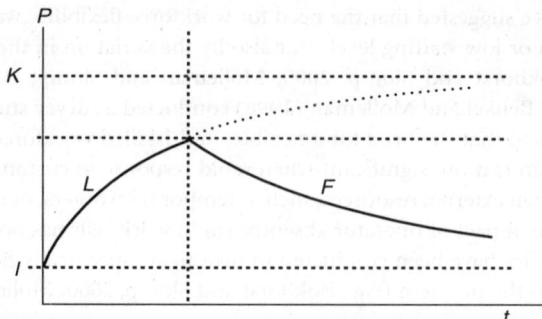

FIGURE 16.2   Model of individual learning and forgetting characteristics.

Shafer et al. (2001) introduced an approach to measuring organizational learning wherein individual worker heterogeneity is modeled. The resulting distribution of individual learning patterns provides a description of organizational learning as well as the associated detail required to parse the worker population into complementary clusters of workers. Nembhard and Uzumeri (2000a) provide an important extension of this approach by incorporating both learning and forgetting into an individual-based model of productivity. In this model, both the learning and the forgetting components were shown to be preferred models among numerous candidate models (Nembhard and Uzumeri, 2000b; Nembhard and Osothsilp, 2001, respectively). To illustrate, the model shown in Figure 16.2 is designed to measure learning and forgetting, where P is a measure of the productivity rate corresponding to t time units. In this model, parameter l represents the initial productivity of the worker, K the steady-state productivity, L the learning rate, and F the forgetting rate.

A potentially significant limitation of many learning/forgetting models in the literature is that they are commonly designed as measurement tools and may not lend themselves to use within other analytical or optimization frameworks.

16.3   Models

This section presents several models that may aid in the understanding of productivity and flexibility in systems with significant job rotation. First, we discuss a math programming-based decision model for assigning and scheduling workers to tasks. The model incorporates the effects of individual worker learning and forgetting for worker-task assignment.

16.3.1   A Worker-Task Assignment Model

Worker-task assignment models can be represented mathematically and solved in a number of different ways, including math programming, enumerative methods (e.g., dynamic programming), and various metaheuristics (tabu search, genetic algorithms, etc.) The following model utilizes math programming and incorporates individual worker learning and forgetting.

We illustrate the construction of a worker-task assignment model with a manufacturing example that incorporates differences among workers' skills and their task-learning rates. The example considers a common type of manufacturing setting in which work is passed from station to station, with intermediate work-in-process buffers between stations and the objective of maximizing system output for a given number of workers over a fixed time interval as shown in Figure 16.3. We make the following assumptions in order to illustrate the construction of a worker-task assignment model using this approach:

There are n workers available to perform the tasks.

There are m sequentially numbered tasks to be completed within the system, and each task is done at a separate workstation. The model can be modified to permit multiple tasks at each workstation.

Time is separated into p periods, and a worker performs only one task during each time period.

There is buffer storage between the workstations. However, the size of this buffer storage can be constrained.

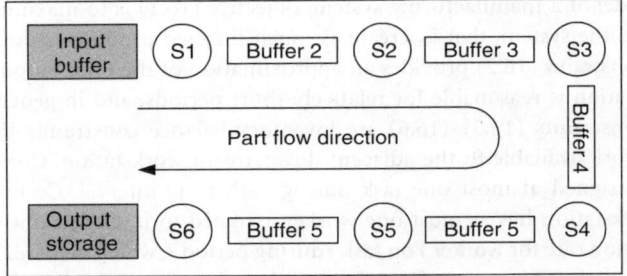

FIGURE 16.3 Serial process configuration.

The rate of learning and forgetting is a function of how much time the worker has spent performing a particular task.

Workers are heterogeneous with respect to initial productivity, learning, and forgetting rates for each task.

Notation

I The set of workers $i = 1, 2, ..., n$

J The set of tasks $j = 1, 2, ..., m$

T The set of time periods $t = 1, 2, ..., p$

$X_{i,j,t}$ Binary variable indicating whether worker i does task j during time t

$P_{i,j,t}$ Productivity if worker i does task j during time t (based on learning/forgetting model)

$O_{i,j,t}$ The output from worker i performing task j for period t

l_{ij} The initial productivity level for worker i on task j (i.e., for the first unit of work)

$K_{i,j}$ Asymptotic learning constant for worker i performing task j (steady-state production rate)

$L_{i,j}, F_{i,j}$ Individual parameters of learning and forgetting, respectively, for worker i on task j

$B_{j,t}$ Buffer (inventory) level at task j at the end of period t

BI_j Beginning inventory for task j, (period 0)

Objective

$$\text{Maximize} \sum_{i=1}^{n} \sum_{t=1}^{p} O_{i,m,t} \tag{16.1}$$

Constraints

$$O_{i,j,t} \le X_{i,j,t} P_{i,j,t} \text{ (period length) } \forall i, \forall j, \forall t \tag{16.2}$$

$$B_{1,1} = BI_1 - \sum_{i=1}^{n} O_{i,1,1} \tag{16.3}$$

$$B_{j,1} = BI_j + \sum_{i=i}^{n} O_{i,j-1,1} - \sum_{i=i}^{n} O_{i,j,1}, \quad j = 2, ..., m \tag{16.4}$$

$$B_{1,t} = B_{1,t-1} - \sum_{i=1}^{n} O_{i,1,t,} \quad t = 2, ..., p \tag{16.5}$$

$$B_{j,t} = B_{j,t-1} + \sum_{i=1}^{n} O_{i,j-1,t} - \sum_{i=1}^{n} O_{i,j,t,} \quad j = 2, ..., m \quad t = 2, ..., p \tag{16.6}$$

$$\sum_{j=1}^{m} X_{i,j,t} \le 1 \; \forall i, \forall t \tag{16.7}$$

$$\sum_{i=1}^{n} X_{i,j,t} \le 1 \; \forall j, \forall t \tag{16.8}$$

$$P_{i,j,t} = l_{i,j} + K_{i,j} \left[1 - \exp\left(-\frac{1}{L_{i,j}} \sum_{k=1}^{t} X_{i,j,k}\right) \right] \exp\left[\frac{1}{F_{i,j}} \left(\sum_{k=1}^{t} X_{i,j,k} - t \right) \right] \forall i, \forall j, \forall t \tag{16.9}$$

In this illustrative model of a manufacturing system, objective (16.1) is to maximize the total output of the last workstation in the system, that is, create as much finished product as possible using n workers over the p periods. Constraint (16.2) provides an approximation of the total output, O, given the period length. The approximation is reasonable for relatively short periods, and in general will underestimate total productivity. Constraints (16.3)–(16.6) are inventory balance constraints that keep track of the number of units of work available to the adjacent downstream workstation. Constraint (16.7) ensures that each worker is assigned at most one task during each time interval. Constraint (16.8) similarly ensures that each workstation has at most one worker assigned to it. Constraint (16.9) is necessary to determine the production rate for worker i on task j during period t, which depends on the worker's experience performing that task. The current formulation is based on log-linear learning and forgetting, and it is assumed that the worker's learning function for a particular task is only related to the worker's initial productivity level for that task and how much time the worker has spent performing that task. Similarly, we assume that the worker's forgetting is only relative to the time spent not performing the task. It is the inclusion of constraint (16.9) that makes this problem nonlinear and this particular formulation a challenge to solve. It is also assumed that productivity, output and inventory are non-negative.

Several challenges present themselves in developing these assignment models. Learning and forgetting are explicitly characterized for each worker and task in this formulation. This introduces nonlinearity into the model but better represents real work environments and the trade-offs between flexibility and production. Thus, the form of the model produces a mixed integer non-linear program (MINLP), which in general is a class of problems difficult to solve. Second, the form of the learning and forgetting model in constraint (16.9) significantly affects the level of difficulty in solving the MINLP. As noted above, learning/forgetting models available in the literature were generally not intended for math programming formulations. Thus, the development of models that better lend themselves to the MINLP formulation is of considerable interest for future research efforts.

16.3.2 Alternate Worker-Task Assignment Models

An assignment model such as the one described may be varied to suit more specific scenarios and variants of the stated problem. For instance, if we assume that the $P_{i,j,t}$ values are constant (i.e., equal to K_{ij}), then the resulting problem becomes an MIP problem, which can be solved far more readily than the MINLP problem. Also, since constraints (16.3)–(16.6) model the precedence requirements of the serial production system, a selective removal of some of these constraints would allow some tasks to be performed in parallel with others.

Alternate objective functions can be investigated using this approach, with the idea that there are three main variables involved in the problem: one, the number of workers available; two, the time available; and three, the number of units to be produced. Thus, the model in the preceding section holds the number of workers and the number of time periods constant, and we maximize the output that is attainable under those constraints. A second objective is to minimize the total number of workers required to produce a predetermined number of units over a fixed time horizon. Such an approach might be appropriate in situations where there is a desire to minimize labor requirements and costs as long as due dates can be met for known demand quantities. Thus, the objective function in Equation (16.1) could be replaced with that in Equation (16.10) combined with the addition of the constraint in Equation (16.11), where M is a large constant and Y_i is a binary variable indicating whether or not worker i is used.

$$\text{Minimize} \sum_{i=1}^{n} Y_i \tag{16.10}$$

$$\sum_{j=1}^{m} \sum_{t=1}^{p} X_{i,j,t} \leq Y_i M \quad \forall i = 1, 2, ..., n \tag{16.11}$$

With a third type of objective, one can minimize the time required for n workers to meet production requirements. That is, if labor resources are limited and demand is known, management may need to determine effective worker-task schedules to minimize the time needed to produce enough to satisfy the demand level. Using a similar approach to that in the preceding case, in Equation (16.12), Y_t is a binary variable indicating that period t is used and the additional constraints found in Equation (16.13) must also be added to the model.

$$\text{Minimize} \sum_{t=1}^{p} Y_t \tag{16.12}$$

$$\sum_{i=1}^{n} \sum_{j=1}^{m} X_{i,j,t} \leq Y_t M \quad \forall t = 1, 2, ..., p \tag{16.13}$$

Using a quadratic loss approach can also limit the number of time periods required to meet a given demand level. This minimizes the elapsed time to produce to a predetermined level. Thus, an alternative objective function, such as that in Equation (16.14), will also serve this purpose:

$$\text{Minimize} \sum_{i=1}^{n} \sum_{j=1}^{m} \sum_{t=1}^{p} t^2 X_{i,j,t} \tag{16.14}$$

As a different interpretation of minimizing the time required for production, we may consider the objective function in Equation (16.15), which will minimize the number of periods workers are assigned to tasks. This will minimize the worker resources by limiting the total number of worker periods used. However, this expression will not necessarily also minimize the overall elapsed time.

$$\text{Minimize} \sum_{i=1}^{n} \sum_{j=1}^{m} \sum_{t=1}^{p} X_{i,j,t} \tag{16.15}$$

Limits on buffer storage at each workstation can be controlled by adding a constraint of the form in Equation (16.16), where $a_{j,j+1}$ represents the maximum number of parts that can be placed in the buffer storage between stations j and $j+1$:

$$B_{j,t} \leq a_{j,j+1} \quad \forall j = 1, 2, ..., m, \quad \forall t = 1, 2, ..., p \tag{16.16}$$

Pay grades based on individual capability might also be incorporated into such a model. In many manufacturing and service settings, worker pay is based on different job grades, which in turn are based on worker skill and task complexity. Workers in higher pay grades can perform tasks that are rated at lower pay grades but the reverse is generally not acceptable. Therefore, an additional objective is to consider the number of workers of each pay grade required for each assignment and to use this information in the cost evaluation for an assignment.

Task-assignment durations can be set for certain tasks where a minimum number of hours is required for workers on each task. This also has the potential to be used to model the requirement that for some tasks, the workers should not rotate too often.

Balancing the workload among the workers is a practical requirement in many work settings. That is, it may not be acceptable to require some workers to perform a greater number of tasks or work for a greater length of time. While it may be feasible to achieve a better objective value without such requirements, the need for certain types of equity is a reality in manufacturing environments. For example, a formulation without requirements of this type might assign one worker to four tasks and another to only one task, which would be unacceptable and impractical.

Work layout design is an important consideration in certain industries. For example, in CM, the question of whether to use I- or U-shaped layouts can be modeled, since in practice these layout designs may produce different results.

Nonadjacent worker assignments may be limited by incorporating travel distance measures into the assignment model in such a manner that the number and frequency of nonadjacent task assignments will be acceptable. A similar approach was taken by Kher and Malhotra (1994), wherein a worker transfer delay was incorporated into the model.

TABLE 16.1 Optimal Station Assignment/Schedule for 2 Workers at 4 Stations over 10 Time Periods[a]

Assignment Based on	Output[b] (units)	Worker[c]	Period									
			1	2	3	4	5	6	7	8	9	10
Steady-State	18.3	1	2	3	3	1	3	4	4	3	4	4
Productivity		2	1	1	2	2	4	2	1	4	2	3
Learning	28.3	1	1	1	3	3	1	3	1	3	3	4
only		2	2	2	4	4	2	4	2	4	4	1
Learning and	36.1	1	1	1	1	4	4	4	4	1	4	4
Forgetting		2	2	2	2	3	3	3	3	4	2	3

[a] To produce each unit of output workers must pass work through each of the 4 stations in order. Stations 1–4 are ordered from most complex (time-consuming) to least complex. Buffers are assumed to have sufficient availability of work initially.

[b] In all three cases the assignment output is evaluated assuming that the workers learn and forget. The assignments themselves are based on the criterion shown in the first column.

[c] Worker 1 has a 15% higher steady-state productivity rate. Worker 2 learns 35% faster (A phenomenon not uncommon in the field, Nembhard and Uzumeri, 2000a).

16.4 Example Application

As an illustrative example of the relevance of including learning and forgetting in this decision problem, Table 16.1 summarizes solutions from the MINLP with objective (16.1) and constraints (16.2)–(16.9) using math programming software (GAMS, CONOPT, and SBB). It illustrates the fact that basing assigned schedules on different assumptions will produce both different worker-task cross-training patterns and different overall productive output. For example, when considering both learning and forgetting in the assignment, we note that worker 1 (the slower learner) cross-trains on only two tasks, as opposed to four tasks under the steady-state (no learning or forgetting) assumption. Further, it is notable that when workers rotate between twice as many jobs during the same time, productivity drops substantially, as evidenced by the drop in productivity from 36.1 units to 18.3 units. This highlights the importance of considering learning and forgetting during worker-task assignment. The differences in output will tend to be smaller over longer planning periods, as workers progress on the learning curve toward steady state. However, as noted previously, many organizations undergo rapid changes in products, technology, and workers, making shorter planning horizons the norm.

While finding provably optimal solutions to our model would prove to be beneficial, in practice it may not be practical to solve the problems exactly owing to the complexity of the computation time. Several heuristic approaches have been developed for similar types of problems (Nembhard, 2001; Tharmmaphornphilas and Norman, 2004; Norman and Bean, 1999; Carnahan et al., 2000).

16.5 Summary and Perspectives

Studies examining the use of cross-training to better understand its impacts on individual learning, forgetting, and productivity in the organization may provide clearer technological bases, allowing organizations to implement cross-training in an informed manner. The need for and the use of manufacturing labor flexibility have increased significantly over recent decades as market competition has become more global. Cross-training is both a potential means for creating and transferring technological knowledge and a common method for obtaining a partially flexible workforce. Organizations have for some time used cross-training as a method for increasing manufacturing flexibility and as a methodology for productivity and quality improvement, in order to permit workers to expand their knowledge toward performing multiple tasks.

Extensions of existing models to a range of process characteristics, operational design conditions, and production measures will further aid managers and researchers in understanding and setting multifunctionality and cross-training levels. For example, absenteeism rates, worker turnover rates, frequencies of product revisions, worker assignment patterns, and the effects of various skill distributions, along with differential cross-training, in which some workers are specialized and others are cross-trained, are of broad interest.

Acknowledgments

A portion of this work was supported by grants from the National Science Foundation, for the first author under SES9986385 and SES0217666, and for the second author under SES0217189.

References

Adler, G.L. and Nanda R., The effects of learning on optimal lot size determination — single product case, *AIIE Trans.*, 6, 14–20, 1974a.

Adler, G.L. and Nanda R., The effects of learning on optimal lot size determination — multiple product case, *AIIE Trans.*, 6, 21–27, 1974b.

Anderson, J.R., Acquisition of cognitive skill, *Psychol. Rev.*, 89, 369–406, 1982.

Anderson, J.R., *Cognitive Psychology and its Implications*, 2nd ed., W.H. Freeman and Company, New York, 1985.

Argote, L., Beckman, S.L., and Epple, E., The persistence and transfer of learning in industrial settings, *Manage. Sci.*, 36, 140–154, 1990.

Argote, L., *Organizational Learning: Creating, Retaining and Transferring Knowledge*, Kluwer, Boston, MA, 1999.

Asher, H., Cost-Quantity Relationships in the Airframe Industry, Rep. No. R291, The Rand Corporation, Santa Monica, CA, 1956.

Askin, R.G. and Huang, Y., Employee Training and Assignment for Facility Reconfiguration, 6th Industrial Engineering Research Conference Proceedings, IIE, 426–431, 1997.

Axsater, S. and Elmaghraby, S.E., A note on EMQ under learning and forgetting, *AIIE Trans.*, 13, 86–90, 1981.

Badiru, A.B., Computational survey of univariate and multivariate learning curve models, *IEEE Trans. Eng. Manage.*, 39, 176–188, 1992.

Bailey, C.D., Forgetting and the learning curve: a laboratory study, *Manage. Sci.*, 35, 340–352, 1989.

Bhaskar, K. and Srinivasan, G., Static and dynamic operator allocation problems in cellular manufacturing systems, *Int. J. Prod. Res.*, 35, 3467–3481, 1997.

Bobrowski, P.M. and Park, P.S., An evaluation of labor assignment rules when workers are not perfectly interchangeable, *J. Oper. Manage.*, 11, 257–268, 1993.

Bokhorst, J. and Slomp, J., Long-Term Allocation of Operators to Machines in Manufacturing Cells, Group Technology/Cellular Manufacturing World Symposium, San Juan, Puerto Rico, 153–158, 2000.

Brusco, M.J. and Johns, T.R., Staffing a multi-skilled workforce with varying levels of perceptivity: an analysis of cross-training policies, *Decision Sci.*, 29, 499–515, 1997.

Brusco, M.J., Johns, T.R., and Reed, J.H., Cross-utilization of two-skilled workforce, *Int. J. Oper. Product. Manage.*, 18, 555–564, 1998.

Buck, J.R. and Cheng, S.W.J., Instructions and feedback effects on speed and accuracy with different learning curve models, *IIE Trans.*, 25, 34–37, 1993.

Buzacott, J.A., The impact of worker differences in production system output, *Int. J. Product. Econ.*, 78, 37–44, 2002.

Campbell, G., Cross-utilization of workers whose capabilities differ, *Manage. Sci.*, 45, 722–732, 1999.

Carnahan, B.J., Redfern, M.S., and Norman, B.A., Designing safe job rotation schedules using optimization and heuristic search, *Ergonomics*, 43, 543–560, 2000.

Caron, G., Hansen, P., and Jaumard, B., The assignment problem with seniority and job priority constraints, *Oper. Res.*, 47, 449–453, 1999.

Carr, G.W., Peacetime cost estimating requires new learning curves, *Aviation*, 45, 76–77, 1946.

Cesani, V.I. and Steudel, H.J., A Classification Scheme for Labor Assignments in Cellular Manufacturing Systems, Group Technology/Cellular Manufacturing World Symposium, San Juan, Puerto Rico, 147–152, 2000a.

Cesani, V.I. and Steudel, H.J., A Model to Quantitatively Describe Labor Assignment Flexibility in Labor Limited Cellular Manufacturing Systems, Group Technology/Cellular Manufacturing World Symposium, San Juan, Puerto Rico, 159–164, 2000b.

Dar-El, E.M., *Human Learning: From Learning Curves to Learning Organizations*, Kluwer, Boston, MA, 2000.

Dar-El, E.M. and Rubinovitz, J., Using learning theory in assembly lines for new products, *Int. J. Product. Econ.*, 25, 103–109, 1991.

Davis, G.S. and Moore, J.S. Manpower allocation for engineering in the product development environment, *Proceedings of the Annual Meeting of the Decision Sciences Institute*, Vol. 2, 1997, pp. 918–920.

Dorroh, J., Gulledge, T.R., and Womer, N.K., Investment in knowledge: a generalization of learning by experience, *Manage. Sci.*, 40, 947–958, 1994.

Ebbinghaus, H., *Memory: A Contribution to Experimental Psychology*, (Trans. Ruger, H. A. and Bussenius, C.E., 1964), Dover, New York, 1885.

Epple, D., Argote, L., and Devadas, R., Organizational learning curves: a method for investigation intra-plant, transfer of knowledge acquired through learning by doing, *Organ. Sci.*, 2, 58–70, 1991.

Farr, M.J., *The Long-Term Retention of Knowledge and Skills: A Cognitive and Instructional Perspective*, Springer, New York, 1987.

Fisk, J.C. and Ballou, D.P., Production lot sizing under a learning effect, *IIE Trans.*, 14, 257, 1982.

Globerson, S., Levin, N., and Shtub, A., The impact of breaks on forgetting when performing a repetitive task, *IIE Trans.*, 21, 376–381, 1989.

Glover, J.H., Manufacturing progress functions I. An alternative model and its comparison with existing functions, *Int. J. Product. Res.*, 4, 279–300, 1966.

Hancock, W.M., The prediction of learning rates for manual operations, *J. Ind. Eng.*, 18, 42–47, 1967.

Karwan, K.R., Mazzola, J.B., and Morey, R.C., Production lot sizing under setup and worker learning, *Nav. Res. Logist.*, 35, 159–179, 1988.

Kher, H.V., Examination of flexibility acquisition policies in dual-resource-constrained job shops with simultaneous worker learning and forgetting effect, *J. Oper. Res. Soc.*, 51, 592–601, 2000.

Kher, H.V. and Malhotra, M.K., Acquiring and operationalizing worker flexibility in dual-resource-constrained job shops with worker transfer delay and learning losses, *Omega*, 22, 521–533, 1994.

Kher, H.V., Malhotra, M.K., Philipoom, P.R., and Fry, T.D., Modeling simultaneous worker learning and forgetting in dual-resource-constrained systems, *Eur. J. Oper. Res.*, 115, 158–172, 1999.

Knecht, G.R., Costing technological growth and generalized learning curves, *Oper. Res. Q.*, 25, 487–491, 1974.

Khoshnevis, B. and Wolfe, P.M., An aggregate production planning model incorporating dynamic productivity. Part I model development, *IIE Trans.*, 15, 111–118, 1983a.

Khoshnevis, B. and Wolfe, P.M., An aggregate production planning model incorporating dynamic productivity. Part II solution methodology and analysis, *IIE Trans.*, 15, 283–291, 1983b.

Levy, F.K., Adaptation in the production process, *Manage. Sci.*, 11, 136–154, 1965.

Malhotra, M.K., Fry, T.D., Kher, H.V., and Donohue, J.M., The impact of learning and labor attrition on worker flexibility in DRC job shops, *Decision Sci.*, 24, 641–662, 1993.

Malhotra, M.K. and Kher, H.V., An evaluation of worker assignment policies in dual resource-constrained job shops with heterogeneous resources and worker transfer delays, *Int. J. Prod. Res.*, 32, 1087–1103, 1994.

Mazur, J.E. and Hastie, R., Learning as accumulation: a reexamination of the learning curve, *Psychol. Bull.*, 85, 1256–1274, 1978.

Mazzola, J.B. and Neebe, A.W., Bottleneck generalized assignment problems, *Eng. Costs Product. Econ.*, 14, 61–66, 1988.

McCreery, J.K. and Krajewski, L.J., Improving performance using workforce flexibility in an assembly environment and learning and forgetting effects, *Int. J. Product. Res.*, 37, 2031–2058, 1999.

McCune, J.C., On the Train Gang, Management Review, American Management Association, October, 57–60, 1994.

Molleman, E. and Slomp, J., Functional flexibility and team performance, *Int. J. Product. Res.*, 37, 1837–1858, 1999.

Nelson, R.T., A simulation of labor efficiency and centralized assignment in a production model, *Manage. Sci.*, 17, 97–106, 1970.

Nembhard, D.A. and Uzumeri, M.V., Experiential learning and forgetting for manual and cognitive tasks, *Int. J. Ind. Ergon.*, 25, 315–326, 2000a.

Nembhard, D.A. and Uzumeri, M.V., An individual-based description of learning within an organization, *IEEE Trans. Eng. Manage.*, 47, 370–378, 2000b.

Nembhard, D.A., A heuristic approach for assigning workers to tasks based on individual learning rate, *Int. J. Product. Res.*, 39, 1955–1968, 2001.

Nembhard, D.A. and Osothsilp, N., An empirical comparative study of models for measuring intermittent forgetting, *IEEE Trans. Eng. Manage.*, 48, 283–291, 2001.

Norman, B.A. and Bean, J.C. A genetic algorithm methodology for complex scheduling problems, *Naval Res. Logist.*, 46, 199–211, 1999.

Norman, B.A., Tharmmaphornphilas, W., Needy, K.L., Bidanda, B., and Warner, R.C., Worker assignment in cellular manufacturing considering technical and human skills, *Int. J. Prod. Res.*, 40, 1479–1492, 2002.

Park, P.S., The examination of worker cross-training in a dual-resource-constrained job shop, *Eur. J. Oper. Res.*, 51, 219–299, 1991.

Pegels, C.C., On startup or learning curves: an expanded view, *AIIE Trans.*, 1, 216–222, 1969.

Pine, B.J., *Mass Customization: The New Frontier in Business Competition*, Harvard Business School Press, Boston, MA, 1993.

Pratsini, E., Camm, J.D., and Raturi, A.S., Effect of process learning on manufacturing schedules, *Comput. Oper. Res.*, 20, 15–24, 1993.

Pratsini, E., Camm, J.D., and Raturi, A.S., Capacitated lot sizing under setup learning, *Eur. J. Oper. Res.*, 72, 545–557, 1994.

Russell, R.S., Huang, P.Y., and Leu, Y., A study of labor allocation strategies in cellular manufacturing, *Decision Sci.*, 22, 594–611, 1991.

Shafer, S.M., Nembhard, D.A., and Uzumeri, M.V., Investigation of learning, forgetting, and worker heterogeneity on assembly line productivity. *Manage. Sci.*, 47, 1639–1653, 2001.

Slomp, J. and Molleman, E., Cross-training policies and performance of teams, *Group Technology/Cellular Manufacturing World Symposium*, San Juan, Puerto Rico, 107–112, 2000.

Smunt, T.L., The impact of worker forgetting on production scheduling, *Int. J. Product. Res.*, 25, 689–701, 1987.

Sule, D.R., A note on production time variation in determining EMQ under influence of learning and forgetting, *AIIE Trans.*, 31, 91, 1981.

Sule, D.R., Effect of learning and forgetting on economic lot size scheduling problem, *Int. J. Product. Res.*, 21, 771–786, 1983.

Tharmmaphornphilas, W. and Norman, B.A., A heuristic search algorithm for stochastic job rotation scheduling, *9th Industrial Engineering Research Conference*, May 21–23, Cleveland, OH, CD-Rom format, 2000.

Tharmmaphornphilas, W. and Norman, B.A., Robust Job Rotation Methodologies to Reduce Worker Injuries, Technical Report 04-03, Department of Industrial Engineering, University of Pittsburgh, 15261, 2004, submitted to *Ann. Oper. Res.*

Van den Beukel and Molleman, E., Multifunctionality: driving and constraining forces, *Hum. Factor. Ergon. Manuf.*, 8, 303–321, 1998.

Warner, R.C., Needy, K.L., and Bidanda, B., Worker assignment in implementing manufacturing cells, *6th Industrial Engineering Research Conference Proceedings*, IIE, 1997, pp. 240–245.

Wisner, J.D. and Pearson, J.N., An exploratory study of the effects of operator relearning in a dual-resource-constrained job shop, *Product. Oper. Manage.*, 2, 55–69, 1993.

Womer, N.K., Learning curves, production rate, and program costs, *Manage. Sci.*, 25, 312–319, 1979.

Yelle, L.E., The learning curve: historical review and comprehensive survey, *Decision Sci.*, 10, 302–328, 1979.

17

Design Issues and Analysis of Experiments in Nanomanufacturing

Harriet Black Nembhard
Pennsylvania State University

Navin Acharya
Pennsylvania State University

Mehmet Aktan
Atatürk University, Turkey

Seong Kim
Pennsylvania State University

17.1 Introduction

The idea of making "nanostructures" that are composed of just one or a few atoms has great appeal, both as a scientific challenge and for practical reasons. In recent years, scientists have learned various techniques for building nanostructures, but they have only just begun to investigate these structures' properties and potential applications (Scientific American, 2002). In a nanoscale environment, the effects of design parameters on product characteristics usually cannot be known purely from phenomenological models; therefore, nanotechnology design often requires data collection and analysis. There is a lot of excellent nanoscience in the literature and in research labs, but there is a gap between nanoscience and nanotechnology. Often, the experimentation used to build up the background of nanoscience, while methodical (e.g., changing one key factor at a time while keeping others constant), is not efficient. Through our interdisciplinary collaborations, we have identified the importance of statistically based design of experiments (DOE) to the research and development of nanomanufacturing. Drawing from these interactions as well as from the literature, this chapter addresses three cases that closely link DOE with the needs of nanomanufacturing.

In particular, Section 17.2 focuses on split-plot designs in a nanoscale milling process. In practice, owing to the physical constraints in the nanomanufacturing process, situations often arise in which it is impossible to execute a completely randomized experiment, and we must resort to a split-plot design. We illustrate such a design along with its usefulness and distinct advantages over fractional factorial designs by using a nanomanufacturing process in which a double-charged arsenic focused ion beam (FIB) is used to mill submicron channels on a gold layer.

Section 17.3 focuses on mixture designs in bonding technology for nanomanufacturing applications. In basic experimental designs, the levels of each factor are independent of the levels of the other factors. In mixture experiments, the levels are not independent because the factors are the components or ingredients of a mixture. We present a mixture design application for bonding with alloy solders based on available data, where the design factors are the proportions of metals in the alloys. We discuss how the results of the analysis can be used to optimize a product characteristic or to obtain a target value of the product characteristic along with considering other characteristics.

Section 17.4 focuses on missing observations in nanotribology. In typical two-level factorial experiments, the design is balanced over every factor, which means that each factor column has an equal number of low and high levels. If all the runs in such a design can be executed, there will be no missing observations. However, in some nanomanufacturing situations, the experimenter is forced to deal with a design having one or more missing observations because the factor combination or molecular structure suggested by a randomized design simply does not exist. We show how to properly handle missing observations using a scenario for a series of nanoscale lubrication experiments for micromechanical (MEMS) devices in order to study the tribological behaviors of combined alcohol molecules.

For each case, we address design issues and advantages and demonstrate the analysis of such experiments using statistical software. In Section 17.5, we conclude the chapter and discuss our perspective for future research. We note that these design issues are rather advanced within the field of DOE, and that by exploring their use in nanomanufacturing, we are also poised to explore related research issues that stem from the iteration between theory and application.

17.2 Split-Plot Designs in Nanoscale Milling

Randomization is one of the key statistical principles of DOE. It refers to the concept that both the allocation of the experimental material and the order in which the individual runs or trials of the experiment are to be performed are randomly determined (Montgomery, 2005). However, owing to the physical constraints in nanomanufacturing applications, situations often arise in which it is impossible to execute a completely randomized 2^k full factorial or a 2^{k-p} fractional factorial design. Potential reasons may be that the manufacturing system is actually a two-stage process, that it is expensive or difficult to change the levels of some of the factors, or that there are physical restrictions on the process. In such cases, the design can be treated as having a split-plot structure or, more specifically, a fractional factorial split-plot (FFSP) structure. (The term "split-plot" comes from agricultural experiments in which a single level of one treatment is applied to a relatively large plot of ground, and all levels of a second treatment are applied to subplots within the large or "whole" plot.) These designs have widely been studied by several authors, including Bisgaard et al. (1995), Box and Jones (1992), Bisgaard (2000), Bingham and Sitter (2001), and Kowalski (2002).

In this section, we will present a split-plot design for investigating the nanoscale milling of submicron channels on a gold layer. In Section 17.2.1, we discuss the critical design issues for the process and show how to establish appropriate split-plot designs for it. In Section 17.2.2, we simulate the experimental output and show how to analyze the output using statistical software. In Section 17.2.3, we discuss our results and recommendations.

17.2.1 Design Issues

Some of the design issues relating to split-plot experiments in the field of nanomanufacturing can be illustrated using an example from Tseng et al. (2004). In this experiment, a double-charged arsenic FIB is used to mill submicron channels on a gold layer. Channels are then milled and the resulting profiles are measured using an atomic force microscope (AFM). The milling is performed by a precise pixel-by-pixel movement using a computer-controlled FIB machine. The process is also known as digital scan and is schematically shown in Figure 17.1.

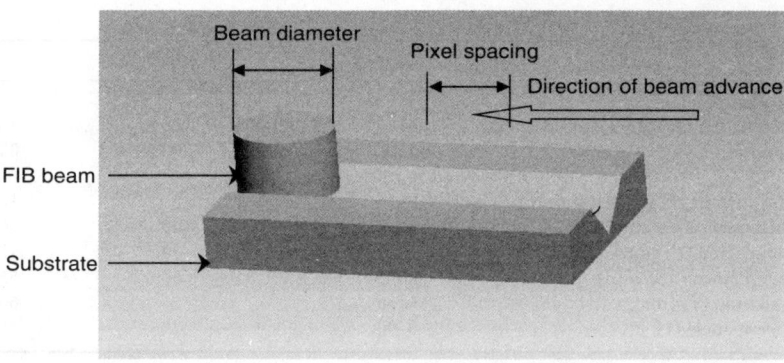

FIGURE 17.1 Schematic of FIB milling.

To mill a submicron channel, the FIB follows a series of adjacent pixels representing the contour of the channel. The amount of time that the beam remains on a given target pixel is called the dwell time (t_d). The distance between the centers of two adjacent pixels is called the pixel spacing (p_s).

The target material for the study is a 125-nm-thick gold layer, a silicon wafer serves as a substrate for the evaporation of the gold layer. The gold layer is coated by a high-vacuum evaporation system with an ion-pumped glass bell jar capable of operating in the range of 6×10^{-7} to 9×10^{-7} torr. Before FIB milling, the surface roughness of the gold films deposited at various speeds and substrate temperatures are studied. The temperature is varied from 25 to 360°C and the deposition speed is varied from 0.02 to 0.06 nm/s. Let us suppose that beam current, differential potential, pixel spacing, beam diameter, dwell time, and incident angle are the design factors that might affect the response. The environmental factors are substrate temperature and deposition speed. Furthermore, let us suppose that the response is the mean milling yield.

We can think of this experiment as a "two-step" process in which the gold films are first deposited on the substrate and then, by varying beam current, differential potential, pixel spacing, beam diameter, dwell time, and incident angle, measurements are recorded. Another way of thinking of it is that factors *A* and *B* affect the gold film deposition and the other factors affect the FIB milling. Since there are eight factors in total, a full factorial design to explore the mean milling yield would require $2^8 = 256$ runs, which would mean 256 different settings of the substrate temperature and deposition speed. However, let us assume that depositing the gold layer is a laborious, time-consuming, and expensive process and should be used minimally. In such a case, even a fractional factorial design would involve many setting changes of factors *A* and *B*. Keeping this in mind, the factors can be broken down into whole plot and subplot factors as shown in Table 17.1, forming a split-plot structure.

With the split-plot structure, we have two whole plot and six subplot factors, each of which is to be investigated at two levels. If we assume that a maximum of 32 runs is feasible for the entire experiment, the most appropriate design for this kind of a study would be a $2^2 \times 2^{6-3}$ FFSP design with 32 runs and only four different settings involving factors *A* and *B*.

Fractionating in FFSP can be classified as two types, namely, confounding within subplots and split-plot confounding. In fractionating within subplots, the generators that specify which combinations of runs to use are chosen from the same plot. Each whole plot is chosen randomly and the subplots are randomized within each whole plot. The design for the example being considered is shown in Table 17.2, for which the generators chosen are $S = PQ$ and $T = QR$.

The defining relation for this design is given by

$$I = PQS = QRT = PRU = PRST = PQTU = QRSU$$

The overall resolution as well as the partial resolutions of the whole plot and subplot design is resolution III, meaning that the interactions between the whole plot and subplots are free of confounding (if three or higher order interactions are assumed to be negligible). However, the main effects of the whole plot

TABLE 17.1   Summary of Factors

Factors	Current Level	Min	Max
Whole plot			
Substrate temperature (A) (°C)	25	25	360
Deposition speed (B) (nm/s)	0.02	0.02	0.06
Subplot			
Beam current (P) (pA)	5	5	10
Differential potential (Q) (kV)	45	45	50
Pixel spacing (R) (nm)	14.5	14.5	15
Beam diameter (S) (nm)	50	50	55
Dwell time (T) (ms)	5–50	5	50
Incident angle (U) (°)	0–90	20	75

TABLE 17.2   Design for Confounding Within Subplots

| | Run | Whole Plot (2^2) | | Subplot (2^{6-3}) | | | | | | Combination |
		A	B	P	Q	R	$S = PQ$	$T = QR$	$U = PR$	
(I)	1	−	−	−	−	−	+	+	+	STU
	2			+	−	−	−	+	−	PT
	3			−	+	−	−	−	+	QU
	4			+	+	−	+	−	−	PQS
	5			−	−	+	+	−	−	RS
	6			+	−	+	−	−	+	PR U
	7			−	+	+	−	+	−	QRT
	8			+	+	+	+	+	+	PQRSTU
a	9	+	−	−	−	−	+	+	+	STU
	10			+	−	−	−	+	−	PT
	11			−	+	−	−	−	+	QU
	12			+	+	−	+	−	−	PQS
	13			−	−	+	+	−	−	RS
	14			+	−	+	−	−	+	PR U
	15			−	+	+	−	+	−	QRT
	16			+	+	+	+	+	+	PQRSTU
b	17	−	+	−	−	−	+	+	+	STU
	18			+	−	−	−	+	−	PT
	19			−	+	−	−	−	+	QU
	20			+	+	−	+	−	−	PQS
	21			−	−	+	+	−	−	RS
	22			+	−	+	−	−	+	PR U
	23			−	+	+	−	+	−	QRT
	24			+	+	+	+	+	+	PQRSTU
ab	25	+	+	−	−	−	+	+	+	STU
	26			+	−	−	−	+	−	PT
	27			−	+	−	−	−	+	QU
	28			+	+	−	+	−	−	PQS
	29			−	−	+	+	−	−	RS
	30			+	−	+	−	−	+	PR U
	31			−	+	+	−	+	−	QRT
	32			+	+	+	+	+	+	PQRSTU

and subplots are confounded with two-factor interactions. This type of design is useful in a robustness study, where the interaction effects between the whole plot and subplot are of main interest, for example, in the study of the environmental effects on the design factors. However, if the objective of the experiment is to estimate the main effects, this design will not be very helpful.

On the other hand, in split-plot confounding, the generators of the subplot involve the whole-plot factors. Once again, randomization is carried out in the same manner as that of confounding within subplots. The corresponding design for this type of confounding with generators $S = APQ$, $T = BQR$, and $U = APR$ is shown in Table 17.3. The corresponding defining relation is given by

$$I = APQS = BQRT = APRU = ABPRST = QRSU = ABPQTU$$

In this case, the main effects of the subplot factors are free of confounding, whereas some of the two-factor interaction terms are confounded with other two-factor interactions. Owing to the alias structure, this type of confounding is more often used in an industry where the subplot contains the design factors of interest. However, this design is not very useful in robustness study experiments. Another important difference between the two types of confounding is that even though the overall resolution remains the same, the partial resolution of the subplot increases by one. In general, it can be said that split-plot confounding possibly increases the partial resolution of the subplot by at least one.

17.2.2 Experimentation and Analysis

Once the design has been chosen to meet the objectives, the experiment is executed and appropriate analysis then needs to be carried out. For the example under consideration, a robustness study is most

TABLE 17.3 Design for Split-Plot Confounding

		Whole Plot (2^2)		Subplot (2^{6-3})						
		A	B	P	Q	R	S = APQ	T = BQR	U = APR	Combination
(I)	1	−	−	−	−	−	−	−	−	(I)
	2	−	−	+	−	−	+	−	+	PS
	3	−	−	−	+	−	+	+	−	QST
	4	−	−	+	+	−	−	+	+	PQT
	5	−	−	−	−	+	−	+	+	RT
	6	−	−	+	−	+	+	+	−	PRST
	7	−	−	−	+	+	+	−	+	QRS
	8	−	−	+	+	+	−	−	−	PQR
a	9	+	−	−	−	−	+	−	+	S
	10	+	−	+	−	−	−	−	−	P
	11	+	−	−	+	−	−	+	+	QT
	12	+	−	+	+	−	+	+	−	PQST
	13	+	−	−	−	+	+	+	−	RST
	14	+	−	+	−	+	−	+	+	PRT
	15	+	−	−	+	+	−	−	+	QR
	16	+	−	+	+	+	+	−	−	PQRS
b	17	−	+	−	−	−	−	+	−	T
	18	−	+	+	−	−	+	+	+	PST
	19	−	+	−	+	−	+	−	−	QS
	20	−	+	+	+	−	−	−	+	PQ
	21	−	+	−	−	+	−	−	+	R
	22	−	+	+	−	+	+	−	−	PRS
	23	−	+	−	+	+	+	+	+	QRST
	24	−	+	+	+	+	−	+	−	PQRT
ab	25	+	+	−	−	−	+	+	+	ST
	26	+	+	+	−	−	−	+	−	PT
	27	+	+	−	+	−	−	−	+	Q
	28	+	+	+	+	−	+	−	−	PQS
	29	+	+	−	−	+	+	−	−	RS
	30	+	+	+	−	+	−	−	+	PR
	31	+	+	−	+	+	−	+	−	QRT
	32	+	+	+	+	+	+	+	+	PQRST

appropriate, wherein the interaction effects between the whole plots and the subplots, along with the main effects of the subplot factors, are our primary interest. Analysis of a split-plot experiment with confounding within subplots is illustrated with the help of a simulated data set. The simulated responses are presented in standard order in Table 17.4.

Owing to the double randomization in split-plot experiments, the effects need to be separated into whole-plot and subplot effects. The whole-plot effects constitute the main effects of the whole-plot factors along with any interactions among them. The subplot effects constitute the main effects of the subplot factors, the interactions among them, and, most importantly, the interactions between the subplot and the whole-plot factors.

The whole-plot and subplot errors are used to test the significance of the whole-plot and subplot effects, respectively. Since there is no replication in this experiment, the whole-plot error is formed by summing the higher order interaction terms of the whole plot; similarly, the subplot error is formed by summing the higher order interactions of the subplots. By doing this, the subplot error now has 13 degrees of freedom. Since there are only two factors in the whole plot, the only way to have an error term for the whole-plot effects is to replicate the experiment. However, as estimation of the main effects of the whole-plot factors is not the primary objective of this robustness study, we can conveniently omit that step.

One way of conducting the analysis of variance (ANOVA) is by using SAS statistical software. The higher order interactions that are pooled to form the error terms are not included in the model statement. SAS automatically treats the terms excluded from the model as error terms. More specifically, wherever a term does not appear in the model, its sum of squares is pooled with the simplest term in the model of which it is a subset. The corresponding output from SAS for the simulated data set is shown in Table 17.5. The plot of the interaction effect of AP is shown in Figure 17.2.

17.2.3 Discussion

From the ANOVA in Table 17.5, it is evident that the effects that significantly affect the response are the beam current (P), dwell time (T), and substrate temperature $\times$ beam current (AP) effects (these effects have a p value less than 0.1). In other words, changing levels of factors P and T, from their low levels to the corresponding high levels, leads to an increased mean milling yield. Furthermore, the interaction plot of effects for factors A and P (refer to Figure 17.2) shows that at the low level of factor A, both levels of factor P produce almost the same yield. In other words, the process is robust to the low substrate temperature, but at the high level of substrate temperature, a significantly larger variation in the yield is observed even though the maximum yield is obtained by maintaining factor P at its high level. In summary, on the basis of this analysis, the recommendations are to use the high values of the beam current (10 pA) and dwell time (50 ms), since the mean yield is better at that level, and to use the low level of the substrate temperature (25°C), since the process is more robust at that level. Of course, follow-up experiments can be similarly conducted in order to confirm and further refine these recommendations.

TABLE 17.4 Simulated Response in Standard Order

Std Ord	Response	Std Ord	Response	Std Ord	Response	Std Ord	Response
1	1.00	9	1.00	17	1.90	25	6.10
2	0.50	10	34.74	18	2.50	26	37.54
3	37.46	11	1.20	19	31.06	27	6.00
4	32.26	12	76.86	20	37.06	28	82.06
5	36.54	13	2.10	21	31.34	29	7.10
6	33.34	14	76.34	22	40.94	30	84.34
7	4.00	15	1.10	23	3.20	31	2.10
8	2.50	16	37.66	24	5.20	32	50.46

TABLE 17.5 SAS® ANOVA Output for the Nanoscale Milling Experiment

Source	DF	Type III SS	Mean Square	F Value	p Value
P	1	6644.162813	6644.162813	83.43	<0.0001
Q	1	5.695312	5.695312	0.07	0.7952
R	1	25.740313	25.740313	0.32	0.5836
S	1	7.125312	7.125312	0.09	0.7716
T	1	5631.257812	5631.257812	70.71	<0.0001
U	1	20.320313	20.320313	0.26	0.6256
P*T	1	797.002812	797.002812	10.01	0.0115
A*P	1	6202.195313	6202.195313	77.88	<0.0001
A*Q	1	0.382812	0.382812	0.00	0.9462
A*R	1	0.227813	0.227813	0.00	0.9585
A*S	1	20.320313	20.320313	0.26	0.6256
A*T	1	275.537813	275.537813	3.46	0.0958
A*U	1	5.695313	5.695313	0.07	0.7952
B*P	1	53.820312	53.820312	0.68	0.4323
B*Q	1	0.137813	0.137813	0.00	0.9677
B*R	1	4.425313	4.425313	0.06	0.8189
B*S	1	4.727812	4.727812	0.06	0.8130
B*T	1	0.227812	0.227812	0.00	0.9585
B*U	1	1.390312	13.390312	0.17	0.6914

Note: Tests of Hypotheses using the Type III MS for $A*B*P*Q*R*S*T*U$ as an error term.

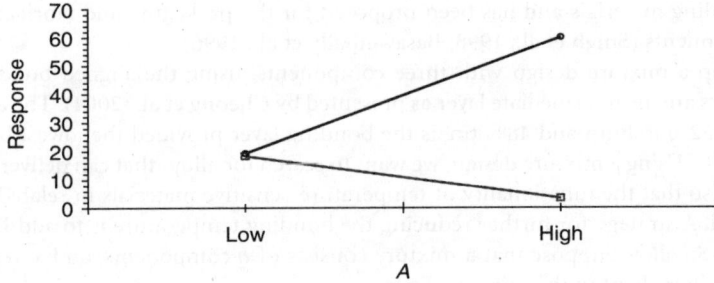

FIGURE 17.2 Interaction plot of effect AP.

In general, we can strongly recommend using split-plot experiments in nanomanufacturing research because they can reduce labor and costs and are often better at detecting the effects of subplot factors. Split-plot designs with the product design factors as the subplot treatments can be used to efficiently estimate the effects of the product design factors and their interactions with the environmental factors. These interactions measure the sensitivity of the product design combinations to the effect of the environmental variables. In designing a split-plot experiment, care should be taken to choose the appropriate type of confounding to best fit the main objectives of the study. In addition, rather than randomly assigning variables to whole plots and subplots, one must examine the complete alias structure of the design and evaluate the best option for the corresponding study.

17.3 Controlling Experimental Levels in Reliable Bonding for Nanotechnology Using Mixture Designs

Mixture experiments are a class of experimental designs in which the design factors are the ingredients or components of a mixture (consequently, their levels are not independent) and the response is a function of the proportions of each ingredient. There is great need for such an experimental arrangement in nanomanufacturing, including the area of bonding technology. It is the efficacy of the bonding technology that enables integration of dissimilar materials such as ceramics, ferroelectrics, polymers, and semi conductors, which are often required for MEMS device fabrication (Cheong et al., 2004). Emerging nanomanufacturing applications require a bonding technology that can provide low processing temperature in order to maintain the functionality of temperature-sensitive materials with, for example, a bond uniformity that is acceptable over an area comparable with the scale of the molecules that need to be manipulated.

In this section, we present a mixture design application for bonding with alloy solders, where the design factors are the proportions of metals in the alloys. We analyze the melting temperature data provided by the Indium Corporation of America (ICA) to investigate the appropriate alloy mixture that results in the lowest processing temperature. In Section 17.3.1, we present a basic discussion on mixture experiments. In Section 17.3.2, we present a mixture design that investigates the solder that minimizes processing temperature. In Section 17.3.3, we provide a summary and discussion.

17.3.1 Design Issues

Several bonding techniques for joining substrates are available, such as anodic bonding (Wei et al., 2003) and silicon–silicon direct bonding (Thompson et al., 2002). Using solders as an intermediate layer is growing in importance, and several techniques for depositing and patterning of solders have been proposed (Lemmerhirt and Wise, 2002; Chan et al., 1992; Cheng et al., 2000; Lee et al., 1991). A number of solder compositions, such as tin, indium, lead, and bismuth, can be used for bonding applications, but using lead is not desirable owing to environmental concerns. Flip-chip technology uses solders as interconnect and bonding materials and has been proposed for the packaging and fabrication of electronic and optical components (Singh et al., 1998; Basavanhally et al., 1996).

We will develop a mixture design with three components, using the aligned bonding technique in which alloy solders are the intermediate layer as presented by Cheong et al. (2004). The authors show that using an alloy of 52% indium and 48% tin as the bonding layer provided the lowest bonding temperature at about 117°C. Using a mixture design, we want to search for alloys that can deliver even lower melting temperatures so that the functionality of temperature-sensitive materials in related products can be better maintained. A strategy for further reducing the bonding temperature is to add Bi as a third component to the In–Sn alloy. Suppose that a mixture consists of q components, and x_i represents the proportion of the ith ingredient in the mixture. Then,

$$x_i \geq 0, \quad i = 1, 2, \ldots, q \tag{17.1}$$

and

$$\sum_{i=1}^{q} x_i = 1 \tag{17.2}$$

Because of the constraint in Equation (17.2), factor levels x_i are not independent; therefore mixture experiments are different from the usual response surface experiments. Figures 17.3 and 17.4 graphically show the constraints in Equations (17.1) and (17.2) for $q = 2$ and $q = 3$ components, respectively. For two components, the feasible space includes all values that satisfy $x_1 + x_2 = 1$, which is the line segment shown in Figure 17.3. With three components, the feasible space requires that $x_1 + x_2 + x_3 = 1$, which is an equilateral triangle, as shown in Figure 17.4. Each of the three vertices in the triangle corresponds to a pure blend, and each of the three sides represents a mixture that has two components (the component labeled on the opposite vertex does not exist in that mixture). Interior points in the triangle represent mixtures in which all three components are present. The experimental region for a mixture problem with q components is a simplex, which is a regularly sided figure with q vertices in $(q-1)$ dimensions (Myers and Montgomery, 1995).

When the q components are combined, the resulting effect is the response y. The expected value of this response, $E(y)$, is therefore a polynomial of a weighted combination of the x_is. In particular, the standard forms of the mixture models widely used are

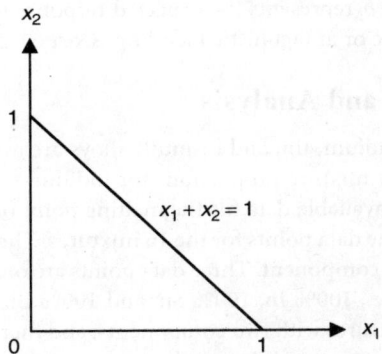

FIGURE 17.3 Constrained factor space for mixtures with $q = 2$ components.

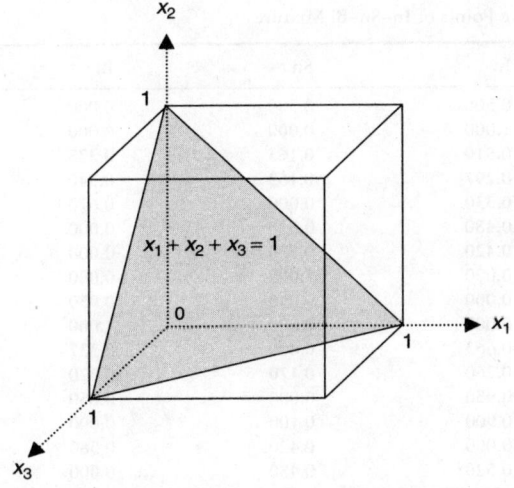

FIGURE 17.4 Constrained factor space for mixtures with $q = 3$ components.

Linear:

$$E(y) = \sum_{i=1}^{q} \beta_i x_i \tag{17.3}$$

Quadratic:

$$E(y) = \sum_{i=1}^{q} \beta_i x_i + \sum_{i<j}^{q} \sum_{i<j}^{q} \beta_{ij} x_i x_j \tag{17.4}$$

Full cubic:

$$E(y) = \sum_{i=1}^{q} \beta_i x_i + \sum_{i<j}^{q} \sum_{i<j}^{q} \beta_{ij} x_i x_j + \sum_{i<j}^{q} \sum_{i<j}^{q} \delta_{ij} x_i x_j (x_i - x_j) + \sum_{i<j<k}^{q} \sum_{i<j<k}^{q} \sum_{i<j<k}^{q} \beta_{ijk} x_i x_j x_k \tag{17.5}$$

Special cubic:

$$E(y) = \sum_{i=1}^{q} \beta_i x_i + \sum_{i<j}^{q} \sum_{i<j}^{q} \beta_{ij} x_i x_j + \sum_{i<j<k}^{q} \sum_{i<j<k}^{q} \sum_{i<j<k}^{q} \beta_{ijk} x_i x_j x_k \tag{17.6}$$

In these models, the parameter b_i represents the expected response to the pure blend, and the parameters b_{ij} represent either synergistic or antagonistic blending (Cornell, 2002).

17.3.2 Experimentation and Analysis

Melting temperatures for some indium, tin, and bismuth alloys are available from the ICA, and we use that data to discover the effects of mixture proportions for indium–tin–bismuth alloys by using a mixture design. Table 17.6 shows the available data for the melting point of In–Sn–Bi mixtures.

Figure 17.5 graphically shows the data points for the 16 mixtures. The gridlines represent a 10% change in proportion from 0 to 1 for each component. Three data points are on the corners of the feasible region. These 3 points are pure blends, i.e., 100% In, 100% Sn, and 100% Bi. There are ten data points on the sides of the triangle, which are mixtures with two components; and there are three data points in the interior of the region, which are mixtures containing all three components, i.e., In, Sn, and Bi.

TABLE 17.6   Melting Points of In–Sn–Bi Mixtures

Data Nos	In	Sn	Bi	Melting Point (°C)
1	0.500	0.500	0.000	125.0
2	1.000	0.000	0.000	156.7
3	0.510	0.165	0.325	60.0
4	0.297	0.163	0.540	81.0
5	0.330	0.000	0.670	109.0
6	0.480	0.520	0.000	131.0
7	0.420	0.580	0.000	145.0
8	0.000	1.000	0.000	232.0
9	0.000	0.050	0.950	251.0
10	0.000	0.000	1.000	271.0
11	0.663	0.000	0.337	72.0
12	0.260	0.170	0.570	79.0
13	0.950	0.000	0.050	150.0
14	0.900	0.100	0.000	151.0
15	0.000	0.420	0.580	138.3
16	0.520	0.480	0.000	118.0

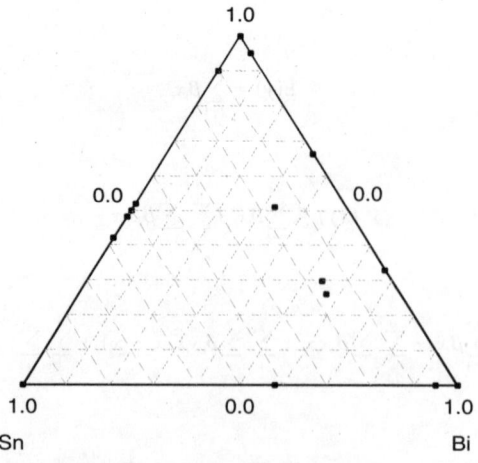

FIGURE 17.5   Data points for the In–Sn–Bi mixtures.

To analyze the experiment, we use Equation (17.4) to fit a second-order nonlinear quadratic mixture polynomial to the data, resulting in

$$\hat{y} = 166.2x_1 + 233.7x_2 + 273.6x_3 - 289.4x_1x_2 - 598.3x_1x_3 - 501.7x_2x_3$$

where x_1, x_2, and x_3 are In, Sn, and Bi proportions in an alloy, respectively, and $0 \leq x_i \leq 1$. This model is a very good representation of the melting temperature response. A summary of the analysis results is given in Table 17.7. The R^2 value of 99.16%, which is close to 1, and the adjusted R^2 value of 98.74%, which is close to the R^2, shows that the model has a good fit to the data. Note that because, $\beta_1 < \beta_2 < \beta_3$, we would conclude that indium (component 1) has the largest main effect (i.e., used alone) of reducing the melting point. (This is also apparent from comparing data values 2, 8, and 9 in Table 17.6) However, because $\hat{\beta}_{12}$, $\hat{\beta}_{13}$, and $\hat{\beta}_{23}$ are negative, blending indium and tin, indium and bismuth, or tin and bismuth

TABLE 17.7 Regression and ANOVA for the Melting Point

Estimated Regression Coefficients for Melting Point					
Term	Coefficient	SE Coefficient	T	P	VIF
In	166.2	4.603	*	*	1.986
Sn	233.7	6.916	*	*	2.297
Bi	273.6	5.276	*	*	1.998
In*Sn	−289.4	22.146	−13.07	0.000	2.655
In*Bi	−598.3	23.088	−25.92	0.000	1.912
Sn*Bi	−501.7	31.829	−15.76	0.000	1.683

Analysis of Variance for Melting Point						
Source	DF	Seq SS	Adj SS	Adj MS	F	P
Regression	5	57903.3	57903.3	11580.7	235.75	0.000
Linear	2	9608.7	12473.2	6236.6	126.96	0.000
Quadratic	3	48294.7	48294.7	16098.2	327.72	0.000
Residual error	10	491.2	491.2	49.1		
Total	15	58394.5				

Note: $S = 7.0087$, PRESS $= 5764.6$; $R^2 = 99.16\%$; $R^2 (\text{adj}) = 98.74\%$

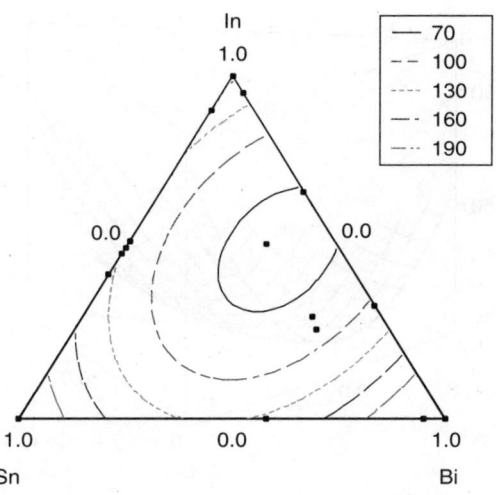

FIGURE 17.6 Contour plot for melting point.

produces lower melting points than would be expected by using any of the components alone. This is an example of "antagonistic" blending effects.

Figure 17.6 is a contour plot of the melting point for the mixtures. In a contour plot, the set of points that yield a particular level of the output is shown with a line. In Figure 17.6, sets of points that yield the melting points of 70, 100, 130, 160, and 190°C are shown. Figure 17.7 is the surface plot for the melting points of the mixtures. The vertical axis shows melting points against the feasible region of In, Sn, and Bi proportions. In effect, Figure 17.6 is a rotated projection of Figure 17.7.

17.3.3   Discussion

From examining Figures 17.6 and 17.7, we observe that if low temperature is desired, a blend of about 50% In–20% Sn–30% Bi should be chosen. We can formally determine the mixture proportions, i.e., x_1, x_2, and x_3, that minimize $\hat{y}$ using nonlinear optimization. The appropriate formulation is

$$\min \ \hat{y} = 166.2x_1 + 233.7x_2 + 273.6x_3 - 289.4x_1x_2 - 598.3x_1x_3 - 501.7x_2x_3$$

subject to

$$x_1 \geq 0, \quad x_2 \geq 0, \quad x_3 \geq 0,$$

$$\sum_{i=1}^{3} x_i = 1$$

The solution is a minimum melting point of 61°C obtained with a mixture of 49.5% In–14% Sn–36.5% Bi. In general, for a mathematical solution, confirmation experiments should be conducted around the estimated optimum point. In our case, we can simply observe that the 51% In–16.5% Sn–32.5% Bi mixture in Table 17.1 has a melting point of 60°C.

For some products, the objective may not be to minimize or maximize an output, but to obtain a desired output. For example, a melting point of 60°C may be too low for the product's operating specification limits, and the desired melting point may be 100°C. Then, contour plots can be used to see the set of mixtures that provide the melting point of 100°C. Since an infinite number of mixtures can satisfy this objective, we can select the mixture by considering other factors, such as cost. Then, from among the infinite number of mixtures that has a melting point of 100°C, we can find the mixture with the lowest cost.

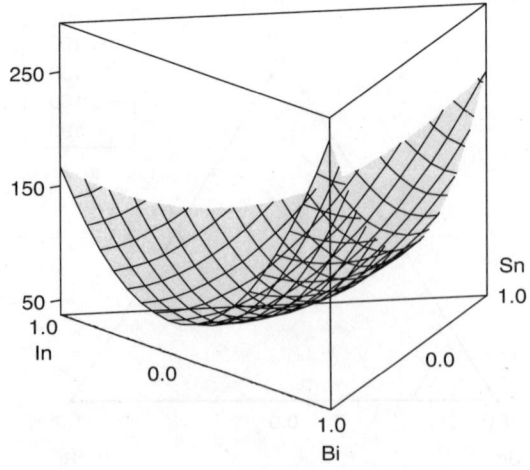

FIGURE 17.7   Surface plot for melting point.

17.4 Missing Observations in Nanotribology Experiments

In any two-level factorial (full or fractional) experiment, the design is balanced over every factor i.e., each factor column has an equal number of low and high levels. If all the runs in such a design can be executed, there will be no missing observations. However, in some nanomanufacturing situations it frequently happens that one or more observations are missing from the data, either because it has to be rejected or because it is impossible to execute the corresponding treatment combination.

In Section 17.4.1, we consider a set of gas-phase nanoscale lubrication experiments for the study of tribological behaviors of C_6 alcohol molecules (alcohols containing six carbon atoms). Since the molecular structures of alcohols only exist in certain combinations, there will be missing values in the experimental design. We use an iterative method due to Shearer (1973) and a noniterative method due to Draper and Stoneman (1964) to estimate the missing data. In Section 17.4.3, we discuss other alternatives that could be used to estimate missing observations.

17.4.1 Design Issues

Strawhecker et al. (2004) investigate the vapor-phase lubrication for MEMS surfaces. More specifically, they report on the effect of saturated alcohol vapors, including ethanol, *n*-propanol, *n*-butanol, and *n*-penatanol, on the tribological properties of a model MEMS surface: the native silicon oxide surface formed on a silicon wafer. We want to design an experiment to extend the study to understand the effect of molecular structure in C_6 alcohol molecules. These molecules have four main structural characteristics of interest, listed as factors *A*, *B*, *C*, and *D* in Table 17.8.

A full factorial design to explore these characteristics would involve $2^4 = 16$ treatment combinations. Table 17.8 only lists those molecules that are physically possible. For example, there exists no molecule that is linear, noncyclic, alken (containing a double bond), and a secondary alcohol (corresponding to treatment combination *ab*). Owing to this physical unavailability of some of the C_6 alcohol molecules, if this 2^4 experiment were to be carried out, only 10 observations could be recorded and we would be left with 6 missing observations (over one third missing). To reduce the proportion of missing observations, we considered a 2^{4-1} design, where there are 24 distinct factor arrangements of the 4 factors and 8 different generator options. In this case, the most efficient combination of factor arrangement and generator choice yields just two missing observations out of a total of eight runs. We will apply estimation techniques to this design.

TABLE 17.8 Factors of Interest and Available Molecules

Molecules	Structural Factors				Treatment
	A "+" Linear "−" Branched	*B* "+" Non "−" Cyclic	*C* "+" Alkyl "−" Alken	*D* "+" Primary Alcohol "−" Secondary OR Tertiary Alcohol	
Phenol	+	−	−	−	*a*
Cyclohexenal	+	−	−	−	*a*
1-Penten-3-ol, 4-methyl	−	+	−	−	*b*
Methylcyclopentanol	−	−	+	−	*c*
Cyclohexanol	+	−	+	−	*ac*
2,3-Methyl-2-butanol	−	+	+	−	*bc*
3-Hexanol	+	+	+	−	*abc*
cis-3-Hexen-1-ol	−	+	−	+	*bd*
trans-3-Hexen-1-ol	+	+	−	+	*abd*
3-Ethyl-1-butanol	−	+	+	+	*bcd*
N-Hexanol	+	+	+	+	*abcd*

Iterative methods for estimating missing data follow the principle that by choosing values that minimize the residual sum of squares, one can obtain the correct least-squares estimates of all estimable parameters as well as the correct residual sum of squares. If there is more than one missing observation, the formula for one missing value is used iteratively by starting with a guessed initial value. These iterative methods, however, have the disadvantages that they involve a great deal of computational work and each design requires a unique formula for a single missing cell. In order to avoid the computational difficulties associated with the iterative methods, many noniterative alternatives have also been proposed.

Shearer's (1973) iterative method for m missing values suggests minimizing the entire residual sum to squares with a two-stage procedure given by

$$y_{\text{miss}}^{i+1} = \sum_{j=1}^{m} b_j^{(i)} x_{\text{miss}j} + \frac{1}{n}\sum_{k=1}^{n} y_k^{(i)}$$

$$b_k^{(i)} = \frac{1}{n}\sum_{j=1}^{n} x_{jk} y_j^{(i)} \quad (k = 1, 2, \ldots, m)$$

where y_{miss} represents the missing observation, x either a "−" or a "+" for the given row and column, and b_k the main effects of the factors. The first equation re-estimates the missing value using the new main effect estimates. This latest estimate (y_{miss}) is used to re-estimate the main effects and the cycle is repeated until convergence is achieved.

Draper and Stoneman's (1964) noniterative method suggests equating a certain number of higher order interactions to 0 to estimate the missing values. The method then is to minimize only the residual sum of squares for the term(s) set to 0 (whereas Shearer's method minimizes all the residual sum of squares). In case of more than one missing observation, Draper and Stoneman suggested setting an equal number of interactions to 0 by making sure that independent equations exist.

17.4.2 Experimentation and Analysis

A typical experimental procedure goes as follows: a silicon wafer substrate is cleaned with a series of organic solvents and then fully dried by blowing with dry nitrogen. A silicon wafer substrate is further cleaned with a UV/ozone treatment to remove any organic residues from the substrate surface. The cleaned substrate is loaded on an AFM sample stage and then placed in an environmental chamber in which the gas-phase composition can be precisely controlled. The chamber is then filled with one of the C_6 alcohol vapors to a given partial pressure, and the friction response is measured with the AFM. These measurements are repeated at different alcohol vapor pressures.

The design shown in Table 17.9 is a basic 2^{4-1} design with $D = ABC$. Let us assume that the third observation is missing and employ the two techniques to estimate the value.

A half-normal probability plot of the factor effects helps to interpret the data. It plots the absolute value of the effect estimates against their cumulative normal probabilities. The effects that are negligible are normally distributed and will tend to fall along a straight line on this plot, whereas the significant effects will have nonzero means and will not lie along the straight line (Montgomery, 2005). In our

TABLE 17.9 Design Matrix with One Assumed Missing Value

Run	A	B	C	D	Treatment	Response
1	−	−	−	−	(I)	0
2	+	+	−	−	ab	3
3	+	−	+	−	ac	0*
4	+	−	−	+	ad	5
5	−	+	+	−	bc	5
6	−	+	−	+	bd	6
7	−	−	+	+	cd	7
8	+	+	+	+	abcd	10

* Assumed to be missing

application of missing observations, the half-normal probability plot can be used to check for procedure bias. If the line in the plot passes through 0, one can conclude that the interactions that were set to 0 were indeed 0 and that no bias has been induced on the other effects.

Figure 17.8a shows the half-normal probability plot when no observations are missing. This can be used as a benchmark to compare the techniques (Daniel, 1959). Using Shearer's iterative method, the value is estimated as 3.99 and Figure 17.8b shows the resulting half-normal plot. Using Draper and Stoneman's method, if we set interaction *ABC* to 0, the half-normal plot in Figure 17.8c shows a bias. This is because *ABC* is aliased with *D*, which is a significant effect. We can go back and re-estimate the missing values by setting *AB* = 0. The half-normal plot in Figure 17.8d does not show any bias and this value of the missing observation (which was 4) can now be used.

Let us now consider that in addition to the third observation, the seventh observation is also missing. Now, we have two assumed missing observations, as shown in Table 17.10. Figure 17.9a represents the normal probability plot when the missing values are estimated using iteration as 3.99 and 6.99. Figure 17.9b shows a plot when interactions *ACD* and *BCD* were set to 0 to obtain the missing values. As a considerable bias is observed, these values are rejected and the effects *BCD* and *AD* are set to 0. The plot for

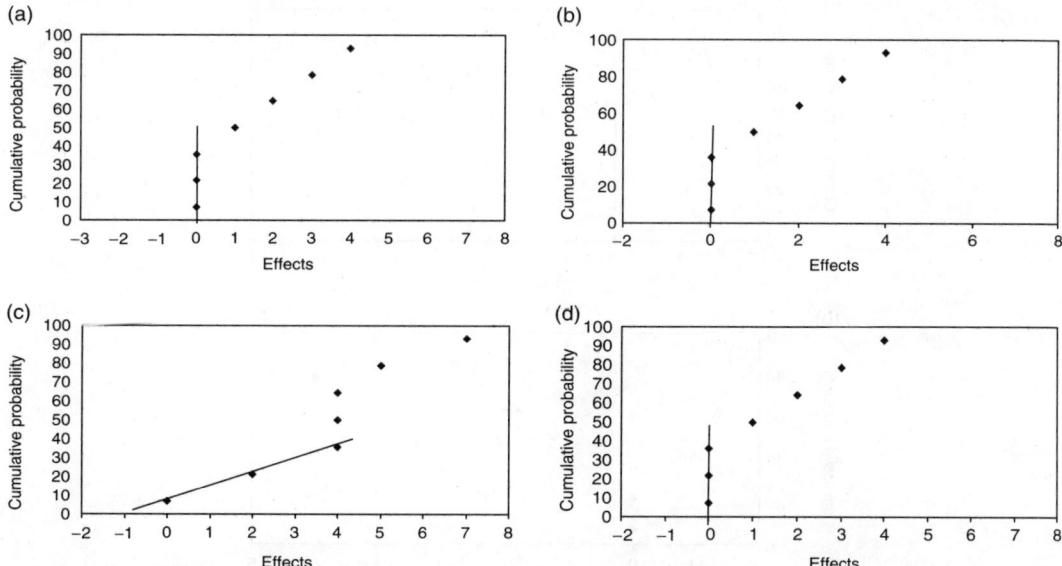

FIGURE 17.8 Normal probability plots for one missing observation using both methods of estimation. Half-normal probability plot of effects: (a) no missing observations; (b) one missing observation from iteration, $y_3 = 3.99$; (c) one missing observation from Draper's method I, setting *ABC* = 0, $y_3 = 20$; (d) one missing observation from Draper's method II, setting *AB* = 0, $y_3 = 4$.

TABLE 17.10 Design Matrix with Two Assumed Missing Values

Run	A	B	C	D	Treatment	Response
1	−1	−1	−1	−1	(I)	0
2	1	1	−1	−1	ab	3
3	1	−1	1	−1	ac	3*
4	1	−1	−1	1	ad	5
5	−1	1	1	−1	bc	5
6	−1	1	−1	1	bd	6
7	−1	−1	1	1	cd	5*
8	1	1	1	1	abcd	10

* Assumed to be missing

the second set is shown in Figure 17.9c, which also represents a bias. We found that using Draper and Stoneman's method, it is impossible to find two effects that yield two independent equations and that do not induce a bias. In this case, the iterative procedure had an advantage over the noniterative one.

17.4.3 Discussion

For the development of appropriate lubricant schemes for various substrate materials, it is very important to understand the effect of the molecular structure of the lubricant molecules adsorbed from the gas phase on tribological properties of each substrate. The DOE study described here helps to investigate the lubrication effects of a series of C_6 alcohols upon adsorption on the silicon oxide surface. If one can fully understand the molecular structure effect in this C_6 series, one can extend this knowledge to other alcohols that have different alkyl chain lengths, for example, C_4 or C_{10} alcohols. The vapor pressure of alcohol is a strong function of alkyl chain length — high for short-chain alcohols and low for long-chain alcohols. In order for the alcohol molecule to be transported effectively through the gas

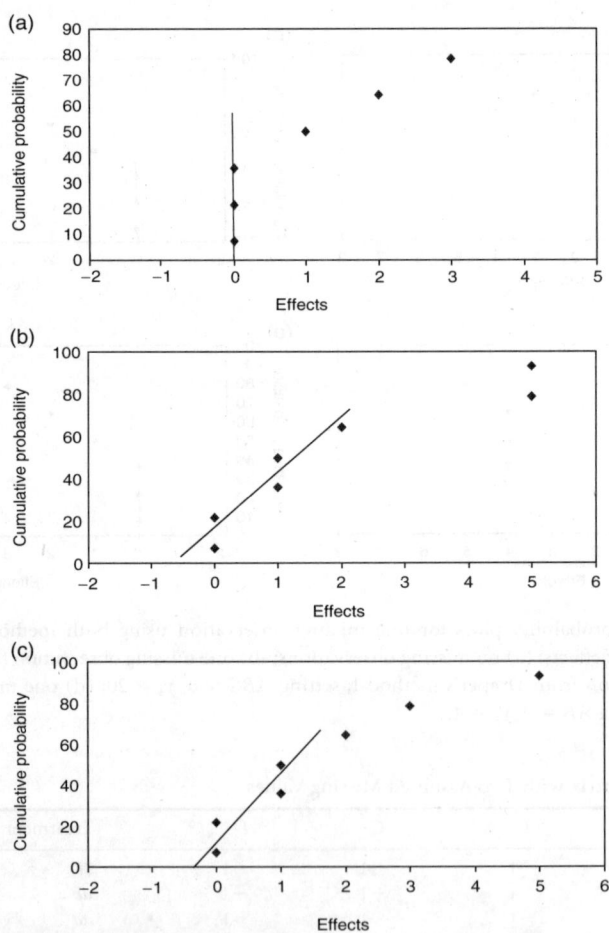

FIGURE 17.9 Normal probability plots for two missing observation using both methods of estimation. Half-normal probability plot of effects: (a) two missing observations from iteration; (b) two missing observations from Draper's method I, setting *ACD* and *BCD*=0, y_3=6, y_7=13; (c) two missing observation from Draper's method II, setting *BCD* and *AD*=0, y_3=2, y_7=9.

phase, the vapor pressure must be in an optimal range. Therefore, one must choose a short-chain alcohol for low-temperature operation of devices and a long-chain alcohol for high-temperature operation. Whenever the alcohol chain length is changed, one must determine the best molecular structure for optimum lubrication performance. In this situation, the knowledge obtained from the C_6 alcohol series can be utilized as a guide. However, it should be noted that the physical availability of alcohols with different molecular structures varies for different chain lengths, and it is not always guaranteed that all the structures will be available. Therefore, development of an algorithm that can sort out and predict the effect of missing components will be essential when a decision is to be made with a limited set of molecular structures.

In addition to Shearer's (1973) and Draper and Stoneman's (1964) techniques discussed above, there are several other techniques in the literature to estimate missing observations. Other iterative methods include those by Healy and Westmacott (1956) and Preece (1971). Other noniterative methods include those by Rubin (1972), John and Prescott (1975), Haseman and Gaylor (1973), and Bulmahn and Chauhan (1994). These techniques are listed in Table 17.11, along with suggestions on the maximum number of observations that can be missing for different full and fractional factorial designs. Note that these numbers are strictly suggestions based on our limited testing.

Although the iterative procedure proposed by Shearer (1973) seems to be extremely useful while dealing with a missing observation in either full or fractional factorial experiments, it may not always be convenient to employ an iterative procedure, especially if the cost of computing is high, as the process sometimes becomes time-consuming. In these cases, noniterative techniques can be employed. One of the major drawbacks of most of the noniterative procedures is their inability to estimate missing observations in fractional factorial experiments. The only two noniterative techniques that have the estimating capability are Draper and Stoneman (1964) and Bulmahn and Chauhan (1994). However, care should be taken when using these procedures, as Draper and Stoneman's method may end up yielding "nonindependent" equations, which would make it impossible to get two missing observations out of two "nonindependent" equations. In Bulmahn and Chauhan's method, the standard deviation of the estimate is a function of the population of the variance, so the larger the variation in the response, the more the estimate deviates from its "true" value.

We also note that Wilkinson (1958) provided an exact least squares approach to missing values in which an ANOVA is used to analyze the data that are actually present by a linear regression. It uses a matrix of independent variables as the "design matrix" of 0s and 1s. However, the structure of the ANOVA design is obscured and, even for modest data sets, the "design matrix" can be large and difficult to handle.

TABLE 17.11 Comparison of Iterative and Noniterative Techniques for Estimation of Missing Observations

		Technique for Estimating Missing Observation(s)							
		Iterative			Noniterative				
Design	Runs	Shearer (1973)	Healy and Westmacott (1956)	Preece (1971)	Draper and Stoneman (1964)	Rubin (1972)	John and Prescott (1975)	Haseman and Gaylor (1973)	Bulmahn and Chauhan (1994)
2^3	8	√ (2)	no info	no info	√ (1)	√ (1)	√ (1)	√ (1)	√ (1)
2^4	16	√ (3)			√ (2)	√ (2)	√ (2)	√ (2)	√ (2)
2^5	32	√ (4)			√ (3)	√ (3)	√ (3)	√ (3)	√ (3)
2^{4-1} or 2^{5-2}	8	√ (2)			√ (1)				√ (1)
2^{5-1} or 2^{6-2}	16	√ (3)			√ (3)				√ (2)
2^{6-1} or 2^{7-2}	32	√ (4)			√ (4)				√ (3)

Note: The number in parentheses indicates the suggested maximum number of missing observations for the design.

17.5 Conclusions and Perspectives

Statistically based DOE techniques have a long history, dating back to the agricultural era, led by the pioneering work of Sir Ronald A. Fisher in the 1920s and early 1930s, and continuing through the industrial era, catalyzed by the work of George E. P. Box in the 1950s, up to the present (see Montgomery [2005] and the references therein). The objective of this chapter was to demonstrate how DOE can be utilized to expedite technical advances as well as to improve fundamental understanding in the "nanotechnology era." Nanotechnology can potentially impact a range of fields, including pharmaceuticals, optics, information storage, sensors, biotechnology, medicine, and microelectronics. We focused on three illustrative cases to show how DOE techniques can be used in nanomanufacturing design and engineering by helping researchers understand the product and process dynamics. There are undoubtedly countless other nanotechnology-related projects that can also benefit from these methods.

In future work, we will use DOE to explore and guide our learning about nano-enabled medical devices. For example, Frecker et al. (2002) suggest a new design method for medical instruments to facilitate minimally invasive surgery. An understanding of nanoparticulate behavior can further move the approach from design to production. We can use DOE in conjunction with topology design to explore the effect of variations in material properties and processing parameters. In a simple case where there are seven factors at two levels, there would be $2^7 = 128$ possible experiments to conduct. Clearly, the traditional approach of exploring one factor at a time while keeping the others constant would be too time-consuming and, more importantly, would likely miss important interactions between the factors. Our approach will be to use a FFSP design to reduce the experimental space. Another important aspect will be to combine nanostructured components, such as nanocrystalline metals with microstructured ceramics materials. This will be an area where mixture designs can be used to discover the relationships between design factors and the product characteristics. Once these relationships are understood, statistical control can be used to monitor and adjust the outcome (Nembhard and Valverde-Ventura, 2003) and to make effective maintenance decisions (Ivy and Nembhard, 2005).

In general, we believe that integrating the types of DOE approaches that we have presented here into the modeling and production for nanomanufacturing research will yield strategic advantages by speeding the research and development cycle, stretching the experimental budget, and helping to create more reliable, robust, and better performing products. We also believe that as advanced nanotechnology applications are explored with DOE, there will be new questions calling for modifications or, perhaps, completely new constructs of experimental designs, which will simultaneously advance the field of DOE.

Acknowledgment

This work was supported in part by NSF Grant No. DMI-0451123 and NSF Grant No. CMS-0408369.

References

Basavanhally, N.R., Brady, M.F., and Buchholz, D.B., Optoelectronic packaging of two-dimensional surface active devices, *IEEE Trans. Components Packag. Manufacturing Technol., Part B-Advanced Packag.*, 19, 107–115, 1996.

Bingham, D. and Sitter, R.S., Design issues in fractional factorial split-plot experiments, *J. Qual. Technol.*, 33, 2–15, 2001.

Bisgaard, S., The design and analysis of $2^{k-p} * 2^{q-r}$ split-plot experiments, *J. Qual. Technol.*, 32, 39–56, 2000.

Bisgaard, S., Fuller, H., and Barrios, E., Two-Level Factorials Run as Split-Plot Experiments, Report #136, Center for Quality and Productivity Improvement, University of Wisconsin-Madison, Madison, WI, 1995.

Box, G. and Jones, S., Split-plot designs for robust product experimentation, *J. Appl. Statistics*, 19, 3–26, 1992.

Bulmahn, B.J. and Chauhan, C.K., Missing data — what do we do now?, *48th Annual Quality Congress Proceedings*, Las Vegas, 1994, pp. 571–576.

Chan, W.K., Yi-Yan, A., and Bhat, R., Tip-chip bonding: solder bonding to the sides of substrates, *Electron. Lett.*, 28, 1730–1732, 1992.

Cheng, Y.T., Lin, L., and Najafi, K., Localized silicon fusion and eutectic bonding for MEMS fabrication and packaging, *J. Microelectromech. Syst.*, 9, 3–8, 2000.

Cheong, J., Goyal, A., Tadigadapa, S., and Rahn, C., Reliable bonding using indium-based solders, in *Reliability, Testing, and Characterization of MEMS/MOEMS III*, Vol. 5343, Tanner, D.M. and Ramesham, R., Eds., Proceedings of SPIE, 2004, pp. 114–120.

Cornell, J., *Experiments with Mixtures*, 3rd ed., Wiley, New York, 2002.

Daniel, C., Use of half-normal plots in interpreting factorial, *Technometrics*, 1, 311–341, 1959.

Draper, N.R. and Stoneman, D.M., Estimating missing values in unreplicated two-level factorial and fractional factorial designs, *Biometrics*, 20, 443–458, 1964.

Frecker, M., Dziedzic, R., and Haluck, R., Design of multifunctional compliant mechanisms for minimally invasive surgery, *Minimally Invasive Therapy and Allied Technologies*, 11, 311–319, 2002.

Haseman, J.K. and Gaylor, D.W., An algorithm for non-iterative estimation of multiple missing values for crossed classifications, *Technometrics*, 15, 631–636, 1973.

Healy, M.J.R. and Westmacott, M., Missing values in experiments analyzed on automatic computers, *Appl. Statist.*, 5, 203–206, 1956.

Indium Corporation of America, *www.indium.com*, accessed January 24, 2005.

Ivy, J.S. and Nembhard, H.B., A modeling approach to maintenance decisions using statistical quality control and optimization, *Qual. Reliability Eng. Int.*, 31(4), 355–366, 2005.

John, J.A. and Prescott, P., Estimating missing values in experiments, *Appl. Statist.*, 24, 190–192, 1975.

Kowalski, S.M., 24 run split-plot experiments for robust parameter design, *J. Qual. Technol.*, 34, 399–410, 2002.

Lee, C.C., Wang, C.Y., and Matijasevic, G.S., A new bonding technology using gold and tin multilayer composite structures, *IEEE Transactions on Components, Hybrids Manuf. Technol.*, 14, 407–412, 1991.

Lemmerhirt, D.F. and Wise, K.D., Field-operable microconnections using automatically-triggered localized solder bonding, *Proceeding of the 15th IEEE International Conference on Micro Electro Mechanical Systems*, 2002, pp. 403–406.

Myers, R.H. and Montgomery, D.C., *Response Surface Methodology*, Wiley, New York, 1995.

Montgomery, D.C., *Design and Analysis of Experiments*, 6th ed., Wiley, New York, 2005.

Nembhard, H.B. and Valverde-Ventura, R., Integrating experimental design and statistical control for quality improvement, *J. Qual. Technol.*, 35, 406–423, 2003.

Preece, D.A., Iterative procedures for missing values in experiments, *Technometrics*, 13, 743–753, 1971.

Rubin, D.B., A non-iterative algorithm for least-squares estimation of missing values in any analysis of variance design, *Appl. Statist.*, 21, 136–141, 1972.

Scientific American, *Understanding Nanotechnology*, Warner Books, 2002.

Shearer, P., Missing data in quantitative designs, *Appl. Statist.*, 22, 135–140, 1973.

Singh, A., Horsley, D.A., Cohn, M.B., Pisano, A.P., and Howe, R.T., Batch transfer of microstructures using flip-chip solder bonding, *J. Microelectromech. Syst.*, 8, 27–33, 1998.

Strawhecker, K., Asay, D.B., and Kim, S.H., Gas-phase lubrication of MEMS devices: using alcohol vapor adsorption isotherm for lubrication of silicon oxides, in *Encyclopedia of Chemical Processing*, Lee, S.K., Ed., Dekker, New York (in press).

Strawhecker, K., Asay, D.B., McKinney, J., and Kim, S.H., Reduction of adhesion and friction of silicon oxide surface in the presence of n-propanol vapor in the gas phase, *Tribology Lett.*, 19, 17–21, 2005.

Thompson, K., Gianchandani, Y.B., Booske, J., and Cooper, R.F., Direct silicon-silicon bonding by electromagnetic induction heating, *J. Microelectromech. Syst.*, 11, 85–292, 2002.

Tseng, A., Insua, I.A., Park, J.S., Li, B., and Vakanas, G.P., Milling of submicron channels on gold layer using double charged arsenic ion beam, *J. Vac. Sci. Technol. B*, 22, 82–89, 2004. •

Wei, J., Xie, H., Nai, M.L., Wong, C.K., and Lee, L.C., Low temperature wafer anodic bonding, *J. Micromech. Microeng.*, 13, 217–222, 2003.

Wilkinson, G.N., Estimation of missing values for the analysis of incomplete data, *Biometrics*, 14, 257–286, 1958.

18

Lean Manufacturing Cell

M. Affan Badar
Indiana State University

Lean principles focus on elimination of waste by applying tools such as just-in-time, level production, standardized work, quality at the source, and continuous improvement. This chapter discusses the application of Lean concepts to designing and organizing a manufacturing cell. A brief description of the system engineering approach as well as a summary of common sources of waste, is also presented. A methodology to organize a cell using a 6S (Sort, Straighten, Shine, Standardize, Safety, and Sustain) program along with a point system that can be designed to account for the 6S, is explained in detail. An audit of the cell will yield a score. Making necessary changes and taking into consideration waste reduction will improve the score. Periodic audits should be conducted to assure continuous improvement.

18.1 Introduction

Lean manufacturing is a set of principles, concepts, and techniques derived from Toyota's just-in-time (JIT) production system (Parks, 2003; Monden, 1998). Just-in-time means achieving the level of production that precisely and flexibly matches customer demand and consists of processes that employ minimal (ideally zero) inventory through a strategy where each operation supplies parts or products to successor operations at the precise time they are demanded. It requires a continuous flow process structure that uses multifunction employees performing only value-added operations. It employs a pull system (kanban system) strategy to meet demand and to limit in-process inventories. This system is known as "Lean" because of its ability to do so much more with fewer resources (space, machines, labor, and materials) than traditional approaches. It uses standardized work practices based on minimal workforce and effort, highest quality, and highest safety in performing each job. It emphasizes building quality into the product rather than inspecting quality through systems that identify and resolve quality problems at their source. Thus, it provides customers with more value and companies with less waste (Durham, 2003). (Waste can be due to any of the following: overproduction, motion, transportation, waiting, set-up time, processing time, inventory, defective products, and underutilized workers [Minty, 1998; Askin and Goldberg, 2002]).

The objectives of Lean manufacturing are (Sobek and Jimmerson, 2003): to use as few resources as possible, to produce the desired amount of product at the highest possible level of quality and in as short a period of time as possible (reducing wait time that occurs as materials wait in queue or in inventory, and decreasing setup time). The key to doing this is to produce in small batch sizes. As inventory levels fall, the cost of defects soars because the system has little slack to absorb them. Thus, great attention is paid to fixing problems if defects occur. Also, work processes must be finely tuned and standardized to achieve predictable processing times and quality. The result is, to the extent possible in a mass-production environment, a system that focuses on individual products made for individual customers.

In recent years, Lean philosophies and practices have been implemented in several manufacturing facilities in the United States and around the world with such success that it is rapidly becoming the dominant manufacturing paradigm. Lean manufacturing integrates simple low-tech tools with advanced production/information technology and unique social/management practices (Sobek and Jimmerson, 2003). This organizational aspect makes it different from total quality management (TQM), which is rooted in meeting customer satisfaction or trying to meet the six-sigma (Elliott, 2003) requirements (3.4 parts per million (ppm) defective or 99.999% perfection). But TQM efforts often do not address organizational systems well and are not responsive to the needs of the employees and management. Integrated with management tools, Lean manufacturing can help ensure the achievement of a company's strategic objectives.

18.1.1 Waste Elimination

Lean manufacturing principles emphasize the elimination of waste from a production system in order to make it more efficient. In reality, however, such waste is generally hidden. Before removing waste, it is important to identify its sources (Minty, 1998; Askin and Goldberg, 2002), which are listed below.

Overproduction: Production of any product involves costs associated with direct material, direct labor, and manufacturing overhead, which can include factory overhead, shop expense, burden, and indirect costs. This means that the quantity produced per period of time from a manufacturing cell should be set to match the demand so that all the items made can be sold. Production should never be set to keep the resources busy. Any amount over the demand is waste as it costs money as well as wearing of machines.

Motion: Motion associated with either human motion or material handling consumes time and energy, and any motion that does not add value is a waste. Therefore, workplace and corresponding processes should be designed to eliminate non-value-added motions and to include ergonomic and safety considerations.

Transportation: This includes the movement between work-cell and storage area. Excessive movements should be minimized. Toolkits can be placed close to the point-of-use (POU). Materials can be stored and oriented in such a way that they can be fed to the cell easily.

Waiting: If a material or work-in-process (WIP) is waiting in queue for a proper machine or worker to be available, this is a kind of waste and causes longer throughput time. Production in small batches with coordinated order processing reduces excessive WIP and cycle time.

Setup time: Every time a tool setup is changed, it requires motion, time, and energy. A workplace should be designed to minimize the number of setups.

Processing time: A production system may consist of value-added and non-value-added activities. Processing time can be reduced by avoiding non-value-added operations.

Inventory: Inventory of finished goods involves costs of space, obsolescence, damage, opportunity cost, and handling. Therefore, excessive inventory should be eliminated.

Defective products: Defective products cause two problems: cost of material and resource, and poor customer satisfaction. Therefore, quality of the system, process, and products should be monitored continually to decrease defective products.

Underutilized workforce: If people working in a manufacturing cell or a production system are not utilized completely, i.e., there is not enough work for all of them, this is also a waste.

18.2   System Engineering Approach

A system engineering approach can help an industry design a manufacturing facility that will be not only-functionally highly productive but also good from the viewpoint of flow of materials or processes. The idea is to consider the manufacturing facility as a system (a set of interacting parts). In the case of the manufacturing facility, different cells are the interacting parts of the system. For any system, adequate stakeholders should be identified and listed (Sawle et al., 2005) — "stakeholders" being persons or organizations that directly have "something at stake" in the creation or operation of the subject system. Examples may include company owners or management, shareholders, employees, the local community, customers, retailers, etc. After the identification and listing of stakeholders, features of the subject system should be examined and enumerated (Sawle et al., 2005). A feature is a behavior of the system that has value to the stakeholders, each feature consisting of one or more feature attributes. For example, for a manufacturing system, production may be a feature and its attributes may include production rate and cost. This should be supplemented with a domain diagram (Sawle et al., 2005) of the system providing a high-level view of the environment in which the system exists and interacts with surrounding subsystems. Functional requirements of the system can be derived from these interactions (Sawle et al., 2005). Further, the feature- and role-attribute mapping will provide an idea of the optimum combination of design attribute values and stakeholder feature attributes (Sawle et al., 2005). This will help decide on feature attributes or parameters to be concentrated on for improvement and investment.

18.3   Lean Manufacturing Cell

In the preceding sections, Lean concepts and system engineering techniques have been described. A manufacturing system and its corresponding cells should be designed considering all these principles in order to efficiently create value for its multiple stakeholders. This requires deep commitment from the top management.

A manufacturing cell consists of a particular group of operations. For instance, a cell may contain hole-punching operations. It is important to note that a cell may be related to other cells in terms of the flow of materials and processes. Thus, each cell and its corresponding operations or processes and sub-processes need to be designed efficiently.

The first step toward good design is to describe and name all the processes and subprocesses, and the second is to measure performance data such as time taken and resources (material, machine, labor, space, etc.) being utilized over a period of time. The third is to analyze the existing process. Why is each process and subprocess necessary? Do all employees follow the same sequence of events prescribed for the process? How much is the process variation? How much is the utilization of each resource? Cause–effect analysis should be performed. Feedback from management, employees, and clients or customers should be obtained. A scoring system that is weighted to assess the importance of each task should be utilized. The fourth step is to propose necessary changes to improve the process. The fifth is to implement the suggested changes for improvements. The sixth is to evaluate (control, standardize, and verify) the new process. Last but not least is continuous improvement. Otherwise, everything will go back to its old state (see Figure 18.1).

A Lean cell needs to be designed to meet customer demands. Depending on the demand, the workforce can be increased or decreased. But operations or tasks within a work-cell must be set in such a way that it can adjust to the workforce level.

18.3.1   Organizing a Cell

Lean concepts can also be implemented to organize a workplace using the 5S philosophy (Parks, 2003; Standard and Davis, 1999; Brian, 2003):

Sort and clear out: Eliminate what is not needed, i.e., remove non-value-added processes or actions that increase the cost but add no value to the product.

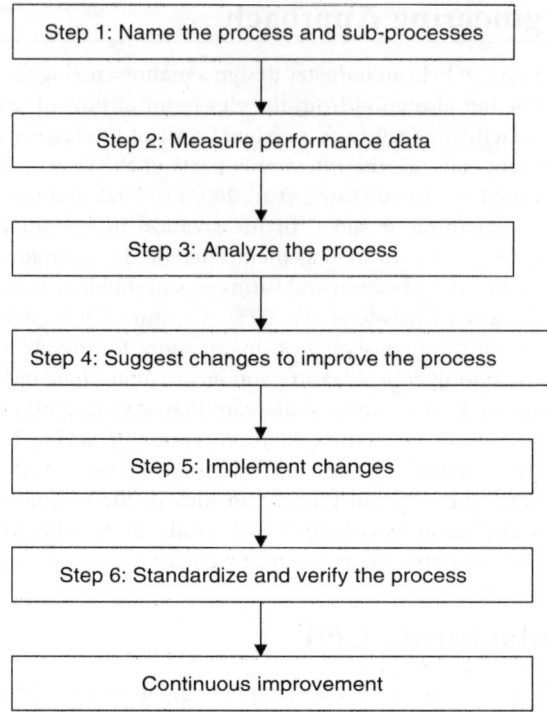

FIGURE 18.1 Steps in designing a process.

Straighten and configure: There should be a place for everything (materials and tools) and everything in its place, so that they are easily accessible.

Shine, scrub, and cleanup: Clean the workplace and look for ways to keep it clean. Clean work environments are more productive and provide workers with less stressful environments.

Safety and self-discipline: Always think safety first. Safety is very important in order to provide workers with satisfactory working conditions.

Standardization: Standardize work methods to eliminate variation and non-value-added time in a process.

Some companies add one more S to make a 6S system (Badar and Johnston, 2004): Sustain or stick to the rules, which includes maintaining and monitoring the above practices.

Badar and Johnston (2004) have investigated and carried out the implementation of Lean concepts or 6S to organize a cell of a local manufacturing company X. It was started with three basic steps (Durham, 2003): walk around, get creative, and look beyond the four walls. The cell consisted of one H5 Cincinnati (a five-axis, CNC-controlled mill) and two omni mills. The procedure is described in the following section.

18.3.2 Procedure

Organization of the workplace is the beginning of Lean application. This can be done using a 6S program: sort, straighten, shine, standardize, safety, and sustain. In Badar and Johnston (2004), a point system was designed to account for each category of 6S: sort, 25 points; straighten, 25 points; shine, 25 points; standardize, 25 points; safety, 30 points; and sustain, 25 points: In addition, 30 points were assigned for the environmental and ISO 14001 consideration, yielding a total of 185 maximum points. Each of the seven categories was further broken down into five subcategories. For each subcategory, 3 to 6 possible points were assigned. Thus, there were 35 subcategories and 185 possible points in total. To ensure the privacy and confidentiality of the company, no further details can be specified.

With this point system in place, the plan was to conduct an audit of the existing workplace; it would be considered acceptable if it scored 148 out of 185 possible points. A score above 164 was needed to classify the cell area in the Green category as per the company standard. The goal was to keep auditing and organizing the cell using Lean concepts so that it could finally be classified into the Green category.

An initial audit of the cell was conducted, and the cell scored 133 points, below the acceptable level. The categories and sub-categories receiving low points were identified. The issue was discussed with the employees to get their input. As outsiders, the authors (Badar and Johnston, 2004) were able to see beyond their wall and give unbiased feedback. This input helped determine what steps were needed, and in what order, for the implementation of the 6S program.

The following steps were taken to organize the cell:

1. The aisles were painted so that no material could be stacked in the aisles. This would also help control the stock on the floor of the cell area better.
2. An attempt was made to investigate what material and tools the machine operators were using and what they were not. Anything that was not being used was removed from the cell.
3. After thinning out the unused items, a place for each item (material and tool) was made and identified, so that anyone working in the cell could readily find what he or she needed. For example, squares were painted on the floor for the inspection cart and the operator's tool chest. This would also help prevent overstocking of materials, as it would be seen by the workers.
4. The floor, machine, and tools were cleaned up.
5. Labels and signs were posted to ensure safety and keep the cell organized.

With the above changes made, an audit was conducted and the cell scored 175 points. This score moved the cell into the Green category. However, periodic audits should be conducted to keep the process of continuous improvement in place.

18.3.3   Simulation Modeling

A simulation model of a manufacturing cell in Arena or other similar software can be developed to provide visual aspects to the complexities of the operations and resources associated with the cell (Thomson and Badar, 2004). The model is a simple yet effective approach to the problem-solving process. It can provide answers that could otherwise not be seen, because the simulation model can be modified and variables changed without affecting the current production. Although the accuracy of the results is as accurate as the data inputted, the Arena software contains an internal randomizer that can provide accurate statistical data.

Simulation can quantify the performance improvements that are expected from implementing the Lean manufacturing philosophy of continuous flow, JIT inventory management, quality at the source, and level production scheduling. Results from the simulation model may be gathered quickly. Detty and Yingling (2000) have used simulation to assist in the decision to apply Lean methods at an existing assembly operation. The manufacturing system in their study comprised multiple identical cells that assembled a finished product from over 80 individual parts. The cells were either fully staffed or shut down as product demand varied. In their study, they built models for the existing system as well as for a new system with the implementation of Lean principles. Their models demonstrated the entire manufacturing system, including the manufacturing processes and the associated warehousing, in-process inventory levels, transportation, and production-control and scheduling systems. Thus, the simulation model helped quantify the benefits of the Lean manufacturing concepts.

18.4   Conclusion

A manufacturing cell designed by employing Lean principles and system-engineering approaches efficiently creates value for its multiple stakeholders. Organization of a cell using the 6S system means that every item will be in its proper place and the workplace will be clean and shining. The 6S philosophy, combined with employee input, common sense, and looking "beyond the walls," can help reduce waste and result in an efficient workplace. Owing to global warming and other environmental problems, an environmental consideration should be made from the beginning and in every aspect of manufacturing. It is to be noted that continuous improvement is an essential concept of Lean philosophy. Therefore, once a cell or a system is designed and organized, it is a must to seek to continuously improve safety, quality, productivity, and the work environment.

References

Askin, R.G. and Goldberg, J.B., *Design and Analysis of Lean Production Systems*, Wiley, New York, 2002.

Badar, M.A. and Johnston, C., Lean application in organizing a manufacturing cell, *Proceedings of the IIE Annual Conference 2004, Solutions*, CD-ROM, Houston, TX, 15–19 May 2004.

Brian, S., Case study in creating a workcell – JIT, TPM and 5S implementation, *IIE 2003 Solutions Conference, Track: Lean Concepts and Applications*, Portland, OR, 17–21 May 2003.

Detty, R.B. and Yingling, J.C., Quantifying benefits of conversion to Lean manufacturing with discrete event simulation: a case study, *Int. J. Prod. Res.*, 38, 429–445, 2000.

Durham, D.R., Lean and green: improving productivity and profitability, *Manuf. Eng.*, 131(3), 16, 2003.

Elliott, G., The race to six sigma, *Ind. Eng.*, 35, 30–34, 2003.

Minty, G., *Production Planning and Controlling*, Goodheart-Willcox, Tinley Park, IL, 1998.

Monden, Y., *Toyota Production System: An Integrated Approach to Just-in-Time*, 3rd ed., Institute of Industrial Engineers, Norcross, GA, 1998.

Parks, C.M., The bare necessities of Lean, *Ind. Eng.*, 35, 39, 2003.

Sawle, S.S., Pondhe, R., Asare, S.A., Badar, M.A., and Schindel, W., Economic analysis of a lawn mower manufacturing system based on systems engineering approach, *IIE Annual Conference 2005, Research Track: Engineering Economics*, Atlanta, GA, 14–18, May 2005.

Sobek, D.K. and Jimmerson, C., Innovating Health Care Delivery Using Toyota Production System Principles, A Proposal to the National Science Foundation Innovation and Organizational Change Program, retrieved on 9/21/03 from http://www.coe.montana.edu/ie/faculty/sobek/IOC_Grant/proposal.htm, 2003.

Standard, C. and Davis, D., *Running Today's Factory: A Proven Strategy for Lean Manufacturing*, Hanser Gardner Publications, Cincinnati, OH, 1999.

Thomson, B.A. and Badar, M.A., Receiving area improvement using Lean concepts and simulation, *Proceedings of the 2004 NAIT Selected Papers*, CD-ROM, Louisville, KY, 19–23 Oct 2004.

19

Cluster Analysis: A Tool for Industrial Engineers

Paul S. Ray
The University of Alabama at Tuscaloosa

H. Aiyappan
The University of Alabama at Tuscaloosa

19.1 Introduction

Classifying objects into similar groups is a task most of us have encountered in our daily lives. A child classifies his toys according to his favorite colors; a real estate agent classifies his homes according to the location and price; and a direct marketer classifies his target according to a variety of geographic, demographic, psychographic, and behavioral attributes. As the field of classification science has become more sophisticated, we have grown to rely more on objective techniques of numerical taxonomy. The advent of high-speed desktop computers having enormous storage capacities has greatly facilitated the applications of advanced statistical tools in business situations. The common term for the class of procedures that are used to isolate component data into groups is "cluster analysis." The application of cluster analysis is prevalent in such diverse areas as (1) psychology for classifying individuals into personality types; (2) chemistry for classifying compounds as per their properties; (3) regional analysis for classifying cities as per their demographic and other variables; and (4) marketing analysis for classifying customers into segments on the basis of their buying behavior and product use.

    Cluster analysis has become a standard tool for marketing research applications. Despite its frequent use, however, there is a general lack of knowledge regarding the clustering methodology, as is apparent from the frequent omission in the literature of the specification of the clustering method used in a case. In the past, cluster analysis has been viewed with skepticism (Frank and Green, 1968; Inglis and Johnson,

1970; Wells, 1975). This skepticism probably arose owing to the confusing array of names for this technique, such as "typology," "classification analysis," "numerical taxonomy," or "Q-analysis". The names differ across disciplines, but they all deal with classification according to some natural relationship. This chapter presents the multivariate statistical techniques popular in providing business solutions; it specifically focuses on a clustering methodology called "*K*-means clustering," and illustrates the use of this type of statistical technique in a marketing situation.

19.2   Multivariate Analysis

Multivariate analysis seeks to examine the statistical properties among three or more variables across multiple subjects. The multivariate statistical techniques that are suitable for segmentation purposes consist of dependence and interdependence techniques. Dependence techniques use one or more types of independent variables to predict or explain a dependent variable. The commonly used dependence techniques for segmentation research include the following (Myers, 1996): automatic interaction detector (AID), chi-square automatic interaction detector (CHAID), and regression and discriminant analysis. Interdependence techniques search for groups of people or items that are found to be similar in terms of one or more sets of basic variables. All variables used are considered to be more or less equal in terms of interest. The most commonly used interdependence techniques for segmentation research include the following: hierarchical clustering, partition clustering, and Q-type factor analysis. The conceptual framework of multivariate methods is given in Figure 19.1 (Doyle, 1977).

The multivariate analytical techniques that are often used to provide business solutions include the following:

- Cluster analysis
- Multiple regression methods
- Multiple discriminant analysis
- Multiple analysis of variance (MANOVA)
- Canonical correlation
- Linear probability models (LOGIT, PROBIT)
- Conjoint analysis
- Structural equation modeling
- Factor analysis
- Multidimensional scaling
- Correspondence analysis

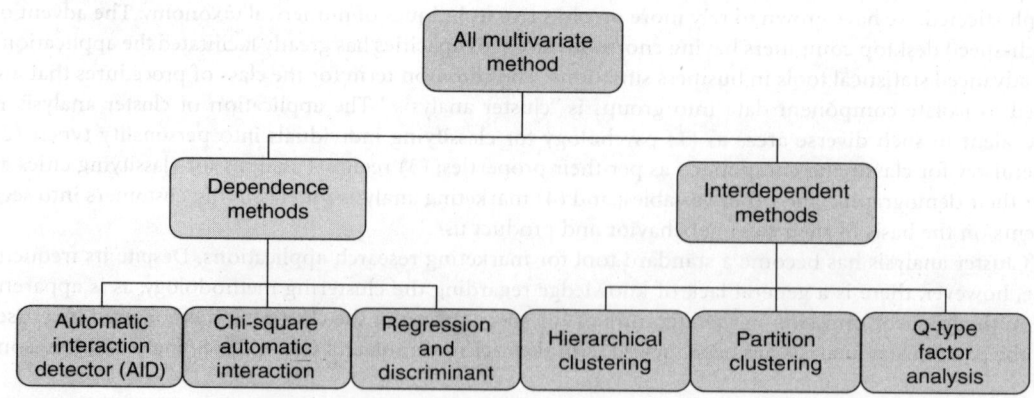

FIGURE 19.1   A conceptual framework of multivariate statistical methods.

19.3 Multivariate Analysis in Clustering

A multivariate analysis is a step beyond univariate analysis and utilizes more than one variable at the same time to explain and divide up the population under study. Variables used as input can be of several forms, and cluster analysis is often a further step from a univariate classification (TargetPro Version 4.5, 2003). Prepackaged approaches to multivariate statistical clustering use a customized version of nonhierarchical cluster analysis known variously as "iterative centroidal relocation" or "K-means clustering." This approach adjusts in multidimensional space the definition of a fixed number of clusters until a criterion involving "sums of squared distances" is minimized. The computer tests a number of different classifications, searches for a set that maximizes the similarity of the objects assigned to the same cluster, and, at the same time, maximizes the statistical distances or differences between individual clusters.

19.4 Multivariable Data Management in Segmentation

Generally, two approaches are used to reduce the amount of data to be analyzed. The first approach is to select on an a priori basis a relatively small number of variables (40 to 50) to represent each dimension that is judged to be most important. The problem with this approach is that a lot of subtlety hidden in the remaining variables is lost. The second approach is called principal component analysis, a statistical process that groups individual variables into separate components of factors and uses these rather than individual variables as a basis for measuring similarities between areas (Wind, 1978). This is an excellent technique for removing distortion caused by taking too many variables from one domain and not enough from another. The disadvantages of using principal components analysis are that it is less effective than using individual variables for building classification systems, and that the level of customization is limited and does not represent the individualistic approach of the end customer.

19.5 Cluster Analysis

19.5.1 Overview

In several statistical procedures, the objects on which measurements made are assumed to be homogeneous. But in cluster analysis, the focus is on the possibility of dividing a set of objects into a number of subsets of objects that display systematic differences. Cluster analysis represents a group of multivariate techniques used primarily to identify similar entities on the basis of the characteristics they possess. It identifies objects that are very similar to other objects in groups or "clusters" with respect to some predetermined selection criteria while the groups exhibit significant differences between them regarding the same criteria. The object cluster then shows high within-cluster homogeneity and high between-cluster heterogeneity. Figure 19.2 illustrates the within- and between-cluster variations.

There is no guideline on what constitutes a cluster, but ultimately it depends on the value judgment of the user. The data are allowed to determine the pattern of clusters inherent in the data. In the literature,

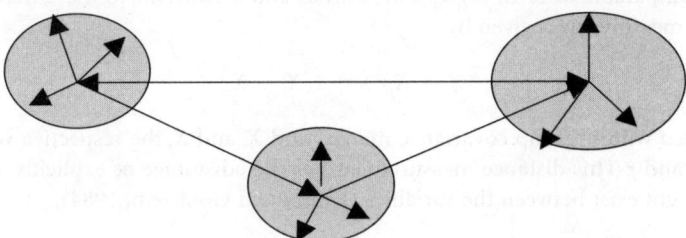

FIGURE 19.2 Within- and between-cluster variations.

an almost endless number of clustering algorithms is found. All of these depend upon high-speed computer capacity and aim to meet some criteria that maximizes the between-cluster variation compared to the within-cluster variation. Many different approaches have been used to measure interobject similarity and numerous algorithms to develop clusters. But until now, there has been no universally agreed-upon and definitive answer to these methods. As such, cluster analysis remains much more of an art than an exact science.

19.5.2 Phases of Clustering Analysis

The process of cluster analysis may be divided into three phases (Hair et al., 1987). These are:

- partitioning
- interpretation
- profiling

19.5.2.1 Partitioning

During the first phase, an appropriate measure is selected for measuring interobject similarity. The proximity or closeness between each pair of objects is used as a measure of similarity. Since distance is a complement of similarity, it is used as a measure of similarity.

19.5.2.1.1 Distance Type Measurement

This type of measurement is possible for quantitative data. The general Minkowski metric for distance measurement is defined by the following equation (Dillon and Goldstein, 1984):

$$d_{ij} = \left\{ \sum_{k=1}^{p} \left| X_{ik} - X_{ij} \right|^r \right\}^{1/r} \tag{19.1}$$

where d_{ij} is the distance between two objects i and j.

When $r = 2$, the Minkowski equation reduces to the familiar Euclidean distance between the objects i and j is given

$$d_{ij} = \left\{ \sum_{k=1}^{p} \left(X_{ik} - X_{jk} \right)^2 \right\}^{1/2} \tag{19.2}$$

When $r = 1$, the equation Minkowski equation reduces to the city-block metric, given by

$$d_{ij} = \sum_{k=1}^{p} \left| X_{ik} - X_{jk} \right| \tag{19.3}$$

Several other options are available in various computer programs. One option is to use the sum of squared differences as a measure of similarity. The raw data are converted to Z scores before computing distances. This step is taken to eliminate the spurious effect of unequal variances of the variables. Another very useful Euclidean distance measure is the Mahalanobis distance. The Mahalanobis D^2 generalized distance measure is comparable to R^2 in regression analysis and is superior to the different versions of the Euclidean distance measures. It is given by

$$(X_i - X_j)'S - 1(X_i - X_j) \tag{19.4}$$

where S is the pooled within-group covariance matrix, and X_i and X_j the respective vectors of measurements on objects i and j. This distance measurement has the advantage of explicitly accounting for any correlations that might exist between the variables (Dillon and Goldstein, 1984).

19.5.2.1.2 Match-Type Measurement

For qualitative data, a match-type or association measure is suitable. Association generally takes the value of 0 to indicate the absence of an attribute and 1 to indicate the presence of an attribute. Two objects or individuals are considered similar if they share common attributes and dissimilar if they do not share common attributes. We can visualize the variables' absence (0) and presence (1) in Table 19.1.

Similarity may be measured by counting the total number of matches (either 0, 0) or (1, 1) between X and Y and dividing the total by the number of attributes (8). The similarity between X and Y in this case is given by

$$\{(\text{Number of } (1, 1)\text{s or } (0, 0)\text{s})/\text{Number of attributes}\} \times 100\% = (5/8) \times 100\% = 62.5\% \quad (19.5)$$

The resulting association table is given in Table 19.2.

Association measure can be computed in several different ways; and unfortunately, this results in different values for the same data sets (Dillon and Goldstein, 1984). Hence, it is essential to assign 1's and 0's based on the importance to the user.

19.5.2.1.3 Clustering Algorithms

The next step is to select a particular type of computational algorithm. The commonly used clustering algorithms are of two types: hierarchical and nonhierarchical. The clustering process strives to maximize between-cluster variability and minimize within-cluster variability. In other words, subjects within the cluster are most similar and each cluster is markedly different from the others. Clustering techniques have been applied to a wide variety of research problems. For example, in the field of medicine, clustering diseases, cures for diseases, or symptoms of diseases can lead to very useful taxonomies (Hartigan and Wong, 1979). In the field of psychiatry, the correct diagnosis of clusters of symptoms such as paranoia, schizophrenia, etc. is essential for successful therapy. In general, whenever one needs to classify a large number of information into manageable piles, cluster analysis is of great utility.

19.5.2.1.4 Hierarchical Clustering

Hierarchical procedures construct tree-like structures. There are basically two types of procedures: agglomerative and divisive. In the agglomerative method, all cases start in their own cluster and are then combined into smaller and smaller numbers of clusters. In other words, all cases start off in the same cluster, and the process commences by dividing the clusters into two groups. The group with the most internal variation, the least homogeneous, gets split in two and now there are three groups, and so on, and the process continues until it can no longer find a statistical justification to continue (Hartigan and Wong, 1979).

19.5.2.1.5 Nonhierarchical Clustering — K-Means Clustering Method

This technique of clustering is gaining popularity for large databases and can be used once agreement is reached with regard to the number of clusters. Nonhierarchical procedures do not involve tree-like construction processes but need to select a cluster center or seed. All objects within a specified distance are included in the resulting cluster. There are three different approaches for nonhierarchical clustering based on sequential threshold, parallel threshold, or optimizing procedures.

TABLE 19.1 Contingency Table of Similarity

Object	1	2	3	4	5	6	7	8
X	0	1	1	0	1	1	1	1
Y	1	1	1	0	0	1	0	1

TABLE 19.2 Association Table for X and Y

	Object 2		Total
	+	−	
Object 1			
+	4	2	6
−	1	1	2
Total	5	3	8

The *K*-means clustering splits a set of objects into a selected number of groups by maximizing between-variations relative to within-variation (Green and Rao, 1969). In general, the *K*-means method will produce exactly *K* different clusters of greatest possible distinction (Sherman and Seth, 1977). In *K*-Means clustering procedure, the value of *K* or the number of clusters has to be decided prior to processing. There appears to be no standard method, but some guidelines are available (Hair et al., 1987). The clustering process may be stopped when the distance between clusters at successive steps exceeds a preselected value. An intuitive number of clusters may be tried, and on the basis of some preselected criteria, the best among the alternatives may be selected. Frequently, a judgment of practicality regarding comprehension and communication becomes the deciding factor.

Some cluster analysis packages, purchasable as "off-the-shelf" software, use segmentation techniques based on a neighborhood-type approach (TargetPro Version 4.5, 2003). These software packages use prepackaged approaches to multivariate statistical clustering that fundamentally follow the same concept, and several of them use a customized version of nonhierarchical cluster analysis, i.e., "*K*-means clustering." This approach consists of testing a number of different classifications and searching for a set of clusters that maximizes the similarity of all the geographic units assigned to the same cluster and, at the same time, maximizes the statistical differences between individual clusters.

19.5.2.2 Interpretation Phase

This phase involves determining the nature of the clusters by examining the criteria used to develop the clusters. One way is to determine the average value of the objects in each cluster for each raw variable and develop average profiles from these data. A cluster may favor one attitude, while the other may favor another. From this analysis, each cluster's attitudes may be evaluated and significant interpretations developed. The interpretations facilitate the assignment of a label that represents the nature of the clusters.

19.5.2.3 Profiling Phase

This phase involves describing the characteristics of each cluster to explain the way they differ in relevant dimensions. The demographics, behavioral patterns, buying habits, consumption characteristics, and other traits relevant to a particular study are usually included in the analysis for profiling. For example, one cluster may represent more affluent and younger customers while the other one may represent older and more conservative persons. This analysis has to focus on the inherent characteristics that differ significantly from those of the other clusters, and are different from the ones used to develop the clusters.

19.6 Testing the Validity of a Clustering Solution

Assessment consists of examining the following:

- Distinctiveness of clusters presented by profiling.
- Optimum number of clusters, depending on a balance between the extent of homogeneity within a cluster and the number of clusters.
- Goodness of fit indicated by high rank-order correlation between the input and the solution output. Because clusters are generated by maximizing the between-cluster sums-of-squares, the usual test of significance of ANOVA (F α, $v1$, $v2$) cannot be conducted in the case of cluster analysis (Dillon et al., 1984). Instead, the maximum value of *F*-statistic among the different alternative groupings is used as an indication of best fit.

19.7 Application of Cluster Analysis in Marketing Management: A Case Study

19.7.1 Cluster Analysis in the Marketing Field

A company's market area is composed of several segments of users, as defined by any number of factors, including industry type, company size, purchase and usage behavior, demographics, psychographics, and sociographic factors. In order to maintain its competitive advantage, a company has to target its market and streamline its communication strategies.

Segmentation is a marketing technique that facilitates targeting a group of customers with specific characteristics (Keller, 1993). By segmenting the market, a company can target product offerings and communications to the appropriate segments for higher overall effectiveness and profitability. Market penetration is defined as the percentage of available target market currently served by the company (Kotler, 1992). Over the past 30 years, encouraged by developments in microchip technology and mass storage devices, the use of computer technology in business applications has expanded rapidly (Freeman, 1991). This trend has permeated marketing data analysis, and a combination of computing power and statistical concepts has played a vital role in advancing the field of marketing management for dealing with large number of variables.

Marketers have to treat different groups or segments differently to derive maximum effectiveness. Cluster analysis is a technique that helps to identify relatively homogenous groups of objects on selected characteristics, using an algorithm that can deal with a large number of objects (Kaufman and Rousseeuw, 1990). But cluster analysis has been traditionally seen as a complex technique, largely the domain of mathematicians (Blamires, 1995). With the advent of mass computing and data storage systems, this technique has now become practical in market segmenting. There are several procedures for clustering, and selection among them has to be made on the basis of each individual case.

19.7.2 Objectives of the Study

The marketing team of a Fortune 500 corporation, with customers all over the continental United States, wanted to analyze and segment the designated marketing areas (DMAs) that it served into specific "clusters" that behaved similarly to each other. The team also had the additional objective of understanding which variables and how the multiple variables affected market penetration among its segments (Ray et al., 2005)

19.7.3 Methodology

The methodology of this research consisted of:

- data reduction by principal components analysis
- market segmentation by K-means cluster analysis of the DMAs based on each variable and penetration using SPSS 11.5 for Windows software
- ANOVA to test the significance of difference between group means for each variable on the DMA penetration

19.7.3.1 Variables

Principal component analysis was conducted to reduce 400 data variables into 29 to make the analysis pragmatic and actionable in the marketing sense. Out of these variables, 18 had usable data in terms of how recently the data had been collected, the source, and the validity of the data. These 18 data variables were:

- Total media spending
- Total amount spent
- All circulation
- Radio advertising
- Newspaper advertising
- April roll fold amount
- Number of data centers
- Number of travelers
- Median home value
- Average household income
- Percentage urban dwellers

- Percentage 55 to 64 years old
- Percentage spent on rent
- Percentage never married
- Percentage with two incomes
- Average liquor consumption
- Average pizza consumption
- Average items per center

19.7.3.2 Data Collection, Storage, Access, and Formats

The variables were reduced using the principal components analysis to factor the hundreds of variables into meaningful variables. The data were obtained in a Visual FoxPro (dBaseIV) format. The original file used for the principal components analysis was data from a leading rental agency that appended information (e.g., demographic and socioeconomic) into the customer file. The dBase file was imported into the SPSS software and stored as SPSS (.SAV) file.

19.7.3.3 Partitioning – Allocation of DMAs to Clusters Using *K*-means Clustering Methodology

The statistical package SPSS 11.5 for Windows had the ability to compute clusters using the *K*-means clustering methodology (Marija, 2002). The requirement for this was to know the variables that needed to be used for creating the SPSS syntax to group DMAs. Each variable was used along with the penetration to create the clusters. The syntax instructs the software to choose a variable and to iterate and classify the cases based on DMA into five clusters that are most disparate based on the principle of maximizing between-group variance and minimizing within-group variance. The number of iterations is set to a maximum of ten, which is a reasonable number to find the cluster solution. The input also requires the determination of the number of clusters for each case and was determined based on heuristics, so that the clusters created could be used for marketing action at a later stage. The syntax also asks to display the cluster membership for each DMA and output the ANOVA table. Each DMA is output as a member belonging to a specific cluster.

19.7.3.4 Results: Interpretation and Profiling of Clusters

On the basis of this study, the DMAs were grouped into clusters with significant differences for 8 of the 18 variables (see Table 19.3). The remaining ten variables did not have significant differences across the DMAs and were not instrumental in the penetration difference, as indicated by the lower *F* values in the ANOVA.

Thus these could not be used for creating valid clusters, which behaved similarly within group and differently across groups with respect to the variable and penetration index. The significant eight variables were:

- Total media spending
- Total amount spent
- All circulation
- April roll fold amount
- Number of centers
- Number of travelers
- Median home value
- Average household income

TABLE 19.3   Variable: Number of Centers

Cluster	No. of DMAs	Penetration Index	No. of Centers Index
1	1	8	2837
2	52	93	51
3	7	101	183
4	42	148	82

The resulting impact for each of the eight significant variables was determined individually. Each report contained the final cluster centers; the ANOVA table, which indicated the *F* values (higher *F* values [typically >1.0] indicated higher differences in variables across DMAs); the number of DMAs in each cluster; and the cluster membership details. The results for the variable "Number of Centers" are given in Table 19.4.

Results for each cluster and representative DMAs were provided for better understanding of the analysis by the client. An example of the results for the variable "Number of Centers" is given below:

- Cluster 1 was an outlier and has only one DMA. There were three centers for a core population of 3223, giving it a high index (2837).
- Cluster 2 had most of the low penetrated DMAs from all regions and the lowest number of centers per core population for any cluster. This fact indicated that increasing the number of centers in these DMAs provided potential for improving penetration.
- Cluster 3 had DMAs where it seemed that increasing centers would not prove as effective because these DMAs typically had a high number of centers per core population.
- Cluster 4 had the majority of high performing DMAs from all regions. These DMAs had high penetration and high index of centers but not as high as the Cluster 3, which suggests that there could be more additions of centers to improve penetration.
- Interestingly, when analyzing a few larger DMAs like Atlanta, Chicago, Houston, and Denver, they were found to be segmented in Cluster 2, which had a low index for centers and a low penetration with this variable. These could be potential markets where increasing the number of centers would improve penetration. The centers were not well represented in these large DMAs.
- It was interesting to note that other large DMAs, such as Charlotte, Orlando, and San Francisco, fell in Cluster 4, which is representative of DMAs that had a higher number of centers and a high penetration index with respect to the variable. The output from ANOVA verified that the variable "Number of Centers" had a significant effect on penetration.

19.7.3.5 Testing Validity of Clustering Solution Using ANOVA

The hypothesis to test is as follows:

$$H_0: \quad \mu_1 = \mu_2 = \mu_3 = \ldots = \mu_r \tag{19.6}$$

$$H_1 = \text{not all are equal } (i = 1 \ldots r) \tag{19.7}$$

There are *r* populations or treatments (DMS in this case). The null hypothesis assumes that all the means across the various DMSs are same for each variable. Alternate hypothesis assumes that at least one DMA differs from the available DMAs in terms of the variables analyzed.

SPSS 11.5 for Windows was used to compute the ANOVA test statistic or *F* value for the difference in the means of the variables across each segment or cluster. A high *F* value (typically > 1) indicates that the means of the variables vary considerably across the clusters and hence is a factor affecting penetration

TABLE 19.4 *F* values for Significant Variable-Index and Penetration-Index

No.	Significant Variable	Variable Index *F* value	Penetration Index *F* value
1	Total media spending	562	33
2	Total amount spent	23,823	42
3	All circulation	105	31
4	April roll fold amount	363	39
5	Number of centers	3,916	40
6	Number of travelers	1,930	62
7	Median home value	206	100
8	All circulation	105	31

across DMAs. The ANOVA table showed F values for each of the eight significant variables and market penetration indexes. These are given in Table 19.4.

Since the F values for all variables are much greater than 1, we conclude that the four clusters are significantly different from each other with respect to the variables and penetration indexes. These high values of F statistic were achieved by the clustering technique.

The F tests were used only for descriptive purposes because the clusters have been chosen to maximize the difference among cases in different clusters. The observed significance levels are not corrected for this and thus cannot be interpreted as tests of the hypothesis that the cluster means are equal. High values of the test statistic F indicate significant difference between the clusters. The variables were assumed to be normally distributed for conducting ANOVA. This assumption was tested by histogram for each variable and was found to be acceptable although in some cases there was a slight skew to the left or right.

19.7.4   Limitations of the Case Study

Owing to the inherent data insufficiencies, 11 factored variables were dropped and the remaining 18 were used for this study. These 18 variables were assumed to be representative of the initial data elements and were used for determining the significant ones that affected penetration. All underlying data were for the year 2001 only.

References

Blamires, C., Segmentation techniques in market research: exploding the mystique surrounding cluster analysis-part 1, *J. Targeting, Measurement Ana. Marketing*, 3, 338, 1995.

Dillon, R.D. and Goldstein, M., *Multivariate Analysis Methods and Applications*, Wiley, New York, 1984.

Doyle, P., In marketing: a review, *J. Bus. Res.*, 5, 235–248, 1977.

Frank, R.E. and Green, P.E., Numerical taxonomy in marketing analysis: a review article, *J. Marketing Res.*, 5, 83–93, 1968.

Freeman P., Using computers to extend analysis and reduce data, *J. Market Res. Soc.*, 33, 127–136, 1991.

Green, P. E. and Rao, V.R., A note on proximity measures and cluster analysis, *J. Marketing Res.*, 6, 359–364, 1969.

Hair, J.F., Anderson, R.E., and Tatham, R.L., *Multivariate Data Analysis*, Macmillan, New York, 1987.

Hartigan, J.A. and Wong, M.A., A K-means clustering algorithm: algorithm AS 136, *Appl. Stat.*, 28, 126–130, 1979.

Inglis, J. and Johnson, D., Some observations on and developments in the analysis of multivariate survey data, *J. Market Res. Soc.*, 12, 75–80, 1970.

Kaufman, L. and Rousseeuw, P.J., *Finding Groups in Data: An Introduction to Cluster Analysis*, Wiley, New York, 1990.

Keller, W.J., Trends in survey data processing, *J. Market Res. Soc.*, 35, 211–219, 1993.

Kotler, P., *Marketing Management*, 8th ed., Prentice-Hall, Englewood Cliffs, NJ, 1992.

Marija, J.N., *SPSS@11.0, Guide to Data Analysis*, Prentice-Hall, Englewood Cliffs, NJ, 2002.

Myers, J.H., *Segmentation and Positioning for Strategic Marketing Decisions*, American Marketing Association, 1996, pp. 30–32.

Ray, P.S., Aiyappan, H., Elam, M.E., and Merritt, T.W., *Int. J. Ind. Eng.*, 12, 125–131, 2005.

Sherman, L. and Seth, J.N., Cluster analysis and its applications in marketing research, in *Multivariate Methods for Market and Survey Research*, Seth, J.N., (Ed.), American Marketing Association, Chicago, IL, 1977, pp. 193–208.

TargetPro Version 4.5, *Users Guide–MapInfo Corporation*, 2003.

Wells, W.D., Psychographics: a critical review, *J. Marketing Res.*, 12, 196–213, 1975.

Wind, Y., Issues and advances in segmentation research, *J. Marketing Res.*, 15, 317–337, 1978.

20

Information Engineering

Teresa Wu
Arizona State University

Jennifer Blackhurst
North Carolina State University

20.1 Information Systems

An information system is an organized combination of people, hardware, software, communication networks, and data resources that collects, transforms, and disseminates information in a system. Typically, information systems have three components, as shown in Figure 20.1: inputs, information processing, and outputs.

Inputs may include a variety of information. In a simple production system, for example, the inputs may be the current inventory levels and shop-floor schedule. Processing takes the data from the inputs and converts them into usable information or outputs. Processing may include application of algorithms to produce an output of a schedule for ordering raw material. Additionally, a feedback loop may be built into the system to improve its operations. Actual orders placed and orders shipped may also be outputs of the feedback system.

In general, information is data associated with the past and the present. The three components of information systems, inputs, information processing, and outputs, are all data-related. A database system is a mechanism used to store data, or information.

This chapter is designed to give the reader an introduction to information systems and how they relate to industrial engineering. Section 20.2 discusses the growing importance of information systems and their relationship to Industrial Engineering. Section 20.3 briefly introduces database systems (including relational database systems and object-oriented database [OODB] systems), followed by a description of

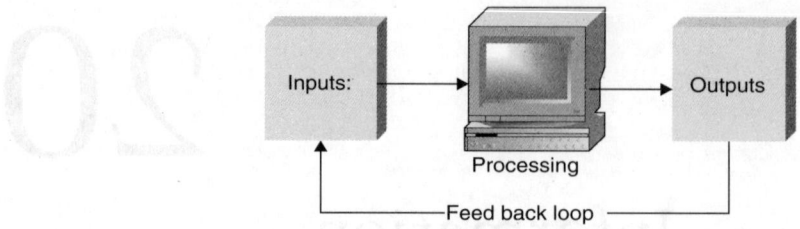

FIGURE 20.1   Three components in an information system.

data representation in Section 20.4. The need for data standards is discussed in Section 20.5. Section 20.6 illustrates some information system applications in industry. Section 20.7 concludes this chapter.

20.2   The Importance of Information Systems

An examination of the importance of information systems to an organization reveals that, the amount that companies invest in such systems is on the rise. The U.S. Department of Commerce Bureau of Economic Analysis reported that less than 5% of capital expenditures of American companies were targeted toward information technology in 1965. The introduction of PCs served as a catalyst and increased this to 15% in the 1980s. By the early 1990s, the amount had reached 30% and had grown to 50% by the end of the decade (Carr, 2003). In a recent information technology spending survey hosted by ITtoolbox (2003), participants were asked to indicate which areas of technology were currently being implemented or had been implemented within the past 6 months. Database applications ranked top among respondents, with 42% of respondents indicating a recent implementation followed by networking technologies (37.5%) and web services (32.4%). All survey participants were also asked about which technologies they plan to implement in the next year. The purpose of this question was to observe future trends in spending and establish which technologies would be most in demand. Of those surveyed, 21% cited business intelligence software as the top technology that they planned to implement during the year, followed by security solutions (20.5%) and web services (19.5%). Clearly, information systems are an integral part of managing a business, and in many cases an effective information system can be a competitive advantage.

While looking at the evolution of the function or role of information systems, a movement toward integration can be noted. Companies need tools to integrate functions to produce the smooth and efficient flows of goods and services. A recent study interviewing supply-chain executives across multiple industries revealed that integration of the supply chain through a comprehensive, real-time, end-to-end system is a tool that companies are lacking. Companies are challenged to improve performance by integrating functions and improving communication between traditional "silos" or typical functional areas (production, marketing, and logistics) that have operated in isolation in the past. Figure 20.2 gives a visual representation of an information system linking functional silos within an organization. An effective information system can integrate separate functions into a single system.

Companies are recognizing the need to seamlessly integrate these functions in order to better manage their facilities and larger supply chains through, in part, the use of information systems.

So, why do information systems matter to industrial engineers? Let us first look at the definition of an industrial engineer: The Institute of Industrial Engineering (IIE, 2004) states that industrial engineering

> "is concerned with the design, improvement, and installation of integrated systems of people, material, information, equipment, and energy. It draws upon specialized knowledge and skills in the mathematical, physical, and social sciences together with the principles and methods of engineering analysis and design to specify, predict, and evaluate the results to be obtained from such systems."

The role of information systems in the field of industrial engineering has grown, making it an integrated component of the field. Industrial engineers need to be able to design, implement, manage, and

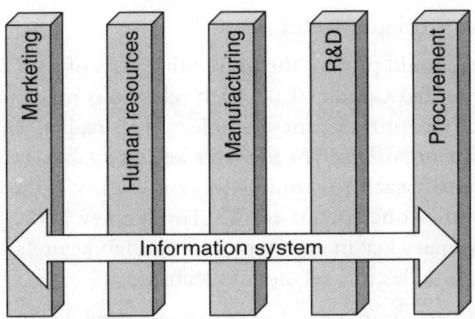

FIGURE 20.2 Information system as a function integrator.

maintain information systems in order to manage complex and dynamic systems. In contrast, let us look at an example of the definition of the industrial engineer from 1971: "Industrial engineering is the engineering approach applied to all factors, including the human factor, involved in production and distribution of products or services" (Maynard, 1971). Clearly, the focus has shifted in more recent times to *systems* and *integration*.

The need to integrate becomes even more apparent as companies expand globally. Management of systems within the four walls of a facility has moved beyond those walls to management of complex and dynamic supply chains, extending across multiple locations, facilities, and international borders. The challenge to manage such systems is complicated when dealing with such additional challenges as currency issues, multiple legacy systems, visibility, and information flows across the entire supply chain, security issues, etc. Effective information systems have become a visibility and management tool to integrate such global systems. This chapter introduces the basic concepts of information systems as they relate to the evolving role of industrial engineers.

20.3 Database Systems

In order to understand information systems, it is first necessary to understand data and database systems. Data is a collection of information. A database is a collection of related data or information. The database management system (DBMS) is a software tool that manages and controls the access to the database (Connolly and Begg, 2005). The DBMS allows the user to store, organize, and manipulate data to better manage complex systems (such as supply chains). The foundation of database systems is built on the hierarchical and network model (Hogan, 1990). Later, the idea that database systems should present the user with a view of data organized as tables called relations enlightens the development of relational databases. Today, the most prevalent type of database is the relational database, which has been applied in many fields, including accounting systems, inventory control, banking, etc. Yet the relational database faces limitations on the data types it can handle. Relational databases can only handle numbers, dates, and characters. Technologies such as the Internet drive the way to store objects such as voice, pictures, video, and text documents (Stephens and Plew, 2003). Objects such as these cannot be stored directly in relational databases, instead, they have to be flattened or parsed and stored in a tabular structure. Object-oriented databases provide a means of storing objects without parsing. The entire object is stored within the database. Therefore, OODBs, congruent with the data defined in objects, have become increasingly popular.

20.3.1 Relational Database

Relational database management systems (RDBMS) have estimated sales of between $15 billion and $20 billion per year, growing at a rate of possibly 25% per year (Connolly and Begg, 2005). There are various commercial relational database products available. The top three makers are IBM, Oracle, and Microsoft.

20.3.1.1   Relational Data Storage Model

The idea that database systems should present the user with a view of data organized as *tables* called *relations* was originally proposed by Ted Codd (1970). Each relation is made up of *attributes*. Attributes are values describing properties of an *entity*, a concrete object in its reality. The connections among two or more sets of entities are called *relationships*. The idea of a key on a table is central to the relational model. The purpose of a key is to identify each row uniquely. Primary key is the attribute (or combination of attributes) that uniquely identifies one row or record. Foreign key is the attribute (or combination of attributes) that appears as a primary key in another table. Foreign key relationships provide the basis for establishing relationships across tables in a relational database.

Example 1   Let us consider a very simple database relational model for a project management system, which can be used for the collaborative product development shown in Figure 20.3. There are two entities within the model: *User* and *Companies*.

- The relation *User,* which keeps the information about the system users, has four attributes: "Name," "Position," "SSN," and "Company Name."
- The relation *Companies,* which records information of company, has three attributes: "Company Name," "Project Name," and "Contact."
- In entity *User,* "SSN" is the primary key which can identify each user; and "Company Name" is the foreign key relating *User* to *Companies* because "Company Name" is the primary key for entity *Companies* as shown in Figure 20.3.

20.3.1.2   Structural Query Language

Based on the relational data model, a relational database can be built where data are stored in tables and all operations on data are either done on the tables themselves or produce other tables as a result of the operation. Several languages have been created to manipulate relational tables. The most common one among them is the ANSI Standard SQL (Structured Query Language). SQL queries are used to perform tasks such as updating data in a database, or retrieving data from a database. Even though there are various versions of SQL software available from different database vendors, the ANSI standard SQL commands can still be used to accomplish most tasks. There are two major groups of SQL commands. One is

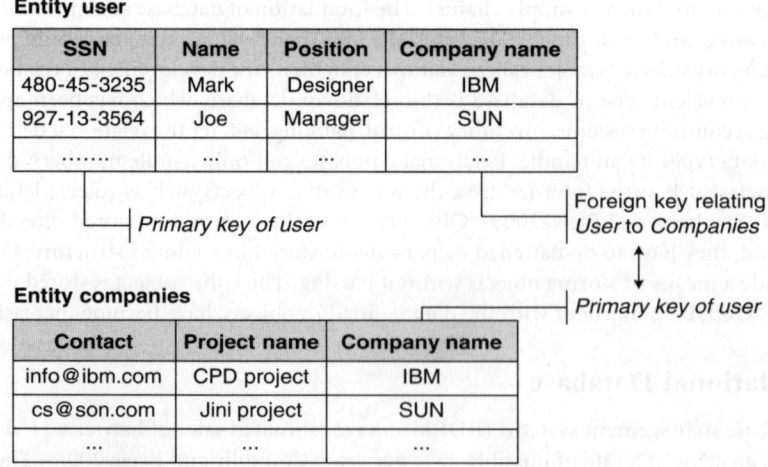

FIGURE 20.3 Relational tables.

to build the database structure, such as create or update table, termed data definition. Another one is to query or input data, termed data manipulation.

Example 2   Suppose we want to implement the relational data model from Example 1. We could use SQL statements to create an empty table named User:

> **CREATE TABLE** User (
>   SSN char (30),
>   Name char (10),
>   Position char (30),
>   Company Name char (20)
>   );

The "Insert" statement is used to insert or add a record into the table. For example:

> **INSERT INTO** User (SSN, Name, Position, Company Name)
> **VALUES** ("480-45-3235", "Mark", "Designer", "IBM");

The "Update" statement is used to update or change records that match specified criteria. For example, the position of system user "Mark" is updated as follows:

> **UPDATE** User **SET** Position= "Manager" **WHERE** Name = "Mark";

The "Select" statement is used to retrieve selected data that matches the criteria specified. For example: the Name and Position of user with SSN (480-45-3235) are retrieved as follows:

> **SELECT** Name, Position **FROM** User **WHERE** SSN = "480-45-3235";

Assume we want to delete table User: The "Drop" statement is used:

> **DROP TABLE** User;

20.3.1.3   Entity-Relationship Model

To ensure that the design of the database precisely presents the data, entity-relationship (ER) models are commonly applied. An ER model is a top-down approach that uses the ER diagram (ERD) to model the real-world information structure by a set of basic objects (entities) and their attributes and relationships among these objects. An ERD maps the tables in the database, details how they are related, and provides information on those attributes of the columns in the tables (Stephens and Plew, 2003). The advantage of ERD is that it is very simple yet very powerful, and any individual can learn how to use it to represent and understand the data requirements. In ERD, each entity is shown as a rectangle labeled with the name of the entity, which is normally a singular noun. The particular properties of entities are called attributes, denoted by ellipses. A relationship is a set of associations between one or more participating entities.

Example 3   Let us consider the previous example presented by an ER model. In the ER model, there are two entity sets and one relationship set as shown in Figure 20.4.

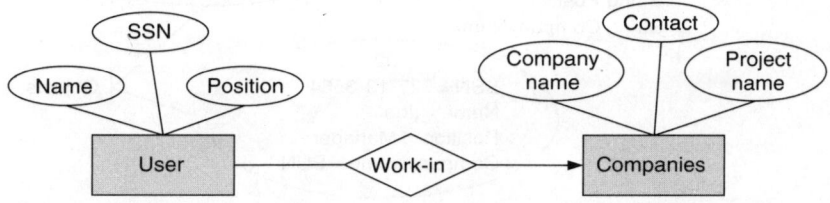

FIGURE 20.4   ER diagram.

The entity set Companies, has three attributes: "Company Name," "Project Name," and "Contact."
  - The entity set *User* has three attributes: "*Name,*" "*Position,*" and "*SSN.*"
  - The relationship *Work-In* relates *User* with *Companies.*

20.3.2 Object-Oriented Database

Today, RDBMS is widely used for traditional business applications. However, existing RDBMS can be inadequate for applications whose requirements are more than what can be addressed by traditional applications. One example is that RDBMS has great difficulty in handling object data. An object is a programming entity with a collection of attributes and functions that are used to operate on the object data. Objects provide an advanced means of storing data within a program. Object-Oriented database was, therefore, developed to work with an object-oriented programming language to store data.

20.3.2.1 Object-Oriented Data Storage Model

In the object-oriented storage model, data are considered as a collection of objects, which are the instances of classes as shown in Figure 20.5. A class is like the blueprint of an object, defined as a group of objects that have similar attributes and unique object identities (IDs). Access to data records/objects follows a systematic approach using the concept of encapsulation, that is, an object contains both the data structure and the set of operations that can be used to manipulate it. Unlike a relational table, where each element is an atomic value such as a string or number, an object can contain another object as its component. In a relational database, real-world entities are fragmented to be stored into many tables. Such fragmentation weakens the representation capability. Object-oriented data models overcome the limitation by giving a perspective that follows from the way human beings view, understand, analyze, and model the universe. Thus, object-oriented models offer great advantages in investigating complex systems and objects.

20.3.2.2 Object Definition Language

Object definition language (ODL) is a proposed standard language for specifying the schema or structure of OODBs. The primary purpose of ODL is to allow object-oriented designs of databases to be written and then translated directly into declarations of an object-oriented database management system. ODL defines the attributes and relationships of types and operations. The syntax of ODL extends the interface definition language of CORBA. Some commonly used keywords are interface, attribute, relationship, and inverse. The keyword "interface" is used to indicate a class; the keyword "attribute" describes a property of an object by associating a value of some simple type, for example string or integer, with that object. The keywords "relationship" and "inverse" declare respective connections between two classes.

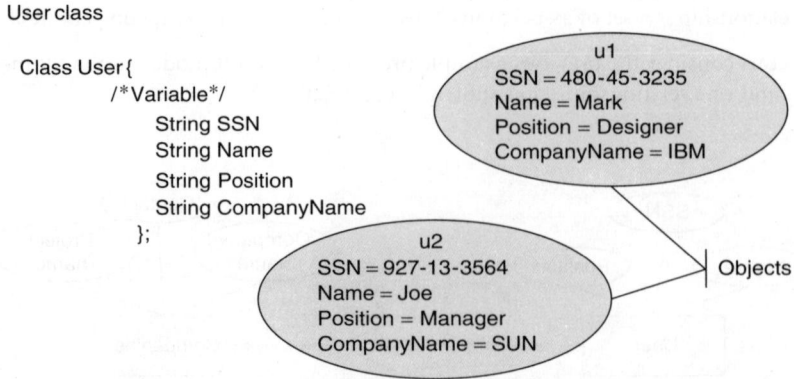

FIGURE 20.5 Object-oriented data model.

```
1.interface User  (Key(SSN)) {

2.       attribute string SSN;

3.       attribute string Name;

4.       attribute string Position;

5.       relationship Companies has ;

6.             inverse Companies :: Work-In

7. }

8. interface Companies (Key(CompanyName){

9.       attribute string CompanyName;

10.      attribute string ProjectName;

11.       attribute string Contact;

12.      relationship Set <User> Work-In

13.             inverse User :: has

14. }
```

FIGURE 20.6   Object-oriented model in ODL.

Example 4   The classes *User* and *Companies* introduced in Example 1 are declared in ODL as shown in Figure 20.6.

For the class *User*, the first attribute shown on line (2) is SSN and has string as its type. Two more attributes are declared on lines (3) and (4). Lines (5) and (6) declare the relationship between User and Companies, i.e., each user works in only one company and each company has many users.

20.3.2.3   Object Query Language

Like SQL, object-oriented query language (OQL) is a standard for database objects querying. Object-oriented query language has an SQL-like notation for the object-oriented paradigm. It aims to provide resolution for object-oriented programming languages, such as C++, Smalltalk, or Java. Objects are thus manipulated both by OQL queries and by the OO programming language.

Example 5   Suppose *u1* is an instance of the class "User." The syntax used to ask for the *Position* of User who has the name "Mark" is as follows:

SELECT u1.Position FROM User u1 WHERE u1.Name = "Mark"

20.4   Web Technology and Data Representation

The World Wide Web (Web) is a powerful platform for the delivery and dissemination of data-centric, interactive applications. Most documents on the Web are currently stored and transmitted in hypertext markup language (HTML). Simplicity is the advantage of HTML, yet it cannot meet the growing needs of users who want to make their documents more attractive and dynamic. In an attempt to satisfy this requirement, extensible markup language (XML) has been, as the standard for data representation and exchange on the web.

20.4.1   Hypertext Markup Language

Hypertext markup language is one of the markup languages for tagging a document so that it can be published on the Web. It is a simple, powerful, platform-independent language (Berners-Lee and Connolly, 1993). Hypertext markup language was originally developed by Tim Berners-Lee. It was standardized in November 1995 and is now commonly referred to as HTML2. The language has evolved and the World Wide Web Consortium (W3C) currently recommends HTML4.01, which has mechanisms for frames, style sheets, scripting, and embedded objects (W3C, 1999). A portion of an HTML document and the corresponding page viewed through a web browser are shown in Figure 20.7.

20.4.2   Extensible Markup Language

It is interesting to note that the growth of Internet application has driven the applications of XML. An XML document is a database only in the strictest sense of the term. That is, it is a collection of data. In many ways, this makes it no different from any other file — after all, all files contain data of some sort. As a "database" format, XML has some advantages. For example, it is self-describing (the markup describes the structure and type names of the data, although not the semantics), it is portable (Unicode), and it can describe data in tree or graph structures. It also has some disadvantages. For example, it is verbose, and access to the data is slow owing to parsing and text conversion.

Example 6   Let us consider the previous example, now presented using XML. In this XML structure, the root element name is Project Management, which could be viewed as the same as the name of a database. There are two types of the supplement User and Companies, which are very like the tables that we used for the database in the previous sections.

```
<?xml version="1.0"?>
<ProjectManagement>
        <User>
    <SSN> 480-45-3235</SSN>
    <Name> Mark</Name>
    <Position> Designer</ Position >
    <CompanyName>IBM</ CompanyName >
```

The code of an HTML file

```
<HTML>

<TITLE> The first html page </TITLE>

<p align="center">

    <B>This is The first HTML page!</B>

</p>

<P align="center">

I can write in <I>Italic</I> or
<B>Bold</B><BR>

</HTML>
```

The display of this HTML file in a browser

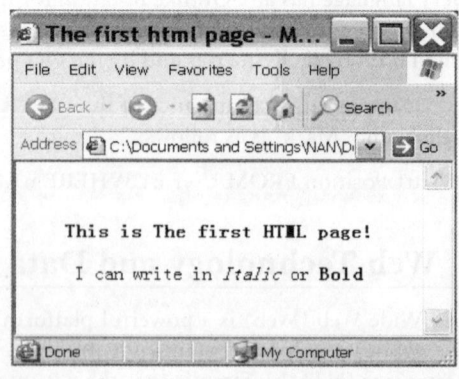

FIGURE 20.7   HTML example.

```
           </ User >
           <User>
              <SSN> 927-13-3564</SSN>
              <Name> Joe</Name>
              <Position> Manager</ Position >
              <CompanyName>SUN</ CompanyName >
           </ User >
           <Companies>
              <CompanyName>IBM</ CompanyName >
              <ProjectName>CDP Project</ ProjectName >
              <Contact>info@ibm.com</ Contact >
        </ Companies >
        <Companies>
              <CompanyName>SUN</ CompanyName >
              <ProjectName>Jini Project</ ProjectName >
              <Contact>cs@sun.com</ Contact >
        </ Companies >
    </ ProjectManagement >
```

Any XML with correct syntax is called a well-formed XML document. The XML syntax can be validated against document type definitions (DTDs) or XML schemas, thus generating a valid XML.

20.4.2.1 XML Document Type Definitions

Document type definition defines the valid syntax of an XML document by specifying the element name, which elements can appear in combination with which other ones, how elements are nested, what attributes are available for each element type, etc. The purpose of a DTD is to define the syntax structure of an XML document.

Example 7 In order to provide a valid XML for the previous example, a possible DTD file is shown as following:

```
<!DOCTYPE ProjectManagement [
<!ELEMENT ProjectManagement (User, Companies)>
<!ELEMENT User (SSN+, Name, Position, CompanyName)>
<!ELEMENT SSN (#PCDATA)>
<!ELEMENT Name (#PCDATA)>
<!ELEMENT Position (#PCDATA)>
<!ELEMENT CompanyName (#PCDATA)>
<!ELEMENT Companies (CompanyName+, ProjectName, Contact)>
<!ELEMENT CompanyName (#PCDATA)>
<!ELEMENT ProjectName (#PCDATA)>
<!ELEMENT Contact (#PCDATA)>
]>
```

As we can see from the example, DTD is not written in XML syntax. In addition, DTD has no support for namespaces and only offers limited data types. The XML schema overcomes the limitations and is more expressive.

20.4.2.2 XML Schema

Unlike DTD, XML schema is an XML-based language. It describes a model that defines the possible arrangement of tags and text in a document. As XML schema documents are XML documents, they can be edited and processed with the same tools that users use to process other XML documents. As opposed to DTD, XML schemas some with a new range of features, including the following. (1) Richer data types, such as Booleans, numbers, dates and times, URLs, integers, decimal numbers, real numbers, intervals of time, etc. (2) User-defined types called *archetypes*. An archetype allows users to define their own named data type. (3) Attribute grouping: it allows the schema author to make this relationship explicit. (4) Refinable archetypes, or "inheritance." A refinable content model is the middle ground: additional elements may be present, but only if the schema defines what they are. (5) Namespace support. (W3C, 1999).

Example 8   Let us rewrite the DTD file in the previous example based on XML schema.

```
<?xml version="1.0" encoding="ISO-8859-1" ?>
<xs:schema xmlns:xs="http://www.w3.org/2001/XMLSchema">
<xs:element name=" ProjectManagement ">
<xs:complexType>
<xs:sequence>
<xs:element name="User" maxOccurs="unbounded">
     <xs:complexType>
     <xs:sequence>
     <xs:element name="SSN " type="xs:string"/>
     <xs:element name="Name" type="xs:string"/>
     <xs:element name="Position" type="xs:string"/>
     <xs:element name="CompanyName" type="xs:string"/>
     </xs:sequence>
     </xs:complexType>
</xs:element>
<xs:element name=" Companies " maxOccurs="unbounded">
     <xs:complexType>
     <xs:sequence>
     <xs:element name=" CompanyName " type="xs:string"/>
     <xs:element name=" ProjectName " type="xs:string"/ >
     <xs:element name=" Contact " type="xs:string"/ >
     </xs:sequence>
     </xs:complexType>
</xs:element>
</xs:sequence>
</xs:complexType>
</xs:element>
</xs:schema>
```

20.4.2.3   XML and Database

Extensible markup language can be used to exchange data that are normally stored in a backend database. There are two major relationships between the XML and the database: (1) mapping between the XML structure and database structure and (2) mapping between the XML file and database tables.

Based on the multitiered architecture of the information system, the mapping between the XML structure to database structure is shown in Figure 20.8.

Example 9   Let us take a look at the following example. As in the example shown below, the root element of the XML data object, Project Management could be mapped to the name of a database. The two types of the supplement User and Companies can be mapped to the tables. The information system's store/query procedures can then be built up based on this mapping.

An XML file can also be directly stored in the form of a database. An XML based data object has to be retrieved and related to the tables in the database. Example 10 illustrates how this is achieved.

Example 10   Let us use the tables shown in Example 1 and map them to XML files.

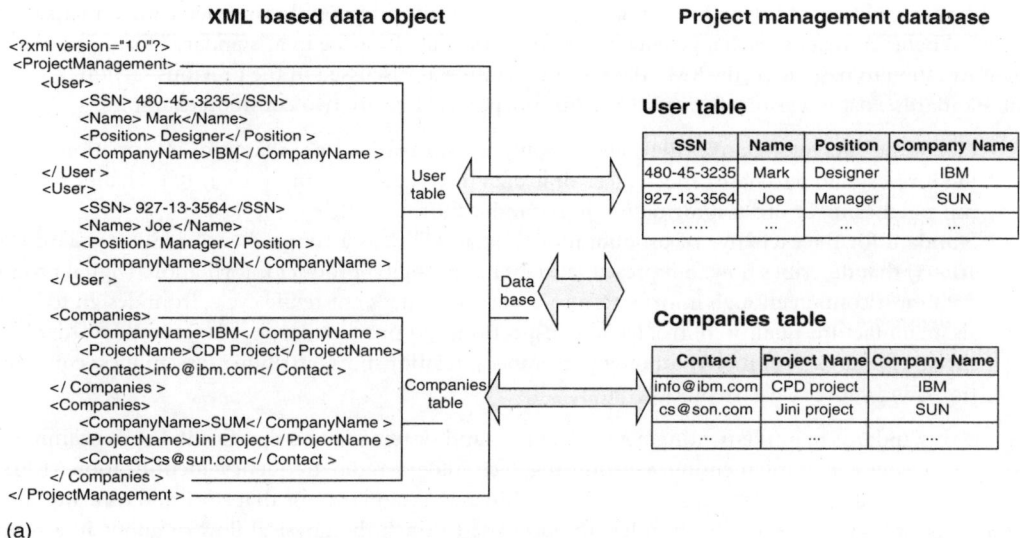

FIGURE 20.8 Mapping between XML structure and database structure. (a) Project management mapped to the name of a database. (b) Project management database.

As the amount of information stored, exchanged, and presented is increased, the ability to intelligently query XML data sources becomes important. There are many existing XML query languages. A relatively simple one is XQL. XQuery is another popular XML query language designed to meet the requirements identified by the W3C XML Query Working Group.

20.5 Industrial Data Standard

The issue of a data standard arises when dealing with data and databases and when using information systems to integrate dispersed entities. The National Institute for Standards and Technology (www.nist.gov), states that "standards are essential elements of information technology-hardware, software, and networks. Standard interfaces, for example, permit disparate devices and applications to communicate and work together. Standards also underpin computer security and information privacy, and they are critical to realizing many widespread benefits that advances in electronic and mobile commerce are anticipated to deliver." Whenever systems exchange data, the form of the data must be in a "standard" form in order for the information to pass correctly. XML data is one example as discussed in the previous section.

Data standards exist in various forms for various purposes. Here are two common examples:

- Electronic data interchange (EDI) allows computer-to-computer exchange of information regarding invoices and purchase orders. Electronic data interchange is simply a way to exchange or transmit purchasing or order information in a standard format.
- Standard for the exchange of product model data (STEP) is a comprehensive ISO standard (ISO 10303) that describes how to represent and exchange digital product information. Digital product data must contain enough information to cover a product's entire life cycle, from design to analysis, manufacture, quality control testing, inspection, and product support functions. In order to do this, STEP covers geometry, topology, tolerances, relationships, attributes, assemblies, configurations, and more (www.steptools.com).

The role of standards in implementing new technology and systems can be significant. Let us examine the current challenges in implementing a promising technology: radio frequency identification (RFID). Radio frequency identification units are tiny microchip and antennae units that store and transmit information. Radio frequency identification has the promise to track the physical flow of goods in a supply chain into an integrated information system. This information can be used to better manage supply chain and material flow. A recent study indicated that companies are developing RFID based on their own standards, which means tags developed by one company may not be readable by another company (Handfield and Dhinagaravel, 2004). Therefore, the lack of a data standard may impede the large-scale implementation of RFID technology. Indeed, the lack of standards is often cited as one of the biggest reasons the technology has not been more widely implemented.

20.6 Information System Applications

20.6.1 Types of Information Systems

For most businesses, there are different requirements for information. Senior managers need information for business planning. Middle management needs more detailed information to monitor and control business activities. Operational employees need information to carry out fundamental duties (Tutor2u, 2004). Thus, various information systems are needed in a company, ranging from lower level transaction processing systems (TPSs), knowledge work systems (KWSs) and office automation systems, management information systems and decision support systems to higher level executive support systems, as may be seen in Table 20.1.

20.6.2 Transaction Processing System

Transaction processing systems are designed to process routine transactions efficiently and accurately. There are basically two types of TPSs: manual and automated. Both are used to process the detailed data and update records about the fundamental business operations. In a TPS, data is processed, classified, sorted, edited based on routine operations, and finally stored. TPS reports are created as the output (shown in Figure 20.9).

20.6.2.1 Advanced Data Exchange for Top of the Tree Bakery Company

Top of the Tree Bakery Company (www.gordonpies.com) is a small fruit pie distributor based in Londonderry, New Hampshire. The company supplies apple, cherry, and other "Gordon's Pies" to major grocery chains located in the northeastern United States. In the past, Top of the Tree relied on paper-based transaction methods such as fax to exchange purchase orders and invoices with its brokers, suppliers, and customers. Such transaction methods cost Top of the Tree huge amounts of time and money. In mid-1999, the company decided to use advanced data exchange (ADX, www.adx.com), a TPS to automate the

TABLE 20.1 Types of Information Systems

System Type	Application Level	Application Users	Examples
TPS	Operational level	Operational employees	Order processing, machine control
KWS	Knowledge level	Employees working on knowledge and data	Computer-aided design, computer-aided manufacture
Office automation system	Knowledge level	Employees working on knoledge and data	Word processing, email systems
MIS	Management level	Middle managers	Inventory control, sales management
Decision support	Management level	Middle managers	Cost analysis, pricing analysis, production scheduling
Executive support system	Strategic level	Senior managers	Long-term operating plan, profit planning, manpower planning

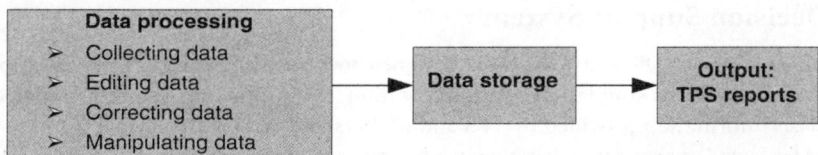

FIGURE 20.9 Transaction processing system.

purchase orders and invoices. With ADX, Top of the Tree conducts EDI-based transactions with several of its major business partners. The company reduced the time it took to send and receive business transactions, saving as much as 13 h per month in personal labor by instituting automated transactions. In its first year using ADX, Top of the Tree reduced its use of paper by 60–80% (Advanced Data Exchange, 2001).

20.6.3 Knowledge Work (Management) System

"Knowledge management (KM) is an effort to increase useful knowledge within the organization. Ways to do this include encouraging communication, offering opportunities to learn, and promoting the sharing of appropriate knowledge artifacts" (McInerney, 2002). Knowledge Management Systems (KMS) are developed to help companies create and share information, in particular, new knowledge and expertise. Knowledge management systems are usually built around systems, which permit efficient distribution of knowledge. For example, documents such as Word and Excel contain knowledge that can be shared by using group collaboration systems such as intranet. Although the TPS and KWS provide different functions and will be used by different kinds of users, their system architectures could be very similar since both of those systems concentrate on the "information" collection, storage, sharing, and display. The structure of a KWS is shown in Figure 20.10.

Mitre Information Infrastructure for Mitre Corporation: Mitre Corp. is an independent, nonprofit company that provides federal agents with system engineering and information technology expertise. In June 1994, Mitre Corp. began to shift the company to using intranet, which provides an accessible corporate knowledge base and intellectual collaboration. In May 1995, Mitre debuted its Mitre Information Infrastructure (MII) company-wide. To date, MII increased the knowledge base; enhanced expertise sharing, in addition, MII enabled Mitre Corp. to save $16.6 million in labor and material costs since 1996, saving $12.8 million in improved staff productivity (Young, 2000).

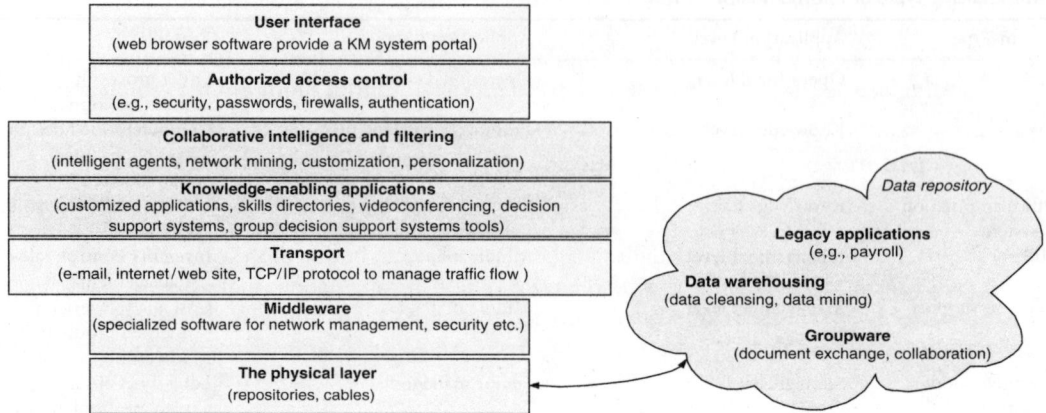

FIGURE 20.10　Knowledge management system.

20.6.4　Decision Support System

Decision support systems (DSSs) are specially designed to help management make decisions by using tools and techniques such as simulation and optimization. DSSs often analyze the options and alternatives on the basis information provided by TPS and KWS (shown in Figure 20.11).

NCR Teradata for Iowa Department of Revenue: In a typical year, the Iowa Department of Revenue processes more than 4 million documents as part of its tax-collection function. In 2003, the state collected in excess of $5.9 billion from individuals and businesses and issued $672 million in refunds (Teradata Staff, 2004). Recognizing the important role of innovation and technology to agency efficiency and strong customer service, the Department of Revenue embarked on a data warehouse journey in November 1999 with its decision to invest in a Teradata® Warehouse. As of December 2003, the State of Iowa has collected more than $34 million in unpaid taxes since deploying its warehouse. According to Teradata, the Teradata enterprise data warehouse has helped Iowa's Department of Revenue realize the following benefits:

- *Optimized revenue and promoted voluntary compliance.* Directly related to this program an additional $35 million in tax revenues has been generated since implementation in early 2000. The program continues to build a strong revenue stream with anticipated revenues of at least $10 million annually.
- *Improved accuracy and greater efficiency.* The integrated database structure enables the matching of data from diverse source systems to perform sophisticated queries with drill-downs to analyze and verify the data to identify and generate improved and more accurate tax gap leads. Value-added data is now available to the state auditors and examiners for each audit lead, providing them with a single view of data to support their work activities. The data warehouse structure supports a repository for the automated movement of data to and from other key legacy systems.
- *Increased management reports.* The data warehouse stores a wealth of data and through the use of query and analysis tools, the department has increased its ability to measure results and obtain the data to support key business questions and further enhance programs and activities.

20.6.5　Management Information System

A management information system (MIS) is mainly concerned with internal sources of information. It organizes and summarizes the data from TPS and provides information for managing an organization based on the generated statistical summaries, exception reports, some analysis and projections, routine decisions such as short-term planning, monitoring internal events, and control of routine work. MIS also provides information for DSSs where information requirement can be identified in advance. Figure 20.12 illustrates the structure of an MIS.

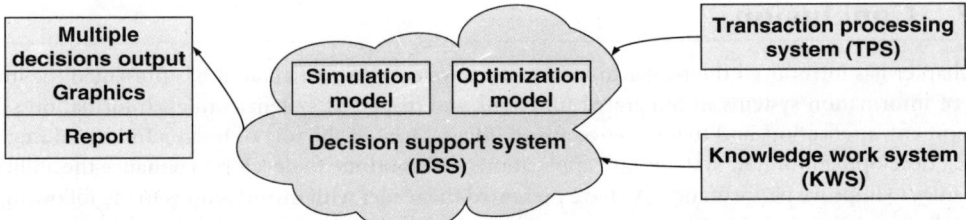

FIGURE 20.11   Decision support system.

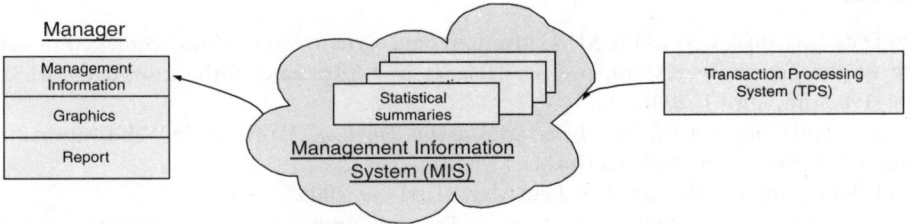

FIGURE 20.12   Management information system.

ShipitSmarter for ING Group: ING Group (www.ing.nl) is a financial institution of Dutch origin with worldwide operations. It has 115,000 employees who offer 60 million customers in 60 countries banking, insurance and asset management, products, and services. ING Group has been using e-marketplace ShipitSmarter (www.shipitsmarter.com) to procure shipping and courier services since 2001. According to Otto Nutzel, a logistics consultant,

> "ShipitSmarter has improved the quality of logistical activities within ING Group enormously. We now have a Management Information System that gives us the information we need. We know precisely what is being shipped every day, by which service provider and at what price. The system has allowed us to reduce the cost of shipping considerably, consolidate our procurement volume and pay for the kilometers calculated on www.locatienet.nl; in other words, we no longer pay for extra kilometers caused by diversions, waiting times or other problems."

20.6.6   Executive Support System

An executive support system (ESS) is designed to help senior managers make strategic decisions. It accepts input from TPS, MIS, DSS, and other information systems and provides easy access to information for monitoring business results and overall business conditions. An ESS integrates customized analysis and presentation of key indicators, status, trends, and exceptions.

20.6.6.1   Executive Viewer for Johnsonville Sausage

Johnsonville Sausage (www.johnsonville.com) has manufactured premium sausage products since 1945. It is the nation's leading brand of bratwurst, Italian Sausage, and fresh breakfast links. In order to provide the external sales force with necessary information to remain competitive, Johnsonville created a data warehouse. Once implemented, the need to provide remote users an easy to use, web-deployed reporting and analysis system was apparent. Johnsonville turned to Temtec (www.temtec.com) for the solution. Executive Viewer from Temtec is a web-based product that is flexible enough to support all levels of users and perform well in an extranet environment. The major benefits of Executive Viewer to Johnsonville Sausage are: reduction of reporting cycles for large ROI, increased reaction time to market trends, and effortless installation and distribution to users, according to Ron Gilson, CIO.

20.7   Conclusion

This chapter has introduced the reader to information systems. In particular, it has presented the importance of information systems in integrated functions and dispersed system entities. Information system development, application, and maintenance are an integral part of the role of today's Industrial Engineer. The fast development of new systems and applications will continue to develop to enhance the abilities of companies to improve performance. We have presented the reader with introductions to the following topics: database systems, data representation, data standards, and distributed information systems. Finally we have presented some illustrative examples of the application of information systems in industry.

References

Advanced Data Exchange, Easy as Pie ADX, advanced data exchange automates business transactions for top of the Tree Bakery Company, available at http://itresearch.forbes.com/detail/RES/987703-102_198.html, April 1, 2001.

Berners-Lee, T. and Connolly, D., *The Hypertext Markup Language*, World Wide Web Consortium, available at http://www.w3.org/MarkUp/MarkUp.html, 1993.

Carr, N., IT doesn't matter, *Harvard Bus. Rev.*, May, 81, 41–49, 2003.

Connolly, T. and Begg, C., *Database Systems: A Practical Approach to Design, Implementation, and Management*, Pearson Education Limited, Harlow, England, 2005.

Handfield, R. and Dhinagaravel, A., Establishing the Business Case for RFID: A status Report and Industry Outlook, Working paper for the Supply Chain Resource Consortium, North Carolina State University, 2004.

Hogan, R., *A Practical Guide to Data Base Design*, Prentice-Hall, Englewood Cliffs, 1990.

Institute of Industrial Engineers, available at http://www.iienet.org/public/articles, 2004.

ITToolbox, *2003 IT Spending Survey*, http://sap.ittoolbox.com/pub/research/spending_survey.htm, 2003.

Maynard, H.B., Ed., *Industrial Engineering Handbook*, McGraw-Hill, New York, 1971.

McInerney, C., Knowledge management and the dynamic nature of knowledge, *J. Am. Soc. Inf. Sci. Technol.*, 53, 1009–1018, 2002.

Stephens, R. and Plew, R., *Sams Teach Yourself Beginning Databases in 24 Hours*, Sams Publishing, Indianapolis, India, 2003.

Ted Codd, E.F., A relational model of data for large shared data banks, *Commun. ACM*, 13, 377–387, 1970.

Teradata Staff, *Closing the Tax Gap in Iowa*, posted at DSSResources.COM, http://dssresources.com/cases/, February, 2004.

Tutor2u, http://www.tutor2u.net/business/ict/intro_information_system_types.htm, 2004.

Young, D., An audit tale, *CIO Magazine*, 46, May, 2000, available at http://www.cio.com/archive/050100/mitre.html

W3C, *HTML 4.01*, World Wide Web Consortium Recommendation 24 December 1999, Available at http://www.w3.org/TR/html4, 1999.

21

Dependability of Computer and Network Systems

Nong Ye
Arizona State University

21.1 Introduction

Industrial Engineering (IE) and computer/network technologies are intertwined in many ways, but usually in the direction of applying computer/network technologies to help solve IE problems. These applications cover a wide range, from design, operation, quality, reliability, user interface, and system needs to other aspects of traditional IE systems, such as those in the manufacturing and service industries. For example, computers and artificial intelligence techniques such as genetic algorithms have been employed as tools to solve optimization problems. Computer databases have been used to manage product, manufacturing, inventory, and enterprise data. Computer-aided design (CAD) and computer-aided manufacturing (CAM) have helped and enhanced product design and manufacturing processes. Networking technologies have facilitated information sharing and integration for collaborative and global enterprise systems. Web-based technologies have enabled marketing and sales via e-commerce. Computer-aided media (e.g., touch and hearing devices) have improved human access to information (especially for people with disabilities in seeing and hearing) and operator performance. In the context of taking computer/network technologies to IE, we use computer/network technologies as tools to assist us in carrying out intensive computation for solutions to optimization problems, information storage and access, data analysis and modeling, and so on.

Although we have greatly benefited from taking computer/network technologies to IE, we have not done much toward using IE to shape computer/network technologies. Work in the direction of applying IE to improve computer/network technologies has been done mostly in the field of human–computer interaction, that is, taking human factors (HF) into account in the design of user interfaces for computer and network systems. Not only can human factors and ergonomics knowledge be applied to enhance the external user interface of computer and network systems (the usability aspect), but other areas of IE knowledge for managing the internal operation of systems can also play a significant role in shaping computer/network technologies to overcome many existing problems in the dependability, performance, and other aspects of computer and network systems.

In this chapter, we present our work in the direction of taking IE to shape computer and network technologies, especially for overcoming problems related to the internal operation of computer and network systems. Our work addresses the lack of service dependability in computer and network systems by using the knowledge and methodologies from operations research (OR)/optimization, quality/reliability engineering, and process control. We hope that by introducing such work we will demonstrate the great benefits for the IE community of putting more effort into this direction — a direction that is still not widely recognized in the IE community but that has a great potential for bringing IE into the core of the computer and telecommunication industries. In Section 21.2, we describe the dependability problem of computer and network systems as well as a subset of the dependability problem related to the security of such systems. In Sections 21.3 and 21.4, we present examples of our work on admission control and job scheduling to achieve the minimization of job waiting time variance for service stability and dependability of computer and network systems. We also discuss our work on process analysis and monitoring to detect cyber attacks for the security of computer and network systems. In Section 21.5, we provide a summary.

21.2   Dependability of Computer and Network Systems

In this section, we first introduce the concepts related to the performance and quality of service (QoS) of computer and network systems. We then highlight the common weaknesses in performance and QoS dependability in computer and network systems and the contributing factors to this problem.

21.2.1   Performance and QoS in Computer and Network Systems

Any system, including a computer or network system, consists of two basic elements: resources and processes (Ye, 2002). A resource provides a service that responds to a request for that service, much as a library provides a book to a reader. The process handles the input and output. For example, a web server is a computer/network resource that provides a response to a request to download information for a web client. A web request presents the input (e.g., the name or link of the web document requested) to the process representing the web request. The process produces the output — in this example, the web document retrieved and sent back to the web client. Another example of a network resource is a router, which provides the service of transmitting data packets (units of data traveling on computer networks) from one location to another. The input to a process representing the request of transmitting the data packet is the data packet itself received at the input port. The output of the process is the data packet sent out through an output port.

On computers and networks, we have information processing resources such as the CPU; the operating system; the spreadsheet, word-processing, and other software programs; information-storage resources such as memory, hard drive, files and databases; and communication resources such as routers, network interface cards, communication cables, and I/O resources (e.g., the keyboard, printer, monitor, and their associated software programs). Multiple processes may request services from a resource at the same or different times. A given resource is understood to be in a certain state at any given time. We are usually interested in three important aspects of the resource state: availability, confidentiality, and integrity (Ye, 2002).

The availability of the resource state reflects the available or usable capacity of a resource in relation to that resource's maximal capacity; the available capacity in turn determines the responsiveness or ability of the resource to respond to service requests. For example, if data packets currently going through an output port of a router use only 20% of the bandwidth capacity of the port (measured by bits transmitted per second[bps]), the router still has the capacity and responsiveness to transmit more data packets through that output port by using the remaining 80% of its bandwidth capacity. Another example can be seen in a router's internal buffer, which holds incoming data packets before transmitting them to an output port. The availability of the router's buffer can be measured by the amount of data packets held in the buffer relative to the buffer's maximum holding capacity.

The second attribute, the integrity of the resource state, is related to the correctness of a process output produced by a resource. For example, if a hacker breaks into a router to make the router send all data packets passing through it to the address of the hacker's computer rather than their original destination addresses, the output of a data packet being sent to an incorrect destination address shows the compromised integrity of the router. The correctness of the resource depends on the internal functioning of the resource. A resource may also produce the correct output for a process but may take a very long time owing to the low availability of the resource, which shows the difference between the availability and integrity attributes of the resource state.

The third attribute of the resource state, confidentiality, is related to the precise amount of output that the resource produces. The confidentiality of the resource also depends on the internal functioning of the resource. For example, if a web server on a computer is compromised by a hacker so that the hacker, through a web request, gets not only the requested web page but also the password file containing passwords for all user accounts on the computer, the output is more than what should have been produced for the web request, and the confidentiality of the computer is compromised. The earlier example of a router sending all data packets to an incorrect destination address also shows the compromised confidentiality of the router, because a correct destination receives 0% rather than 100% of its data.

The above examples demonstrate that a change of the resource state has, as well, an important impact on the output performance of a process. There are three attributes of the output performance, corresponding to the availability, integrity and confidentiality attributes of the resource state. They are, respectively, timeliness, precision, and accuracy. Timeliness measures how long it takes to produce the output. Precision measures how much output is produced, as related to the quantity of the output. Accuracy measures the correctness of the output, related to the quality of the output. For example, when a system's CPU — an information processing resource — services a process, the availability attribute of the CPU state changes because less CPU time becomes available. The availability state of the CPU in turn affects the timeliness attribute of the output performance for all processes sharing the CPU time. Activities in a computer or network system consist of user activities that initiate processes and receive services as well as resource operations that provide services. User activities to initiate processes change the state of resources, and resource state changes in turn produce impact on the output performance of processes. Hence, in a computer or network system there exist both resource-process interactions and activity-state-performance interactions.

The output performance of processes determines the QoS received by processes (Ye, 2002). Hence, the timeliness, precision, and accuracy attributes of the output performance are also the attributes of QoS. We often see such measures as delay, response time, and jitter (reflecting the variance of delay) used to describe the timeliness attribute of QoS; throughput, bandwidth, and loss rate (e.g., packet drop rate) represent the precision attribute of QoS; and error rate describes the accuracy attribute of QoS.

21.2.2 Contributing Factors to Lack of Dependability in Computer and Network Systems

Computers and networks are currently not dependable in terms of QoS. With regard to the timeliness of QoS and availability of resource, for example, the response time to the same web request may vary according to when the web request is issued. No guarantee is provided as to when an email message will be received after it is sent. Denial of service (DoS) attacks have happened to a number of commercial web sites, rendering web servers unavailable to provide services and support business transactions. With regard to the accuracy of QoS and the integrity of a resource, we have seen cases in which the appearance of U.S. Government web sites was modified by hackers. With regard to the precision of QoS and the confidentiality of a resource, both identifiable and unidentifiable incidents of stolen computer documents and information have occurred. Further, the integrity of individuals' email programs and the confidentiality of their email address books have been compromised, permitting unsolicited messages to be sent to people in their email address books.

The lack of QoS dependability in computer and network systems can be attributed to two major causes: design drawbacks and cyber attacks. Although the phenomenon of cyber attacks has become known to the general public (Skoudis, 2002), the design drawbacks that compromise QoS dependability in computer and network systems (even when no cyber attacks are present) are less obvious. Currently, most computer and network systems are designed using the best-effort model with the First-In-First-Out (FIFO) method for job scheduling and no admission control (Gevros, et al., 2001).

For example, with FIFO, a router services data packets on the basis of their arrival times, so a given data packet waits in a queuing buffer while the router is providing services to data packets that arrived before it. Since there is no admission control, all the arriving data packets are admitted into the router, which holds them in the queuing buffer. If a data packet arrives but the buffer is full, it is dropped by the router. Consequently, the waiting time of a data packet in the queuing buffer is variable because it depends on the number of data packets that arrive ahead of it, and that number varies over time. Thus, the timeliness of the service that a data packet receives from the router is not stable. If a data packet is dropped because the queuing buffer is full, the output— transmitting the data packet — is zero, thus affecting the precision of service that the data packet receives from the router. Hence, the FIFO protocol and the lack of admission control in the best effort model for the router lead to a lack of QoS stability or dependability, not to mention the possibility of cyber attacks on the router causing other effects on QoS stability and dependability.

The lack of QoS dependability in computers and networks makes them a vulnerable platform, an unsettling prospect considering how much an increasing number of business transactions, critical missions, and other important activities depend on those computers and networks. QoS dependability is a "must" for next-generation computers, networks, and the information infrastructure. The following sections show examples from our work of using IE to move computer and network technologies toward achieving the goal of QoS dependability in computers and networks.

21.3   Engineering for QoS Dependability

In this section, we present three examples from our work that address QoS dependability. The first example draws on HF data as well as computer and network technology data to determine QoS metrics for various computer and network applications. The second example advances the admission control method and the optimization method for job scheduling in order to minimize the variability of waiting times for computer and network jobs and thus to achieve service stability and dependability in the timeliness of QoS. The third example applies a conventional feedback-control method to reduce the variability and thus stabilize the job waiting time for greater QoS stability and dependability. Our other related work for QoS dependability can be found in the following publications: Park et al. (2004), Wu et al. (2005), Ye et al. (2003a), Ye et al. (2005a), Ye et al. (2005b), Ye and Farley (in press-a), Ye et al. (in review-a), Ye et al. (in press-a).

21.3.1   QoS Metrics Based on Data from Human Factors and Computer/Network Technology

Various computer and network applications have different service features and different QoS requirements. For example, email applications have a low bandwidth requirement, and are more sensitive to data loss than to delay or jitter, but a network application involving mainly voice data such as voice-over IP, while it also has a low bandwidth requirement, is more sensitive to delay and jitter than to data loss. To assure QoS dependability, we must be clear about the QoS requirements of various computer and network applications that specify the desired QoS. If no QoS requirements are provided and attached to applications, any level of QoS should be considered acceptable since no expectation of QoS is set. Hence, to assure QoS dependability on computers and networks, we must first understand and specify the QoS requirements of various computer and network applications.

Human factors, an area of IE, is one of the key factors in determining QoS metrics since HF reveals how users perceive QoS from computers and networks to be acceptable. For example, as recommended by the International Telecommunication Unit G.114, a phone-to-phone delay should be no more than 150 msec to allow for appropriate and easy understanding (Chen, et al., 2003). Outside of this limit, the user will become annoyed and find the service unacceptable. This HF datum is applicable both to voice-over IP applications and to other applications involving voice transmission. Computer and network technology is another key factor that needs to be taken into account in determining QoS metrics. For example, when considering the delay requirement of the voice-over IP application, we need to take into account the bandwidth capacity of routers and communication cables on the path of voice data transmission. The bandwidth capacity of a Cisco 12000 Series Internet router goes up to 10G bps, whereas a Cisco 7100 Series Internet Router supports 140 Mbps.

Since HF data and computer/network technology data exist in separate fields, they need to be put together in the context of various computer and network applications to determine quantitative QoS metrics. In our previous work (Chen et al., 2003), we first classified a number of common computer and network applications on the basis of their technology attributes of time dependence (real-time vs. non-real-time) and symmetry (symmetric vs. asymmetric in volumes of request and response data). These applications included web browsing, email, file transfer, telnet, Internet relay chat, audio broadcasting, video broadcasting, interactive audio on demand, interactive video on demand, telemetry, teleconferencing, videophone, and voice-over IP. We then extracted and evaluated HF data and computer/network technology data related to those applications and put together the QoS metrics for each of these applications as the starting point for the computer and network industry to establish QoS requirement standards. Table 21.1 shows QoS metrics (Chen et al. 2003) that we established for audio broadcasting, which is a real-time, asymmetric application on computer networks. Details of this work and QoS metrics for other computer and network applications can be found in Chen et al. (2003).

21.3.2   Admission Control and Job Scheduling through Optimization

In general, the time that a job takes, in a computer or network system, to receive service from a resource includes the waiting time and the processing time. The processing time is the time the resource takes to process the job after removing it from a waiting queue. The size of the job usually determines its processing time, whereas the waiting time depends on admission control, which determines whether or not to admit a job into the system, and job scheduling, which determines the order in which a resource will service jobs admitted into the system.

Let us first look at admission control and its effect on the job waiting time. We indicated in Section 21.2.2 that a lack of admission control results in variable waiting times for the jobs in a computer or network system, which in turn leads to variable response times, delays, and jitters of those jobs, and unstable or undependable timeliness of QoS. To reduce the variance of job waiting times, we have developed an admission control method (Ye et al., in review-b) that we call batch scheduled admission control (BSAC).

A computer or network system with BSAC decides whether or not to admit an incoming job into the system, thereby also controlling job admission based on this decision. With the use of the BSAC method, the system admits dynamically arriving jobs in batches. Many criteria can be considered to determine the size of each batch. For example, the number of jobs in the batch can be used to define its size if the processing

TABLE 21.1   QoS Metrics for Audio Broadcasting

Timeliness			Precision			Accuracy
Response time (sec)	Delay (msec)	Jitter (msec)	Data rate (bps)	Bandwidth (bps)	Loss rate	Error rate
2–5	<150	<100	56–64 K	60–80 K	<0.1%	<0.1%

time does not vary much among jobs. The total load of the processing times of all the jobs in the batch can also be used to define the batch size if there is a large variance in the processing times of jobs. Take the example of a router. We may define a batch size of 19,531 small (64 byte) data packets (jobs) or limit the batch size to a 10 MB total load, which both translate to approximately 1 sec total processing (data transmission) time load for the router output with the bandwidth capacity of 10 Mbps. We recommend defining the batch size by the total load of the processing times of all the jobs in the batch to bound or limit the waiting time of any individual job so that it is not greater than the time required to process the full batch.

At a given time, the system maintains two batches, a current batch and a waiting batch. The resource in the system takes the jobs in the current batch one by one for service according to their scheduled order. During the processing of the jobs in the current batch, arriving jobs, if admitted, are placed in the waiting batch. For an incoming job, the system admits the job if adding it to the waiting batch does not produce a batch whose length exceeds the batch size; otherwise, it rejects the job. The system may advise the sender of a rejected job to resubmit it later or send the job to another system with a similar resource. When the resource has finished processing all the jobs in the current batch, the jobs in the waiting batch are moved into the current batch for processing.

The BSAC method reduces the variance in job waiting times and thus improves the waiting time stability because BSAC bounds and stabilizes the total load of the jobs in each batch, and at any given time a bounded batch is processed for service. Each job's waiting time includes the waiting time in the waiting batch as well as the waiting time in the current batch, since each job in the current batch still needs to take its turn for service by the resource. The waiting times of the jobs in the current batch are different, but are bounded owing to the bounded batch size. Hence, the waiting time of each job in a computer or network system with BSAC is bounded and stabilized, as we demonstrate in our test results (Ye et al. in review-b).

The jobs in the current batch must follow a scheduled order to receive the service from the resource. The job scheduling problem for minimizing the variance of the jobs' waiting times is called the waiting time variance (WTV) problem. We formulate the WTV problem as a quadratic integer programming problem. The decision variable is s_{ij}'s, for $i = 1, 2, ...n$ and $j = 1, 2, ...n$, representing a job sequence as well as the position of each job in the job sequence. The binary integer variable, s_{ij}, takes the value of 1 if job j is scheduled at position i, and 0 otherwise. There are n positions in the job sequence since there are n jobs. The job to be scheduled and thus processed first is placed at position 1, the job to be scheduled second is placed at position 2, and so on. The processing time of job j is p_j, which is given. The waiting time of the job at position i is w_i.

Objective function:

$$\text{minimize} \quad \frac{1}{n-1} \sum_{k=1}^{n} \left(w_k - \frac{1}{n} \sum_{i=1}^{n} w_i \right)^2 \tag{21.1}$$

$$\text{subject to} \quad \sum_{j=1}^{n} s_{ij} = 1, \forall j, \quad i = 1, ..., n, \tag{21.2}$$

$$\sum_{i=1}^{n} s_{ij} = 1, \forall j \quad j = 1, ..., n, \tag{21.3}$$

$$s_{ij} = 0 \text{ or } 1, \quad i, j = 1, ..., n, \tag{21.4}$$

$$w_1 = 0, \tag{21.5}$$

$$w_i = w_{i-1} + \sum_{j=1}^{n} s_{i-1,j} {}^* p_j, \quad i = 2, ..., n. \tag{21.6}$$

The objective function is to minimize the sample variance of the waiting times of n jobs. Equation (21.2) describes the constraint that there can be only one job to be assigned to each position. Equation (21.3) indicates that one job can be placed at only one position. Equation (21.4) gives the integer constraint. The waiting time of the first job to be processed is 0, which is given in Eq. (21.5). Equation (21.6) defines the waiting time of the job at position i ($i \geq 2$), which is the waiting time of the job at position $i-1$ plus the processing time of the job at position $i-1$. Since the WTV problem is NP-hard, we develop two job scheduling methods, called Verified Spiral (VS)[*] and Balanced Spiral (BS), to schedule jobs in a given batch for minimizing the variance of the jobs' waiting times.

On the basis of the work by Schrage (1975) and Hall and Kubiak (1991), in an optimal sequence for the WTV problem, the first longest job is scheduled last, the second longest is first, and the third longest is next to last; in a dual optimal sequence, the first longest job is scheduled last, the second longest is next to last, and the third longest is first. Several studies (Merten and Muller, 1972; Eilon and Chowdhury, 1977; Mittenthala et al. 1995) have proved the V-shape property of optimal job sequences for the WTV problem. This V-shape property of a job sequence states that the jobs before the job with the shortest processing time are scheduled in descending order of their processing times, and the jobs after the job with the shortest processing time are scheduled in ascending order of their processing times.

Eilon and Chowdhury (1977) developed two job scheduling methods for the WTV problem based on the V-shape property and Schrage's conjecture about the placement of the first four longest jobs in the optimal sequences of the WTV problem. We modify their methods by first incorporating Hall and Kubiak's (1991) proof about the placement of only the first three longest jobs; and by enhancing the spiral placement of the remaining jobs by verifying whether to place the next job before the tail or after the head of the job sequence, depending on which position produces a smaller variance of the waiting times for the jobs already in the job sequence, as follows.

Suppose an arbitrary job set $p = \{p_1, p_2, \ldots, p_n\}$ for a batch needs to be scheduled for a single resource. Assume that the jobs are numbered such that $p_1 \leq p_2 \leq \cdots \leq p_n$.

1. To start, first place job p_n in the last position, job p_{n-1} in the last-but-one position, and then job p_{n-2} in the first position. Now the right side of the job sequence has p_{n-1} and p_n, or $R = \{p_{n-1}, p_n\}$. The left side of the job sequence has p_{n-2} or $L = \{p_{n-2}\}$. Place the smallest job, p_1, in between L and R. The job pool has the remaining jobs, $\{p_2, \ldots, p_{n-3}\}$.
2. Remove the longest job from the job pool, place the job either before R or after L and then add the job to either R or L, depending on which position produces a smaller variance of subschedule in the job sequence so far (placing the job at the position producing the smaller variance of waiting times).
3. Repeat Step 2 until the job pool is empty.

The VS method requires more computation than the method developed by Eilon and Chowdhury (1977) owing to the verification of the waiting time variances for the two possible positions of placing a job from the job pool.

To reduce the computational cost associated with the VS method, the BS method replaces the verification of the waiting time variances for two possible positions of placing a job in VS. This replacement is by simply maintaining the balance of L and R in the total processing time of jobs in L and R while placing a job from the job pool. The steps for the replacement are as follows:

1. To start, first place job p_n in the last position, and job p_{n-1} in the last-but-one position, and then job p_{n-2} in the first position. Let $L = \{p_{n-2}\}$ and $R = \{p_{n-1}\}$. Note that p_n is not included in R. We denote the sum of the processing times of the jobs in L and R as SUM_L and SUM_R, respectively. The job pool has the remaining jobs $\{p_1, p_2, \ldots, p_{n-3}\}$.
2. Remove the largest job from the job pool. If $SUM\_L < SUM\_R$, place the job after L, add the job to L, and update SUM_L; or if $SUM\_L \geq SUM\_R$, place the job before R, add the job to R, and update SUM_R.
3. Repeat Step 2 until the job pool is empty.

---

[*] A patent has been filed by Arizona State University for the verified spiral method under the name of YELF (YE, Li, and Farley) Spiral.

By testing and comparing VS and BS with the four other job scheduling methods (FIFO, shortest processing time first, and the two methods developed by Eilon and Chowdhury [1977]), we show that VS consistently produces the best performance for the WTV problems tested in our study (Ye et al. in review-a), followed by BS, with less computational cost which may be considered for practical use on computers and networks owing to its computational efficiency and comparable performance with that of VS.

21.3.3 Feedback Control for Stabilizing Job Waiting Time

Conventional process control methods such as feedback control can also help in stabilizing the job waiting time for QoS dependability. In Yang et al. (in press), the proportional-integral-differential (PID) control is employed to develop a QoS model for a router as shown in Figure 21.1. We compare and test the QoS model of a router with a differentiated service (DiffServ) model of a router.

The DiffServ model considers two classes of service: high- and low-priority service. Data packets arrive at the router with the mark of their service class. The DiffServ model aims at providing service differentiation between two classes of data packets to assure QoS to high-priority data packets, and providing the best-effort service to low-priority data packets only after satisfying the QoS need of high-priority data packets. There are two input ports and one output port in the router model. At each input port, the admission control mechanism applies the token bucket model (Almquist, 1992; Blake, et al., 1998) to high-priority data packets in order to control the admission of high-priority data packets into the router. Admission control is applied to high-priority data packets to assure the availability of the resource in the router to provide the premium service to the admitted high-priority data packets. There is no admission control for low-priority data packets because (1) low-priority data packets are served only when there are no high-priority data packets waiting for the resource in the router and (2) QoS of low-priority data packets is not of concern and is provided on the best-effort basis.

In the token bucket model, admission control is determined by two parameters: token rate r and bucket depth p. Token rate r dictates the long-term rate of the admitted traffic, and bucket depth p defines the maximum burst amount of the admitted traffic. Any incoming data packets that cause the token rate and

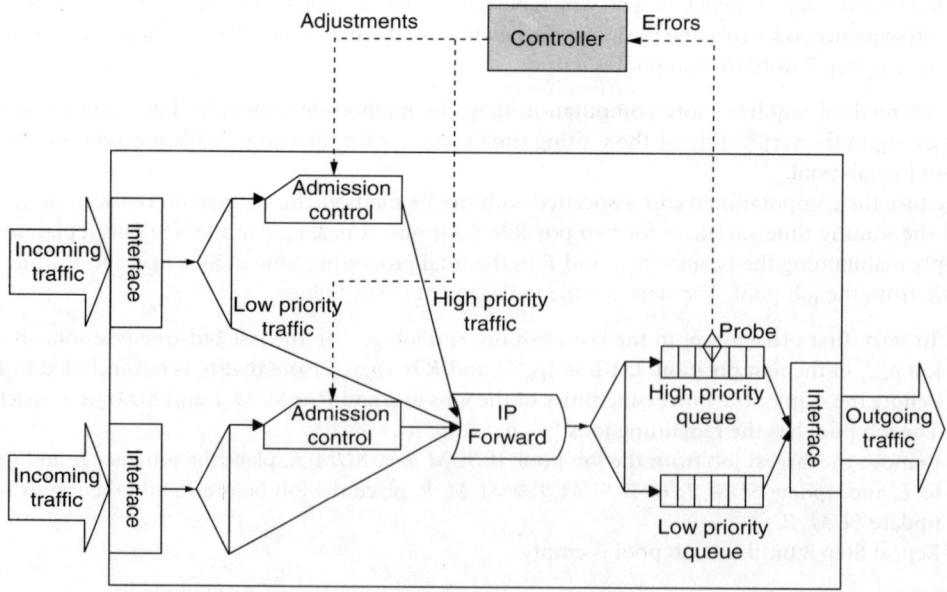

FIGURE 21.1 A QoS model of a router with PID feedback control (From Yang, Z., Ye, N., Lai, Y.-C., *Qual. Reliab. Eng. Int.*, in press).

the bucket depth of the admitted traffic to exceed r and p, respectively, are not admitted into the router and are dropped by the admission control mechanism. The admission control mechanism in the DiffServ model of a router uses a fixed token rate, r, and a fixed bucket depth, p.

There are two queuing buffers, a high- and a low-priority queuing buffer, for the output port of the router to keep the admitted data packets in before their transmission through the output port. The sizes of the two queuing buffers are predetermined and fixed based on the traffic characterization and bandwidth allocation between the two classes of data packets. The admitted high-priority data packets are placed into the high-priority queuing buffer to form a queue, and, similarly, the incoming low-priority data packets are placed into the low-priority queuing buffer to form a queue. The capacity of the high-priority queuing buffer is usually small in order to bind the delay of the high-priority data packets. The output port first serves the data packets in the high-priority queue, as long as the queue is not empty. The output port serves the data packets in the low-priority queue only when the high-priority queue is empty. Hence, the two separate queuing buffers together provide the mechanism in the basic QoS model of a router to enforce service differentiation between the two classes of data packets. For each queue, the FIFO queuing discipline is applied to determine the order of serving data packets in the queue. If there is not enough space in each queuing buffer to take in a data packet, the data packet is dropped.

In the QoS model with the PID feedback control, the feedback control loop is applied to the high-priority queuing buffer because the QoS of low-priority data packets is not of concern and is provided on a best-effort basis. The QoS model of the router monitors the usage state of the high-priority queuing buffer and accordingly adjusts the token rate adaptively in order to prevent packet drop and transmission delay, and thus, a compromised QoS. The usage state of the high-priority queuing buffer is measured by the total length of data packets (queue length) in the buffer in relation to its size or capacity. If at any given time we keep the total length of data packets in the high-priority queuing buffer within a certain range (e.g., less than the size of the high-priority queuing buffer by at least the maximum length of one data packet), packet drop and loss will not occur. By setting an upper bound of the queue length, packet drop and loss can be prevented and transmission delay can be bounded since transmission delay is correlated with the queue length. A detailed description of the QoS model with the PID feedback control can be found in Yang et al. (in press), along with the testing results of the DiffServ model and the PID-based QoS model for purposes of comparison.

21.4 Engineering for Computer and Network Security

Layers of defense can be set up against cyber attacks on computers and networks through prevention, detection, and reaction (Ye et al., 2001a). Firewalls, authentication, and cryptography are some examples of the prevention mechanisms. Although attack prevention can increase the difficulty of launching successful attacks, it cannot completely block skilled and determined attackers with considerable resources from breaking into computers and networks. The purpose of attack detection is to capture cyber attacks while they are acting on a computer and network system. Successful reaction to a detected attack includes stopping the attack in progress, diagnosing the origin and path of the attack, assessing the impact of the attack on the system state and performance, correcting the vulnerability of the system, and returning the system to normal operation. Our work focuses on cyber attack detection.

Existing techniques for cyber attack detection fall into three general approaches: signature recognition, anomaly detection, and attack-norm separation (Ye et al., 2001b; Ye and Farley, in press-b). Signature recognition techniques match the observed system data against the signatures of known attackers and signal an attack when there is a match. For a subject (user, file, privileged program, computer, network, etc.) of interest, anomaly detection techniques establish a profile of the subject's normal behavior (norm profile), compare the observed behavior of the subject with its norm profile, and signal attacks when the subject's observed behavior deviates significantly from its norm profile. Attack-norm separation techniques consider cyber attack data as cyber signals and normal operation data as cyber noises. When an attack occurs on a computer and network system, there are usually normal user activities going on in the system at the same time. Hence, the data collected from the system contains a mixture of both cyber attack data (cyber signals) and normal use data (cyber noises). To detect cyber signals mixed with cyber noises, attack-norm separation techniques define both the model of norm data (cyber noise) and the model of attack data (cyber signal), use the model of cyber noise to separate or filter out cyber noise data from the observed system data, and then use the model of cyber signal to check the presence of cyber signal in the remaining of the observed system data after noise separation or filtering. Therefore, signature recognition employs only knowledge of cyber attacks (cyber signals), whereas anomaly detection employs only knowledge of normal use (cyber noises). Attack-norm separation employs knowledge of both cyber attack data and normal use data for detection accuracy and efficiency.

In this section, we first describe the χ^2—square distance monitoring method (CSDM) as an anomaly detection technique that we have developed to detect cyber attacks based on the principle of multivariate statistical process control (SPC). We then present a System-Fault-Risk framework to analyze cyber attacks for deriving data, features and characteristics of activity, and state and performance data in the attack condition — cyber attack data for the attack-norm separation approach. Our other related work, such as the application of exponentially weighted moving average (EWMA) control charts, Markov chain models and data mining techniques to cyber attack detection, can be found in Emran and Ye (2001); Ye et al. (2002a), Ye and Chen (2003), Ye et al. (2004, 2002b), Li and Ye (2001, 2002), Ye and Li (2002), and Ye (2001, 2003).

21.4.1 χ^2-Square Distance Monitoring Method for Cyber Attack Detection

Before we introduce the CSDM method for cyber attack detection, we describe the observed system data that the CSDM method monitors to detect cyber attacks. In most existing work on cyber attack detection, two sources of data are widely used: network traffic data and computer audit/log data (Ye et al., 2001b). Network traffic data contain data packets traveling on computer networks. Computer audit/log data capture events occurring on computers.

We now show an example of computer audit data that we used in some of our studies (Ye and Chen, in review, Borror, et al., in review, Ye et al., 2001b, Ye and Chen, 2001, Li and Ye, 2001, 2002, in press, Emran and Ye, 2002, Ye and Li, 2002, Ye et al., 2002b, 2002c, 2004) to illustrate how the CSDM method monitors such data to detect cyber attacks. The audit data come from a Sun SPARC 10 workstation using the Solaris UNIX operating system. The Solaris operating system has an auditing facility, called the Basic Security Module (BSM), which monitors and records the events related to the security of a system.

A BSM audit record for each event contains a variety of information, including the event type, user ID, group ID, process ID, session ID, the system object accessed, and so on. We extract and use the event type from the record of each audit event. There are 284 different types of BSM audit events from Solaris 2.5. Hence, activities on a computer are captured through a continuous stream of audit events, each of which is represented by the event type. To determine whether or not the current event at time t is part of an attack, we examine the events in the recent past of the current event (including the current event) to obtain the EWMA (exponentially weighted moving average) smoothed frequencies of the 284 event types in this recent past of the current event and represent the smoothed event frequencies by a vector $(X_1, X_2, \ldots, X_{284})$, where X_i represents the smoothed frequency of event type i. Specifically, the EWMA data smoothing method works as follows (Ye et al., 2001):

$$X_i(t) = \begin{cases} \lambda^*1 + (1 - \lambda)\,^*X_i(t-1) & \text{if the audit event at time } t \text{ belongs to the } i\text{th event type} \quad (21.7) \\ \lambda^*0 + (1 - \lambda)\,^*X_i(t-1) & \text{if the audit event at time } t \text{ is different from the } i\text{th event type} \quad (21.8) \end{cases}$$

$X_i(t)$ is the observed value of the ith variable in the vector of an observation at time t, λ is a smoothing constant that determines k or the decay rate, and $i = 1, \ldots, 284$. The most recent observation at time t receives a weight of λ, the observation at time $t-1$ receives a weight of $\lambda(1-\lambda)$, and the observation at time $t-k$ receives a weight of $\lambda(1-\lambda)^k$. Hence, $X_i(t)$ represents an exponentially decaying count of event type i, measuring the intensity of event type i in the recent past of the current event. A multivariate observation, $(X_1, X_2, \ldots, X_{284})$, represents the intensity distribution of various event types. In our study, we initialize $X_i(0)$ to 0 for $i = 1, \ldots, 284$, and set λ to 0.3. Hence, for each audit event in an event stream from the computer, we obtain a vector, $(X_1, \ldots, X_{284})$.

A conventional multivariate SPC method such as Hotelling's T^2 control chart (Ryan, 2000; Montgomery, 2001) can be applied to monitoring $(X_1, ..., X_{284})$ for anomaly detection. We first take a look at the application of Hotelling's T^2 control chart to cyber attack detection (Ye et al., 2001b, 2002c, Ye and Chen, 2001; Emran and Ye, 2002). Let $X = (X_1, X_2, ..., X_p)'$ denote an observation of p measures on a process or system at time t. Assume that when the process is operating normally (in control), the population of X follows a multivariate normal distribution with the mean vector μ and the covariance matrix Σ. When using a data sample of size n, the sample mean vector $\overline{X}$ and the sample covariance matrix S are used to estimate μ and Σ, where

$$\overline{X} = (\overline{X}_1, \overline{X}_2, ..., \overline{X}_p) \tag{21.9}$$

$$S = \frac{1}{n-1}\sum_{i=1}^{n}(X_i - \overline{X})(X_i - \overline{X})' \tag{21.10}$$

Hotelling's T^2 statistic for an observation X_i is determined by the following:

$$T^2 = (X_i - \overline{X})'S^{-1}(X_i - \overline{X}) \tag{21.11}$$

A large value of T^2 indicates a large deviation of the observation X_i from the in-control population. We can obtain a transformed value of the T^2 statistic, $(n(n-p)/p(n+1)(n-1))T^2$ which follows an F distribution with p and $n–p$ degrees of freedom. If the transformed value of the T^2 statistic is greater than the tabulated F value for a given level of significance, α, then we reject the null hypothesis that the process is in control (normal) and thus signal that the process is out of control (anomalous).

An out-of-control anomaly from Hotelling's T^2 control chart can be caused by a shift from the in-control mean vector (mean shift), a departure from the in-control covariance structure or variable relationships (counterrelationship), or combinations of the two situations. Hotelling's T^2 control chart detects both mean shifts and counterrelationships.

Hotelling's T^2 control chart requires the computation of the inverse of the variance–covariance matrix. For a large number of process variables (e.g., 284 variables for cyber attack detection), it is impossible to keep this large matrix in the computer memory for efficient computation. For many real-world data sets, it is often difficult to compute the inverse of the variance–covariance matrix owing to the colinearity problem among variables. Hence, the inverse of the variance–covariance matrix in Hotelling's T^2 control chart presents high computational cost and difficulty for complex process data with a large number of monitored variables.

Since Hotelling's T^2 statistic essentially represents a statistical distance of an observation from a statistical profile of an in-control process, we develop the following χ^2—square distance (Ye et al., 2001b, 2002c; Ye and Chen, 2001; Emran and Ye, 2002) to overcome the problems with Hotelling's T^2 statistic:

$$X^2 = \sum_{i=1}^{p}\frac{(X_i - \overline{X}_i)^2}{\overline{X}_i} \tag{21.12}$$

This test statistic measures the distance of a data point from the center of a data population. When the p variables are independent and p is large (usually > 30), the X^2 statistic follows approximately a normal distribution according to the central limit theorem, regardless of what distribution each of the p variables follows. Using a sample of X^2 values, the mean and standard deviation of the X^2 population can be estimated from the sample mean $\overline{X^2}$ and the sample standard deviation S_{X^2}. The in-control limits to detect out-of-control anomalies are usually set to 3-sigma control limits as determined by $[\overline{X^2} - 3S_{X^2}, \overline{X^2} + 3S_{X^2}]$. Since we are interested in detecting significantly large X^2 values for cyber attack detection, we need to set only the upper control limit to $\overline{X^2} + 3S_{X^2}$ as a signal threshold. That is, if X^2 for an observation is greater than $\overline{X^2} + 3S_{X^2}$, we signal an anomaly.

When applying the CSDM method to cyber attack detection, we use the training data of an event stream obtained in the normal operation condition to estimate $\overline{X}$ in Eq. (21.12) and then compute the value of X^2 for each event in the testing data obtained in the attack condition to detect cyber attacks. Our testing results show comparable performance of Hotelling's T^2 control chart and the CSDM method in accuracy of detecting cyber attacks (Ye et al., 2001b, 2002c; Ye and Chen, 2001; Emran and Ye, 2002).

Our further studies (Ye et al. 2003b, in press-b) on these two multivariate SPC methods indicate that the CSDM method performs either better than or as well as Hotelling's T^2 control chart in detecting the out-of-control anomalies of mean shifts, counterrelationships, and their combinations for multivariate process data with uncorrelated and normally distributed data variables, auto-correlated and normally distributed data variables, and nonnormally distributed data variables without correlation or autocorrelation, regardless of application fields. Hotelling's T^2 control chart has been shown (Mason, et al., 1997; Mason and Young, 1999) to be sensitive to the normality assumption, and hence it is not able to perform as well as the CSDM method for nonnormally distributed process data. Only for multivariate process data with correlated and normally distributed data variables, Hotelling's T^2 control chart is superior to the CSDM method for detecting the mean shifts and counterrelationships. However, for correlated and normally distributed process data, it is computationally efficient to discover underlying independent variables from a large number of original data variables by using techniques such as principal component analysis and then applying the computationally efficient CSDM method to monitor a few uncorrelated variables. On the basis of those findings and the advantages of the CSDM method in computational efficiency and scalability, the CSDM method merits serious consideration as a solid, viable multivariate SPC technique for monitoring large-scale, complex process data (e.g., computer and network data for cyber attack detection), especially when uncorrelated, auto-correlated, or nonnormally distributed variables are present.

21.4.2 System-Fault-Risk Framework to Analyze Data, Features, and Characteristics

Because the attack-norm separation approach to cyber attack detection employs knowledge of both the cyber attack data signal and the cyber norm data (Ye and Farley, in press-b; Ye and Chen, in review), the attack-norm separation approach has an advantage in detection accuracy over the anomaly detection approach and the signature recognition approach, which employ knowledge of the cyber norm data or the cyber attack data only. The attack-norm separation approach requires a clear scientific understanding of three elements: data, features, and characteristics, for building sensor models of detecting cyber attacks (cyber signals). Figure 21.2 illustrates these three elements along with a sensor model of detecting cyber attacks. In the figure, raw data (e.g., network traffic data) collected from computers and networks go through data processing to obtain the desired data (e.g., the intensity ratio of packets for the web server to all packets for all network applications) from which the feature is extracted using a feature extraction method (e.g., an arithmetic calculation of the sample average). The sensor model incorporates the characteristics of both the cyber attack data and the cyber norm data (e.g., a step increase of the feature from cyber norm data to cyber attack data) and monitors the feature to detect the characteristics of the cyber attack data mixed with the cyber norm data to decide if an cyber attack is present.

On the basis of the fault modeling and risk assessment theories in combination with the system model of resource-process interactions and activity-state-performance interactions described in Section 2.1, we have developed the system-fault-risk framework (Ye et al., in review-c) to assist in the discovery of data, features and characteristics of the cyber attack and norm data. Fault modeling and risk assessment are important parts of reliability engineering that we study in IE. A fault has a propagation effect in a system, involving activity-state-performance interactions, as discussed in Section 21.2.1. In general, three factors contribute to any risk: asset, vulnerability, and threat (Fisch and White, 2000). Computer and network resources are assets. These assets have vulnerabilities that can be exploited by hackers who may be disposed to launch cyber attacks. Figure 21.3 shows the SFR framework. Table 21.2 gives some examples of cyber attacks analyzed using the SFR framework. The analysis and classification of cyber attacks based on the SFR framework will help to develop a set of data, features, characteristics, and sensor models to detect a group of similar cyber attacks.

In fault modeling, a fault such as a cyber attack can be modeled in a cause-effect chain or network of activity, state change, and performance impact, all occurring in the system during the fault effect propagation. An

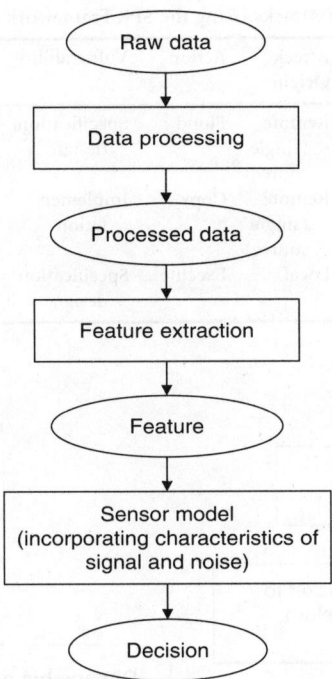

FIGURE 21.2 Data, features, characteristics, and sensor models in the attack-norm separation approach to cyber attack detection.

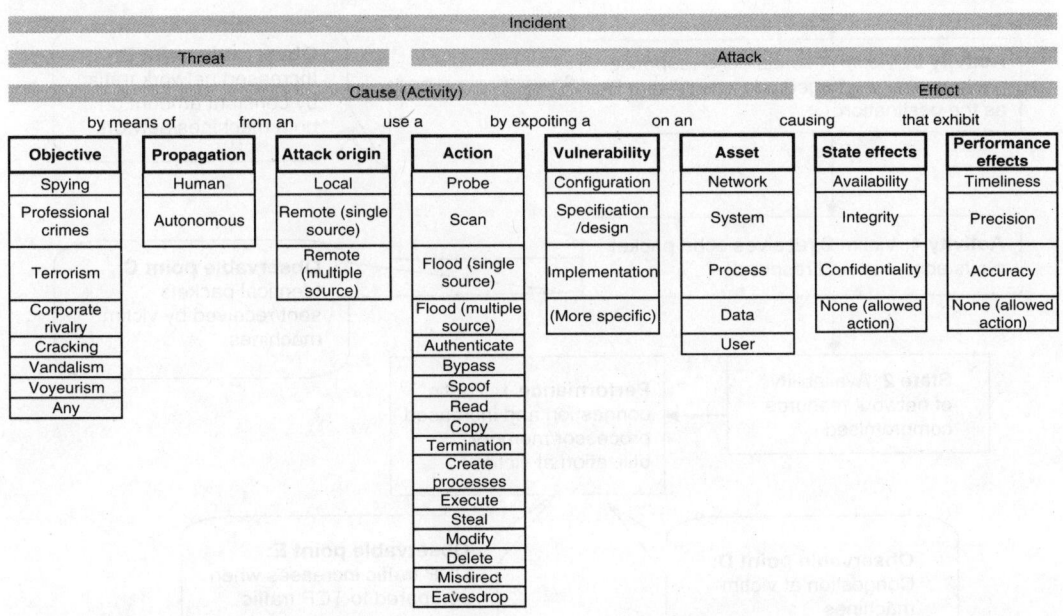

FIGURE 21.3 The System-Fault-Risk framework for attack analysis and classification (From Ye, N., Newman, C., and Farley, T., Information Systems Frontiers, in review).

TABLE 21.2   Analysis of Selected Cyber Attacks Using the SFR Framework

Attack Name	Objective	Propaga-tion	Attack Origin	Action	Vulnerability	Asset	State Effect	Performance Effect
UDP storm	Any	Human	Remote (single source)	Flood	Specification/design	Network	Availability	Timeliness
Slammer worm phase I	Cracking	Autono-mous	Remote (single source)	Copy	Implemen-tation	Process	Integrity	Accuracy
Slammer worm phase II	Cracking	Autono-mous	Local	Execute	Specification/design	System	Integrity	Accuracy

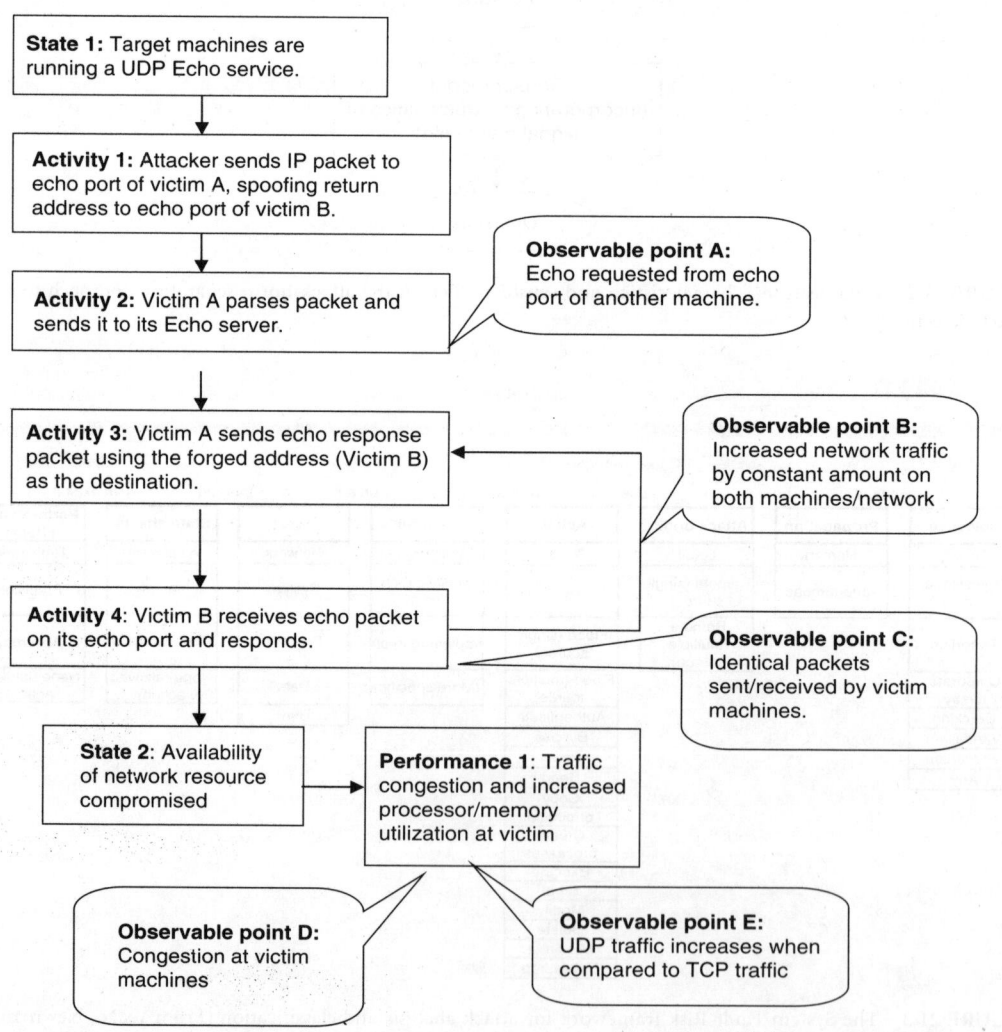

FIGURE 21.4.   Cause-effect chain of a UDP storm attack on computer networks. (From Ye, N., Newman, C., and Farley, T., *Information Systems Frontiers*, in review).

attacker's activities cause the state change of computer and network resources, which in turn produces performance impact (e.g., performance error of processes or degraded QoS). For example, the cause–effect chain of a UDP Storm DoS attack is shown in Figure 21.4. The UDP Storm attack is a DoS attack that causes network congestion and slowdown. When a connection is established between two UDP services, each of which produces output, these two services can produce a very high number of packets that can lead to a denial of service on the machine(s) where the services are offered. By examining each step of the cause-effect chain for the attack, we analyze the activities, state changes or performance impacts involved in each step to identify the activity, state, and performance data that are useful to detect the attack. We then extract the feature and characteristic of the feature that enables the distinction of the cyber attack data from the cyber norm data. Table 21.3 gives a description of the observation points derived from the cause-effect chain of the UDP storm attack. Table 21.4 shows the data, features, and characteristics for these observation points that enable the distinction of the cyber attack data from the cyber norm data. In Table 21.4, the location of L1 represents the victim computer. The location of L2 represents the router on the network of the victim machine.

21.5 Summary

This chapter presents some examples of our work, which introduces IE principles into the design and operation control of computer and network systems. Specifically, we demonstrate the great potential of using IE to shape computer and network technologies through our research work, which employs HF, optimization/OR, and Quality/Reliability engineering to design computer and network systems for QoS dependability and to develop sensor models for cyber attack detection.

TABLE 21.3 The Description of Observation Points for the UDP Storm Attack

OBS	Description
A	Echo requested from echo port of another machine.
B	Increased network traffic by constant amount on both machines/network
C	Identical packets sent/received by victim machines.
D	Congestion at victim machines
E	UDP traffic increases when compared to TCP traffic

Source: From Harish, B., M.S. thesis, Department of Computer Science and Engineering, Arizona State University, 2004.

TABLE 21.4 Data, Features, and Characteristics for the Observation Points of the UDP Storm Attack

OBS	Location	Data	Feature	Characteristic
A	L1	SRC port and DEST port of received UCP packet	Individual observation	Has value 8 (echo port)
	L2	SRC port and DEST port of received UCP packet on network	Individual observation	Has value 8 (echo port)
B	L2	IP packets received/sec with DEST IP having victims IP address.	Exponentially weighted moving average	Step increases in both victims by same amount
C	L1	Similarity scores of pair-wise observation of consecutive IP packet contents	Exponentially weighted moving average	Step Increase
D	L1	IP packets received/sec from performance log	Exponentially weighted moving average	Step Increase
	L2	IP packets received/sec to victim IP	Exponentially weighted moving average	Step Increase
E	L1	Ratio of UDP packets/sec to TCP packets/sec	Exponentially weighted moving average	Step increase
	L2	Ratio of UDP packets/sec to TCP packets/sec to victim IP	Exponentially weighted moving average	Step increase

Source: From Harish, B., M.S. thesis, Department of Computer Science and Engineering, Arizona State University, 2004.

Acknowledgments

This material is based upon work supported by the Department of Defense (DoD) and the Air Force Office of Scientific Research (AFOSR) under Grant No. F49620-01-1-0317; the Air Force Research Laboratory (AFRL) and Advanced Research and Development Activity (ARDA) under Contract No. F30602-03-C-0233; the Air Force Office of Scientific Research (AFOSR) under Grant No. F49620-03-1-0109 and Grant No. F49620-98-1-0257; AFRL under Grant No. F30602-01-1-0510 and Grant No. F30602-98-2-0005; the Defense Advanced Research Project Agency (DARPA) and AFRL under Grant No. F30602-99-1-0506; and gifts from Symantec Corporation. Any opinions, findings and conclusions, or recommendations expressed in this material are those of the author(s) and do not necessarily reflect the views of DoD, AFOSR, AFRL, ARDA, DARPA, or Symantec Corporation.

References

Almquist, P., Type of service in the Internet Protocol suite. Request for Comments 1349, Internet Engineering Task Force, July 1992. Available: http://www.ietf.org/rfc.html.

Blake, S., Black, D., Carlson, M., Davies, E., Wang, Z., and Weiss, W., An architecture for differentiated service. Request for Comments (Informational) 2475, Internet Engineering Task Force, Dec. 1998. Available: http://www.ietf.org/rfc.html.

Borror, C., Ye, N., Montgomery, D., and Chen, Q., CUSUM control chart for detecting computer intrusions, *Quality and Reliability Engineering International*, (in review).

Chen, Y., Farley, T., and Ye, N., QoS requirements of network applications on the Internet, *Inf. Knowledge Sys. Manage.*, 4, 55–76, 2003.

Eilon, S. and Chowdhury, I.D., Minimizing waiting time variance in the single machine problem, *Manage. Sci.*, 23, 567–574, 1977.

Emran, S.M. and Ye, N., Robustness of chi-square and Canberra techniques in detecting intrusions into information systems, *Qual. Reliability Eng. Int.*, 18, 19–28, 2002.

Emran, S.M. and Ye, N., "A system architecture for computer intrusion detection." *Inf. Knowledge Sys. Manage.*, 2, 271–290, 2001.

Fisch, E.A. and White, G.B., *Secure Computers and Networks: Analysis, Design and Implementation*, CRC Press, Boca Raton, FL, 2000.

Gevros, P., Crowcorft, J., Kirstein, P., and Bhatti, S., Congestion control mechanisms and the best effort service model, *IEEE Network*, 15 (3), 16–26, 2001.

Hall, N.G. and Kubiak, W., Proof of a conjecture of Schrage about the completion time variance problem, *Oper. Res. Lett.*, 10, 467–472, 1991.

Harish, B., Cyber Attack Profiling Using Cause Effect Networks, M.S. thesis, Department of Computer Science and Engineering, Arizona State University, 2004.

Li, X. and Ye, N., Detection of cyber attacks from network traffic data with mixed variables. *IEEE Trans. Systems, Man, Cybernet.* (in press).

Li, X. and Ye, N., Grid- and dummy-cluster-based learning of normal and intrusive clusters for computer intrusion detection. *Qual. Reliability Eng. Int.*, 18, 231–242, 2002.

Li, X. and Ye, N., Decision tree classifiers for computer intrusion detection. *J. Parallel Distributed Comput. Pract.*, 4, 179–190, 2001.

Mason, R.L. and Young, J.C., Improving the sensitivity of the T^2 statistic in multivariate process control. *J. Qual. Tech.*, 31, 155–164, 1999.

Mason, R.L., Champ, C.W., Tracy, N.D., Wierda, S.J., and Young, J.C., Assessment of multivariate process control techniques. *J. Qual. Tech.*, 29, 29, 1997.

Merten, A.G. and Muller, M.E., Variance minimization in single machine sequencing problems. *M. Sci.*, 18, 518–528, 1972.

Mittenthala, J., Raghavacharia, M., and Rana, A.I., V- and Λ-shaped properties for optimal single machine schedules for a class of non-separable penalty functions, *Eur. J. Oper. Res.*, 86, 262–269, 1995.

Montgomery, D.C., *Introduction to Statistical Quality Control*, Wiley, New York, 2001.

Park, K., Lai, Y.-C., and Ye, N., Characterization of weighted complex networks, *Phys. Rev. E*, 70, 026109-1–026109-4, 2004.

Ryan, T.P., *Statistical Methods for Quality Improvement*, Wiley, New York, 2000.

Schrage, L., Minimizing the time-in-system variance for a finite jobset. *Manag. Sci.*, 21, 540–543, 1975.

Skoudis, E., *ounter Hack*, Prentice-Hall, Upper Saddle River, NJ, 2002.

Wu, T., Ye, N., and Zhang, D., Comparison of distributed methods for resource allocation, *Int. J. Prod. Res.*, 43 (3), 515–536, 2005.

Yang, Z., Ye, N., and Lai, Y.-C., QoS model of a router with feedback control, *Qual. Reliab. Eng. Int.*, (in press).

Ye, N., QoS-centric stateful resource management in information systems, *Inf. Sys. Frontiers*, 4, 149–160, 2002.

Ye, N., Robust intrusion tolerance for information systems. *Inf. M. Compu. Security*, 9, 38–43, 2001.

Ye, N., Mining computer and network security data, in *The Handbook of Data Mining*, Ye, N., Ed., Lawrence Erlbaum Associates, Mahwah, NJ, 2003, pp. 617–636.

Ye, N. and Chen, Q., An anomaly detection technique based on a chi-square statistic for detecting intrusions into information systems. *Qual. Reliability Eng. Int.*, 17(2), 105–112, 2001.

Ye, N. and Chen, Q., Computer intrusion detection through EWMA for auto-correlated and uncorrelated data. *IEEE Trans. Reliability*, 52, 73–82, 2003.

Ye, N., Chen, Q., and Borror, C., EWMA forecast of normal system activity for computer intrusion detection, *IEEE Transactions on Reliability*, 53(4), 557–566, 2004.

Ye, N., Ehiabor, T., and Zhang, Y., First-order versus high-order stochastic models for computer intrusion detection, *Qual. Reliability Eng. Int.*, 18, 243–250, 2002b.

Ye, N. and Farley, T., Attack-norm separation – A new approach to cyber attack detection, *IEEE Compu.*, (in press-b).

Ye, N. and Farley, T., Information sharing and control in homogeneous and heterogeneous supply networks, *Int. J. Model. Simulation*, (in press-a).

Ye, N., Farley, T., and Aswath, D., Data measures and collection points to detect traffic changes on large-scale computer networks, *Inf. Knowledge Sys. Manage.*, (in press-a).

Ye, N., Giordano, J., and Feldman, J., A process control approach to cyber attack detection, *Commn. ACM.*, 44, 76–82, 2001a.

Ye, N., Borror, C., and Parmar, D., Scalable chi square distance versus conventional statistical distance for process monitoring with uncorrelated data variables, *Qual. Reliability Eng. Int.,* 19, 2003b, 505–515.

Ye, N., Borror, C., and Zhang, Y., EWMA techniques for computer intrusion detection through anomalous changes in event intensity. *Qual. Reliability Eng. Int.*, 18, 443–451, 2002a.

Ye, N. and Chen, Q., Attack-norm separation for detecting attack-induced quality problems on computers and networks, Quality and Reliability Engineering International, (in review).

Ye, N., Lai, Y.-C., and Farley, T., Dependable information infrastructures as complex adaptive systems, *Sys. Eng.*, 6, 225–237, 2003a.

Ye, N. and Li, X., A scalable, incremental learning algorithm for classification problems. *Compu. Ind. Eng. J.*, 43, 677–692, 2002.

Ye, N., Newman, C., and Farley, T., A System-Fault-Risk framework for cyber attack classification, *Information Systems Frontiers*, (in review-c).

Ye, N., Parmar, D., and Borror, C.M., A hybrid SPC method for monitoring large-scale, complex process data, *Qual. Reliability Eng. Int.*, (in press-b).

Ye, N., Yang, Z., and Lai, Y.-C., Enhancing router QoS through job scheduling with weighted shortest processing time—adjusted, *Compu. Oper. Res.*, 32(9), 2255–2269, 2005b.

Ye, N., Zhang, Y., and Borror, C.M., Robustness of the Markov-chain model for cyber-attack detection, *IEEE Trans. Reliability*, 53, 116–123, 2004.

Ye, N., Gel, E., Li, L., Farley, T., and Lai, Y.-C., Web-server QoS models: Applying scheduling rules from production planning, *Compu. Oper. Res.*, 32(5), 1147–1164, 2005a.

Ye, N., Emran, S.M., Chen, Q., and Vilbert, S., Multivariate statistical analysis of audit trails for host-based intrusion detection, *IEEE Trans. Compu.*, 51, 810–820, 2002c.

Ye, N., Harish, B., Li, X., and Farley, T., Batch scheduled admission control for service dependability of computer and network systems, *Information, Knowledge, Systems Management,* (in review-b).

Ye, N., Li, X., Chen, Q., Emran, S.M., and Xu, M., Probabilistic techniques for intrusion detection based on computer audit data. *IEEE Trans. Sys. Man Cybern.,* 31, 266–274, 2001b.

Ye, N., Li, X., Farley, T., and Xu, X., Job scheduling methods to reduce the variance of job waiting times on computers and networks. *Computers & Operations Research,* (in review-a).

PART V

NEW TECHNOLOGIES

PART V

NEW TECHNOLOGIES

22

Optimization Problems in Applied Sciences: From Classical through Stochastic to Intelligent Metaheuristic Approaches

Oye Ibidapo-Obe
University of Lagos, Nigeria

Sunday Asaolu
University of Lagos, Nigeria

22.1 Introduction

The desire for optimality (perfection) is inherent in humans. The search for extremes inspires engineers, scientists, mathematicians, and the rest of the human race to excel. Human development is motivated by faith and the zeal to excel. Faith is very fuzzy and fuzziness implies unpredictability. The goal of optimization theory is the creation of reliable methods to catch the extremum of a function by an intelligent arrangement of its evaluations (measurements). Optimization is vitally important for modern scientific investigations, engineering, and planning that incorporate optimization at every step of the complicated decision-making process.

Science makes living worth our while as it searches for ways and means of ensuring that life in the broader sense is more abundant. The methodology for knitting all these variables together is very intricate: this is the realm of mathematical analysis and, specifically, modeling. The language of nature is mathematics, while that of change is differential equations and probability, hence stochasticity. The use of mathematical models to support decision-making continues to make a great impact on both the public and private sectors' policies; it is the power of modeling that allows us to play God.

Optimization theory guides searches through valleys and through hills toward the peaks; it fights the curse of dimensionality and models evolution gambling and other human passions. The optimizing algorithms themselves are mathematical models of intellectual and intuitive decision-making.

Everyone who has studied calculus knows that an extremum of a smooth function is reached at a stationary point where its gradient vanishes. Some may also remember the Weierstrass theorem, which proclaims that the minimum and the maximum of a function in a closed finite domain do exist.

Optimization encompasses such areas as the theory of ordinary maxima and minima, the calculus of variations, linear and nonlinear programming, dynamic programming, maximum principles, discrete and continuous games, and differential games of varying degrees of complexity (Luenberger, 1984).

22.2   Optimization, Control, and Modeling

In this work, the methodical development of the optimization problem is examined in relation to applications in applied sciences from classical techniques through stochastic approaches to contemporary and recent methods of intelligent metaheuristics. Indeed, computational techniques have grown from classical techniques, which afford closed form solutions, to approximation and search techniques, and more recently to intelligent search techniques. Such intelligent heuristics include Tabu Search, Simulated Annealing (SA), Fuzzy systems, Neural Networks and Genetic Algorithms, among a myriad of evolving modern computational intelligent techniques. Furthermore, important applications of these heuristics to the evolution of self-organizing adaptive systems such as modern economic models, transportation, and mobile robots do exist (Gill et al., 1983; Glover and Macmillan, 1986).

Central to all problems in optimization theory are the concepts of payoff, controllers or players, and system and information sets. In order to define what one means by a solution to an optimization problem, the concept of payoff must be defined and the controllers must be identified. If there is only one person on whose decision the outcome of some particular process depends and the outcome can be described by a single quantity, then the meaning of payoff (and hence a solution to the optimization problem) and controller or player is clear.

The simplest, of course, is the problem of parameter optimization, which includes the classical theory of maxima and minima, linear and nonlinear programming. In parameter optimization, there is one (deterministic or probabilistic) criterion, one controller, one complete information set, and the system state described by static equations and inequalities in the form of linear or nonlinear algebraic or difference equations.

On the next rung of complexity are optimization problems of dynamic systems where the state is defined by ordinary or partial differential equations. These can be thought of as limiting cases of multistage (static) parameter optimization problems where the time increment between steps tends to 0. In this class, developed extensively in recent years as *optimal control*, we encounter the classical calculus of variation problems and their extension through various maximum and optimality principles, i.e., Pontryagin's minimum principle and the dynamic programming principle. We are still concerned with one criterion, controller and information set, but have added dimension in that the problem is dynamic and might be deterministic or stochastic.

The next level would introduce two controllers (players) with a single conflicting criterion. Here we encounter elementary or finite matrix game theory, where the controls and payoff are continuous functions. Each player at this level has complete information regarding the payoff for each strategy but may or may not have knowledge of his opponent's strategy. Such games are known as zero-sum games since the sum of the payoffs to each player for each move is zero: what one player gains, the other loses. If the order in which the players act does not matter, i.e., the minimum and maximum of the payoff are equal (this minimax is called the "value" of the game and is unique), the optimal strategies of the players are unaffected by knowledge or lack of knowledge of each other's strategy. A solution to this game involves the value and at least one optimal strategy for each player.

Next, we can consider extensions to dynamic systems where the state is governed by differential equations; we have one conflicting criterion and two players, i.e., zero-sum, two-player differential games. The information available to each player may be complete or incomplete. In cases with complete information, the finite game concept of a solution is directly applicable. For the incomplete information case, it is reasonable to expect mixed strategies to form the solution, but not much is known of solution methods or whether a solution always exists in the finite game theoretic sense.

In the last group or uppermost rung of the hierarchy, we identify a class of optimization problems where the concept of a solution is far from clear. To this class belong multiple criteria, n-person games with complete or incomplete information, non-zero-sum wherein either or both players may lose or gain simultaneously.

22.3 The Classical Methods

"*What's new?*" is an interesting and broadening eternal question but one which, if pursued exclusively, results only in an endless parade of trivia and fashion, the silt of tomorrow. I would like, instead, to be concerned with the question "*What is best?*" a question that cuts deeply rather than broadly, a question whose answers tend to move the silt downstream (Pirsig, 1974). Mathematical optimization is the formal title given to the branch of computational science that seeks to answer the question "What is best?" for problems in which the quality of the answer can be expressed as a numerical value. Such problems arise in all areas of mathematics; the physical, chemical, and biological sciences; engineering; architecture; economics; and management. The range of techniques available to solve them is nearly as wide (Gray et al., 1997).

Definitions. The goal of an optimization problem can be formulated as follows: find the combination of parameters (independent variables) that optimize a given quantity, possibly subject to some restrictions on the allowed parameter ranges. The quantity to be optimized (maximized or minimized) is termed the *objective function*, the parameters that may be changed in the quest for the optimum are called *control or decision variables*, and the restrictions on allowed parameter values are known as *constraints*. A maximum of a function f is a minimum of $-f$. The general optimization problem may be stated mathematically as

$$\text{minimize} \quad f(X), \quad X = (x_1, x_2, \ldots, x_n)^T$$

$$\text{Subject to} \quad C_i(X) = 0, \quad i = 1, 2, \ldots, m' \tag{22.1}$$

$$C_i(X) > 0, \quad i = m' + 1, m' + 2, \ldots, m$$

where $f(X)$ is the objective function, X the column vector of the n independent variables, and $C_i(X)$ the set of constraints. Constraint equations of the form $C_i(X) = 0$ are termed equality constraints and those of the form $C_i(X) > 0$ are inequality constraints. Taken together, $f(X)$ and $C_i(X)$ are known as the problem functions (Table 22.1 and Table 22.2).

Optimality Conditions. Before continuing to consider individual optimization algorithms, we describe the conditions that hold at the optimal sought.

The strict definition of the global optimum X^* of $f(X)$ is that

$$f(X^*) < f(Y) \quad \forall Y \in V(X), \quad Y \neq X^* \tag{22.2}$$

TABLE 22.1   Optimization Problem Classifications

Characteristics	Property	Classification
No. of decision variables	One	Univariate
	More than one	Multivariate
Types of decision variables	Continuous real numbers	Continuous
	Integers	Discrete
	Both continuous real numbers and integers	Mixed integer
	Integers in permutation	Combinatorial
Objective functions	Linear functions of decision variables	Linear
	Quadratic functions of decision variables	Quadratic
	Other nonlinear functions of decision variables	Nonlinear
Problem formulation	Subject to constraints	Constrained
	Not subject to constraints	Unconstrained
Decision variable realization within the optimization model	Exact	Deterministic
	Subject to random variation	Stochastic
	Subject to fuzzy uncertainty	Fuzzy
	Subject to both random variation and fuzzy uncertainty	Fuzzy-stochastic

TABLE 22.2   Typical Applications

Field	Problem	Classification
Nuclear engineering	In-core nuclear fuel management	Nonlinear
		Constrained
		Multivariate
		Combinatorial
Computational chemistry	Energy minimization for 3D structure prediction	Nonlinear
		Unconstrained
		Multivariate
		Continuous
Computational chemistry and biology	Distance geometry	Nonlinear
		Constrained
		Multivariate
		Continuous

where $V(X)$ is the set of feasible values of the control variables X. Obviously, for an unconstrained problem, $V(X)$ is infinitely large.

A point Y^* is a strong local minimum of $f(X)$ if

$$f(Y^*) < f(Y) \quad \forall Y \in N(Y^*, \eta), \quad Y \neq Y^* \tag{22.3}$$

where (Y^*, η) is defined as the set of feasible points contained in the neighborhood of Y, i.e., within some arbitrary small distance of Y. For Y^* to be a weak local minimum, only an equality needs to be satisfied:

$$f(Y^*) \leq f(Y) \quad \forall Y \in N(Y^*, \eta), \quad Y \neq Y^* \tag{22.4}$$

More useful definitions, i.e., more easily identified optimality conditions, can be provided if $f(X)$ is a smooth function with continuous first and second derivatives for all feasible X. Then a point X^* is a stationary point of $f(X)$ if

$$g(X^*) = 0 \tag{22.5}$$

where $g(X)$ is the gradient of $f(X)$. This first derivative vector $\wedge f(x)$ has components given by

$$g_i(X) = \frac{\partial f(X)}{\partial x_i} \tag{22.6}$$

The point X is also a strong local minimum of $f(X)$ if the Hessian matrix $H(X)$, the symmetric matrix of second derivatives with components

$$H_{ij}(X) = \frac{\partial^2 f(X)}{\partial x_i \partial x_j} \tag{22.7}$$

is positive-definite at X^*, i.e., if

$$u^T H(X^*) u > 0 \quad \forall u \neq 0 \tag{22.8}$$

This condition is a generalization of convexity, or positive curvature to higher dimensions. Figure 22.1 illustrates the different types of stationary points for unconstrained univariate functions.

As shown in Figure 22.2, the situation is slightly more complex for constrained optimization problems. The presence of a constraint boundary in Figure 22.2, in the form of a simple bound on the permitted values of the control variable, can cause the global minimum to be an extreme value, an *extremum* (i.e., an end point), rather than a true stationary point. Some methods of treating constraints transform the optimization problem into an equivalent unconstrained one, with a different objective function (Figure 22.3).

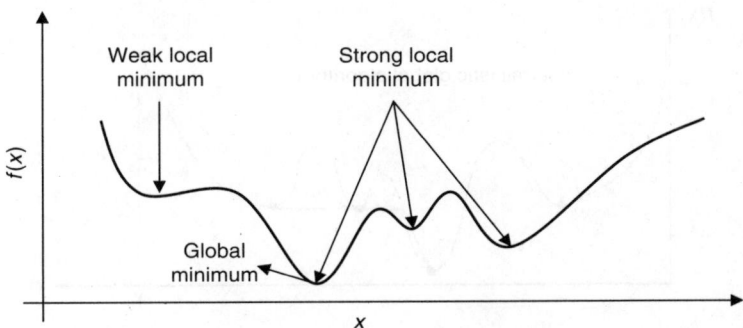

FIGURE 22.1 Types of minima for unconstrained optimization problems.

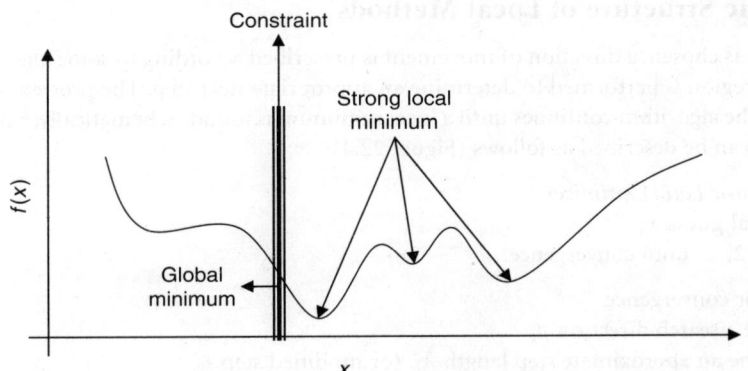

FIGURE 22.2 Types of minima for constrained optimization problems.

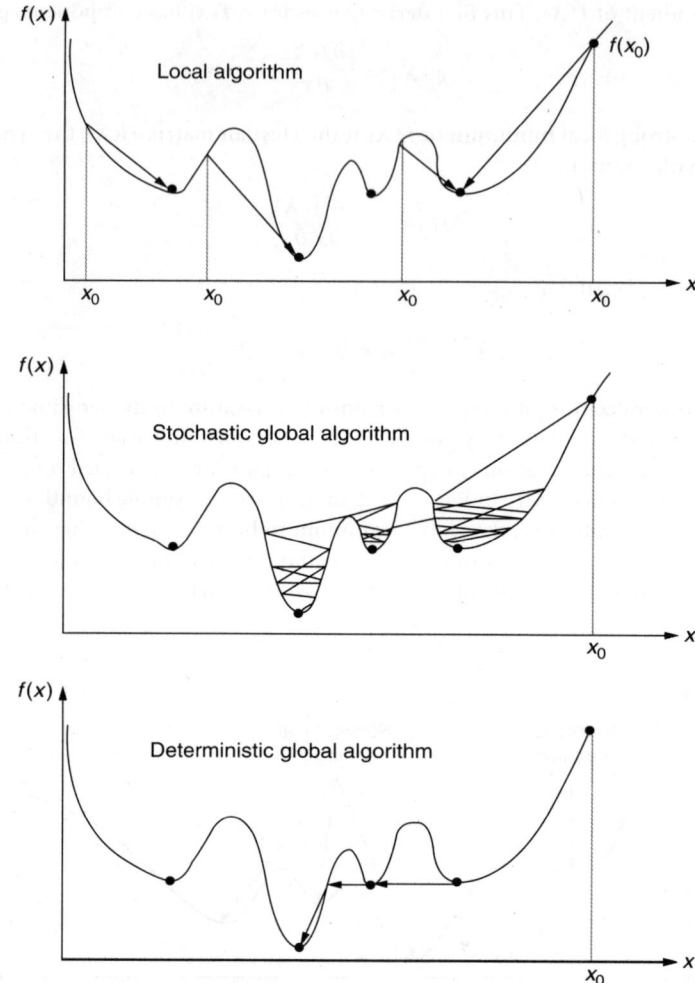

FIGURE 22.3   Types of structure of local and global minimization algorithms.

22.3.1   Basic Structure of Local Methods

A starting point is chosen, a direction of movement is prescribed according to some algorithm, and a line search or trust region is performed to determine an appropriate next step. The process is repeated at the new point and the algorithm continues until a local minimum is found. Schematically, a model local minimizer method can be described as follows (Figure 22.4):

Algorithm 3.1 Basic Local Optimizer
Supply an initial guess x_0
   For $k = 0, 1, 2, \ldots$ until convergence,

   1 Test x_k for convergence
   2 Calculate a search direction p_k
   3 Determine an approximate step length λ_k (or modified step s_k)
   4 Set x_{k+1} to $x_k + \lambda_k p_k$ (or $x_k + s_k$)

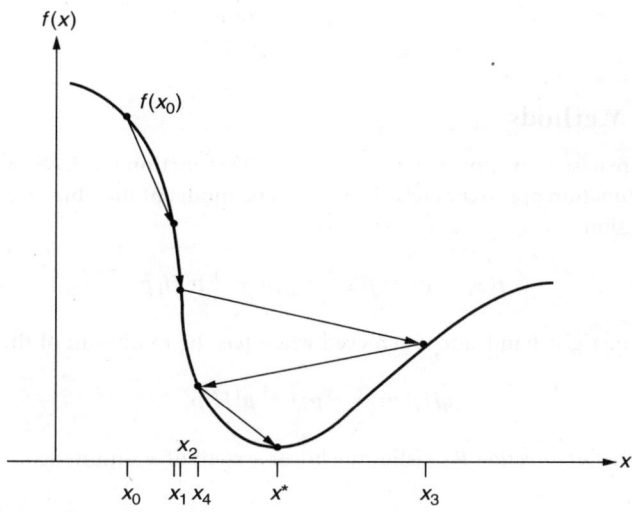

FIGURE 22.4 The descent structure of local minimization algorithms.

22.3.2 Descent Directions

It is reasonable to choose a search vector p that will be a descent direction, i.e., a direction leading to function value reduction. A descent direction P is defined as one along which the directional derivative is negative:

$$g(X)^T p < 0 \tag{22.9}$$

when we write the approximation

$$f(X + \lambda p) \approx f(X) + \lambda g(X)^T p \tag{22.10}$$

we see that the negativity of the right-hand side guarantees that a lower function value can be found along p for sufficiently small λ.

Steepest descent. Steepest descent (SD) is one of the oldest and simplest methods. At each iteration of SD, the search direction is taken as $-g_k$, the negative gradient of the objective function at x_k. Recall that a descent direction p_k satisfies $g_k^T p_k < 0$. The simplest way to guarantee the negativity of this inner product is to choose $p_k = -g_k$. This choice also minimizes the inner product $-g_k^T p$ for unit-length vectors and, thus gives rise to the name steepest descent. Steepest descent is simple to implement and requires modest storage, $O(n)$. However, progress toward a minimum may be very slow, especially near a solution.

Conjugate Gradient. The first iteration in conjugate gradient (CG) is the same as in SD, but successive directions are constructed so that they form a set of mutually conjugate vectors with respect to the (positive-definite) Hessian $\mathbf{A}$ of a general convex quadratic function $qA(X)$.

Algorithm 3.2.2 CG method to solve $AX = -b$

1. Set $r_0 = -(Ax_0 + b)$, $d_0 = r_0$
2. For $k = 0, 1, 2, \ldots$ until r is sufficiently small, compute:

$$\lambda_k = r_k^T r_k / d_k^T A d_k$$
$$x_{k+1} = x_k + \lambda_k d_k$$
$$r_{k+1} = r_k - \lambda_k d_k$$

$$\beta_k = r_{k+1}^T r_{k+1}/r_k^T r_k$$
$$d_{k+1} = r_{k+1} + \beta_k d_k$$

22.3.3 Newton Methods

All Newton methods are based on approximating the objective function locally by a quadratic model and then minimizing that function approximately. The quadratic model of the objective function f at x_k along p is given by the expansion

$$f(x_k + p) \approx f(x_k) + g_k^T p + \tfrac{1}{2} p^T H_k p \tag{22.11}$$

The minimization of the right-hand side is achieved when p is the minimum of the quadratic function:

$$qH_k(p) = g_k^T p + \tfrac{1}{2} p^T H_k p \tag{22.12}$$

Alternatively, such a *Newton direction P* satisfies the linear system of n simultaneous equations, known as the *Newton equation*

$$H_k p = -g_k \tag{22.13}$$

In the "classic" Newton method, the Newton direction is used to update each previous iterate by the formula $x_{k+1} = x_k + p_k$, until convergence. For the one-dimensional version of Newton's method for solving a nonlinear equation $f(X) = 0$

$$x_{k+1} = x_k - f(x_k)/f'(x_k) \tag{22.14}$$

The analogous iteration process for minimizing $f(X)$ is

$$x_{k+1} = x_k - f'(x_k)/f''(x_k) \tag{22.15}$$

Newton variants are constructed by combining various strategies for the individual components above.

Algorithm: Modified Newton
   For $k = 0, 1, 2, \ldots$, until convergence, given x_0:

1. Test x_k for convergence.
2. Compute a descent direction p_k so that $\|H_k p_k + g_k\| \le \eta_k \|g_k\|$,

where η_k controls the accuracy of the solution and some symmetric matrix $\overline{H}_k$ may represent H_k.

3. Compute a step length λ so that for $x_{k+1} = x_k + \lambda p_k$,

$$f(x_{k+1}) \le f(x_k) + \alpha \lambda g_k^T p$$
$$|g_{k+1}^T p_k| \le \beta |g_k^T p_k|$$

   with $0 < \alpha < \beta < 1$.
2 Set $x_{k+1} = x_k + \lambda p_k$.

Newton variants are constructed by combining various strategies for the individual components.

22.4 The Stochastic Approach

An object that changes randomly in both time and space is said to be *stochastic*. A basic characteristic of applied optimization problems treated in engineering is the fact that the data for these problems, e.g., parameters of the material (yield stress, allowable stresses, moment capacities, and specific gravity), external loadings, manufacturing errors, cost factors, etc., are not known at the planning stage and have to be considered to be random variables with a certain probability distribution; in addition, there is always some uncertainty in the mathematical modeling of practical problems. Typical problems of this type are:

- The limit (collapse) load analysis and the elastic or plastic design of mechanical structures represented mathematically by means of:
  - › The equilibrium equation
  - › Hooke's law
  - › The member displacement equation
- Optimal trajectory planning of robots by offline programming such that the control strategy, based on the optimal open loop control, causes only low online correction expenses. Here, the underlying mechanical system is described by the kinematics and the dynamic equation, and the optimal velocity profile and configuration variables are determined for fixed model parameters by a certain variation problem.

Since the (online) correction of a decision, e.g., the decision on the design of a mechanical structure or on the selection of a velocity profile after the realization/observation of the random data, might be highly expensive and time-consuming, the pre-existing statistical information about the underlying probability mechanism generating the random data should be taken into account during the planning phase. By applying stochastic programming instead of ordinary mathematical programming methods, the original optimization problem with random data is replaced using appropriate decision criteria by a deterministic substitute problem, e.g.,

- Using a chance-constrained programming approach, the objective function is replaced by its mean value, and the random constraints are replaced by chance constraints.
- Evaluating the violation of the random constraints by means of penalty functions, a weighted sum of the expectation of the primary objective function and the total expected penalty costs are minimized subject to the remaining deterministic constraints, e.g., box constraints or the mean value of the objective function is minimized subject to constraints involving upper bounds for the expected penalty cost arising from violations of the original constraints.

A main problem in the solution of these problems is the numerical computation and differentiation of risk functions.

22.4.1 Stochastic Approximation

While some nonclassical optimization techniques are able to optimize on discontinuous objective functions, they are unable to do so when complexity of the data becomes very large. In this case the complexity of the system requires that the objective function be estimated. Furthermore, the models that are used to estimate the objective function may be stochastic owing to the dynamic and random nature of the system and processes.

The basic idea behind the stochastic approximation method is the gradient descent method (Wasan, 1969 and Marti, 1996). Here, the decision variable is varied in small increments and the impact of this variation (measured by the gradient) is used to determine the direction of the next step. The magnitude of the step is controlled such that when the perturbations in the system are small, the steps are larger, and vice versa. Stochastic approximation algorithms based on various techniques have been developed recently. They have been applied to both continuous and discrete objective functions.

22.4.2 General Stochastic Control Problem

The control of a random, dynamic system in some optimal fashion using imperfect measurement data is the general problem. It also constitutes a problem about which it is very difficult to obtain any meaningful insights. Although feedback is used in order to compensate for unmodeled errors and inputs, most controllers are designed and analyzed in a deterministic context. The control inputs for the system generally must be based on imperfect observations of some of the variables that describe the system. The control policy that is utilized must be based on *a priori* knowledge of the system characteristics on the time history of the input variables.

The mathematical model of the system is described by a nonlinear difference equation

$$x_{k+1} - f(x_k, H_k) + w_k, \quad k = 0, 1, ..., N \tag{22.16}$$

The noise *w* has been assumed to be additive primarily for reasons of convenience. The state *x* is *n*-dimensional and the input *u* is *p*-dimensional. In general, a probabilistic model for the initial state x_0 and for the plant W_k is assumed to be known except for some unknown parameters. With rare exceptions these variables are regarded as having a Gaussian distribution such that

$$E[x_0] = U_0, \quad E[w_k] = 0, \quad \forall k \tag{22.17}$$

$$E[(x_0 - H_0)(x_0 - U_0)] = M_0, \quad E[x_0 w_k^T] = 0, \quad \forall k \tag{22.18}$$

$$E[w_k w_k^T] = O_k \delta_k \tag{22.19}$$

Thus, the plant noise sequence is white and independent of the initial state.

The measurement system is described by a nonlinear algebraic relation to the state. The *m*-dimensional measurement vector is given by

$$z_k = h_k(x_k) + v_k, \quad k = 0, 1, ..., N \tag{22.20}$$

The noise *v* is considered to be additive for reasons of convenience. It is assumed to be a zero-mean, white Gaussian sequence that is independent of the initial state and the plant noise sequence.

$$E[v_k] = 0, \quad \forall k \tag{22.21}$$

$$E[v_k v_i^T] = R_k \delta_{ki}, \quad \forall k \tag{22.22}$$

$$E[v_k x_0^T] = 0, \quad \forall k \tag{22.23}$$

$$E[v_k w_j^T] = 0, \quad \forall k, j \tag{22.24}$$

The equations above provide the mathematical description of the system. It is this part of the complete system that represents the physical system that must be controlled. The structure of the controller, of course, depends on the exact form of the system model equations $f(\bullet, \bullet)$ and $h(\bullet)$.

The behavior of the system is controlled through the input signals u_k, which are introduced at each sampling time t_k. The manner in which the controls are generated can be accomplished in a limitless number of ways. Certainly, the controls are constrained by the objectives that are defined for the control action and by the restrictions on the control and state variables themselves. Generally, there will be more than one control policy that satisfies the system constraints and achieves the prescribed objectives. Then it is reasonable to attempt to select the control policy from among all these admissible policies that is "best" according to some well-defined performance measure. Optimal stochastic control theory is concerned with the determination of the best admissible control policy for the given system.

The following performance index is assumed:

$$J_0 = E \sum_{i=0}^{N-1} w_i(x_{i+1}, u_i) \tag{22.25}$$

Notice that the summation $E \sum_{i=0}^{N-1} w_i(x_{i+1}, u_i)$ is a random variable. Consequently, it is appropriate to consider its minimization; instead it is mapped into a deterministic quantity by considering its expected value.

22.5 The Intelligent Heuristic Models

There have been vast and continuing advances in optimal solution techniques for intelligent systems in the last two decades. The heuristic methods offer a very viable approach; however, the design and implementation of a problem-specific heuristic can be a long and expensive process, and the result is often domain-dependent and not flexible enough to deal with changes that may occur over time. Hence, considerable interest is focused on general heuristic techniques that can be applied to a variety of different combinatorial problems (Rayward-Smith, 1995). This has yielded a new generation of intelligent heuristic techniques such as Tabu Search and SA, and Evolutionary Algorithms such as GAs and Neural Networks (Haykin, 1999).

22.5.1 Heuristics

Heuristics are the knowledge used to make good judgments or strategies, tricks, or "rules of thumb" used to simplify the solution of problems. They include "trial and error" (experience-based) knowledge and intelligent guesses/procedures for domain-specific problem solving. They are particularly suitable for ill-defined or poorly posed problems and poor models, such as when there are incomplete data. Heuristics play an important role in such strategies because of the exponential nature of most problems. They help to reduce the number of alternatives from an exponential number to a polynomial number and thereby render a solution obtainable in a tolerable amount of time.

22.5.2 Intelligent Systems

Intelligence is the ability to acquire, understand, and apply knowledge or the ability to exercise thought or reasons. It also embodies knowledge and accomplishments, both conscious and unconscious, which animate beings have acquired and achieved through study and experience. Artificially intelligent systems are thus machines and coded programs aimed at mimicking such accomplishments and knowledge. Systems have been designed to perform many types of intelligent tasks. These can be physical systems such as robots; mathematical computational systems such as scheduling systems that solve diverse tasks; systems used in planning complex strategies for military and for business; system used in medicine, diagnosis and control of diseases, and so on.

22.5.3 The General Search Paradigm

The General Search Algorithm is of the form:

General Search
Objective is to maximize $f(x), x \in U$
X, Y, Z: multiset of solutions $\subset U$
Initialize (X);
While not finish (X) do

> **Begin**
> $Y := \text{select }(X)$
> $Z := \text{create }(Y)$
> $X := \text{merge}(X, Y, Z)$
> **End**

where X is the initial pool of one or more potential solutions to the problem. Since X may contain multiple copies of some solutions, it is more appropriately called a multiset. Y is a selection from X and Z is created from Y.

When a new solution is *created* either initially or by using the operator "*create*," the function value, $f(X)$ is applied to determine the value of the solution. X is reconstructed from the penultimate pool of X, Y and Z by the operator "*merge*." The process is repeated until the pool X is deemed satisfactory.

22.5.4   Integrated Heuristics

Modern approaches to local searches have incorporated varying degrees of intelligibility. The contribution of intelligent search techniques should not be solely viewed in terms of improved performance alone as the traditional systems engineers or analysts expect, even though that would be very desirable. The trust of the contribution of intelligent search techniques should be in terms of *improved intelligibility, flexibility,* and *transparency of these emerging computational techniques.* A synergy between intelligibility and performance is normally of utmost importance in assessing the efficiency of an intelligent heuristic.

22.5.5   Tabu Search

Tabu Search is one successful variant of the neighborhood search paradigm and is designed to avoid the problem of becoming trapped in a local optimum.

    The Tabu Search paradigm is as below:

Tabu Search
Objective is to maximize $f(x)$, $x \in U$
X, Z: multiset of solutions $\subset U$
Tabu set of rules of type $U \rightarrow$ {true, false}
Initialize (X);
Initialize (Tabu); {very often to Φ}
While not finishing (X) do

                Begin
                Z: = create $(X,$ Tabu)
                Tabu = update(Tabu)
                X: = merge(X, Z)
                End

The difficulty in Tabu Search is in constructing the set of rules. Considerable expertise and experimentation is required to construct the rules and to ensure their dynamic nature is correctly controlled. If the expertise is available, the resulting search can be efficient. Aspiration criteria are often included to prevent the Tabu Search from being too restrictive. These criteria are rules that say that certain moves are to be preferred over others. Some form of expert rules may also serve as Tabu Search rules. Each rule may have an associated weight, negative if Tabu and positive if an aspiration. The combined set of rules thus associates a weight with each neighbor. A large positive weight suggests it is a desirable move, while a large negative weight suggests it can be discounted. Tabu Search has found applications to real-world problems such as packing and scheduling problems (flow shop problems, employee scheduling problem, machine scheduling, etc.); traveling salesman; vehicle routing, and telecommunications.

22.5.6   Simulated Annealing

Simulated annealing exploits an analogy between the way in which a metal cools and freezes into a minimum energy crystalline structure (the annealing process) and the search for a minimum in a more general system (Ingber, 1993 and Siarry et al., 1997).

    The SA is essentially a local search technique in which a move to an inferior solution is allowed with a probability that decreases as the process progresses, according to some Bolzmann-type distribution. The inspiration for the SA approach is the law of thermodynamics, which states that at temperature t, the probability of an increase in energy of magnitude, δE, is given by

$$P(\delta E) = \exp(-\delta E / kt) \tag{22.26}$$

where k is the physical constant known as the Bolzmann constant. The equation is applicable to a system that is cooling until it converges to a "frozen" state. The system is perturbed from its current state and the resulting energy change δE is calculated. If the energy has decreased, the system moves to the new state, otherwise the new state is only accepted with the probability given above. The cycle can be repeated for a number of iterations at each temperature and subsequently reduced, and the number of cycles repeated for the new lower temperature. This whole process is repeated until the system freezes to its steady state (Figure 22.5).

We can associate the potential solutions of an optimization problem with the system states, the cost of a solution corresponds to the concept of energy and moving to any neighbor corresponds to a change of state. A simple version of the SA paradigm is of the form:

Genetic Algorithms are search techniques (Cadenas and Jimenez, 1994) based on an abstracted model of Darwinian evolution (Darwin, 1859). Fixed-length strings represent solutions over some alphabet. Each such string is thought of as a "chromosome." The value of the solution then represents the "fitness" of the chromosome. The concept of "survival of the fittest" is then used to allow better solutions to combine to produce offspring (Davis, 1991). The Genetic Algorithm paradigm closely follows the same search concept exploited in Tabu Search and SA. The paradigm is:

Genetic Algorithm
Objective is to maximize $f(x)$, $x \in U$
X, Y, Z: multiset of solutions $\subset U$
Initialize (X);
While not finish (X) do

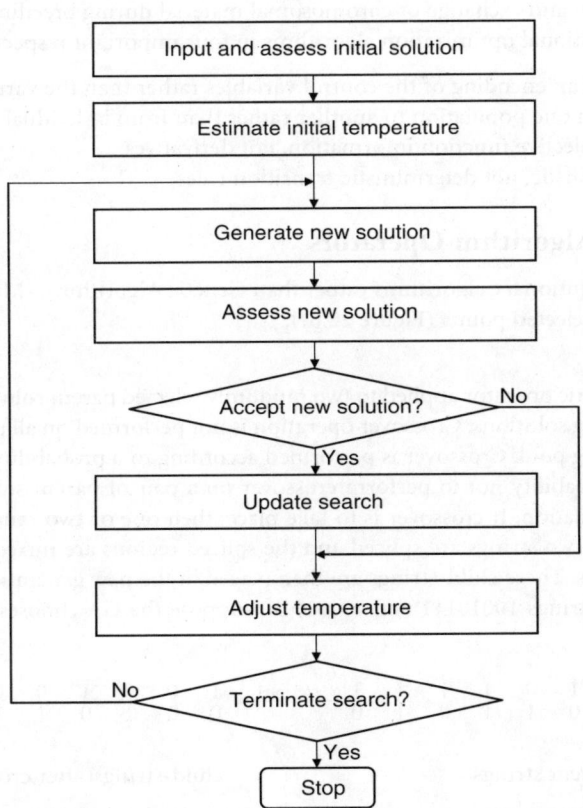

FIGURE 22.5   The structure of the SA algorithm.

> **Begin**
> $Y: = \text{select}(X)$
> $Z: = \text{create}(Y)$
> $X: = \text{merge}(X, Y, Z)$
> **End**

Here the operator's *select*, *create* and *merge* correspond to the operator's select, reproduction, and crossover. The iteration loop of a basic GA is as follows

Procedure GA

> **Begin**
> Generate initial population, $P(0)$; $t = 0$;
> Evaluate Chromosomes in $P(0)$;
> **Repeat**
> $t = t + 1$'
> Select $P(t)$ from $P(t-1)$;
> Recombine chromosomes in $P(t)$ using genetic operators;
> Evaluate Chromosomes in $P(t)$;
> **Until** termination condition is satisfied;
> **End**

In a natural evolution, each species searches for beneficial adaptations in an ever-changing environment. As species evolve, these new attributes are encoded in the chromosomes of individual members. This information does change by random mutation, but the real driving force behind evolutionary development is the combination and exchange of chromosomal material during breeding.

GAs differ from traditional optimization algorithms in four important respects:

- They work using an encoding of the control variables rather than the variables themselves.
- They search from one population to another rather than from individual to individual.
- They use only objective function information, not derivatives.
- They use probabilistic, not deterministic transition rules.

22.5.7   Genetic Algorithm Operators

These are known as evolutionary algorithms rather than Genetic Algorithms. (Mutation: bit-wise change in strings at randomly selected points (Figure 22.6)).

Examples
Crossover: This is a generic operator applied to two randomly selected parent solutions in the mating pool to generate two offspring solutions. Crossover operation is not performed on all pairs of parent solutions selected from the mating pool. Crossover is performed according to a probability, p (usually $p = 0.6$). If GA decides by this probability not to perform crossover on a pair of parent solutions, they are simply copied to the new population. If crossover is to take place, then one or two random splicing points are chosen in a string. The two strings are spliced and the spliced regions are mixed together to create two (potentially) new strings. These child-strings are then placed in the new generation.

For example, using strings 10010111 and 00101010, suppose the GA chooses at random to perform crossover at point 5

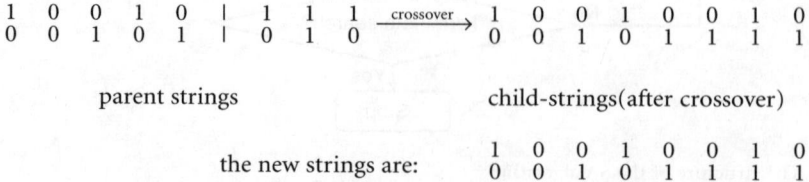

parent strings

child-strings(after crossover)

the new strings are:

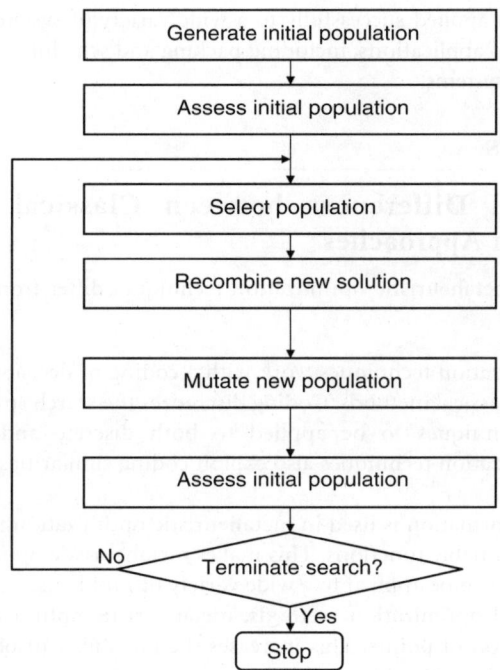

FIGURE 22.6 The basic structure of a genetic algorithm.

For two crossover points at points 2 and 6, the crossing looks like

$$
\begin{array}{cc}
1\ \ 0 \mid 0\ \ 1\ \ 0\ \ 1 \mid 1\ \ 1 \\
0\ \ 0 \mid 1\ \ 0\ \ 1\ \ 0 \mid 1\ \ 0
\end{array}
\xrightarrow{\text{crossover}}
\begin{array}{cc}
1\ \ 0 \mid 1\ \ 0\ \ 1\ \ 0 \mid 1\ \ 1 \\
0\ \ 0 \mid 0\ \ 1\ \ 0\ \ 1 \mid 1\ \ 0
\end{array}
$$

parent strings   child-strings(after crossover)

the new strings are:
$$
\begin{array}{cccccccc}
1 & 0 & 1 & 0 & 1 & 0 & 1 & 1 \\
0 & 0 & 0 & 1 & 0 & 1 & 1 & 0
\end{array}
$$

Mutation: Selection and crossover alone can obviously generate a staggering amount of differing strings. However, depending on the initial population chosen, there may not be enough variety of strings to ensure that GA sees the entire problem space, in which event GA may find itself converging on strings that are not close to the optimum it seeks owing to a bad initial population. Some of these problems may be overcome by introducing a mutation operator into the GA. The GA has a mutation probability, m, which dictates the frequency at which mutation occurs. Mutation can be performed either during selection or crossover (although crossover is more usual). For each string, GA checks if it should perform a mutation. If it should, it randomly changes the element value to a new one. In a binary string, 1s are changed to 01s and 0s to 1s. For example, given the string 10101111, if the GA determines to mutate this string at point 3 the 1 in that position is changed to 0, i.e.,

$$10101111 \xrightarrow{\text{mutuate}} 10001111$$

The mutation probability is kept as low as possible (usually about 0.01) as a high mutation rate will destroy fit strings and degenerate the GA algorithm into a random walk, with all the associated problems.

Genetic Algorithm has been applied successfully to a wide variety of systems. We can only highlight a small subset of the myriad of applications, including packing and scheduling, design of engineering systems, robots, and transport systems.

22.6 Conclusions

22.6.1 Fundamental Differences between Classical and Metaheuristic Optimization Approaches

For a number of reasons, metaheuristic optimization techniques differ from classical search and optimization methods.

1. Metaheuristic optimization techniques work with a coding of decision variables, a matter that is very uncommon in classical methods. Coding discretizes the search spaces and allows metaheuristic optimization techniques to be applied to both discrete and discontinuous problems. Metaheuristic optimization techniques also exploit coding similarities to make faster and parallel searches.
2. Since no gradient information is used in metaheuristic optimization techniques, they can also be applied to nondifferentiable functions. This makes metaheuristic optimization techniques robust in the sense that they can be applied to a wide variety of problems.
3. Unlike many classical optimization methods, metaheuristic optimization techniques, like GAs, work with a population of points. This increases the possibility of obtaining the global optimal solution even in ill-behaved problems.
4. Metaheuristic optimization techniques use probabilistic transition rules instead of fixed rules. For example, in early GA iterations, this randomness in GA operators makes the search unbiased toward any particular region in the search space and has an effect, i.e., not making a hasty, wrong decision. Use of stochastic transition rules also increases the chance of recovering from a mistake.
5. Metaheuristic optimization techniques use only payoff information to guide them through the problem space. Many search techniques need a variety of information to guide them. Hill-climbing methods require derivatives, for example. The only information a metaheuristic needs is some measure of fitness about a point in the space (sometimes known as objective function value). Once the metaheuristic knows the current measure of "goodness" about a point, it can use this to continue searching for the optimum.
6. Metaheuristic optimization techniques allow procedure-based function declaration. Most classical search methods do not permit such declarations. Thus, where procedures of optimization need to be declared, metaheuristic optimization techniques prove a better optimization tool.

The coding of decision variables in metaheuristic optimization techniques also makes it proficient in solving optimization problems involving equality and inequality constraints.

22.6.2 Some In-House Applications of Heuristics to Intelligent Systems

At the University of Lagos, some of these techniques have been successfully applied to model and simulate some intelligent systems. These include the application of GAs to bimodal transport scheduling problem, (Ibidapo-Obe and Ogunwolu, 2001). In this application, GA is applied using a bilevel programming approach to obtain transit schedules for a bimodal transfer station in an urban transit network with multiple dispatching stations. The arrival rates of passengers are captured as fuzzy numbers in order to confer intelligibility in the system. The model results in comparably better schedules with better levels of service in the transit network (in terms of total waiting and transfer times for passengers) than is obtained otherwise.

Another application is an intelligent path planner for autonomous mobile robots (Asaolu, 2002). For this robotic motion planner, a real-time optimal path planner was developed for autonomous mobile robots navigation. Metaheuristic optimization techniques were used to guide the robot within a workspace with both static and roving obstacles. With the introduction of arbitrarily moving obstacles, the dynamic obstacle avoidance problem was recast into a dynamic graph-search. The instantaneous graph is made of the connected edges of the rectangles boxing the ellipses swept out by the robot, goal, and moving obstacles. This was easily transversed in an optimal fashion. We are presently also looking at the problems of machine translation as another application of intelligent metaheuristic optimization techniques.

22.6.3 High-Performance Optimization Programming

Future developments in the field of optimization will undoubtedly be influenced by recent interest and rapid developments in new technologies — powerful vector and parallel machines. Indeed, their exploitation for algorithm design and solution of "grand challenge" applications is expected to bring new advances in many fields, including computational chemistry and computational fluid dynamics. Supercomputers can provide speedup over traditional architectures by optimizing both scalar and vector computations. This can be accomplished by pipelining data as well as offering special hardware instructions for calculating intrinsic functions (e.g., $\exp(x)$, $\sqrt{x}$), arithmetic, and array operations. In addition, parallel computers can execute several operations concurrently. Multiple instructions can be specified for multiple data streams in multiple instruction multiple data (MIMD) designs, whereas the same instructions can be applied to multiple data streams in single instruction multiple data (SIMD) prototypes. Communication among processors is crucial for efficient algorithm design so that the full parallel apparatus is exploited. These issues will only increase in significance as massively parallel networks enter into regular use. In general, one of the first steps in optimizing codes for these architectures is implementation of standard basic linear algebra subroutines. These routines are continuously being improved, expanded, and adapted optimally to more machines performing operations such as dot products ($x^T y$) and vector manipulations ($ax + y$), as well as matrix/vector and matrix/matrix operations.

Specific strategies for optimization algorithms have been quite recent and are not yet unified. For parallel computers, natural improvements may involve the following ideas:

(1) Performing multiple minimization procedures concurrently from different starting points
(2) Evaluating function and derivatives concurrently at different points (e.g., for a finite-difference approximation of gradient or Hessian or for an improved line search)
(3) Performing matrix operations or decompositions in parallel for special structured systems (e.g., Cholesky factorizations of block-band preconditioned)

With increased computer storage and speed, the feasible methods for the solution of very large (e.g., $O(10^5)$ or more variable) nonlinear optimization problems arising in important applications (macromolecular structure, meteorology, and economics) will undoubtedly expand considerably and make possible the solution of larger and far more complex problems in all fields of science and engineering.

References

Asaolu, O.S., An Intelligent Path Planner for Autonomous Mobile Robots, Ph.D. thesis, University of Lagos, Lagos, Nigeria, 2002.

Cadenas, M. and Jimenez, F., Genetic algorithm for the multi-objective solid transportation problem: a fuzzy approach, *in International Symposium on Automotive Technology and Automation, Proceedings for the Dedicated Conferences on Mechatronics and Supercomputing Applications in the Transportation Industries*, Aachen, Germany, 1994, pp. 327–334.

Darwin, C., *On The Origin of Species*, 1st ed., Harvard University Press, Cambridge, MA, (facsimile - 1859).

Davis, L., *Handbook of Genetic Algorithms*, Van Nostrand Reinhold, New York, 1991.

Gill, P.E., Murray, W., and Wright, M.H., *Practical Optimization*, Academic Press, New York, 1983.

Glover, F. and Macrnillan, C., The general employee scheduling problem: An integration of management science and artificial intelligence, *Comput. Oper. Res.*, 15, 563–593, 1986.

Gray, P., Hart, W., Painton, L., Phillips, C., Trahan, M., and Wagner, J., A Survey of Global Optimization Methods, Sandia National Laboratories Albuquerque, NM, http://www.cs.sandia.gov/opt/survey/, 1997.

Haykin, S., *Neural Networks: A Comprehensive Foundation*, 2nd ed., Prentice-Hall International Inc., New York, 1999, ISBN 0-13-908385-5.

Ibidapo-Obe, O. and Ogunwolu, F.O., An Optimal Scheduling of a Bi-modal Urban Transit System Using Genetic Algorithms, Unpublished work, Department of Systems Engineering, University of Lagos, Lagos, Nigeria, 2001.

Ingber, L.A., Simulated annealing: Practice versus theory, *J. Math. Comput. Model.*, 18(11), 29–57, 1993.

Luenberger, D.G., *Linear and Nonlinear Programming*, 2nd ed., Addison-Wesley, Reading, MA, 1984.

Marti, K., Stochastic optimization methods in engineering, in *System Modeling and Optimization*, Dolezal, J. and Fiedler, J., Eds., Chapman & Hall, London, New York, 1996.

Pirsig, R.M., *Zen and the Art Motorcyle Maintenance*, Bantam Books, Toronto, 1974, pp. 92–96.

Rayward-Smith, V.J., *Applications of Modern Heuristic Methods*, Alfred Walter Limited Publishers in association with UNICOM, 1995, pp. 145–156.

Siarry, P., Berthiau, G., Durdin, F., and Haussy, J., Enhanced simulated annealing for globally minimizing functions of many-continuous variables, *ACM Trans. Math. Software*, 23, 209–228, 1997.

Wasan, M.T., *Stochastic Approximation*, Cambridge University Press, Cambridge, 1969.

23

An Architecture for the Design of Industrial Information Systems

Richard E. Billo
Oregon State University

J. David Porter
Oregon State University

Richard J. Puerzer
Hofstra University

23.1 Introduction

Industrial information systems have been incorporated into almost all facets of industry in recent years, including warehouse management (Kirzner, 2002; Obstgarten, 2002; Porter et al., 2004), shop floor control (Chausse et al., 2000; Campbell, 2001; Manthou and Vlachopoulou, 2001), health care (Puerzer, 1997; Quinn, 1997; Parker, 2002), transportation (Gilmore, 1998; Rose, 2002), and retail (Strozniak, 2002). In manufacturing, these products typically perform large and diverse functions such as job or lot tracking, labor tracking, receiving, inventory management, statistical process control, EDI transactions, maintenance management, cost accounting, and scheduling. Often they are extended beyond the shop floor and warehouse to include functionality for financial accounting and purchasing. They are often called manufacturing execution systems (MES), warehouse management systems (WMS), or enterprise resource planning (ERP) systems.

With advances in technology, including those related to computer processing speeds, bar code technology, database management systems, and distributed data architectures, there is also an increasing number of issues that must be addressed by potential users of industrial information systems. If the team responsible for the design and implementation of such a system does not have an understanding of these issues and how to address them from the onset of the project, the selection of the product and its ultimate implementation can be time-consuming, expensive, and frustrating for all parties involved.

Such issues can be grouped into two major categories: *technology* and *function*. With respect to technology, a potential user of industrial software has a variety of choices. These choices fall into three general categories: computer choices (e.g., portable data terminal, laptop, desktop PC, workstation, mainframe), automatic data capture (ADC) choices (e.g., bar code scanner technologies, bar code printers, and bar code symbologies), and network choices (e.g., wired vs. wireless, bus topologies, ring topologies). Within each of these categories, there remains a myriad of other technology choices that must be made. For example, if the team decides to use a wireless network for industrial data capture and processing, they must decide whether it should use narrow band or spread spectrum communication technology. If they choose spread spectrum, they must still make a decision as to what data rate standard is the best wireless transmission mechanism (Pahlavan et al., 1995).

In addition to technology choices, the team is also faced with the task of matching the functional capabilities available in each product with the actual operational capabilities of their own manufacturing system. When investigating each product under consideration, the team learns that every product varies greatly in functionality and complexity, and few, if any, match the actual operation of the customer's information processes.

This variety results from the fact that no two manufacturing or warehouse systems operate in exactly the same way. For example, even though some manufacturing shop floors may operate consistently under some very general principles (e.g., pull vs. push, manufacturing cells vs. job shop production vs. mass production), in reality, every manufacturing system has its own rules and guidelines for how product is to be scheduled, manufactured, moved, and stored throughout the process. Most commercial products available for industrial environments are at one of two extremes. Either they are severely limited in their capabilities and features due to an assumption that such functions as shop floor control and warehouse management in a facility will be carried out in a similar manner as the software itself; or they are extremely complex to implement in an attempt to flexibly meet the demands of a large and varied customer base.

The consequences of this situation are projects that are often late to implementation, exceed budget, and typically fall far short of users' expectations. Given the potential for these problems, how does a team go about the systematic selection and implementation of an industrial information system that will maximize the likelihood of success for their manufacturing environment, and still have a product that is of reasonable cost and can be implemented in a timely manner?

There is little recent work on methodologies for maximizing success of the design of such systems in industrial environments. Although much work has been done describing scheduling algorithms, SPC, manufacturing computing architectures, and architectures for business-to-business (B2B) transactions, little literature is available describing approaches for selection of one technology over another, determining user requirements, customizing these applications for a particular customer's industrial system, or implementing such systems in reasonable time spans.

23.2   Purpose

The purpose of this chapter is to present a framework and tools to aid a team in addressing the issues critical for the successful design or selection of effective industrial information systems. We define an effective industrial information system as one that provides the actual product information or work order information requested by a user, provides such information in a user-friendly format, provides the information to the user in a timely manner, and ensures that the information presented to the user is accurate. Such a product can be implemented in a reasonable time frame and is cost-effective.

23.3 Zachman Framework for Information Systems Design

The most effective framework we have found for designing industrial information systems is the Zachman framework for enterprise systems architecture (Sowa and Zachman, 1992; Zachman, 1997, 1999). For general information system development efforts, the Zachman framework has become the *de facto* standard used by many successful organizations (Perkins, 1997). The framework is a logical structure for classifying and organizing the descriptive representations of an enterprise information system as it proceeds through various design and development steps. It was derived from analogous structures, found in the older disciplines of architecture and manufacturing, that classify and organize the design artifacts created over the process of designing and manufacturing such complex physical products as buildings or airplanes (Zachman, 1996).

In brief, the framework is a taxonomy with 15 cells organized into three columns and five rows (see Figure 23.1). Columns correspond to the three major subsystems comprising an information system: data, process, and network. Rows correspond to the different representations of the information as it proceeds through its development cycle. Items inside the cells represent tools that can be used to model the different systems as they proceed through their design cycle.

The Zachman framework is considered to be a comprehensive, logical structure for descriptive representations of any information system. In this chapter, we will describe not only the steps that comprise the framework, but also some useful modern modeling tools that we have found to be successful for designing and implementing industrial information systems specifically for manufacturing and warehouse environments.

Throughout this chapter, we provide excerpts from several actual projects that illustrate how the architecture and tools were used. One application is the development of a wireless job-tracking system for an extrusion-honing manufacturing process. Another is the design and implementation of a wireless inventory-tracking system for a warehouse distribution center.

23.4 Previous Research

Several years ago, we undertook this line of research owing to the lack of comprehensive modern tools to aid in the selection, design, and implementation of material-tracking systems appropriate for manufacturing environments. The exception was the early integrated computer aided manufacturing (ICAM) project work sponsored by the US Air Force (ICAM, 1981), which addressed modeling techniques for industrial environments. This work led to the development of early process and data modeling tools such as integrated definition version 0 ($IDEF_0$) and integrated definition version 1 ($IDEF_1$), respectively.

Along with $IDEF_0$ and $IDEF_1$, other general data-flow diagramming and data-modeling tools were developed in this same timeframe. Some of the more widespread tools included structured analysis (DeMarco, 1979), structured systems analysis (Gane and Sarson, 1979), structured development (Ward and Mellor, 1985), entity relationship modeling (Chen, 1982), and semantic hierarchy modeling (Brodie and Ridjanovic, 1984). These tools were used primarily by computer scientists and information technologists and have since been incorporated into modern computer aided software engineering (CASE) systems. Over time, modern relational database modeling tools (Date, 2003; Elmasri and Navathe, 2004) and object-oriented tools (Loomis, 1995; Rumbaugh, Blaha, Premerlani, Eddy, and Lorensen, 1991) have been developed to model information systems, in general. These tools have been gathered into a collective suite of modeling tools termed the unified modeling language (UML) (Booch et al., 1999). However, UML includes no comprehensive framework for the usage of these tools.

According to Sowa and Zachman, (1992), each of these techniques is specialized for a different purpose. By concentrating on one aspect of the information systems development effort, each technique loses sight of the overall information system and how it relates to the enterprise and its surrounding environment. No framework existed to organize and classify the application of these tools for a comprehensive development and implementation effort.

In our research experience with industrial clients attempting to select, design or implement material tracking systems, we found Sowa and Zachman's, (1992) statements to be quite true. When teams tried to use these tools independent of a comprehensive design framework, the project efforts often failed owing

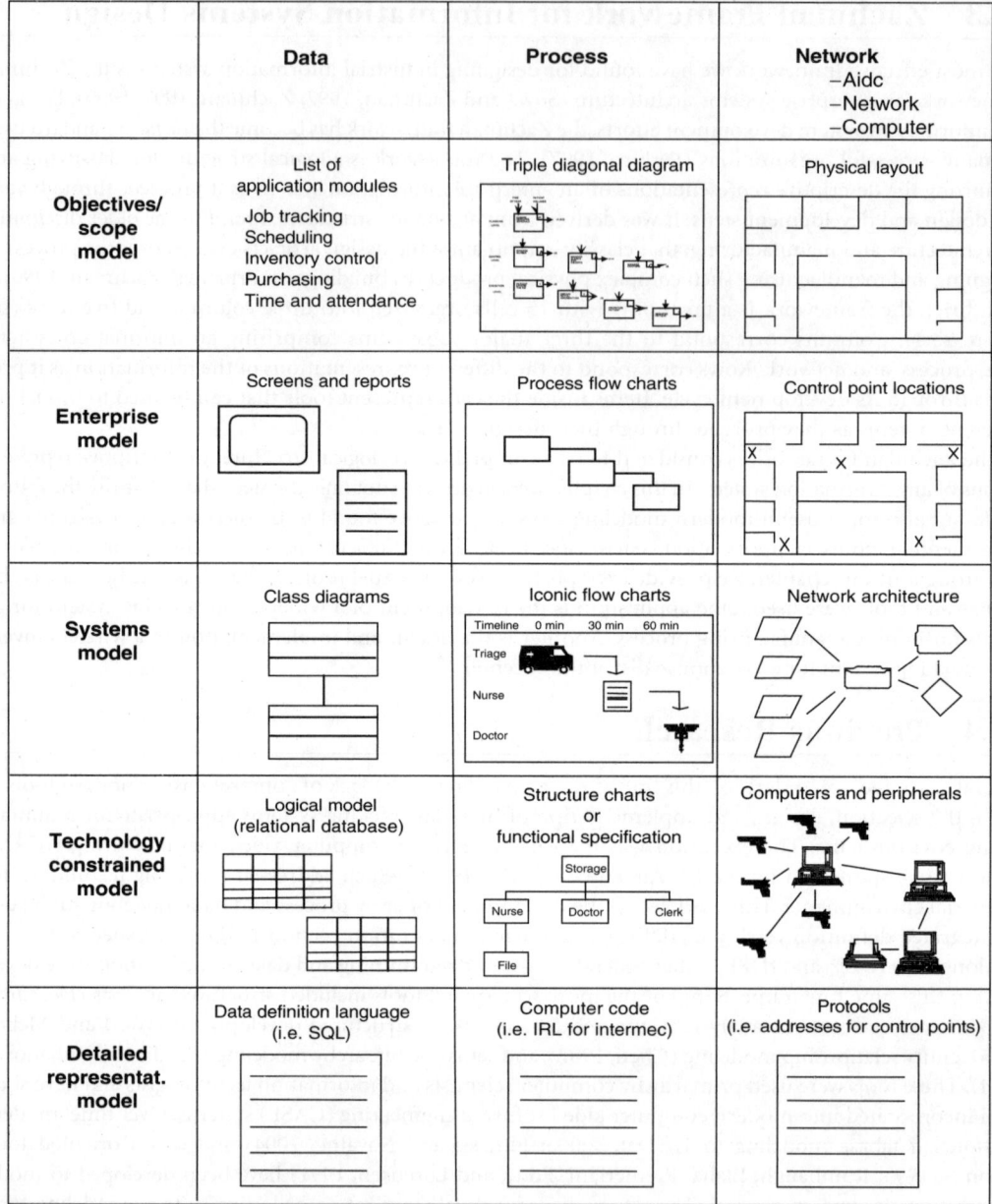

FIGURE 23.1   The Zachman framework for enterprise systems architecture.

to an overreliance on the usage of a single tool without consideration of the need for a framework (Billo et al., 1994). The Zachman framework was designed not to replace modern CASE tools such as UML, but to organize and classify them so that they can be used more appropriately.

23.5   Overview of the Framework

The Zachman framework for enterprise systems architecture is general framework that can be used for designing any information system. It is necessary in order to effectively design and control the integration of all of the needed components into an application environment. Such a framework is used to identify the necessary data the system will convey, the important functions and processes of the system, how the system should communicate, and how the system should be physically distributed and constructed. It provides users with specific perspectives of the environment in which they are implementing an information system and allows for the systematic analysis of planning for the implementation process.

The Zachman framework attempts to emulate the engineering discipline as it applies to information systems development. The framework looks at three components of the proposed information system from five different design phases. Cells within the framework consist of tools used in the design and implementation process.

With respect to industrial information systems, the columns of the matrix list the major components of the system. The *Data* column (the "what") of the matrix focuses on the identification and modeling of the data structures to be collected, manipulated, and stored by the system. The *Process* column (the "how") of the matrix focuses on the identification and modeling of the detailed functional capabilities of the application in support of the industrial process. The *Network* column (the "where") of the matrix focuses on the particular data capture technology, computer architecture, and network technology that the tracking application will eventually utilize. Because the data, process, and network are so tightly integrated, the framework assumes they will be modeled concurrently.

The rows of the matrix list the various steps that are typically carried out in the implementation of the industrial information system. These steps include the objectives/scope, enterprise, systems, technology constrained, and detailed representations models. The *Objectives/Scope* model provides an enterprise view of the system allowing for a determination of the gross level of effort required to implement the system. This model serves to place boundaries on the upcoming project and helps to identify critical modules. The *enterprise* model is a view from the outlook of the system's owner and provides insight into the user's requirements of the system. The *systems* model is the view of the system designers, providing a design of the system free from technological or equipment constraints. The *technology constrained* model looks at the system from the perspective of the system implementers and serves to convert the design to a format suitable for particular choices, such as radio frequency data communication tools and relational databases. Finally, the *detailed representations* model focuses on the actual programming and implementation required in the development of the system.

The framework is used by applying the tools defined in each of the matrix cells. These tools are supposed to provide the analyses necessary to gain all necessary information for the identification of requirements as well as for the selection and implementation of the information system appropriate for the particular application at hand.

Many of the problems that hinder the successful implementation of industrial information systems can be attributed to the failure to strictly follow such a strategy. Many modern industrial information system project efforts, as well as older "legacy" information systems currently in use by manufacturers, fail by skipping the first four rows of the development effort and going directly to Step 5. Often, organizations do not create models of their requirements or information flows. According to Zachman (1996), "those models that are not explicitly produced are implicitly assumed by default, and a lot of assumptions ... have become erroneous." In addition, Row 1 and Row 2 models can change as quickly as management changes its mind. Unfortunately, like architectural buildings, Row 5 implementations are "poured in concrete" through software development, making changes expensive, difficult, and time-consuming once the programming has been completed. By procuring or developing new industrial information systems without proceeding with the appropriate design steps, we are only creating new legacy systems where the maintenance of such systems claims the majority of information technology resources.

23.6 Tools Comprising the Framework

By using the Zachman framework as a guide, a description of each of the steps for designing and implementing an industrial information system is provided in the following sections. We have supplemented the architecture with modeling tools and techniques that we have found to be quite effective for designing large integrated industrial information systems.

23.6.1   Step 1: Objectives/Scope Models: Facilitate a Focus Group

The first step in developing an industrial information system is to determine the scope and magnitude of the system. A mechanism must be found to identify and prioritize particular modules of the system appropriate for early implementation, thus keeping the effort at a manageable level. In most current systems development efforts, this task is very poorly performed. Typically, such requirements are solicited from a single stakeholder such as a vice president or production manager interested in the system, without involving other eventual users of the system. Soliciting the scope and objectives only from a single stakeholder, rather than from representatives from all parties that will be impacted by the system, often causes serious problems later in the project. Typical problems that often occur include project implementation delays due to modifications of software to meet others' requirements, later reprioritization of important modules, and mismatch among user requirements, often resulting in poorly written application software.

To properly identify the objectives and scope of the project, a focus group consisting of representative users should be brought together. This group should consist of production management, warehouse and shop floor operators, representatives from sales/order entry, engineering staff, purchasing, and so forth. The facilitator of the focus group solicits input from focus group members to identify all of the important modules required of the information system. A single high-priority module is selected for implementation. Other modules are then ranked for later implementation. We try to scope a project so that a single module is completely implemented every 3 to 6 months. If a module cannot be completely and successfully implemented in that time, then the scope is narrowed further until the time span can be successfully met. If the company wishes a larger portion or an additional module to be implemented, and they have the financial resources available to them, then we respond by adding personnel to concurrently develop an additional module.

Of critical importance to the successful completion of this step is to understand that the system must provide information to aid management in supporting the industrial process. Therefore, any tools that are developed to model the industrial information system must be structured to incorporate the process flow of the company, whether that be a manufacturing process flow, a warehouse shipping-and-receiving process, or an accounts-payable process. The reasoning is that information drawn from the information system is typically used to aid management personnel in decisions about the industrial process.

Two different approaches can be used to scope the project, depending on the size and complexity of the industrial system that must be supported. If the information system is to be designed for a small- or medium-sized facility with only a moderate number of industrial processes, then scoping the project is a simple task. For this task, the focus group is charged with the task of generating a prioritized list of modules needed for the system. This begins with each member listing the modules of most importance. Conflicts often arise when many different modules are quickly needed. However, through proper facilitation, participants can be encouraged to debate their needs among each other and to come to a consensus on a prioritized list. In many industrial companies, this list often results in a consensus for a receiving system, as this is the location where product first enters the facility. At other times, order entry or purchasing ranks high on the list, as the focus group sees the beginning of the process at the order entry desk or through the material-procurement process.

For the data description at the scope level of the Zachman framework, the focus group first selects the highest ranked module from the process description. They then identify a prioritized list of reports that they would like to see generated from the module. For example, if job-tracking ranks high on the process list, then high ranking reports may include travelers, process histories, and work-in-process status.

The network description at the scope level should tell the locations *where* the information system will be collected, processed, and used. These will be locations where data collection will ultimately be done, where reports will be generated, or where peripherals such as bar code scanners or printers will be located. For industrial information systems, we have found that the tool that best serves this purpose is a schematic of the physical layout of the manufacturing facility where the information system will be used (see Figure 23.2). For example, if job-tracking ranks high on the process description, then the focus group

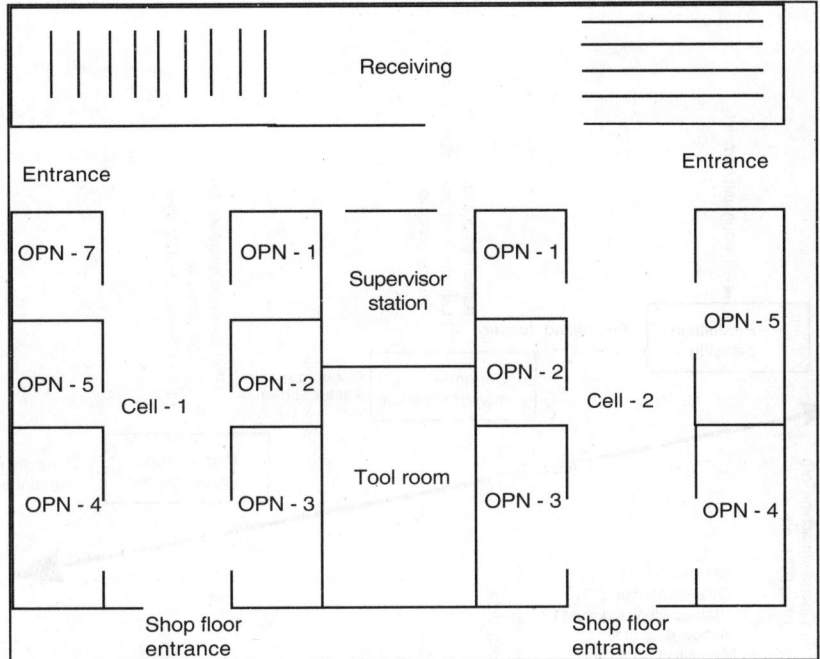

FIGURE 23.2   Facility layout for scoping network requirements.

must make a decision as to which organizations will first receive computer equipment and peripherals. In the development of the wireless job-tracking system, the focus team decided that the system needed to track the product from the receiving area, through the manufacturing shop floor operations, and into the shipping area. A facility layout of these three areas was drawn and served as a starting point for the network architecture design that was subsequently designed and implemented.

If the manufacturer is a large company such as a shipbuilder or automobile assembly plant, then a more structured approach is needed to identify and prioritize software functional modules and information. For such large manufacturing systems, manufacturing processes are often numerous, large, and complex. As a result, the information system requirements for such systems are also diverse and complex. Recently, the authors developed an architecture for a new naval ship repair facility. This facility was designed to repair over 700,000 individual ship components through 33 different group technology part families (Bidanda et al., 1998). For such large manufacturing systems, a popular and useful tool found to aid in definition of the scope of the project was the *triple diagonal* diagram (Shunk, 1992; Billo et al., 1994). The triple diagonal diagram provides a systematic method to identify and prioritize industrial application modules as well as to identify critical points for ADC and reporting. The triple diagonal provides a "big picture" of integration that shows, on a single diagram, how the industrial information system must be designed in order to allow the customer to control the process at hand.

The triple diagonal comprises three levels: the *execution level*, which models the flow of material or people that the information system must track; the *control level*, which models the control mechanisms governing the flow of material or people; and the *planning level*, which models the long-term planning functions that regulate the control systems. Figure 23.3 illustrates a portion of a triple diagonal to track machined parts as they proceed through the extrusion-honing process that served as the case. Details describing the development of a triple diagonal diagram can be found in other recent writings (Shunk et al., 1987, 1992).

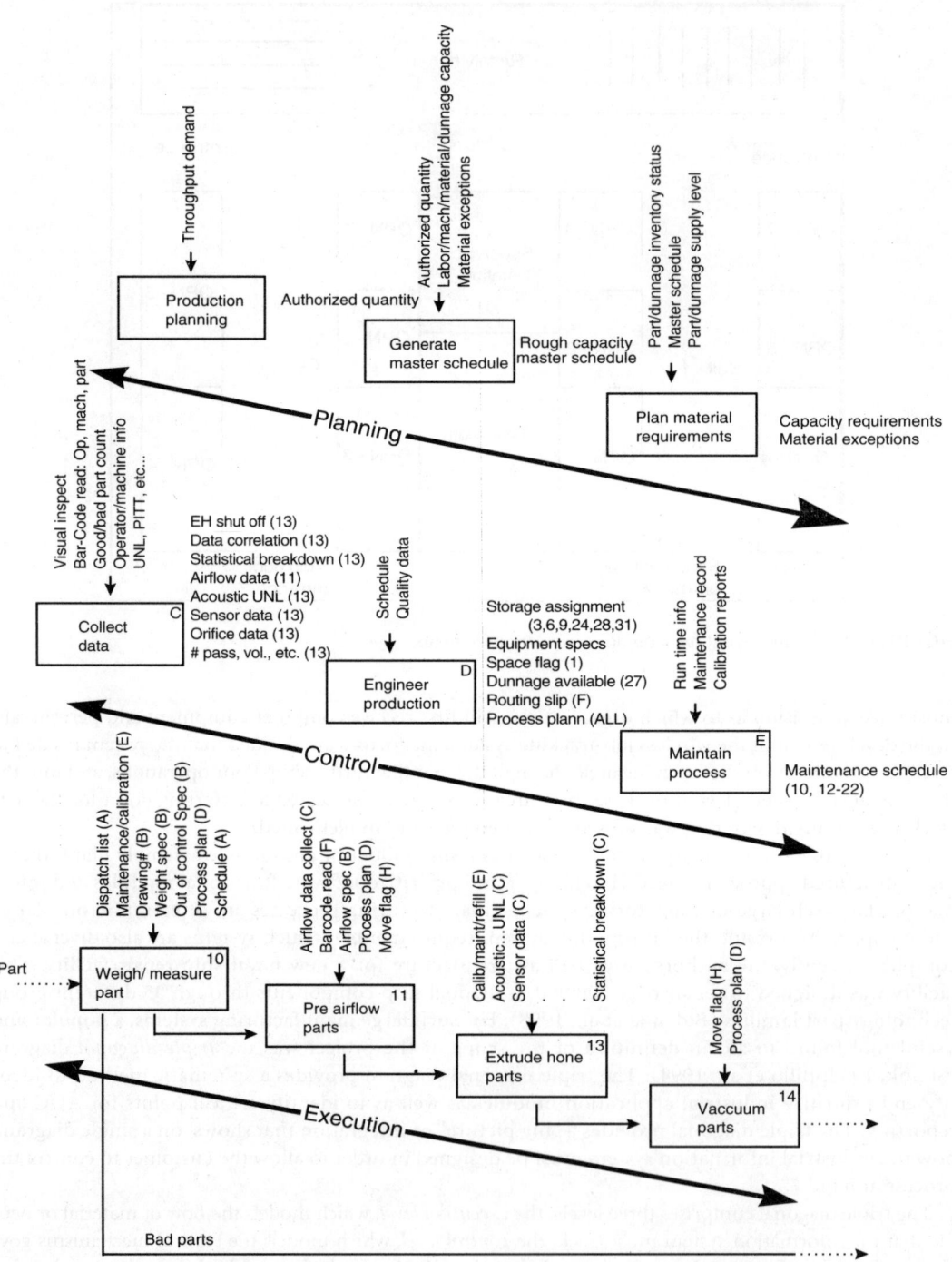

FIGURE 23.3   A triple diagonal diagram.

TABLE 23.1   Application Modules and Reports Resulting from Triple Diagonal

Modules	Data Flows
Material movement	Routings
	Component part KanBan
	Control item KanBan
Scheduling	Layouts/routings
	Material availability
	Bill of material
	Depot overhaul factor
	Master schedule
	Weekly finite schedule
	Asset invoice
	Daily dispatch list
	Transfer of accountability
WIP tracking	Material availability
	Individual production counts
	Individual scrap counts
	Part attribute data
	Received/issued part counts
	Hazardous material levels
	Material requests
	Move scrap ticket
	New parts attributes
	WIP status reports

Once the triple diagonal diagram is completed, critical functional modules and reports necessary to track and control the process can easily be derived by the focus group. Similar data flows providing inputs to the execution level functions are grouped into modules to identify the general business applications comprising the information system. Table 23.1 displays an example of the lists resulting from this process. The modules represent the process description for the scope level of the framework, while the data flows represent the important reports for each module.

23.6.2   Step 2: Enterprise Models: Develop the Required Forms, Reports, and Locations

Once the scope of the project has been identified and the highest priority module is selected for either design (if custom development is to be done) or further specification (if a commercial product is to be procured), then a series of tasks must be undertaken to specifically determine the business needs that the information system must meet. This step is reflected in the Enterprise phase of the framework. For industrial systems, tools designed for each cell of the framework must be easy to communicate to the user. Therefore, such tools as ER diagrams or traditional data-flow diagrams (which convey information to the information systems designer better than to the user from whom requirements are being solicited) *should be avoided* at this level of the project. The goal is to develop models for this level of the framework that physically appear similar to the desired end-product of the project.

For the enterprise level data description, an efficient way to extract business needs from users is to solicit their ideas on critical information they feel the system should collect and reports that should result as output from the system. This solicitation of requirements is not merely a listing of entity types. Working with users, the design team must actually sketch sample reports and input screens. This exercise forces the users to think not only about what information they feel they need from the industrial information system, but also about the way the information should be presented. This effort, if performed early in the design cycle, significantly reduces the number of revisions when the system is finally implemented. This idea came about by first determining that the user's interaction with the information system will be only at two points: data collection (input) and information reporting (output). Therefore, a technique was needed that would

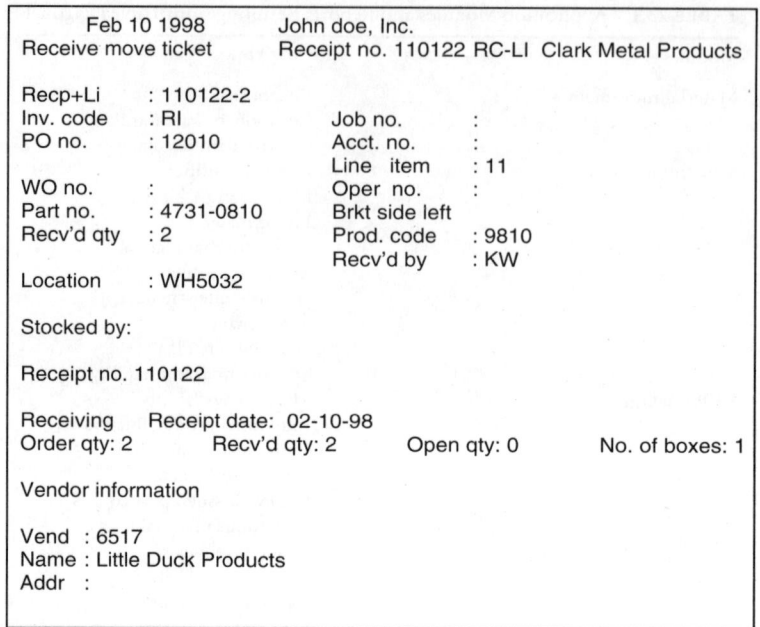

FIGURE 23.4   Current receiving move ticket.

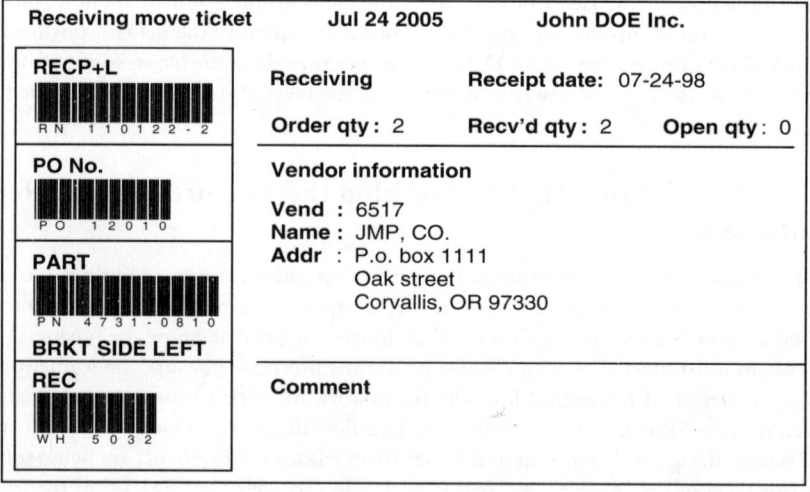

FIGURE 23.5   Proposed receiving move ticket.

mimic the information resulting from these two interface points. Figure 23.4 and Figure 23.5 illustrate an example of this technique. Figure 23.4 illustrates the existing Receive Move Ticket for a warehouse Receiving operation. In discussions with warehouse personnel, much of the information in this figure was unused, was entered manually, and had no method for ADC. Figure 23.5 illustrates a redesigned Receive Move Ticket for the warehouse receiving operation. In this redesigned form, the user wished to simplify his receiving and storage process through the incorporation of a bar code symbol onto the Move Ticket. In addition, the redesigned form incorporated only the data that were actually used by warehouse personnel.

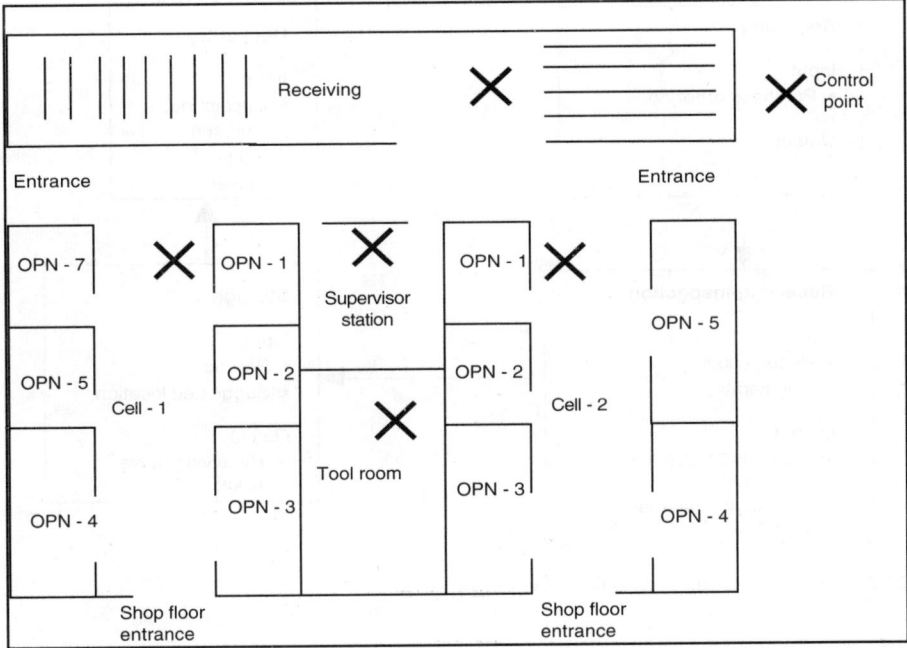

FIGURE 23.6   Control point diagram.

This Move Ticket served as a *visual target* for the application programmers of how the end product should actually appear.

For the enterprise level network description, the design team must begin to identify the control points — physical location of data entry and reporting points for the information system. The control points are typical locations for data capture scanning technology, computer clients, bar code printers, etc., and serve as the next step in the development of a local area network for the information system. Control points can be mapped directly onto the facility layout of the proposed area, as illustrated in Figure 23.6.

For the wireless work order tracking system, users decided that data must be collected within each manufacturing cell of the facility as well as the shipping/receiving area in the warehouse. Each cell operator would be given one wireless bar code reader that they could take from operation to operation for work order scanning.

The enterprise level process model describes the functional needs of the business managers and staff that the system must support. Once again, the tools used for this task center around the process flow of the industrial system. Users are best able to determine the needed requirements of the information system by associating them with the detailed steps necessary in carrying out the process. Analysis of non-value added activities associated with this process leads to the design of a new process that is lean and automated to the greatest extent possible.

Modifications of UML activity diagrams (Booch et al., 1999) work well for this task. Figure 23.7 and Figure 23.8 illustrate an example of a receiving process in a manufacturing facility. In Figure 23.7, the boxes in the illustration describe each step in the receiving process. The lists below each box describe the information that is input or output from each process step. In the analysis of this Receiving process, it was learned that the Receive Move Tickets were collected (but not entered) at the data entry desk for a 24-h period. This information was then entered into the computer by manually keying it into the system. In addition, input data at the preceding operations was either manually written onto forms or keyed manually into the system. These inefficiencies caused the database to be out of date with respect to received material; also, it often had errors because of periodic mistakes in data entry. As a

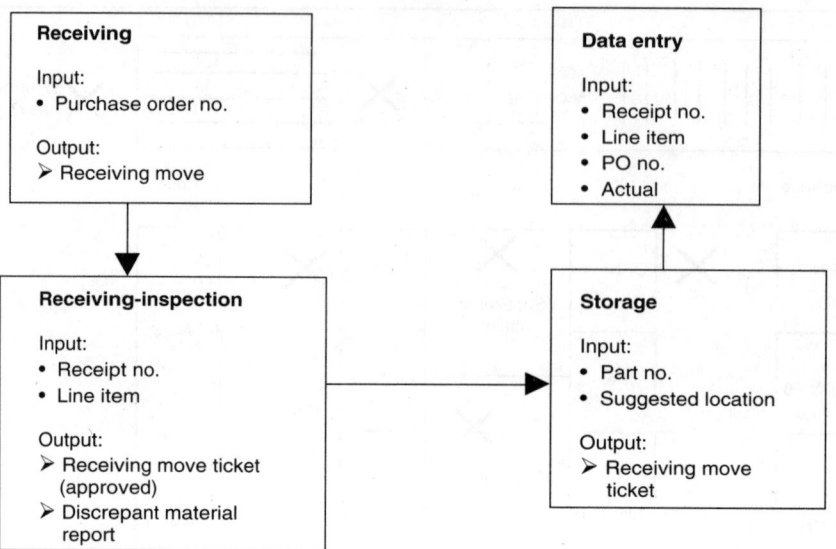

FIGURE 23.7   Current receiving process flow diagram.

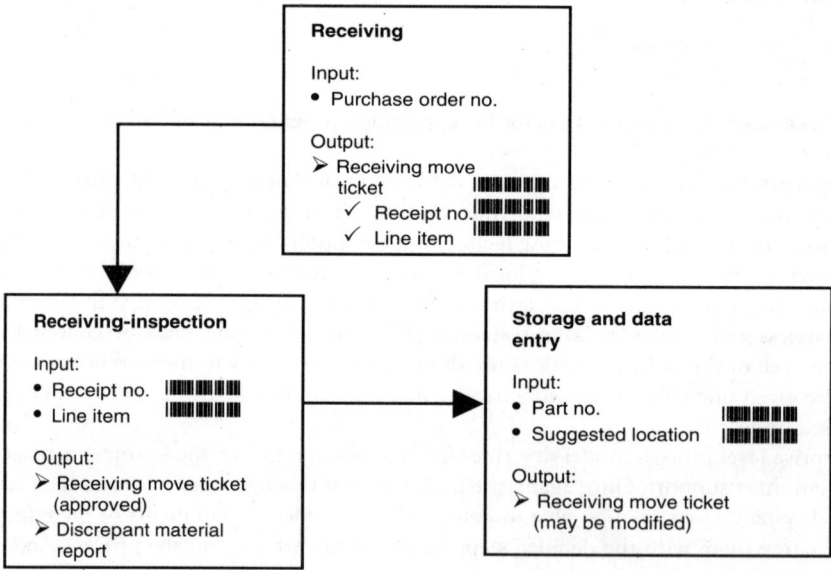

FIGURE 23.8   Proposed receiving process flow diagram.

result, manufacturing personnel often made incorrect decisions on scheduling product for production, and sales personnel were unable to make or keep commitments on sales orders owing to the lack of timeliness and to errors in the data.

Figure 23.8 illustrates the newly designed receiving process. Corresponding to the redesigned forms illustrated above, many input entities were automatically captured via bar code symbols on a real-time basis. This made the final manual data entry operation unnecessary, allowing such advantages as a reduction in labor, a timely and up-to-date database, and an elimination of data errors.

23.6.3 Step 3: Systems Models: Use Application Appropriate Modeling Tools

This phase of the design is intended to provide models of the industrial information system that are independent of any technological or equipment constraints. This point of the project is purely in the domain of the system designers. Tools and language used in this phase of the project are those commonly used by database designers, network engineers, specialists in ADC, and functional experts in the particular applications to be developed.

The systems data level is equivalent to conceptual design as described in the ANSI three schema architecture for database design (Tsichritzis and Klug, 1978). In this stage, information identified from screens and reports drafted from the previous step is extracted and placed into a data model such as the well-known ER model (Chen, 1982; Elmasri and Navathe, 2004), or UML class diagrams (Booch et. al., 1999). Figure 23.9 displays a small portion of the class diagram for the wireless shop floor tracking system in support of the extrusion honing process. In the figure, this model supports both the job-tracking module and provides data in support of the mathematical expressions needed to monitor the extrusion honing machine tool.

At the systems process level, functional models of the application at hand must first be developed before software can be written. Just as a building cannot be designed unless an architectural blueprint is first completed to act as a guide to the builders and to solicit input and feedback from owners, no information

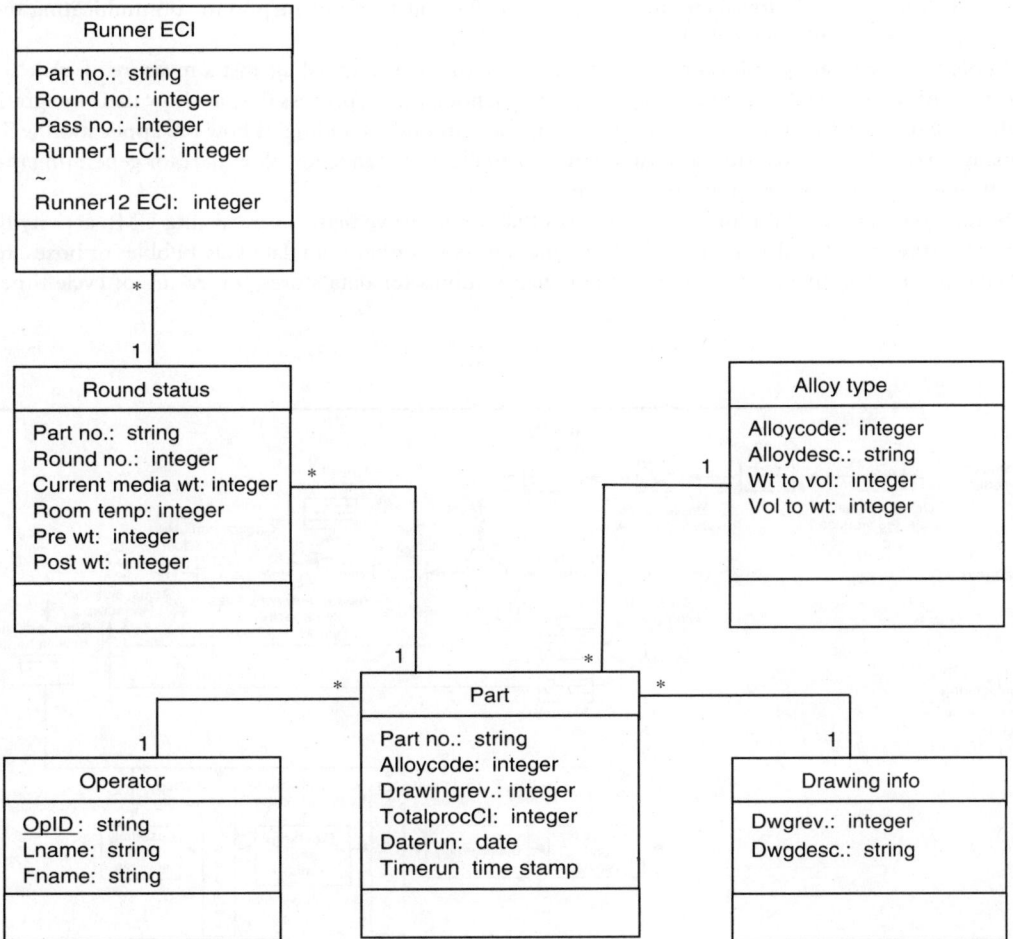

FIGURE 23.9 UML class diagram.

system can be developed without there being first a functional model of how it will operate. For these reasons, functional application experts must now be added to the project team. Their role is to identify the correct application theory, and then to customize this theory for the particular organization under study. In addition, it is at this point that the design team should be trained on the concepts and theories of the application at hand, so that they can take part in identifying and customizing the theory for their particular manufacturing system. This is the case for all industrial information system modules, including work order tracking, inventory management, SPC, financials, purchasing, and so forth. Without such models, the chances of a mismatch between a manufacturer's actual operating requirements and the functionality of any procured or developed software are very high.

There are two types of tools needed to complete descriptions of the three cells of the systems process model: (1) tools to describe operation of the application and (2) tools to describe the integration of the application into the manufacturing process. We do not attempt to provide a formula to represent the particular application, as it will vary on the basis of the application type. For example, a scheduling application may best be modeled through mathematical expressions; whereas a cost accounting application will best be modeled through identification and definition of company cost drivers, activities, overhead allocations, etc.

Integration of the application into the manufacturing process is best achieved through the use of flow-charting techniques. Because the system data collection, processing, and reporting are at least one step removed from the user's first-hand knowledge, *iconic flowcharts* can be helpful in communicating these processes (Martin, 1988; Hostick et al., 1991; Billo et al., 1994).

Iconic flowcharts are graphical models of business processes mapped against a measure of cycle time and areas of responsibility. They are mapped to the manufacturing process flow to show their integration with the manufacturing process, giving the design team an understanding of how the application will be actually carried out on a day-to-day basis. Figure 23.10 illustrates an iconic flowchart for generating manufacturing schedules for an engine overhaul process.

Iconic flowcharts contain four major features that are improvements over existing UML activity diagrams or other structured analysis tools. These include icons rather than data flow bubbles or boxes, representations of documents and reports rather than symbols for data stores, a measure of cycle time to

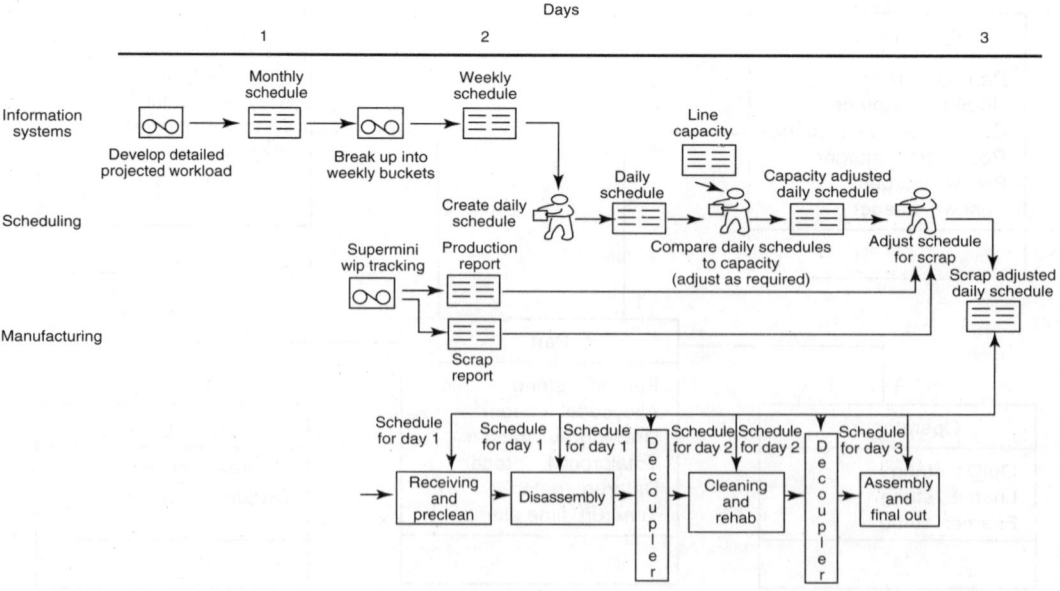

FIGURE 23.10   Iconic flowchart.

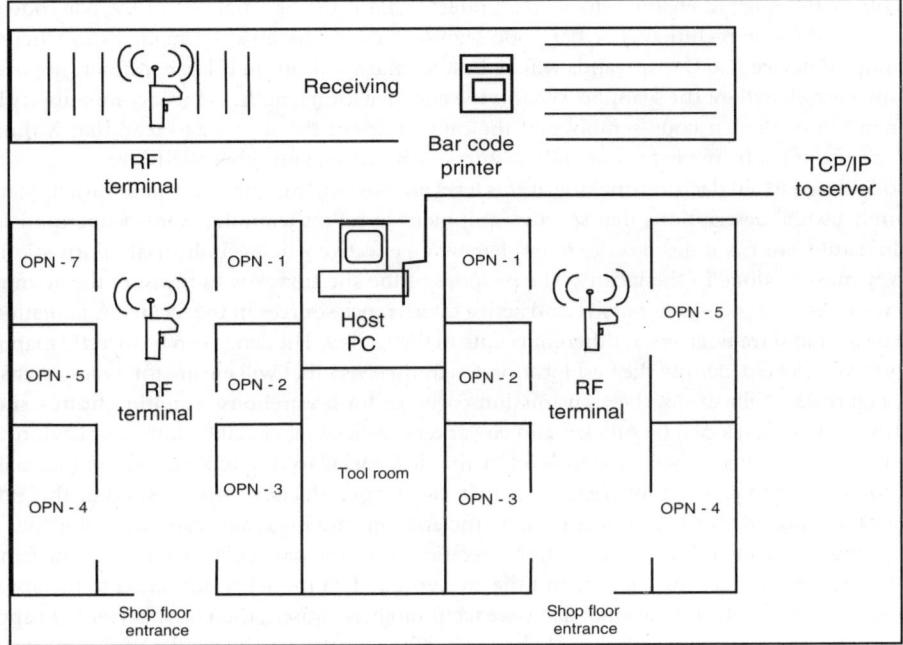

FIGURE 23.11   Overlay of material tracking system on facility layout.

determine elapsed time for completion of the information process, and depiction of areas of responsibility for different portions of the information flow (Rosch and Mervis, 1975; Mervis and Pani, 1980).

At the systems network level, the first draft of a computer architecture is designed. This architecture will consist of the selection of the generic ADC technology, computer technology, and network connections. In this model, the design is once again placed directly on the shop floor layout (see Figure 23.11). This model is kept free of any technological constraints that may be imposed by selection of any particular vendor product. For the material tracking system for the extrusion-honing process, a wireless data collection system was selected as the technology of choice. For this system, users decided that data must be collected within each manufacturing cell of the facility as well as the shipping/receiving area. Each cell operator would be given one wireless bar code reader that they could take from operation to operation within the cell for job scanning. This was deemed more cost-effective than having a traditional wire-bound scanner at each operation in the cell.

23.6.4   Step 4: Models: Select Technologies and Complete Designs

This is the most critical juncture of the project, for it is at this point that the design team makes the decision to either build a custom information system or to procure a commercial system. Once again, it is necessary to have experts available in each of the three facets of the architecture to aid in these detailed design steps. Technical knowledge of the advantages and constraints of specific functional applications, ADC technologies, computer architectures, and database technologies are addressed. Although the decisions at this step are numerous, varied, and complex, they are more manageable because of the compartmentalization that the Zachman model requires.

For the technology constrained network level, a variety of decisions must be made. For example, consider some of the following questions that must be answered concerning the ADC technology for the information system. If bar codes are going to be used, then which symbology is most appropriate for the manufacturing setting? What is the optimal X dimension (nominal width of the narrowest bar) of the particular symbology to ensure reliable read rates? What is the optimal size of the bar code symbol? Should a check digit be included? For bar code scanners, decisions must be made as to whether laser, charged coupled device (CCD), or wands will be best for the application. If laser scanners are used, what is the optimal wavelength of the scanner? What is the effective focal length? Is there compatibility between the X dimension of the bar code symbol and the capabilities of the scanner to read that X dimension? Similar decisions must be made for bar code printers, computers, and other hardware.

The models used to aid decision-making at this level are obviously going to be quite varied. Many models are simply textual descriptions that specify equipment to reflect team decisions. Some models can be pictorial in nature. For example, in order to implement an effective wireless industrial information system, a site survey must be done in the facility. The purpose of the site survey is to measure the attenuation of the radio waves as they encounter passive and active interference sources in the facility. Attenuation affects the range of the radio transceivers at different points in the facility. The range can be directly mapped onto the facility layout to help identify the best location for transceivers that will ensure total radio transmission coverage, Figure 23.12 illustrates the transmission coverage for a warehouse resulting from a site survey optimization system developed by Adickes and coworkers (Adickes et al., 2002; Billo and Layton, 2004).

At the technology constrained process level, if the decision is to develop the system internally, then standard software engineering tools (e.g., structure charts, pseudocode) will be used at this point and throughout the remainder of the project so that efficient software programs can be developed.

At the technology constrained data level, the decision as to the particular database management system (DBMS) technology is now chosen, and the conceptual data model is translated to the appropriate logical model. For example, if relational database technology is chosen, then formal rules are applied for migrating keys, and tables are normalized. The formats for displaying the results of this migration task can be many, including the application of appropriate relational data models such as $IDEF_{1X}$, UML class diagrams, or a text-based data dictionary that contains a description of each table and its attributes.

For procurement of commercial systems, three deliverables will be sent to potential vendors for bid: a *data specification* of the reports, screens, data, and the database proposed for the project (e.g., class diagrams,

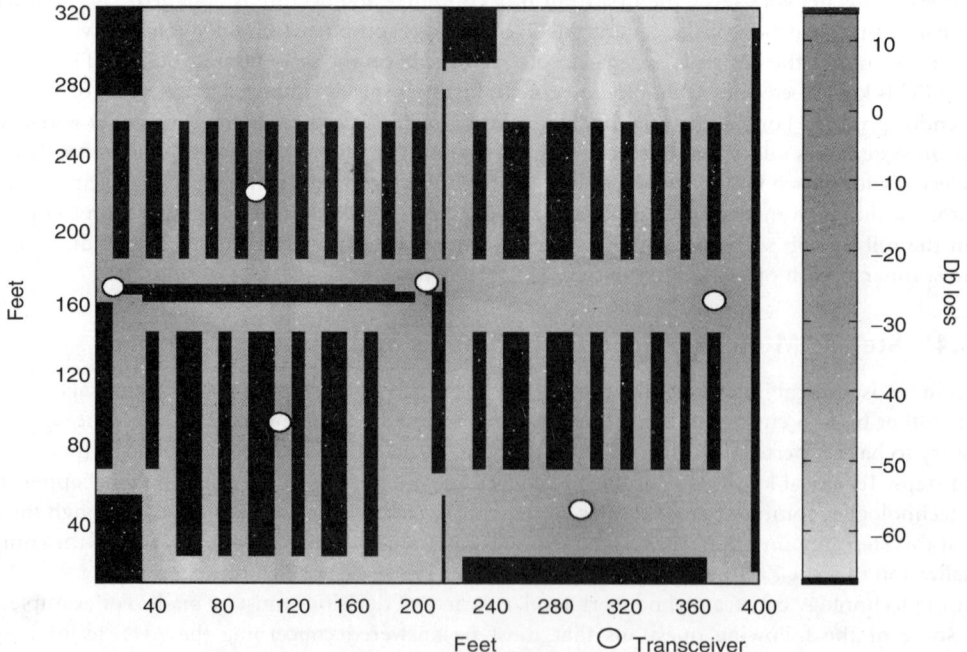

FIGURE 23.12   Overlay of site survey results on facility layout.

data dictionaries, sample reports, sample screens); a *functional specification* describing each module and the functional capabilities expected of each module; and an *architecture specification* describing the data capture technology proposed (e.g., laser scanners, thermal transfer printers) and the network protocol (e.g., IEEE 802.11 b/g for wireless communication), along with network diagrams showing data-collection points, computer architecture, site survey results, etc.

For the data-collection scanning technology, the design team will often devise a series of internal tests to compare the read rates and reliability of different bar code scanners under consideration. Evaluation copies of these instruments are often provided by the vendors for such purposes. For the wireless tracking system we have described, various types of bar code readers (wands, lasers, pen-based, etc.) were demonstrated to the cell operators. The operators came to a consensus on the technology they felt would be most productive and easiest to use in their setting. Particular vendor products were then tested to determine which yielded the fastest scanning times.

23.6.5 Step 5: Detailed Representation Models: Develop Software Programs and Rapid Prototype

This final step of the project is the actual coding of the software. If a commercial system has been selected, then training on its detailed functionality must be conducted. In addition, this product must be customized to meet the unique needs of the particular company. If the previous design steps have not been completed in a thorough manner, this effort can be a time-consuming and expensive task. A poor match between the capabilities of a commercial software product and the actual requirements of the company will require extensive modifications of the commercial product. The costs for such modifications can far exceed the initial purchase price of the product itself.

If a system is to be developed, the detailed data model cell consists of data definition language statements (e.g., SQL statements) for creation of database tables or type definitions. The process description consists of actual computer programs; and the network description consists of programs configuring specific computers and peripherals with individual node addresses.

As a final step in the detailed representations level of the framework, we recommend that every application module be subjected to a short testing period in a limited area in the industrial setting. Usually, the testing period can be limited from 2 to 5 days of actual usage. Programmers and engineers from the design team should be close by to help users with operation of the system and to obtain firsthand feedback on needed modifications to the system. Such a testing period has many benefits for the project. Users have an opportunity to provide firsthand feedback to developers on the strengths of the information system, and things they would like to see changed. If there are any errors in the system, they will be quickly caught.

23.7 Case Study

The tools described above were utilized in the development of a shop floor tracking system for a company opening its new 150,000 ft^2 extrusion-honing manufacturing facility. The company was awarded a large subcontract with a major U.S. automotive manufacturer. Specifically, it was contracted to apply a series of extrusion-honing operations to polish the inside passageways of engine intake manifolds and heads. As part of the quality reporting requirements, automated tracking of the status and quality of each manifold and head was required through the entire manufacturing process, from point of warehouse receipt of unpolished packaged components on pallets through final shipments of completed product. Parts were to be tracked through each and every shop floor operation including all inspection operations. Information was to be entered by operators using handheld radio frequency terminals, which, in turn, automatically and on a real-time basis, communicated with a central computer that uploaded the information to the database. Data from this database were used for generating part counts, process times, and audit trails that accompanied the final product shipments to the customer. In addition to the above requirements, the tracking system had to be seamlessly integrated with the sensors monitoring each machine tool to provide real-time process monitoring and control.

TABLE 23.2   Tools and Products Developed for Wireless Tracking System

Data	Process	Architecture
Triple diagonal model	Data collection strategy	Shop floor cell layout
Data model	Part count strategy	Control points
MS Access database	Bar code generation software	Hardware identification
		—Computers
		—RF equipment
		—Bar codeprinters
File structure for data upload	RF collection software	RF site survey
	Data upload/processing software	Rapid prototype/test
	Report generation software	

Table 23.2 lists many of the tools and products that resulted from this project. After proceeding through the Zachman model using the described tools through the technology model level, the users felt it was more cost-effective to design a work-order tracking system rather than to procure a commercial system. The primary reason for this decision was the necessity for seamless integration with process monitor/control data provided by the extrusion-honing machine tools.

Because of the proprietary nature of the details of the extrusion-honing process, not all of the models and tools of the project and process can be revealed. However, several of the design tools used have been illustrated throughout this report. For example, Figure 23.3 illustrates the triple diagonal diagram that was developed to scope the project and to identify initial requirements with the company's managers and supervisors. The focus group used this diagram as a guide to aid them in selection of the part tracking system as the primary focus of the development effort. Once this diagram was completed, the focus group only needed about an hour to identify and prioritize major modules. The diagram also helped to scope the beginning and ending points of the tracking process to be incorporated into the material tracking software.

Figure 23.5 shows a sample of the detailed data requirements identified in the receiving process. This diagram helped the project team identify the critical data elements that had to be stored in the database for the receiving application. It also served as an impetus for redesigning Receiving Move Tickets (Figure 23.4) and process travelers that contained bar code symbols.

Figure 23.9 illustrates the data model developed to identify the structure of the machine tool process model data. This model allowed the project team to integrate the product tracking data with the machine tool quality data. This model actually illustrates only the machine tool monitor data that were stored in the database. What is not shown are the proprietary mathematical operations developed by a team of senior fluid engineers for the calculation of such parameters as extrusion media viscosity, media loss as a function of machine tool passes, and so forth. These models were ultimately incorporated into the process monitoring software for the machine tools, with resulting data integrated with the component tracking data. In addition, neural nets models were developed to serve in adaptive control of the extrusion honing process.

Finally, Table 23.3 illustrates one of the reports generated by the system showing the process flow of several components. This report was derived from an initial sketch of the report that was developed in the early stages of the framework process.

Because the design steps had been modularized, the first usable prototype of the tracking software and completion of the programming of required reports was completed in 4 months. The entire project was completed within 6 months, using only one engineer from our research center and two members of the company project team. However, it should be noted that this team did not work in isolation. The success of the project was due not only to the usage of the tools and framework, but also to the countless interactive design discussions and feedback sessions with users of the proposed system as well as experts in fluid engineering and process control. The cost of the project, including labor and equipment, was approximately $55,000. As a comparison, bids for commercial software began at $100,000 for software modules, with approximately an additional $120,000 estimated for labor costs to customize and implement the

TABLE 23.3 Sample Part Operations Summary Report Manifolds Operations Tracking Report

Part Number	Operation Performed	Operator	Date	Time
11112	Preclean	00000123554437	09-18-19	11:36:0
	Preinspect	00000123554437	09-18-19	11:36:1
	Preairflow	00000123554437	09-18-19	11:36:2
	Extrude hone	00000123554437	09-18-19	11:38:0
	Part Vacuum	00000694813232	09-18-19	11:59:3
	Part Wash	00000694813232	09-18-19	11:59:4
	Postairflow	00000694813232	09-18-19	12:00:0
12548	Preclean	00000123554437	09-18-19	11:54:2
22222	Part label	00000123554437	09-18-19	11:36:3
	Postairflow	00000123554437	09-18-19	11:36:5

product for the facility. At the conclusion of the project, the industry and government sponsor expressed their own satisfaction with the project's success, the user satisfaction with the material tracking system itself, and the team's efficiency and timeliness in the product design and implementation.

23.8 Concluding Remarks

The Zachman framework is one of the best-known architectures for implementing information systems today. However, successful usage of this technological approach for industrial settings requires the use of separate modeling tools at each phase of the architecture. In the present work, we have supplemented the framework with tools and techniques appropriate for the design of industrial information systems.

The utilization of the tools and techniques within the context of the Zachman framework has proven invaluable in developing cost-effective, high-quality, and timely manufacturing and warehouse tracking applications. When viewed alone, these methods and tools appear simplistic, even naive, as they are often no more than maps, drawings, and schematics. The contribution of this work is not only in the development of the individual tools per se, but also in their proper placement into the framework. When taken together, each tool builds on the knowledge obtained from earlier tools used in the framework.

We have used the Zachman model and supporting tools for over 15 years. The framework has been used in developing dozens of different industrial applications including shop floor control, activity-based costing, purchasing, inventory and process control, and warehouse and maintenance management. We have also successfully used the model and tools for such nonmanufacturing applications as aircraft inspection reporting, capital project tracking for package delivery companies, and patient tracking for hospital emergency departments (Puerzer, 1997). It has been successfully applied to businesses of all sizes, from start-up manufacturing firms to large government overhaul and repair facilities. The most useful feature of the framework is that it compartmentalizes the decision-making and design of industrial projects. This allows project scheduling to be more precise, ensures that specific design tasks are completed thoroughly, and keeps the project within established budget guidelines.

References

Adickes, M.D., Billo, R.E., Norman, B.A., Banerjee, S., Nnaja, B., and Rajgopal, J., Optimization of Indoor Wireless Communication Network Layouts, *IIE Trans.*, 34, 823–836, 2002.

Bidanda, B., Billo, R., Boucher, R., Canning, A., Collier, D., Dickson, C., Gallagher, T., Jessop, D., and O'Reilly, D., Designing and implementing large scale cellular manufacturing systems using group technology principles, *Industrial Engineering Solutions '98 Conference Proceedings*, Norcross, GA, Institute of Industrial Engineers, May 10–13, 189–196, 1998.

Billo, R.E. and Layton, W., A Report of a Site Survey for Radio Frequency Communication, Oregon State University, Technical Report, June 2004.

Billo, R.E., Rucker, R., and Paul, B.K., Three rapid and effective requirements definition modeling tools: evolving technology for manufacturing system investigations, *Int. J. Compu. Integrated Manuf.*, 7, 186–199, 1994.

Booch, G., Rumbaugh, G.J., and Jacobson, I., *The Unified Modeling Language User Guide*, Addison-Wesley, Reading, MA, 1999.

Brodie, M.L. and Ridjanovic, D., A strict database transaction design methodology, in *On Conceptual Modelling: Perspectives From Artificial Intelligence, Databases, and Programming Languages*, Brodie, M.L., Mylopoulos, J., and Schmidt, J.W., Eds., Springer-Verlag, New York, 1984.

Campbell, K., barcode tracks progress, *Packag. Dig.*, 38, 2, 2001.

Chausse, S., Landry, S., Pasin, F., and Fortier, S., Anatomy of a kanban: a case study, *Prod. Inventory Manage. J.*, 41, 11–16, 2000.

Chen, P.P., Applications of the entity-relationship model, in *Lecture Notes in Computer Science*, Goos, G. and Hartmanis, J., Eds., Springer-Verlag, New York, 1982.

Date, C.J., *An Introduction to Database Systems*, 8th. ed., Addison Wesley, New York, 2003.

DeMarco, T., *Structured Analysis and System Specification*, Prentice-Hall, Englewood Cliffs, NJ, 1979.

Elmasri, R. and Navathe, S., *Fundamentals of Database Systems*, 4th ed., Benjamin/Cummins, Redwood City, CA, 2004.

Gane, C. and Sarson, T., *Structured Systems Analysis*, Prentice-Hall, Englewood Cliffs, NJ, 1979.

Gilmore, D., Transportation logistics, *ID Sys.*, 18, 1998, http://www.scs-mag.com/reader/1998_04/vio0498.htm

Hostick, C.J., Billo, R.E, and Rucker, R., Making the most of structured analysis in manufacturing information system design: application of icons and cycle time, *Compu. Ind.*, 16, 267–278, 1991.

ICAM Architecture Part II – Volume IV – Function Modeling (IDEF0), Technical Report AFWAL-TR-81-4023,s Materials Laboratory, Air Force Wright Aeronautical Laboratories, Air Force Systems Command, Wright Patterson Air Force Base, OH, June, 1981.

Kirzner, K., Turning inventory into profits, *Frontline Solutions*, 3, 17–20, 2002.

Loomis, M.E.S., *Object Databases: The Essentials*, Addison-Wesley, New York, 1995.

Manthou, V. and Vlachopoulou, M., Bar-code technology for inventory and marketing management systems: a model for its development and implementation, *Int. J. Prod. Econ.*, 71, 157–164, 2001.

Martin, C.F., *User-Centered Requirements Analysis*, Prentice-Hall, Englewood Cliffs, NJ, 1988.

Mervis, C. and Pani, J., Acquisition of basic object categories, *Cogn. Psychol.*, 12, 523–553, 1980.

Obstgarten, M., Towards paperless aerospace manufacture: 2-D data matrix coding for jet-engine parts, *Adv. Imaging*, 15, 12–14, 2002.

Pahlavan, K. and Levesque, A.H., *Wireless Information Networks*, Wiley, New York, 1995, pp. 359–417.

Parker, C., AHA outlines to FDA effectiveness, utility of barcode labels in hospitals, *AHA News*, 38, 6, 2002.

Perkins, A., Implementing the Zachman framework for enterprise architecture, http://www.ozemail.com.au/~visible/papers/Zachman.html, 1997.

Porter, J.D., Billo, R.E., and Mickle, M.H., A standard test protocol for evaluation of radio frequency identification systems for supply chain applications, *J. Manuf. Sys.*, 23, 46–55, 2004.

Puerzer, R.J., *A Patient Tracking and Control System for use in the Emergency Department*, Ph.D. thesis, University of Pittsburgh, Pittsburgh, PA, 1997.

Quinn, P., Major new study to explore health care's use of AIDC, *ID Systems 17*, 11, November, 86, 1997.

Rosch, E. and Mervis, C., Family resemblances: studies in the internal structure of categories, *Cogn. Psychol.*, 7, 573–605, 1975.

Rose, G., Providing premium carpool parking using a low-tech ITS initiative, *Inst. Transp. Eng. J.*, 72, 32–36, 2002.

Rumbaugh, J., Blaha, M., Premerlani, W., Eddy, F., and Lorensen, W., *Object-Oriented Modeling and Design*, Prentice-Hall, Englewood Cliffs, NJ, 1991.

Shunk, D.L., *Design and Development*, Business One/Irwin, Homewood, IL, 1992.

Shunk, D., Sullivan, B., and Cahill, J., Making the most of IDEF modeling – the triple diagonal concept, *CIM Rev.*, 12–17, 1987.

Shunk, D.L., Paul, B.K., and Billo, R.E., Managing technology through effective user needs analysis: a federal government case study, in *Management of Technology III*, Khalil, T.M. and Bayraktar, B.A., Eds., Institute of Industrial Engineers, Norcross, GA, 1992.

Sowa, J.F. and Zachman, J.A., Extending and formalizing the framework for information systems architecture, *IBM Sys. J.*, 31, 3, 1992.

Strozniak, P., Replenishment Heaven, *Frontline Solutions*, 3, 22–26, 2002 .

Tsichritzis, D. and Klug, A., Eds., *The ANSI/X3/SPARC DBMS Framework*, AFIPS Press, Montvale, NJ, 1978.

Ward, P.T. and Mellor, S.J.,*Structured Development for Real-Time Systems: Introduction & Tools*, Yourdon Press, New York, 1985.

Zachman, J.A., Concepts of the framework for enterprise architecture http://www.ozemail.com.au/~visible/papers/zachman3.htm, 1997.

Zachman, J.A., A framework for information systems architecture, *IBM Sys. J.*, 38, 2 & 3, 454 – 470, 1999.

Zachman, J.A., Enterprise architecture and legacy systems, http://www.ozemail.com.au/~visible/papers/zachman1.htm, 1996.

24

Fuzzy Group Decision-Making

David Ben-Arieh
Kansas State University

Zhifeng Chen
Kansas State University

Group decision-making is an essential activity in many domains, including such important ones as the financial, engineering, and medical fields. In group decision-making, we basically solicit opinions from experts and combine these judgments into a coherent group decision. Experts typically express their opinions both quantitatively and qualitatively.

Experts often cannot express judgment in accurate numerical terms and use, instead, linguistic labels or what are called "fuzzy" preferences. The use of linguistic labels makes expert judgment more reliable and informative. This chapter presents a review of group decision-making methods with emphasis on using fuzzy preference relations and linguistic labels, explores the various methods to gather individual opinions into a group decision, and shows how to calculate a consensus level that represents the degree of consistency and agreement between the experts.

We discuss the benefits and limitations of the various methods and provide numerous examples.

24.1　Introduction

Decision-making, as a specialized field of operations research (OR), is the process of specifying a problem or opportunity, identifying alternatives and criteria, evaluating alternatives, and selecting a preferred alternative from among the possible ones. Typically, there are three kinds of decision-making approaches:

1. Multiple criteria decision-making (MCDM)
2. Multiple objective decision-making
3. Group decision-making

24.1.1　Multiple Criteria Decision-Making

Multiple criteria decision-making is one of the most widely used methods in the decision-making area (Hwang and Yoon, 1981). The objective of MCDM is to select the best alternative from several mutually exclusive alternatives based on their general performance regarding various criteria (or attributes) chosen by the decision-maker. Depending on the type and characteristics of the problem, a number of MCDM methods have been developed, such as the *simple additive weighting method*, the analytical hierarchical process (AHP) method, various outranking methods, max–min methods, and lexicographic methods. Introduced by Thomas Saaty in the early 1970s, AHP has gained wide popularity and acceptance in decision-making. The AHP method supports a hierarchical structure of the problem and uses pair-wise comparison of all objects and alternative solutions. A lexicographic method is appropriate for solving problems in which the weight relationship among criteria is dominant and noncompensatory (Liu and Chi, 1995).

24.1.2　Multiple Objective Decision-Making

In multiple objective decision-making, the decision-maker wants to attain more than one objective or goal in choosing a course of action while satisfying the constraints imposed by environment, processes, and resources. This problem is often referred to as a vector maximum problem (VMP). There are two approaches for solving VMPs (Hwang and Masud, 1979). The first approach is to optimize one of the objectives while appending the other objectives to a constraint set so that the optimal solution will satisfy these objectives at least up to a predetermined level. This method requires the decision-maker to rank the objectives in order of importance. The preferred solution obtained by this method is one that maximizes the objectives, starting with the most important and proceeding according to the order of importance of the objectives.

The second approach is to optimize a superobjective function created by multiplying each objective function with a suitable weight and then by adding them together. One well-known approach in this category is goal programming, which requires the decision-maker to set goals for each desired objective. A preferred solution is then defined as the one that minimizes the deviations from the set goals.

24.1.3　Group Decision-Making

Group decision-making has gained prominence owing to the complexity of modern-day decisions, which involve complex, social, economical, technological, political, and many other critical domains. Many times, a group of experts needs to make a decision that represents individual opinions and yet is mutually agreeable.

Such group decisions usually involve multiple criteria with multiple attributes. Clearly, the complexity of MCDM encourages group decisions as a way to combine interdisciplinary skills and improve the management of the decision process. The theory and practice of multiple objectives and multiple-attribute decision-making for a single decision-maker has been studied extensively over the past 30 years. However, extending this methodology to group decision-making is not simple, because of the complexity introduced by the conflicting views of decision-makers and the varying significance or weight of those views in the decision-making process.

Moreover, the problem of group decision-making is complicated by several additional factors. Usually, one expects a decision-making model to follow a precise mathematical model because such a model can bring consistency and precision to the decision generated. Human decision-makers, however, are often reluctant to follow a decision generated by a formal model unless they are confident about the model's assumptions and methods. The input to such a decision model often cannot be quantified precisely, thus reducing the perceived accuracy of the model. Intuitively, the optimization of the group decision, in a mathematical sense, is contradictory to the concept of consensus and a group agreement.

The benefits of group decision-making, however, are quite numerous, justifying the additional effort required. Some of the benefits are

1. *Better learning.* Groups are better than individuals at understanding problems.
2. *Accountability.* People are held accountable by themselves and others for decisions in which they participate.
3. *Fact screening.* Groups are better than individuals at catching errors.
4. *More knowledge.* A group has more information (knowledge) than any one member. Groups can combine this knowledge to create new knowledge. More creative alternatives for problem solving can be generated, and better solutions can be derived (by group stimulation, for example).
5. *Synergy.* The problem-solving process may generate better synergy and communication among the parties involved.
6. *Creativity.* Working in a group may stimulate the creativity of the participants.
7. *Commitment.* Group members often have their egos invested in the decision, so they will be more committed to the solution.
8. *Risk propensity is balanced.* Groups moderate the instincts of high-risk takers and encourage conservatives or those who are averse to risk.

Generally, there are three basic approaches toward group decision-making (Hwang and Lin, 1987):

1. *Game theory.* This approach implies a conflict or competition between or among the decision-makers.
2. *Social choice theory.* This approach represents voting mechanisms that allow the majority to express a choice.
3. *Group decision using expert judgment.* This approach deals with integrating the preferences of several experts into a coherent and just group position.

24.1.3.1 Game Theory

Game theory can be defined as the study of mathematical models of conflict and cooperation between intelligent and rational decision-makers (Myerson, 1991). Modern game theory gained prominence after the work of Von Neumann in 1928 and later (Von Neumann and Morgenstern, 1944). Game theory became an important field during World War II and the Cold War that followed, culminating with the famous Nash Equilibrium. The objective of the games as a decision tool is to maximize some utility function for all decision-makers under uncertainty. Since this technique does not explicitly accommodate multiple criteria for selection of alternatives, however, we will not consider it in this chapter.

24.1.3.2 Social Choice Theory

Social choice theory deals with MCDM, since this methodology considers the votes of many individuals as the instrument for choosing a preferred candidate or alternative. The candidates can exhibit many characteristics, such as honesty, wisdom, and experience, as the criteria to be evaluated. The complexity of this seemingly simple problem of voting can be illustrated by the following example: a committee of nine people need to select an office-holder from three candidates, a, b, and c. The votes that rank the candidates are as follows:

- three votes have the order a, b, c.
- three votes agree on the order b, c, a.
- two votes have the preference of c, b, a.
- one vote prefers the order c, a, b.

Even a cursory review of the results reveals that each candidate received three votes as the preferred option, resulting in an inconclusive choice.

The theory of social choice was studied extensively with notable theories such as Arrow's Impossibility Theorem (Arrow, 1963, Arrow and Raynaud, 1986). This type of decision-making is based on the ranking of choices by the individual voters, while the scores that each decision-maker gives to each criterion of each alternative are not considered explicitly. Therefore, this methodology is less suitable for MCDM, in which each criterion in each alternative is carefully weighed by the decision-makers.

24.1.3.3 Expert Judgment Approach

Within the expert judgment approach, there are two minor styles, team decision and group decision (terminology based on Zimmermann, 1987). These styles differ in the degree of disagreement that the experts are allowed to have while constructing the common decision.

Generally, expert judgment methods can be divided into the following categories:

- *Methods of generating ideas*. These methods include brainstorming in verbal or written forms.
- *Methods of polling ideas*. These methods produce quick estimates of the preferences of the experts. Surveys, the Delphi method, and conferencing are implementations of polling ideas.
- *Simulation models*. These models include cognitive maps and the SPAN method (successive proportional additive network also known as social participatory allocative network).

There is a vast amount of literature on this topic, and this chapter provides only the most basic review in order to provide the background for the more detailed discussion on fuzzy group decision-making. A good review of the general MCDM field can be found in Triantaphyllou (2000).

The essence of group decision-making can be summarized as follows: there is a set of options and a set of individuals (experts) who provide their preferences over the set of options. The problem is to find an option (or a set of options) that is most acceptable to the group of experts. Such a solution entertains the concept of *majority* that is further explored below.

This chapter explores the application of fuzzy logic toward generating a group decision using expert judgment. As a part of the decision evaluation, the chapter explains the concept of *consensus* and develops various measures of that property.

The motivation behind using fuzzy sets in group decision-making comes from several sources:

1. The available information about the true state of nature lacks evidence, and thus, the representation of such pieces of knowledge by a probability function is not possible.
2. The user preferences are assessed by means of linguistic terms instead of numerical values. In many cases, these terms are subjective and inconsistent.
3. The decision-maker's objectives, criteria, or preferences are vaguely established and cannot induce a crisp relation.

24.2 Background

When trying to make a group decision, it is not always reasonable to expect all experts to choose the same option. Thus, a necessary ingredient in the group decision process is the issue of *majority*. Such a majority can be *strict*, as in "more than half" or "more than 75%." The majority could also be *soft*, as in "most" or "a large majority."

Handling the softer, qualitative terms leads the decision model toward using linguistic quantifiers based on fuzzy logic. In order to express softer terms, the following standard fuzzy set notation is used.

A *fuzzy set A* in X, $A \subseteq X$, is defined using a membership function $\mu_A: X \rightarrow (0.1); \mu_A \in (0, 1)$ is the grade of membership of X in A. This membership varies from full membership to nonmembership, through all intermediate values.

Linguistically quantified statements belong to the category of soft statements defining the degree of agreement among the experts. These statements, essential in everyday life, can be represented in general as

Qy's are F

where Q is a linguistic quantifier (such as "most"), y belongs to a set of objects (such as experts), and F is a verb property (such as convinced) (Kacprzyk et al., 1992). In addition to this definition, it is possible to add more information regarding the importance of the experts in the quantified statement. The *importance B* can be added, resulting in the statement:

QBy's are F

Such a statement can represent the understanding that "*Most* of the *important* experts are *convinced*." This statement can support the group decision that "alternative A is superior" or a similar decision outcome.

There are two types of linguistic quantifiers: proportional and absolute, as will be further discussed in Section 24.5.2. One approach toward group decision-making problems is to find the degree of truth of such a quantified statement.

A group decision-making process typically consists of generating a list of relevant ideas, screening out the poor ideas, and modifying the more promising ones to fit current goals. Idea generation can take the form of unstructured and structured group procedures. Popular unstructured group processes are interactive group brainstorming (group brainstorming) and nominal group brainstorming (solitary brainstorming). Structured group processes include Delphi and nominal group techniques (NGT) (Hwang and Lin, 1987).

24.2.1 The Fuzzy Group Decision-Making Process

In a fuzzy environment, the group decision-making problem can be solved in four steps (Herrera et al., 1997). First, one should unify the evaluations from each expert. The second step is to aggregate the opinions of all group members to a final score for each alternative. This score is usually a fuzzy number or a linguistic label, which is used to order the alternatives. The third step is to rank the linguistic labels or fuzzy numbers and select the preferred alternatives based on this order. Finally, the decision manager assesses the consensus level and the individual contribution to the group decision. The following procedure describes these steps in more detail:

1. Expressing fuzzy preference of alternatives
2. Aggregating individual preferences into a group decision
3. Ranking alternatives
4. Measuring group consensus

Before we explore the details of fuzzy group decision-making, the next section presents direct fuzzy group decision-making methods that were extended from the crisp methodologies.

24.3 Direct Fuzzy Group Decision Methods

24.3.1 Fuzzy Delphi

The Delphi approach uses expert opinion surveys with three special features: anonymous response, iteration and controlled feedback, and statistical group response. The number of iterations of Delphi questionnaires may vary from three to five, depending on the degree of agreement and the amount of additional information being sought or obtained. Generally, the first questionnaire asks individuals to respond to a broad question. Each subsequent questionnaire is built upon responses to the preceding questionnaire. The process stops when consensus has been approached among participants, or when sufficient information exchange has been obtained. Thus, one of the most attractive properties of this approach is the ability to gather and evaluate information from a group of experts without requiring a face-to-face meeting.

The Delphi approach typically involves three different groups: decision-makers, staff, and experts (Hwang and Lin, 1987). Decision-makers are responsible for the outcome of the Delphi study. A work group of five to nine members, composed of both staff and decision-makers, develops and analyzes all questionnaires, evaluates collected data, and revises the questionnaires if necessary. The staff group is directed by a coordinator who should have experience in designing and conducting the Delphi method and is familiar with the problem area. The staff coordinator's duties also involve supervising a support staff in typing, mailing questionnaires, receiving and processing results, and scheduling meetings. Respondents are recognized as experts on the problem.

The Delphi method is suitable for decision domains:

- Where expertise is subjective and inputs likely to be judgmental
- Where problems are large, complex, and multidisciplinary and considerable uncertainties exist
- Where there is the possibility of unexpected breakthroughs
- Where causal models cannot be built or validated
- Where time frames are particularly long
- Where opinions are required from a large group, and anonymity is preferred

One of the weaknesses of the Delphi method is that it requires repetitive surveys of the experts to allow the evaluations to converge. The cost of this method rapidly increases with repetitive surveys, especially in large and complicated problems (Ishikawa, 1993).

The fuzzy Delphi method is applied to alleviate this problem. Using fuzzy numbers or linguistic labels for evaluating the experts' opinions allows a faster convergence to an agreeable group decision. An example presented in Cheng (1999) uses linguistic terms to express the experts' responses.

24.3.2 Fuzzy Analytic Hierarchy Process (Fuzzy AHP)

In some instances, decision problems are hard to conceptualize or even clearly define. The AHP was formulated to support the decision-maker in these situations.

Analytical Hierarchical Process is based on the following two steps: structuring the decision as a hierarchical model, and then using pairwise comparison of all criteria and alternatives to find the calculated weight of the criteria and the score of each alternative. This approach allows decision-makers to examine complex problems in a detailed rational manner. The hierarchical representation helps in dealing with large systems, which are usually complex in nature. The decisions are made one level at a time, from the bottom-up, to more clearly aggregate strategic levels.

The advantages of AHP include highly structured and more easily understood models and consistent decision-making.

The disadvantages of AHP arise mainly from the decision-maker, who has to make many pairwise comparisons to reach a decision while possibly using subjective preferences.

The fuzzy AHP approach uses the concepts of fuzzy set theory for evaluation of alternatives and defining the weights of criteria. Shamsuzzaman et al. (2003) integrated fuzzy sets and the AHP for selecting the best-ranked flexible manufacturing system from a number of feasible alternatives. Fuzzy sets are employed to recognize the selection criteria as linguistic variables rather than numerical ones. The AHP is used to determine the weights of the selection criteria in accordance with their relative importance.

24.4 Expressing Fuzzy Preference of Alternatives

There are four ways in which experts can express their opinions.

24.4.1 Preference Ordering of the Alternatives

An expert here gives his/her preferences of alternatives as an individual preference ordering, $O^k = \{o^k(1), ..., o^k(n)\}$, where $o^k(i)$ is a permutation function over the index set $\{1, ..., n\}$ showing the place of alternative i in the sequence (Chiclana et al., 1998; Seo and Sakawa, 1985). Therefore, according to this point of view, an ordered vector of alternatives, from best to worst, is given. For example, an expert may give a four-alternative evaluation as $O = \{1, 3, 4, 2\}$, which means alternative A_1 is the best, A_4 is in the second place ($O(4) = 2$), A_2 is in third place, and alternative A_3 is last. An alternative representation of the same preference vector is $(1, 4, 2, 3)$.

24.4.2 Fuzzy Preference Relation

In this case, the expert's preferences on alternatives are described by a fuzzy preference relation P^k, with membership function $\mu_{p^k}(A_i, A_j) = P_{ij}^k$ which denotes the preference degree or intensity of alternative A_i over A_j. Here $P_{ij}^k = 1/2$ indicates indifference between A_i and A_j, $P_{ij}^k = 1$ indicates that A_i is unanimously preferred to A_j, and $P_{ij}^k > 1/2$ indicates that A_i is preferred to A_j. It is usual to assume that $P_{ij}^k + P_{ji}^k = 1$ and $P_{ii}^k = 1/2$ (Orlovsky, 1978; Tanino, 1990). For instance, the fuzzy preference relation to four alternatives can be

$$P = \begin{bmatrix} 0.5 & 0.55 & 1.0 & 0.25 \\ 0.45 & 0.5 & 0.6 & 0.2 \\ 0 & 0.4 & 0.5 & 0.95 \\ 0.75 & 0.8 & 0.05 & 0.5 \end{bmatrix}$$

where $P_{12} = 0.55 > 1/2$ means that A_1 is slightly preferred to A_2 while $P_{13} = 1.0$ indicates that A_1 is unanimously preferred to A_3.

24.4.3 Multiplicative Preference Relation

In this case, the preferences of alternatives of expert E_k are described by a positive preference relation $A^k = (a_{ij}^k)$, where a_{ij}^k indicates a ratio of the preference intensity of alternative A_i to that of A_j, i.e., it is interpreted as A_i is a_{ij}^k times as good as A_j. Saaty (1980) suggests using a scale of 1 to 9, where $a_{ij}^k = 1$ indicates indifference between A_i and A_j and $a_{ij}^k = 9$ indicates that A_i is unanimously preferred to A_j. An example of multiplicative preference relation can be expressed as

$$A = \begin{bmatrix} 1 & 2 & 9 & 1/5 \\ 1/2 & 1 & 4 & 1/7 \\ 1/9 & 1/4 & 1 & 8 \\ 5 & 7 & 1/8 & 1 \end{bmatrix}$$

One can observe that the preference matrix has the property of a multiplicative reciprocity relationship (i.e., $a_{ij}^k \cdot a_{ji}^k = 1$).

24.4.4 Utility Function

In this case, the expert provides the preferences as a set of n utility values, $U^k = \{u_i^k, i = 1,.....,n\}$, where $u_i^k \in [0,1]$ represents the utility evaluation given by expert E_k to alternative A_i (Luce and Suppes, 1965; Tanino, 1990). For example, the utility function of four alternatives can be $U = \{0.7, 0.4, 0.2, 0.6\}$.

24.4.5 Transformation between the Various Representations

Since the experts may provide their preferences in different ways, it is necessary to convert the various representations to a unified form. A common transformation between the various preferences is as follows (Chiclana et al., 1998):

$$P_{ij}^k = \frac{1}{2}\left(1 + \frac{o_j^k - o_i^k}{n-1}\right) \tag{24.1}$$

$$P_{ij}^k = \frac{(u_i^k)^2}{(u_i^k)^2 + (u_j^k)^2} \tag{24.2}$$

$$P_{ij}^k = \frac{1}{2}(1 + \log_9 a_{ij}^k) \tag{24.3}$$

These transformations allow all the experts' judgments to be converted into fuzzy preference relations.

24.5 Aggregating Individual Preferences into a Group Decision

Multi-criteria decision making and Multi-Expert-MCDM (ME-MCDM) are two rich and well-studied problem-solving approaches usually aimed at the ranking of alternatives (see, for example, Triantaphyllou, 2000). Both approaches aggregate scores given by an expert to each alternative in correspondence with selected criteria into one score, which represents the overall performance of that alternative. This solution approach allows a ranking of the alternatives, with the most preferred one ranked at the top.

The aggregation in both MCDM and ME-MCDM is usually done by an averaging function, or by using the maximum or minimum value among the scores assigned to that alternative. Yager (1988) introduced the ordered weighted averaging (OWA) aggregation method, in which linguistic quantifiers are used in the aggregation function. This approach allows decision-making based on linguistic requirements such as "choose the best alternative on the basis of *most* of the criteria," or based on "*all* of the experts."

24.5.1 Ordered Weighted Averaging Operators

In 1988, Yager introduced the OWA aggregation method. This is a unique and special approach that allows direct use of linguistic quantifiers. This method is defined in the following way:

Definition: An aggregation operator $F: I^n \rightarrow I$ is called an OWA operator of dimension n if it has a weighting vector $W = (w_1 \cdot w_2 \cdot .... \cdot w_n)$ such that $w_i \in (0, 1)$ and $\sum_{i=1}^{n} w_i = 1$, and where $F(a_1, a_2,.....,a_n) = \sum_{j=1}^{n} w_j b_j$, where b_j is the jth largest of the a_i.

The OWA operator has four important properties, allowing it to be considered as a mean operator (Yager, 1988, 1996). These properties are: commutativity, monotonicity, idempotency, and boundedness.

A key step of this aggregation is the reordering of the arguments a_i in descending order so that the weight w_j is associated with the ordered position of the argument. The weight itself can represent either the importance of the criteria or the effect of a linguistic quantifier, as described next.

24.5.2 Linguistic Quantifiers

According to Zadeh (1983), linguistic quantifiers $Q(r)$ can be viewed as linguistic probability, which determines the degree that the concept Q has been satisfied by r. In exploring this concept, Zadeh also proposed the concepts of absolute and relative or proportional quantifiers. The absolute quantifier represents the linguistic terms that relate to an absolute count, such as "at least five" and "more than ten." The relative or proportional quantifier represents the term containing the proportion r, where r belongs to the unit interval. Examples of relative quantifiers are "at least 0.5" and "more than 0.3," as well as "many" and "few." Yager (1991) categorized the relative quantifiers into three categories.

1. *Regular monotonically non-decreasing.* As previously mentioned, the quantifier $Q(r)$ can be perceived as the degree that the concept Q has been satisfied by r. In this type of quantifier, as more criteria are satisfied, the higher the value of the quantifier. Examples for this type of quantifier are "most," "all," "more than α," "there exists," and "at least α." This type of quantifier has the following properties:

$$Q(0) = 0$$

$$Q(1) = 1$$

$$\text{If } r_1 > r_2 \text{ then } Q(r_1) \geq Q(r_2)$$

2. *Regular monotonically non-increasing.* These quantifiers are used to express linguistic terms such as "few," "less than α," "not all," and "none" in which the quantifier prefers fewer criteria to be satisfied. Such a quantifier has these properties:

$$Q(0) = 1$$

$$Q(1) = 0$$

$$\text{If } r_1 < r_2 \text{ then } Q(r_1) \geq Q(r_2)$$

3. *Regular unimodal.* These quantifiers are used to express linguistic terms such as "about α" or "close to α," which implies that the maximum satisfaction is achieved when exactly α is satisfied.

24.5.3 Aggregation Functions

There are numerous approaches to the aggregation of the various scores of the criteria (to a combined score for each alternative) or the scores of the various experts (to a combined group score). A popular approach uses quasi-arithmetic means as a family of algebraic aggregation methods (Smolíková and Wachowiak, 2002). This family of means is defined as

$$F_\alpha(x) = h^{-1}\left[\frac{1}{n}\sum_{i=1}^{n} h(x_i) \right], \quad x \in I^n \tag{24.4}$$

where h is a continuous, strictly monotonic function (h^{-1} is the inverse function of h). This general function can be divided into four types of means:

1. *Root power or generalized mean:*
   Let $h(x) = x^\alpha$ then $h^{-1}(x) = x^{1/\alpha}$

$$F_\alpha(x) = \left(\frac{1}{n}\sum_{i=1}^{n} x_i^\alpha \right)^{1/\alpha}, \quad x \in I^n \tag{24.5}$$

2. *Geometric mean:*
   For $\alpha \to 0$,

$$F_0(x) = \lim_{\alpha \to 0} F_\alpha(x) = \sqrt[n]{\prod_{i=1}^{n} x_i} \qquad (24.6)$$

3. *Harmonic mean:*
   For $\alpha = -1$,

$$F_{-1}(x) = \frac{n}{\sum_{i=1}^{n} 1/x_i} \qquad (24.7)$$

4. *Arithmetic mean:*
   For $\alpha = 1$,

$$F_1(x) = \frac{1}{n} \sum_{i=1}^{n} x_i \qquad (24.8)$$

and for every $x \in I^n$, $F_{-1}(x) \le F_0(x) \le F_1(x)$.

These four types of aggregation functions can be naturally expanded into weighted functions.

24.5.4 Weights Calculation

Since the OWA aggregation method requires a set of weights w_i, these weights have a profound effect on the solution (the ranking of the alternatives in order of preference). One approach for generating the weights has been proposed in Yager (1993a, 1996) for the regular, monotonically nondecreasing quantifiers. Using this approach, the weights are calculated using

$$w_i = Q\left(\frac{i}{n}\right) - Q\left(\frac{i-1}{n}\right), \quad i = 1, \ldots, n \qquad (24.9)$$

Calculating the weights for the regular monotonically nonincreasing quantifiers is based on the fact that these quantifiers are the antonyms to the regular, monotonically nondecreasing quantifiers. The generated weights have the following properties:

$$\sum w_i = 1$$

$$w_i = [0, 1]$$

Example (Some Basic Quantifiers)

1. "All"

This quantifier is also defined as the logical "AND" quantifier and can be represented as (Kacprzyk and Yager, 1984; Yager, 1988, 1993a, 1996)

$$Q_*\left(\frac{i}{n}\right) = \begin{cases} 0 & \text{for } \dfrac{i}{n} < 1 \\ 1 & \text{for } \dfrac{i}{n} = 1 \end{cases} \qquad (24.10)$$

This representation shows that the satisfaction is a step function achieved only when all the criteria are included (as expected).

2. "There exists"

This quantifier is equivalent to the term "at least one" and can be represented as

$$w_i = \begin{cases} 1 & \text{for } i = 1 \\ 0 & \text{for } i \neq 1 \end{cases}$$

The level of satisfaction $Q(r)$ is presented in Figure 24.1.

Thus, this quantifier exhibits complete satisfaction when one criterion is included.

3. "More than α"

The term "more than α" can be represented using the weights shown in Figure 24.2 (e.g., Wang and Lin, 2003).

All these weights are used in the OWA process to generate the overall score of each alternative. This is done using the ordered scores of the criteria of each alternative.

Example (From Yager, 1996)

Take two alternatives, A_1 and A_2, evaluated based on four criteria, C_1 to C_4. The importance associated with the criteria are $V_1 = 1$, $V_2 = 0.6$, $V_3 = 0.5$, and $V_4 = 0.9$. The decision-maker estimates the level of satisfaction of each criterion in each alternative given the following scores:

$$C_1(A_1) = 0.7, \ C_2(A_1) = 1.0 \ C_3(A_1) = 0.5 \ C_4(A_1) = 0.6$$

$$C_1(A_2) = 0.6 \ C_2(A_2) = 0.3 \ C_3(A_2) = 0.9 \ C_4(A_2) = 1.0$$

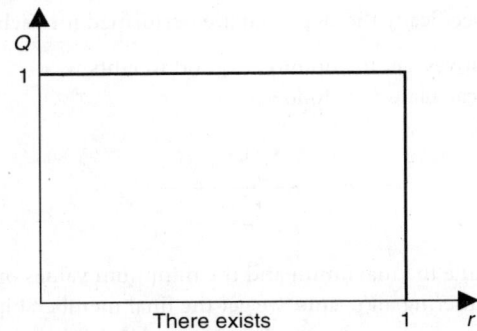

FIGURE 24.1 Satisfaction level of the "There exists" linguistic quantifier.

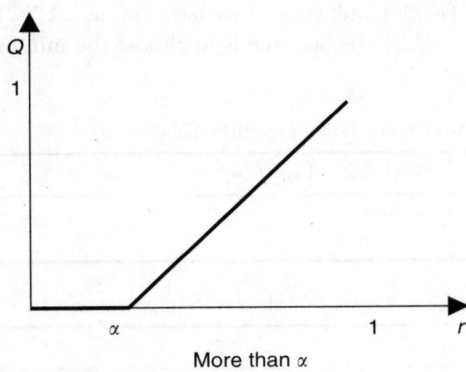

FIGURE 24.2 Satisfaction level of the "More than α" linguistic quantifier.

Using the quantifier "most" with a value function $Q(r)=r^2$, the score of each alternative is calculated to be $A_1= 0.609$ and $A_2 = 0.567$, showing that A_1 is the preferred alternative.

24.5.4.1 Aggregation in Fuzzy Domain

Typically, aggregation in the fuzzy domain is more complicated than in crisp domains. Generally, there are two different approaches for aggregation and comparison of linguistic values:

1. Using associated membership functions
2. Direct computations on labels

These approaches are detailed below.

24.5.5 Using Associated Membership Functions

Most methods use the associated membership functions. Such methods include Baas and Kwakernaak's rating and ranking algorithm (1977) and the fuzzy compromise programming method (Prodanovic and Simonovic, 2003), among others.

24.5.5.1 Rating and Ranking Algorithm

Baas and Kwakernaak's rating algorithm assumed that all the alternatives in the choice set can be characterized by a number of properties (criteria), and that information is available to assign weights to these properties. The method basically consists of computing fuzzy weighted ratings for each alternative and comparing these ratings. This method aggregates fuzzy scores and fuzzy weights at different α-cut levels with their associated membership functions. For each α-cut, we determine the maximum and the minimum of the weighted scores. By gathering all α-cuts, one can get the final membership function of the aggregated result. More specifically the steps that are performed for each expert are:

1. For each pair of alternatives, get the opinions x_{ij} and weights w_j
2. For α-cut from 0 to 1, calculate U as follows:

$$U = \frac{\sum_{j=1}^{n} w_j x_{ij}}{\sum_{j=1}^{n} w_j} \qquad (24.12)$$

3. For each α-cut, determine the maximum and the minimum values of U
4. For each alternative, gathering all α-cuts, we get the final membership function

Example   For an alternative, two experts have the evaluations using linguistic labels, shown in Table 24.1. The membership functions associated with the four linguistic labels are defined as shown in Figure 24.3 and Figure 24.4.

Since there are two pairs of (w_1, x_1) and, (w_2, x_2) we have $(w_1^L, w_1^R, x_1^L, x_1^R)$ and $(w_2^L, w_2^R, x_2^L, x_2^R)$, so there are $16\,U$ values from $(U_1, U_2, ..., U_{16})$. The last step is to choose the minimum and the maximum from these values for each α-cut level.

TABLE 24.1   Evaluations from Two Experts Using Lignuistic Labels

Expert	Weighting of Expert ($\tilde{w}_j$)	Rating of Alternative (x_j)
E_1	Medium	Very good
E_2	Very important	Good

TABLE 24.2   Example of MCDM Using Linguistic Labels with Importance

Criteria	C_1	C_2	C_3	C_4	C_5	C_6
Importance	P	VH	VH	M	L	L
Score	H	M	L	P	VH	P

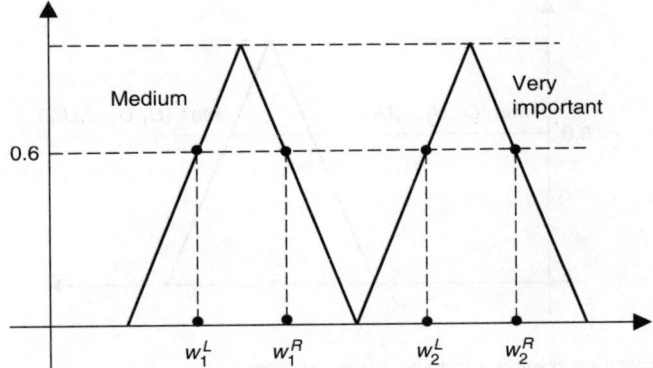

FIGURE 24.3   Membership values of weight of experts 1 and 2.

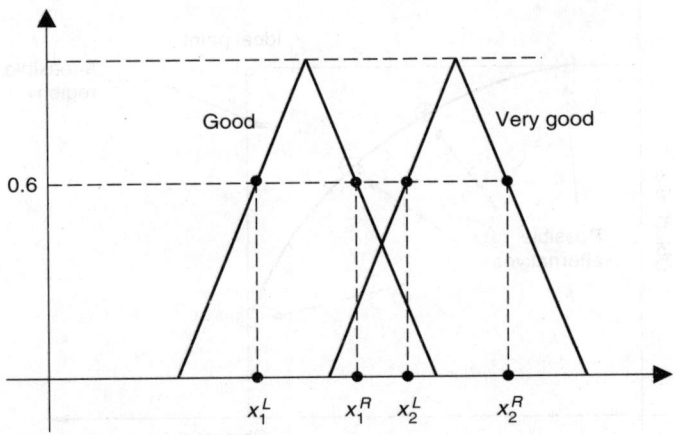

FIGURE 24.4   Membership values of rating of the alternative by experts 1 and 2.

The aggregated score of both experts, considering their importance, is depicted in Figure 24.5. The figure highlights the point of $\alpha = 0.6$.

24.5.5.2   Fuzzy Compromise Programming

Using fuzzy compromise programming, the decision-maker evaluates each alternative according to its distance from an ideal value (Prodanovic and Somonovic, 2003). Thus, for each alternative, the decision-maker sums the distances of each criterion from the ideal value, as shown in Figure 24.6. This sum represents the value of the alternative and is used to compare the alternatives. The distance calculation can be Euclidean or more generally presented as

$$D_j = \left[\sum_{z=1}^{t} \left\{ W_s^p \left(\frac{f_z^* - f_z}{f_z^* - f_z^-} \right)^p \right\} \right]^{1/p} \tag{24.13}$$

The equation represents the distance for alternative j using t criteria $z = 1, ..., t$. The values f^* and f^- represent the positive and negative ideal values, while f_z is the actual value of criterion z. The weight of each criterion is represented by w_z, while p is a parameter.

Fuzzy compromise programming considers all input parameters as fuzzy sets, not just criteria values. This approach benefits from using fuzzy sets in representation of the various parameters, which ensures

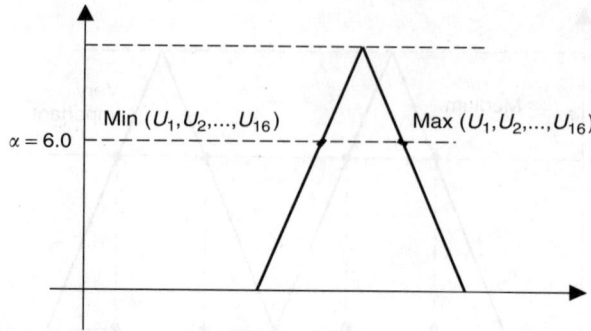

FIGURE 24.5   The aggregation result is a triangular fuzzy number.

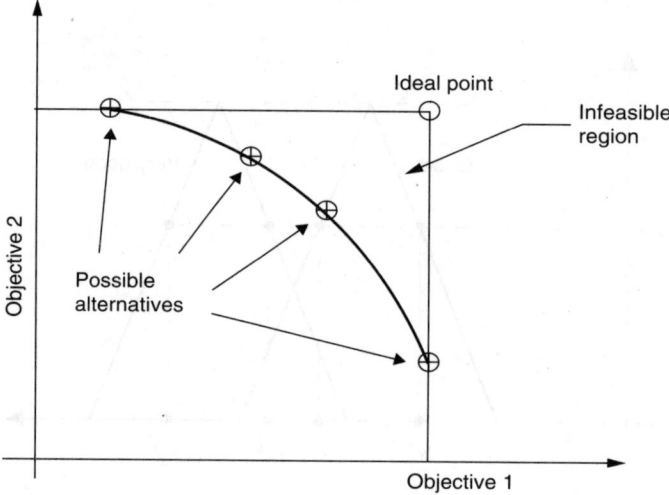

FIGURE 24.6   An illustration of compromise programming.

that the model uses as much of the relevant information as possible. The more certain the expert is in a particular parameter value, the less fuzziness is assigned to the fuzzy number, resulting in a more focused solution. The downside of this approach is that the distance measures are also fuzzy, also requiring a heavy computational load, again based on α-cut values.

24.5.6   Direct Computations on Labels

Zadeh (1983) introduced the concept of the linguistic variable as a term having a linguistic value (a name such as "low", "medium") associated with a fuzzy membership value. Usually this variable is a member of a set $S_T = \{s_0, s_1, \ldots, s_T\}$, where T is an even number.

The properties of a linguistic label set are

- Ordered, $s_i > s_j$, if $i > j$
- Negation operator, $\text{Neg}(S_i) = S_j$, such that $j = T - i$
- Maximization operator, $\text{MAX}(s_i, s_j) = s_i$, if $s_i > s_j$
- Minimization operator, $\text{MIN}(s_i, s_j) = s_j$, if $s_i > s_j$
- Symmetric, s_i and s_{T-i+2} have symmetrical meaning with regard to the middle label $s_{T/2+1}$

As an example, Herrera et al. (1997) uses a nine linguistic labels set S = {I, EU, VLC, SC, IM, MC, ML, EL, C} with their respective associated semantic meaning and fuzzy values:

C	Certain	$(1, 1, 0, 0)$
EL	Extremely Likely	$(0.98, 0.99, 0.05, 0.01)$
ML	Most Likely	$(0.78, 0.92, 0.06, 0.05)$
MC	Meaningful Chance	$(0.63, 0.80, 0.05, 0.06)$
IM	It May	$(0.41, 0.58, 0.09, 0.07)$
SC	Small Chance	$(0.22, 0.36, 0.05, 0.06)$
VLC	Very Low Chance	$(0.1, 0.18, 0.06, 0.05)$
EU	Extremely Unlikely	$(0.01, 0.02, 0.01, 0.05)$
I	Impossible	$(0, 0, 0, 0)$

This approach calculates the aggregated opinion of the various experts directly as a label representing the group opinion. There are few methods that use this approach, but two are presented below. Both approaches can accommodate the expert opinion as a label, and the experts' weights (importance) as a fractional number.

24.5.6.1 Linguistic OWA Operator

This method is based on the OWA (Yager, 1988) and the *convex combination of linguistic labels* (Delgado et al., 1993). The idea is that the combination resulting from two linguistic labels should be an element in the set S. So, given s_i, $s_j \in S$ and $i, j \in (0, T)$, the linguistic OWA (LOWA) method finds an index k in the set S representing a single resulting label.

As an example of this approach, three experts E_1, E_2, and E_3 are evaluating an alternative A. Each chooses a linguistic label from the set S to express his/her opinion. Let us use the same nine linguistic labels set defined above S= {I, EU, VLC, SC, IM, MC, ML, EL, C}. Suppose the labels that the experts choose are X = {s_1, s_5, s_7}. The aggregate value of these three linguistic labels is the score of the alternative under consideration. Also, the experts have weights of, w_5 = 0.5, w_1 = 0.125, and w_7 = 0.375. Using the LOWA algorithm, the aggregated opinion of the three experts is simply s_5.

24.5.6.2 Fuzzy Linguistic OWA Operator

Based on the LOWA method, Ben-Arieh and Chen (2004) presented a new aggregation operator denoted as fuzzy linguistic OWA (FLOWA). In the FLOWA approach, the final result is not a single label but a range of labels, each with a membership function showing the degree of belief that the group has in the label. This approach considers the weight of each expert as an indication of the strength of the expert's belief. These weights are then linearly spread among the labels included in the aggregation. This is demonstrated in Figure 24.7.

For the same example with three experts E_1, E_2, and E_3, the final result shown in Figure 24.8 is a fuzzy set {$0/s_0$, $0.0357/s_1$, $0.0893/s_2$, $0.1429/s_3$, $0.1964/s_4$, $0.25/s_5$, $0.1706/s_6$, $0.1071/s_7$, $0/s_8$}. The result of the aggregation indicates that the linguistic label s_5=MC has the highest possibility as the aggregation result.

24.5.6.3 Induced Ordered Weighted Averaging (IOWA)

When both scores and weights are not crisp numbers, the OWA methodology can be extended to include weights (Yager, 1998). The IOWA method allows aggregation of the scores considering their respective importance. Using this approach and the OWA terminology, the aggregated value of each alternative is defined as the OWA value using modified weights as presented below.

$$a^* = F(b_1, b_2, ...b_m)$$

where

$$a^* = F(w_j, b_j)$$

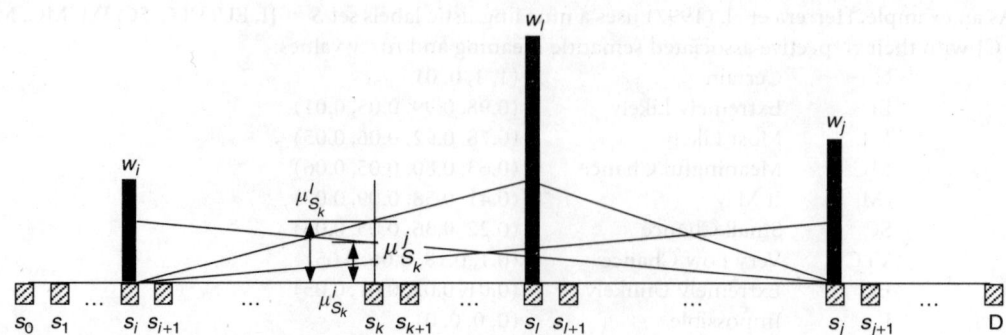

FIGURE 24.7   The concept of FLOWA.

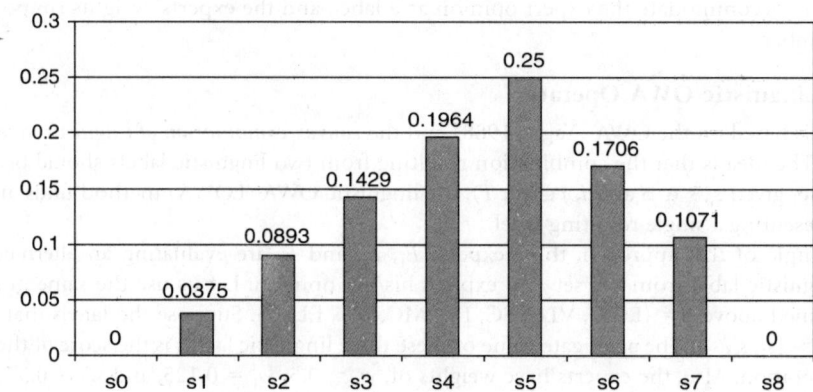

FIGURE 24.8   Example of aggregate three linguistic labels.

$$b_j = G(u_j, a_j)$$

$$G(u,a) = \frac{\text{high}(\alpha)\,G_{\max}(u,a) + \text{medium}(\alpha)\,G_{\text{avg}}(u,a) + \text{low}(\alpha)\,G_{\min}(u,a)}{\text{high}(\alpha) + \text{medium}(\alpha) + \text{low}(\alpha)}$$

In this case, a_j represents the scores and u_j is the importance of the jth criteria. α is the *orness* used to measure the degree of optimism of a decision-maker (Yager, 1988). $G(u, a)$ is a transformation function that depends on the following rules defined by the decision-maker or analyst.

      If the degree of *orness* is high, $G(u, a)$ is $G_{\max}(u, a)$

      If the degree of *orness* is medium, $G(u, a)$ is $G_{\text{avg}}(u, a)$

      If the degree of *orness* is low, $G(u, a)$ is $G_{\min}(u, a)$

Example   (Yager, 1998)

Define the high, medium, and low as the functions of *orness* α as

$$\text{High}(\alpha) = \begin{cases} 2\alpha - 1, & \alpha \geq 0.5 \\ 0, & \alpha \leq 0.5 \end{cases}$$

$$\text{Medium}(\alpha) = \begin{cases} 2\alpha, & \alpha \leq 0.5 \\ 2 - 2\alpha, & \alpha \geq 0.5 \end{cases}$$

$$\text{Low}(\alpha) = \begin{cases} -2\alpha+1, & \alpha \leq 0.5 \\ 0, & \alpha \geq 0.5 \end{cases}$$

Since, in this special case, $\text{High}(\alpha)+\text{Medium}(\alpha)+\text{Low}(\alpha)=1$, we have

$$G(u,a) = \frac{\text{high}(\alpha)G_{\max}(u,a) + \text{medium}(\alpha)G_{\text{avgx}}(u,a) + \text{low}(\alpha)G_{\min}(u,a)}{\text{high}(\alpha)+\text{medium}(\alpha)+\text{low}(\alpha)}$$

$$= \text{high}(\alpha)G_{\max}(u,a) + \text{medium}(\alpha)G_{\text{avg}}(u,a) + \text{low}(\alpha)G_{\min}(u,a)$$

By defining $G_{\max} = ua$, $G_{\min} = \bar{u} + ua$, $G_{\text{avg}}=\left(n\big/\sum_{j=1}^{n} u_j\right)ua$, the final transformation function is

$$G(u, a) = \begin{bmatrix} (-2\alpha +1)(1-u + ua)+2\alpha\left(\dfrac{n}{T}ua\right), & \alpha \leq 0.5 \\ \\ (2\alpha -1)ua +(2-2\alpha)\left(\dfrac{n}{T}ua\right), & \alpha \geq 0.5 \end{bmatrix}$$

Given four criteria with the following score and importance: $(u,a) = (0.7, 0.8), (1, 0.7), (0.5, 1.0), (0.3, 0.9)$. The weights required for linguistic quantification are: $W = (0.4, 0.3, 0.2, 0.1)$. We can calculate the *orness* to be $\alpha = 0.67 > 0.5$, then, the final aggregated result is: $a^* = F(0.78, 0.98, 0.7, 0.38)=F_W(0.98, 0.78, 0.7, 0.38)=0.98\cdot0.4+0.78\cdot0.3+0.7\cdot0.2+0.38\cdot0.1=0.8$.

24.5.6.3.1 *Yager's All/and/min Aggregation*

Yager's group MCDM evaluation process (1993b) is a two-stage process. In the first stage, individual experts are asked to provide an evaluation of the alternatives. The evaluation consists of a rating for each alternative on the basis of each of the criteria. Each of the criteria may have a different level of importance. The values to be used for the evaluation of the ratings and importance will be drawn from a linguistic scale, which makes it easier for the evaluator to provide the information. The second stage performs the aggregation of the individual evaluations to obtain an overall linguistic value for each alternative. Implicit in this linguistic scale are two operators, the maximum and minimum of any two scores, as discussed earlier. The aggregated score of each alternative with j criteria is simply defined as

$$A_{ik} = \text{Min}_j \left[\text{NEG}(I(q_j)) \vee A_{ik}(q_j)\right] \tag{24.14}$$

where $I(q_j)$ is the importance of criteria j, A_{ik} the opinion of expert k, and $\vee$ indicates the max operation. Negation of a linguistic term is calculated as

$$\text{Neg}(S_i) = S_{l-i+1} \text{ if we have a scale of } l \text{ items} \tag{24.15}$$

Example

$$\begin{aligned} A_{jk} &= \text{Min}[\text{NEG}(P) \vee H, \text{NEG}(VH) \vee M, \text{NEG}(VH) \vee L,...,\text{NEG}(L) \vee P] \\ &= \text{Min}[N \vee H, VL \vee M, VL \vee L, M \vee P, H \vee VH, H \vee P] \\ &= \text{Min}[H, M, L, P, VH, P] \\ &= L \end{aligned}$$

24.6 Ranking Alternatives

Ranking fuzzy numbers is an essential activity in many applications and has been widely reported in the literature. Good summaries of fuzzy numbers ranking methods can be found in books (e.g., Chen and

Hwang, 1992), and papers (e.g., Lee and Li, 1988; Chang and Lee, 1994; Dubois and Prade, 1999; Lee-Kwang and Lee, 1999). A good ranking method should satisfy the following four criteria (Yuan, 1991): allow presentation of fuzzy preference, represent rational preference ordering, be robust, and be efficient.

24.6.1   Ranking Based on Defuzzification

Basically, fuzzy sets outranking methods can be classified into two categories. The first category is based on defuzzification. A simple method for ranking fuzzy intervals consists of turning each fuzzy set into a precise crisp value (also referred to as utility value) and then using the ranking induced by these values. The advantage of this approach is that it provides a complete ranking. Usually, a method that follows this approach is easy and fast to calculate, which explains its popularity. The main disadvantage of this approach is that defuzzification results in a loss of information regarding the existing uncertainty.

Lee and Li (1988) proposed a ranking method based on the center of gravity and the associated standard deviation of the fuzzy numbers. Using this approach, fuzzy numbers are ranked on the basis of two criteria: the fuzzy mean (calculated as center of gravity) and the fuzzy spread. This method is justified by imitating human intuition, which favors a fuzzy number with a higher mean value and, at the same time, a lower spread. The fuzzy mean is defined as

$$\bar{x}_u (A) = \frac{\int_{S(A)} x \mu_A(x)\,dx}{\int_{S(A)} \mu_A(x)\,dx} \tag{24.16}$$

where $S(A)$ is the support of fuzzy number A. The standard deviation is calculated as

$$\sigma_u(A) = \left[\frac{\int_{S(A)} x^2 \mu_A(x)\,dx}{\int_{S(A)} \mu_A(x)\,dx} - [\bar{x}_u(A)]^2 \right]^{1/2} \tag{24.17}$$

Assuming that the mean values and spreads are calculated for two fuzzy numbers A and B, the rules for ranking are shown in Table 24.3.

Example   Given three fuzzy sets,

$$A = \{0/s_0, 0/s_1, 0.0833/s_2, 0.1667/s_3, 0.2500/s_4, 0.1875/s_5, 0.1250/s_6, 0.0625/s_7, 0/s_8, 0/s_9\}$$
$$B = \{0/s_0, 0/s_1, 0.0625/s_2, 0.1250/s_3, 0.1875/s_4, 0.2500/s_5, 0.1667/s_6, 0.0833/s_7, 0/s_8, 0/s_9\}$$
$$C = \{0/s_0, 0/s_1, 0/s_2, 0.25/s_3, 0.208/s_4, 0.167/s_5, 0.125/s_6, 0.0833/s_7, 0.0417/s_8, 0/s_9\}$$

The results of the comparison are presented in Table 24.4.

The ranking results show that, since $\bar{x}(U_B) = \bar{x}(U_C) > \bar{x}(U_A)$, therefore B > A, C > A. While $\bar{x}(U_B) = \bar{x}(U_C)$, but $\sigma(U_B) < \sigma(U_C)$ thus B > C. Finally, we conclude that B > C > A.

TABLE 24.3   Ranking Rules for Fuzzy Mean and Spread Method

Relation of $\bar{x}(A)$ and $\bar{x}(B)$	Relation of $\sigma(A)$ and $\sigma(B)$	Ranking Order
$\bar{x}(A) > \bar{x}(B)$	—	A>B
$\bar{x}(A) = \bar{x}(B)$	$\sigma(A) < \sigma(B)$	A>B

TABLE 24.4   Ranking Rules for Fuzzy Mean and Spread Method

	A	B	C
$\bar{x}(U_i)$	3.792	4.083	4.083
$\sigma(U_i)$	1.925	2.01	2.08

24.6.2 Ranking Based on Linguistic Evaluation

The other approach uses fuzzy relations to compare pairs of fuzzy sets, and then constructs a relationship that produces a linguistic description of the comparison. The ordering results are something like "fuzzy number A is slightly better than fuzzy number B."

Some methods retrieve the areas under the membership function as the fuzzy preference relation of the fuzzy sets. For example, when comparing two fuzzy numbers A and B, we can rank two fuzzy numbers by comparing the Hamming distances between each of A and B and Max (A, B) to decide which is the greatest.

In Figure 24.9, the Hamming distances for A and B are: $H_A(A, \max(A, B)) = S_1 + S_2$ and $H_B(B, \max(A, B)) = S_3 + S_4$. If H_A is larger than H_B, then the corresponding fuzzy number A is larger than B.

Chang and Lee (1994) use a measure called overall existence ranking index (OERI) to estimate the similarity between two convex fuzzy numbers as

$$\text{OERI}(A) = \int_0^1 \omega(\alpha) \, [\chi_1(\alpha) \, \mu_{AL}^{-1}(\alpha) + \chi_2(\alpha) \, \mu_{AR}^{-1}(\alpha)] \, d\alpha \qquad (24.18)$$

where $\chi_1(\alpha)$ and $\chi_2(\alpha)$ are subjective weighting functions indicating neutral, optimistic, and pessimistic preferences of the decision-maker, with the restriction that $\chi_1(\alpha) + \chi_2(\alpha) = 1$. Parameter $\omega(\alpha)$ is used to specify weights, representing the decision-maker's preference. $\mu_{AL}^{-1}(\alpha)$ represents an inverse of the left part and $\mu_{AR}^{-1}(\alpha)$ the inverse of the right part of the membership function. Note that linear and nonlinear functions for the subjective-type weighting are possible, thus giving the user more control in the ranking.

Both methodologies have advantages and disadvantages. It is argued that defuzzification methods lose some information by reducing the analysis to a single crisp number. This methodology, on the other hand, produces a consistent ranking of all fuzzy sets considered. As Yuan (1991) points out, using linguistic comparison methods may not always produce total ordering among all alternatives based on pairwise comparison of fuzzy preference relations.

24.7 Measuring Group Consensus

Ness and Hoffman (1998) define consensus as: "a decision that has been reached when most members of the team agree on a clear option and the few who oppose it think they have had a reasonable opportunity to influence that choice. All team members agree to support the decision."

The expression of concerns and conflicting ideas is considered desirable and important. When a group creates an atmosphere that nurtures and supports disagreement without hostility and fear, it builds a foundation for stronger, more creative decisions. Consensus is viewed as a pathway to a true group decision. Sharing opinions prior to reaching a decision, as is done in jury settings, clearly reduces the effective

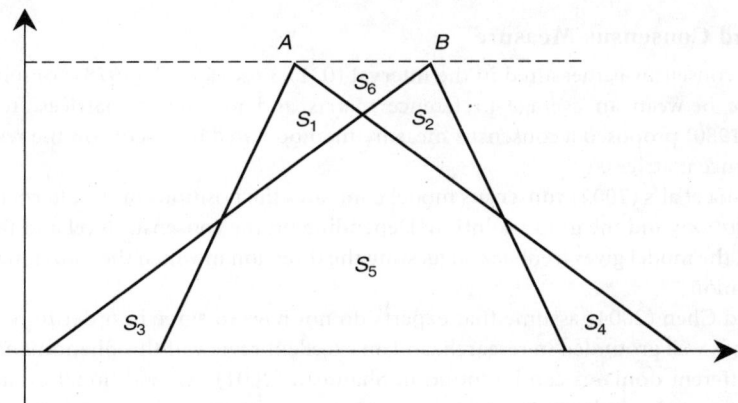

FIGURE 24.9 Distance between two fuzzy numbers.

number of independent voices. In some situations, discussion is not allowed. For example, during figure skating competitions, judges are expressly forbidden from interacting.

24.7.1 Difficulties in Generating Consensus

Hypothetically, agreement with other experts is a necessary characteristic of an expert. Common wisdom claims that experts in a given field should agree with each other. If opinions do not match, then some of the members of this set of experts must not be functioning at the appropriate level. However, in practice, a consensus among experts implies that the expert community has largely solved the problems of the domain. In that case, each individual expert is getting the correct answer, usually with the aid of well-developed technology; therefore, their answers agree (Weiss and Shanteau, in press). In reality, however, disagreement among experts is inevitable and even useful. Moreover, one might argue that too much interindividual agreement is a signal that the problem is trivial and scarcely worthy of an expert.

Consensus makes it possible for a group to reach a final decision that all group members can support despite their differing opinions. True expertise is characterized by the following properties:

- The domains in which experts work are very complex. A single optimal solution does not exist.
- A distinction can be made among the different levels of decisions made by experts. Experts might disagree at one level, but agree at another.
- Despite the assumption made by many researchers, experts are seldom asked to make single-outcome decisions. The job of the expert is to clarify alternatives and describe possible outcomes for clients.
- Experts generally work in dynamic situations with frequent updating. Thus, the problems faced by experts are unpredictable, with evolving constraints.
- Experts work in realms where the basic science is still evolving.

24.7.2 Consensus Measure Methods

The ultimate goal of a procedure in group decision-making, in the context considered here, is to obtain an agreement between the experts as to the choice of a proper decision, i.e., to reach consensus (Kacprzyk et al., 1992). Initially, the group may be far from consensus. However, it can be expected that during the decision-making process, the opinions of its members will converge. Consensus is not to be enforced or obtained through negotiations or a bargaining process, but is expected to emerge after exchanges of opinions among the experts.

The study of consensus has become a major area of research in group decision-making. Generally, there are two ways to represent consensus (Bordogna et al., 1997):

24.7.2.1 Hard Consensus Measure

In this category, consensus is measured in the interval (0, 1). Bezdek et al. (1978) computed a consensus as the difference between an average preference matrix and preference matrices from each expert. Spillman et al. (1980) proposed a consensus measure method based on α-cuts on the respective individual fuzzy preference matrices.

Herrera-Viedma et al.'s (2002) consensus model compares the positions of the alternatives on the basis of individual solutions and the group solution. Depending on the consensus level and the offset of individual solutions, the model gives feedback suggesting the direction in which the individual experts should change their opinion.

Ben-Arieh and Chen (2004) assume that experts do not have to agree in order to reach a consensus. This assumption is well grounded in research, and an excellent review of this phenomenon of expert disagreement in different domains can be found in Shanteau (2001). An additional example for such an expert decision is, again, judging figure skating. In this case, the judges, who are carefully trained experts, evaluate the performance using very well-defined performance guidelines based on uniform criteria. In such judging, there are no expectations that all experts will eventually converge to an agreement. On the contrary, the experts are expected to produce diversified opinion, and the usual procedure is to eliminate the high- and low-extreme opinions (assign a weight of 0) and average the rest (assign a weight of $n-2$).

In the model presented by Ben-Arieh and Chen (2004), the degree of importance of each expert is considered in calculating the consensus. Moreover, once the consensus is calculated, the experts with a more extreme opinion will lose some of their weight (credibility, influence, etc.). The experts, however, need not modify their opinions to achieve consensus.

$$C_i = \sum_{k=1}^{q}\left[\left(1-\frac{|O^G_{A_i}-O^{Ek}_{A_i}|}{n-1}\right)\times u_k\right] \tag{24.19}$$

$$C_G = \frac{1}{p}\sum_{i=1}^{p}C_{[i]} \tag{24.20}$$

where $[i]$ represents the alternative ranked in ith position, C_G the group consensus of the final solution, C_i the consensus level of the ith alternative achieved by the group, $O^G_{A_i}$ he index of the ith alternative in the group's selection order, $O^E_{A_i}$ the order of the ith alternative based on the kth expert, u_k the importance of the kth expert's opinion, q the number of experts, and n the number of alternatives.

This definition compares each individual solution presented by an expert with the group solution generated separately.

This "hard" interpretation of consensus is sometimes counterintuitive, since one may be fully satisfied (consensus $= 1$) even in the case of agreement only among *most* of the experts, but not *all* (Bordogna et al., 1997).

24.7.2.2 Soft Consensus Measure

In this category, the consensus is not measured by a crisp number but by a linguistic label such as "most." This soft consensus is actually a linguistic quantifier helping to aggregate the evaluations from experts. Fedrizzi (1990) proposed a consensus measure based on dissimilarity between the preference relations. The procedure computes a "soft" degree of consensus, which is a numeric value assessing the truth of a statement such as "most pairs of experts agree on almost all the alternatives."

Similarly, Bordogna et al. (1997) uses the statement, "Most of the experts agree on alternative Ax," which is interpreted as "Most of the experts agree with most of the other experts on alternative Ax." For example, q experts express their overall evaluations to n alternatives, as summarized in Table 24.5.

A linguistic degree of consensus among the experts' overall performances is computed for each alternative. A procedure to evaluate the consensus degree among Q experts for each alternative (Q is a quantifier identifying a fuzzy majority) is as follows:

(a) For each alternative, pairwise comparisons of experts' overall performance labels produce the degree of agreement between pairs of experts. A matrix of $q \times q$ is then constructed for each alternative. An element $Ag(E_i, E_j)$ is the linguistic label, which expresses the closeness between the overall performance labels of expert E_i and E_j.

TABLE 24.5 q Experts Express their Evaluations to n alternatives Using Linguistic Labels

Overall Performance Values	E_1	E_2	...	E_q
A_1	High	...	...	...
...	...	...	O_{ix}	Perfect
A_n	...	Low	...	...

$$Ag(E_i,E_j) = Neg(d(O_{ix},O_{jx})) \tag{24.21}$$

where O_{ix} denotes the linguistic overall performance label of expert E_i on alternative Ax. The d function is a difference operator of linguistic labels in the same scale S: $d(s_i,s_j) = s_r$ with $r = |i-j|$. This process is depicted in Table 24.5.

(b) For each expert E_i (a row of the matrix $q \times q$), $A_g(E_i, E_j)$, $i \neq j$ are pooled to obtain an indication of the agreement $A_g(E_i)$ of expert E_i with respect to Q other experts. This is shown in the last column of Table 24.6.

(c) The values $Ag(E_i)$ are finally aggregated to compute the truth of the sentence "Q experts agree on alternative Ax." This value is stored in the bottom-right cell of Table 24.6.

24.7.3 Procedure to Reach the Consensus (see The Ball Foundation website)

Here is a general procedure for helping to reach consensus among experts.

(a) Identify areas of agreement.

(b) Clearly state differences.

- State positions and perspectives as neutrally as possible.
- Do not associate positions with people. The differences are between alternative valid solutions or ideas, not between people.
- Summarize concerns and list them.

(c) Fully explore differences.

- Explore each perspective and clarify.
- Involve everyone in the discussion. Avoid a one-on-one debate.
- Look for the "third way": make suggestions or modifications, or create a new solution.

(d) Reach closure.

(d) Articulate the decision.

- Ask people if they feel they have had the opportunity to fully express their opinions.
- Obtain a sense of the group. (Possible approaches include "go rounds" and "straw polls," or the consensus indicator tool. When using the consensus indicator, if people respond with two or less, then repeat steps (a)–(c) until you can take another poll.)
- At this point, poll each person, asking, "Do you agree with and will you support this decision?"

24.7.3.1 Some Guidelines for Reaching Consensus (Crow, NPD Solution)

(a) Make sure everyone is heard and feels listened to. Avoid arguing for one's own position. Present each expert's position as clearly as possible. Listen to other team members' reactions and com-

TABLE 24.6 Using Linguistic Quantifier Q to Aggregate q Experts' Evaluations

A_i	E_1	E_2	$\cdots$	E_q	$Ag(E_i)$ from OWA_Q
E_1	High	$\cdots$	$\cdots$	$\cdots$	—
E_2	$\cdots$	$\cdots$	$\cdots$	Medium	—
$\cdots$	$\cdots$	$\cdots$	$Ag(E_i, E_j)$	$\cdots$	—
E_q	$\cdots$	Low	$\cdots$	$\cdots$	
		Q E_i agree on alternative Ax			Final

ments to assess their understanding of each other's position. Consider their reactions and comments carefully before pressing one's own point of view further.

(b) Do not assume that someone must win and someone must lose when a discussion reaches a stalemate. Instead, look for the next most acceptable alternative for all parties. Try to think creatively. Explore what possibilities would exist if certain constraints were removed.

(c) Do not change your mind simply to avoid conflict, to reach agreement, or to maintain harmony. When agreement seems to come too quickly or easily, be suspicious. Explore the reasons and be sure that everyone accepts the solution for basically similar or complementary reasons. Yield only to positions that have objective or logically sound foundations or merits.

(d) Avoid conflict-reducing techniques such as majority vote, averaging, coin toss, or bargaining. When dissenting members finally agree, do not feel that they have to be rewarded or accommodated by having their own way on some later point.

(e) Differences of opinion are natural and expected. Seek them out, value them, and try to involve everyone in the decision process. Disagreements can improve the group's decision. With a wider range of information and opinions, there is a greater chance that the group will hit upon a more feasible or satisfactory solution.

24.8 Conclusions

The area of decision-making is a very fertile domain with numerous methods and approaches. This large area can be divided into multiple objective, multiple criteria, and group decision-making. Although group decision-making includes multicriteria decisions, group decision methods are mostly concerned with aggregation of the individual opinions, while pure multicriteria decisions emphasize the necessary compromise between the conflicting criteria. Group decision methods can be divided into game theoretic, social choice approach, and expert judgments methods. This chapter focuses mostly on the last approach, narrowing the field by using fuzzy set theory as a decision aid.

This chapter has presented some direct approaches to fuzzy group decision-making that are extensions of crisp methodologies, such as fuzzy AHP and fuzzy Delphi. The issue of expressing individual preferences numerically is described in detail. Four methods for capturing individual opinions are presented with a formula that converts the various forms into one unified representation.

Next, this chapter has presented various methods for aggregating individual opinions into a group opinion. The more interesting approaches use linguistic quantifiers as descriptors for the aggregation. For example, such aggregation can use the phrase, "the decision supported by *most* experts," which gives group decision-making a much broader semantics.

The chapter has also given methods for ranking fuzzy numbers, thus allowing ranking of alternatives with fuzzy scores. This ranking ultimately supports the group decision by pointing to the alternative that is ranked in first place.

This chapter has also discussed the complex issue of consensus in group decision-making. Contrary to common belief, experts do not necessarily converge in unanimous agreement. Thus, a group decision may be supported by the experts only to a limited extent. This extent can be measured using various consensus measures. Some group decision methods try to "close the loop" by identifying the experts that deviate the most from the group and influencing them to change their preferences. Other methods weigh the opinions of the experts on the basis of their deviation from the group opinion.

References

Arrow, K.J., *Social Choice and Individual Values*, 2nd ed., Wiley, New York, 1963.

Arrow, K.J. and Raynaud H., *Social Choice and Multicriterion Decision-making*, The MIT Press, Cambridge, MA, 1986.

Baas, S.M. and Kwakernaak, H., Rating and ranking of multiple-aspect alternatives using fuzzy sets, *Automatica*, 13, 47–58, 1977

Ben-Arieh, D. and Chen, Z., A new linguistic labels aggregation and consensus in group decision-making, *Conference of IERC 2004*, Houston, TX, May 15–19, 2004.

Bezdek, J., Spillman, B., and Spillman, R., A fuzzy relation space for group decision theory, *Fuzzy Sets Syst.*, 1, 255–268, 1978.

Bordogna, G., Fedrizzi, M., and Pasi, G., Linguistic modeling of consensus in group decision-making based on owa operators, *IEEE Trans. Syst., Man, Cybern. A: Syst. Hum.*, 27, 126–133, 1997.

Chang, P.-T. and Lee, E.S., Ranking of fuzzy sets based on the concept of existence, *Comput. Math. Appl.*, 27, 1–21, 1994.

Chen, S.-J. and Hwang, C.-L., *Fuzzy Multiple Attribute Decision-Making*, Springer, Berlin, 1992.

Cheng, C., Simple fuzzy group decision-making method, *IEEE Int. Conf. Fuzzy Syst.*, 2, II-910–II-915, 1999.

Chiclana, F., Herrera, F., and Herrera-Viedma, E., Integrating three representation models in fuzzy multipurpose decision-making based on fuzzy preference relations, *Fuzzy Sets Syst.*, 97, 33–48, 1998.

Crow, K., New product development solutions, http://www.npd-solutions.com/consensus.html

Delgado, M., Verdegay, J.L., and Vila, M.A., On aggregation operations of linguistic labels, *Int. J. Intelligent Syst.*, 8, 351–370, 1993.

Dubois, D. and Prade, H., A unified view of ranking techniques for fuzzy numbers, *IEEE International Fuzzy Systems Conference Proceedings*, August 22–25, Seoul, Korea, 1999.

Fedrizzi, M., 1990, On a consensus measure in a group MCDM problem, *Multiperson Decision-Making Models Using Fuzzy Sets and Possibility Theory*, Kluwer Academic Publishers, Dordrecht, pp. 231–241.

Herrera, F., Herrera-Viedma, E., and Verdegay, J.L., Rational consensus model in group decision-making using linguistic assessments, *Fuzzy Sets Syst.*, 88, 31–49, 1997.

Herrera-Viedma, E., Herrera, F., and Chiclana, F., A consensus model for multiperson decision-making with different preference structures, *IEEE Trans. Syst., Man, Cybern. A: Syst. Hum.*, 32, 394–402, 2002.

Hwang, C.L. and Lin, M.-J., *Group Decision-making Under Multiple Criteria*, Lecture Notes in Economics and Mathematical Systems, Vol. 281, Springer, Berlin, 1987.

Hwang, C.L. and Masud, A.S.M., *Multiple Objective Decision-making — Methods and Applications, A State of the Art Survey*, Springer, Berlin, 1979.

Hwang, C.L. and Yoon, K., *Multiple Attribute Decision-making — Methods and Applications, A State of the Art Survey*, Springer, Berlin, 1981.

Ishikawa, A., The new fuzzy Delphi methods: economization of GDS (group decision support), System Sciences, 1993, *Proceeding of the Twenty-Sixth Hawaii International Conference*, Vol. 4, pp. 255–264, 1993.

Kacprzyk, J., Fedrizzi, M., and Nurmi, H., Fuzzy logic with linguistic quantifiers in group decision-making, in *An Introduction to Fuzzy Logic Applications in Intelligent Systems*, Yager, R.R. and Zadeh, L.A., Eds., Kluwer Academic Publishers, Dordrecht, 1992.

Kacprzyk, J. and Yager, R.R., "Softer" optimization and control models via fuzzy linguistic quantifiers. *Inf. Sci.*, 34, 157–178, 1984.

Lee, E.S. and Li, R.-J., Comparison of fuzzy numbers based on the probability measure of fuzzy events, *Comput. Math. Appl.*, 15, 887–896, 1988.

Lee-Kwang, H. and Lee, J., Method for ranking fuzzy numbers and its application to decision-making, *IEEE Trans Fuzzy Syst.*, 7, 677–685, 1999.

Liu, S.-Y. and Chi, S.-C., Fuzzy multiple attribute decision-making approach using modified lexicographic method, *Proceedings of the IEEE International Conference on Systems, Man and Cybernetics*, Vol. 1, Vancouver, BC, Canada, 1995, pp. 19–24.

Luce, R.D. and Suppes, P., 'Preferences' utility and subject probability, in *Handbook of Mathematical Psychology*, Vol. III, Luce, R.D. et al., Eds., Wiley, New York, pp. 249–410.

Myerson, R.B., *Game Theory*, Harvard University Press, Cambridge, MA, 1991.

Ness, J. and Hoffman, C., *Putting Sense into Consensus: Solving the Puzzle of Making Team Decisions*, VISTA Associates, Tacoma, WA, 1998.

Orlovsky, S.A., Decision-making with a fuzzy preference relation, *Fuzzy Sets Syst.*, 1, 155–167, 1978.

Prodanovic, P. and Simonovic, S.P., Fuzzy compromise programming for group decision-making, *IEEE Trans. Syst., Man, Cyber. A*, 33, 358–365, 2003.

Saaty, Th.L., *The Analytic Hierarchy Process*, McGraw-Hill, New York, 1980.

Seo, F. and Sakawa, M., Fuzzy multiattribute utility analysis for collective choice, *IEEE Trans. Syst. Man, Cybern.*, SMC-15, 45–53, 1985.

Shamsuzzaman, M., Sharif, Ullah A.M.M., and Bohez, Erik L.J., Applying linguistic criteria in FMS selection: fuzzy-set-AHP approach, *Integrated Manuf. Syst.*, 14(3), 247–254, 2003.

Shanteau, J., What does it mean when experts disagree, in *Naturalistic Decision-making*, Klein, G. and Salas, E., Eds., Lawrence Erbaum Associates, Hillsdale, NJ, 2001.

Smolíková, R. and Wachowiak, M.P., Aggregation operators for selection problems, *Fuzzy Sets Syst.*, 131, 23–34, 2002.

Spillman, B., Spillman, R., and Bezdek, J., A fuzzy analysis of consensus in small groups, in *Fuzzy Sets Theory and Applications to Policy, Analysis and Information Systems*, Wang, P.P. and Chang, S.K., Eds., Plenum, New York, 1980, pp. 291–308.

Tanino, T., On group decision-making under fuzzy preferences, in *Multiperson Decision-making Using Fuzzy Sets and Possibility Theory*, Kacprzyk, J. and Fedrizzi, M., Eds., Kluwer, Norwell, MA, 1990, pp. 172–185.

Triantaphyllou, E., *Multi-Criteria Decision-making Methods: A Comparative Study*, Kluwer Academic Publishers, Dordrecht, 2000.

Von Neumann, J. and Morgenstern, O., *Theory of Games and Economic Behavior*, Princeton University Press, Dordrecht, 1944.

Wang, J. and Lin, Y.I., A fuzzy multicriteria group decision-making approach to select configuration items for software development. *Fuzzy Sets Syst.*, 134, 343–363, 2003.

Weiss, D.J. and Shanteau, J., The vice of consensus and the virtue of consistency, in *Psychological investigations of Competent Decision Making*, Shanteau, J. and Johnson, P., Eds., Cambridge, UK: Cambridge University Press, 2004, pp. 226–240.

Yager, R.R., On ordered weighted averaging aggregation operators in multicriteria decision-making, *IEEE Trans. Syst., Man, Cyber. A*, 18, 183–190, 1988.

Yager, R.R., Connectives and quantifiers in fuzzy sets, *Fuzzy Sets Syst.*, 40, 39–76, 1991.

Yager, R.R., Families of OWA operators. *Fuzzy Sets and Syst.*, 59, 125–148, 1993a.

Yager, R.R., Non-numeric multi-criteria multi-person decision-making, *Group Decision and Negotiation*, 2, 81–93, 1993b.

Yager, R.R., Quantifier guided aggregation using OWA operators, *Int. J. Intelligent Syst.*, 11, 49–73, 1996.

Yager, R.R., Including importances in OWA aggregations using fuzzy systems modeling, *Fuzzy Systems, IEEE Trans.*, 6, 286–294.

Yuan, Y., Criteria for evaluating fuzzy ranking methods, *Fuzzy Sets Syst.*, 44, 139–157, 1991.

Zadeh, L., A computational approach to fuzzy quantifiers in natural languages, *Computing Math. Appl.*, 9, 149–184, 1983.

Zimmermann, H.-J., *Fuzzy Sets, Decision-making and Expert Systems*, Kluwer Academic Publishers, Boston, MA, 1987.

The Ball Foundation, http://www.ballfoundation.org/ei/tools/consensus/steps.html

25

Introduction to Applications of Fuzzy Set Theory in Industrial Engineering

Pamela R. McCauley-Bell
University of Central Florida

Lesia L. Crumpton-Young
University of Central Florida

25.1 Introduction

A primary issue in the development of realistic expert systems, human factors models, simulation systems, and other industrial engineering systems is the management of uncertainty. Fuzzy set theory (FST) is a proven technique for handling uncertainty in many of these application areas. In the early development of fuzzy logic concepts, Zadeh (1965) felt that humans had the ability to analyze imprecise concepts that were not thoroughly understood. However, this ability to analyze these concepts encounters difficulties when one is attempting to represent this type of knowledge numerically.

This problem of accurate knowledge representation is commonly encountered by experts in any specific field of knowledge. Zadeh felt that the ability to consider this type of "imprecise" knowledge was a key attribute in the human thinking process, and because it could not be represented numerically, he introduced FST. In FST, the sets are not defined as sharply as in ordinary set theory. Conversely, FST does not restrict variable conditions to the rigid guidelines of ordinary set theory. Instead, FST deals with the imprecision associated with many variables by permitting the grade of membership to be defined over the interval [0,1]. The grade of membership expresses the degree of possibility to which a particular element belongs to a fuzzy set.

Fuzzy set theory is a tool that has illustrated encouraging results in bridging the gap between natural language and computer language. It is capable of handling the ambiguity associated with linguistic terms and can, in turn, translate the linguistic terms into fuzzy values that can be interpreted by the computer language. The nature of FST makes it useful for handling a variety of imprecise or inexact cognitive conditions, which may arise owing to a number of situations.

As previously stated, FST does not sharply define sets as is traditionally done in ordinary set theory. Ordinary set theory is governed by binary principles such that a variable either belongs to a set, which would indicate a membership value of 1, or it does not belong to the set and maintains a membership value of 0. In contrast, FST does not restrict variable conditions to these rigid guidelines. Instead, FST deals with the imprecision associated with many variables by permitting a grade of membership to be defined over the interval [0,1]. The grade of membership expresses the degree of strength with which a particular element belongs to a fuzzy set, Thus allowing a more realistic modeling of variable status.

25.2 Fundamentals of Fuzzy Set Theory

The nature of FST makes it useful for handling a variety of imprecise or inexact conditions. To differentiate between these situations or classes of problems associated with the use of FST, the uncertainty can be divided into categories within which most problems assessed by FST can be partitioned. These categories include generality, ambiguity, and vagueness.

1. Generality is the condition in which a given variable or level can apply to a number of different states. A variety of situations can be characterized in this manner, where the defined universe is not just a point. For example, FST was used in a previous study with nurses to specify the level of fatigue associated with performing specific job demands (Crumpton, McCauley-Bell and Soh, 1997).
2. Ambiguity is the use of FST to describe a condition in which more than one distinguishable sub-concept can simultaneously exist. An example of ambiguity representation with FST is presented in a research study with the construction industry where fuzzy variables were used to describe the overall awkward posture of the worker's hand, wrist, and arm while performing his job tasks (Dortch and Trombly, 1990).
3. Vagueness is the use of FST to present those cases whose precise boundaries are not well defined. The boundaries in this case may be described as nonprecise or noncrisp. This is particularly relevant when attempting to define linguistic variables. For example, when evaluating the impact of a risk factor, such as awkward joint posture, on injury development, the exact degree of deviation that produces a particular risk level is not clearly defined and is thus considered vague.

All of these types of fuzziness (i.e., generality, ambiguity, and vagueness) are present in real-world applications and can be represented mathematically by a fuzzy set.

25.3 Mathematics of Fuzzy Set Theory

A fuzzy set is a class of objects with a continuum of grades of membership defined for a given interval. Such a set is characterized by a membership function, which assigns a degree of membership ranging between 0 and 1 to each object. To understand the mathematical definition of fuzzy sets, consider a finite set of objects X.

1. Define the finite set as

$$X = x_1, x_2, \ldots, x_n \tag{25.1}$$

where x_i are elements in the set X. Each element, x_i, has a particular membership value, μ_i, which represents its grade of membership in a fuzzy set. The set of membership values associated with the fuzzy set occurs along the continuum $[0, 1]$. A fuzzy set A can thus be represented as a linear combination of the following form:

$$A = \mu_1(x_1), \mu_2(x_2), \ldots, \mu_n(x_n) \tag{25.2}$$

A fuzzy set could also be expressed as a vector, a table, or a standard function whose parameters can be adjusted to fit a given system. The interval over which a fuzzy set applies, known as a universe of discourse U, is thus characterized by a membership function that associates each element x_i of X with a degree of membership μ_i.

2. The membership values, μ_i, of each element in the fuzzy set A may be normalized such that they all are represented by values over a desired range. In fuzzy sets, this typically means representing the potential outcomes of the elements over the interval $[0,1]$. A fuzzy set is considered "normal" if the maximum value of the elements is 1 and the minimum value is 0. On the basis of the previous numerical basis for a fuzzy set, a graphical representation is created to further illustrate the progression of the set from one state to another.

To understand the mathematical definition of fuzzy sets, consider a set of objects A defined over a sample space X.

1. Consider a finite set defined as

$$X = x_1, x_2, \ldots, x_n \tag{25.3}$$

where the grade of membership of x_i in A is defined over the interval $[0, 1]$. Each element will have a particular membership value μ_i. A membership function may be expressed as a vector, a table, or a standard function whose parameters can be adjusted to fit a given system. Given the membership values, μ_i, the set A can be represented as a fuzzy set with the linear combination of the following:

$$A = \mu_1(x_1), \mu_2(x_2), \ldots, \mu_n(x_n) \tag{25.4}$$

Thus, a fuzzy set A of a universe of discourse U, is characterized by a membership function that associates each element x_i of X with a degree of membership (u_i). A fuzzy set A is considered as the union of its constituent singletons (King, 1988).

2. Normalization of a set simply means normalizing the values of the outcomes such that they are all represented over a desired range. In fuzzy sets this typically means representing the potential outcomes on the interval $[0, 1]$. Thus, all potential values of the set occur over the interval $[0, 1]$. A fuzzy set A is normal if the maximum value of the membership function $\mu(x) = 1$.

3. Two fuzzy sets are said to be equal if and only if for any x in U,

$$\mu_A(x) = \mu_B(x) \tag{25.5}$$

Fuzzy sets share many of the properties of conventional sets. Some of the properties that apply to fuzzy sets include the following:

Equality:

$$A = B \quad \text{iff} \quad \mu_A(x) = \mu_B(x), \forall x \in X \tag{25.6}$$

Containment:

$$A \subseteq B \quad \text{iff} \quad \mu_A(x) \le \mu_B(x), \forall x \in X \tag{25.7}$$

Standard Intersection:

$$\mu_{A \cap B}(x) = \min[\mu_A(x), \mu_B(x)] \tag{25.8}$$

Standard Union:

$$\mu_{A \cup B}(x) = \max[\mu_A(x), \mu_B(x)] \tag{25.9}$$

Standard Complement:

$$\mu_A(x) = 1 - \mu_A(x) \tag{25.10}$$

Concentration:

$$\mu_A(x) = [\mu_A(x)]^2 \tag{25.11}$$

Dilation:

$$\mu_A(x) = [\mu_A(x)]^{1/2} \tag{25.12}$$

Commutative property:

$$A \cap B = B \cap A \tag{25.13}$$

$$A \cup B = B \cup A \tag{25.14}$$

These operations on fuzzy sets are very similar to standard sets. Other operation properties that hold for fuzzy sets include the commutative, distributive, associative, and idempotence properties, as well as DeMorgan's Law.

Distributive property:

$$A \cup (B \cap C) = (A \cup B) \cap (A \cup C) \tag{25.15}$$

$$A \cap (B \cup C) = (A \cap B) \cup (A \cap C) \tag{25.16}$$

Associative property:

$$(A \cup B) \cup C = A \cup (B \cup C) \tag{25.17}$$

$$(A \cap B) \cap C = A \cap (B \cap C) \tag{25.18}$$

Idempotence property:

$$A \cap A = A \tag{25.19}$$

$$A \cup A = A \tag{25.20}$$

DeMorgan's law:

$$\mu_{(A \cap B)'}(x) = \mu_{(A' \cup B')}(x) \tag{25.21}$$

$$\mu_{(A \cup B)'}(x) = \mu_{(A' \cap B')}(x) \tag{25.22}$$

These operations become useful when manipulating fuzzy variables, particularly when two or more sets of variables are involved.

25.4 Membership Functions

Several geometric mapping functions have been developed, including S, π, trapezoidal, and triangular-shaped functions. All of these functions have utility in characterizing the environments of particular interests (Cox, 1994). In ordinary (i.e., crisp) subsets, a phenomenon is represented by a characteristic function. The characteristic function is associated with a set, S, which is represented as a binary mapping function

$$\mu_s: X \rightarrow [0, 1] \tag{25.23}$$

such that for any element x in the universe,

$$\mu_s(x) = 1 \quad \text{if } x \text{ is a member of } S \text{ and}$$

$$\mu_s(x) = 0 \quad \text{if } x \text{ is not a member of } S$$

Figure 25.1 illustrates the mapping of a characteristic function for a crisp set.

The process of translating a crisp set to a fuzzy set is known as fuzzification. In order to be "fuzzified," the real-world characteristics of a phenomenon must be mapped to a fuzzy function. The goal of this function is to map a subjective and ambiguous real-world phenomenon, X, into a membership domain, for example, [0,1]. This mapping function is a graphical representation of an element as it passes throughout a continuum (i.e., nonbinary) set of potential membership values. In other words, the mapping function provides a means to view the progression of the changes in the state of a given variable. Thus, this representation is referred to as a membership function for the fuzzy set. The term

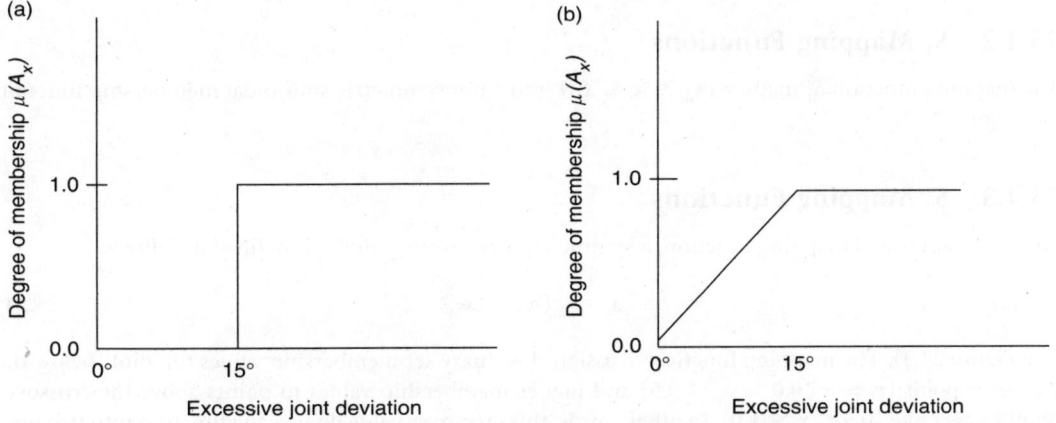

FIGURE 25.1 (a) Crisp set for excessive joint deviation. (b) Fuzzy membership function for exvessive joint deviation.

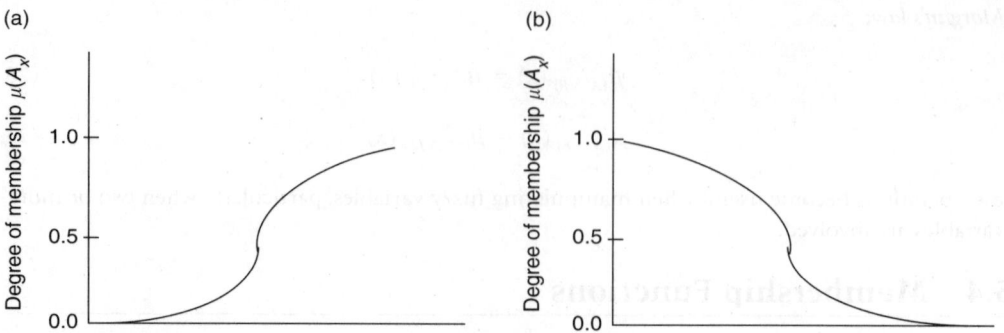

FIGURE 25.2   (a) Growth S-shaped membership function. (b) Decline S-shaped membership function.

"membership function" emphasizes the previously stated premise of fuzzy sets: for a fuzzy set A, each x value within the set has an associated $\mu(x)$ value that indicates the degree to which x is a member of the set A.

Membership functions are a characteristic of the data set under analysis and can take on many forms. Given the primary geometric mapping functions of S, π, trapezoidal, and triangular-shaped functions, the sinusoidal mapping functions, which include the S and π shapes (Gupta et al., 1988), are the most frequently implemented. For all S and π mapping functions discussed below, consider the fuzzy phenomenon X, defined over a real (i.e., nonnegative) interval $[x_m, x_M]$, where x_m and x_M correspond to the lower and upper bounds of the set X, respectively.

25.4.1   S-Shaped (Sigmoid/Logistic) Mapping Functions

These mapping functions are termed "S" because they are shaped like the letter S. The curves that comprise the S mapping functions may be referred to as growth and decline curves (Cox, 1994). The growth S-curve set moves from no membership at its extreme left-hand side, to complete membership at its extreme right-hand side. The declining S-curve behaves in just the opposite manner, beginning with complete membership at its extreme left-hand side, and progressing to 0 membership at its extreme right-hand side. Two common S-shaped curves are the S_1 and S_2 membership functions. Each of these functions has utility in the representation of fuzzy elements. An illustration and example of each function are provided in Figure 25.2.

25.4.2 S_1 Mapping Functions

The mapping function S_1 maps x_i ($x_m \leq x_i \leq x_M$) into a nonsymmetric sinusoidal membership function (see Figure 25.2).

25.4.3 S_2 Mapping Functions

For the S_2 sinusoidal mapping function, a symmetrical crossover point x_c is defined as follows:

$$x_c = \tfrac{1}{2}(x_m + x_M) \tag{25.24}$$

(see Figure 25.3). The mapping function S_2 assigns low fuzzy set membership values to points below the crossover point $[x_i \leq x_c; 0.0 \leq x_i \leq 0.5]$ and higher membership values to points above the crossover point $[x_i \geq x_c;$ i.e., $0.5 \leq x_i \leq 1.0]$. In other words, this crossover value divides the function into two portions and is located at the center point of the curve.

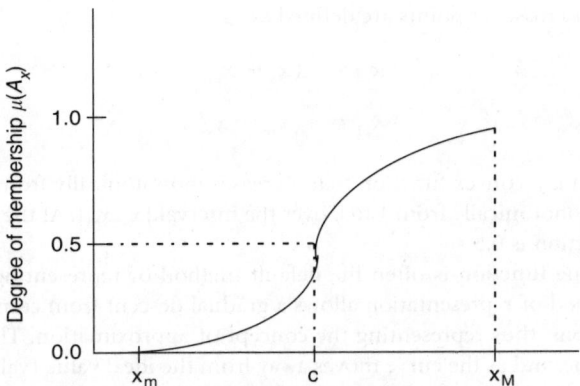

FIGURE 25.3   Growth S-shaped membership function with crossover point (c).

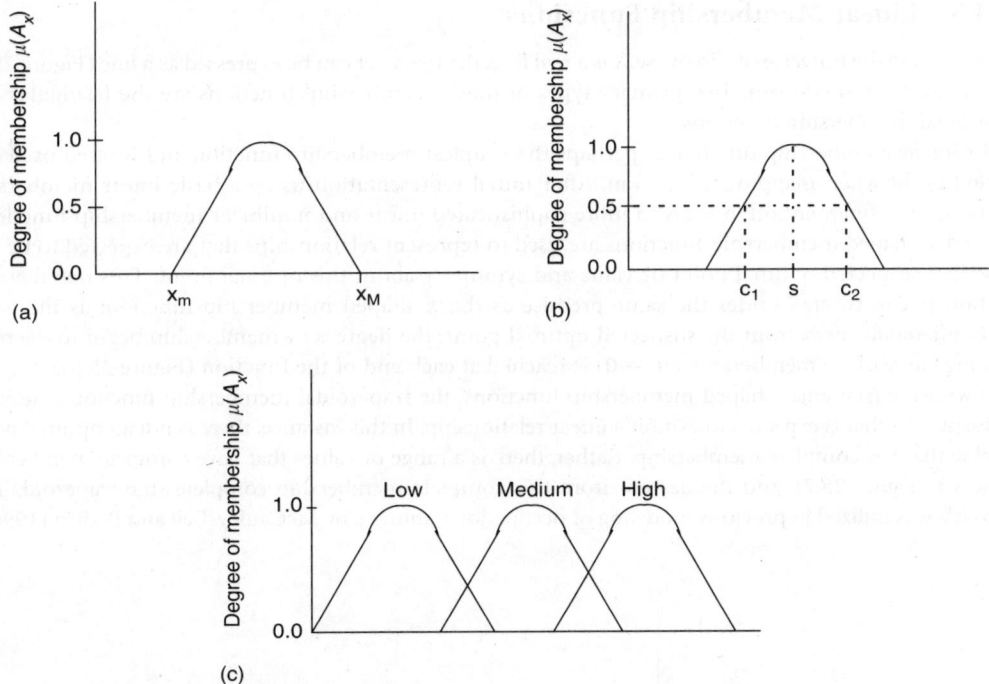

FIGURE 25.4   (a) π-Shaped membership function. (b) π-Shaped membership function with crossover points. (c) Overlapping π-shaped membership function.

25.4.4 π Mapping Functions

The π mapping functions are so named because they approximately simulate the shape of the Greek letter π. Figure 25.4 illustrates a symmetrical π mapping function for a fuzzy phenomenon X over the interval $[x_m, x_M]$; where x_m represents the minimum value for the curve and x_M the maximum value in the given curve. For this mapping function, the symmetrical point x_S is defined as the midpoint of x_m and x_M,

$$x_s = \tfrac{1}{2}(x_m + x_M) \tag{25.25}$$

and the lower and upper crossover points are defined as

$$x_{c2} = \tfrac{1}{2}(x_s + x_M) \tag{25.26}$$

$$x_{c1} = \tfrac{1}{2}(x_m + x_s) \tag{25.27}$$

The π mapping function is a convex function that increases monotonically from 0 to 1 over the interval $[x_m, x_s]$ and decreases monotonically from 1 to 0 over the interval $[x_s, x_M]$. At the crossover points x_{c1} and x_{c2}, the value of the function is 0.5.

The π shaped mapping function is often the default method of representing a fuzzy variable (Cox, 1994) because this method of representation allows a gradual descent from complete membership for a number in both directions, thus representing the concept of approximation. The symmetric π curve is centered on a single value, and as the curve moves away from the ideal value (value with complete membership), the degrees of membership begin to taper off until the curve reaches a point of no membership, where $\mu_s = 0$.

25.4.5   Linear Membership Functions

In cases where the universe of discourse X is a real line, the fuzzy set can be expressed as a line (Figure 25.5) or as some functional form. Two primary types of linear membership functions are the triangular and trapezoidal membership functions.

The linear membership function is perhaps the simplest membership function and is often used as a starting point when mapping a function. After initial representation using a basic linear membership function, set refinement often leads to more sophisticated linear and nonlinear membership functions. Triangular-shaped membership functions are used to represent relationships that are expected to be linear with a suspected optimal point or value and symmetry about this optimal point. This membership function is constructed under the same premise as the π-shaped membership function: as the value moves bilaterally away from the suspected optimal point, the degrees of membership begin to decrease until the value of no membership ($\mu_s = 0$) is reached at each end of the function (Figure 25.6).

As with the triangular-shaped membership functions, the trapezoidal membership function is used to represent a set that is expected to exhibit a linear relationship. In this instance, there is not an optimal point or value that has complete membership. Rather, there is a range of values that have complete membership in the set (Figure 25.7), and the descent from this complete membership completes the trapezoid. This approach was utilized in previous modeling of occupational injuries by McCauley-Bell and Badiru (1996a).

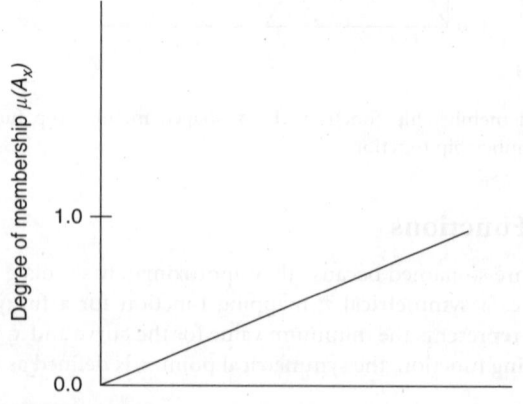

FIGURE 25.5   Linear membership function.

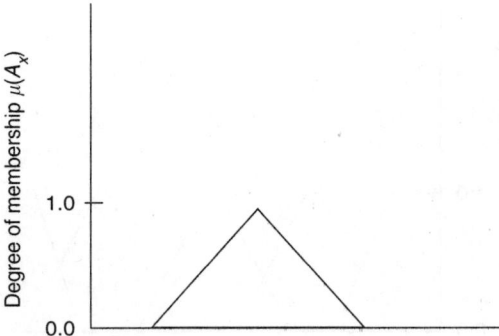

FIGURE 25.6 Triangular membership function.

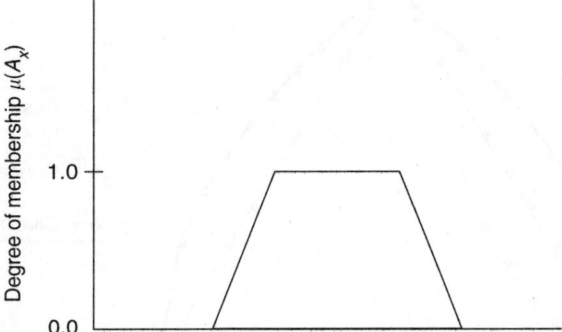

FIGURE 25.7 Trapezoidal shaped membership function.

Fuzzy sets may also take on a combination of triangular and trapezoidal membership functions. For instance, in process-control systems, variables are decomposed into overlapping arrays of triangular-shaped membership functions. The endpoints of these variables represent regions that begin and end in complete membership for a given set. These outer membership functions are often expressed as "shouldered" sets, and they appear as bisected trapezoids (Figure 25.8).

The development of an appropriate membership function is critical for effective representation and modeling of a fuzzy set. It is generally possible to represent virtually any domain through a membership function because such functions may assume a variety of different shapes and forms to accommodate a given data set. Irregular and uniquely shaped membership functions can also be developed to represent a fuzzy set in unusual cases. The previously mentioned membership functions, however, will be useful for graphically representing most fuzzy sets.

25.4.6 Hedges

Hedging in FST is the process of using adjectives to modify the set of values associated with a membership function. For example, consider the linguistic variable "short," represented by the variable x. A hedge would be the addition of a linguistic term such as "very" or "somewhat" to modify the linguistic term such that the two created linguistic variables are "very short" and "somewhat short." A mathematical operation is then performed on all values of the set $\mu(x)$, resulting in new membership functions for the hedges (Figure 25.9). This operation may also be performed on a single value within a membership function.

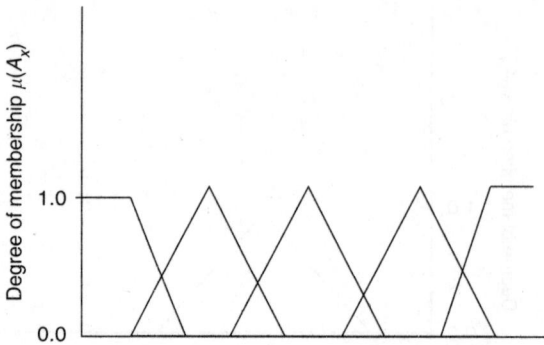

FIGURE 25.8   Triangular shaped membership function with trapezoidal shoulders.

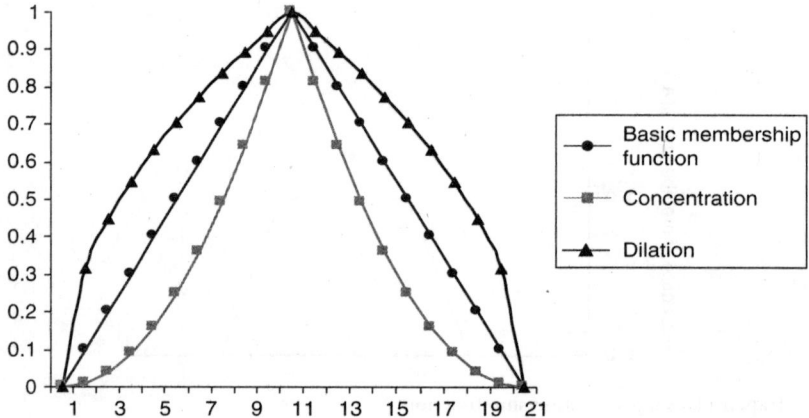

FIGURE 25.9   Concentration and dilation of a membership function.

25.5   Fuzziness vs Probability

The proponents of FST must deal with a fundamental issue: with probability theory available to characterize uncertainty, what is the added utility of FST? Although randomness and fuzziness share many similarities, there are fundamental differences that set them apart. Both systems describe uncertainty with numbers in the unit interval [0,1]. Thus, both systems numerically represent uncertainty. Both systems also combine sets and propositions associatively, commutatively, and distributively.

A key distinction, however, is how the systems jointly treat a set A and its complement. Classical set theory states that the intersection of a set A and its complement A^c is the null set. This is represented as

$$A \cap A^c = \varnothing \tag{25.28}$$

and by probability theory:

$$P(A \cap A^c) = P(\varnothing) = 0 \tag{25.29}$$

Fuzzy set theory, however, states that the intersection of a set A and its complement A^c may have events in common. This is represented as:

$$A \cap A^c \neq \varnothing \tag{25.30}$$

For example, consider a fuzzy set A {1,2,3,4,5}, which contains a set of numbers that are close to the number 5.

$$A : \{3/0.5, 4/0.8, 5/1\}$$

Each element has a particular membership value, which represents its grade of membership in the fuzzy set A (i.e., the membership grade of the number 3 is 0.5, of 4 is 0.8, and of 5 is 1). The complement of A, A^c, contains a set of numbers that are not close to 5.

$$A^c : \{3/0.3, 2/0.5, 1/0.7\}$$

In traditional set theory, the numbers 3, 4, and 5 could not be a member of the set A^c since they are in the set A. Note, however, that in the above example both sets A and A^c contain the number 3. The number 3 belongs to the set A with a membership grade of 0.5 and to the set A^c with a membership grade of 0.3. Thus, there is a distinct difference in what the numeric values for fuzzy sets and probabilistic sets are capable of representing.

In order to understand fuzziness as an alternative to randomness for describing uncertainty, consider the following. Fuzziness describes event ambiguity. It measures the degree to which an event occurs, not whether it occurs. Randomness, on the other hand, describes the uncertainty of event occurrence. Thus, whether or not an event occurs is "randomness" and the degree to which it occurs is "fuzziness." For example, the uncertainty associated with the outcome for a roll of a die has a certain probability associated with it. This event does not represent ambiguity but rather uncertainty of event occurrence. Once the die is rolled, the need for the probability due to the lack of knowledge concerning the future outcome dissipates. Now consider the term *tall* when used to describe the height of a man. The ambiguity associated with this event lies in differentiating where tall begins and ends — or the degree to which the person is tall. This type of ambiguity is characteristic of fuzziness because there does not exist a specific height at which we consider a person to be tall or not tall. Instead, there is a progression, which is more appropriately represented by linguistic terms (i.e., tall, somewhat tall, not tall). Thus, in contrast to the probabilistic representation of uncertainty, no addition of information is useful in removing the ambiguity associated with the boundaries for describing the variables "tall" and "not tall." It is in such situations that the concept of fuzziness becomes useful.

25.6 Fuzzification

The process of converting a crisp variable into a fuzzy variable is defined as fuzzification. This process involves the determination of factor existence over a universe of discourse.

Subsequently, each level of a variable's existence is assigned a degree of membership. This is often accomplished by providing an equation to express the membership function throughout the universe of discourse. In the absence of a known set of factors to define an equation, empirical evidence can be used. Generally, after an empirical model is validated, an equation or specific criteria are defined to represent the variable existence throughout the entire universe of discourse.

25.6.1 Fuzzification Example

To further explain the process of fuzzification, consider a task evaluation where the amount of repetition associated with task performance is the variable being fuzzified. The fuzzy set will be defined as "extreme repetition." The values within the universe of discourse represent the capabilities of the machinery and individual in task performance. In this case, let the variable x represent repetition level and $\mu(x)$ will represent the respective membership value for the variable x. For the task being evaluated, the variable x, repetition level, can range from 15 to 40 units within the given time interval. For simplification, the fuzzification will be performed for values at 5-unit intervals. Thus, each of these values of repetition will have a corresponding level of membership in the set.

25.6.2 Fuzzification of Variable:

$$[x, \mu(x)]$$

where x is number of repetitions, and $\mu(x)$ degree of membership for variable x.

$$[15, 0.50] \quad [30, 0.90]$$

$$[20, 0.70] \quad [35, 0.95]$$

$$[25, 0.80] \quad [40, 1.00]$$

Thus, the value of 15 units has a degree of membership of 0.5 in the set "extreme repetition," where as the value of 40 units has a degree of membership of 1.00 in the same set.

25.6.3 Aggregation

Aggregation is the process of combining various parameters to describe a given condition. For example, in an effort to define the degree of risks in an occupational setting, one may determine that six variables are critical in the development of risks. Second, if fuzzy modeling is performed, these six variables would then be fuzzified. Upon application of the fuzzy model to a problem, the goal may then be to determine the aggregated impact of the variables. Thus, there exists a need to understand the holistic impact of all fuzzy variables. The process of measuring the impact of various fuzzy variables on a given condition is known as fuzzy aggregation.

25.6.4 Aggregation Operations

Aggregation operations are a variety of methods designed to represent resultant fuzzy sets through inter-action between two or more fuzzy sets. These approaches include previously mentioned operations such as union and intersection. When desiring approaches that consider the degrees of memberships for all elements of a fuzzy set, mathematical functions are often utilized.

Some of the most commonly used tools for aggregation are the union and intersection. A variety of approaches are used to obtain the union and intersection of fuzzy sets; however, the standard union and standard intersection are taken given the fuzzy sets' maximum and minimum values, respectively. The fol-lowing example illustrates the standard approaches to the fuzzy union and intersections.

25.6.5 Union and Intersection: Max/Min Example

Obtain the fuzzy set that represents the standard union and standard intersection of the following sets:

$$A: [a, 0.7], [b, 0.3], [c, 0.6]$$

$$B: [a, 0.5], [b, 0.2], [c, 0.8]$$

$$A \cup B = \{\max (a, b, c): [a, 0.7], [b, 0.3], [c, 0.8]\}$$

$$A \cap B = \{\min (a, b, c): [a, 0.5], [b, 0.2], [c, 0.6]\}$$

Other aggregation techniques that are similar to the process of defuzzification are often desired to deter-mine a resultant impact of two or more fuzzy variables. These aggregation approaches include the use of various arithmetic operations, linear functions, and nonlinear functions. Weighted summations of func-tions are often used to aggregate various variables impacting a particular environment. For example, con-sider the variable condition defined by parameters $w_1, w_2, \ldots, w_n$, where each w represents a fuzzy variable that contributes to the resultant condition, R. These values may be aggregated using the arithmetic mean or a weighted mean. A weighted mean is generally preferred when all variables are not expected to have equal impact on the representation of the resulting condition. Thus, the following linear function is representative of an approach that can be used to obtain a weighted representation for fuzzy variables:

$$R = a_1 w_1 + a_2 w_2 + \ldots + a_n w_n \tag{25.31}$$

25.6.6 Defuzzification

The purpose of defuzzification is to convert fuzzy conclusions, obtained from inferencing, into a single real number. This single number summarizes the constraint imposed on possible values of the output variable by the fuzzy set. Three of the most commonly used defuzzification methods include the centroid or center-of-area method, the center-of-maxima method, and the mean-of-maxima method.

25.6.6.1 Centroid Method

The centroid method provides a single numerical output by selecting that value within the range of the variable v for which the area under the graph of membership function C is divided into two equal subareas. This defuzzified value is represented as $d_{CA}(C)$. This is calculated for the variable x as follows:

$$d_{CA}(C) = \frac{\int_{-c}^{c} C(x) x \, dx}{\int_{-c}^{c} C(x) \, dx} \tag{25.32}$$

For the discrete case, in which C is defined on a finite universal set $\{x_1, x_2, \ldots, x_n\}$, the formula is

$$d_{CA}(C) = \frac{\sum_{k=1}^{n} C(x_k) x_k}{\sum_{k=1}^{n} C(x_k)} \tag{25.33}$$

If $d_{CA}(C)$ is not equal to any value in the universal set, the value closest to it is taken. Thus, the value obtained from the previous function can be interpreted as the expected value of variable v.

25.6.6.2 Center of Maxima Method

The center of maxima method basically takes the average of the range for the variable. Thus, $d_{CM}(C)$ is defined as the average of the smallest value and the largest value for v for which $C(x)$ is the height, $H(C)$, of C. Therefore, $d_{CM}(C)$ is defined as

$$d_{CM}(C) = \frac{\inf M + \sup M}{2} \tag{25.34}$$

where $\inf M$ is the minimal membership value among the continuous functions and $\sup M$ the maximal membership value among the continuous functions

$$M = \{x \in [-c, c] \mid C(x) = h(C)\}$$

For the discrete case,

$$d_{CM}(C) = \frac{\min\{x_k \mid x_k \in M\} + \max\{x_k \mid x_k \in M\}}{2} \tag{25.35}$$

where

$$M = \{x_k \mid C(x_k) = h(C)\}$$

25.6.6.3　Mean of Maxima Method

The mean of maxima method is usually defined only for the discrete case. The defuzzified value, $d_{MM}(C)$, is the average of all values in the crisp set M, defined by

$$d_{MM}(C) = \frac{\sum_{x_k \in M} x_k}{|M|} \tag{25.36}$$

In the case of a continuous set, when M is given by the previous equation, $d_{MM}(C)$ may be calculated using the arithmetic average of mean values of all intervals contained in M, including intervals of length 0.

25.7　Fuzzy Linear Regression

Fuzzy linear regression has been explored by several researchers in different domains. Tanaka, Uejima, and Asai (1982) have introduced an LP-based regression method using a linear fuzzy model with symmetric triangular fuzzy parameters. This approach has been applied to forecasting in uncertain environments by Heshmaty and Kandel (1985). Tanaka (1987) extended symmetric triangular fuzzy parameters to more general fuzzy numbers defined by an L-shaped function. Celmins (1987a) and Diamond (1988) have developed several models for fuzzy least squares fitting from the viewpoint of least-squares regression. Tanaka et al. (1992) conducted LP formulation of regression analysis with a linear interval model. This new approach has not been applied in practice. It is the tool that is used in this study.

In general, fuzzy linear regression analysis has the following advantages:

1. It is based on possibility theory, which describes the ambiguity of events or the degree to which they occur.
2. It is effective for small data sets.
3. It can accommodate the inaccuracy and distortion introduced by linearization. The linear regression model itself contains vagueness, which is interpreted by a regression band, not a regression line.

The theory of fuzzy sets represents an attempt to construct a conceptual framework for a systematic treatment of vagueness and uncertainty due to fuzziness in both quantitative and qualitative ways (Zadeh, 1965). The fuzzy set approach has been widely applied to decision-making and specifically to multicriterion decision-making (Juang, 1988; Eldukair and Ayyub, 1992; Knosala and Pedrycz, 1992; Lasek, 1992).

Fuzzy regression techniques may provide another choice, which makes it possible to measure the "goodness" of the relationship between y and x on the basis of a small data set. The purpose of the technique is to find the regression model, which minimizes a measure of vagueness of y. This measure of "vagueness" replaces the least-squares criterion used in statistical regression analysis.

A fuzzy linear regression model is given by Tanaka et al. (1992):

$$y = A_0 + A_1 x_1 + A_2 x_2 + \ldots + A_n x_n = Ax \tag{25.37}$$

where A_i is a fuzzy parameter consisting of ordered pair $(\alpha_i, c_i)_L$. The center of the fuzzy parameters α_i denotes the center value for the ith regression coefficient, and the width of the ith fuzzy parameter coefficient is c_i, also known as the fuzziness of its parameter. L is a representation associated with a membership function. In fuzzy regression analysis the fuzzy coefficient is A_i ($i = 1, 2, \ldots, m$). The values of the fuzzy linear regression model can be obtained from the interval parameters $\{\alpha_i, c_i\}$, where α_i is the center and c_i the width of the interval (Tanaka et al., 1992).

The following LP model is formulated to obtain the above interval parameters (Tanaka et al., 1992): suppose that m input–output pairs (x_{ij}, y_i) for the input–output relationship of a system are given, where x_{ij} is n-dimensional risk factors input values $(x_{i1}, x_{i2}, \ldots, x_{in})$ for the ith subject, and y_i is an output of CTD level of the subject. The index, i, corresponds to the subjects from 1 to m. Then, the following LP formulation can be derived:

$$\text{Min } y_{1c} + y_{2c} + \ldots + y_{mc}$$

$$\text{subject to } y_i \in Y_i \quad i = 1, 2, \ldots, m \tag{25.38}$$

$$c_j \geq 0, j = 0, 1, \ldots, n$$

By simplifying fuzzy coefficients A_i to interval coefficients A_i, and from Eqs. (25.3) and (25.4), the above LP problem can be reformulated as follows (Tanaka and Ishibuchi, 1992):

$$\text{Min } J(c) = \sum_{i=1}^{m} (c_0 + c_1|x_{i1}| + \ldots + c_n|x_{in}|)$$

subject to

$$\alpha_0 + \sum_{j=1}^{n} \alpha_j x_{ij} - c_0 - \sum_{j=1}^{n} c_j|x_{ij}| \leq y_i$$

$$\alpha_0 \sum_{j=1}^{n} \alpha_j x_{ij} + c_0 + \sum_{j=1}^{n} c_j|x_{ij}| \leq y_i \tag{25.39}$$

$$c_0, c_j \geq 0$$

$$i = 0, 1, \ldots, m, j = 0, 1, \ldots, n$$

The objective of the above formulation is to minimize the sum of the width $J(c)$ subject to the linear interval model and should include all the given data. Based on the above LP problem, the interval coefficient $A_i = \{\alpha_i, c_i\}$ of the interval model is determined as the optimal solution of the above LP problem. It can be proven that there are always optimal solutions in the above LP problem (Tanaka et al., 1992).

In fuzzy linear regression, the degree of fitting, h, for a fuzzy linear system corresponds to R^2 for traditional regression. In this case, h is an index that indicates that observation y_i is contained by fuzzy estimation Y_i with more than h degree. According to h-level set constraints (Tanaka et al., 1992):

$$y_i \in [Y_i(x)]_h \tag{25.40}$$

$$[Y_i(x)] \, h = \{Y_i | \mu_{Y_i}(y_i) \geq h\} \, 0 \leq h < 1 \tag{25.41}$$

the fuzzy coefficients $A_i = \{\alpha_i, c_i\}_L$ of the fuzzy linear regression model can be derived as

$$\alpha_i^* = \alpha_i, \quad i = 1, 2, \ldots, n \tag{25.42}$$

$$c_i^* = c_i/(1 - h), \quad 0 \leq h < 1 \tag{25.43}$$

According to Eq. (25.9), the relationship between c_0 and $c_{0.5}$ is $2c_{i\,(h=0)}^* = c_{i\,(h=0.5)}^*$, which means that twice the width from the given data is obtained when $h = 0.5$.

Therefore, the fuzzy parameters A_i of fuzzy linear regression models can be obtained:

$$[A_i]h = \{\alpha_i^*\, c_i^*\}_L \tag{25.44}$$

There is a trade-off between h and $J(h)$. h values used in previous research vary widely, ranging from 0 to 0.9 (Moskowitz and Kim, 1993). Tanaka and Watada (1988) suggest that the selection of the h value be based on the sample size of the data set. Set $h=0$ when the data set is sufficiently large; set h at a higher value when the data set becomes smaller compared with some ideal size. Savic and Pedrycz (1992) recommended that the h value should not exceed 0.9. Bardossy et al. (1990) suggested that selection of h value be dependent on the decision-maker's belief in the model, generally recommending an h value between 0.5 and 0.7. The hypothetical decision-maker prefers the improvement of fuzzy threshold h at the expense of the fuzziness, $J(c)$, of the model. The higher the level of h, the fuzzier the estimated coefficient (i.e., the broader the triangular fuzzy number). However, high values of h could immediately cause the lack of specificity of the model. In some literature (Heshmathy and Kandel, 1985; Tanaka et al., 1982), $h=0.5$ is used consistently. What makes this technique attractive in fuzzy environments is dependent upon an h value because the h value can be subjectively selected by a decision-maker as an input to the model. On assessing the h value in fuzzy linear regression, Moskowitz and Kim (1993) proposed two approaches — an analytic approach and a search approach — to assess an h value that incorporates the decision-maker's belief. These are still in development. In general, the fuzzy output of the model should "cover" all of the data points to a certain degree of h, which makes the model as specific as possible.

25.8 Fuzzy Set Theory in Industrial Engineering and Human Factors/Ergonomics

The variability of system components in the modeling of an industrial engineering environment often limits the utility of traditional uncertainty management tools such as statistics. For this reason, FST has found application in a variety of these situations.

In an overview of mathematical modeling tools employed in ergonomics/human factors applications, Evans and Karwowski (1986) highlight the fact that fuzziness, vagueness, and imprecision exist in all aspects of the "man/machine/environment system." The fuzziness, vagueness, or imprecision can be related to three primary attributes: (1) the inability of researchers to obtain and process information relating to the behavior of the system, (2) variation associated with individuals and their responses to the system, and (3) subjective perceptions of the systems and human information processing.

Karwowski and Mital (1986) also discuss the applicability of FST in ergonomics/human factors research. Fuzzy set theory can be used to minimize the mismatch between human capabilities and job task requirements. Various applications, including control tasks, analytical tasks, manual materials handling tasks, and fault diagnosis, have employed FST with success.

McCauley-Bell et al. (1999) developed a fuzzy linear regression model to determine the level of risk associated with developing a work-related musculoskeletal disorder (WMSD) of the forearm and hand in office environments. Several data collection techniques were employed to collect information on the risk factors of forearm and hand WMSDs in the office. Based on these techniques, a total of 30 risk factors were identified and classified into four groups (joint deviation risk factors, anthropometric risk factors, personal risk factors, and task-related risk factors). Using the analytical hierarchy process, weights were assigned to each risk factor based on expert opinions in the area of WMSDs. The level of risk was determined using seven linguistic categories ranging from "nothing" to "very high risk." Numeric representations of the level of risk were developed based on the output of the model.

Jiang and Smith (1986) utilized FST to determine the relationship between an operator's psychophysical capabilities for individual and combined manual material handling (MMH) activities. Four individual and three combined MMH activities were performed for a duration of 1 h under varying lifting frequencies. The tasks involved the three most common MMH activities in industry: lifting, lowering, and carrying. The dependent variable included psychophysical capacity and body weight. Body weight was included as part of the dependent variable because it has been shown that body weight affects muscular strength, which in turn will affect the individual's perception regarding the maximum amount of weight that he or she can handle safely. The results of the fuzzy analysis found that with respect to predictability, the models performed well. It was also found that physiological responses for the combined MMH activities were limiting factors in the MMH capacities for faster frequencies.

In similar studies, Karwowski (1983) and Karwowski and Ayoub (1984) utilized FST in predicting stress associated with MMH tasks. The researchers investigated the relationship between biomechanical, physiological, and psychological acceptability limits during lifting tasks. Results from these studies found that combining biomechanical and physiological acceptability measures closely paralleled the individual's perceived psychological acceptability limits.

Another application by Hafez et al. (1986) examined the effects of heat stress on individuals discomfort during job-task performance. On the basis of previous research regarding acceptable and unacceptable thermal environments for office or sedentary work, a fuzzy model was developed. Following the same procedure, various work-intensity levels were investigated, resulting in another fuzzy model. The resulting models can then be used to estimate the amount of discomfort an individual will experience under various environmental temperatures and levels of work intensity.

25.9 Representative Applications of Fuzzy Modeling of Occupational Injuries

Although FST and occupational injury representation are compatible in many respects, the application of this technology to ergonomics and the modeling injury risk has been limited. Karwowski and Mital (1986) produced a volume designed to highlight fuzzy applications to human factors and ergonomics. This book is entitled *Applications of Fuzzy Set Theory in Human Factors*. This volume hosted work that had applied fuzzy modeling to the representation of occupational back-injury risk levels and material handling tasks as well as other topic areas. Some of the topics in this book include "Towards an Algorithmic/Procedural Human Consistency of Decision Support Systems: A Fuzzy Logic Approach"; "A Fuzzy Method for the Modeling of HCI in Information Retrieval Tasks"; "Techniques for Fitting Fuzzy Connective and Logical Operators to Human Judgement Data in Design and Evaluation of Man–Machine Systems"; and "Dealing With the Vagueness of Natural Languages in Man–Machine Communication." Although the projects presented in this volume were useful and of sound methodology, they did not produce widespread application of fuzzy modeling in human factors and ergonomics problems.

Additional applications of fuzzy modeling to ergonomics were applied by McCauley-Bell and Badiru (1997a and b), McCauley-Bell and Crumpton (1997), McCauley-Bell et al. (1998), and Crumpton et al. (1996). Crumpton, McCauley-Bell, and Soh (1997) applied fuzzy modeling techniques to develop a model for evaluating fatigue using data from several estimators of fatigue. Some of the fatigue estimators included in this model were choice reaction time, heart rate, perceived exertion, and skin temperature. These research projects included application of fuzzy modeling to various occupational settings including risks of nursing injuries and an array of cumulative trauma disorders. Sufficient time has not yet elapsed to judge the impact of this work on the proliferation of fuzzy modeling in human factors/ergonomics. Nonetheless, there still exists a need to provide the occupational injury, human factors, and ergonomics communities with practical, comprehensive methodologies to produce reliable and quantitative models of occupational risks using the techniques associated with FST.

25.10 Conclusion

The underlying premise in FST is the graded membership concept, which recognizes that variables differ in their levels of association with a response variable. The grade of membership is highly dependent on the data under consideration and is defined by a membership function. The function is used to map each variable along an interval. Fuzzy set theory not only allows for the determination of the presence or absence of a concept, but also assists in determining the exact degree or level of a particular concept.

As can be seen by this brief description of FST, the possible applications and uses of this technique within the various areas of industrial engineering are as diverse as they are in the ergonomics/human factors area. In addition, FST provides a unique method for describing the human–machine relationships often considered, such as injury and illness prevention and control, quantification of mental and physical work capacities, descriptions of physiological responses to work and stress, etc. The use of FST has been shown to provide more accurate and reliable representations of these systems than traditional techniques.

Glossary

α-cut and strong α-cut – the crisp set that contains all the elements of the universal set XX whose membership grades in A are greater than or equal to (or only greater than for strong α-cut) the specified value of α.

Cardinality – the number of members of a finite set A.

Crossover point – the element or area within the membership function U at which its membership function is 0.5.

Fuzzy Set – any set that allows its members to have different grades of membership (membership function) in the interval $[0,1]$.

Fuzzy Singleton – a fuzzy set whose support is a single point in U with a membership function of one.

Support – the crisp set of all points in the Universe of Discourse U such that the membership function of F is nonzero.

Supremum – maximal value in a fuzzy set.

t-norms – often referred to as "fuzzy ands," these are tools to combine fuzzy sets.

There are many ways to compute "and." The two most common are:

1. Zadeh – $\min(\mu_A(x), \mu_B(x))$ computes the "and" by taking the minimum of the two (or more) membership values. This is the most common definition of the fuzzy "and";
2. Product – $\mu_A(x)$ times $\mu_B(x)$. This technique computes the fuzzy "and" by multiplying the two membership values.

Triangular Conorms (*t-conorms*) – often referred to as "fuzzy or." The fuzzy "or" can be obtained by either of the following two techniques:

1. Zadeh – $\max(\mu_A(x), \mu_B(x))$. This technique computes the fuzzy "or" by taking the maximum of the two (or more) membership values. This is the most common method of computing the fuzzy "or."
2. Product – $\mu_A(x) + \mu_B(x) - \mu_A(x)\mu_B(x)$. This technique uses the difference between the sum of the two (or more) membership values and the product of the membership values.

Universe of Discourse – the range of all possible values for an input to a fuzzy system.

References

Afifi, A.A. and Clarke, V., *Computer Aided Multivariate Analysis*, 2nd ed., Van Nostrand Reinhold, New York, 1990.

Anon, IBM and IBM compatible keyboards. Technology Strategy Paper, Honeywell Inc. *REPSIG Newsletter Ergonomics Society Aust.*, 4–5, 1986

Armstrong, T.J. and Chaffin, D.B., Carpal tunnel syndrome and selected personal attributes, *J. Occup. Med.*, 21, 481–486, 1979.

Armstrong, T.J. and Lifshitz, Y., Evaluation and design of jobs for control of cumulative trauma disorders, *Ergonomic Interventions to Prevent Musculoskeletal Injuries in Industry*, Chelsea, Lewis Publishers, Inc., 1987

Armstrong, T.J., Radwin, R.G., Hansen, D.J., and Kennedy, K.W., Repetitive trauma disorders: job evaluation and design, *Hum. Factors*, 28, 325–336, 1986.

Armstrong, T.J., Castelli, W., Evans, F., and Diaz-Perez, R., Some histological changes in carpal tunnel contents and their biomechanical implications. *J. Occupa. Med.*, 26, 197–201, 1984.

Baker, E. and Ehrenberg, R., Preventing the work-related carpal tunnel syndrome: physician reporting and diagnostic criteria, *Ann. Intern. Med.*, 112, 317–319, 1990.

Bleecker, M.L., Bohlman, M., and Moreland, R., Carpal tunnel syndrome: role of carpal canal size. *Neurology*, 35, 1599–1604, 1985.

Bobbins, H., Anatomical structure of the study of the median nerve in the carpal tunnel and etiologies of carpal tunnel syndrome, *J. Bone Joint Surg.*, 45, 953–966, 1963.

Borg, G. and Noble, B.J., Perceived exertion, *Exercise and Sport Sciences Reviews*, Vol. 2, Academic Press, New York, 1974.

Burnette, J., CTD-123: A Cumulative Trauma Disorder Risk Assessment Model, Unpublished Doctoral Dissertation, North Carolina State University, Department of Industrial Engineering, 1989.

Calisto, G.W., Jiang, B.C., and Cheng, S.H., A checklist for carpal tunnel syndrome, *Proceedings of the Human Factors Society-30th Annual Meeting*, Santa Monica, CA, 1986, pp. 1438–1442.

Celmins, A., Least squares model fitting to fuzzy vector data, *Fuzzy Sets Syst.*, 22, 245–269, 1987a.

Celmins, A., Multidimensional least-squares fitting to fuzzy models, *Math. Modeling*, 9, 669–690, 1987b.

Chao, E.Y., Opgrande, J.D., and Axmear, F.E., Three dimensional force analysis of finger joints in selected isometric hand functions, *J. Biomech.*, 9, 387–396, 1976.

Cox, E., *The Fuzzy Systems Handbook: A Practitioner's Guide to Building, Using, and Maintaining Fuzzy Systems*, Academic Press, New York, 1994.

Crumpton, L.L. and Congleton, J.J., Use of risk factors commonly associated with carpal tunnel syndrome to model median nerve conduction, *Adv. Ind. Ergonomics Safety*, VI, 511–514, 1994a.

Crumpton, L.L. and Congleton, J.J., An evaluation of the relationship between subjective symptoms and objective testing used to assess carpal tunnel syndrome, *Adv. Ind. Ergonomics Safety*, VI, 515, 1994b.

Crumpton, L. and Congleton, J., Methods of assessing carpal tunnel syndrome: an investigation of accuracy and usability, *Proceedings of the Third Annual Institute of Industrial Engineering Conference*, 1994.

Crumpton, L., McCauley-Bell, P., and Soh, T., Modeling overall level of fatigue using linguistic modeling and fuzzy set theory, *Appl. Ergonomics*, 1997, submitted.

Diamond, P., Fuzzy least squares, *Inf. Sci.*, 46, 141–157, 1988.

Dortch, H. and Trombly, C., The effects of education on hand use with industrial workers in repetitive jobs, *Am. J. Occup. Ther.* 3, 777–782, 1990.

Eastman, K., *Ergonomic Design for People at Work*, Vol. 1, Van Nostrand Reinhold, New York, 1986, pp. 28–34, 333–341.

Eiser, H., Subjective scale of force for a large number of group, *J. Exp. Psychol.*, 64, 253–257, 1962.

Eldukair, Z.A. and Ayyub, B.M., Multi-attribute fuzzy decisions in construction strategies, *Fuzzy Sets and Syst.*, 46, 155–165, 1992.

Elmer-Dewitt, P., A royal pain in the wrist, *Time*, 144(17), 60–62, 1994.

Gellman, H., Gelberman, R.H., and Tan, A.M., Carpal tunnel syndrome: an evaluation of the provocative diagnosis tests, *J. Bone Joint Surg.*, 68(5), 735–737, 1986.

Gharpuray, M.M., Tanaka, H., and Fan, L.T., Fuzzy linear regression analysis of cellulose hydrolysis, *Chem. Eng. Commun.*, 41, 299–314, 1986.

Goldstein, S.A., Biomechanical Aspects of Cumulative Trauma to Tendons and Tendon Sheaths, Unpublished Doctoral Dissertation, The University of Michigan, 1–124, 1991.

Goldstein, S.A., Armstrong, T.J., Chaffin, D.B., and Matthews, L.S., Analysis of cumulative strain in tendons and tendon sheaths, *J. Biomec.*, 20, 1–6, 1987.

Gregory, B., An ounce of prevention is worth a pound of cure, *Inside Seminole*, 9, 1994.

Heshmaty, B. and Kandel, A., Fuzzy linear regression and its applications to forecasting in uncertain environment, *Fuzzy Sets Syst.*, 15, 159–191, 1985.

Hudock, S.D. and Keran, C.M., *Risk profile of cumulative trauma disorders of the arm and hand in the U.S. mining industry*, Bureau of Mines Information Circular, United States Department of the Interior, 1–5, 1992.

Ishibuchi, H., Fuzzy regression analysis. *Japanese J. Fuzzy Theory Syst.*, 4, 137–148, 1992.

Ishibuchi, H. and Tanaka, H., Identification of fuzzy parameters by interval regression models, *Electron. Commun. Japan, Part 3*, 73, 19–27, 1990.

Jetzer, T., Use of vibration testing in the early evaluation of workers with carpal tunnel syndrome, *J. Occup. Med.*, 33, 117–120, 1991.

Kandel, A. and Heshmaty, B., Using fuzzy linear regression as a forecasting tool in intelligent systems, in *Fuzzy Logic in Knowledge-Based Systems, Decision and Control*, Elsevier, Amsterdam, The Netherlands, 1988, pp. 361–366.

Karwowski, W. and Mital, A., Eds., Applications of fuzzy set theory in human factors, *Advances in Human Factors/Ergonomics*, Vol. 6, Elsevier, Amsterdam, 1986, pp. 232–237.

Karwowski, W., Evans, G.W., and Wilhelm, M.R., Fuzzy modeling of risk factors for industrial accident prevention: some empirical results, in *Applications of Fuzzy Sets Methodologies in Industrial Engineering*, Elsevier, New York, 1989, pp. 141–154.

Keyserling, W.M., Stetson, D.S., Silverstein, B.A., and Brouwer, M.L., A checklist for evaluating ergonomics risk factors associated with upper extremity cumulative trauma disorders, *Ergonomics*, 36(7), 807–831, 1993.

Kilbom, A. and Persson, J., Work technique and its consequences for musculoskeletal disorders, *Ergonomics*, 30, 273–279, 1987.

Kroemer, K.H.E., Cumulative trauma disorders: their recognition and ergonomics measures to avoid them, *Appl. Ergonomics*, 20, 274–280, 1989.

Kuorinka, I.A., and Jonsson, B., Standardized nordic questionnaire for the analysis of musculoskeletal symptom, *Appl. Ergonomics*, 18, 233–237, 1987.

Lifshitz, Y. and Armstrong, T.J., A design checklist for control and prediction of cumulative trauma disorder in intensive manual jobs, *Proceedings of the Human Factors Society-30th Annual Meeting*, Santa Monica, The Society, 1986, pp. 837–841.

McCauley-Bell, P., A Study of Job Related Physical Performance Tests, Unpublished Thesis, Industrial Engineering Department, University of Oklahoma, Norman, Oklahoma, 1991.

McCauley-Bell, P., A Fuzzy Linguistic Artificial Intelligence Model for Assessing Risks of Cumulative Trauma Disorders of the Forearm and Hand, Published Doctoral Dissertation, Industrial Engineering Department, University of Oklahoma, Norman, Oklahoma, 1993.

McCauley-Bell, P. and Badiru, A., Fuzzy modeling and analytic hierarchy processing to quantify risk levels associated with occupational injuries part I: the development of fuzzy linguistic risk levels, *IEEE Trans. Fuzzy Syst.*, 4, 124–131, 1996a.

McCauley-Bell, P. and Badiru, A., Fuzzy modeling and analytic hierarchy processing as a means to quantify risk levels associated with occupational injuries part II: the development of a fuzzy rule-based model for the prediction of injury, *IEEE Trans. Fuzzy Syst.*, 4, 132–138, 1996b.

McCauley-Bell, P. and Wang, H., Fuzzy linear regression models for assessing risks of cumulative trauma disorders, *Fuzzy Sets Syst.*, 92, 95–105, 1997.

McKenzie, F., Storment, J., Van Hook, P., and Armstrong, T., A program for control of repetitive trauma disorders, *Am. Ind. Hyg. Assoc. J.*, 46, 674–677, 1985.

Meagher, S., Pre-employment anthropometric profiling, in *Advances in Industrial Ergonomics and Safety II*, Taylor & Francis, New York, 1990, pp. 375–379.

Moore, A., Wells, R., and Raaney, D., Quantifying exposure in occupational manual tasks with cumulative trauma disorder potential, *Ergonomics*, 34, 1433–1453, 1991.

Moskowitz, H. and Kim, K., On assessing the H value in fuzzy linear regression, *Fuzzy Sets Syst.*, 58, 303–327, 1993.

Nahorski, Z., Regression analysis: a perspective, *Fuzzy Regression Analysis*, Omnitech Press, Warsaw and Physica-Verlag, Heidelberg, 1992, pp. 2–13.

Nubar, Y. and Contini, R., A minimal principle in biomechanics, *B. Math. Biophys.*, 23, 377, 1961.

OSHA, Ergonomics program management guidelines for meatpacking plants, U.S. Document, 1–21, 1990.

Owen, R.D., Carpal tunnel syndrome: a products liability prospective, *Ergonomics*, 37, 449–476, 1994.

Park, D., A Risk Assessment Model of Hand Cumulative Trauma Disorders, Unpublished Doctoral Dissertation, The Pennsylvania State University, PA, 1–288, 1993.

Parker, K.G. and Imbus, H.R., *Cumulative Trauma Disorders Current Issues and Ergonomic Solution: A Systems Approach*, Lewis Publishers, Ann Arbor, 1992, pp. 31–67.

Peters, G., Fuzzy linear regression with fuzzy intervals, *Fuzzy Sets Syst.*, 63, 45–55, 1994.

Phalen, G.S., The carpal tunnel syndrome: seventeen years' experience in diagnosis and treatment of 654 hands, *J. Bone Joint Surg.*, 48, 211–228, 1966.

Putz-Anderson, V., *Cumulative Trauma Disorders: A Manual for Musculoskeletal Disease of the Upper Limbs*, Taylor and Francis, London, 1988, pp. 1–70.

Ramezani, R. and Duckstein, L., Fuzzy regression analysis of the effect of university research on regional technologies, in *Fuzzy Regression Analysis*, Kacprzyk, J. and Fedrizzi, M., Eds., Omnitech Press, Warsaw and Physica-Verlag, Heidelberg, 1992, pp. 237–263.

Rose, M.J., Keyboard operating postures and actuation force: implications for muscle over-use, *Appl. Ergonomics*, June, 198–203, 1991.

Saaty, T., *The Analytic Hierarchy Process*, McGraw-Hill, New York, 1980.

Sakawa, M. and Yano, H., Fuzzy linear regression and its applications, *Fuzzy Regression analysis*, Omnitech Press, Warsaw and Physica-Verlag, Heidelberg, 1992, pp. 61–79.

Sauter, S., Schleifer, L., and Knutson, S., Work posture, work station design, and musculoskeletal discomfort in a VDT data entry task, *Hum. Factors*, 33, 151–167, 1991.

Savic, D.A. and Pedrycz, W., Fuzzy linear regression models: construction and evaluation, *Fuzzy Regression Anal.*, Omnitech Press, Warsaw and Physica-Verlag, Heidelberg, 1992, pp. 91–100.

Schoenmarklin, R.W., Marras, W.S., and Leurgans, S.E., Industrial wrist motions and incidence of hand/wrist cumulative trauma disorders, *Ergonomics*, 37, 1449–1459, 1994.

Silverstein, B., The Prevalence of Upper Extremity of Cumulative Trauma Disorders in Industry, Doctoral Dissertation, The University of Michigan, Ann Arbor, MI, 1985.

Smith, E.M., Sonstegard, D.A., and Anderson, W.H., Contribution of flexor tendons to the carpal tunnel syndrome, *Arch. Phys. Med. Medical Rehabil.*, 58, 379–385, 1977.

Smith, M.J. and Zehel, D.J., Cumulative trauma injuries control efforts in select Wisconsin manufacturing plants, in *Advances in Industrial Ergonomics and Safety II*, Das, B. Ed., Elsevier Science Publishers, Amsterdam, 1990, pp. 259–265.

Snyder, D., McNeese, M., Zaff, B., and Gomes, M., Knowledge Acquisition of Tactical Air-to-Ground Mission Information Using Concept Mapping, Unpublished Research Report, Wright-Patterson Air Force Base, Air Force Institute of Technology, Ohio, 1992.

Sommerich, C.M., McGlothlin, J.D., and Marras, W.S., Occupational risk factors associated with soft tissues disorders of the shoulder: a review of recent investigations in the literature, *Ergonomics*, 36, 697–717, 1993.

Steward, J., Safety Engineer, Seagate Technology, personal communication, 1991.

Swanson, A.B., Swanson, G.G., and Goran-Hagert, C., Evaluation of impairment of hand function, in *Rehabilitation of the Hand: Surgery and Therapy*, Morsby, St Louis, 1990, pp. 109–138.

Tabourn, S., Cumulative trauma disorders: an ergonomics intervention, *Advances in Industrial Ergonomics and Safety II*, North-Holland, Elsevier, The Netherlands, 1990, pp. 277–284.

Tanaka, H., Fuzzy data analysis by possibilistic linear models, *Fuzzy Sets Syst.*, 24, 363–375, 1987.

Tanaka, H. and Ishibuchi, H., Possibilistic regression analysis based on linear programming, in *Fuzzy Regression Analysis*, Kacprzyk, J. and Fedrizzi, M., Eds., Omnitech Press, Warsaw and Physica-Verlag, Heidelberg, 1992, pp. 47–60.

Tanaka, H. and Watada, J., Possibilistic linear systems and their application to the linear regression model, *Fuzzy Sets Syst.*, 27, 275–289, 1988.

Tanaka, H., Uejima, S., and Asai, K., Fuzzy linear regression model, *Appl. Syst. Cybern.*, VI, 2933–2939, 1980.

Tanaka, H., Uejima, S., and Asai, K., Linear regression analysis with fuzzy model, *IEEE Trans. Syst., Man, Cybern.*, SMC-12, 903–906, 1982.

Terano, T., Asai, K., and Sugeno, M., *Fuzzy Systems Theory and its Applications*, Academic Press, San Diego, CA., 1992, pp. 69–84.

Wang, H., Fuzzy Linear Regression Models for Cumulative Trauma Disorders of the Hand and Forearm in Office Environments, Unpublished Masters Thesis, The University of Central Florida, Department of Industrial Engineering and Management Systems, Orlando, Florida, 1995.

Zadeh, L., Fuzzy sets, *Inf. Control*, 8, 338–353, 1965.

26

Maintenance Management in the 21st Century

S.A. Oke
University of Lagos

26.1 Introduction

Maintenance is a combination of all technical, administrative, and managerial actions during the life cycle of an item intended to keep it in or restore it to a state in which it can perform the required function (Komonen, 2002) (see Figure 26.1). Traditionally, maintenance has been perceived as an expense account with performance measures developed to track direct costs or surrogates such as the headcount of tradesmen and the total duration of forced outages during a specified period. Fortunately, this perception is changing (Tsang, 1998; Kumar and Liyanage, 2001, 2002a; Kutucuoglu et al., 2001b). In the 21st century, plant maintenance has evolved as a major area in the business environment and is viewed as a value-adding function instead of a "bottomless pit of expenses" (Kaplan and Norton, 1992). The role of plant maintenance in the success of business is crucial in view of the increased international competition, market globalization, and the demand for profitability and performance by the stakeholders in business (Labib et al., 1998; Liyanage and Kumar, 2001b; Al-Najjar and Alsyouf, 2004). Today, maintenance is acknowledged as a major contributor to the performance and profitability of business organizations (Arts et al., 1998; Tsang et al., 1999; Oke, 2005). Maintenance managers therefore explore every opportunity to

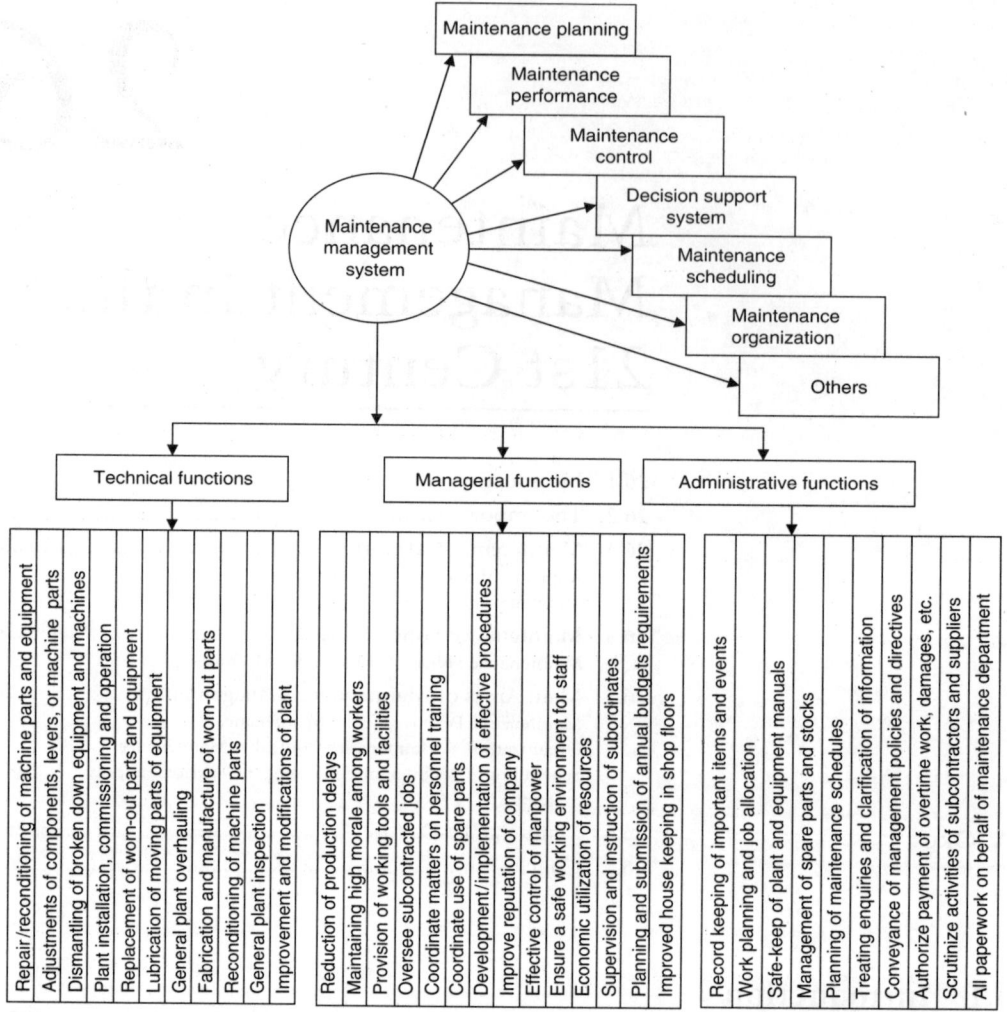

FIGURE 26.1   The maintenance management system.

improve on profitability and performance and achieve cost savings for the organization (Al-Najjar and Alsyouf, 2004). A major concern has been the issue of what organizational structure ought to be adopted for the maintenance system: should it be centralized or decentralized? Such a policy should offer significant savings as well (HajShirmohammadi and Wedley, 2004).

The maintenance organization is confronted with a wide range of challenges that include quality improvement, reduced lead times, set up time and cost reductions, capacity expansion, managing complex technology and innovation, improving the reliability of systems, and related environmental issues (Kaplan and Norton, 1992; Dwight, 1994, 1999; De Groote, 1995; Cooke and Paulsen, 1997; Duffua and Raouff, 1997; Chan et al., 2001). However, trends suggest that many maintenance organizations are adopting total productive maintenance, which is aimed at the total participation of plant personnel in maintenance decisions and cost savings (Nakajima, 1988, 1989; HajShirmohammadi and Wedley, 2004). The challenges of intense international competition and market globalization have placed enormous pressure on maintenance system to improve efficiency and reduce operational costs (Hemu, 2000). These challenges have forced maintenance managers to adopt tools, methods, and concepts that could stimulate performance growth and minimize errors, and to utilize resources effectively toward making the organization a "world-class manufacturing" or a "high-performance manufacturing" plant.

Maintenance information is an essential resource for setting and meeting maintenance management objectives and plays a vital role within and outside the maintenance organization. The need for adequate maintenance information is motivated by the following four factors: (1) an increasing amount of information is available and data and information are required on hand and to be accessible in real-time for decision-making (Labib, 2004); (2) data lifetime is diminishing as a result of shop-floor realities (Labib, 2004); (3) the way data is being accessed has changed (Labib, 2004); and (4) it helps in building knowledge and in measurement of the overall performance of the organization. The computerized maintenance management system (CMMS) is now a central component of many companies' maintenance departments, and it offers support on a variety of levels in the organizational hierarchy (Labib, 2004). Indeed, a CMMS is a means of achieving world-class maintenance, as it offers a platform for decision analysis and thereby acts as a guide to management (Labib, 1998; Fernandez et al., 2003).

Consequently, maintenance information systems must contain modules that can provide management with value-added information necessary for decision support and decision-making. Computerized maintenance management systems are computer-based software programs used to control work activities and resources, as well as to monitor and report work execution. Computerized maintenance management systems are tools for data capture and data analysis. However, they should also offer the capability to provide maintenance management with a facility for decision analysis (Bamber et al., 2003).

With the current advancements in technology, a number of contributions have been made by the development and application of sophisticated techniques that have enhanced the quality of decision-making using CMMSs. Data mining and web data management are two new concepts that are gradually finding applications in maintenance management. The utilization of data-mining applications in maintenance has helped in the discovery, selection, and development of core knowledge in the management of large maintenance databases hitherto unknown. Data-mining applications in maintenance encourage adequate and systematic database analysis for correct maintenance management decisions. With today's sophistication in data mining, evaluation and interpretation of relational, transactional, or multimedia databases in maintenance are much easier than before. We can classify, summarize, predict, describe, and contrast maintenance data characteristics in a manufacturing milieu for efficient maintenance data management and high productivity. With the emergence of the Internet and World Wide Web (WWW), numerous communication problems in maintenance have been solved. This new communication mechanism quickly distributes information to distant and multiple locations (HajShirmohammadi and Wedley, 2004). This challenges many firms to reassess how they organize and manage their resources (Blanchard, 1997). This is particularly important where information exchange is very vital. Examples of such systems are multinationals, which have manufacturing plants scattered all over the world where the exchange of resources takes place.

With the users connected to the Internet, information exchange is possible since plants can communicate with one another through high-speed data transmission paths. With the emergence of the Internet and the web data management, significant improvements have been made by the maintenance organization in terms of updating information on equipment management, equipment manufacturer directory services, and security systems in maintenance data. The web data system has tremendously assisted the maintenance organization to source for highly skilled manpower needed for maintenance management, training and retraining of the maintenance workforce, location of equipment specialists and manufacturers, and the exchange of maintenance professionals across the globe.

26.2 The Importance of Maintenance

Industrial maintenance has two essential objectives: (1) a high availability of production equipment and (2) low maintenance costs (Komonen, 2002). However, a strong factor militating against the achievement of these objectives is the nature and intensity of equipment failures in plants. Since system failure can lead to costly stoppages of an organization's operation, which may result in low human, material, and equipment utilization, the occurrence of failure must therefore be reduced or eliminated. An organization can have its customers build confidence in it by having uninterrupted flow in operations. Thus, maintenance ensures system sustenance by avoiding factors that can disrupt effective productivity, such as machine breakdown and its several attendant consequences. In order to carry out effective maintenance activities, the team players must be dedicated, committed, unflagging, and focused on achieving good maintenance practices. Not only are engineers and technicians are involved, but also every other employee, especially those involved in production and having physical contact with equipment. Thus, maintenance is not only important for these reasons, but its successful implementation also leads to maximum capacity utilization, improved product quality, customer satisfaction, adequate equipment life span, among other benefits. Equipment does not have to finally breakdown before maintenance is carried out. Implementing a good maintenance policy prevents system failures and leads to high productivity (Vineyard et al., 2000).

26.3   Maintenance Categories

In practice, failure of equipment could be partial or total. Even with the current sophistication of equipment automation, failure is still a common phenomenon that generates serious consideration of standard maintenance practices. Nonetheless, the basic categories of maintenance necessary for the control of equipment failures are traditionally divided into three main groups: (1) preventive maintenance (PM) (condition monitoring, condition-based actions, and scheduled maintenance); (2) corrective/breakdown maintenance (BM); and (3) improvement maintenance (Komonen, 2002). Breakdown maintenance is repair work carried out on equipment only when a failure has occurred. Preventive maintenance is carried out to keep equipment in good working state. It is deployed before failure occurs, thereby reducing the likelihood of failure. It prevents breakdown through repeated inspection and servicing.

26.3.1   Implementing Preventive Maintenance

Preventive maintenance is basically of two types: (1) a control necessitating machine stoppages and (2) a tool change, which results in machine stoppage. In order to attain constant system functioning, it is necessary to be aware of when a system is susceptible to failure or when it requires servicing. This PM will enable these faults to be forestalled before failure occurs. Failure can occur at any time in the life of equipment. It is known as infant mortality when failure occurs during the initial stages of equipment use. Thus, manufacturers guide against this initial "setback" through various tests before sales. More often than not, these initial failures are the results of improper usage by the user. Thus, operators must be adequately trained in the handling of equipment and process. Once the machine is operating normally, we can then determine the MTBF (mean-time-between-failure) distribution. Any departure from this distribution is an indication that a fault may be developing; PM can then be carried out. The MTBF distribution is an indication of how economical maintenance will be. The narrower the distribution, the more expensive is maintenance. Sometimes, the cost of PM may roughly equal the cost of repair of breakdown. In this case, equipment may be left to break down before repair work is done. Occasionally, the cost of PM may be low, even though there is a wide departure from the distribution. Minor breakdowns must not be overlooked, as they may result in major problems. These minor faults should be tackled with readily available tools.

Figure 26.2 shows the relationship between PM cost and BM cost under normal and inflation conditions. The operation managers should find the balance between the two costs. Utilizing inventory, personnel, and money in PM can reduce breakdown experienced by a firm. But the increase in PM costs will be more than the decrease in BM costs at a point, leading to an increase in total cost (TC). Beyond this optimal point, it could be better to allow breakdowns to occur before repair is carried out. This analysis does not consider the full costs of breakdown and neglects many costs not associated with the breakdown. This negligence does not nullify the validity of our assumption. Here, we will consider two costs usually neglected: (i) cost of inventory, which compensates for downtime and (ii) low employee morale resulting

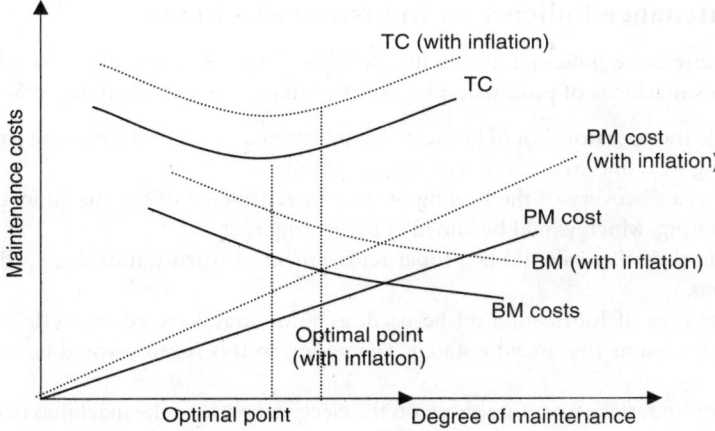

FIGURE 26.2 Maintenance costs (including total, preventive, and breakdown). *Note*: preventive maintenance – PM; breakdown maintenance – BM; and total cost – TC.

from downtime. In theory, it is possible to find the optimal point of maintenance activity considering downtime and its associated cost. Here, the distribution of breakdown probabilities, maintenance cost, and number of repairs over a period must be precisely computed.

26.3.2 Breakdown Maintenance

Breakdown maintenance is sometimes referred to as emergency maintenance. It is maintenance strategy on equipment that is allowed to fail before it is repaired. Here, efforts are made to restore the equipment back to operational mode in order to avoid serious consequences that may result from the breakdown of the equipment. Such consequences may take the dimension of safety, economic losses, or excessive idle time. When equipment breaks down, it may pose safety and environmental risks to workers if it produces fumes that may be injurious to the health of workers or excessive noise that could damage the hearing mechanism of human beings. Other consequences may include high production loss, which would result in economic losses for the company. Consequently, the maintenance manager must restore the facility to its operating condition immediately.

26.4 Maintenance Policy

These are guidelines for decision-making. All organizations have such policies, which vary from place to place (Vineyard et al., 2000). However, policies are not static but dynamic. They are reviewed periodically depending on changes both within and outside the company. For instance, changes in the management of the organization may result in changes in maintenance policies since the new head may have a different orientation, technical know-how, experience, and drive from the previous chief executive. Also, changes in government policies affect the maintenance policies of the organization. For instance, ban on importation of spare parts into a country would affect maintenance policies regarding just-in-time purchases from companies located abroad. The result of utilizing a maintenance policy can be assessed prior to its implementation by using a simulation model. Thus, future breakdowns can be guided against by using simulation techniques to evaluate the effect of changing components before actual failure occurs. Also, the maintenance department can determine their staff strength by using this model to compare the costs of machine downtime with additional labor. Advanced applications involve the use of computerized simulation models to determine when to engage in turn-around maintenance.

26.4.1 Maintenance Policies for Industrial Machines

Integration of maintenance policies into the industrial settings for continuity is based on the intrinsic study of the various machines of production for effective maintenance. The policy may stipulate that:

(i) Once a week, the operator should lubricate the points of contact of the moving parts. This is aimed at preventing wear and maintaining cooling condition.

(ii) Once there is a discovery of the peeling of the protective coat of the machine, there should be a proper recoating, which would be aimed at preventing rust.

(iii) As the need arises, there should be proper replacement of worn-out or damaged parts by recommended ones.

(iv) A particular type of lubricating oil be used, as some may have adverse effect on the machine among other reasons that may be stated. The policy, in this regard, would be ensuring compatibility.

(v) The operator undertake a routine check on the electrical parts of the machines in order to (1) forestall any potential occurrence of a fault and (2) to ensure proper connections for proper machine operation. A breakdown due to electrical faults would invariably slow down production.

(vi) The operator of the machine should see that quick attention is paid to the machine during the machine operation. The noise from the operating machine may also be a herald of an adverse damage of the machine that may require cost-consuming resources to restore if not quickly attended to.

(vii) A specified limit to the period for which a machine is run be set in order to avoid, for instance, unwanted internal heat generation, which could lead to damage of internal components of the machine.

(viii) Total overhauling of the machines for general servicing at a set time be encouraged. This could be arranged in such a way that the machine off-period will not interfere negatively with the demand at hand. Excesses are usually produced at this period to meet consumers' demand.

However, each of the maintenance guidelines highlighted in the foregoing may be categorized either as a preventive or corrective measure taken against the failure of the machine.

26.5 Main Areas of Maintenance Management

As shown in Figure 26.3, the main areas of maintenance management include maintenance performance, maintenance control, maintenance planning, decision support system (DSS), maintenance scheduling, and maintenance organization, among others (Coetzee, 1997, 1999).

26.5.1 Maintenance Performance

To perform effectively on his job, the maintenance manager should be well versed in performance measurement (Pintelon and Van Puyvelde, 1997; Kutucuoglu et al., 2001a). Measures such as productivity, efficiency, and effectiveness, quality, quality of working life, innovation, and profitability should be regularly used to assess the performance of the system and that of the subsystem within the maintenance function (Oluleye et al., 1997; Ljungberg, 1998; Bamber et al., 2003). There exist several methods for maintenance performance (Blanchard, 1997; Maggard, and Rhyne, 1992).

The maintenance manager should be able to assess the productivity levels of individual staff within the maintenance function. When the productivity of a staff person declines as a result of age or the monotonous nature of the work done by this person, the engineering manager could redeploy him to another section where higher productivity may be attained. At this point, he should plan for a successor of this staff who has worked in the maintenance function for a significant period. Thus, he must ensure that all that the incumbent knows is transferred to the successor before the incumbent retires, is redeployed, or leaves the organization.

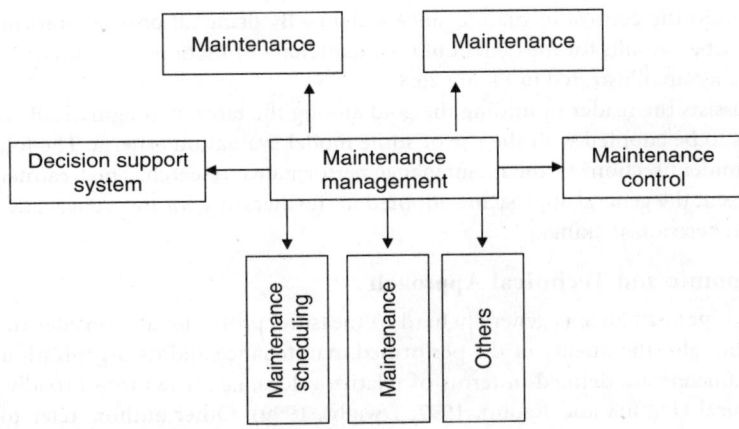

FIGURE 26.3 The main areas of maintenance mangement.

The maintenance manager must have a detailed understanding of how to measure the performance of all the resources under him, both human and nonhuman. Knowledge of the outcome of such measures will help him to control the maintenance function for the improved profitability of the organization. He should understand that controlling the maintenance department means that he directs the affairs of the maintenance function to conform to some set of specifications and standards. Several performance measures are used in practice, such as value-based approach, balanced scorecard (BSC) methodology, composite formulation approach, and others (Perry and Starr, 2001). If these models are developed, there is a need to perform a sensitivity analysis in order to determine the degree of responsiveness of the model variables to changes in their values (Tsang, 1998). We may then proceed to determine the robustness of our model.

The body of knowledge on maintenance performance is both quantitatively and qualitatively based. The quantitative approaches are economic and technical, value-based, BSC, system audit, strategic, composite formulation, statistical, and partial maintenance productivity indices (Oke and Oluleye, 1999; Paulsen, 1997). The qualitative approaches include Luck's method, among others (Luck, 1965). Qualitative approaches are adopted in view of the inherent limitations of measuring a complex function such as maintenance effectively through quantitative models. Very often, maintenance decision-makers come to the best conclusion using heuristics backed up by qualitative assessment and supported by quantitative measures. Applications are found in manufacturing (i.e., steel), mining, transportation, utilities, petrochemical, etc. (Kumar and Liyanage, 2000b).

Advances on maintenance performance are continually being made in this area due to competition in the international market requiring fast and efficient maintenance systems. Maintenance performance measurement, analysis, and control were probably developed during the Second World War, since enormous maintenance activities occurred at that time owing to the need to maintain war equipment. Since the 1960s, some published papers have appeared on maintenance performance. These vary from reviews to rigorous mathematical analyses of specific problems. Unfortunately, the maintenance practitioner is not very well assisted in finding practical ways of controlling the maintenance organization's output to properly serve the company's bottom line. A lot of the "analysis paralysis" does not bring him nearer to being able to measure his maintenance performance in a practical, usable way. The work by Luck, Priel, and a few others offers the more practical approach to this problem (Komonen, 2002; Luck, 1965; Priel, 1974).

Documentation in the form of texts or research papers received very little attention until relatively recently (i.e., during the second half of the 1990s, when maintenance performance achieved recognition as a subject worthy of academic study in courses, learned journals, and articles [Kincaid, 1994]). Maintenance performance attempts to provide a systematic and rational approach to the fundamental problems involved in the control of maintenance systems. By using all possible information, decision-makers achieve the best results for the benefit of the maintenance function. The important areas of maintenance performance are illustrated in Figure 26.4.

In Table 26.1 assists the reader in finding the gold among the tares by pragmatically identifying profitable approaches to be adopted with the use of some model evaluation criteria. The following is a brief summary of prominent sections of the maintenance performance research. The treatment methods presented here represent the general approaches adopted in the literature for theoretical and practical analysis, modeling, and decision-making.

26.5.1.1 Economic and Technical Approach

Since maintenance performance is generally hard to measure, professionals consider not only quantifiable parameters but also the quality of the performed maintenance and its organization. Consequently, performance parameters are defined in terms of relative values, i.e., true ratios broadly defined as economic and technical (Duffua and Raouff, 1997; Dwight, 1999). Other authors refer to these ratios as financial and nonfinancial ratios, respectively (Kumar and Liyanage, 2000a). Economic ratios are elements that allow the follow up of the evolution of internal results and certain comparisons between maintenance services of similar plants. Technical ratios are elements that give the maintenance manager the means of following the technical performance of the installation (Duffua and Raouff, 1997). In particular, economic ratios are linked to maintenance cost.

26.5.1.2 Strategic Approach

The strategic approach relates to strategic issues of acquisition, improvement, replacement, and disposal of physical assets (Tsang, 1998; Jooste and Page, 2004). This approach to maintenance reflects the maintenance organization's conception of its intended long-term goal. In adopting this approach to the measurement of maintenance performance, the maintenance strategy is linked to the corporate strategy. Three common examples of maintenance strategies that relate to maintenance performance measurement are maximization of asset utilization, improvement of responsiveness of maintenance to customer needs, and development of core competencies in all maintenance areas. The success of the strategic maintenance performance management approach is built on the stakeholders' understanding of the compelling need for

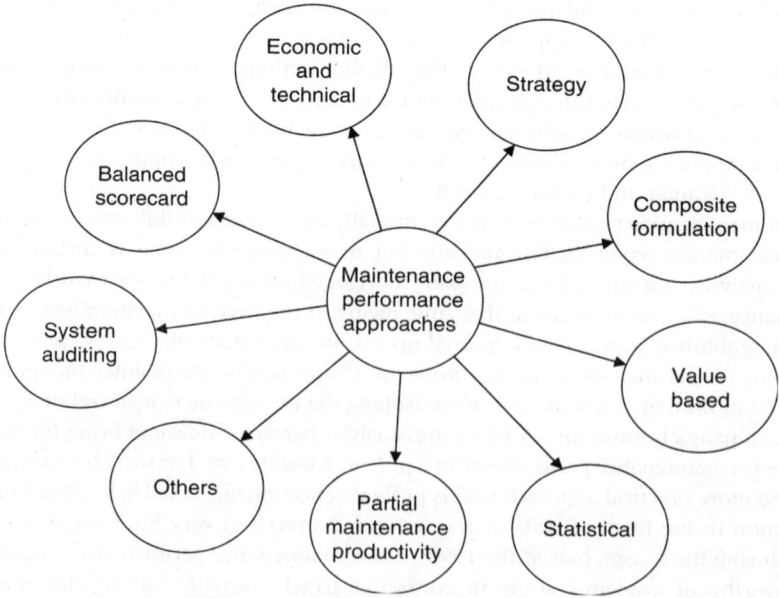

FIGURE 26.4 Maintenance performance approaches.

change, a shared vision of future performance, commitment, deployment of the needed resources, feedback of progress made, constant communication, and a reinforcing reward and recognition system.

26.5.1.3   System Auditing Approach

System auditing approach focuses on the system and its attributes rather than specific outcomes. It concentrates on the maintenance system itself as contrasted with quantifying its inputs and outputs. It is anticipated that the results from such an approach yield a level of accuracy that is compatible with the information normally available about real performance. The systems audit approach is represented by a transformation coefficient or coefficient of influence and the correlation between the performance of a particular system element or activity and the organization's success. This is built up according to the relationship between the failure attributes, addressed by system activity and business success, and potential of the system activity to influence technical system performance relative to the failure attributes.

26.5.1.4   Statistical Approach

This approach is based on the use of subsurvival functions for the analysis of censored reliability life data for the evaluation of maintenance performance (Perry and Starr, 2001). The interest is to separate the effects of maintenance from the intrinsic reliability of the components as expressed in the naked failure rate that would be observed if no maintenance were performed. Incidents are classified as "corrective" and "preventive" maintenance.

26.5.1.5   The Composite Formulation Method

The composite formulation method employs the following detailed steps for measuring performance and instituting a corrective action needed for improvement (Oke and Oluleye, 1999). First, performance measures are obtained and ranked in order of decreasing importance. The best and worst performance for each measure is considered over a chosen span of time. Appropriate utility values of 1.0 and 0 are then assigned respectively. Utility curves are developed using the allotted values for each measure.

26.5.1.6   Partial Maintenance Productivity Approach

The approach here is input–output based (Lofsten, 2000). Partial factor productivity is defined as the ratio of total output to single input. Total output is expressed as the *availability* and *amount of production*. The approach is based on the idea that good productivity is achieved when the total maintenance costs and downtime costs are reduced to a minimum level for the state of production system, i.e., the minimum production rate is fulfilled. The method involves first carrying out an object analysis of each individual maintenance object. The next stage consists of carrying out a resource allocation analysis. Here, one chooses a combination of measures from each object that uses the available limited resources.

26.5.1.7   The Value-Based Approach (Liyanage and Kumar, 2002b, 2002c, 2003)

This takes into account the impact of maintenance activities on the future value of the organization. The value-based approach emphasizes the value rather than the cost of maintenance in the emerging business environment, extending our understanding beyond pure financial implications of maintenance. Mathematical models that could assist in calculating the performance of an organization when the value of the maintenance function is considered have been developed (Liyanage and Kumar, 2001c, 2002a).

26.5.1.8   The Balanced Scorecard

The BSC is a vehicle that translates a business unit's mission and strategy into a set of objectives with quantitative measures built around four perspectives: financial, customer, internal process, and learning and growth (Tsang, 1998). Kaplan and Norton (1992) introduced it as a template for the balanced presentation of analysis. The BSC has been implemented in a number of major corporations in the engineering, construction, microelectronics, and computer industries (Kaplan and Norton, 1996; Kumar and Liyanage, 2000a). Experiences in these pioneering organizations indicate that the scorecard will have the greatest impact on business performance only if it is used to drive a change process. The development of a BSC also engenders the emergence of a strategic management system that links long-term strategic objectives to short-term actions (Liyanage and Kumar, 2000a, 2000b, 2001a).

26.5.1.9 Criteria for the Evaluation of Maintenance Performance Approaches

The following are the criteria used to evaluate the various approaches to maintenance performance covered in this work (see Table 26.1):

- *Training required by user:* This attribute refers to the level of training that the user of model is expected to have acquired to enable the proper usage of the model. It is assumed that the user must have taken courses in areas such as mathematics, physics, programming, and statistics in order to understand the procedure and application of the model. The user must have had this training in an organized training environment. As part of the training, the user must have been assessed at a variety of levels to determine the level of competence attained. Thus, in using the model, this criterion measures whether the training required is high or low.
- *Accuracy of model:* The model is expected to give the correct output result since the right data would be fed into the model. Thus, the correct input data must yield the correct output at all times. The model should be relied upon for accuracy and consistency while processing the input data. In evaluating the input data, the model must give the corresponding and correct output information required. This implies that the processing capability of the model should be strong enough to accept required data input and then release the correct information needed to achieve a particular task.
- *Clarity of model procedure :* The model must not show any ambiguity. It must be precisely clear and distinct. The various steps involved in evaluation must not be cloudy. Clarity in how to follow the procedure and interpret the results is of paramount importance.
- *Number of constraints of model :* The constraints or limitations of the model must be well defined. The condition under which the model will not function must be clearly spelled out. The boundary of operation of the model should be mapped out, and the parameters both defined and undefined with respect to the model should be stated. The number of limitations to the functioning of the model should be enumerated. These constraints are the factors or parameters that will not allow the model to function outside its boundary. The models are thus demobilized or limited in scope by these constraints. Therefore, we will consider the number of such constraints available in each model.
- *Correctness of data input :* It is clear that "garbage in, garbage out" holds for any sequential-factored data system model. For the result of model evaluation to be correct, the input data must — as a matter of necessity — be correct. To avoid error right from the initial stage, the model user must ensure the correctness of the data input and, therefore, of the data computed. The values must be checked for homogeneity in units, readings, computations, etc.
- *Assessment ability by user :* The model user must be able to assess or evaluate the performance of the model in terms of its operation and results. More importantly, the user must know how to assess, analyze, and interpret the end result of computation to discover hidden knowledge embedded in the operation of the model.
- *Mathematical soundness of model :* In the derivation and usage of the model, the mathematics involved in the various steps must be clear and sound. Starting from the initial stage, the step-by-step development of model mathematics and its attributes must establish accuracy or soundness. The building blocks of topics (such as probability, series and summation, and inequality) must be based on logical reasoning and rational usage. The need for sound mathematics in the development and results generated by the model cannot be overemphasized.
- *Logical soundness of model:* The logic inherent in the development of the model must be sound. The model must logically be arrived at and effectively utilized.
- *Experience requirement of user :* This defines the level of experience required by the user of the model. Experience is measured in terms of proper model analysis based on usage of similar models for analysis. A new user will have to go the extra length to carry out the evaluation task. Therefore, the level of experience required by the user is a necessary criterion in proper analysis of model results.
- *Nonambiguity of Input data:* The data input into the model must be distinct. It must not be ambiguous or misrepresented. It must show clarity and uniqueness.

TABLE 26.1 An Evaluation of the Various Approaches Used in the Measurement of Maintenance Performance

S/N	Criteria for Evaluating Various Approaches	Economic and Technical	Strategic	System Auditing	Statistical Functions	Composite Formulation	Partial Maintenance Productivity	Value-Based/Balanced Scorecard
					Category of Approach			
1.	Training required by user	3	2	2	4	1	•	2
2.	Accuracy of model	2	3	2	Δ	3		
3.	Clarity of model procedure		•	3	•	3		
4.	Number of constraints of model	4	4	2	3	3		4
5.	Correctness of data input			2	4	4	2	3
6.	Assessment ability by user	3			Δ			3
7.	Mathematical soundness of model	X	•			•		X
8.	Logical soundness of model	X			•	•		X
9.	Experience requirement of user	3	2	2	1	2	3	2
10.	Nonambiguity of input data		2	2	3	•	2	2
11.	Practicality of model	3	2	3	Δ	•	2	2
12.	Effectiveness of model	2	•	3	4	•	2	2
13.	Correct interpretation of result by user		2		2	4	4	
14.	Number of input values needed	2	•	4	4	4	4	3
15.	Number of steps	4	4	4	4	3	4	3
16.	Does it involve making comparisons?	4	4	4	4	4	5	3
17.	Computational complexity	2	Δ	2	4	4	4	3
18.	Capabilities of model		3		Δ	3	3	2
19.	Faulty design of model	5	Δ	4	•	3	4	4
20.	Inherent weaknesses of model	4	4	4	2	2	3	4
21.	Wide utilization of model		Δ		Δ	Δ	3	1
22.	Sensitivity of model parameters	2	4		2	•	•	Δ
23.	Consistency of model parameters		3		4	•	•	Δ
24.	Presentation of model				Δ	Δ	Δ	Δ
25.	Readily availability of input data	Δ		Δ	Δ	Δ	Δ	

Key 1: Very Good — X; Good — X; Average — •; Poor — ; Very poor — Δ; *Key 2:* Very high — 1; High — 2; Average — 3; Low — 4; Very low — 5.

- *Practicality of model* : The model must be easy to use in practice and must be readily applicable in real-life situations. Real attributes of performance must be displayed in the working of the model. Thus, the model has an immediate significant influence on systems operations and the final result.
- *Effectiveness of model* : The model must be able to produce the correct result that is wanted or intended. It should display expected and reliable output information by properly coordinating its processing of data. The difference in values generated by model to the actual correct values indicates the degree of error and it is a measure of model effectiveness.
- *Correct interpretation of result by user*: The way the user views, understands, and explains the model procedure and results will affect the application of the model. The user must ensure correct and adequate analysis and evaluation of model results.
- *Number of input values needed* : The model's input values should be minimal. The input values should be as few and as concise as possible. In a working environment, the lesser the input parameters, the less cumbersome the computation, and the higher the functionality of user and model.
- *Number of steps*: The different steps or stages involve in the model procedure should also be as few as possible. The link from one step to the next should also be logical.
- *Does it involve making comparison?*: Models should be carefully examined with others to reveal similar or different patterns. This comparison would show hidden knowledge concerning the result and how interrelated the models are.
- *Computational complexity*: The mathematical computations must not be too complicated or clumsy as to cause difficulty in understanding and usage. Clarity and simplicity should be a prominent feature of the model.
- *Capabilities of model*: The model should be able to carry out the task of computing, processing, and presenting the input data to give accurate output data required. It should cover a wide range of input data with few or no exceptions.
- *Faulty design of model*: The model is designed to correctly work and be free from error. Faulty design can lead to poor evaluation and processing of data.
- *Inherent weaknesses of model*: The weakness embedded in a model right from its design stage constitutes a stumbling block toward its effective performance. This basic fault or weakness limits the application of the model.
- *Wide utilization of model*: Use of the model in wide areas of application is a pointer to its effectiveness. It must be limited by few or no constraints. A good model is able to accurately process a wide range of input data.
- *Sensitivity of model parameters*: This implies that parameters input into the model must respond to small changes in factors producing them.
- *Consistency of model parameters*: Input parameters should develop in a consistent manner. Their derivation should be regular.
- *Presentation of model* : The model procedure and result presentation should be made in a clear manner devoid of any obscurity. Results should not be liable to many interpretations.
- *Ready availability of input data*: The data input should be readily available at all times needed. It should not be difficult to get.

26.5.1.10 Implications for Researchers and Managers

The future of maintenance performance has promising opportunities for system modeling, specific applications in industries, and general surveys. System modeling will experience an explosion with a stream of new modelers joining the camp. Prominent tools in operations research that have been profitably applied in other areas may also be successfully employed in maintenance performance research. For example, operations research involves the application of scientific methods to the management of maintenance performance problems that involve complex systems of people, machinery, material, money, and information. Investigation of operations research problems in maintenance performance will seek to produce an understanding of maintenance management problems and to develop models, which will enable consequences of maintenance decisions to be investigated. The prediction of optimal maintenance performance models, their calibration, sensitivity analysis, and application should add knowledge to research on maintenance performance. The purpose of the model would be to provide a quantitative description of the interactions that occur between component variables.

26.5.2 Maintenance Organization

The structure of the maintenance system must be such that it allows for unhindered information flow in order to efficiently detect and rectify faults, and for smooth, efficient, and productive operations. Maintenance organization defines a laid-down structure of working relationships in an enterprise for achieving effective maintenance decision-making and the means of transmitting such decisions into timely remedial actions. It involves grouping or assignment of activities with a provision for authority delegation and coordination. The structuring of a maintenance organization in line with a system that supports problem and process analysis and also efficient solution integration is crucial in attaining plant economy and equipment effectiveness. We shall briefly examine some desirable characteristics of an effective maintenance organization, a clearly defined structure in which the division of authority is either functional or geographical. A fusion of the two is also possible to avoid confusion or conflicts. Each personnel must be aware of his duties, procedures, and policies in maintenance operations. Moreover, efficient utilization of manpower involves an optimal number of subordinates. Although this optimal number depends on the capacity of the maintenance function, the emphasis must be on skill development. The maintenance managers must not be burdened with on-the-job supervision at the expense of time for planning and administering both technical and administrative functions. Delays must be avoided with minimum levels of decision centers. These reduced levels avoid needless overhead cost. Maintenance organization structure should be such that team cooperation is encouraged.

This will foster an operating environment that encourages high morale, stimulates interest, and increases productivity among the maintenance personnel. A maintenance organization structure should be flexible enough to permit change when the need arises. With emerging technologies across the globe ushering in improvements in techniques and methodologies in maintenance practices, it is imperative that an effective maintenance structure must incorporate these innovations into their operations in order to deliver quality control of maintenance operations. It is well known that jobs are contracted out when it is no longer economical for the department to implement some maintenance functions. Thus, the changing nature of competition in industry not only demands flexibility in operations, but also interaction with other players inside and outside the sector to keep abreast of latest developments in products, processes, and applications. The organization of maintenance must respond promptly to the slightest anomaly in operations. It must have an efficient feedback or information flow system that will detect fault early and utilize the mechanism in place to rectify the fault. A remarkable trend in developing nations shows a marked difference between government-owned enterprises and the organized private sector in terms of maintenance organization. Multinationals operating in the bottling subsector such as Coca Cola are reputed to implement high-maintenance functionality.

Communication between the production and maintenance team is usually high such that the maintenance manager quickly recognizes any sign of irregularity and takes prompt action to solve the problem. This is in sharp contrast with public enterprises where a fault could lead to idle time lasting days (or more) before amendment is made. A lot of bureaucracies hinder efficient problem-solving in these organizations, resulting in low performance. The administrative part of maintenance is also notable. A secretary or typist engaged in the paper work is making important contributions. In typical hierarchy of a maintenance organization, plant engineering is at the top, assisted by maintenance engineers who are immediately above the maintenance supervisor, foremen, and technician/craftsmen, in that order.

26.5.3 Maintenance Control

The need for a control mechanism is evident as a firm increases in complexity (Priel, 1974). The implementation of control measures in maintenance management aims to achieve efficiency and effectiveness in the operational coordination of demand and resources. To achieve and ensure the applications of directives in all maintenance functions, maintenance managers are saddled with the responsibility of communicating policies and decisions taken throughout the department. This serves as a means of control against aberrations that could undermine maintenance standards. Control in maintenance can be measured in both monetary and operational units. Control involves information gathering on conduct of operations, comparing this information with planned results or targets, and taking appropriate corrective measures that will establish mastery over situations. This characterizes the essentials of a control system. A control procedure should highlight simplicity and clarity. Except for a large organization with complex operations, the simpler a control mechanism is, the more useful. Since continuous equipment utilization depends on maintenance functions, we must focus on "controlling" the effects of the service on the equipment being serviced (Priel, 1974).

Thus, the organization and performance and costing of this service is very important and must be coordinated to achieve optimum functionality by striking a balance between expected benefits and cost. Such benefits include effective plant operation, extended equipment life, reduced idle time, and reduced accidents, among others. Costs cover materials, spare parts, tools, overheads, etc. The implementation of control measures in maintenance management aims to arrive at a balance between maintenance expenditure and accruing benefits. Large spending on maintenance may not translate into increase or any appreciable benefits if the control system is malfunctioning. Effective control monitors and analyzes data, investigates aberrations, and implements corrective steps to rectify irregularities noticed. With this experience, such deviations will be controlled when next it develops. The accuracy of data collected also determines the quality of control. In order to facilitate the work of analysis, we must have answers to such questions as: Are the data readily available? What data are we looking for? Where do we locate them? Control data are expressed in terms of figures and values. These figures are derived from the maintenance functions being carried out. To be carefully monitored are the accuracy and consistency of details transmitted and recorded. Upon discovery where the "weakness" lies, corrective and preventive actions must be executed to deal with the anomalies. Also data must meet the expectations for meaningful interpretation. Control functions should be continuous and improve on its effectiveness, efficiency, flexibility, and adaptability. Since new information develops constantly during maintenance, feedback and progress monitoring are *sine qua non* in maintenance control.

A major factor affecting the production process in any organization is the coordination of maintenance management functions to reduce the probability of failure and ensure proper functioning of production systems. The changing nature of competition in the global market requires a maintenance policy implementation that is able to support and sustain intricate production processes through effective, efficient, flexible, and adaptable transformation of material, energy, and power to the desired product. Maintenance control anticipates unavoidable short-term variations in normal operations, eliminates them and ensures a return to the long-term goal of continual improvement, thereby increasing the organization's ability to meet customers' satisfaction. Maintenance control focuses on achieving the intended performance of technical systems with respect to effective and efficient execution of specific activities. This aim is in line with the objectives of production and the organization as a whole. Maintenance control is affected by (i) complexity of systems and operation, (ii) uncertainty in the problem arising, and (iii) flexibility in adapting to changing demand.

Maintenance control involves the measures taken to ensure that the maintenance of a system is standard, efficient, and as safe (while performing) as possible. A good grip on the maintenance procedure and policy of an organization will always keep it above all organizations, especially in terms of profit-making (Lofsten, 1999; Kumar and Liyanage, 2002b). Adequate and proper control of maintenance in any organization will definitely lead to an all-round efficiency of the maintenance procedure/policy undertaken by the company. The control function, in the maintenance department, is an important routine toward getting to the destination of high profit for the organization, while gratification of customers' satisfaction is not compromised. A typical maintenance control function consists of inputs and outputs, which are compared, while deviation is kept to the minimum.

26.5.4   Maintenance Scheduling

The maintenance scheduling literature has a variety of applications. These diverse applications include aircraft maintenance, process industry, vehicle fleet maintenance, railway track maintenance, power generation, pavement maintenance, highway maintenance, refinery, and production facilities. An attempt to codify the maintenance scheduling literature portrays a line of inquiry that has grown in volume and in depth. Generally, important areas of maintenance where scheduling principles have been applied are the main industries of oil and chemical railways, transport companies airlines, steel works, and discrete part manufacturing, as well as public sector applications such as electricity generation, defense, and infrastructure (e.g., roads). Maintenance scheduling research dominates a large portion of contemporary studies in operations management. It is a broad stream with two main currents: methods and applications (see Figure 26.5).

26.5.5   Maintenance Planning

This refers to the daily organization and implementation of maintenance tasks, including the provision of resources to carry out these tasks. It defines the practical steps to be taken to prevent, reduce, or eliminate failures, detect and diagnose faults, and repair or correct the effects of usage on facilities in the most economical manner in the given circumstances. For planning purposes, maintenance managers must study guidelines in manuals that would include the assembly and operation guide, the likeable maintenance approach to adopt when a characteristic fault is developed, etc. The manufacturers' recommendations may vary from equipment to equipment. Planning must ·be structured to prevent failure where maintenance is undertaken at regular time intervals. Planning should respond to unexpected failures, which may require emergency maintenance. In order to gain more insight into maintenance planning, we utilize the following approaches as a basis for planning:

- *Period-dependent activities*: Here, planning involves anticipating problems and making necessary adjustments in good time to either prevent failure or replace deteriorated parts. It may require regular inspections, tests, and repairs, e.g., changing machine elements, painting, etc.
- *Equipment condition-dependent activities*: The maintenance function is dependent on a set of conditions before implementation. A part to be replaced or repaired must be justified.

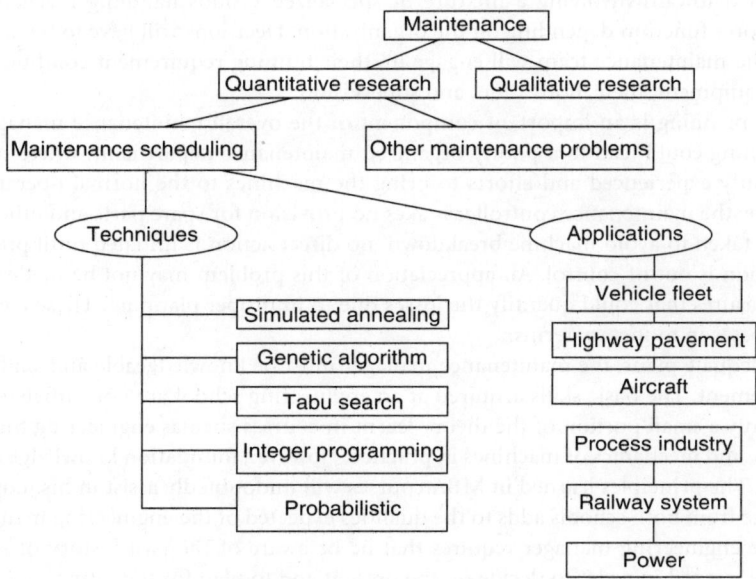

FIGURE 26.5   Methods and applications of maintenance scheduling.

- *Operation of equipment to failure*: Where it is difficult to establish or predict failure, the equipment may be left to fail especially when the consequence of failure is insignificant or minimal, e.g., replacing light bulbs.

It is obvious that sound recording of information is indispensable in planning maintenance operations. The documentation of previous activities and solutions recording problems encountered will increase the ease with which similar problems are tackled in future. This previous knowledge will greatly improve the planning process; thus, efficient maintenance planning supports continuous improvement programs. This resulting enhancement will lead to high effectiveness and efficiency in the organization of the maintenance activities. A poorly organized maintenance department is characterized by inadequate record keeping, although it is not compulsory to record every single activity. For instance, in a food and beverage industry where baking of snacks is implemented, the maintenance procedures in the repair of the oven used in baking cannot be fully documented in the event of breakdown. Another approach to maintenance planning activities involves the coordination of the following maintenance functions: management procedures, technological practice, personnel management, and performance controls. The coordination of these varied functions will enhance the operating efficiency of maintenance implemented. Although very few maintenance departments are able to fully cover these areas, there is always room for improvement since techniques are not always fully utilized. Another approach is to ask the salient questions when planning for maintenance: "what is to be done and why", "how the work is to be done and when", and "where the job is and who is to do it".

In the first instance, planning must recognize what constitutes maintenance work, its functions and expectations. Basically, the work to be done covers all activities that assist in keeping plant and equipment in good condition. These activities include cleaning, adjustments, repairs, inspection, overhauls, lubrication, replacement of parts, calibration, refurbishment, renewal, painting, and even staff training, among others. The varied reasons for embarking on maintenance include preventing failures, detecting faults, improving performance, counteracting deterioration, minimizing wear and tear, repairing breakdowns, etc. In planning, therefore, the integration of these activities to reach high performance in its operations is paramount. In another instance, responsibilities of the key personnel must be clearly defined to avoid disputes that may rear their ugly head. Thus, right from inception, the various job descriptions must be clearly spelled out, starting from the top echelons such as the plant engineer and maintenance supervisors, to lower cadre foremen and craft workers. The internal structure of large maintenance departments is usually planned toward involving a mixture of specialized groups handling mechanical, electrical, building, and stores function depending on the organization. Decisions will have to be made concerning the work that the maintenance team will engage in, their training requirement condition, amount and complexity of equipment, their motivation, and supervision.

Maintenance planning is an important component of the overall maintenance management system. Ineffective planning could lead to a poorly organized maintenance department, where machine breakdown is frequently experienced and efforts to bring the machines to the normal operating conditions seem futile. Here, the maintenance controller makes no provision for spare parts and other resources; no precautions are taken to avoid machine breakdown, no direct action is initiated until production stops, when the situation is out of control. An appreciation of this problem may not be made unless we have value-based measures that could quantify the losses due to improper planning. These could be in terms of costing the losses in monetary terms.

In making adequate plans, the maintenance manager must be knowledgeable and skilful in engineering and management. The basic skills acquired at an engineering school are not sufficient. For example, he may apply only a small portion of the theory learnt in courses such as engineering thermodynamics, fluid mechanics, and mechanics of machines in practice. However, foundation knowledge of such courses will surely help. The principles learned in MBA courses will undoubtedly assist in his judgment. A combined knowledge from both schools adds to the qualities expected of the engineering manager. The planning task of the engineering manager requires that he be aware of the past history of the engineering function that he learned in order to decide on the present and to plan for the future.

- *Planning for budgets:* Another area that must be planned for is budgeting. Finance is a single factor, and if unavailable, it could paralyse the activities of the maintenance function. At this point, no matter how much effort is put into the maintenance activities of the organization, intangible results will be obtained. A budget is a prelaid plan of most specific expenditure for a set period. Planning for a budget can be compared to preparing to execute laid-down plans of expenditure over a definite period. Such preparations involve strategic planning of the best (mostly step-by-step) procedure or set of procedures, to follow all outlines of the budget to the letter.
- *Planning for manpower:* Every organization has a set target to meet in a specified period of time. To meet this target, the company would have to make adequate plans of how much manpower, depending on the resources readily available for this purpose, can be acquired or employed.
- *Preventive maintenance planning:* Prevention is better than cure; the reason being that an item that is well preserved tends to perform far better than one that is not preserved. The worst scenario, for every company, is bound to occur. But this can be prevented by proper maintenance activities, aimed specifically at preventing this. How it will be undertaken, performed, organized, directed, and supervised will depend on the effectiveness of whatever plan the organization has for PM.

26.5.6 Decision Support Systems

In the maintenance-engineering field, DSSs refer to a category of computerized information systems that support the day-to-day running of maintenance activities. They also ensure a high-quality organizational decision-making process. A properly designed DSS should be interactive. It should be a software-based system developed to help the maintenance manager in capturing, analyzing, and presenting valuable maintenance information. Usually, information is obtained from raw maintenance data, maintenance service documents, personal experience, and knowledge of the manager who engages in the collection of data. Maintenance models are also used to identify and solve problems and make decisions in DSSs. Decision support systems could be used to

(1) access historical records of downtime, production hours, energy costs, etc. Therefore, relationship data analysis could be performed on such historical records with data warehouses developed for them; and
(2) compare the results between one week and the next of the following:
   (a) down-time hours, breakdown, scrap due to malfunctions, and stoppages;
   (b) strength of maintenance workforce, recruitment and departures, vacancies, attendants and tardiness, absence from training;
   (c) maintenance/production hours ratio, total maintenance expenditure, average cost of maintenance hour, cost of spares and supplies issued, annual expenditure, total departmental clocked hours, total directly applied hours, manpower utilization ratio, total direct and indirect payroll, downtime hours of all plants, cost of downtime, etc.;
   (d) performance data, which are of two types: (i) job class, which includes emergencies, plant development, repairs and overhauls, routine services and (ii) type of hours, which includes regular hours, overtime hours, idle/waiting time, indirect services, supervision, clerical work, service to production;
   (e) control ratios including maintenance cost per kWh, % utilization of hours, % planned work, total weekdays, cumulative expenditure, cost of maintenance hour, and variance from budget.

26.5.6.1 Computerized Maintenance Management Systems

We have partially introduced the concept of CMMSs in the introductory section of this chapter. What follows is a brief extension of the discussion. The major features of CMMSs include the processing of maintenance data to yield useful information on which management decisions are based. This information can be analyzed or evaluated with respect to previous results such that performance over a period of time can be assessed. Managers will therefore find it convenient in making use of the available data to plan for present and future goals. There is also the advantage of printing out this information in hard copy at any desired period, e.g., balances on all stock items can be made. Other useful information, such as work done in the maintenance department and its TCs, lists of jobs worked on during a period, and inventory taking, can be presented in a manner that will greatly ease both the technical and administrative task of maintenance. Through networking of the systems, information can be passed efficiently between the maintenance department and the organization's management. The CMMSs can be used to analyze budgets, downtime, supplies, screening of applicants, etc. The database programs in CMMS systems contain structured data on workers' names; job titles; daily; weekly or yearly schedules; etc. The data processed on CMMSs can be stored permanently or retrieved much faster for future use or modifications. Also, CMMSs can be used in maintenance planning and scheduling, coordinating people, and controlling resources and costs of maintenance functions. It can also be used in such areas as the analysis of a week's activity and budget proposals.

26.6   Maintenance Culture

Generally, the fundamental reason for establishing maintenance is to prevent the occurrence of failure. However, the approach employed to meet maintenance demands may vary depending on the economic, geographical, and technical background of the organization in view. The variation, nonetheless, does not run contrary to a common trend for maintenance that spans the entirety of similar organizations. For instance, Companies X and Y are paint manufacturing industries established in a developing country. Owing to the close proximity in location of these two, the same maintenance guideline against corrosion of machines may be implemented in the companies since they face relatively similar environmental conditions. On the other hand, there may be a conspicuous gulf in the approach if there is a huge difference in the financial strength of the two companies. Equally, the type of machine system installed may be the sole determinant of the type of maintenance policy that would be considered, regardless of other underlying factors. This may be attributed to paying strict attention to the manufacturer's guide for sophisticated maintenance systems. In this ever-increasing corporate world both in age and advancement, modern technology has ushered in better platforms for research into the optimization of industrial maintenance applications. And hence, maintenance culture adopts different approaches to research for new results. On the basis of profit-making, with improved quality of productivity in mind, stakeholders have sought to review their maintenance policies at intervals to enhance the incorporation of research results for further development. Alternatively, assessment of the maintenance policies from time to time may be tailored to reducing the cost of integrating a new maintenance system into the existing one.

26.7   Conclusion

This chapter presents a conceptual viewpoint of maintenance in the 21st century. Drawing on the scientific definition of maintenance, this chapter presents the wide range of technical, administrative, and managerial actions in which maintenance engineers are expected to be involved (see Figure 26.1). Further, the chapter presents two main categories of maintenance — preventive and breakdown — and also considers the effect of inflation on the TC incurred during the maintenance process (see Figure 26.2). The maintenance management system is then viewed from the perspectives of maintenance performance, maintenance control, maintenance planning, maintenance scheduling, DSS, maintenance organization, etc. The maintenance performance approaches are briefly introduced. The approaches considered are limited to quantitative measures, which includes economic and technical, strategic, composite formulation, value-based, statistical, partial maintenance productivity, system auditing, and BSC approaches, etc. On the basis of observations from the literature, 25 criteria were developed and used to evaluate these approaches in order to assist the reader to pragmatically identify profitable approaches to be adopted in practice. The result of this evaluation is shown in Table 26.1. In all, this chapter presents a new conceptual view of maintenance management in the current century.

References

Al-Najjar, B. and Alsyouf, I., Enhancing a company's profitability and competitiveness using integrated vibration-based maintenance: a case study, *Eur. J. Oper. Res.*, 157, 643–657, 2004.

Arts, R.H., Knapp, G.M., and Mann, L. Jr., Some aspects of measuring maintenance performance in the process industry, *J. Qual. Maint. Eng.*, 4, 6–11, 1998.

Bamber, C.J., Castka, P., Sharp, J.M., and Motara, Y., Cross-functional team working for overall equipment effectiveness (OEE), *J. Qual. Maint. Eng.*, 9, 223–238, 2003.

Blanchard, S.B., An enhanced approach for implementing total productive maintenance in the manufacturing environment, *J. Qual. Maint. Eng.*, 3, 69–80, 1997.

Coetzee, J.L., *Maintenance*, Maintenance Publishers, Pretoria, Republic of South Africa, chap. 11, 1997.

Coetzee, J.L., A holistic approach to the maintenance problem, *J. Qual. Maint. Eng.*, 5, 276–281, 1999.

Chan, K.T., Lee, R.T.H., and Burnett, J., Maintenance performance: a case study of hospitality engineering systems, *Facilities*, 19, 494–504, 2001.

Cooke, R. and Paulsen, J., Concepts for measuring maintenance performance and methods for analysing competing failure modes, *Reliability Eng. Syst. Saf.*, 55, 135–141, 1997.

De Groote, P., Maintenance performance analysis: a practical approach, *J. Qual. Maint. Eng.*, 1, 4–24, 1995.

Duffua, S.O. and Raouff, A., Continuous maintenance productivity improvement using structured audit, *Int. J. Ind. Eng.*, 3, 151–160, 1997.

Dwight, R., Concepts for measuring maintenance performance, in *New Developments in Maintenance: An International View*, Martin, H.H., Ed., IFRIM, Eindhoven, 1994, pp. 109 – 125.

Dwight, R., Searching for real maintenance performance measures, *J. Qual. Maint. Eng.*, 5, 258–275, 1999.

Fernandez, O., Labib, A.W., Walmsley, R., and Petty, D.J., A decision support maintenance management system: development and implementation, *Int. J. Qual. Reliability Manage.*, 20, 965–979, 2003.

HajShirmohammadi, A. and Wedley, W.C., Maintenance management: an AHP application for centralization/decentralization, *J. Qual. Maint. Eng.*, 10, 16–25, 2004.

Hemu, M., Using benchmark data effectively sustainable maintenance performance targets in our industry, *J. Maint. Asset Manage.*, 15 (July/August), 2000.

Jooste, J.L. and Page, D.C., A performance management model for physical asset management, *South African J. Ind. Eng.*, 15, 45–66, 2004.

Kaplan, R.S. and Norton, D.P., The balance scorecard measures that drive performance, *Harv. Bus. Rev.*, January–February, 71–79, 1992.

Kaplan, R.S. and Norton, D.P., Using the balance scorecard as a strategic management system, *Harv. Bus. Rev.*, January–February, 75–84, 1996.

Kincaid, D.G., Measuring performance in facility management, *Facilities*, 12, 17–20, 1994.

Komonen, K., A cost model of industrial maintenance for profitability analysis and benchmarking, *Int. J. Prod. Econ.*, 79, 5–31, 2002.

Kumar, U. and Liyanage, J., A strategically balanced measurement system for maintenance process: some foundational issues for a development method (Part II), *Maint. Oper.*, 4, 147–155, 2000a.

Kumar, U. and Liyanage, J., Value based maintenance: An agenda to streamline operations and maintenance performance to improve process profitability, *The 2nd European Petrochemical Technology Conference (EPTC-2000)*, Prague, Czech Republic, 2000b.

Kumar, U. and Liyanage, J., The process of maintenance performance management in an integrated plant environment: setting the foundation for a new perspective, *Eng. Prod. Maint.*, 4, 37–47, 2001.

Kumar, U. and Liyanage, J., A value based working algorithm to manage maintenance performance: key learning points from the oil & gas industry about an integrated production asset, *New Eng. J.*, 4, 5–7, 2002a.

Kumar, U. and Liyanage, J., Adjusting maintenance policy to business conditions: value-based maintenance performance measurement, *Proceedings of the International Foundation for Research in Maintenance, Maintenance Management and Modeling Conference*, Paper No. 20, Vaxjo, Sweden, 2002b.

Kutucuoglu, K.Y., Hamali, J., Frani, Z., and Sharp, J.M., A framework for managing maintenance using performance measurement systems, *Int. J. Operat. Prod. Manage.*, 21, 173–194, 2001a.

Kutucuoglu, K.Y., Hamali, J., Irani, Z., and Sharp, J.M., A framework for managing maintenance using performance measurement systems, *Int. J. Oper. Prod. Manage.*, 21, 173–194, 2001b.

Labib, A.W., World-class manufacturing using a computerised maintenance management system, *J. Qual. Maint. Eng.*, 14, 66–75, 1998.

Labib, A.W., A decision analysis model for maintenance policy using a CMMS, *J. Qual. Maint. Eng.*, 10, 191–202, 2004.

Labib, A.W., O'Conor, R.F., and Williams, G.B., An effective maintenance system using the analytical hierarchy process, *Integr. Manuf. Syst.*, 9, 87–98, 1998.

Liyanage, J.P. and Kumar, U., A strategically balanced measurement system for maintenance process: Some foundational issues for a development method (Part I), *Maint. Oper.*, 4, 79–87, 2000a.

Liyanage, J.P. and Kumar, U., Measuring maintenance process performance using the balanced scorecard, *The 15th European Maintenance Congress: Euromaintenance 2000*, 2000b, pp. 25–32.

Liyanage, J.P. and Kumar, U., VBM & BSC: an adaptive performance measurement system using the balanced scorecard, *J. Maint. Asset Manage.*, 16, 12–21, 2001a.

Liyanage, J.P. and Kumar, U., In search of performance excellence in asset maintenance: performance indicators, knowledge management & process intelligence, *Maint. J.*, 14, 54–64, 2001b.

Liyanage, J.P. and Kumar, U., Value based maintenance performance diagnostics: An architecture to measure maintenance performance in petroleum assets, *The International Conference of Maintenance Societies (ICOMS-2001)*, Paper no. 050, Melbourne, Australia, 2001c.

Liyanage, J.P. and Kumar, U., A value-based working algorithm to manage maintenance performance: key learning points from oil & gas industry about an integrated production asset, *New Eng. J.*, 4, 5–7, 2002a.

Liyanage, J.P. and Kumar, U., Value-based maintenance performance management for the petroleum industry, *Proceedings of the 4th International Conference on Quality, Reliability, Maintenance (QRM-2002)*, 2002b, pp. 113–116.

Liyanage, J.P. and Kumar, U., Value-based management of maintenance performance: the best practice for the 21st century maintenance based on experiences & learning from oil & gas industry, *Proceedings of the 16th European Maintenance Congress: Euromaintenance 2002*, Helsinki, Finland, 2002c, pp. 29–36.

Liyanage, J.P. and Kumar, U., Towards a value-based view on operations and maintenance performance management, *J. Qual. Maint. Eng.*, 9, 333–350, 2003.

Ljungberg, O., Measurement of overall equipment effectiveness as a basis for TPM activities, *Int. J. Oper. Prod.*, 18, 495–507, 1998.

Lofsten, H., Management of industrial maintenance — economic evaluation of maintenance policies, *Int. J. Oper. Prod.*, 19, 716–737, 1999.

Lofsten, H., Measuring maintenance performance: in search for a maintenance productivity index, *Int. J. Prod. Econ.*, 63, 47–58, 2000.

Luck, W.S., Now you can really measure maintenance performance, *Factory Management and Maintenance*, Vol. 114, McGraw-Hill, New York, 1965, pp. 81–86.

Maggard, B.N. and Rhyne, D.M. Total productive maintenance: a timely integration of production and maintenance, *Prod. Invent. Manage. J.*, Fourth Quarter, 6–11, 1992.

Nakajima, S., *An Introduction to TPM*, Production Press, Cambridge, MA, 1988.

Nakajima, S., *TPM Development Programme*, Production Press, Cambridge, MA, 1989.

Oke, S.A., An analytical model for the optimisation of maintenance profitability, *Int. J. Prod. Perform. Manage.*, 54, 113–136, 2005.

Oke, S.A. and Oluleye, A.E., A template for composing maintenance performance measures, *Conference Proceedings of the Nigerian Institute of Industrial Engineers*, 1999, pp. 114–125.

Oluleye, A.E., Tade, A.O., and Olajire, K.A., A schema for assessing maintenance effectiveness, *7th International Management of Industrial Reliability and Cost Effectiveness Symposium*, University of Exeter, United Kingdom, 1997, pp. 331–341.

Paulsen, J., Cooke, R., and Nyman, R., Comparative evaluation of maintenance performance using sub-survival functions, *Reliability Eng. Syst. Saf.*, 58, 157–163, 1997.

Perry, D. and Starr, A.G. Introducing value-based maintenance, in *Condition Monitoring and Diagnostic Engineering Management*, Starr, A.G. and Rao, R.B.K.N., Eds., Elsevier Science, Amsterdam, 2001.

Pintelon, L. and Van Puyvelde, F., Maintenance performance reporting systems: some experiences, *J. Qual. Maint. Eng.*, 3, 4–15, 1997.

Priel, V.Z., *Systematic Maintenance Organization*, MacDonald & Evans Ltd., London, 1974.

Tsang, A.H.C., A strategic approach to managing maintenance performance, *J. Qual. Maint. Eng.*, 4, 87–94, 1998.

Tsang, A.H.C., Jardine, A.K.S., and Kolodny, H., Measuring maintenance performance: a holistic approach, *Int. J. Oper. Prod. Manage.*, 19, 691–715, 1999.

Vineyard, M., Amoako-Gyampah, K., and Meredith, J.R., An evaluation of maintenance policies for flexible manufacturing systems, *Int. J. Oper. Prod. Manage.*, 20, 409–426, 2000.

27

Ranking Irregularities When Evaluating Alternatives by Using Some Multi-Criteria Decision Analysis Methods

Xiaoting Wang
Louisiana State University

Evangelos Triantaphyllou
Louisiana State University

27.1 Introduction to Multi-Criteria Decision Analysis

People make decisions almost every day and everywhere. Normally, individuals seldom need to use sophisticated decision-making tools when making their decisions. But in many fields of engineering, business, government, and science, where decisions are often either worth millions of dollars or may have a significant impact on the welfare of society, decision-making problems are usually complex and anything but easy. In such settings, powerful decision analysis and decision-making tools must be built and used to help decision-makers make better choices.

There are many decision-making tools in the literature. Some focus on inventory control, investment selection, scheduling, etc. Among them, multi-criteria decision analysis (MCDA) is one of the most widely used decision methodologies. Multi-criteria decision analysis can help to improve the quality of decisions by making decision-making more explicit, rational, and efficient.

A typical problem in MCDA is the task of ranking a finite set of decision alternatives, each of which is explicitly described in terms of different characteristics (often called attributes, decision criteria, or objectives) that have to be taken into account simultaneously. Usually, an MCDA method aims at one of the following four goals, or "problematics" (Roy, 1985; Jacquet-Lagreze and Siskos, 2001):

Problematic 1: Find the best alternative.
Problematic 2: Group the alternatives into well-defined classes.
Problematic 3: Rank the alternatives in order of total preference.
Problematic 4: Describe how well each alternative meets all the criteria simultaneously.

Many interesting aspects of MCDA theory and practice are discussed in Hobbs (1986), Hobbs et al. (1992, 2000), Stewart (1992), Triantaphyllou (2000), and Zanakis et al. (1995, 1998).

Another term that is used interchangeably with MCDA is multi-criteria decision-making (MCDM). It should be stated here that the term MCDM is also used to mean finding the best alternative in a continuous setting.

Although different MCDA methods follow different procedures, almost all of them share the following common essentials: a finite set of alternatives and a finite set of decision criteria. Each alternative is described by how well it meets each one of the decision criteria. If a given criterion refers to a qualitative aspect of the alternatives, then the alternatives may be described in relative or qualitative terms regarding that criterion. If the criterion is easily quantifiable, then the alternatives may be described in absolute terms regarding that criterion. Meanwhile, the criteria may be associated with weights of importance.

For example, in the hypothetical problem of selecting the best car among three candidate cars, say cars A, B or C, the decision criteria may be price, mileage per gallon, and the physical attractiveness of the shape of a car. That is, we have three criteria. Of these three criteria, the first two are easy to quantify, as one may know the exact price value of each car and also the exact fuel consumption. On the other hand, expressing the alternatives in terms of the last criterion might be trickier as that criterion is a qualitative one. In such cases one may use percentages expressing how much a given car is more desirable than another car.

The above data can also be viewed as the entries of a *decision matrix*. The rows of such a matrix correspond to the alternatives of the problem, the columns to the decision criteria. The a_{ij} element of a decision matrix represents the performance value of the *i-th* alternative in terms of the *j-th* criterion. The typical decision matrix can be represented as in Figure 27.1 (observe that the criteria weights are depicted in this matrix as the w_j parameters). Data for MCDA problems can be determined by direct observation (if they are easily quantifiable) or by indirect means if they are qualitative (Triantaphyllou et al., 1994), as we have demonstrated in the previous car selection example.

From the early developments of the MCDA theories in the 1950s and 1960s, a plethora of MCDA methods have been developed in the literature, and new contributions are continuously coming forth in this area. There are many ways to classify the already existing MCDA methods. One of the ways is to classify

	Criteria			
	C_1	C_2	...	C_n
	$(W_1$	W_2	...	$W_n)$
Alternatives				
A_1	a_{11}	a_{12}	...	a_{1n}
A_2	a_{21}	a_{22}	...	a_{2n}
	.	.		.
	.	.		.
	.	.		.
A_m	a_{m1}	a_{m2}	...	a_{mn}

FIGURE 27.1   Structure of a typical decision matrix.

MCDA methods according to the type of data they use. Thus, we have deterministic, stochastic, or fuzzy MCDA methods (Triantaphyllou, 2000). Another way of classifying MCDA methods is according to the number of decision-makers involved in the decision process. Hence, we have single decision-maker MCDA methods and group decision-making MCDA methods. For some representative articles in this area, see George et al. (1992), Hackman and Kaplan (1974), and DeSanctis and Gallupe (1987). For a comprehensive presentation of some critical issues in group decision-making, the interested reader may want to consult the papers regularly published in the journal *Group Decision Making*. In this chapter we concentrate on single decision-maker deterministic MCDA methods that attempt to find the best alternative, subject to a finite number of decision criteria.

This chapter is organized as follows. Section 27.2 presents some well-known MCDA methods. Applications of MCDA methods in different engineering fields are described in the Section 27.3. The Section 27.4 discusses various ranking issues that emerge when evaluating alternatives by using different MCDA methods. Finally, some concluding comments are presented in the last section.

27.2 Some MCDA Methods

Among the numerous MCDA methods, there are several prominent families that have enjoyed a wide acceptance in the academic world and have also been used in many real-world applications. Each of these methods has its own characteristics, background logic, and application areas. In the following sections, we will give a brief description of some of them.

27.2.1 The Analytic Hierarchy Process and Some of Its Variants

The analytic hierarchy process (AHP) method was developed by Thomas Saaty (1980, 1994). It is a powerful decision-making process that can help people set priorities and choose the best options by reducing complex decision problems to a system of hierarchies. Since its inception, it has evolved into several different variants and has been widely used to solve a broad range of multi-criteria decision problems. Its applications can be found in business, industry, government, and the military.

27.2.1.1 The Analytic Hierarchy Process

The AHP method uses the pairwise comparison and eigenvector methods to determine the a_{ij} values and also the criteria weights w_j. Details of the pairwise comparison and eigenvector methods can be found in Saaty (1980, 1994). In this method, a_{ij} represents the relative value of alternative A_i when it is considered in terms of criterion C_j. In the original AHP method, the a_{ij} values of the decision matrix need to be normalized vertically. That is, the elements of each column in the decision matrix should add up to 1. In this way, values with various units of measurement can be transformed into dimensionless ones. If all the criteria express some type of benefit, according to the original AHP method, the best alternative is the one that satisfies the following expression:

$$P^*_{AHP} = \max_i P_i = \max_i \sum_{j=1}^{n} a_{ij} w_j, \quad \text{for } i = 1, 2, 3, \ldots, m \tag{27.1}$$

From the above formula, we can see that the original AHP method uses an additive expression to determine the final priorities of the alternatives in terms of all the criteria simultaneously. Next, we consider the revised AHP, which is an additive variant of the original AHP method.

27.2.1.2 The Revised Analytic Hierarchy Process

The revised AHP model was proposed by Belton and Gear (1983) after they had found a case of ranking abnormality that occurred when the original AHP model was used. In their case, the original AHP was used to rank three alternatives in a simple test problem. Then a fourth alternative, identical to one of the three alternatives, was introduced in the original decision problem without changing any other data. The ranking of the original three alternatives was changed after the revised problem was ranked again by the same method. Later, this ranking abnormality was defined as a rank reversal. According to Belton and Gear, the root of this inconsistency is the fact that the sum of relative values of the alternatives for each criterion is 1. So instead of having the relative values of the alternatives sum up to 1, they proposed to divide each relative value by the maximum value of the relative values. According to this variant, the a_{ij} values of the decision matrix need to be normalized by dividing the elements of each column in the decision matrix by the largest value in that column. As before, the best alternative is given again by the additive formula (27.1), but now the normalization is different.

$$P^*_{\text{Revised}-\text{AHP}} = \max_i P_i = \max_i \sum_{j=1}^{n} a_{ij}w_j, \quad \text{for } i = 1, 2, 3, \ldots, m$$

The revised AHP was sharply criticized by Saaty (1990). After many debates and a heated discussion (see Dyer, 1990a and b; Saaty, 1983, 1987, 1990; Harker and Vargas, 1990), Saaty accepted this variant, which is now also called the *ideal mode* AHP (Saaty, 1994). However, even earlier, the revised AHP method was found to suffer from some other ranking problems even without the introduction of identical alternatives (Triantaphyllou and Mann, 1989). In that study and also in (Triantaphyllou, 2000, 2001), it was found that most of the problematic situations in the AHP methods are caused by the required normalization (either by dividing by the sum of the elements or by the maximum value in a vector) and the use of an additive formula on the data of the decision matrix for deriving the final preference values of the alternatives. However, in the core step of one of the MCDA methods known as the weighted product model (WPM) (Bridgeman, 1922; Miller and Starr, 1969), the use of an additive formula is avoided by using a multiplicative expression. This ushered in the development of a multiplicative version of the AHP method, known as the multiplicative AHP.

27.2.1.3 The Multiplicative Analytic Hierarchy Process

The use of multiplicative formulas in deriving the relative priorities in decision-making is not new (Lootsma, 1991). A critical development appears to be the use of multiplicative formulations when one aggregates the performance values a_{ij} with the criteria weights w_j. In the WPM method, each alternative is compared with others in terms of a number of ratios, one for each criterion. Each ratio is raised to the power of the relative weight of the corresponding criterion. Generally, the following formula is used (Bridgeman, 1922; Miller and Starr, 1969) in order to compare two alternatives A_K and A_L:

$$R\left(\frac{A_K}{A_L}\right) = \prod_{j=1}^{n} \left(\frac{a_{Kj}}{a_{Lj}}\right)^{w_j} \tag{27.2}$$

If $R(A_K/A_L) \geq 1$, then A_K is more desirable than A_L (for the maximization case). Then the best alternative is the one that is better than or at least equal to all other alternatives.

Based on the WPM method, Barzilai and Lootsma (1994) and Lootsma (1999) proposed the multiplicative version of the AHP method. This method was further analyzed in Triantaphyllou (2000, 2001). According to this method, the relative performance values a_{ij} and criteria weights w_j are not processed according to formula (27.1), but the WPM formula (27.2) is used instead.

Furthermore, one can use a variant of formula (27.2) to compute preference values of the alternatives that in turn, can be used to rank them. The preference values can be computed as follows:

$$P_{i,\ multi-AHP} = \prod_{j=1}^{n} (a_{ij})^{w_j} \tag{27.3}$$

Note that if $P_i > P_j$, then $P_i/P_j > 1$, or equivalently, $P_i - P_j > 0$. That is, two alternatives A_i and A_j can be compared in terms of their preference values P_i and P_j by forming the ratios or, equivalently, the differences of their preference values.

From formula (27.2), we can see that not only was the use of an additive formula avoided in the multiplicative AHP, but the negative effects of normalization can also be eliminated by using the multiplicative formula. These properties of the multiplicative AHP are demonstrated theoretically in Triantaphyllou (2000). In that study, it was also proved that most of the ranking irregularities that occurred when the additive variants of the AHP method were used would not occur with the multiplicative AHP method.

27.2.2 The ELECTRE Methods

Another prominent role in MCDA methods is played by the ELECTRE approach and its derivatives. This approach was first introduced in Benayoun et al. (1966). The main idea of this method is the proper utilization of what are called "outranking relations" to rank a set of alternatives. The ELECTRE approach uses the data within the decision problems along with some additional threshold values to measure the degree to which each alternative outranks all others. Since the introduction of the first ELECTRE method, a number of variants have been proposed. Today two widely used versions are ELECTRE II (Roy and Bertier, 1971, 1973) and ELECTRE III (Roy, 1978) methods. Since the ELECTRE approach is more complicated than the AHP approach, the process of ELECTRE II is described next to provide a simple introduction of its logic.

The ELECTRE methods are based on the evaluation of two indices, the *concordance index* and the *discordance index*, defined for each pair of alternatives. The concordance index for a pair of alternatives a and b measures the strength of the hypothesis that alternative a is at least as good as alternative b. The discordance index measures the strength of evidence against this hypothesis (Belton and Stewart, 2001). There are no unique measures of concordance and discordance indices.

In ELECTRE II, the concordance index $C(a, b)$ for each pair of alternatives (a, b) is defined as follows:

$$C(a, b) = \frac{\sum_{i \in Q(a, b)} w_i}{\sum_{i=1}^{m} w_i}$$

where $Q(a, b)$ is the set of criteria for which alternative a is equal or preferred to (i.e., at least as good as) alternative b and w_i the weight of the ith criterion. One can see that the concordance index is the proportion of the criteria weights allocated to those criteria for which a is equal to or preferred to b. The discordance index $D(a, b)$ for each pair (a, b) is defined as follows:

$$D(a, b) = \frac{\max_{j} [g_j(b) - g_j(a)]}{\delta}$$

where $\delta = \max_j |g_j(b) - g_j(a)|$ (i.e., the maximum difference on any criterion). This formula can only be used when the scores for different criteria are comparable. After computing the concordance and discordance indices for each pair of alternatives, two outranking relations are built between the alternatives by comparing the indices with two pairs of threshold values. They are referred to as the "strong" and "weak" outranking relations.

We define (C^*, D^*) as the concordance and discordance thresholds for the strong outranking relation and (C^-, D^-) as the concordance and discordance thresholds for the weak outranking relation where $C^* > C^-$ and $D^* < D^-$. Then the outranking relations will be built according to the following rules:

(1) If $C(a, b) \geq C^*, D(a, b) \leq D^*$ and $C(a, b) \geq C(b, a)$, then alternative a is regarded as strongly outranking alternative b.

(2) If $C(a, b) \geq C^-, D(a, b) \leq D^-$ and $C(a, b) \geq C(b, a)$, then alternative a is regarded as weakly outranking alternative b.

The values of (C^*, D^*) and (C^-, D^-) are decided by the decision-makers for a particular outranking relation. These threshold values may be varied to give more or less severe outranking relations; the higher the value of C^* and the lower the value of D^*, the more severe (i.e., stronger) the outranking relation is. That is, the more difficult it is for one alternative to outrank another (Belton and Stewart, 2001). After establishing the strong and weak outranking relations between the alternatives, the *descending* and *ascending distillation* processes are applied to the outranking relations to get two preorders of the alternatives. Next by combining the two preorders together, the overall ranking of the alternatives is determined. For a detailed description of the distillation processes, we refer interested readers to Belton and Stewart (2001) and Rogers et al. (1999).

Compared with the simple process and precise data requirement of the AHP methods, ELECTRE methods apply more complicated algorithms to deal with the complex and imprecise information from the decision problems and use these algorithms to rank the alternatives. ELECTRE algorithms look reliable and neat. People believe that the process of this approach could lead to an explicit and logical ranking of the alternatives. However, this may not always be the case. This point is further explored in section 27.4.

27.2.3 Utility or Value Functions

In contrast with the above approaches, there is another type of analysis that is based on value functions. These methods use a number of trade-off determinations that form what is known as utility or value functions (Kirkwood, 1997). The utility or value functions attempt to model mathematically a decision-maker's preference structure by a utility function (if the problem is stochastic) or a value function (if the problem is deterministic), and these functions are next used to identify a preferred solution (Al-Rashdan et al., 1999).

The functions attempt to map changes of values of performance of the alternatives in terms of a given criterion into a dimensionless value. Some key assumptions are made in the process of transferring changes in values into these dimensionless quantities (Kirkwood, 1997). The roots of this type of analysis can be found in Edwards (1977), Edwards and Barron (1994), Edwards and Newman (1986), and Dyer and Sarin (1979).

27.3 Some Applications of MCDA in Engineering

Multi-criteria decision analysis methods have long been used in many areas of real-life applications, especially in the engineering world. For example, the ELECTRE methods have been widely used in civil and environmental engineering (Hobbs and Meier, 2000; Zavadskas et al., 2004). Some related projects include water resources planning (Raj, 1995); wastewater or solid waste management (Hokkanen and Salminen, 1997; Rogers and Bruen, 1999); site selection for the disposal of nuclear waste (nuclear waste management); and highway design selection. Multi-criteria decision analysis methods have also been the main tools that are used to solve many kinds of environmental decision-making problems by the U.S. Department of Energy's Environmental Management in the National Research Council. Hobbs and Meier (2000) have presented an extensive study on the applications of MCDA methods in energy and environmental decision-making.

Multi-criteria decision analysis methods also play a significant role in financial engineering. Its applications within this area have covered many important issues, including venture-capital investment, business failure risk, assessment of granting credit and investments, and portfolio management. Zopounidis and Doumpos (2000) offer a detailed description about the applications of some MCDA methods in financial engineering and how to combine those methods with techniques such as expert systems and artificial intelligence technologies to address decision problems in financial engineering.

Industrial engineering is another field where MCDA methods are studied intensively and used extensively. One of the most important contributions of industrial engineering is in assisting people to make sound decisions with appropriate, scientific decision-making tools. Triantaphyllou and Evans (1999) coedited an issue of the journal *Computers and Industrial Engineering*, which focused on some vital MCDA issues in industrial engineering, including facility layout and location problems, maintenance-related decision-making, process planning, production planning, and some theoretical issues about MCDA methods in industrial engineering.

Other engineering applications of MCDA include the use of decision analysis in integrated manufacturing (Putrus, 1990), in flexible manufacturing systems (Wabalickis, 1988), and material selection (Liao, 1996). It is impossible to give an exhaustive review of the applications of MCDA methods in engineering, which has accumulated a vast literature in the past quarter century. It should be clear from the above enumeration that efficient scientific decision-making methods have played and are playing an important and indispensable role in many decision-making activities related to engineering.

27.4 Ranking Irregularities When Evaluating Alternatives in MCDA

We have seen that many methods have been proposed to analyze and solve MCDM problems in various fields. However, an important topic in the MCDM area is that often, different MCDA methods may yield different answers to exactly the same problem. Sometimes, ranking irregularities may occur in such well-known MCDA methods as, for example, the AHP method.

27.4.1 Ranking Irregularities When the Additive Variants of the AHP Method Are Used

The AHP method has been widely used in many real-life decision problems. Thousands of AHP applications have been reported in edited volumes and books (e.g., Golden et al., 1989; Saaty and Vargas, 2000) and on websites (e.g., www.expertchoice.com). However, the AHP method has also been criticized by many researchers for some of its problems. One key problem is rank reversals. Belton and Gear (1983) first described the problem of rank reversals with the AHP. The example of rank reversal that they provide (please refer to Section 27.2) demonstrated that the ranking of alternatives may be affected by the addition (or deletion) of nonoptimal alternatives. This phenomenon has inspired some doubts about the reliability and validity of the original AHP method.

After the first report, some other types of ranking irregularities with the original AHP method were also found. Dyer and Wendell (1985) studied rank reversals when the AHP was used and near copies were considered in the decision problem. Triantaphyllou (2000) reported another type of rank reversal with the additive AHP methods, in which the indication of the optimal alternative may change when one of the nonoptimal alternatives is replaced by a worse one. Next, Triantaphyllou (2001) reported two new cases of ranking irregularities when the additive AHP methods are used. One is that the ranking of the alternatives may be different when all the alternatives are compared two at a time and also simultaneously. Another case is that the ranking of the alternatives may not follow the transitivity property when the alternatives are compared two at a time.

As we know, the MCDA problems usually involve the ranking of a finite set of alternatives in terms of a finite number of decision criteria. Such criteria may often be in conflict with each other. That is, an MCDA problem may involve both benefit and cost criteria at the same time. How to deal with conflicting criteria is another factor that may also cause some ranking irregularities. In Triantaphyllou and Baig (2005), it was found that some ranking irregularities occurred with some additive MCDA methods (which include the additive variants of the AHP method) when two different approaches for dealing with conflicting criteria are used. The two approaches are the benefit-to-cost ratio approach and the benefit-minus-cost approach. It was demonstrated that when the two approaches for aggregating conflicting criteria into two groups are used on the same problem, even when using the same additive MCDA method, one may derive very different rankings of the alternatives. Furthermore, an extensive empirical study revealed that this situation might occur rather intensively in random test problems. The only methods that are immune to these ranking irregularities are two multiplicative MCDA methods: the weighted product model (WPM) and the multiplicative AHP.

Many researchers have also put a lot of effort into explaining the reasons behind the rank reversals and studying how to avoid them. Belton and Gear (1983) proposed the revised AHP method in order to preserve the ranking of the alternatives in the presence of identical alternatives. Saaty (1987) pointed out that rank reversals were due to the inclusion of duplicates of the alternatives. So he suggested that people should avoid the introduction of similar or identical alternatives. However, other cases were later found in which rank reversal occurred without the introduction of identical alternatives (Triantaphyllou, 2000, 2001). Dyer (1990a) indicated that the sum to unity normalization of priorities makes each one dependent on the set of alternatives being compared. He also claimed that the resulting individual priorities are thus arbitrary, as arbitrary sets of alternatives may be considered in the decision problem. Stam and Silva (1997) revealed that if the relative preference statements about alternatives were represented by judgment intervals (i.e., the pairwise preference judgments are uncertain [stochastic]), rather than single values, then the rankings resulting from the traditional AHP analysis based on the single judgment values may be reversed and will therefore be incorrect. On the basis of this statement, they developed some multivariate statistical techniques to obtain both point estimates and confidence intervals for the occurrence of certain types of rank reversal probabilities with the AHP method. Yue et al. (2004) introduced a grouping method based on direct comparisons between all alternatives. Their method divides the alternatives into groups in such a way that a dominant relationship exists between groups but not among alternatives within each group, and a rank reversal will not happen between ranking groups. This method can be used in situations where just a group ranking is desired. The above references are just a sample of the research that has been conducted on ranking problems when evaluating alternatives by using various MCDA methods. It is evident that many of these ranking problems have not been fully explained. That means that disputes and studies about this important topic are still ongoing in the MCDA area, and more studies are needed.

27.4.2 Some Test Criteria for Evaluating MCDA Methods

Most of the past research studies on ranking irregularities have concentrated on the AHP method. There are very few studies that explore the reliability and validity of the other MCDA methods. Does that mean that decision-makers can trust the other MCDA methods without questioning the validity of their answers? The answer is "No." Usually, decision-makers undertake some kind of sensitivity analysis to examine how the decision results will be affected by changes in some of the uncertain data in a decision problem. For example, is the ranking of the alternatives stable or easily changeable under a different set of criteria weights? By this process, decision analysts may better understand a decision problem.

However, another intriguing problem with decision-making methods is that different methods may often yield different answers (rankings) when they are faced with exactly the same decision-making problem (numerical data). Thus, the issue of evaluating the relative performance of different MCDA methods is naturally raised. This, in turn, raises the question of how one can evaluate the performance of different MCDA methods. Since for some problems, it may be practically impossible to know which one is the best alternative, some kind of testing procedure has to be employed. The above subjects, along with some other related issues, have been discussed in detail in Triantaphyllou and Mann (1989) and Triantaphyllou (2000, 2001). In these studies, three test criteria were established to test the relative performance of different MCDA methods. These test criteria are as follows:

Test Criterion #1: *An effective MCDA method should not change the indication of the best alternative when a nonoptimal alternative is replaced by another worse alternative (given that the relative importance of each decision criterion remains unchanged).*

Suppose that an MCDA method has ranked a set of alternatives in a particular way. Next, suppose that a nonoptimal alternative, say A_k, is replaced by another alternative, say A'_k, which is less desirable than A_k. Then, the indication of the best alternative should not change when the alternatives are ranked again by the same method. The same should also be true for the relative rankings of the rest of the unchanged alternatives.

Test Criterion #2: *The rankings of alternatives by an effective MCDA method should follow the transitivity property.*

Suppose that an MCDA method has ranked a set of alternatives of a decision problem in some way. Next, suppose that this problem is decomposed into a set of smaller problems, each defined by two alternatives at a time and the same number of criteria as in the original problem. Then, all the rankings that are derived from the smaller problems should satisfy the transitivity property. That is, if alternative A_1 is better than alternative A_2, and alternative A_2 is better than alternative A_3, then one should also expect that alternative A_1 is better than alternative A_3.

The third test criterion is similar to the previous one, but now one tests for the agreement between the smaller problems and the original, undecomposed problem.

Test Criterion #3: *For the same decision problem and when using the same MCDA method, after combining the rankings of the smaller problems into which an MCDA problem is decomposed, the new overall ranking of the alternatives should be identical to the original overall ranking of the undecomposed problem.*

As before, suppose that an MCDA problem is decomposed into a set of smaller problems, each defined on two alternatives and the original decision criteria. Next, suppose that the rankings of the smaller problems follow the transitivity property. Then, when the rankings of the smaller problems are all combined together, the overall ranking of the alternatives should be identical to the original ranking before the problem was decomposed.

27.4.3 Ranking Irregularities When the ELECTRE Methods Are Used

The performance of some ELECTRE methods was tested in terms of the previous three test criteria in Wang and Triantaphyllou (2004, 2006). During these experiments, the three test criteria were used to evaluate the performance of TOPSIS (Hwang and Yoon, 1981), ELECTRE II, and the ELECTRE III methods. In these tests, each one of these three methods failed in terms of each one of the three test criteria. This revealed that the same kinds of ranking irregularities that occurred when the additive AHP methods were used also occurred when the ELECTRE methods were used.

For a deeper understanding of these ranking irregularities, a computational experiment was undertaken by Wang and Triantaphyllou (2004, 2006). The experimental results demonstrated that the ranking irregularities were fairly significant in the simulated decision problems. For instance, in terms of test criterion #1, the ranking reversal rate was about 20% with the increase of the number of criteria from 3 to 21 for the ELECTRE III method. Sometimes, the best alternatives will become the second-best or even lower than that. In terms of test criterion #2, with the increase of the number of alternatives from 3 to 21, the frequency of violating the transitivity property tended to be 100%. Among the decision problems that followed the transitivity property, it was also very likely that the overall ranking of the alternatives from the smaller problems was partially or completely different from the original overall ranking of the undecomposed problem.

Although the computational results have revealed that these three types of ranking irregularities occurred frequently in simulated decision problems, ten real-life cases selected randomly from the literature were also studied in order to better understand this situation. The results of this study indicated that the rates of these ranking irregularities were also rather high in these real-life cases. For example, six out of ten cases failed test criterion #1. The rankings of nine out of ten case studies did not follow the transitivity property. The one case in which the rankings from the smaller problems did not violate the transitivity property failed to pass test criterion #3.

This is the first time in the literature that rank reversals have been reported with the ELECTRE methods. These findings can be viewed as a wake-up call to people that the methods they have already been using are not as reliable as they may have expected. More reliable decision-making methods are needed to help people make better decisions.

27.5 Conclusion and Future Research Directions

From the above ranking problems with the AHP and the ELECTRE methods, it can be seen that it is hard to accept an MCDA method as being accurate all the time, although such methods may play a critical role in many real-life problems. The research work in Wang and Triantaphyllou (2004, 2006) complements previous studies and reveals that even more MCDA methods suffer from ranking irregularities. The ELECTRE methods are widely used today in practice. However, the ranking irregularities should function as a warning not to accept ELECTRE's recommendations without questioning their validity. Previous and current research indicates that the above ranking irregularities tend to occur when the alternatives appear to be very close to each other. If, on the other hand, the alternatives are very distinct from each other, then it is less likely that these ranking irregularities will occur. However, one needs a more powerful MCDA method when alternatives are closely related to each other. In Section 27.3 it has been shown how widely MCDA methods have been used in various engineering fields. Decisions in those areas are often worth millions or even billions of dollars and have a great influence on the economy and welfare of society. Thus, when evaluating alternatives by different MCDA methods, ranking problems are worth a great deal of attention.

As it has been mentioned previously in Triantaphyllou (2000, 2001), it is demonstrable that the multiplicative AHP is immune to all of the above ranking irregularities. This means the multiplicative AHP can pass all the previous three test criteria. Of course, that does not mean it is perfect. It has been found that it may suffer from some other ranking problems (Triantaphyllou and Mann, 1989). This method uses a multiplicative formula to compute the final priorities of the alternatives. The multiplicative formula can help it to avoid the distortion from any kind of normalization and also some arbitrary effects introduced by the additive formulas. Thus, an intriguing task for the future is to try to see if a new MCDA method can be designed that combines the good qualities from the multiplicative AHP and some other MCDA methods and is also immune from any ranking problems. Another direction for future research is to discover more test criteria against which existing and future MCDA methods can be evaluated. Clearly, this is a fascinating area of research and is of paramount significance to both researchers and practitioners in the MCDM field.

References

Al-Rashdan, D., Al-Kloub, B., Dean, A., and Al-Shemmeri, T., Environmental impact assessment and ranking the environmental projects in Jordan, *Eur. J. Oper. Res.*, 118, 30–45, 1999.

Barzilai, J. and Lootsma, F.A., Power relations and group aggregation in the multiplicative AHP and SMART, *Proceedings of the Third International Symposium on the AHP*, George Washington University, Washington, DC, 1994, pp. 157–168.

Belton, V. and Gear, A.E., On a shortcoming of Saaty's method of analytic hierarchies, *Omega*, 13, 143–144, 1983.

Belton, V. and Stewart, T.J., Outranking methods, *Multiple Criteria Decision Analysis: An Integrated Approach*, Kluwer Academic Publishers, Boston, MA, 2001, Chap. 8.

Benayoun, R., Roy, B., and Sussman, N., Manual de Reference du Programme Electre, Note De Synthese et Formaton, 25, Direction Scientifque SEMA, Paris, France, 1966.

Bridgeman, P.W., *Dimensionless Analysis*, Yale University Press, New Haven, CT, 1922.

DeSanctis, G. and Gallupe, R.B., A foundation for the study of group decision support systems, *Manage. Sci.*, 33, 589–609, 1987.

Dyer, J.S., Remarks on the analytic hierarchy process, *Manage. Sci.*, 36, 249–258, 1990a.

Dyer, J.S., A clarification of remarks on the analytic hierarchy process, *Manage. Sci.*, 36, 274–275, 1990b.

Dyer, J. S. and Sarin, R.K., Measurable multiattribute value functions, *Oper. Res.*, 27, 810–822, 1979.

Dyer, J.S. and Wendell, R.E., A Critique of the Analytic Hierarchy Process, Technical Report 84/85-4-24, Department of Management, the University of Texas at Austin, Austin, TX, 1985.

Edwards, W., How to use multiattribute utility measurement for social decision making, *IEEE Trans. Man Syst. Cyber.*, SMC-7, 1977, 326–340.

Edwards, W. and Barron, F.H., SMARTS and SMARTER: Improved simple methods for multiattribute utility measurement, *Organ. Behav. Hum. Decis. Process.*, 60, 306–325, 1994.

Edwards, W. and Newman, J.R., Multiattribute evaluation, in *Judgment and Decision Making: An Interdisciplinary Reader*, Arkes, H.R. and Hammond, K.R., Eds., Cambridge University Press, Cambridge, U.K., 1986, pp. 13–37.

George, J.F., Dennis, A.R., and Nunamaker, J.F., An experimental investigation of facilitation in an EMS decision room, *Group Decis. Negot.*, 1, 57–70, 1992.

Golden B., Wasil, E., and Harker, P., *The Analytic Hierarchy Process: Applications and Studies*, Springer, Berlin, German, 1989.

Hackman, J.R. and R.E. Kaplan, Interventions into group process: An approach to improving the effectiveness of groups, *Decis. Sci.*, 5, 459–480, 1974.

Harker, P.T. and Vargas, L.G., Reply to 'Remarks on the Analytic Hierarchy Process', *Manage. Sci.*, 36, 269–273, 1990.

Hobbs, B.F., What can we learn from experiments in multiobjective decision analysis, *IEEE Trans. Syst. Manage. Cyber.*, 16, 384–394, 1986.

Hobbs, B.F., Chankong, V., Hamadeh, W., and Stakhiv, E., Does Choice of Multi-Criteria Method Matter? An Experiment in Water Resource Planning, *Water Resour. Res.*, 28, 1767–1779, 1992.

Hobbs, B.F. and Meier, P., *Energy Decisions and the Environment: A Guide to the Use of Multi-Criteria Methods*, Kluwer Academic Publishers, Boston, MA, 2000.

Hokkanen, J. and Salminen, P., Choosing a solid waste management system using multi-criteria decision analysis, *Eur. J. Oper. Res.*, 98, 19–36, 1997.

Hwang, C.L. and Yoon, K., *Multiple Attribute Decision Making: Methods and Applications*, Springer-Verlag, New York, 1981.

Jacquet-Lagreze, E. and Siskos, Y., Preference disaggregation: 20 years of MCDA experience, invited review, *Eur. J. Oper. Res.*, 130, 233–245, 2001.

Kirkwood, C.W., *Strategic Decision Making: Multiobjective Decision Analysis with Spreadsheets*, Duxbury Press, Belmont, CA, 1997 (ISBN: 0534516920).

Liao, T.W., A fuzzy multi-criteria decision making method for material selection, *J. Manuf. Syst.*, 15, 1–12, 1996.

Lootsma, F.A., Scale Sensitivity and Rank Preservation in a Multiplicative Variant of the AHP and SMART, Technical Report 91-67, Delft University of Technology, Delft, the Netherlands, 1991.

Lootsma, F.A., *Multi-Criteria Decision Analysis via Ratio and Difference Judgment*, Applied Optimization Series, 29, Kluwer Academic Publishers, Dordrecht, The Netherlands, 1999.

Miller, D.W. and Starr, M.K., *Executive Decisions and Operations Research*, Prentice-Hall, Englewood Cliffs, NJ, 1969.

Putrus, P., Accounting for intangibles in integrated manufacturing (nonfinancial justification based on the analytical hierarchy process), *Inform. Strategy*, 6, 25–30, 1990.

Raj, P.A., Multi-criteria methods in river basin planning — A case study, *Water Sci. Technol.*, 31, 261–272, 1995.

Rogers, M.G. and Bruen, M.P., Applying ELECTRE to an option choice problem within an environmental appraisal – three case studies from the Republic of Ireland, in *Advances in Decision Analysis*, Meskens, N. and Roubens, M., Eds., Kluwer Academic Publishers, Dordrecht, The Netherlands, 1999, Chap. 10.

Rogers, M.G., Bruen, M.P., and Maystre, L.-Y., *The Electre Methodology, Electre and Decision Support*, Kluwer Academic Publishers, Boston, MA, 1999, Chap. 3.

Roy, B., ELECTRE III: Un algorithme de classements fonde sur une representation floue des preference en presence de criteres multiples, *Cahiers CERO*, 20, 3–24, 1978.

Roy, B., *Methodologie Multicritiere d'Aide a la Decision*, Econometrica, Paris, France, 1985.

Roy, B. and Bertier, P., La methode ELECTRE II: Une methode de classement en presence de critteres multiples, SEMA (Metra International), Direction Scientifique, Note de Travail No. 142, Paris, 1971, 25pp.

Roy, B. and Bertier, P., La methode ELECTRE II: Une methode au media-planning, in *Operational Research 1972*, Ross, M. Ed., North-Holland Publishing Company, Amsterdam, 1973, pp. 291–302.

Saaty, T.L., *The Analytic Hierarchy Process*, McGraw-Hill, New York, 1980.

Saaty, T.L., Axiomatic foundations of the analytic hierarchy process, *Manage. Sci.*, 32, 841–855, 1983.

Saaty, T.L., Rank generation, preservation, and reversal in the analytic hierarchy process, *Dec. Sci.*, 18, 157–177, 1987.

Saaty, T.L., An exposition of the AHP in reply to the paper remarks on the analytic hierarchy process, *Manage. Sci.*, 36, 259–268, 1990.

Saaty, T.L., *Fundamentals of Decision Making and Priority Theory with the AHP*, RWS Publications, Pittsburgh, PA, 1994.

Stam, A. and Silva, A.P.D., Stochastic judgments in the AHP: The measurement of rank reversal probabilities, *Dec. Sci.*, 28, 655–688, 1997.

Saaty, T.L. and Vargas, L., *Models, Concepts and Applications of the Analytic Hierarchy Process*, Kluwer Academic Publishers, Boston, MA, 2000.

Stewart, T.J., A critical survey of the status of multiple criteria decision making theory and practice, *OMEGA*, 20, 569–586, 1992.

Triantaphyllou, E., *Multi-Criteria Decision Making Methods: A Comparative Study*, Kluwer Academic Publishers, Boston, MA, 2000.

Triantaphyllou, E., Two new cases of rank reversals when the AHP and some of its additive variants are used that do not occur with the multiplicative AHP, *Multi-Criteria Dec. Anal.*, 10(May), 11–25, 2001.

Triantaphyllou, E. and Baig, K. The impact of aggregating benefit and cost criteria in four MCDA methods, *IEEE Trans. Eng. Manage.*, 52, 213–226, 2005.

Triantaphyllou, E. and Evans, G.W., Eds., Multi-criteria decision making in industrial engineering, *J. Comput. Ind. Eng.*, 37, 1999 (special issue).

Triantaphyllou, E., Lootsma, F.A., Pardalos, P.M., and Mann, S.H., On the evaluation and application of different scales for quantifying pairwise comparisons in fuzzy sets, *J. Multi-Criteria Dec. Anal.*, 3, 133–155, 1994.

Triantaphyllou, E. and Mann, S.H., An examination of the effectiveness of multi-dimensional decision-making methods: a decision-making paradox, *Int. J. Dec. Support Syst.*, 5, 303–312, 1989.

Wabalickis, R.N., Justincation of FMS with the analytic hierarchy process, *J. Manuf. Syst.*, 17, 175–182, 1988.

Wang, X. and Triantaphyllou, E., Some ranking irregularities when the ELECTRE method is used for decision-making, *Proceedings of the 2004 IIE Annual Conference*, Houston, TX, May 2004.

Wang, X. and Triantaphyllou, E., Ranking irregularities when evaluating alternatives by using some ELECTRE methods, under review, *Omega*, 2006.

Yue, J., Chen, B., and Wang, M., Generating Ranking Groups in the Analytical Hierarchy Process, Working paper, 2004.

Zanakis, S., Mandakovic, T., Gupta, S.K., Sahay, S., and Hong, S., A review of program evaluation and fund allocation methods within the service and government sectors, *Socio Econ. Plan. Sci.*, 29, 59–79, 1995.

Zanakis, S., Solomon, A., Wishart, N., and Dublish, S., Multi-attribute decision making: a comparison of select methods, *Eur. J. Oper. Res.*, 107, 507–529, 1998.

Zavadskas, E.K., Ustinovičius, L., and Stasiulionis, A., Multicriteria valuation of commercial construction projects for investment purposes, *J. Civil Eng. Manage.*, X, 151–166, 2004.

Zopounidis, C. and Doumpos, M., *Intelligent Decision Aiding Systems Based on Multiple Criteria for Financial Engineering*, Kluwer Academic Publishers, Boston, MA, 2000.

28

e-Design Systems

Bartholomew O. Nnaji
University of Pittsburgh

Yan Wang
University of Pittsburgh

Kyoung-Yun Kim
Wayne State University

28.1 Introduction

Today, businesses face many challenges owing to the growth of globalization. This is true in multiple domains. For instance, manufacturing enterprises must adapt to meet the worldwide availability of technology, capital, information, and labor. Faster change in market demand drives faster obsolescence of established products. Global marketing competition makes manufacturers more conscious of quality, cost, and time-to-market. This distributed economic and technological environment poses a challenge: how to manage collaborative engineering, that is, how to let engineers collaborate globally during the product-development period. As the Internet has evolved in recent years, it has had an enormous impact on a whole spectrum of industries. The application of network technologies in manufacturing is indispensable because manufacturers face numerous challenges in the practice of collaborative design: lack of information from suppliers and working partners; incompleteness and inconsistency of product information/knowledge within the collaborating group; and incapability of processing information/data from other parties due to the problem of interoperability. Hence, collaborative design tools are needed to improve collaboration among distributed design groups, enhance knowledge sharing, and assist in better decision-making. There now exist many computational tools in those different areas. However, there are many problems that prevent such tools from working together transparently and seamlessly without human intervention. Problems mostly arise from the lack of common communication protocols, such as different computer-aided design (CAD) data formats, different computer operating systems, and different programming languages.

This chapter presents an innovative *e*-design paradigm and a model for service-oriented collaboration. This design paradigm aims at seamless and dynamic integration of distributed design objects and engineering tools over the Internet. *e*-Design involves conceptualizing, designing, and realizing a product using tools that allow for the interoperability of remote and heterogeneous systems, collaboration among remote supply-chain and multidisciplinary product-design team stakeholders, and virtual testing and validation of a product in a secure, Internet-based information infrastructure.

An *e*-design system is an integrated product-development environment that allows customers, suppliers, engineers, sales personnel, and other stakeholders to participate in product life cycle management while simultaneously shortening product-development time and cost. This integration should be realized by using a service-oriented infrastructure. Service provides functional use for a person, an application program, or another service in the system, which is the core for integration of engineering tools. Various computational engineering tools make certain services available to other design participants in a network-based distributed environment. Service-oriented infrastructure allows design stakeholders to *use* engineering services, but not *own* engineering tools. Instead of the traditional client–server relationship, peer-to-peer relations exist among service providers. The services that are provided by different engineering tools are published by a service manager, and are available within the distributed environment.

28.2 Design Process

A conventional design process is shown in Figure 28.1. The first phase starts with the recognition that a customer has a need for a product, thus triggering initial design. This recognition can take the form of

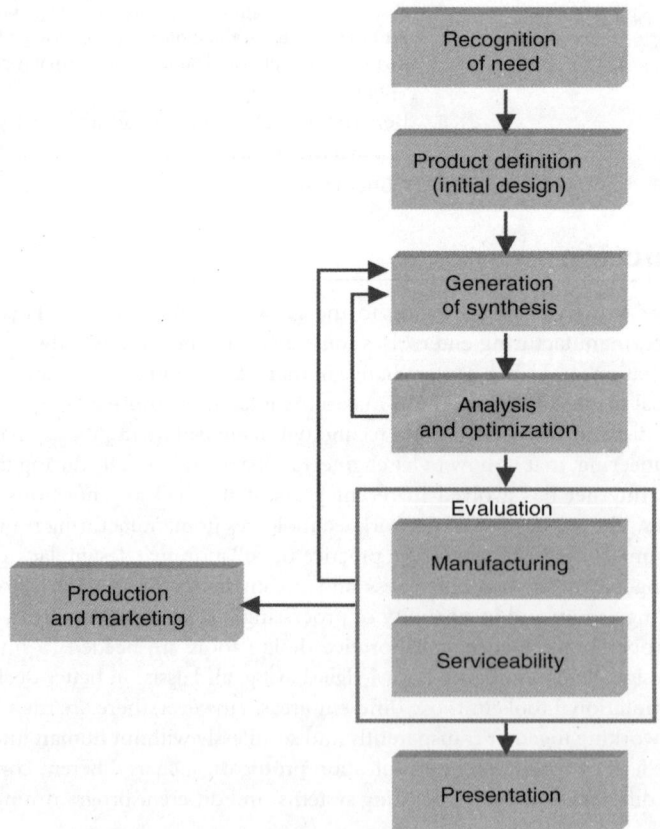

FIGURE 28.1 Traditional design processes.

the discovery of a dysfunctional system that must be redesigned, or a realization that the market needs a certain new product.

The second phase is the problem definition. This is the specification of the entity that is to be designed. It includes the determination of the physical and functional characteristics of the product, and its operating principles and service life. It can also involve the gathering of data about costs, legal requirements and standards, manufacturing, and quality and serviceability requirements.

The third phase is the conception phase, which involves new product generation or the synthesis of alternative designs. This is a very creative activity. It is often believed that this phase of the design process is the most crucial and is where the designer's creativity is employed. The design synthesis is normally tied to the fourth phase, the analysis phase, since the representation of a concept would normally be subjected to analysis, which can result in improved design, depending on the constraints of the analysis. This process can be repeated several times until the design is optimized.

The fifth phase is the evaluation, which may use defined specifications such as standard engineering and manufacturing practices, whereby the design is evaluated in order to ensure nonviolation of the established constraints at the problem definition stage. The evaluation systems must include manufacturability and serviceability evaluations. In the past, evaluation could require the fabrication and testing of a prototype model to determine its performance, quality, life, and so on. But with advanced computer technology, it is now often possible to model and test prototypes on the computer.

Finally, the design is presented as an engineering document, and in a modern manufacturing environment, the design data can also be directly transferred to a process planning system for product manufacture.

28.2.1 Design Hierarchy

Complex designs, such as aircraft, are built by many teams, each of which may yet have smaller teams. For instance, an aircraft is made up of thousands of components, one of which is the jet engine. The engine is normally designed by a team of engineers who would have the responsibility for individual segments of the engine. The designs produced by various teams would have to be assembled in order to test the final product. This test does not preclude the tests that individual teams must have performed on their specific components in order to certify them as completed.

It is clear that because of the component generate-and-test as well as the system generate-and-test nature of design, there are two discernible approaches to design. These are the bottom-up approach and the top-down approach. In the bottom-up approach, the designer proceeds from detailed design of the parts to the system assembly of the product. In the top-down approach, the designer makes a global decision about the product before considering the detailed parts. For the design of manufacturing processes, similar approaches are taken. In employing modern information technology in design, the desired approach can be quite cumbersome, especially if cost and time-to-market must be significantly reduced.

28.2.2 *e*-Design Paradigm

Industries now understand that the best way to reduce life-cycle costs is to evolve a more effective product-development paradigm using the Internet and web-based technologies. However, there remains a gap between these current industry market demands and current product-development paradigms. Today's CAD systems do not accommodate conceptual design or the direct imposition of multidisciplinary preferences and constraints such as functionality, manufacturability, safety, ergonomics, and material property. The existing CAD systems require that a product developer possess all the design analysis tools in-house, making it impractical to employ all the needed and newest tools. While *view* and *edit* functions on a product can be accomplished at remote locations in some advanced CAD systems, there is no platform that allows a customer at a remote location to participate in the design of the product through direct imposition of preferences and multidisciplinary constraints. This is particularly hindered by the fact that there is no mechanism for creating form from product specifications. The emergence of the Internet has offered technologies that can allow the creation of a design that will leapfrog product-development into an effective, high-quality design paradigm that significantly reduces development time and costs and achieves effective high-quality products.

The design of a product requires concurrent availability of dozens of technical supports from various engineering and nonengineering fields, just a few of which are drawing, material, manufacturing process, quality, marketing, maintenance, government regulations. The Internet provides an opportunity for these engineering tools to work together and utilize these services optimally. To connect these "islands of automation," universally accepted protocols are needed. This alternative design paradigm is service-oriented collaborative design, which incorporates different engineering services (assembly, manufacturing, finite element analysis [FEA], optimization, ergonomics, etc.) and makes them available for automatic transactions in product development. Each of these CAD and computer-aided engineering tools can be hooked up to an open *e*-design platform over the Internet and can provide certain services that result in a successful distributed product-development environment.

28.3 Efforts toward the Modernization of the Design Process

The advent of the Internet and World Wide Web (WWW) has introduced a new wave of research on collaborative product-development environments. Comprehensive reviews have been conducted by Yang and Xue (2003) and Fuh and Li (2005). There are two major topical areas in this field. One is how to manage product life-cycle information effectively within a distributed enterprise environment. The other is on network-centric concurrent design and manufacturing, which concentrates on new product design and a manufacturing methodology facilitated by network technologies.

In the first focus area, research topics comprise the integration of product and process information, both temporally and spatially. The product information for the whole life cycle needs to be storable and retrievable enterprise-wide. The accessibility, security, and integrity of information are major concerns. By merging the processes of design documentation and design data management through linking CAD drawings with external network-accessible relational databases, integrated geometric information and related documentation can be shared enterprise-wide (e.g., Maxfield et al., 1995; Dong and Agogino, 1998; Roy and Kodkani, 1999; Huang and Mak, 1999; Kan et al., 2001; Xue and Xu, 2003). This category of research utilizes existing network protocols to achieve enterprise-wide communication. Some research has focused on agent-based communication methodology over networks (e.g., Sriram and Logcher, 1993; Kumar et al., 1994; Huang and Mak, 2000). Those researchers considered the following research issues relating to collaborative design systems: multimedia engineering documentation, message and annotation organization, negotiation/constraint management, design, visualization, interfaces, and web communication and navigation among agents.

In the second focus area, research is more focused on efficient tools for product design and manufacturing collaboration using networked computers in a distributed environment. The importance of design collaboration has gained the attention of industry as well (e.g., the National Science Foundation (NSF) Workshop, 2000; FIPER; OneSpace; Windchill; Smarteam; Teamcenter). Meanwhile, the integration of distributed environment for product designers and manufacturers has been studied by many academic research groups (e.g., Wagner et al., 1997; Kao and Lin, 1998; Chui and Wright, 1999; Larson and Cheng, 2000; Qiang et al., 2001). Some research utilizes middleware technologies for communication in the areas of feature modeling, feature recognition, and design composition (e.g., Han and Requicha, 1998; Lee et al., 1999; Abrahamson et al., 2000).

Instead of looking at various engineering tools from the viewpoint of traditional computer hardware and software, *e*-design focuses on the service implications of those tools from a more abstract level. This approach assures the openness of the collaborative engineering environment. The Internet is no longer a simple network of computers; in fact, it is more than possible that the concept of "computer" will vanish in the future. From an applications perspective, the Internet is a network of potential services. Functional views of services at different levels need to be clearly defined to form an interoperable, Internet-based, distributed engineering environment. The information supply chain (from customers to product vendors and makers) is required to deal with many information issues, including the following: (1) life-cycle needs; (2) protection and security; (3) seamless sharing of information across international boundaries; (4) storage/retrieval and data-mining strategies; (5) creation of a knowledge depository; (6) classification in the depository (proprietary, public, and shared) along with the means to deposit information and knowledge into corporate memory; (7) maintaining and representing the interpretation of information for use by downstream applications and processes; (8) information interpretation for consistency; and (9) a record of the reasoning process.

In this chapter, a service-oriented product engineering information infrastructure is described that will make a collaborative design environment cost- and time-effective. In a service-oriented *e*-design environment, customers, designers, manufacturers, suppliers, and other stakeholders can participate in the early stages of product design so as to reduce the new product development cycle time and cost. *e*-Design allows for lean data exchange, instantaneous remote constraint imposition, remote service provider invocation, and active customer participation through direct preference imposition.

Four key areas in an *e*-design paradigm include: *enabling information infrastructure* that will enable interoperability for design tools and for collaboration; *conceptual design tools and design process models* focusing on function-based design; *life cycle, collaborative, multidisciplinary design* for product life cycle; and *virtual prototyping and simulation tools.*

28.4   Service-Oriented Architecture Concept

In service-oriented architecture (SOA), any service can be integrated and shared with components of the architecture. To utilize this architecture, services should be specified from the functional aspect of service providers. To make an existing tool available online or to build a new tool for such a system, services associated with the tool should be defined explicitly. The service transaction among service providers, service consumers, and the service manager within an *e*-design system is illustrated in Figure 28.2. Once a service is registered with an administrative manager, it is then available within the legitimate domain. This is a service publication process. When a service consumer within the system needs a service, the consumer will request a lookup service from the service manager. If the service is available, the service consumer can request it from the service provider with the aid of the service manager. Most importantly, this service triangular relationship can be built dynamically at any time. The service consumer does not know the name, the location, or even the way to invoke the service from the service provider. The collaboration between engineering tools is established and executed based on the characteristics of services that can be provided.

Service publication and lookup are the primary services provided by the service manager. As depicted in Figure 28.3, service publication for service providers includes name publication, catalog publication, and implementation publication. A name publication service is similar to the white-page service provided by telephone companies. Catalog publication service is similar to the yellow-page service: the name and the functional description of the service are published. Implementation publication service is the procedure

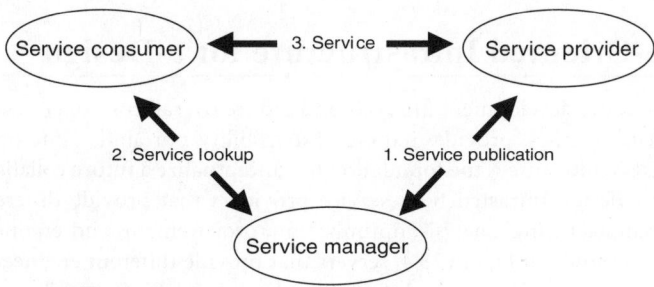

FIGURE 28.2   Service triangular relationship.

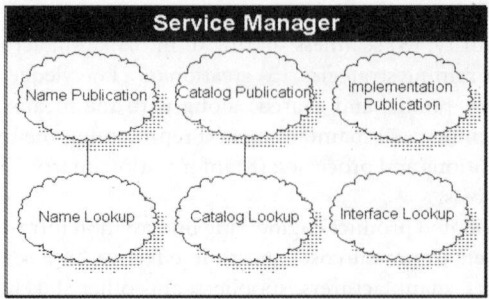

FIGURE 28.3   Services provided by service manager.

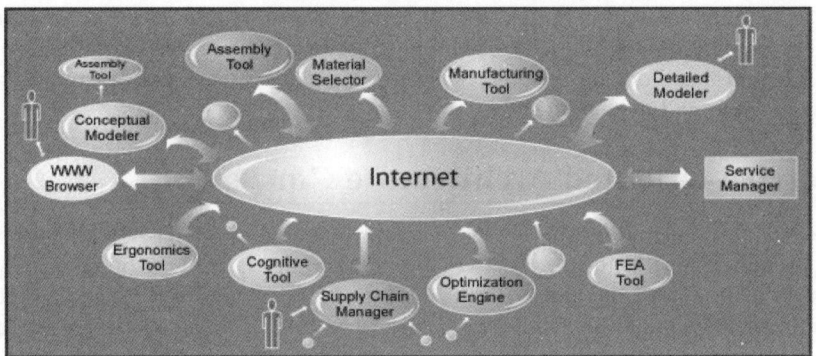

FIGURE 28.4   *e*-Design system architecture.

by which the service provider makes its implementation and invocation of services public so that clients can invoke the service at run-time. Service lookup for service consumers includes name, catalog, and interface lookups. Name lookup service is provided so that consumers can locate the service providers by the service names. Catalog lookup service is for those consumers who need certain services according to their needs and specifications but do not know the names of the providers. Interface lookup service provides a way for consumers to check the protocols of how to invoke the service.

A cost-effective collaborative design environment should consider several issues: (1) security (which includes access control, identification, authentication, and auditing); (2) concurrency and consistency; (3) heterogeneity and transparency (which includes transparency of access, location, performance, and scaling); (4) interprocess communication and naming (which separates physical and logical names to preserve scalability and transparency); and (5) scalability and resource sharing and management. It is also desirable to reduce the coupling and dependency of data, control information, and administrative information.

28.5   Service-Oriented Infrastructure for *e*-Design

In order to shorten product development life cycle and reduce overall cost, openness and ease of collaboration are needed. The openness provides required extensibility, portability, interoperability, and scalability. Service-oriented architecture is the foundation to conceptualize a future collaborative development environment. In the *e*-design infrastructure, service providers that provide different services such as drafting, assembly, manufacturing, analysis, optimization, procurement, and ergonomics can be developed independently. As shown in Figure 28.4, servers that provide different engineering services (which are represented by nodes) are linked by the Internet. Each node in this network may require or provide certain engineering services. Thus, it may be a client or a server for different services, depending on whether it is the recipient or the provider of such a service. The system is open for future expansion and extension. Plug-and-play (PnP) is an important consideration of this structure.

Figure 28.5 illustrates an *e*-Design demonstration system. This system was developed by researchers at the NSF Center for *e*-design and named *Pegasus* (Nnaji et al., 2004b), based on the information infrastructure. The infrastructure includes three major components (i.e., *e*-design service management, *e*-tools, and participating *e*-design stakeholders called *e*-designers). These components are integrated through the Internet and share their resources. The *e*-design service management components provide administrative services (i.e., brokerage service). Each *e*-tool has its own services defined. During service transaction, an *e*-tool may request multiple transactions in collaboration with other *e*-tools. The *e*-design stakeholder (e.g., an enterprise) can have its own collaboration network. The intra enterprise collaboration can be managed by the *e*-design architecture. In addition, the architecture can realize enterprise-to-enterprise collaboration. This service transaction chain should be transparent to the stakeholders. Different components of the *e*-design infrastructure are described in the following sections.

28.5.1 *e*-Design Service Management

The service management builds a bridge between stakeholders and *e*-tools, in which engineering services are defined, queried, dispatched, and protected according to real-time requirements. The management tasks can be performed by a central service provider as well as by distributed service managers. The required services provided by this management role include:

- *Service brokerage.* This allows a transparent and extendable service transaction between stakeholders and *e*-tools, where distributed and specialized computational tools can be developed and interconnected in a modular way and dedicated *e*-tools can serve more effectively.

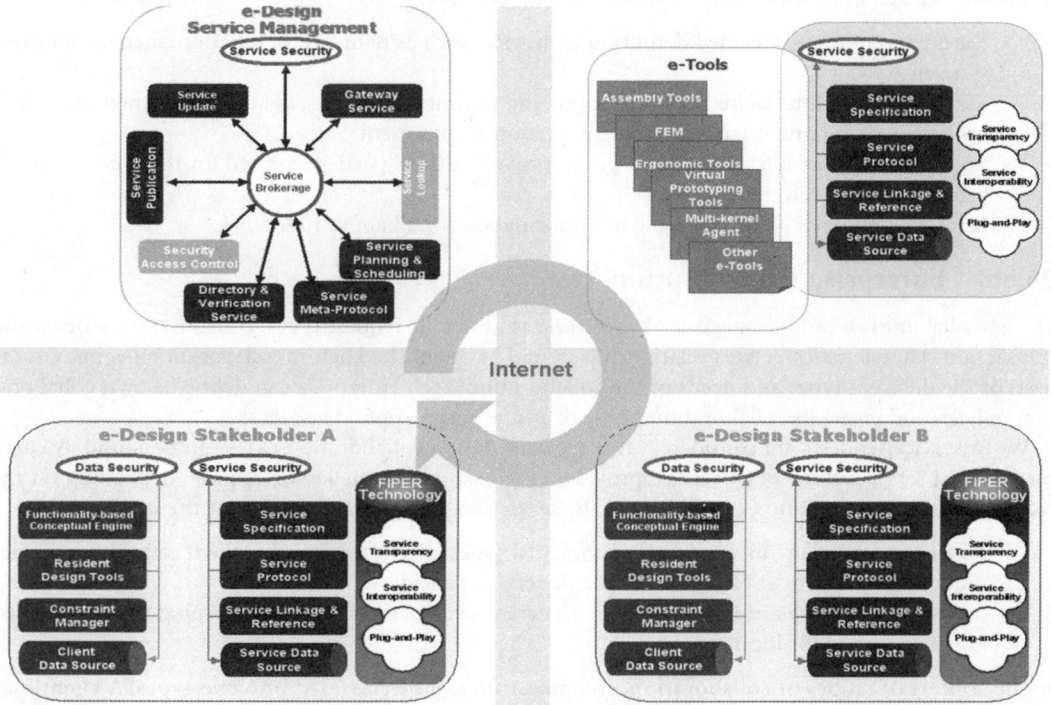

FIGURE 28.5 Service-oriented *e*-design information infrastructure.

- *Service publication.* This allows for *e*-tools or third-party agents to publish engineering services for legitimate clients so that open information flow can assist community communication.
- *Service lookup.* This is a directory service provided for stakeholders or third-party agents to query and retrieve engineering service meta-information.
- *Service subscription.* This provides different levels of client access to service and allows ease of service customization.
- *Security and trust management.* This ensures a secured system for service brokerage and secured information transaction so that collaborators can have a trustworthy and accountable environment.
- *Service certification.* This ensures security and quality of services by independent certification authorities for *e*-tools as well as brokers.
- *Service planning and scheduling.* These perform distributed resource management such as differentiation of services and distribution of jobs among service providers with identical services, and throughput and service cycle time.
- *Service update.* This is service version control and maintains consistency for interface and implementation.
- *Protocol publication* (service meta-protocol). This is a service provided for service brokers. It provides meta-information about service protocols such that brokers can look up and update as necessary.
- *Financial accounting management.* This allows financial compensation transactions for *e*-tools being monitored and managed in the pay-per-use service model.

28.5.2 *e*-Tools

e-Tools include any hardware and software resources providing engineering services. Examples are engineering solvers (e.g., FEA and computational fluid dynamics solvers), information database/knowledge base (e.g., material library, component library), web servers (e.g., ontology server), intelligent agents (e.g., design and assembly advisors), as well as computing servers (e.g., clusters, supercomputer grids). To be seamlessly utilized as *e*-tools, the tools should be implemented in accordance with the following:

- *Service specification.* Detailed definition of service, such as name, type, function, metrics, and version, etc.
- *Service protocol.* Interfacing protocol, input and output parameters, detailed implementation, and brokerage requirement (such as implementation requirement).
- *Service linkage and reference.* Other service providers/third parties required for the service; should be specified and referred.
- *Service data source.* Necessary data and information for all above functions.

28.5.3 Enterprise Collaboration

Design collaboration can be categorized according to different request types, collaboration scopes, and transaction characteristics. A trust relationship should be established before collaboration begins. On the basis of the different types and needs of the collaboration, each enterprise can define its own collaboration policies and create its collaboration network at one agreed-upon trust level.

When service requests are considered, two types of design collaboration cases can be found. A client may request services directly, or service providers themselves may make the request, depending on the requirements of performance, complexity of the service, as well as the frequency of the service:

- *Service request*: For an extremely complex or specialized service, the client submits a service request and receives a result from available service providers.
- *Service provider request*: For a simple or recurrent service, tools (e.g., applets, plug-ins, and agents) are downloaded for local use.

In the context of scopes of collaboration, collaboration can be classified into two types. A client may request services within the enterprise or across the boundary of the enterprise based on the availability of service.

- *Intra-enterprise collaboration*: Collaboration within an enterprise, where the firm's collaboration policy is easy to implement. As shown in Figure 28.6, collaboration networks normally exist within an enterprise.
- *Inter-enterprise collaboration*: Collaboration among enterprises, where collaboration policy should be strictly followed and trust and contract management are required. As a result, more overhead is involved. This collaboration needs to be pursued on the basis of trust policies in a trust track.

A client may request services with different patterns of service cycles and transactions. This affects the specifications, protocols, and performance requirements of services when collaboration is defined.

- *Single-cycle transaction*: In one cycle, final service results can be obtained.
- *Multiple cycle transaction* requires multiple cycles and may use a service manager multiple times.

28.5.4 Enabling Technologies Support

Beyond the SOA, the *e*-design infrastructure should also support enabling technologies to meet the requirements from various design and analysis aspects in collaborative product development, as listed in Table 28.1.

The general information infrastructure should support product data and information management throughout the complete product life cycle. As shown in Figure 28.7, an interoperable, scalable, and secure information flow channel from product conceptualization to physical realization are critical in the *e*-design environment.

28.6 Collaborative Scenarios

To enable the openness of an integrated system that has good extensibility, portability, interoperability, and scalability, SOA is the basic requirement. Based on SOA, engineering services in the networks are available to *e*-design stakeholders based on needs. Engineering services can become pay-per-use utilities, and users are free of software and hardware installation, update, and maintenance. Therefore, engineering tasks can be simplified, and product development cycle time can be reduced. Figure 28.8 illustrates a procedure to make the services available for clients. When a company wants to make its service available for *e*-design stakeholders, the company needs to register its service by providing related information (e.g., company information, service name and category, and transaction protocols). An *e*-design participant or

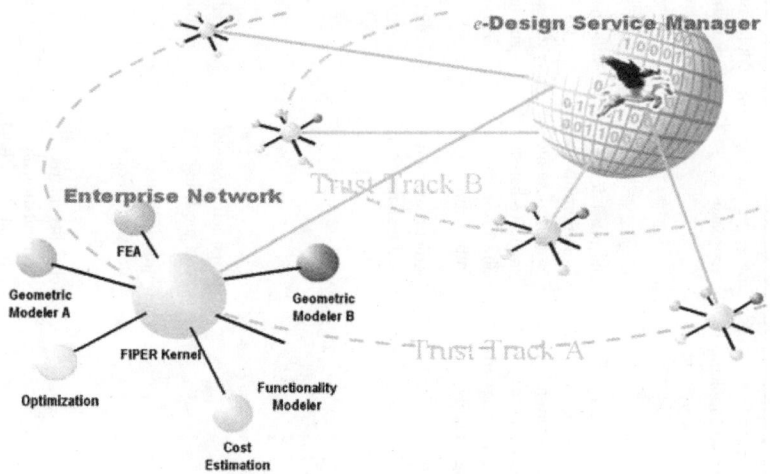

FIGURE 28.6 Intra- and inter-enterprise collaboration.

TABLE 28.1 The Enabling Technologies in Collaborative Product Development

Inter-Operability from CAD to CAD, from CAD to CAE, from CAE to CAE	Multidisciplinary Design in Product Life-cycle Management	Multidisciplinary Constraints and Preferences Capture	Lean Product Data Management, Instant Access and Visualization	Proactive Analysis and Transparency	Virtual Simulation and Prototyping	Security and Trust Management
Standards and protocols supplement	System engineering approach	Multidisciplinary constraints and preferences representation and management	Lean product data modeling	Capacity for process of analysis, etc., "behind the scene" at remote locations without setup	Development of physics-based models	Security and trust modeling
Parameters and constraints representation, nongeometric constraints capturing design intents	Conceptual design (functionality-based, ergonomics- and cognitive-based) direct constraint imposition	Multidisciplinary constraints representation and multi-objective decision-making	Subscription-based hybrid data modeling	High level modeling knowledge capturing in engineering analysis modeling	Virtual reality, product, environment models for system level simulation	Trust for service for distributed, enterprise-wide e-business networks
	Conflict resolution and management	Conflict resolution	Distributed data linkage mode	Domain specific analysis knowledge modeling and ontology, model reusability, adaptability, and interoperability	Models for simulation-based design under uncertainty	Trust issues concerning honestly, openness, reliability, competence and benevolence
	Design activity based cost modeling	Material representation methodology	Multi-views with different levels of details	Open system architecture	Virtual collaboration and sharing	Trust-support infrastructure
	Multi-attributes decision models				VAA, design, and knowledge capturing	
	Reliability matrix, computationally efficient, high-fidelity predictive models				Simulation-based acquisition	
					Real-time visualization	

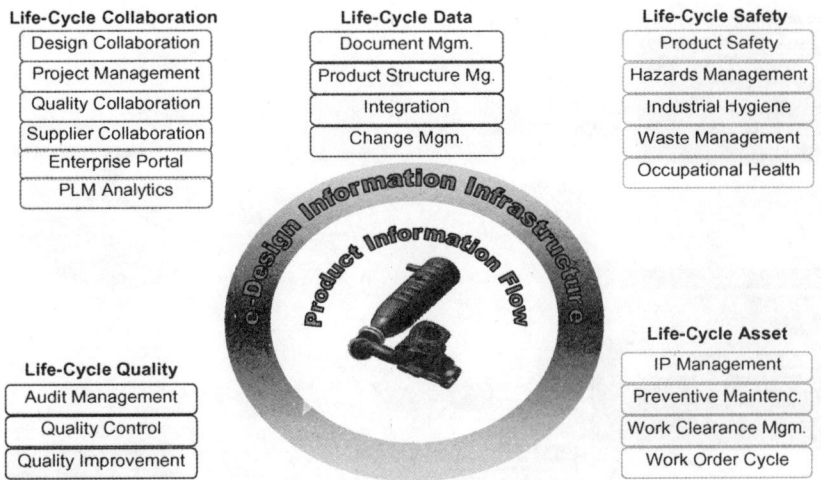

FIGURE 28.7 Product information life-cycle support.

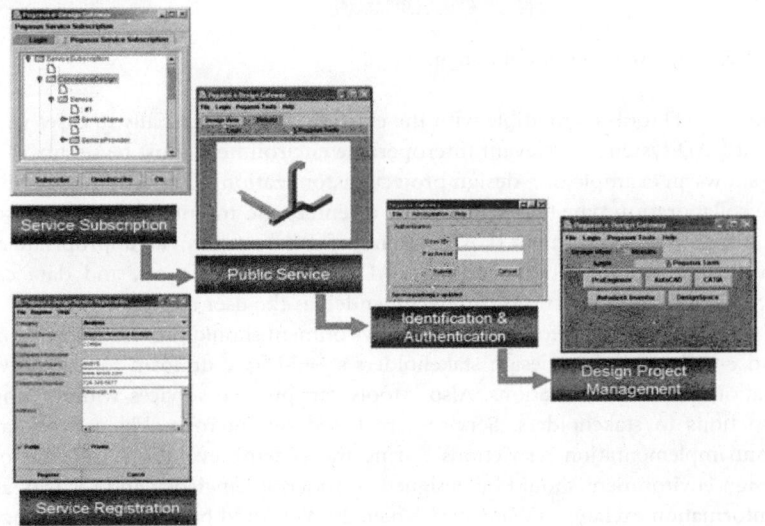

FIGURE 28.8 *e*-Design service set-up.

stakeholder needs to subscribe to a service. Through this service subscription process, services are classified into visible or invisible services. A set of services can be assembled into a service package, such as conceptual design and analysis package. All *e*-design participants can access a public service that provides fundamental functions, such as a graphics viewer. However, to use the subscribed services, the participants must go through the identification and authentication process. The subscribed service will be displayed through a common interface called the *e-design gateway*.

The *e*-design gateway provides a customizable environment for a participant's design project, as is illustrated in Figure 28.9. Each enterprise or participant may require a different design environment because every project is different. For example, if a design project is generative in nature and is to conceptualize design specifications without having a previous design, the project will require conceptual design tools, such as a functionality-based design tool. If the project is based on an existing design, this

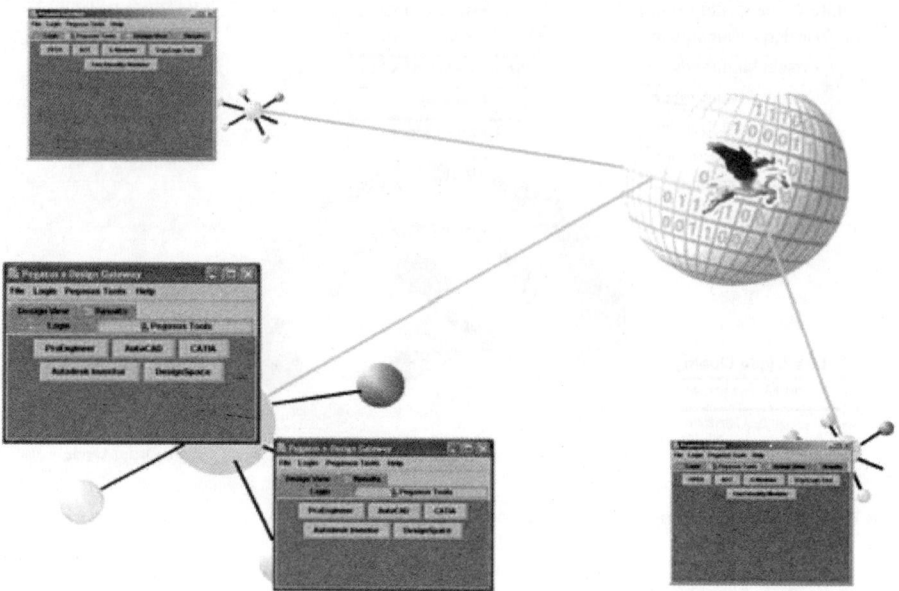

FIGURE 28.9   *e*-Design gateway and customization.

project will require CAD tools compatible with the existing design. Especially in cases where the designs are from different CAD systems, a relevant interoperable environment must be set up.

Figure 28.10 shows an example of *e*-design project customization. By specifying the project type (new or existing), the collaboration type (enterprise-wide or enterprise-to-enterprise), the existence of an initial model, and the system environment (CAD systems of collaborators), an appropriate *e*-design system environment including interoperability among CAD tools, analysis tools, and data can be custom-designed. Additional tools and functions can be appended as the user requests.

For different collaboration scenarios, an *e*-design environment should provide transparent views to both stakeholders and *e*-tools. That is, *e*-design stakeholders should have uniform functional views of *e*-tools without notification of physical locations. Also *e*-tools can provide services without imposing system-dependent restrictions to stakeholders. Services are based on interoperable network and application protocols without implementation restrictions during the system's evolution. Information flow process within the *e*-design environment should be designed considering efficiency and security as well as transparency. Lean information exchange (Wang and Nnaji, 2004) should be supported among heterogeneous tools considering the limitation of communication bandwidth. Protection of intellectual property (Wang et al., 2004a, 2004b) is vital for building a trustworthy cyberspace for product development.

28.7   Case Study: Virtual Assembly Analysis for e-Design and Realization Environment

In this section, an innovative virtual assembly analysis (VAA) process is introduced as a case study of an *e*-design and realization environment. Unlike the current sequential design process used for verifying an assembly design (AsD), the VAA process integrates AsD and analysis in collaborative product-design. In addition, VAA components are developed to predict the various effects of joining in the actual AsD stage. The information obtained from the VAA process can guide designers to make appropriate design decisions in the early stages of AsD. Virtual assembly analysis helps the designer to generate an AsD for joining and can eliminate the time-consuming feedback processes between the AsD process and the assembly analysis process. Previous work has largely focused on assembly modeling and assembly process planning,

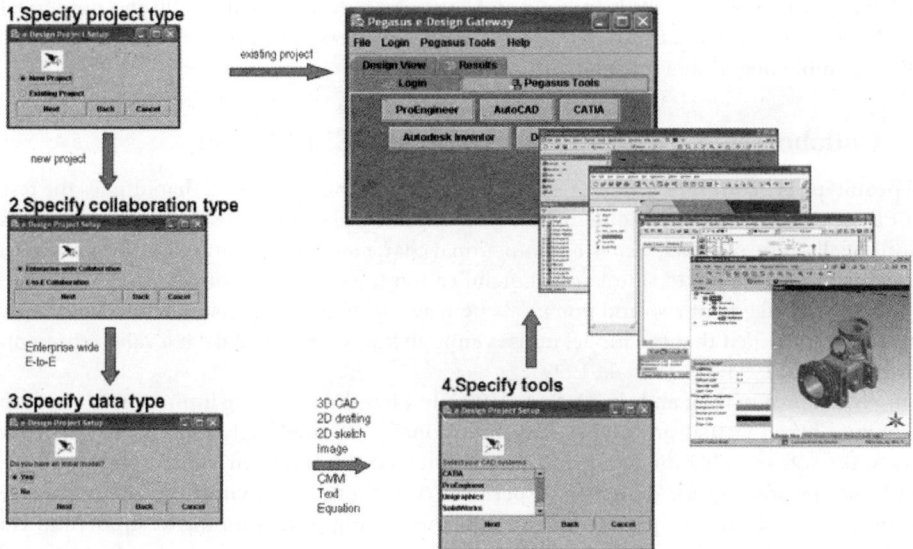

FIGURE 28.10 *e*-Design project customization.

without considering the assembly operations and their effects in a collaborative design environment. The developed VAA framework provides a concurrent environment for designers to predict physical effects transparently and remotely. The captured physical effects of assembly operations provide information critical to realizing an Internet-based collaborative AsD environment.

28.7.1 Assembly and Joining Operations

An assembly is a collection of manufactured parts brought together by assembly operations to perform one or more of several primary functions. Assembly operation is defined as the process or series of acts involved in the actual realization of an assembly. Joining finalizes the assembly operation and generates joints. Messler (1993) divided the primary functions into three categories: structural, mechanical, and electrical. To enable material and structural optimization, determining appropriate joining methods and joint design is critical. This can provide additional benefits in terms of assemblability, joinability, and quality and damage tolerance by changing properties along a potential crack path and disrupting and arresting crack propagation. Local joints should be compatible with the overall structure design. For example, if a deformation effect of a weld joint on a metal frame is propagated onto an automobile windshield area, it can result in a fitting distortion problem between the window and the metal frame (Nnaji et al., 2004a). Understanding the physical effects of each assembly operation is very important for generating an appropriate joint design for an assembly.

This section discusses the role of VAA in an *e*-design environment by using selected joining processes (e.g., arc welding and riveting). In the arc-welding process, owing to the highly localized transient heat input from arc welding, considerable residual stresses and deformations, such as welding distortion, welding shrinkage, and welding warping, occur during heating and cooling in the welding cycle (Masubuchi, 1980; Moon and Na, 1997). A common method of permanent or semipermanent mechanical joining is riveting. Thousands of rivets may be used in the construction and assembly of many objects, such as airplanes, ships, and automobiles. Installing a rivet consists of placing a rivet in a hole and deforming the end of its shank by upsetting (heading). Sufficient compressive elastic energy must be stored in the components to ensure that the rivet is placed in tension by stress relaxation when the compressive forging pressure is released. The rivet design should be determined by considering the required strength of the assembled joint, the required ductility of the rivet material, and the control of the forging process. There has been much research done to study welding and riveting operations analytically and numerically (e.g., Tsai et al., 1999; Rahman et al., 2000). However, these researchers have not presented a methodology to analyze the assembly operations transparently and concurrently.

28.7.2 Collaborative Virtual Prototyping and Simulation

A virtual prototype is a computerized representation of assembly components that implies the testing and analysis of three dimensional (3D) solid models on computing platforms. It should provide access to representations of all of the physical, visual, and functional characteristics of the actual physical device. It can be subjected to virtual testing to simulate full qualification tests prior to manufacture. This has the benefits of reducing management risk and potential engineering changes associated with a new manufacturing environment, provided that the model utilizes enough knowledge to make it a valid representation of the real world (Pratt, 1994; Chua et al., 1999).

Current virtual prototyping and simulation (VP and S) has the following limitations: (1) complex and accurate physics-based virtual prototypes have traditionally required high-end and expensive workstations to host the software; (2) modeling and simulation software has been located on stand-alone computers with no opportunity for designers to perform collaboration activities; and (3) virtual testing is typically unavailable or must be performed through painstaking programming of simulation characteristics of a virtual prototype. In addition, collaborators are limited as to how much information they can process at once.

To enable true collaborative VP and S, a virtual prototype should be generated with consideration of various product life-cycle aspects and should be shared among distributed design collaborators. To generate a robust virtual prototype, an understanding of assembly geometry and its physical effects is a prerequisite. However, current solid modelers and simulation software are not really advantageous for driving a robust virtual prototype since they provide incomplete product definitions and are not able to act according to the semantics of the information. The reason for this is that traditionally the geometric model was at the center of product models. Of course, geometry is of primary importance in AsD, but the morphological characteristics are consequences of the principle physical processes (e.g., deformation effects of welding) and the design intentions (e.g., joint intent) (Kim et al., 2005a).

28.7.3 Service-Oriented Collaborative Virtual Prototyping and Simulation

An SOA provides a fundamental infrastructure for collaborative VP and S. For collaborative VP and S, the virtual prototype should be generated after consideration of various product life cycle aspects (e.g., functionality, manufacturing, assembly/joining, safety, ergonomics, material, packaging and shipping, and disassembly/recycling) and should be shared among distributed/remote design collaborators. Current virtual prototype models cannot fully capture multidisciplinary data/information from collaborators or engineering tools in a distributed and heterogeneous collaboration environment. Also, to represent the quality of a product, the virtual prototype should be capable of capturing realistic manufacturing situations. Figure 28.11 illustrates the concept of collaborative VP and S and virtual model generation based on SOA. As shown in the figure, the information about different aspects of a virtual prototype model can be linked so that *e*-design collaborators/*e*-tools of various disciplines can access and manipulate it. The collaborative virtual prototyping architecture provides methods for sharing data/information between different collaborators/*e*-tools, allowing concurrent access to a virtual prototype model and moving objects through the Internet.

28.7.4 Virtual Assembly Analysis

The current design practice and analysis for verifying a design concept is usually performed after selecting a final design concept. Prediction of various effects corresponding to specified assembly processes in up-front design is critical to understanding the performance of an assembly. Virtual assembly analysis is a transparent and remote virtual simulation and testing paradigm utilized in a service-oriented collaborative environment. A VAA tool embedded in the AsD process can be used to represent an assembly and imply the physical effects of a joint. Figure 28.12 illustrates the concept of VAA. A designer who participates in service-oriented collaborative design can request analysis services through the Internet/intranet. An analysis service provider solves the analysis problem requested and provides the results to the designer. This VAA process is embedded in the distributed AsD environment and can guide designers to make appropriate design decisions. It generates an AsD for joining and eliminates the time-consuming feedback processes between the AsD and analysis processes.

28.7.4.1 Service-Oriented VAA Architecture and Components

To realize VAA in an SOA, an appropriate VAA service triangular relationship should be developed. In this service triangular relationship, each analysis service provider has its own service defined and published by the service manager. For example, an AsD engine and VAA tool provide the services of assembly functional specification, engineering relations construction, and design presentation to end-users. Many third-party analysis solvers can serve as the analysis service provider; the ANSYS solver provides the services of structural nonlinearities, heat transfer, dynamics, electromagnetic analyses, etc. During the process of service, one service provider may require services from other service providers. The first service provider will then send a service request to the providers who provide these additional services. This service chain action should be transparent to the end consumer. For instance, when a design engineer completes the design of two parts, he/she may want to build an assembly model based on the part models. The detailed modeler then calls the assembly procedure. When the assembly model is finished, the design engineer may want to do further mechanical analysis of the assembled parts by calling the service of an FEA tool through the VAA tool. The locations of various service providers are not known until run-time; the relation between the service consumers and the service providers is built dynamically. Figure 28.13

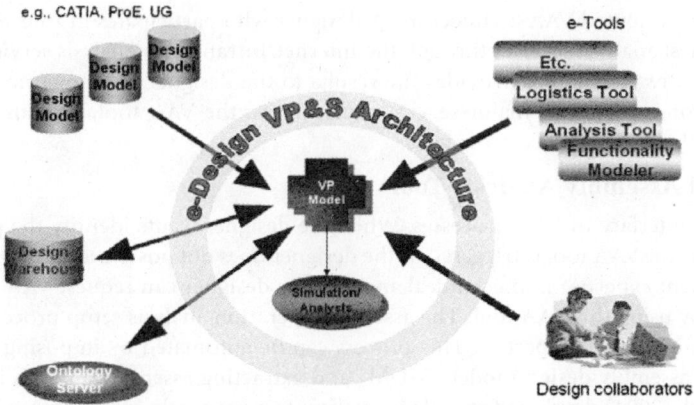

FIGURE 28.11 Collaborative virtual prototype model generation.

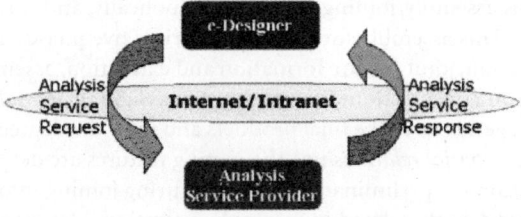

FIGURE 28.12 Virtual assembly analysis.

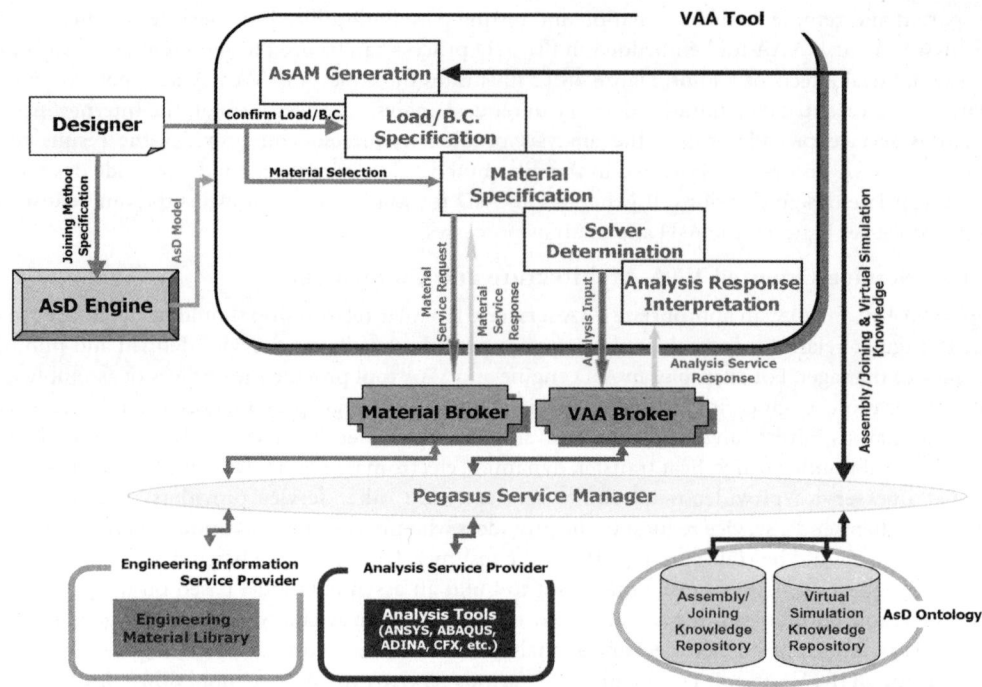

FIGURE 28.13   Service-oriented VAA architecture.

illustrates the service-oriented VAA architecture. A designer who participates in Internet-based collaborative AsD can request analysis services through the Internet/Intranet. An analysis service provider solves the analysis problem requested and provides the results to the designer. As shown in Figure 28.13, the VAA architecture consists of four major service components: the VAA tool, Pegasus service manager, *e*-design service brokers, and service providers.

28.7.4.2   Virtual Assembly Analysis Tool

The VAA tool is an interface of VAA processes. When the designer wants identify the physical effects of the specified joining, the VAA tool is triggered. If the designer does not possess any analysis tools in house, and/or has insufficient expertise in mechanical analysis, the designer can request VAA services remotely and transparently by using this VAA tool. The assembly operation analysis setup process is cumbersome and requires a certain level of expertise. This process can be automated by imposing assembly/joining information on an assembly design model (AsDM) and extracting assembly analysis information from the AsDM. Kim et al. (2004) developed an AsD formalism to persistently capture assembly/joining information in collaborative AsD. The following subsection briefly describes their AsD formalism.

28.7.4.3   Assembly Design Formalism and Assembly Design Model Generation

The AsD formalism specifies assembly/joining relations symbolically, and it is used as the mechanism to perform product AsD tasks. This assembly formalism comprises five phases: spatial relationship specification, mating feature extraction, joint feature formation and extraction, assembly feature formation, and assembly-engineering relation construction. By interactively assigning *spatial relationships*, the designer can assemble components together to make final products and infer the degrees of freedom remaining on the components. In assigning *spatial relationships*, the mating features are defined and extracted from the parts. Mating feature extraction is a preliminary step for capturing joining information. This process provides geometric information directly related to assembly operation. However, the mating feature is not sufficient to represent a joining operation. The joint feature captures the information of actual joining operations. The designer can specify specific joining methods and constraints, such as welding conditions and fixture locations in joint features. After joint features are generated, assembly features are formatted. The purpose of assembly feature formation is to group the mating features and joint features together and thus integrate the data embedded at the component design stage with new assembly information for subsequent processes such as assembly violation detection and process planning. Having designated spatial relationships, mating features, and joint features, the system can then trace back to the component design stage to determine from which design features these mating features originate and what their design specifications are. From the generated assembly features, assembly-engineering relations, including assembly/joining relations, are automatically extracted, and mating bonds (MBs) are generated. An MB is a data structure representing a mating pair and its mating conditions. Assembly-engineering relations of the entire assembly are constructed on the basis of assembly features after the spatial relationships and joining relationships between components are specified. The MBs and an assembly relation model are used to represent the engineering relationships of the entire structure. A detailed description can be found in Kim et al. (2004).

In later publications, Kim et al. (2005b, 2005c) extended the AsD formalism to represent AsD joining information and constraints explicitly using ontology. Their approach is to represent all AsD concepts explicitly and in a universally acceptable manner. By relating concepts through ontology technology rather than just defining data syntax (e.g., via XML), assembly/joining concepts can be captured in their entirety and extended as necessary. Furthermore, the higher semantic richness of ontologies allows computers to infer additional assembly/joining knowledge and make that knowledge available to AsD decision-makers. By using ontology technology, AsD constraints can be represented in a standard manner regardless of geometry file formats. Such representation will significantly improve integrated and collaborative assembly development processes, including VP and S. Lastly, given that knowledge is captured in a standard way through the use of ontology, it can also be retrieved, shared, and reused during collaboration. This AsD ontology forms an assembly knowledge repository.

From the AsD model, the VAA tool automatically generates an assembly analysis model (AsAM) that includes the analysis variables, such as environmental variables, loading/boundary conditions, and material properties.

28.7.4.4 Assembly Analysis Model Generation

To integrate AsD and assembly operation analysis, the AsDMs should be translated to an AsAM. To perform VAA, the assembly/joining information necessary to assembly operation analysis can be extracted using an assembly-analysis solution model (AASM) to explain physical phenomena based upon the assembly/joining information. The AASM is an implantation of mapping functions (Ω) of AsD and joining analysis. It translates an AsDM to an AsAM: $_{AsDM}\Omega_{AsAM}$. Information essential for an assembly analysis is extracted from the AsD model, which is generated by the AsD formalism.

The material of the assembly components from the joint features is translated to the material property for AsAM. Through this mapping, the material property for the specified material name is automatically assigned from a material library. If the resident material library does not have the information about the specified material, the material service can be invoked through the SOA; this way, a designer does not need to hold all material information in-house.

Heat input on a weldline, which is essential information to perform a weld analysis, can be calculated based upon assembly operation information, such as welding conditions (e.g., amperage, voltage, and welding speed) and material properties. Deposition of weld metal is simulated by defining the weld elements at elevated weld deposition temperatures. All other nodes are defined at the ambient temperature as the initial temperature field.

According to the assembly model information and assembly-engineering information, additional geometric features, such as a weld bead for welded joints and a rivet for riveted joints, can be generated for detailed joint modeling. This detailed joint modeling provides realistic representation for engineering analyses. The geometric configuration of the joint geometric features can be determined automatically from the assembly model information and assembly-engineering information. For rivet joints, the designer specifies the location, head type, and radius of the rivet, and the AsD model contains the information. For welded joints, the cross-sectional area of the weld bead can be determined from the existing theoretical relationships between the welding condition and the material properties imposed in the AsD model.

28.7.4.5 Pegasus Service Manager

The Pegasus service manager collaborates with third-party analysis servers (service providers) such as ANSYS to achieve the VAA process. Currently, the VAA architecture is implemented by common object request broker architecture (CORBA) (Siegel, 2000), which is an architecture and specification for creating, distributing, and managing distributed program objects in a network. It allows programs at different locations and developed by different vendors to communicate in a network through an "interface broker." An object request broker (ORB) acts as a "broker" between a client request for a service from a distributed object or component and the completion of that request. The ORB allows a client to request services from a server program or object without having to understand where the server is in a distributed network or what the interface to the server program looks like. CORBA serves as a bond to integrate the whole system and provides good features of openness for collaborative computation. The components in the distributed system have peer-to-peer relationships with each other. From the end-users' point of view, distributing application components between clients and servers does not change the look and feel of any single application, meaning that the system provides end-users with a single system image.

28.7.4.6 *e*-Design Brokers

e-Design brokers handle service invocation and service result conveyance through SOA. The brokers reside in local sites. Each client, such as the VAA tool, and each service provider need the brokers to request or register service. The VAA tool can request the services by invoking these service brokers with relevant service inputs, such as analysis input files and material names. It minimizes the code modification of a service requesting system and provides PnP capability. Before the VAA process, the analysis service providers register their service through an *e*-design broker at each server site. When VAA service is requested by an *e*-designer, the VAA tool sends a request with an analysis input to the *e*-design broker at the client site, and the *e*-design broker conveys the request to the Pegasus service manager. After an analysis result is obtained from the analysis service provider, the Pegasus service manager informs the client's *e*-design broker and conveys the result to the *e*-designer.

28.7.4.7 VAA Service Providers

The Pegasus service manager and the service providers play key roles in the VAA service chain management. The Pegasus service manager allocates service resources according to service consumers' demand and service providers' capability and capacity, while service providers respond with the requested service.

In this case study, two types of service providers are considered: material service providers and analysis service providers. A specialized material service provider can provide the material properties, which are usually too cumbersome to store in the assembly designer's site. The *e*-designer can request certain material properties from the engineering material service provider by specifying the material name or certain material specifications. Any available engineering material library can provide relevant material properties to the client. To perform VAA to predict the physical effects of the joining, FEA tools such as ANSYS, ADINA, and ABAQUS can provide various FEA services. Generally, FEA tools allow certain command-based external analysis inputs. Depending upon the FEA tools and analysis types, different sets of commands and analysis procedures are needed. Appropriate analysis procedures, including specific analysis commands, can be provided from available analysis service providers through an analysis procedure service. In this case study, typical analysis procedures considering the characteristics of joining methods are investigated, and appropriate analysis procedures are predetermined. Analysis service providers provide analysis procedure templates based upon the analysis procedures. The researchers at the NSF Center for *e*-design are developing the assembly analysis knowledge repositories based on ontology technology, and these repositories will be integrated into AsAM.

28.7.5 Implementations

This subsection describes VAA procedures and implemented VAA system components. This VAA process predicts the physical effects of joining processes in which the VAA tool is embedded into AsD processes in collaborative product design environments.

VAA for assembly operations requires specific analysis methodology and procedures. As a case study, a thermo structural analysis is used to understand the thermal and structural behavior of the welding operation. In addition, structural analysis is employed to predict structural phenomena of the riveting operation. To enable VAA for specific joining processes, proper analysis procedures must be preinvestigated and built into an analysis procedure library.

In this case study, the VAA tool is implemented in the ANSYS Workbench environment of ANSYS, Inc. The ANSYS solver is employed as the analysis service provider. Engineering material information is represented in XML format in the material database. The Pegasus service manager is implemented in Java. e-Design brokers are implemented in C++. IONA's ORBacus implementation of CORBA is used in the service architecture.

Figure 28.14 illustrates the transaction flow of services for VAA. Detailed processes are described below. The numbers in the figure stand for the index of each process.

Step 1: e-Designers can exchange product data, such as AsD models, and select assembly components through the product data sharing (PDS) service (①).

Step 2: The selected assembly components are loaded in an AsD engine to generate joints (②). The system integrator, e-designer 1, can specify joining methods on the assembly (③).

Step 3: When the e-designer wants to know physical effects of the specified joining, the VAA tool is triggered and a newly generated AsD model is sent to the VAA tool (④). From the AsD model, the VAA tool extracts analysis information and generates an AsAM. The designer can add additional loading and boundary conditions (⑤).

Step 4: If the material specified in the AsD model does not exist in a local database, the material property is obtained from remote material libraries though the SOA. The designer can also request a certain

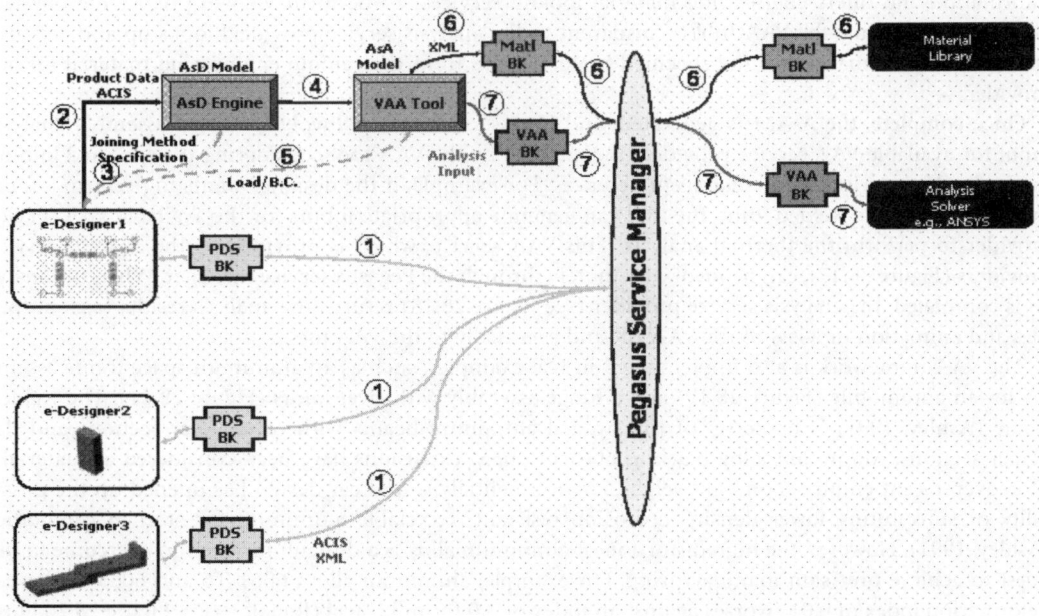

FIGURE 28.14   Service transactions in VAA.

material to be entered in the VAA tool. The VAA tool dynamically requests the service by invoking the material service broker (Mtl BK) with relevant material information (⑥).

Step 5: Once the VAA inputs are ready, the VAA tool invokes the VAA service broker (VAA BK) with the VAA input. When the analysis is completed, the analysis service provider returns the analysis results to the VAA tool (⑦).

As shown in Figure 28.14, PDS, material, and VAA services are accomplished through service brokers (i.e., PDS, material, and VAA brokers). These service brokers at the user's site handle service invocation and service result conveyance through the SOA. The VAA tool can request the services by invoking these service brokers with relevant service inputs, such as analysis input files and material names. Figure 28.14 also illustrates how the service brokers are used in the service architecture. For example, *e*-designers can exchange product data, such as AsD models, through PDS service and select assembly components by using the PDS service.

The developed VAA tool (see Figure 28.15(a)) is used as an interface to capture assembly and joining specifications. When the designer wants to know the physical effects of the joining, the VAA tool is triggered to interpret the AsD model. From the AsD models, the VAA tool automatically generates an AsAM including the analysis variables, such as environmental variables (e.g., as convection and fixed support); loading condition (e.g., given temperature and force/pressure); and material properties such as Young's modulus, specific heat, and thermal expansion coefficient (see Figure 28.15(b)). The joining parameters (e.g., welding conditions and welding speed) are extracted from the AsD model, and relevant analysis variables are obtained and assigned to the AsAM. For example, the degrees of freedom at fixture locations are restricted as fixed supports. Temperature at the specified weld seam is estimated from the welding condition. Through this analysis setup process, the designer can impose additional analysis constraints on the AsAM in the VAA tool.

The Pegasus service manager allocates service resources according to service consumers' demand and service providers' capability and capacity. Figure 28.16 shows an implementation of the Pegasus service manager.

A specialized material service provider can provide the material properties, which are usually too cumbersome to store in the assembly designer's site. Here, an engineering material service provider has this information and offers engineering material lookup services. To perform VAA to predict the physical effects of the joining, the VAA tool (transparent to the analysis service provider) looks up and acquires the material information on the specified material type from the remote engineering material service provider.

Once complete, the AsAM is generated, and VAA service can be invoked. Virtual assembly analysis input for available VAA service providers is generated by the VAA tool considering the specified joining method's characteristics and analysis preferences. For example, if the designer wants to perform a thermal analysis for the welded joint, the tool can generate appropriate inputs for the available VAA service provider to perform the thermal analysis. This VAA service, which is provided by the analysis service providers (Figure 28.17), can be invoked remotely through the Pegasus service manager. When the analysis is completed, the analysis service provider (see Figure 28.18) returns the analysis results (e.g., output files and animation movies) to the VAA BK and eventually to the VAA tool.

The VAA framework is also implemented in real examples, such as an aluminum space frame assembly for an automobile (Figure 28.19) and a hinge assembly with rivet joints (Figure 28.21). The welded frame (Ashley, 1991) is made up of thin-walled aluminum beams with rectangular sections and flat planer sections. Aluminum alloy (such as 6061 or 6063) extrusions have been considered as materials. Moreover, recent emphasis on lightweight, environmentally sound car design has opened up the possibility of substituting lower density, corrosion-resistant, recyclable aluminum for steel in car bodies (Ashley, 1991). However, the high distortion of aluminum alloy is a difficult problem to overcome in the quest to achieve precision manufacturing. For example, aluminum alloys 6061-T6 and 6063-T6 have a deformation index of 0.01 (worse) against an index of 1.0 for mild steel (Radaj, 1992). Figure 28.20 illustrates a simple car frame model in the VAA tool and the VAA result. The result clearly shows deformation of this structure and stresses concentrated at the welded joint. Deformation beyond allowable tolerance

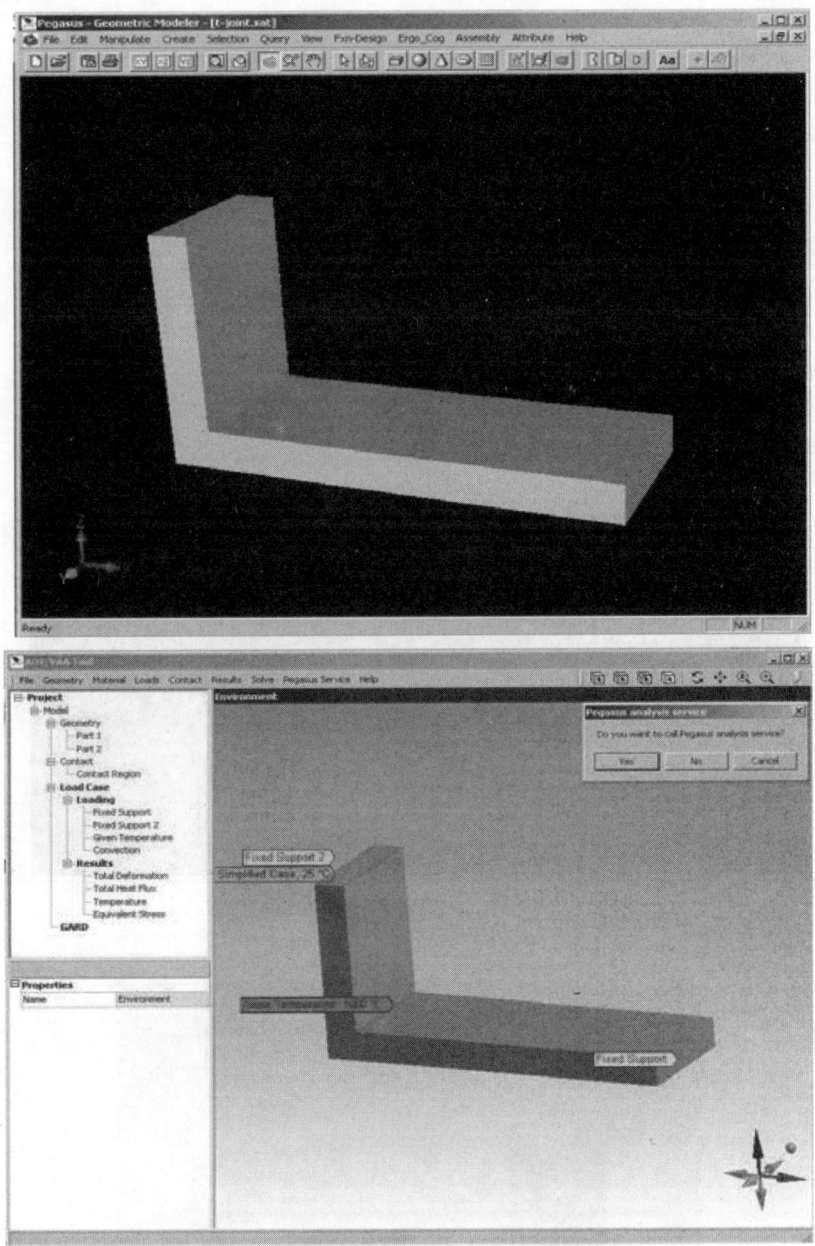

FIGURE 28.15   Assembly models for VAA. (a) AsD model. (b) AsAM.

will be indicated easily. Based on this result, the designer can make a decision on whether this joining method is feasible within this nominal geometry.

As another illustration, a hinge assembly with rivet joints is used. The material of the hinge is structural steel. Figure 28.21 shows a structural VAA result for the hinge joint. Based upon this result, although stresses are concentrated on the top component and the stress affects one of the rivets, the designer can clearly see that this joint is robust in this specific test environment.

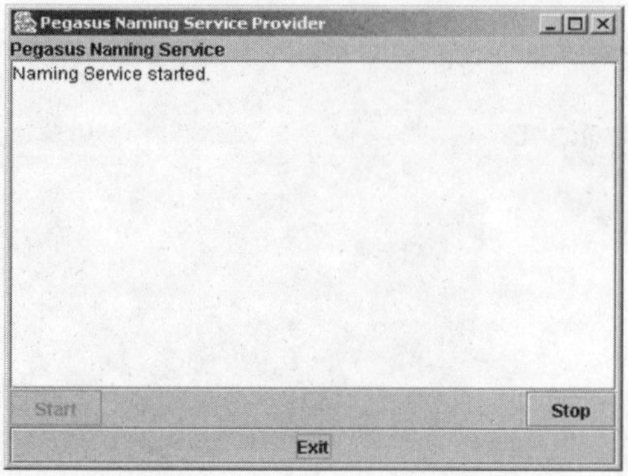

FIGURE 28.16 Pegasus service manager.

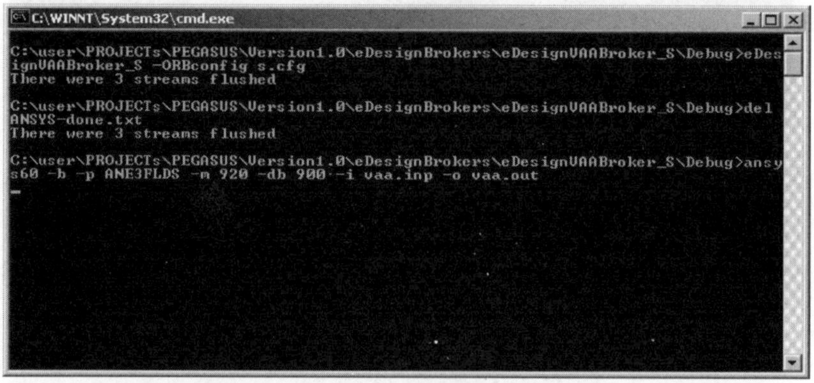

FIGURE 28.17 VAA service provider.

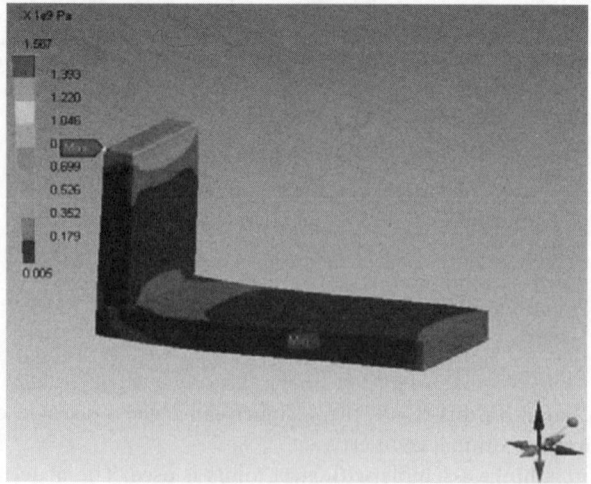

FIGURE 28.18 Equivalent stress and deformation obtained from VAA service.

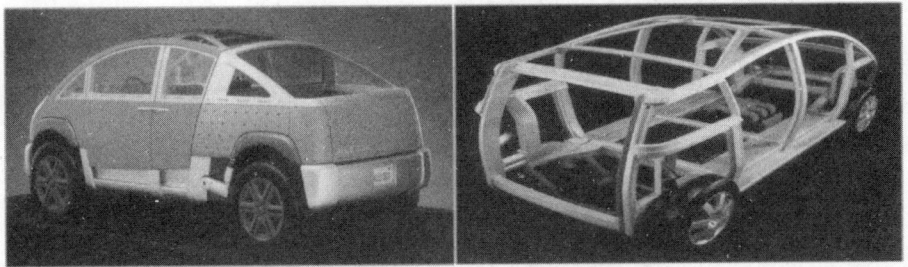

FIGURE 28.19 Aluminum concept car and body frames (Buchholz, 1999).

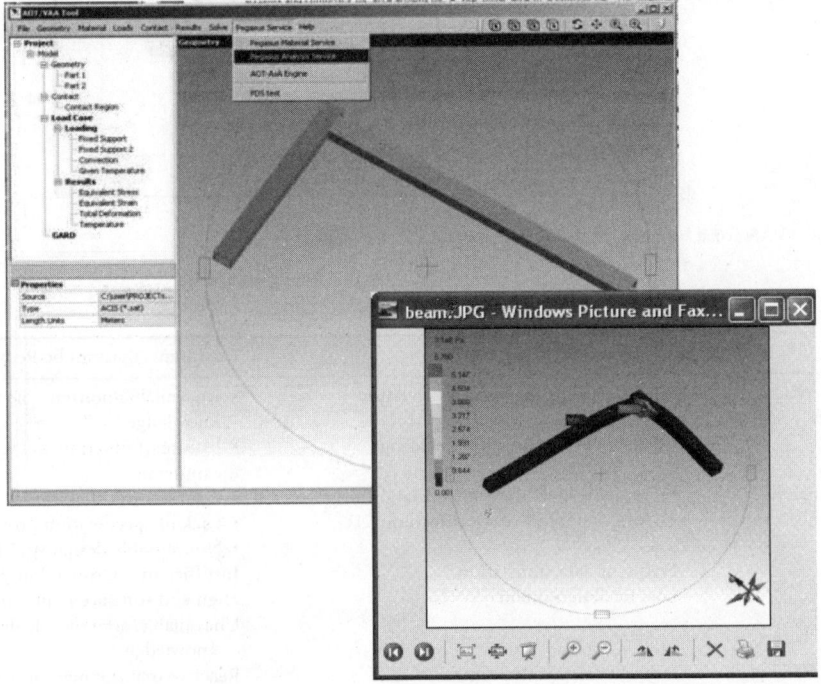

FIGURE 28.20 VAA for a welded extruded frame.

28.8 Benefits of *e*-Design for Product Life-Cycle Management

The goal of *e*-design is to meet customers' requirements in product mass customization. Service-oriented *e*-design infrastructure provides opportunities to manufacturers to reduce cost and time-to-market, as listed in Table 28.2.

The *e*-design infrastructure allows a dynamic engineering environment to be established for integrated product development. Manufacturers can outsource engineering services to specialized service providers in a virtual environment. Traditional computational tools for engineering activities such as design, analysis, simulation, optimization, and data management, can become utilities that are readily accessible for enterprises of different sizes. The infrastructure will become indispensable for the newly networked engineering and business paradigm.

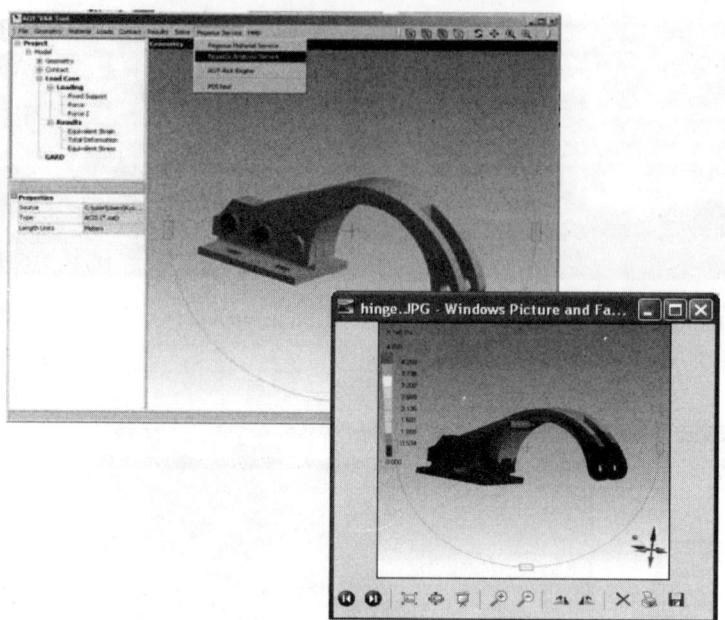

FIGURE 28.21   VAA for a hinge with three rivets.

TABLE 28.2   Costs Needed to be Reduced in Product Life-Cycle Activities

Activities	Time that can be Saved	Cost Items that can be Reduced
Concept development Market analysis	• Lack of tools for conceptualization • Lack of communication/sharing	• Unavailable/nonreusable design knowledge • Failure of functionality requirement capturing
Requirement analysis Specifications	• Frequent specification translation • Human process of specification	• Conflicts in collaboration • Lack of specification protocol • Nonreusable design specification
Design	Frequent data translation Loss of design intent	Insufficient interoperability High end software requirement Unavailable/nonreusable design knowledge
Test	Long simulation/test time Long resource/knowledge acquisition time	Reactive management of design failures Generation of physical prototypes High end software/hardware requirement
Full production	Long Manufacturing preparation time	Reactive management of production failure
Marketing and distribution Maintenance Recycling	• Lack of collaboration • Lack of knowledge transfer	• Lack of design information/intent • Lack of environmental consciousness

28.9   Conclusion

This chapter has presented the overall structure of the new collaborative design paradigm, *e*-design, which enables the seamless and dynamic integration of distributed design objects and engineering services over the Internet. The concept of SOA was described for the future engineering design environment, which allows cost-effective distributed design collaboration. The openness of such infrastructure provides required extensibility, portability, interoperability, and scalability for a ubiquitous engineering service environment, which transparently integrates overall product life-cycle management activities. This chapter further detailed the requirements for openness and for enabling technologies to realize the *e*-design environment. The subscription-based service chain structure that allows for customizable engineering service utility was presented. Functional modules to ensure transparency of service transactions were also shown. Finally, the *e*-design concept was illustrated by VAA with prototypes of service manager, transaction agents, and several service providers.

References

Abrahamson, S., Wallace, D., Senin, N., and Sferro, P., Integrated design in a service marketplace, *Comput. Aided Des.*, 32, 97–107, 2000.

Ashley, S., Contour: the shape of cars to come? *Mech. Eng.*, 113, 36–44, 1991.

Buchholz, K., Alcoa shows aluminum association its concept vehicle, *Automot. Eng. Int.*, 107, 53–55, 1999.

Chua, C.K., Teh, S.H, and Gay, R.K.L., Rapid prototyping versus virtual prototyping in product design and manufacturing, *Int. J. Adv. Manuf. Technol.*, 15, 597–603, 1999.

Chui, W.H. and Wright, P.K., A WWW computer integrated manufacturing environment for rapid prototyping and education, *Int. J. Comput. Integr. Manuf.*, 12, 54–60, 1999.

Dong, A. and Agogino, A.M., Managing design information in enterprise-wide CAD using 'smart drawings', *Comput. Aided Des.*, 30, 425–435, 1998.

FIPER, Engineous Software, http://www.engineous.com.

Fuh, J.Y.H. and Li, W.D., Advances in collaborative CAD: The-state-of-the art, *Comput. Aided Des.*, 37, 571–581, 2005.

Han, J.H. and Requicha, A.A.G., Modeler-independent feature recognition in a distributed environment, *Comput. Aided Des.*, 30, 453–463, 1998.

Huang, G.Q. and Mak, K.L., Design for manufacturing and assembly on the Internet, *Comput. Ind.*, 38, 17–30, 1999.

Huang, G.Q. and Mak, K.L., WeBid: A web-based framework to support early supplier involvement in new product development, *Robot. Comput. Integr. Manuf.*, 16, 169–179, 2000.

IONA, http://www.iona.com.

Kan, H.Y., Duffy, V.G., and Su, C.J., An internet virtual reality collaborative environment for effective product design, *Comput. Ind.*, 45, 197–213, 2001.

Kao, Y.C. and Lin, G.C.I., Development of a collaborative CAD/CAM system, *Robot. Comput. Integr. Manuf.*, 14, 55–68, 1998.

Kim, K.Y., Manley, D. G., Nnaji, B. O., and Lovell, M. R., Framework and technology for virtual assembly design and analysis, *IIE Annual Conference (IERC)*, Atlanta, GA, May 14–18, 2005a.

Kim, K.Y., Manley, D. G., and Yang, H. J., The role of ontology in collaborative virtual prototyping, *2005 IIE Annual Conference (IERC)*, Atlanta, GA, May 14–18, 2005b.

Kim, K.Y., Manley, D.G., Yang, H.J., and Nnaji, B.O., Ontology-based virtual assembly model for collaborative virtual prototyping and simulation, *International Symposium on Collaborative Technologies and Systems (CTS 2005)*, Saint Louis, MO, May 15–19, 2005c.

Kim, K.Y., Wang, Y., Muogboh, O.S., and Nnaji, B.O., Design formalism for collaborative assembly design, *Comput. Aided Des.*, 36, 849–871, 2004 (the special issue on Distributed CAD).

Kumar, V., Glicksman, J., and Kramer, G.A., A SHAREd web to support design teams, *IEEE Proceedings of the Third Workshop on Enabling Technologies: Infrastructure for collaborative Enterprises*, April 17–19, Morgantown, WV, 1994, pp. 178–182.

Larson, J. and Cheng, H.H., Object-oriented cam design through the internet, *J. Intell. Manuf.*, 11, 515–534, 2000.

Lee, J.Y., Han, S.B., Kim, H., and Park, S.B., Network-centric feature-based modeling, *IEEE Proceedings of the Seventh Pacific Conference on Computer Graphics and Applications*, October 5–7, Seoul, Korea, 1999, pp. 280–289.

Masubuchi, K., *Analysis of Welded Structures: Residual Stresses, Distortion, and their Consequences*, Pergamon Press Inc., New York, 1980, pp. 60–236.

Maxfield, J., Fernando, T., and Dew, P., A distributed virtual environment for concurrent engineering, *IEEE Proceedings on Virtual Reality Annual International Symposium*, March 1–15, Research Triangle Park, NC, 1995, pp. 162–170.

Messler, R. W., *Joining of Advanced Materials*, Butterworth-Heinemann, Oxford, 1993.

Moon, H.S. and Na, S.J., Optimum design based on mathematical model and neural network to predict weld parameters for fillet joints, *J. Manuf. Syst.*, 16, 13–23, 1997.

Nnaji, B.O., Gupta, D., and Kim, K.Y., Welding distortion minimization for an aluminum alloy extruded beam structure using a 2D model, *ASME J. Manuf. Sci. Eng.*, 126, 52–63, 2004a.

Nnaji, B.O, Wang, Y., and Kim, K.Y., Cost-effective product realization – Service-oriented architecture for integrated product life-cycle management, *The 7th IFAC Symposium on Cost Oriented Automation*, June 7–9, 2004, Gatineau/Ottawa, Canada, 2004b.

NSF Workshop, National Science Foundation Workshop on e-Product Design and Realization for Mechanically Engineered Products, University of Pittsburgh, Pittsburgh, PA, October 19–20, 2000.

OneSpace, *CoCreate Corporate*, http://www.cocreate.com.

Pratt, M. J., Virtual prototypes and product models in mechanical engineering, *Proceedinds IFIP WG 5.10 on Virtual Environments and their Applications and Virtual Prototyping*, 1994, pp. 113–128.

Qiang, L., Zhang, Y.F., and Nee, A.Y.C., A distributive and collaborative concurrent product design system through the WWW/Internet, *Int. J. Adv. Manuf. Technol.*, 17, 315–322, 2001.

Radaj, D., *Heat Effects of Welding*, Springer, New York, 1992, pp. 21–23.

Rahman, A. and Bakuckas, J.G., Boundary correction factors for elliptical surface cracks emanating from countersunk rivet holes, *AIAA J.*, 38, 2171–2175, 2000.

Roy, U. and Kodkani, S.S., Product modeling within the framework of the World Wide Web, *IIE Trans.*, 31, 667–677, 1999.

Siegel, J., *CORBA 3: Fundamentals and Programming*, 2nd ed., Wiley, New York, 2000.

Smarteam, *Dassault Systems*, http://www.3ds.com.

Sriram, D. and Logcher, R., The MIT dice project, *IEEE Comput.*, 26, 64–65, 1993.

Teamcenter, *EDS*, http://www.eds.com.

Tsai, C.L., Park, S.C., and Cheng, W.T., Welding distortion of a thin-plate panel structure, *Weld. J.*, 78, 156–165, 1999.

Wagner, R., Castanotto, G., and Goldberg, K., FixtureNet: Interactive computer-aided design via the world wide web, *Int. J. Hum. Comput. Stud.*, 46, 773–788, 1997.

Wang, Y., Ajoku, P.N., and Nnaji, B.O., Scheduled role-based distributed data access control model for data sharing in collaborative design, *Proceedings of the International Symposium on Collaborative Technologies and Systems (CTS2004)*, January 18–23, 2004, San Diego, CA, 2004a, pp. 191–196.

Wang, Y., Ajoku, P.N., and Nnaji, B.O., Distributed data access control for lean product information sharing in collaborative design, *Proceedings of the ASME International Design Engineering Technical Conferences & The Computer and Information in Engineering Conference)*, September 28–October 3, 2004, Salt Lake City, UT, Paper No. DETC2004/CIE-57748, 2004b.

Wang, Y. and Nnaji, B.O., UL-PML: Constraint-enabled distributed design data model, *Int. J. Prod. Res.*, 42, 3743–3763, 2004.

Windchill, *Parametric Technology Corporate*, http://www.ptc.com.

Xue, D. and Xu, Y., Web-based distributed system and database modeling for concurrent design, *Comput. Aided Des.*, 35, 433–452, 2003.

Yang, H. and Xue, D., Recent research on developing web-based manufacturing systems: A review, *Int. J. Prod. Res.*, 41, 3601–3629, 2003.

PART VI

GENERAL APPLICATIONS

29

Generating User Requirements in Project Management

David Ben-Arieh
Kansas State University

Zhifeng Chen
Kansas State University

29.1 Introduction

The management of projects consists of a concentrated effort to accomplish a task that is unique and has a well-defined duration and cost. Most projects have specialized requirements and constraints dictated by the end-user, a fact that makes the clarification of user requirements crucial to the success of any project.

Generating the user requirements has to be a carefully managed process aimed at generating complete, correct, unambiguous, and nonredundant user requirements. This chapter reviews the various methodologies developed in order to generate these requirements. In addition, it presents the classification of projects into eight types and recommends a preferred method for generating user requirements for each project type.

The specification of user requirements is an important part of project management. If user requirements are inadequately or incorrectly defined, the failure to grasp user expectations can only result in the failure of the whole project. As a part of the project requirements, the user usually defines the project objectives, the criteria with which to evaluate the project's success, and constraints such as cost, weight, or schedule. Many times, the user can also clarify the need for the project and delineate the expected major milestones, in addition to the user requirements (Kerzner, 2004).

The correct definition of user requirements is crucial to the success of any project, since these requirements support the definition of the activities and events included within the scope of each project, elucidate the relationships between the activities and events, and describe the attributes of the activities and events. Moreover, the user requirements are necessary for the development of validation criteria for the project.

In effect, the requirements outline the user's expectations of what the system must do. However, the user requirements do not specify the way the system will actually be implemented. *What* is to be done and *how* to do it are not always easy to separate.

Generating the user requirements is the first step in the process of implementing a project. The project life cycle is composed of five stages: conceptual design, advanced development, detailed design, execution, and termination (Shtub et al., 1994). Extracting the user requirements has to take place before the detailed design in this life-cycle description. The Project Management Institute (PMI) defines the first stage in the life cycle of a project as the "initial" or "assessment" phase, which includes the generation of user specifications (PMBok® Guide, 2000). This approach recognizes the fact that the user requirements may shift or evolve with time, and may thus require active management. This requires good communication between the user groups and the project management team, which, in turn, requires established communication channels and a common vocabulary and terminology, as well as a communication protocol between the user and the project management.

In many cases, where the system is very complex, the user requirements may be generated via an iterative process. In such cases, the requirements as received initially are incomplete owing to a lack of knowledge, and are gradually defined. A prototyping process is one approach that allows the user to provide detailed feedback to the project management. Examples of such prototypes might be wind-tunnel models or architecturally scaled models. Prototypes are usually classified into three types: evolutionary, incremental, and throwaway prototypes. The evolutionary model allows communication with the user in order to extract more details as the prototype evolves. The incremental approach takes more pronounced steps toward defining the user requirements. This is useful when a project consists of several subsystems; each increment can then define the requirements for a subsystem. The throwaway approach develops a prototype that is used to confirm the user expectations and then discarded. An example in this category could be a mechanism that is demonstrated to the user via a rapid prototyping machine. This machine quickly and inexpensively converts the design — as understood by the project team — to a physical model, which then is used to confirm that the model agrees with the user intentions.

In real life, it is difficult for project managers to gather complete and accurate system requirements. This difficulty is due to various factors, such as:

- Constraints on humans as information processors and problem solvers; this view is supported by Simon's bounded rationality theory (Simon, 1957).
- The variety and complexity of the information requirements.
- The complex interaction required between the users and the project management.
- Specifying the requirements correctly so that both the user and the project management agree on what are important and challenging tasks.

Failure to reach such an agreement is termed "expectations failure" (Lyytinen and Hirschheim, 1987). This failure mode joins the other three failure types: correspondence failure (the system does not match the goals); process failure (the project overruns the schedule or budget); and interaction failure (the delivered system is not used as intended).

The user requirements need to be documented and formally agreed upon by the user and the project management. The suggested contents of a user-requirements document are as follows:

- *System environment.* The objectives of the system and their degrees of significance as well as the various constraints with their importance should be clearly specified.
- *Conceived subsystems.* For example, the Mars Rover project included the rover itself, the deploying spacecraft, and the landing parachute and inflatable balls subsystems; these should also be clearly specified and agreed upon.

- Planned activities and responses of two types: normal and unplanned or failure mode activities and responses.
- System constraints, such as reliability and human factors.
- Evolution of the project and a reporting system.

Ideally, the iterative process of generating the user requirements is frozen after a few iterations (Rowen, 1990).

In many instances, the project specifications have to be derived from a group of users. This is usually accomplished using two approaches: focus groups and decision support tools (Bias et al., 1993). The process of obtaining user requirements is often concluded with a formal agreement on the project objectives (also termed a project charter by PMI).

According to Davis (1982), there are two levels of requirements for a project. The first one is the overall systems level (also termed the Master Plan). This plan defines the various applications, the boundary and interface between the applications, and the order of producing the applications. The second level is the detailed level of the application. At the application level, the manager needs to define the activities of each application and the design and implementation requirements.

The two levels of requirements can be viewed as a recursive system, or as independent methods being applied to each level separately. Another pair of requirements represents two basic types of application requirements: social (behavioral) and technical. The social requirements specify job design objectives, work organization, individual role assumptions, responsibility assumptions, and organizational policies. The technical requirements specify the outputs, inputs, and process for each activity.

To summarize, the main desired attributes of the requirements are that they should be *correct, useful, complete, consistence, unambiguous, accessible, modifiable, traceable, annotated, verifiable, and reusable* (Gueguen and Chlique, 1993).

29.2 General Approaches Toward Generating User Requirements

Project managers use different fundamental approaches to specify the various aspects of the user requirements. Davis (1982) describes four basic approaches toward requirements specifications:

(1) Direct interaction with the user (termed "asking" strategy)
(2) Deriving requirements from existing projects
(3) Synthesis from previous knowledge
(4) Discovering by experimentation

29.2.1 Direct Interaction with User

In its pure form, the project management team obtains the user requirements exclusively from user groups by directly asking for the requirements. This method is useful only under the assumption that the user has an unbiased and complete view of the requirements, a condition that holds only in the case of very stable and conforming projects.

This approach toward generating user requirements can be implemented using a variety of methods, such as:

(1) *Structured questions.* This method, which is appropriate for a well-defined project that is very well understood by the management team, uses multiple-choice questions.
(2) *Open questions.* Here, the possible responses from the user are not known in advance or cannot be narrowed down to few options. This approach allows the user to specify opinions and qualitative descriptions. This approach requires the project management team to be able to interpret the user responses and formulate the requirements.
(3) *Brainstorming.* This method allows a wider variety of opinions and suggestions and an open flow of ideas. It is useful for nonconventional projects in which new ideas need to be generated by the user.

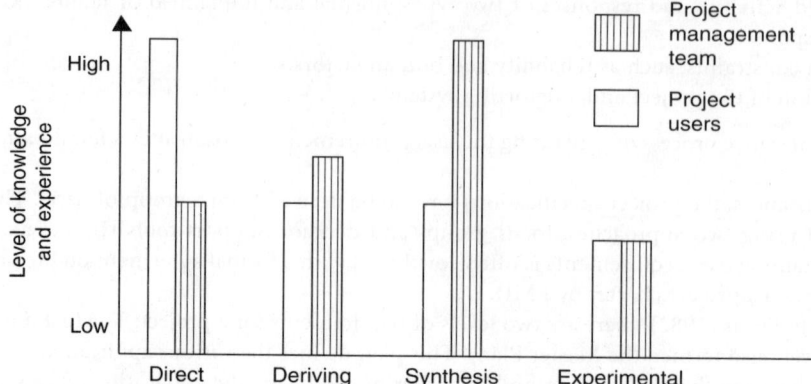

FIGURE 29.1   Relative estimated level of knowledge and experience required by project management team and project users.

(4) *Guided brainstorming.* This more limited version of (3) allows the users to define criteria for selection of a preferred option. This approach is useful when the user and project management have knowledge about the project but still need to prioritize and select some preferred options.

(5) *Group consensus.* This approach, which includes the Delphi method, allows the users to quantify their expectations regarding major project parameters. Such an approach is useful when the user groups need to develop the requirements in more detail, including quantitative estimation of some important parameters.

The *direct interaction* approach requires the highest level of user knowledge and experience, as described in Figure 29.1.

29.2.2   Deriving Requirements from Existing Projects

In this method, requirements are derived from past experience with the same type of project. This approach includes the "case-based reasoning (CBR)" methodology, in which a database of old cases is used to generate pertinent information regarding new projects (see Section 29.4.10). This approach is of limited usefulness when there is a lack of past experience or a bias or shift in the approach of management.

29.2.3   Synthesis from Previous Knowledge

This approach may overcome some of the limitations stated above. Having knowledge about similar projects can be useful in deriving the user requirements. This can be done using various types of analyses of the existing systems (Davis, 1982), such as:

- *Normative analysis.* This analysis uses prescribed or normative requirements. This approach requires tailoring the normative requirements to meet nonstandard expectations.
- *Objectives-based analysis.* This approach concentrates on the objectives of the organization in order to derive the requirements. An example is an objective to improve profitability that will define cost requirements for a project.
- *Critical factor analysis.* An example is the critical success factors, which define the requirements based on the factors that are critical to the success of the organization (Rockart, 1979).
- *Process Analysis.* In this case, the requirements can be derived by observing the process that the project helps to manage or control.
- *Input–process–output analysis.* This is similar to process analysis, with the exception that it analyzes the input to each activity and its outputs. These in turn help define the requirements of the project. Tools such as $IDEF_0$ are useful in this case.

29.2.4   Experimental Approach

This iterative discovery method starts by capturing some initial requirements and implementing a system to provide those requirements. As the users analyze the system, they generate additional requirements. This approach is also described as prototyping or heuristic development (Berrisford and Wetherbe, 1979). This method is useful when there is no well-defined model of the activities, the experience of the users is insufficient to define the requirements, or the users' expectations are evolving.

29.2.5   Information Requirements

The four approaches toward generating user requirements assume different levels of knowledge and maturity of the project objectives. Clearly, the experimental approach intends to develop this knowledge during the requirements-generation process, while direct derivation by the user assumes a high level of stability. This is demonstrated in Figure 29.1(Davis, 1982).

Figure 29.1 compares the level of information and domain knowledge required by the project management and users for the basic four approaches. The figure shows that the lowest amount of knowledge of both users and managers is required for the *experimental* approach. The *direct* strategy requires a high level of user expertise, while the *synthesis* approach requires a high level of management knowledge.

29.3   Uncertainty and Risks in Generating User Requirements

In the selection of strategies for extracting user requirements, the uncertainty of the project and its requirements needs to be considered. The uncertainty of the requirements can come from several sources: uncertainty with respect to the existence and stability of the requirements, uncertainty due to users' inability to specify the requirements, and uncertainty due to the project manager's inability to elicit requirements and evaluate their correctness and completeness.

The sources of the first uncertainty can be:

- Lack of well-understood procedures to implement the activity, confused objectives, and unclear organization and poorly defined operating procedures
- Lack of stability in structure and operation of the system
- A large number of users with different expectations

The second type of uncertainty can be caused by:

- An overly complex system.
- A large number of users with various levels of participation and feelings of ownership in specifying requirements.
- Various types of users providing the requirements. For example, clerical staff can provide detailed low-level requirements but not overall content. Management will have the opposite tendency.

The third type of uncertainty is related to the level of training, experience, and ability of the project manager. The same characteristics of the applications that affect users also affect the manager's performance.

One can use more than one strategy to cope with different components of user requirements or one can employ one of the strategies as the primary one and supplement it with other determination strategies.

Some risks involved with generating the user requirements are presented in Table 29.1 below.

29.4   User Requirements Specification Tools

This section describes the various tools used to generate and develop user requirements. These tools are a part of the four general approaches described above.

TABLE 29.1   Representation of Risks to Generating User Requirements

	Risk type	Severity Level			
		Highest	High	Moderate	Low
1	Requirements identification	No defined process or model to identify requirements	Model or process exists but is not proven	Model or process established and being implemented	Model or process established and implemented successfully
2	Requirements complexity	Requirements are very complex and require interaction	Requirements are complex with limited interaction	Requirements are typical with some interaction required	Requirements are established based on similar projects
3	Requirements volatility	Requirements constantly change or grow	Significant amount of equirements change or growth	Changes occurring or likely to occur in some critical requirements	Requirements are very stable with little change activity
4	Requirements verifiability	Verification methods do not exist for most requirements	Verification methods do not exist for some critical requirements	Verification methods exist for most requirements	Verification methods exist for almost all requirements

29.4.1   States Method

In the states approach, the requirements are represented through a series of actions that move the system within a state space (Duke and Harrison, 1995). Thus, the requirements are defined by three structures: the state space defining the possible states of the system, a family of operations that defines how the system can move from one state to another, and a set of initial states. Functional refinement allows decomposition of each state and actions into a more detailed description. Clearly, this approach is very appropriate for an information system project.

Also, Gueguen and Chlique (1993) use state charts (Harel, 1988) to represent requirements for control of a system with prescribed dynamic behavior.

29.4.2   Task Analysis

An alternative approach presented by Duke and Harrison (1995) is the representation of requirements by a list of tasks to be performed. This method creates a model of the project by naming significant features that are necessary to complete the tasks involved with the project.

Because task descriptions are usually hierarchical, a project can be represented by a task hierarchy of several levels. At the higher levels, tasks in the work domain are themselves abstractions of the functionality expected of the system. These high-level tasks place constraints on the order in which operations can be performed. At the lower levels, the tasks represent the operations that define the detailed functionality of the project.

29.4.3   Waterfall and the Quality Function Deployment Model

The process of requirements development often has to bridge several cultures of users. Two models of requirements developments that are taken from the software engineering world are the *waterfall model* and the *quality function deployment (QFD) model*, in which the voice of the customer defines the requirements (Boehm et al., 1998). However, it can happen that the customers have no idea of the cost or difficulty of a requirement and may define an infeasible requirement as a statement of need. A partial solution to this problem is the stakeholder "win–win" approach in which all "win" conditions are covered by agreements and no unresolved issues remain. An effective approach for achieving a win–win compromise is prototyping concurrently with a negotiation process. In this case, the prototype can demonstrate that even a relaxed set of requirements will be able to satisfy the customer.

29.4.4   Team-Based Approach

The team-based approach for requirements specifications includes some of the following methods:

- *Graphical approaches.* The graphical issue-based information systems (gIBIS) application (Conklin and Begeman, 1988) is derived from a specific method, IBIS. This approach is based on the concept that the design process for complex problems is fundamentally a conversation among designers, customers, and implementers, in which they bring their respective expertise and viewpoints to the resolution of design issues. This graphical implementation uses color and a high-speed database server to facilitate building and browsing the application over a network by the various teams.
- *Viewpoints approach* (Finkelstein et al., 1992). A viewpoint can be thought of as a combination of the idea of an "actor," a "knowledge source," a "role," or "agent" in the development process and the idea of a "view" or "perspective," which an actor maintains.

Each viewpoint includes the following components:

- A *representation style*, the scheme and notation by which the viewpoint expresses its view of the project.
- A *work plan*, describing the process by which the specification can be built.
- A *work record*, an account of the history and current state of the development.

The project development team develops and shares the viewpoint, thus defining the project requirements. This can be achieved in various ways:

- *Team-based goal-oriented approach.* This approach is demonstrated by the implementation called GRAIL (Dardenne et al., 1991). This goal-directed acquisition approach aims to support the elaboration of a requirements model that is guaranteed to satisfy the goal of the clients and records how that satisfaction is realized. The concept of a goal is a central component of this model.
- *Participatory design (PD) and joint application design (JAD)* (Carmel et al., 1993). These are well-known methodologies for activating user involvement and user participation. Both JAD and PD focus on facilitated interactions between users and designers and are employed for eliciting and refining ideas.

Joint application design is intended to accelerate the design of information systems and promote comprehensive, high-quality results, while PD seeks to accelerate the social context of the workplace and promote workers' control over their work and their project environment. Joint application design emphasizes structure and agenda, while PD practices mutual reciprocal learning in which users and designers teach one another about work practices and technical possibilities through "joint experiences."

29.4.5   Spiral Model

The spiral model is currently gaining popularity over the traditional waterfall development model. The spiral model is a risk-driven approach. The process steps are determined by the need to resolve high-risk situations — ones that are most likely to fail. This approach contrasts with traditional document-driven approaches (the states model), where the process steps are determined by the type of document that has to be submitted. In this model, the project gradually evolves following a testing or analysis phase after the completion of each main task. This development and confirmation continues through the project's evolution as the system's functionality is gradually extended. This model, which is also supported by the PMI, is depicted in Figure 29.2.

29.4.6   Design Cycles

This approach is suitable for design-oriented projects (Wheelwright and Clark, 1992; Shenhar, 1988). During the project development, the project is successively refined using design cycles. Each cycle helps one to understand the project better and refines the requirements together with product development.

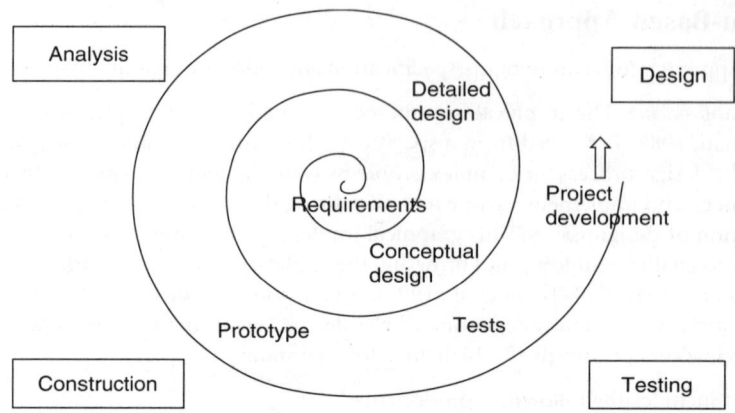

FIGURE 29.2   The spiral model.

This process continues until the product is sufficiently defined and the process reaches a "design freeze" stage. After that stage, the user requirements, together with the product, are sufficiently defined and the project proceeds with very little (or sometimes no) changes allowed. In effect, the design cycles model is similar to the spiral model except that circles can have a different set of activities in each circle.

29.4.7   General Analysis of Requirements

There are many domain-specific methods to describe user specifications. Boehm et al. (1998) specify factors that complicate development of requirements (*complicators*) and factors with the opposite effect (*simplifiers*). In the domain of project management, simplifiers include facts such as:

- A project scope that fits within the user's scope of authority
- Proposed requirement that reduces organizational friction and infrastructure clashes
- No hidden costs or complications
- Tasks that are well-defined, understood, and experienced by users
- A single criterion for defining success

Cheong and Jarzabet (1998) suggested adding a suffix in order to distinguish between mandatory and optional requirements.

29.4.8   Scenarios

Another approach for generating and representing requirements is using scenarios (Some et al., 1995). A scenario is a description of the interaction between a system and its environment in a restricted situation. This description consists of a succession of operations constrained by timing requirements. A scenario can be represented as a sequence of operations and time of occurrence that depends on conditions in the system and the environment. In such cases, scenarios are represented as sequences of events, each with pre- or postconditions.

Dearden and Howard (1998) demonstrate this approach using techniques drawn from scenarios-based design, soft system methodology, and QFD. They divide the process into four phases: initial hypothesis, modeling the context, analyzing the models, and identifying and recording user needs. Each phase of this method is intended to develop the system analyst's understanding of the project and its context.

29.4.9   Goal-Driven Approach

Another approach toward specifying project requirements (at any level) is the goal-driven approach (Lee and Xue, 1999). In this approach, the project manager identifies the various users of the systems (the same user can have several roles). Goals have three facets: competence, view, and content. The competence facet defines the goals as "rigid" or "soft": rigid goals have to be satisfied completely, while soft goals are simply desirable. The view facet defines whether a goal is user- or system-specific. A user-specific goal defines what the user needs to be able to accomplish, and a system-specific goal defines the requirements from the overall system (such as response time). The content facet divides requirements into functions and nonfunctional requirements. The nonfunctional requirements often define constraints on the system.

29.4.10　Case-Based Reasoning

Case-based reasoning is a methodology that utilizes old cases in order to derive solutions to new problems (Kolodner, 1991, 1993). A CBR system uses an indexed database of old cases and comparison schemes. When a new problem arises, it is compared with the problems in the database, and the one with the most similarity is extracted from the database. The database also stores the solutions used in the past and thus the solution of the selected problem is retrieved. This solution is modified to fit the new problem and is then adopted.

This approach can be used to extract the requirements from similar projects performed in the past. The drawback of this approach is the need to catalog old projects in a database with their requirements and to develop the similarity measures required by the CBR methodology. This setup is only justified in a company that performs a large number of diverse projects.

Projects that are very structured (usually a construction-type project) and with little innovation are suitable for CBR analysis.

29.5　Project Classification and Selection of Appropriate Method

In order to be able to recommend the most appropriate method for generating user requirements, we first need to look at the classification of projects. This classification allows us to match the best method with the right type of project.

A project can be classified in many different ways. One of the more obvious is by project type, i.e., whether it is an in-house R&D, small construction, large construction, aerospace/defense, MIS, or engineering project (Kerzner, 1995). A similar classification of projects can be done based on the following seven dimensions: size, length, experience required, visibility of upper-level management, location, available resources, and unique aspects of the project (Kerzner, 1995). This is a more nebulous classification system and does not provide a clear assignment of projects into groups. Another classification system of projects is two-dimensional, based on the level of technical uncertainty and project complexity. The

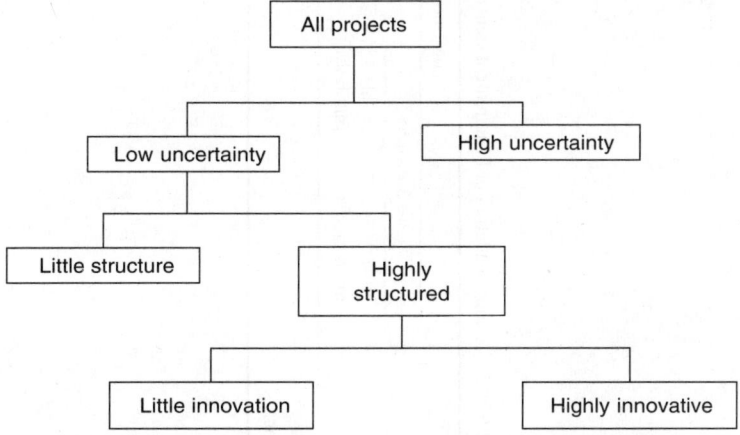

FIGURE 29.3　Project classification.

TABLE 29.2 Recommendation of Requirement Generation Method

| | Low Uncertainty | | | | High Uncertainty | | | |
| | Little Structure | | Highly Structured | | Little Structure | | Highly Structured | |
	Little Innovation	Highly Innovative	Little Innovation	Highly Innovative	Little Innovation	Highly Innovative	Little Innovation	Highly Innovative
Primary	3	4	1	9	5, 7	6	3	7
Secondary	1, 2	7, 8	2	8, 10	6	5	2	4

uncertainty axis is divided into four levels: low uncertainty (low-tech), medium, high, and super-high technical uncertainty (corresponding to super-high-tech projects). The project complexity axis is divided into three ranges: assembly-level project, systems-level project, and an array of systems-level projects (Shenhar, 2001). Thus, this system allows one to classify projects into 12 types.

Projects can also be classified on the following basis of the five factors (Shenhar, 1998): project style, structure, tools, processes, and people. However, such a classification system is fuzzy and does not allow the crisp assignment of projects into their corresponding groups. Another project classification system can be based on technology employed, depending on whether the projects use existing technology or innovative applications of technology. The innovative usage of technology can itself be further broken down into incremental innovation, radical, modular or architectural changes (Henderson and Clark, 1990).

In order to select the most appropriate method for generating user requirements, this paper presents a project analysis method based on project uncertainty, level of structure and level of innovation. Each category consists of low and high levels. This classification is presented in Figure 29.3, which, however, does not actually show the complete tree of all combinations.

Using the project dichotomy, Table 29.2 represents the method most suitable for each type of project: the primary method and the secondary methods. The table also clearly shows the eight possible types of projects.

The numbers in the table refer to the section of this chapter describing the preferred method; so for example method number 1 is the states-based method (described in Section 29.4.1). The numbers in the first row represent the primary method recommended, while the second row represents the secondary method recommended.

29.6 Conclusion

Understanding user requirements is an essential part of project management. Projects are usually complex and nonrepetitive endeavors largely defined by the user. Thus, generating correct and complete user requirements is necessary for the success of the project.

This chapter starts with a presentation of strategies used by the project manager to specify the user requirements. These strategies define approaches that the project manager can take to manage the project and communicate with project staff. The chapter also reviews the information requirements for a successful generation of user requirements and lists common uncertainty and risk factors. Next, the chapter presents ten methods of developing user requirements. These methods vary from simple states or task analysis to more complex iterative methods such as the "waterfall" or the "spiral" methods. The chapter then presents a classification system of projects that helps assign the preferred method to the appropriate project type. This system is based on level of uncertainty, level of structure of the project, and level of innovation. On the basis of this system, we provide recommendations regarding the primary and secondary methods.

The chapter is intended for the researcher as well as for the practitioner and provides a comprehensive literature review for the researcher and concrete recommendations for the practitioner.

References

Berrisford, T.R., and Wetherbe, J.C., Heuristic development: A redesign of system design, *MIS Q.*, 3, 11–19, 1979.

Bias, R.G., Lanzetta, T.M., and Scanlon, J., Consensus requirements: Low and high tech methods for discerning system requirements from groups of users, *Proceedings of the IEEE International Conference on Systems, Man and Cybernetics*, Vol. 1, 1993, pp. 191–196.

Boehm, B., Abi-Anton, M., Port, D., Kwan, J., and Lynch, A., Requirements Engineering, Expectations Management, and the Two Cultures, Internal report, University of Southern California, http://sunset.usc.edu/publications/TECHRPTS/1998/usccse98-518, 1998.

Carmel, E., Whitaker, R., and George, J., PD and joint application design: A transatlantic comparison, *Commun. ACM*, June, 36(6), 40–48, 1993.

Cheong, Y.C., and Jarzabek, S., Modeling variant user requirements in domain engineering for reuse, in *Information Modeling and Knowledge Bases,* Jaakkola, H., Kangassalo, H., and Kawaguchi, E., Eds., IOS Press, Netherlands, 1998, pp. 220–234, ISSN: 0922-6389.

Conklin, J., and Begeman, M. *gIBIS* : A hypertext tool for exploratory policy discussion, *ACM Trans. OIS,* 6(4), 303–331, 1988.

Dardenne, A., Fickas, S., and van Lamsweerde, A., Goal-directed concept acquisition in requirement elicitation, *Proceedings IWSSD 6, IEEE,* 1991, pp. 14–21.

Davis, G. B., Strategies for information requirements determination, *IBM Syst. J.,* 21, 4–30, 1982.

Dearden, A. and Howard, S. Capturing user requirements and priorities for innovative interactive systems, *Australian Computer Human Interaction Conference Proceedings,* November 29–December 12, Adelaide, Australia, 1998, pp. 160–167.

Duke, D.J., and Harrison, M.D., Mapping user requirements to implementations, *Software Eng. J.,* 10, 13–20, 1995.

Finkelstein A., Kramer, J., Nusibeh, B., Finkelstein, L., and Goedicke, M., Viewpoints: A framework for integrating multiple perspectives in system development, *Int. J. Software Eng. Knowledge Eng.,* 31–58, 1992.

Gueguen, H. and Chlique, P., A method for users' requirements specification for control of hybrid systems, *Proceedings of International Conference on Systems, Man and Cybernetics,* Vol. 2, 1993, pp. 687–691.

Harel, D., On visual formalisms, *Commn. ACM,* 31, 514–531, 1988.

Henderson, R.M. and Clark, K.B., Architectural innovation: The reconfiguration of existing product technologies and the failure of established firms, *Admin. Sci. Q.,* 35, 9–30, 1990.

Kerzner, H., *Project Management, a System Approach to Planning, Scheduling and Controlling,* 5th ed., Van Nostrand Reinhold, New York, 1995.

Kerzner, H., *Project Management, a System Approach to Planning, Scheduling and Controlling,* 8th ed., Wiley, New York, 2004.

Kolodner, J., Improving human decision making through case-based decision aiding, *AI Mag.,* 12, 52–68, 1991.

Kolodner, J., *Case Based Reasoning,* Morgan Kaufmann, San Mateo, CA, 1993.

Lee, J. and Xue, N.-L., Analyzing user requirements by use cases: A goal-driven approach, *IEEE Software,* 16, 92–101, 1999

Lyytinen, K. and Hirschheim, R., Information systems failures: A survey and classification of empirical literature, *Oxford Surv. Inf. Technol.,* 4, 257–309, 1987.

PMBok® Guide, *A Guide to the Project Management Body of Knowledge,* 2000 Edition, Project Management Institute, Newton Square, PA, 2000.

Rockart, J.F., Critical success factors, *Harvard Bus. Rev.,* 57, 81–91, 1979.

Shenhar, A. J., From theory to practice: Toward a typology of project-management styles, *IEEE Trans. Eng. Manage.,* 45, 33–48, 1988.

Shenhar, A. J., One size does not fit all projects: exploring classical contingency domain, *Manage. Sci.,* 47, 393–414, 2001.

Shtub, A., Bard, J.F., and Globerson, S., *Project Management, Engineering Technology and Implementation,* Prentice-Hall, Englewood Cliffs, NJ, 1994.

Simon, H., *Models of Man,* Wiley, New York, 1957.

Some, S., Dssouli, R., and Vaucher, J., From scenarios to timed automata: Building specifications from users' requirements, *Proceedings of Software Engineering Conference,* Asia Pacific, 1995, pp. 48–57.

Wheelwright, S.C. and Clark, K.B., *Revolutionizing Product Development,* Free Press, New York, 1992.

30

Learning and Forgetting Models and Their Applications

Mohamad Y. Jaber
Ryerson University, Toronto, Canada

30.1 Overview

Many experts believe the only sustainable advantage an organization will have in the future is its ability to learn faster than its competitors (Kapp, 1999). This competitive advantage can be achieved by transforming the organization into a learning organization. The measure of how fast organizations learn is captured by the learning-curve theory.

Learning curves have been receiving increasing attention by researchers and practitioners for almost seven decades. Cunningham (1980) presented 15 examples from U.S. industries. He wrote (p. 48): "Companies that have neglected the learning-curve principles fall prey to more aggressive manufacturers." With examples from aerospace, electronics, shipbuilding, construction, and defense sectors, Steven (1999, p. 64) wrote on learning curves: "They possibly will be used more widely in the future due to the demand for sophisticated high-technology systems, and the increasing interest in refurbishment to extend asset life."

The learning curve can describe group as well as individual performance, and the groups can comprise both direct and indirect labor. Technological progress is a kind of learning. The industrial learning curve thus embraces more than the increasing skill of an individual by repetition of a simple operation. Instead, it describes a more complex organism — the collective efforts of many people, some in line and others in staff positions, but all aiming to accomplish a common task progressively better. This may be why the learning phenomenon has been variously named as *start-up curves* (Baloff, 1970), *progress functions* (Glover, 1966), and *improvement curves* (Steedman, 1970). In this chapter, the term "learning curve" is used to denote this characteristic learning pattern.

Researchers and practitioners have unanimously agreed the power-form learning curve to be by far the most widely used model that depicts the learning phenomenon (Yelle, 1979). However, a full understanding of the forgetting phenomenon and the form of the curve that best describes it has not yet been reached.

In this chapter, we shed light on past and recent developments in modeling the learning–forgetting process. In doing so, we will attempt to be comprehensive but concise.

30.2 The Learning Phenomenon

Early investigations of learning focused on the behavior of individual subjects. These investigations revealed that the time required to perform a task declined at a decreasing rate as experience with the task increased (Thorndike, 1898; Thurstone, 1919). Such behavior was experimentally recorded, with its data fitted with an equation that adequately described the trend line and the scattered points around it. The first attempt made to formulate relations between learning variables in quantitative form was by Wright (1936), and resulted in the theory of the "learning curve."

30.2.1 Psychology of Learning

Even though psychologists have been conducting research on the learning phenomenon for decades, it is still difficult to define. There is considerable agreement among psychologists that learning includes the trend of improvement in performance that comes about as a result of practice. Hovland (1951) founded the definition that learning is the change in performance associated with practice and not explicable on the basis of fatigue, artifacts of measurement, or of receptor and effector changes as the most suitable. The *learning process* passes mainly through three phases, which are: acquisition, training, and retention of learning (Hovland, 1951). *Acquisition* of learning involves a series of trials, each based on retention and influenced by the conditions under which acquisition takes place. Learning or acquiring a knowledge or skill is influenced by three factors, which are: (1) the *length* of the material learned, (2) the *meaningfulness* of the material learned, and (3) the *difficulty* of the material learned. *Meaningfulness* is an obvious factor affecting ease of learning, where material that makes sense is learned more easily than nonsense material. *Difficulty* of the material affects the ability to learn. Easy topics or tasks are learned rapidly while, as the material increases in complexity, the time required to learn will increase. The factors of length and meaningfulness affect the time required for learning as well.

In most verbal and motor learning, the materials and movements are learned in a serial manner. Ebbinghaus (in Hergenhahn, 1988), unlike the exponents of traditional schools of psychology, emancipated psychology from philosophy by demonstrating that the *higher mental processes* of learning and memory could be studied experimentally. He conducted an experiment on serial learning to study the degree of association between adjacent as well as remote items learned. Ebbinghaus emphasized the law of frequency as an important principle of association. The law of frequency states that the more frequently humans experience a verbal or motor skill, the more easily the material learned is recalled. Practice makes perfect, due to the fact that memory is strengthened through repetition.

Evidence for the importance of motivation in human learning comes from the immense number of studies showing that learning is facilitated when *this* or *that* motivation is employed with adults or children (Hergenhahn, 1988).

Age differences, *sex* differences, and differences in *mental ability* are the most extensively studied factors affecting the human ability of acquiring knowledge or a skill (Hovland, 1951). Hovland (1951) reported that learning is transferable even under highly controlled conditions. The learning of a new task is not independent of previous learning but is built upon previous acquisitions. This is shown in our everyday life, where our old habits interfere with the formation of new ones. How previous learning affects new learning has customarily been dealt with under the term "transfer of training."

The effect of previous learning may either improve (*positive transfer*) or retard (*negative transfer*) the new learning. Negative transfer of learning, *forgetting*, occurs when there is not enough similarity between the conditions under which the student studied and the conditions under which he or she is tested (Hergenhahn, 1988). Hovland (1951) reported that when old learning interferes with new learning, forgetting occurs. He further discussed various factors producing loss in retention (forgetting) with the passage of time. This factor in forgetting is the alternation of the stimulating conditions from the time of learning to that of the measurement of retention. This means that forgetting occurs when some of the stimuli present during the original learning are no longer present during recall. All the works reviewed in Section 30.2 assume that forgetting occurs with the passage of time. Yet there has been no model developed for industrial settings that considers forgetting as a result of interference. This is a potential area of future research.

30.2.2 Learning-Curve Theory

The earliest learning-curve representation is a geometric progression that expresses the decreasing cost required to accomplish any repetitive operation. The theory states that as the total quantity of units produced doubles, the cost per unit declines by some constant percentage. Wright's (1936) learning curve (WLC), as illustrated in Figure 30.1, is a power-function formulation that is represented as

$$T_x = T_1 x^{-b} \tag{30.1}$$

where T_x is the time to produce the xth unit, T_1 the time to produce the first unit, x the production count, and b the learning-curve exponent. In practice, the b parameter value is often replaced by another index number that has had more intuitive appeal.

This index is referred to as the "learning rate" (LR), which occurs each time the production output is doubled:

$$LR = \frac{T_{2x}}{T_x} = \frac{T_1 (2x)^{-b}}{T_1 x^{-b}} = 2^{-b} \tag{30.2}$$

The time to produce x units, $t(x)$, is given as

$$t(x) = \sum_{i=1}^{x} T_1 x^{-b} \cong \int_0^x T_1 x^{-b} \, dx = \frac{T_1}{1-b} x^{1-b} \tag{30.3}$$

For example, consider a production process that has an 80% learning rate ($b = 0.3219$; if an 80% learning curve is equivalent to a 20% [100 − 80 = 20%] progress function) where the time to produce the first

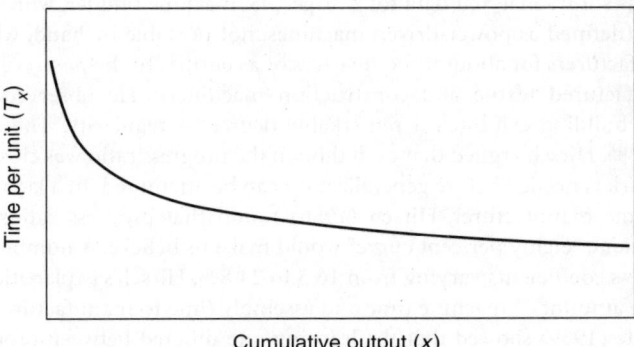

FIGURE 30.1 Wright's learning curve.

unit is 60 min ($T_1 = 60$), then the time to produce the 50th unit is $T_{50}=60(50)^{-0.3219}=17.03$ min, and the time to produce all 50 units is $t(50) = [60/(1-0.3219)](50)^{1-0.3219}=1255.82$ min. How much time would it take to produce the next 50 units? To answer this question, we need to define the first 50 units as the cumulative experience, $u = 50$, and write Eq. (30.3) as

$$t(x) = \int_{u+1/2}^{x+u+1/2} T_1 x^{-b}\, dx = \frac{T_1}{1-b} \left\{ (x+u+1/2)^{1-b} - (u+1/2)^{1-b} \right\}$$ (30.4)

and the time to produce the next 50 units is $t_2(50) = 753.53$. Logarithmic transformation of the model demonstrates the linearity of the start-up phase when plotted on log coordinates as

$$\log (T_x) = \log(T_1) - b\log(x)$$ (30.5)

A learning curve is usually drawn on log–log paper simply because in a straight-line form it is much easier to project. At the same time it must be remembered that on logarithmic coordinates, time and cumulative output are greatly compressed.

30.2.3 Is There a Universal Learning Rate?

Wright (1936) reported an 80% learning rate. Many researchers who investigated the application of learning curves to industrial engineering problems indisputably tend to assume their learning curves to follow the 80% learning rate. Conversely, there has been enough evidence to support that learning rates vary among products and across organizations.

Hammer (1954) explained the rate of progress by a number of during-production factors, such as tooling, methods, design changes, management, volume changes, quality improvements, incentive pay plans, and operator learning. Most of these factors are related to *process* design, which is the predominant factor that determines the progress rate for a particular product or assembly (Nadler and Smith, 1963). Nadler and Smith (1963) define "process design" as the operations performed on one type of standard factory equipment, or operations performed by an operator with a particular skill with special training to become proficient at a specific task. Examples are milling, drilling, grinding, welding, punch press, and precision subassembly operations, which are performed in most factories. This theory is supported by McCampbell and McQueen (1956), who reason that the learning (progress) function is based on (1) the human element, or man's ability to learn and improve, and the elimination of nonproductive activities through repetition, (2) the natural process of improvements of methods by the operator and his foreman, (3) more efficient material handling, transportation, pallet moving, and so forth, and (4) a smaller percentage owed to scrap, rejection, and rework. They suggested that a completely automatic operation such as an automatic screw machine should have a 100% curve; and a complicated hand assembly might be as low as 60%. Wertmann (1959) suggested a similar approach to that of McCampbell and McQueen (1956) for estimating the learning rate.

Hirsch (1952),in his study, analyzed data for a large U.S. machine builder, which has been one of the largest machine-tool (defined as power-driven machines, not portable by hand, which cut, hammer, or squeeze metal) manufacturers for about three-quarters of a century. In the post-World War II period, the company also manufactured textile and construction machinery. He observed that the company's progress in machine building exhibited a remarkable degree of regularity. The progress ratio varied between 16.5 and 20.8%. Hirsch argued that even though the progress ratio was close to 20% (80% learning curve), further work is needed before generalization can be attempted. In a later study on eight products made by the same manufacturer, Hirsch (1956) found that progress ratios are less uniform in magnitude than the name "eighty per cent curve" would make us believe. A number of empirical studies have produced progress coefficients varying from 16.5 to 24.8%. Hirsch's explanation of this variation is mainly because of the amount of machine time and assembly time to manufacture each product.

Conway and Schultz (1959) showed that the learning rate differed between products, manufacturing facilities, and industries. They experienced one product which ceased to progress in one industry but continued to progress when transferred to another industry. Also, they experienced that one product had a learning rate of 85%, whereas another company producing the same product had a learning rate of 80%.

Alchian (1963) found that fitting learning curves to the aggregate past performance of a single manufacturing facility to predict the future could result in a significant margin of error. This study was significant in airframe manufacturers, which had been operating on the assumption of an 80% curve. This assumption did not take into consideration the margin of error or the difference between airframe types. The progress ratio for the same wartime period (World War II) for the production of Liberty ships, Victory ships, tankers, and standard cargo vessels ranged from 16 to 22% (84 to 78% learning curve).

Hirschmann (1964) said that there is no such thing as a universal curve that fits all learning. He related it to the great variations in level at which a curve starts, i.e., the cost of the first unit. This is simply because of the different ranges of complexity of items. Operations performed by people have steeper slopes than those performed by machines. He referred to aircraft manufacture, where three-quarters of the direct labor input is assembly and one-quarter machine work, and an 80% learning curve is found. If the ratio of assembly to machine work is 50–50, the learning curve is about 85%. If the ratio is one-quarter assembly and three-quarters machine work, the operation is largely machine-paced, and the curve is about 90%.

In a study conducted on 28 separate cases of new products and new process start-ups occurring in five separate companies in four different industries, Baloff (1966) found that the learning slope varied widely. In a later study, Baloff (1967) described the results of an empirical approach to estimating the learning-curve parameters using manufacturing experience and experimental studies in group learning. The primary focus of the paper was on estimating the learning slope, given a reliable measure of the time to produce the first unit.

Cunningham (1980) collected learning rates reported in 15 different U.S. industries over the period 1860–1978. The learning rates ranged from 95% in electric power generation to 60% in semiconductor manufacturing. Dutton and Thomas (1984) showed the distribution of learning rates for 108 firms, with a mean of 80.11%. The learning rates for 93 firms fall in the range of 71 to 89%. Dar-El (2000, pp. 58–60) collected learning rates quoted for specific types of work that ranged from 68 to 95%. Thus, one may reasonably conclude that a universal learning rate does not exist, and understanding the reasons why learning rates vary is a major challenge for research (Argote and Epple, 1990).

30.2.4 Learning-Curve Models

Several authors have debated the form of the learning curve. The traditional representation of the improvement, experience, or learning phenomenon has been the model introduced by Wright (1936). In addition, there are several continuous or smooth curves which have been used to fit experience data. Among these models are the Stanford B curve, the DeJong curve, and the "S" curve. The Stanford B curve has a shallow slope of the early units by assuming that the equivalent of B units have already been experienced (Carlson, 1973):

$$T_x = T_1(x+B)^{-b} \qquad (30.6)$$

where B is a constant, which may be expressed as the number of units theoretically produced prior to the first unit accepted. The DeJong curve (1957) considers that many operations consist of an incompressible component which is incorporated to the equation

$$T_x = T_1(M+(1-M)x^{-b}) \qquad (30.7)$$

where M ($0 \leq M \leq 1$) is the factor of incompressibility, and the other parameters are the same as defined in Eq. (30.1). DeJong (1957) found that $M = 0.25$ for assembly operations, ranging upward to unity for operations which contained a large number of machine-controlled times. The DeJong curve function reduces to that of Wright's mathematical expression, when the job is fully manual ($M = 0$) and to $T_x = T_1$ when the job is fully automated ($M = 1$). Because the Stanford B has a better fit for the early part of the curve, and the DeJong the latter of the curve, the temptation is to combine these two methods to yield an S-curve (Carlson, 1973):

$$T_x = T_1(M+(1-M)(x+B)^{-b})$$ (30.8)

Without any assumed form, it is necessary to estimate or preselect the incompressibility factor M, equivalent experience unit B, and even the expected learning exponent b. Glover (1966) derived a model for the learning of individuals that applies to learning of all kinds of industry. He confirmed that the model that best fits the data is of the form

$$\Sigma y_i + C = a(\Sigma x_i)^m$$ (30.9)

where y_i and x_i are interchangeable in the sense that either may represent time or quantity, depending on the format devised, C is a "work commencement" factor, a the time of the first cycle, and m the index of the curve equal to $1-b$. The factor C was seldom required, i.e., it was always nearly zero. It was only when general data were obtained from records of companies and exact starting times could not be guaranteed, that a value had occasionally been assigned to C in order to obtain a good fit of the data.

Thomopoulos and Lehman (1969) developed a learning curve for mixed-model assembly lines. They started with the formulation of the "two-model" assembled on one line case, then the "three-model," to end up with the general mixed-model learning curve of the form

$$f(x) = T_x = T_1 x^{-b} Q$$ (30.10)

where Q represents, mathematically, the added increase in assembly time required to process these units over an equivalent single model line. Thomopoulos and Lehman's (1969) model predicts the time to assemble the xth unit under the assumption that the production up to the xth unit contains the same relative proportions of models designated for total production runs. Thomopoulos and Lehman (1969) advocate that the mixed-model learning curve described in Eq. (30.10) helps managers gain a better understanding of their assembly processes. They suggested that potential applications could include the comparison of learning costs on single and mixed-model lines, and the selection of models to mixed-model lines, which tend to minimize the percentage increase in assembly time. Thomopoulos and Lehman (1969) based their study on the learning-curve function presented by Wright (1936).

Levy (1965) presented a new type of learning function useful in describing how firms improve their performance in new processes. The function, termed as the "adaptation function," is shown to be capable of being related to variables that may affect a firm's rate of learning. He based his study on the traditional learning-curve model presented by Wright (1936) to reach a curve of the form

$$R(x) = P - (P - x^b/T_1)$$ (30.11)

where $R(x)$ is the rate of production after j units have been produced, P the maximum rate of output the firm would like to achieve, and j the cumulative output. For $R(x)$ to approach P, he multiplied the bracketed expression by the damping factor $e^{-\lambda x}$, rearranged terms, and obtained

$$R(x) = P(1-e^{-\lambda x}) + (x^{-b}/T_1)e^{-\lambda x}$$ (30.12)

Levy (1965) also showed that the adaptation function had normative implications for firms in their formal training and equipment replacement decisions.

Goel and Bucknell (1972) at national cash register (NCR) corporation undertook the project of determining a consistent scientific approach in order to arrive at a solution to the problem of assigning a number of days or weeks to learn various manufacturing operations, with the learning curve described by

$$T_x = (T_1 - T_s/1.45)LR^{\log(x)/\log 2} + T_s/1.45 \tag{30.13}$$

where T_s is the standard time to produce one unit, T_j the time to produce the jth unit, and LR the learning rate. No subsequent researcher reported Goel and Bucknell's learning curve.

The above models were selected in particular because they plot the time required to produce the xth unit against cumulative output. Such a relationship facilitates the representation of labor time required as a function of the total units produced.

Several authors have developed mathematical functions that represent particular learning processes. Among these learning models is the time-constant model (Bevis et al., 1970), whose modified version is found in Hackett (1983) to have the form

$$y_i = Y_c + Y_f(1 - e^{-t/\tau}) \tag{30.14}$$

where y_i is rate of production at time t of cycle i, Y_c the initial rate of production, Y_f the difference in the rate of output between the initial rate of production, Y_c, and the maximum rate, Y_m, t_i time (days), and τ the time constant for a particular curve. Other models of the same family are found, for example, in Wiltshire (1967), Cherrington and Towill (1980), and Towill (1976). Refer to Yelle (1979), Belkaoui (1986), and Badiru (1992) for a review of learning-curve models.

30.2.5   Which Learning-Curve Model to Use?

Kilbridge (1959) performed an experiment to establish a system for predicting learning time on routine clerical operations. The experiment was conducted in the clerical department of a mail-order house to determine the form of the learning curves, and a system was then devised for predetermining the parameters of these curves from job content. The experiment showed that these curves follow the pattern described by Wright (1936).

Buck et al. (1976) suggested that desirable features for learning-curve models include: (1) flexibility to fit a wide variety of known learning behavior, (2) easy and accurate parameter estimation from small quantities of data and various methods of data collection, (3) parameter estimation methods which are relatively insensitive to noise in the collected data, (4) forecast accuracy which is insensitive to parameter estimation precision, and (5) the availability of auxiliary techniques for computing the sequence sums, averages, costs, etc., that are needed in the various roles which learning curves serve in production/quality control and economic analysis. Buck et al. (1976) presented a discrete exponential model for use as a learning curve. This model appears to offer many of the features cited above. Their paper describes a particular learning-curve model that was previously proposed and empirically demonstrated by Pegels (1969). Pegels' (1969) discrete exponential model was of the form

$$T_x = \alpha T_1^{x-1} + \beta \tag{30.15}$$

where T_x, T_1 and x are defined as for the Wright model and α and β are empirically based parameters. Pegels (1969) concluded that his proposed model was slightly more difficult to apply than the Wright's power model, which has the advantage of simplicity of use. For review of exponential-form learning curves, refer to Bevis et al. (1970) and Lerch and Buck (1975).

Towill (1982) identified a number of patterns in the basic data, each of which is an important source of information, where he argued that deriving a model that copes with these patterns simultaneously is a difficult problem. Towill (1982) identified three main sources of prediction error: (1) there are errors due to natural fluctuations in performance, with the fluctuations random or deterministic, such as sinusoidal oscillation; (2) deterministic errors usually vary more slowly and include plateaus, for which there may well be physiological, psychological, or environmental causes; and (3) a complete description of experimental data is only achieved by taking account of modeling errors. That is, the form of the model may not permit adequate description of the trend line.

Hackett's (1983) research investigated the effectiveness of training at British Telecommunications. With the learning curve, models selected for assessment had a format that allowed rate of production to be plotted against time. He concluded that the best model could not be selected purely on the basis of statistical tests alone. However, Hackett (1983) further added that, given very large quantities of learning data, this might be possible.

Globerson (1980) investigated the influence of three groups of variables on the chosen model: working conditions, type of instruction, and previous experience. He used three learning models in his analysis, which are the power model of Wright (1936), the exponential model of Bevis et al. (1970), and the linear model of Hancock (1967). The overall findings of this research show that out of the three possible models analyzed, the power model is probably to be preferred, since it is at least as good as the others for all ranges, and sometimes even better. This conclusion supports the historical precedence of choosing the power form as the model that depicts the learning phenomenon.

Naim (1993) examined the interrelationship between the diffusion of the product into the marketplace and the start-up of industrial systems. His aim was to enhance the practical scope of the suite of Industrial Dynamics models to develop *least-squares error* (LSE)-based algorithms for the "S"-shaped and Ripple learning-curve models. Naim realized that in some cases the "S"-or Ripple-curve models yielded better fits, in LSE terms, although more data were required to estimate the parameters of these complex models. He found that a rough usable estimation may be obtained sooner and therefore more usefully with the Time Constant model. In general, a simpler model with fewer parameters is easier and requires less data. This finding is consistent with that of Towill (1982), who suggested that a simple model derived on the back of an envelope can be more profitable for management than a sophisticated computerized model. However, and as demonstrated above, the form of the learning curve has been debated by many researchers and practitioners. By far, the WLC is the most widely used and accepted, where it has been found to fit empirical data quite well (Yelle, 1979; Lieberman, 1987).

30.2.6 The Plateauing Phenomenon

When does the learning process cease? Crossman (1959) claimed that this process continues even after 10 million repetitions. Conway and Schultz (1959) showed that the learning rate differs between products, manufacturing facilities, and industries. They experienced one product which ceased to progress in one industry but continued to progress when transferred to another industry. Baloff (1966, 1970) concluded that plateauing is much more likely to occur in machine-intensive industries than it is in labor-intensive industries. Corlett and Morcombe (1970) related plateauing to either consolidating what has already been learned before making further progress, or to forgetting. Yelle (1980, p. 317) stated that two general conclusions can be drawn from Baloff's studies. One possible explanation for plateauing could be strongly associated with labor ceasing to learn. Secondly, plateauing could be associated with the management's unwillingness to invest additional capital in order to generate the technological improvements necessary for the continuation of the learning process. Thirdly, Hirschmann (1964) makes the point that skepticism, on the part of the management, that improvement can continue may in itself be a barrier to its continuance. This position is supported by Conway and Schultz (1959). On these explanations for the plateauing phenomenon, Li and Rajagopalan (1998a, p. 148) wrote: "There does not appear to be strong empirical evidence to either support or contradict these hypotheses." They attributed plateauing to depreciation in knowledge.

Previous to these works, DeJong (1957) was the first to alter WLC by introducing a third parameter (factor of compressibility) to force the WLC to plateau. DeJong's factor of compressibility embodies the assumption that every job contains two components, one subject to improvement with an elasticity (learning exponent) that does not vary between jobs and the other subject to no improvement. In a subsequent paper, DeJong (1961) has suggested that when shop floor organization, work scheduling, jigs used, etc., remain constant, cost decrease will be bound below by a positive limit, and in his formula with positive factor of compressibility, but when they are subject to change, cost decrease will be bound below only by zero (Steedman, 1970, p. 194). This simply means that the factor of compressibility would be meaningful in a constant, rather than in a changing manufacturing environment. In a real-world setting where shop floor organization, work scheduling, jigs used, etc., will vary, the WLC might be more appropriate where the factor of compressibility is zero. Dar-El et al. (1995b, p. 275) stated that there is no scientific work reported in literature to help in determining values for the number of repetitions needed to attain the "standard time" (associated with the factor of compressibility in the case of DeJong [1957]). He further added that DeJong (1957) assumes a value of 1000 repetitions to reach standard time; however, there is no basis for his assumption. On DeJong's factor of compressibility, Badiru (1992, p. 179) wrote: "Regrettably, no significant published data is available on whether or not DeJong's model has been successfully used to account for the degree of automation in any given operation. With the increasing move towards automation in industry, this certainly is a topic for urgent research." On the same issue, Dar-El (2000, p. 38) wrote: "How would M (the factor of compressibility) be determined? Even DeJong isn't much help on this point." He further stated (p. 38): "Regrettably, no significant field data is available to support DeJong's model." Recently, Jaber and Guiffrida (2004) modified WLC with their analytical study suggesting that plateauing might be attributed to problems in quality.

From the above studies one can deduce that there is no tangible consensus among researchers as to what causes learning curves to plateau. This remains an open research question.

30.2.7   Is Cumulative Output a Good Measure of Learning?

Perhaps the most conventional learning-curve model (Wright, 1936) is of a power form where intensive hours per unit decreases as the cumulative number of units produced increases. Wright's model has been found to fit empirical data quite well (Yelle, 1979; Lieberman, 1987). Some researchers have raised the issue of whether cumulative production alone could be used as a proxy to detect improvement. Lieberman (1987) stated that several studies have suggested that learning is a function of time rather than cumulative output. Fine (1986) pointed out that quality-related activities unveil inefficiencies in the production process, and suggested the use of cumulative output of good units as a proxy to measure knowledge or, alternatively, improvement. Globerson and Levin (1995) recommended the use of equivalent number of units, which is equal to the sum of finished product and in-process inventory units, as a proxy to measure experience (knowledge).

Implicit in the conventional learning-curve model (Wright, 1936) is the assumption that knowledge acquired (measured in units of output) through learning by doing does not depreciate (Epple et al., 1991). Several empirical studies have refuted such a claim. For example, Argote et al. (1990) suggested that the conventional measure of learning, cumulative output, significantly overstates the persistence of learning (knowledge depreciation). They found evidence of depreciation of knowledge during breaks to be a more important predictor of current production than cumulative output. Epple et al. (1991) demonstrated how a conventional learning curve could be generalized to investigate factors responsible for organizational learning. They contrasted their results on intra-plant transfer of knowledge in automotive production to results on inter-plant transfer of knowledge in shipbuilding (Argote et al., 1990). The findings of Epple et al.'s (1991) on persistence of learning in automotive production were viewed as an interesting supplement to those found in shipbuilding. Argote (1993) concluded that learning captured by the traditional learning curve is a combination of employees, organizational systems, and outside actors.

Recently, some researchers have raised the need to understand how learning occurs in organizations. Zangwill and Kantor (1998) proposed a learning-curve model that accounts, in addition to learning-by-doing, for learning through continuous improvement efforts. This model enables management to observe what techniques are causing greater improvement and can therefore be used to accelerate the production process. In a subsequent paper, Zangwill and Kantor (2000) extended their earlier work by attempting to articulate a conceptual framework for the learning curve. Contrary to the previous, almost exclusive employment of the learning curve as forecasting curve, they employed what they called the learning-cycle concept, which provides information on which improvement efforts fail and which succeed. Hatch and Mowery (1998) indicated that the improvement of manufacturing performance through learning is not an exogenous result of output expansion but is influenced primarily by the systematic allocation of engineering intensive to problem-solving activities, that is, learning is subject to managerial discretion and control. Lapré et al. (2000) derived a learning-curve model for quality improvement in a dynamic production model. Terwiesch and Bohn (2001) concentrated in their study on deliberate learning through experiments using the production process as a laboratory.

They emphasized that such experiments are essential for diagnosing problems and testing proposed solutions and process improvements. However, these researchers have criticized the traditional approach to modeling performance as a function of cumulative production, but they have never proposed that cumulative production should be excluded. This remains an open research question to be investigated by researchers in future studies.

30.2.8 Quality-Learning Relationship

Abernathy et al. (1981) and Fine (1986) related the question of why some organizations learn faster than their competitors to product quality.

Some researchers have called for accounting for quality when modeling the learning curve. For example, Levy (1965) is believed to be the first to capture the linkage between quality and learning when he introduced the concept of autonomous learning (learning by doing) and induced learning (learning from continuous improvement efforts) that was later extended by Dutton and Thomas (1984).

Fine (1986) pointed out that quality-related activities can lead to the discovery of inefficiencies in the production process, thus providing an opportunity for learning. Tapiero (1987) established a linkage between quality control and the production learning process, in which he proposed a model to determine optimal quality control policies with inspection providing an opportunity to learn about the process. Garvin (1988), with examples from the Japanese manufacturing organizations, discussed the common sources of improvement for quality and productivity. Lower quality implies higher scrap and rework, which in turn means wasted material, intensive, equipment time, and other resources. Chand (1989), who considered the impact of learning on process quality, was unable to specify the shape of the quality learning curve. Koulamas (1992) hypothesized that during the production process, a small number of quality-related problems are encountered and that these decrease with time due to learning effects. Badiru (1995a) defined quality in terms of the loss passed on to the customer. He further added that learning affects worker performance, which ultimately can affect product quality. Li and Rajagopalan (1997) empirically studied the impact of quality on learning by addressing the following questions: (1) how well does the cumulative output of defective or good units explain the learning-curve effect? (2) Do defective units explain learning-curve effects better than good units? (3) How should cumulative experience be represented in the learning-curve model when the quality level may have an impact on learning effects? The results of Li and Rajagopalan (1997) supported the findings of Fine (1986) that learning is the bridge between quality improvement and productivity increase. This corroborates the observation of Deming (1982) that quality and productivity are not to be traded off against each other; instead, productivity increases follow from quality improvement efforts. Li and Rajagopalan (1998b) presented a model based on economic tradeoffs to provide analytical support for the continuous improvement philosophy. They explicitly modeled knowledge creation as the result of both autonomous and induced learning and considered the impact of knowledge on both production costs and quality.

As previously mentioned, investigating the learning–quality relationship has been attracting the attention of many researchers. Among the recent works are those of Vits and Gelders (2002), Lapré and Van Wassenhove (2003), Franceschini and Galetto (2003), and Allwood and Lee (2004). These works could be described as attempts at modeling the "Quality Learning Curve," but still more research is required in this area. The only analytical model that has a strong theoretical foundation is that of Jaber and Guiffrida (2004). Unfortunately, this model has not been validated empirically.

The learning–quality relationship has also attracted the attention of researchers in the service industry. For example, see the works of Waldman et al. (2003) and Ernst (2003), who investigated the learning-curve theory and the quality of service in healthcare systems.

30.2.9   Some Recent Trends in Learning-Curve Research

The last 20 years showed new trends in learning-curve research. Some of these works were geared toward a better understanding of the learning process and are surveyed below.

Globerson and Seidmann (1988) investigated the behavioral pattern of individuals when performing a typical industrial task, with their study aimed at investigating the interacting effects of imposed goals, expressed by a dictated learning pace and the individual's manual performance. Their results revealed that such imposition has an adverse effect when the imposed pace seems to be too difficult. However, subjects can outperform their learning curves if they are motivated to do so by other techniques such as an amplified incentive scheme.

Globerson and Millen (1989) investigated the interaction between "group technology" and learning curves, and the changes that should be introduced into learning-curve models to fit an environment where multiple products are processed across some identical operations. They developed a model allowing for the explicit consideration of shared learning resulting from the application of group technology principles.

Adler (1990) explored the forms of learning that characterize the evolution of productivity performance at Hi-Tech, a multinational, multiplant electronics firm. The results from this study suggested that three forms of shared learning are critical to modeling manufacturing productivity improvements: (1) sharing across the development/manufacturing interface, (2) sharing between the primary location and plants that started up later, and (3) the ongoing sharing among plants after start-up.

Dar-El et al. (1995a) noted the problems associated with using the traditional learning curve in industrial settings where learning can occur at different speeds in different phases. They developed and validated the dual-phase learning model (DPLM), which expresses learning as a combination of cognitive and motor skills. A natural result of this finding is that the learning constant for tasks involving cognitive and motor learning is not "constant," as is assumed in the classical learning curve of Wright (1936).

Mazzola and McCardle (1997) generalized the classical, deterministic learning curve of Wright (1936) to a stochastic setting. An interesting outcome of this analysis is the discovery that faster learners do not necessarily produce more.

Bailey and Gupta (1999) investigated whether human judgment can be of value to users of industrial learning curves, either alone or in conjunction with statistical models. Experimental results indicate substantial potential for human judgment to improve predictive accuracy in the industrial learning-curve context.

Smunt (1999) attempted to resolve the confusion in the literature concerning the appropriate use of either a unit or cumulative average learning curve for cost projections. He provided guidelines for the use of these learning-curve models in production research and cost estimation. Waterworth (2000) reported work similar to that of Smunt (1999).

Schilling et al. (2003) attempted to provide answers to the following questions: (1) is the learning rate maximized through specialization? (2) or does variation, related or unrelated, enhance the learning process? They found that there are no significant differences in the rates of learning under the conditions of specialization and unrelated variation. Schilling et al.'s results yield important implications for how work should be organized and for future research into the learning process.

Jaber and Guiffrida (2004) extended upon WLC (Wright, 1936) by assuming imperfect production. A composite learning curve, which is the sum of two learning curves was developed. The first learning curve describes the reduction in time for each additional unit produced, where the second learning curve described the reduction in time for each additional defective unit reworked. The composite learning model was found to have three behavioral patterns: concave, plateau, and monotonically decreasing.

These patterns provided valuable managerial insights. A convex behavior may caution managers not to speed up production without also improving the quality of the process. This finding is in harmony with that of Hatch and Mowery (1998), who observed that some new processes experience yield declines following their introduction, which may reflect the impact of rapid expansion in production volumes before a new process is fully stabilized or characterized. A plateau behavior may provide managers with an indication as to when an additional investment in training or new technology might be required to break the plateau barrier. This suggests that the plateauing usually observed in learning-curve data might be attributed to problems in quality. Finally, a monotonically decreasing behavior of the quality learning curve could occur because the time to rework a defective item becomes insignificant. This might be the reason why the WLC, being observed in an aircraft manufacturing facility where the cost of producing a defective plane is extremely high, assumed that all units produced are of good quality.

There is a plethora of work on learning curves; however, a complete understanding of how the learning process occurs is still unattainable. The coming years may bring interesting, if not tantalizing, findings in this research subject.

30.3 The Forgetting Phenomenon

The collapse of trade barriers among nations has transformed the market to a global and competitive one. This change in the market environment has imposed tremendous pressures on companies to deliver quality products at competitive prices in a short period of time. Shorter product life cycle is becoming the norm, requiring companies to reduce the time from concept to market and also requiring a better responsiveness to market changes. To cope with these pressures, manufacturing companies have been trying to be responsive, efficient, and flexible. As a result, the occurrence of worker learning and forgetting effects is becoming quite common in such manufacturing environments (Wisner and Siferd, 1995; Wisner, 1996).

"Learning is the essence of progress, forgetting is the root of regression" (Badiru, 1994). Although there is almost unanimous agreement by scientists and practitioners on the form of the learning curve presented by Wright (1936), as discussed in Section 30.1, scientists and practitioners have not yet developed a full understanding of the behavior and factors affecting the forgetting process.

Contrary to the plethora of literature on learning curves, there is a paucity of literature on forgetting curves. This paucity of research has been attributed probably to the practical difficulties involved in obtaining data concerning the level of forgetting as a function of time. (Globerson et al., 1989).

Many researchers in the field have attempted to model the forgetting process mathematically, experimentally, and empirically. In this section, we will review these models and shed light on potential ones.

30.3.1 What Causes Forgetting?

There is enough empirical evidence that knowledge depreciation (forgetting) occurs in organizations (e.g., Argote et al., 1990; Argote, 1993). Argote et al. (1990) reported that it was not clear why knowledge depreciates. They reported that knowledge could depreciate because of of worker turnover, as workers leave and take their knowledge with them. Knowledge may also depreciate due to changes in products or processes that make previous knowledge obsolete. Argote et al. (1990) further suggested that depreciation in knowledge could occur if organizational records were lost or became difficult to access. The work of Argote et al. (1990) was based on a data set from shipyards, which was used in a number of learning-curve studies. Further evidence of knowledge depreciation was also reported in pizza stores (Darr et al., 1995) and in a truck plant (Epple et al., 1996).

As discussed in Section 30.2.1, psychologists have reported that forgetting occurs in any of the following situations: (1) when there is not enough similarity between the conditions of encoding and retention of material learned, (2) when old learning interferes with new learning, and (3) when there is an interruption in the learning process for a period of time (production break). In industrial engineering literature, production breaks are viewed as the main cause of forgetting. For example, Hancock (1967) indicated that very short breaks have no effect on learning, whereas longer breaks retard the learning process. Anderlohr (1969) and Cochran (1973) agreed that the length of the production break has a direct effect on the degree to which humans forget. However, they also attributed forgetting to other factors such as the availability of the same personnel, tooling, and methods. Production breaks are typical today in organizations where product and process variety are high (Boone and Ganeshan, 2001), yet there has been no model developed for industrial settings that considers forgetting as a result of factors other than production breaks. This is a potential area of future research.

30.3.2   The Form of the Forgetting Curve: Power or Exponential?

The behavior of the knowledge decay (forgetting) function is frequently assumed to be of the same form as the standard progress function except that the forgetting rate is negative whereas the learning rate is positive (Badiru, 1994). This is consistent with the suggestion of Globerson et al. (1989) that learning and forgetting may be considered as mirror images of each other. As discussed in the previous section, a common assumption among these models is that the length of the interruption period is the primary factor in the deterioration of performance.

The debate over whether forgetting is really a power law, as opposed to exponential, has attracted the attention of several researchers. Wixted and Ebbesen (1991) conducted three experiments to identify the nature of the forgetting function. Two experiments involved human subjects and one involved pigeons. The first experiment involved the encoding and retrieval of a list of words, while the second involved face recognition followed by a retention interval. In their study, six potential forgetting functions were investigated. These functions are the linear, exponential, hyperbolic, logarithmic, power, and exp-power. The results of Wixted and Ebbesen (1991) showed that the exponential function, which describes the behavior of many natural processes, did not improve much over the linear function. The hyperbolic function (Mazur and Hastie, 1978) did not fit well. However, the exp-power function reformed reasonably well, although it was outperformed by the power function. In a later study, Wixted and Ebbesen (1997), conducting an analysis of the forgetting functions of individual participants, confirmed their earlier findings, i.e., a forgetting function is better described by a power function than by an exponential function. Refer to the works of Sikström (1999) and Heathcote et al. (2000).

Wright (1936) is believed by many researchers to be the first to investigate learning in an industrial setting. However, to industrial engineers, the forgetting phenomenon is fairly new. It was not until the 1960s that the first attempt was made to investigate the opposite phenomenon, i.e., forgetting. Researchers in both disciplines unanimously agree that the learning process is negatively affected when subjects are interrupted for a significant period of time, over which some of the knowledge acquired in earlier learning sessions is lost. Industrial engineers have not debated the form of the forgetting curves to the extent that psychologists did. In general, they assumed that the forgetting curve is a mirror image of the learning curve (Globerson et al., 1989), i.e., of a power form, except that it is upward rather than downward. That is, the suggested form of the forgetting curve is similar to Eq. (30.1) and is given as

$$\hat{T}_x = \hat{T}_1 x^f \tag{30.16}$$

where $\hat{T}_x$ is the time for the xth repetition of lost experience on the forgetting curve, $\hat{T}_1$ the intercept of the forgetting curve, and f the forgetting exponent. Figure 30.2 illustrates such behavior of Eqs. (30.1) and (30.16).

Existing studies addressing forgetting in industrial settings can be categorized into three groups. One set of researchers has focused on modeling forgetting mathematically. Other researchers have focused on modeling forgetting experimentally using data collected from laboratory experiments performed by students. Finally, some researchers have modeled forgetting curves using empirical data from real-life settings.

30.3.3   Mathematical Models of Forgetting

Unlike industrial learning curves, which have been studied for almost seven decades, a full understanding of the behavior and factors affecting the forgetting process has not yet been developed. The earliest

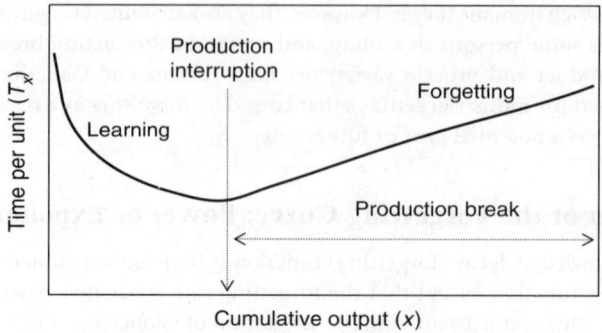

FIGURE 30.2   The learning–forgetting process.

attempts by industrial engineers to model the forgetting functions, Hoffman (1968) and later Adler and Nanda (1974), presented refined mathematical techniques for incorporating the effects of production breaks into production planning and control models. However, these models do not conform to the learning–forgetting relationship illustrated in Figure 30.2. Thus, those models conforming to the relationship described in Figure 30.2 are discussed. These models are the variable regression variable forgetting model (VRVF; Carlson and Rowe, 1976), the variable regression invariant forgetting model (VRIF; Elmaghraby, 1990), the learn–forget curve model (LFCM; Jaber and Bonney, 1996), and the power integration diffusion (PID) model (Sikström and Jaber, 2002).

How do the VRVF, VRIF, and LFCM differ from each other? Jaber and Bonney (1997) addressed this question by conducting a numerical comparison of the three models. Both the VRVF and VRIF use a fixed and externally specified forgetting rate (exponent), f; however, the VRVF uses a variable intercept for the forgetting curve, $\hat{T}_1$, whereas the latter uses a fixed intercept for the forgetting curve. Contrary to these models, the LFCM uses both a variable forgetting exponent and a variable intercept for the forgetting curve. All three models hypothesize two important relationships between learning and forgetting. The first is that when total forgetting occurs, the time to process reverts to the time required to process the first unit with no prior experience. The second relationship is that the performance time on the learning curve equals that on the forgetting curve at the point of interruption. Jaber and Bonney (1997) show that the VRIF satisfies the first relationship but not the second, and that the VRVF satisfies the second relationship but not the first. The LFCM satisfied both relationships. This implies that the LFCM satisfies the characteristic that learning and forgetting are mirror images of each other (Jaber et al., 2003). Also, the LFCM was found to fit the time estimates to the experimental (laboratory) data of Globerson et al. (1989) with less than 1% error. Jaber and Sikström (2004a) empirically compared the LFCM to available empirical data from Nembhard and Osothsilp (2001). The LFCM fitted the data well.

In a recent paper, Jaber et al. (2003) identified seven characteristics of forgetting that should be considered by a learning–forgetting model, and evaluated existing learning–forgetting models to assess the degree to which they incorporate these characteristics. The characteristics of forgetting that were inferred from laboratory and empirical studies are the following: (1) the amount of experience gained before interruption occurs in the learning process influences the level of forgetting; (2) the length of the interruption interval influences the level of forgetting; (3) the relearning rate is the same as the original learning rate, where relearning is faster after a break; (4) the power function is appropriate for capturing forgetting; (5) learning and forgetting are mirror images of each other; that is, the learning curve can be summarized by a power function that increases with learning, and forgetting can be summarized with a power function that decays with time; (6) the level of forgetting is positively related to the rate that worker learns; that is, workers who learn rapidly also tend to forget rapidly; and (7) the nature of the task being performed, that is, whether the task is cognitive or manual. Jaber et al. (2003) found that the LFCM handles (1) through (6), whereas (7) was not tested.

Recently, the LFCM was subjected to two extensions. First, Jaber and Kher (2002) developed the dual-phase learning–forgetting model (DPLFM) to predict task times in an industrial setting. Combining the DPLM (Dar-El et al., 1995a) and the LFCM forms the DPLFM. The DPLM expresses learning as a combination of cognitive and motor skills. The results of Jaber and Kher (2002) imply that a worker who learns rapidly also tends to forget rapidly. This behavior of the model concurs with the empirical findings reported in Nembhard (2000) and Nembhard and Uzumeri (2000) in industrial settings. Their results also showed that workers' learning tasks that are dominated by motor elements are less susceptible to forgetting, while workers' learning tasks that have a greater cognitive content (i.e., complex tasks) are more susceptible to forgetting. Second, Jaber et al. (2003) extended the LFCM by accounting for job similarity factor. The new model was investigated in a job shop where a worker who is being trained on two or three similar tasks is likely to experience relatively less forgetting as compared to the workers being trained on very dissimilar tasks. Jaber et al. (2003) found that with increasing similarity, the importance of upfront training and transfer policy decline. This result has important implications for work environments in which extensive worker flexibility is desirable for performing a set of closely related tasks. Third, Jaber and Kher (2004) corrected the assumption of the LFCM that the time for total forgetting is invariant of the experience gained prior to interruption. This was done by incorporating the findings of Hewitt et al. (1992) into the LFCM.

The model suggested by Sikström and Jaber (2002) combines three basic findings, namely, that single memory traces decay according to a power function of the retention interval, that memory traces can be combined by integration, and that the time to produce a unit can be described by a diffusion process on the memory trace. This model is referred to as the PID model. The basic idea of PID is that every time a task is performed, a memory trace is formed. The strength of this trace decays as a power function over time. When the same task is repeated, an aggregated memory trace can then be found by integrating the strength of the memory trace over the time interval that the task is repeated. The integral of a power function is a power function. The time it takes to perform a task is determined by "a diffusion" process in which the strength of the memory constitutes the signal. Unlike the VRVF, VRIF, and LFCM, the PID uses a single parameter to capture both learning and forgetting. This also makes the model relatively easy and the predictions from the model straightforward. Furthermore, the assumptions of power-function decay, integration, and a diffusion process are plausible and have empirical support.

The LFCM and the PID models will be discussed further in a later section as potential models for capturing the learning–forgetting process in industrial settings.

30.3.4 Experimental Models of Forgetting

A second group of researchers performed laboratory experiments resulting in a better understanding of the learning–forgetting relationship. Globerson et al. (1989) indicated that the degree of forgetting is a function of the level of experience gained prior to interruption and the length of interruption. Bailey (1989) concluded that the forgetting rate is a function of the amount learned and that with the passage of time, the relearning rate is not correlated with the learning rate, and the learning rate is highly correlated with the time taken to complete the first unit. Shtub et al. (1993) used the forgetting model and the data of Globerson et al. (1989) to partially validate Bailey's hypothesis. Bailey used Erector set assembly and disassembly as his experimental tasks, whereas Globerson et al. (1989) and Shtub et al. (1993) simulated the operations of a data-entry office in a microcomputer laboratory. All three studies concur that forgetting is influenced by the amount learned prior to the interruption as well as the length of the interruption interval in the learning process. Thus for a given level of experience, a worker interrupted for a shorter duration will likely forget less than a worker interrupted for a longer duration. Similarly for a given interval length, a worker with greater experience will tend to forget less than a less-experienced worker. Bailey (1989) and Shtub et al. (1993) also concluded that forgetting is not a function of the worker's learning rate. Globerson et al. (1989), Sparks and Yearout (1990), and Shtub et al. (1993) supported the finding that the relearning rate is the same as the original learning rate. This has been debated by several researchers, with partial support in Bailey and McIntyre (1997). However, Hewitt et al. (1992) indicated a lack of correlation between the original learning rate and the relearning rate.

Dar-El et al. (1995a) suggested that learning an industrial task consists of two parallel phases: cognitve and motor. They collected data from an experimental apparatus of assembling electric matrix board and electronic components that involved one complex task and another simple one. Dar-El et al. (1995a) concluded that the learning decreases as experience is gained. In a subsequent paper, Dar-El et al. (1995b) defined forgetting as a consequence of a specific sub-task reappearing in the next cycle after a whole cycle time of other activities is completed. Dar-El et al. (1995b) empirically determined forgetting as a function of the learning constant and interruption length. Arzi and Shtub (1997), who compared learning and forgetting of cognitive and motor tasks, concluded that forgetting intensity is affected only by the rate of learning before the break. This finding is in contradiction with those of Bailey (1989) and Shtub et al. (1993). Finally, Bailey and McIntyre (2003) developed and tested the parameter prediction models (PPMs) to predict relearning curve parameters following a production break. The potential usefulness of PPMs is that they can provide a means of estimating a post-break production time well before post-break data become available. Although the learning rate is less predictable, Bailey and McIntyre (2003) found that the best PPM showed that it was related to the original learning rate and the length of the break. Their study could be considered as a correction of an earlier study by Bailey (1989) who was unable to correlate the relearning rate to the original learning rate before a break.

The above studies indicate that a full understanding of the factors that affect the learning–forgetting process is not yet attainable. More research is required in this area.

30.3.5 Empirical Forgetting Models

According to Globerson and Levin (1987; p. 81), "One of the major reasons that forgetting has not been studied is probably due to the practical difficulties in obtaining data concerning the level of forgetting as a function of time." This is reflected in the limited number of studies that investigate the learning and forgetting relationship using empirical data. These studies are those of Badiru (1995a) and Nembhard and Uzumeri (2000).

Badiru (1995a) presented an approach for multivariate learning-curve models that accounts for periods of learning and forgetting. Badiru's empirical data consist of 48 monthly observations. There are four independent variables: (1) production level, (2) number of workers, (3) number of hours of production downtime, and (4) the number of hours of production rework. The independent variable is the average production cost per unit. Badiru (1995a) concluded that the average unit cost would be underestimated if the effect of downtime hours were not considered. Thus, a multivariate model is more accurate in representing the learning and forgetting relationship. Unfortunately, Badiru's study (1995a) neither specified a forgetting model, nor did it investigate intensive performance.

Nembhard and Uzumeri (2000) pointed out an important limitation of experimentally derived models, which is the fact that these models assume a single production break and that the time the break occurs is known. They remarked that in real production systems, multiple breaks will occur and they will occur sporadically. Nembhard and Uzumeri (2000) addressed two research questions in examining patterns of worker variation with respect to learning and forgetting. First, how are patterns of learning behavior distributed among a workforce? Second, how is forgetting related to the other learning characteristics, and to what extent is this relation specific to the type of task involved (manual or cognitive)? The data for the manual task consist of performance measures taken at frequent intervals in a textile manufacturing plant. The data for the procedural task were collected on the final-assembly test-inspection stations for units of an automotive electronics component. In their paper, Nembhard and Uzumeri (2000) presented a model of learning and forgetting that is an extension of an earlier model (Mazur and Hastie, 1978; Uzumeri and Nembhard, 1998), shown to be useful for the tasks involved in their study. In their paper, Nembhard and Uzumeri (2000) introduced a concept called the "recency" of experiential learning. They defined the recency measure, R_x, as how recently an individual's practice was obtained of cumulative production x. The recency model (RC) is considered to be a modification of the three-parameter hyperbolic learning functions of Mazur and Hastie (1978). The results of Nembhard and Uzumeri (2000) indicate that despite the differences between the tasks and whether the worker was new or being retrained, workers who learn more gradually tended to reach a higher steady-state rate of productivity. In addition, workers who learned more rapidly also tended to forget more rapidly during breaks in production. Nembhard and Uzumeri (2000) recommended that workers who learn gradually should be scheduled to long production runs and could also be cross-trained on secondary tasks. However, they recommended that workers with fast learning should be scheduled to short production runs and should not be cross-trained on secondary tasks.

In a follow-up paper, using the procedural data set in Nembhard and Uzumeri (2000), Nembhard and Osothsilp (2001) compared 14 different published models of forgetting breaks. There were 11 statistical models; seven GLS models (Globerson et al., 1989), the GNE model (Globerson et al., 1998), the exponential model and the S-shaped model (Globerson and Levin, 1987), and the RC model (Nembhard and Uzumeri, 2000). There were three deterministic models, the VRVF model (Carlson and Row, 1976), the VRIF model (Elmaghraby, 1990), and the LFCM model (Jaber and Bonney, 1996). They concluded that the RC model performed the best, in terms of efficiency, stability, and balance, of these models. Jaber and Sikström (2004a) contradicted the findings of Nembhard and Osothsilp (2001). Nembhard and Osothsilp (2001) found that the LFCM showed the largest deviation from empirical data. Jaber and Sikström (2004a) demonstrated that the poor performance of the LFCM in the study of Nembhard and Osothsilp (2001) might be attributed to an error on their part when fitting the LFCM to their empirical data. Both the LFCM and the RC models will be discussed in further detail in the next section.

Nembhard (2000) examined the effects of task complexity and experience on parameters of individual learning and forgetting. In this study, three attributes of task complexity and experience were addressed: the method, machine, and material employed. The data used were collected, consisting of performance measures taken from a large number of learning/forgetting episodes. The study took place at a textile manufacturing plant. Nembhard (2000) used the RC model in that study (Uzumeri and Nembhard, 1998; Nembhard and Uzumeri, 2000). The results from this study indicated that complexity significantly affects learning/forgetting parameters and that the effects depend on the experience of the workers. Nembhard (2000) also found that task complexity and experience were found to be useful in predicting individual learning and forgetting characteristics.

Using the same data set as in Nembhard (2000) and the RC model, Nembhard and Osothsilp (2002) examined the effects of task complexity on the distribution of individual learning and forgetting parameters. The results from this study indicated that task complexity significantly affects the variance of individual learning rates, forgetting rates, and steady-state productivity rates, where the variability of these parameters among individuals increase as task complexity increases. Nembhard and Osothsilp (2002) adopted the approach of Goldman and Hart (1965) that identifies tasks with longer assembly times indicate a higher information content and greater complexity. However, this may seem to be a simplistic measure of task complexity. This issue requires further research.

30.3.6 Potential Models

As aforementioned, there are three learning and forgetting models that have promising applications in industrial settings. These models are the LFCM (Jaber and Bonney, 1996), RC (Nembhard and Uzumeri, 2000), and PID (Sikström and Jaber, 2002) models. All three models assume that learning conforms to that of Wright (1936) as described in Section 30.1. The mathematics of these models are presented below.

30.3.6.1 Mathematics of the Learn-Forget Curve Model

Jaber and Bonney (1996) suggest that the forgetting curve exponent could be computed as

$$f_i = \frac{b(1-b)\log(u_i+n_i)}{\log(1+D/t(u_i+n_i))} \tag{30.17}$$

where $0 \leq f_i \leq 1$, n_i is the number of units produced in cycle i up to the point of interruption, D the break time to which total forgetting occurs, and u_i the number of units remembered at the beginning of cycle i

from producing x_{i-1} in previous $i-1$ cycles (note that in i production cycles there are $i-1$ production breaks), where $x_{i-1} = \sum_{j=1}^{i-1} n_j$ and $0 < u_i < x_{i-1}$. That is, if the learning process is interrupted at a time of length D, then performance reverts to a threshold value, usually equivalent to T_1. Let $t(u_i + n_i)$ denote the time to produce $u_i + n_i$ units (equivalent units of cumulative production accumulated by end of cycle i), and b the learning-curve constant in Eq. (30.1). $t(u_i + n_i)$ is computed from Eq. (30.1) as

$$t(u_i+n_i) = \sum_{x=1}^{n_i} T_1(u_i+x)^{-b} \cong \int_0^{u_i+n_i} T_1 x^{-b}\, dx = \frac{T_1}{1-b}\,(u_i+n_i)^{1-b} \tag{30.18}$$

The number of units produced at the beginning of cycle $i + 1$ is given from Jaber and Bonney (1996) as

$$u_{i+1} = (u_i+n_i)^{(1+f_i/b)} y_i^{-f_i/b} \tag{30.19}$$

where $u_1 = 0$, and y_i the number of units that would have been accumulated, if production was not ceased for d_i units of time. y_i is computed from Eq. (30.18) as

$$y_i = \left\{ \frac{1-b}{T_1}[t(u_i+n_i)+d_i] \right\}^{1/(1-b)} \tag{30.20}$$

When total forgetting occurs we have $u_{i+1} = 0$. However, from Eq. (30.4), $u_{i+1} \to 0$ as $y_i \to +\infty$; or alternatively, as $d_i \to +\infty$, where all the other parameters in Eq. (30.4) are of nonzero positive values. Thus, we deduce that total forgetting occurs only when d_i holds a very large value. This does not necessarily contradict the assumption of finite value of D at which total forgetting occurs. Doing so, $u_{i+1} < 1$ when $d_i = D$, and it plateaus at zero for increasing values of $d_i > D$. However, assuming total forgetting will occur might not seem unrealistic given that McKenna and Glendon (1985) and Anderlohr (1969) empirically reported such findings. The intercept of the forgetting curve could be determined as

$$\hat{T}_{1i} = T_1(u_i+n_i)^{-(b+f_i)} \tag{30.21}$$

The time to produce the first unit in cycle i could then be predicted from Eq. (30.1) as

$$\tilde{T}_{1i}^{\text{LFCM}} = T_1(u_i+1)^{-b} \tag{30.22}$$

30.3.6.2 Mathematics of the Recency Model

The RC model has the capability of capturing multiple breaks. Nembhard and Uzumeri (2000) modified the three hyperbolic learning functions of Mazur and Hastie (1978) by introducing a measure they termed recency of experiential learning, R. For each unit of cumulative production x, Nembhard and Uzumeri (2000) determined the corresponding recency measure, R_x, by computing the ratio of the average elapsed time to the elapsed time of the most recent unit produced. Nembhard and Osothsilp (2001) suggested that R_x could be computed as

$$R_x = 2\frac{\sum_{i=1}^{x}(t_i-t_0)}{x(t_x-t_0)} \tag{30.23}$$

where x is the accumulated number of produced units, t_x the time when units x are produced, t_0 the time when the first unit is produced, t_i the time when unit i is produced, and $R_x \in (1,2)$. Altering Eq. (30.1), the performance of the first unit after a break could be computed as

$$\tilde{T}_{1i}^{RC} = T_1(xR_x^\alpha)^{-b} \tag{30.24}$$

where α is a fitted parameter that represents the degree to which the individual forgets the task. However, Nembhard and Uzumeri (2000) and Nembhard and Osothsilp (2001) did not provide evidence to how Eq. (30.8) was developed, or the factors affecting α.

30.3.6.3 Mathematics of the Power Integration Diffusion Model

The PID model of Sikström and Jaber (2002) advocates that each time a task is performed, a memory trace is formed. The strength of this trace decays as a power function over time. For identical repetitions of a task, an aggregated memory trace could be found by integrating the strength of the memory trace over the time interval of the repeated task. The integral of the power-function memory trace is a power function. The time it takes to perform a task is determined by a "diffusion" process where the strength of the memory constitutes the signal. The strength of the memory trace follows a power function of the retention interval since training was given. That is, the strength of a memory trace (at which t time units has passed between learning and forgetting) encoded during a short time interval of (dt) is

$$S'(t) = S_0 t^{-a} dt \tag{30.25}$$

where a is the forgetting parameter, $a \in (0, 1)$, and S_0 a scaling parameter > 0 (compare with the parameter in other models that represents the time to produce the first unit).

The strength of a memory trace encoded for an extended time period is $S(t_{e,1}, t_{e,2})$, where $t_{e,1}$ time units passed since the start of encoding of unit e and $t_{e,2}$ time units passed since the end of encoding of unit e and $t_{e,1} > t_{e,2}$. This memory strength can be calculated by the integral over the time of encoding

$$S(t_{e,1}, t_{e,2}) = \int_{t_{e,1}}^{t_{e,2}} S'(t) dt = \frac{S_0}{1-a} \left[t_{e,2}^{1-a} - t_{e,1}^{1-a} \right] \tag{30.26}$$

The strength of the memory trace following encoding during N time intervals is the sum over these intervals, and it is determined from Eq. (30.26) as

$$S(t_{e,1}, t_{e,2}) = \frac{S_0}{1-a} \sum_{e=1}^{N} \left[t_{e,2}^{1-a} - t_{e,1}^{1-a} \right] \tag{30.27}$$

The time to produce a unit is the inverse of the memory strength, and is given from Eq. (30.27). The start time of the diffusion process constitutes a constant (t_0) that is added to the total time to produce a unit

$$T(t_r) = S(t_{e,1}, t_{e,2})^{-1} + t_0 = \frac{1-a}{S_0} \left\{ \sum_{e=1}^{N} \left[t_{e,2}^{1-a} - t_{e,1}^{1-a} \right] \right\}^{-1} + t_0 \tag{30.28}$$

$$= S_0' \left\{ \sum_{e=1}^{N} \left[t_{e,2}^{a'} - t_{e,1}^{a'} \right] \right\}^{-1} + t_0$$

where $S_0' = \dfrac{1-a}{S_0}$, is a rescaling of S_0, and $a' = 1 - a$, $a' \in (0, 1)$, is and a rescaling of a. The rescaling of the parameters is introduced for convenience to simplify the final expression.

30.3.6.4 Numerical Comparison of the Models

Jaber and Sikström (2004b) conducted a numerical study to investigate and discuss the three models described above. Their results indicate that for a moderate learning scenario (where the learning rate classifies a task as being more cognitive than motor), it might be difficult to differentiate between the three models. As learning becomes faster (as a task becomes highly cognitive) the predictions of the LFCM and the RC models are below those of PID. Conversely, for slower learning scenarios (as a task becomes highly motor) the predictions of the LFCM and the RC models are above those of PID. However, and for both cases, the predictions of LFCM were considered, on the average, to be closer to those of PID than the predictions of the RC model. Numerical results for the PID and LFCM further suggest that as learning becomes slower, forgetting becomes faster. This result is inconsistent with that of the RC model, which suggests that fast (slow) learners forget faster (slower). The RC model also showed that regardless of the length and frequency of the production break, forgetting is most significant in earlier than in later cycles where cumulative production is larger. These findings are inconsistent with Jost's law of forgetting, but they are consistent with those of the PID and the LFCM, where the length and frequency of the production break is significant in all cycles.

The results of Jaber and Sikström (2004b) also indicate that the PID and RC models, and the PID and the LFCM models, could best be differentiated for cases characterized by high initial processing times, long production breaks, and for tasks that are identified as being more motor than cognitive. Under those circumstances the deviations between the models were at a maximum. Finally, the three models were investigated for the phenomenon of plateauing. Results indicate that PID could be a better model at capturing plateauing than the RC and LFCM models.

In summary, these results suggest possible ways to differentiate between models that capture learning and forgetting. Based on these results, Jaber and Sikström (2004b) suggest that these models should further be investigated in different industrial settings before concluding which model of the three proposed, LFCM, RC, or PID, most accurately describes the learning–forgetting process.

30.4 Applications to Industrial Engineering Problems

The application of the WLC to industrial engineering and management problems has received increasing attention by researchers in the field over the years. For example, and not limited to, budgeting (Summers and Welsch, 1970), lot sizing (Wortham and Mayyasi, 1972), break-even analysis (McIntyre, 1977), line balancing (Dar-El and Rubinovitz, 1991), bidding and planning (Reis, 1991), resource allocation decisions (Badiru, 1995b), lot-splitting (Eynan and Li, 1997), purchasing (Sinclair, 1999), single-machine scheduling (Biskup, 1999), teleportation (Bukchin et al., 2002), supply chain management (Macher and Mowery, 2003), and cellular manufacturing (Kannan and Jensen, 2004). However, the simultaneous application of learning and forgetting has been limited, to the best of the author's knowledge, to lot sizing, dual resource constrained (DRC) job shops, and project management. Thus, in this section, the application of the learning–forgetting process to these three areas will be briefly discussed.

30.4.1 Lot-Sizing Problem

Keachie and Fontana (1966) are believed to be the first to investigate the effect of both learning and forgetting in production on the lot-sizing problem. They assumed an economic order quantity model (EOQ) situation with two extreme cases: (1) either a full transmission of learning between cycles, i.e., no forgetting, (2) or total loss of learning, i.e., forgetting. The research problem has caught the attention of several researchers since then. However, up to the 1980s, the modeling of forgetting has been a simplistic one, and the impact of the length of the production break was not taken into account. The only works that account for the length of the production break when modeling the lot-size problem are those of Elmaghraby (1990) and Jaber and Bonney (1996). For a nearly comprehensive review, readers may refer to Jaber and Bonney (1999). Results from this research advocate that with learning it is recommended to produce in smaller lots more frequently. However, when forgetting is accounted for, it is recommended to produce in larger lots less frequently. In either case, the lot-size quantity is bounded by the economic manufacture quantity (EMQ) model, and the EOQ quantities, where $EMQ = \sqrt{2Kr/h(1-r/p)}$ and $EOQ = \sqrt{2Kr/h}$, with K being the set-up cost, r the demand rate, p the production rate without learning effects, and h the holding cost per unit.

Some researchers extended the EMQ/EOQ models with learning and forgetting in several directions. Jaber and Abboud (2001) investigated the simultaneous effect of learning and forgetting in production when production breaks occur randomly. Chiu et al. (2003) studied the deterministic time-varying demand with learning and forgetting in set-ups and production considered simultaneously. Jaber and Bonney (2003) studied the effects that learning and forgetting have in set-ups and product quality on the lot-sizing problem, but they did not consider learning and forgetting in production.

However, the above models were not investigated in a supply chain management context. This might be a plausible immediate extension.

30.4.2 Dual Resource Constrained Job Shop

The DRC system is a job shop where both machines and intensive are limiting resources (Nelson, 1967). The introduction of the intensive constraint opened up a new level of design parameters, such as the quantity and quality of the intensive force. Nelson (1967) noted that as the quantity decreased, the degrees of possible worker flexibility increased, and therefore the quality of the workers increased. Conversely, as the quantity of the workforce increased, the amount of possible flexibility decreased, and therefore the quality of the workforce decreased.

The benefits of a flexible workforce, wherein workers are cross-trained to operate machines in functionally different departments, have been well documented in practice (Wisner and Siferd, 1995) and in the DRC literature (e.g., Treleven, 1989; Hottenstein and Bowman, 1998). Assigning workers to different departments is useful in alleviating the detrimental effects of bottlenecks resulting from machine breakdowns, product type changes, or external demand changes. Worker transfers in DRC systems emulate interruptions in the work environment. As a result, the relevancy of the learning and forgetting effect on DRC systems research is obvious.

Kher et al. (1999) were the first to address both learning and forgetting effects in a DRC system context with intensive attrition. They assumed that the forgetting follows the VRVF model developed by Carlson and Rowe (1976). McCreery and Krajewski (1999) examined worker deployment issues in the presence of the effects of task variety and task complexity on the performance of a DRC system with learning and forgetting effects. The forgetting model assumed was a simple linear function of the length of the interruption break. Shafer et al. (2001) examined the effect of learning and forgetting on assembly-line performance when the intensive force is heterogeneous. They assumed that the forgetting follows the RC model (Nembhard and Uzumeri, 2000). Jaber et al. (2003) extended upon the findings of Kher et al. (1999) by incorporating the LFCM of learning and forgetting of Jaber and Bonney (1996) instead of the VRVF model of Carlson and Rowe (1976). Jaber et al. (2003) also included the effect of the degree of job similarity in the LFCM model.

The major findings from the aforementioned articles are summarized herein. Kher et al. (1999) concluded that in the presence of high forgetting and attrition rates, workers do not even achieve their standard processing time efficiency. This suggests that in such situations, managers should focus on reducing the attrition rates of their workers rather than providing them with additional training. McCreery and Krajewski (1999) found that when both task complexity and product variety increase, cross-training should increase, but deployment should be restricted. On the other hand, when task complexity and product variety are low, a moderate amount of cross-training with a flexible deployment of the workforce is best. Shafer et al.'s (2001) results suggest, contrary to intuition, that the productivity of the workforce increases as the variability of the learning and forgetting parameters of the workers increase. In other words, the productivity of the system is greater if workers are modeled as having unique learning–forgetting distributions as compared to assuming a fixed distribution across workers. Jaber et al. (2003) found that with increasing task similarity, the importance of upfront training and transfer policy decline. This result has important implications for work environments in which extensive worker flexibility is desirable for performing a set of closely related tasks. However, there is a common consensus among these works that cross-training workers in more than three tasks worsens system performance.

There are ample possible extensions to the research in this area. For example, worker learning and forgetting and flexibility of the workforce in DRC systems could be investigated in the presence of machine flexibility (referred to as group technology). Another example is to account for process quality and reworks in DRC systems.

30.4.3 Project Management

Earlier research on learning in project management environments has been limited to multiple repetitions of a single project (e.g., Shtub, 1991; Teplitz and Amor, 1993). However, limited work has been done to account for the effect of forgetting. Ash and Smith-Daniels (1999) investigated the impact of learning, forgetting, and relearning (LFR) on project completion time when preemption is allowed in multiproject development environments. Their forgetting function is time-based, similar in form to that in Eq. (30.2). Ash and Smith-Daniels (1999) concluded that LFR cycle effects are significant for both the flow time and the mean resource utilization performance measures. Lam et al. (2001) explored the learning and forgetting phenomena that exist in repetitive construction operations, and their influence on project productivity. Lam et al. (2001) assumed that the forgetting function is S-shaped, as described in Globerson and Levin (1987). They emphasized that understanding the phenomenon of the learning and forgetting effect can enable the main contractor to better forecast the progress of different work tasks and have the resources and materials delivered on site in a "just-in-time" manner that can alleviate site congestion, which is a common problem in confined areas.

30.5   Summary and Conclusions

In this chapter, we have presented a survey of literature on the learning and forgetting process with applications to industrial engineering issues. While doing so, we have also suggested directions of possible future research.

The above survey suggests that although there is almost unanimous agreement by scientists and practitioners on the form of the learning curve presented by Wright (1936), scientists and practitioners have not yet developed a full understanding of the behavior and factors affecting the forgetting process. The paucity of empirical data on learning and forgetting makes it more difficult to attain such an understanding. This suggests that a close cooperation between the industry and academia is needed to encounter this problem. On this issue, Elmaghraby (1990, p. 208) wrote: "We fear that unless such empirical validation is undertaken, these, and other models shall take their place in literature as exercises in armchair philosophizing, not as 'doing science' in the field of production systems."

Some other pressing research questions that need to be addressed by researchers and practitioners are: (1) how learning and forgetting interact, (2) whether cumulative production alone is an adequate measure of organizational knowledge contributing to performance improvement, and (3) how knowledge transfers among members in a group, and across different groups in an organization.

The above survey indicates that the application of the learning–forgetting process has been limited to the lot-sizing problem, DRC systems, and project management. However, the author believes that there are ample opportunities to extend application of the learning–forgetting process to other industrial engineering areas, e.g., concurrent engineering, supply chain management, and virtual manufacturing.

References

Abernathy, W.J., Clark, K.B., and Kantorow, A.M., The new industrial competition, *Harv. Bus. Rev.*, 59, 68–81, 1981.

Adler, P.S., Shared learning, *Manage. Sci.*, 36, 938–957, 1990.

Adler, G.L. and Nanda, R., The effects of learning on optimal lot determination — single product case, *AIIE Trans.*, 6, 14 –20, 1974.

Alchian, A.A., Reliability of progress curves in airframe production, *Econometrica*, 31, 679–693, 1963.

Allwood, J.M. and Lee, W.L., The impact of job rotation on problem solving skills, *Int. J. Prod. Res.*, 42, 865–881, 2004.

Anderlohr, G., What production breaks cost, *Ind. Eng.*, 20, 34–36, 1969.

Argote, L., Group and organizational learning curves: individual, system and environmental components, *Br. J. Soc. Psychol.*, 32, 31–51, 1993.

Argote, L., Beckman, S.L., and Epple, D., The persistence and transfer of learning in industrial settings, *Manage. Sci.*, 36, 140–154, 1990.

Argote, L. and Epple, D., Learning curves in manufacturing, *Science*, 247, 920–924, 1990.

Arzi, Y. and Shtub, A., Learning and forgetting in mental and mechanical tasks: a comparative study, *IIE Trans.*, 29, 759–768, 1997.

Ash, R. and Smith-Daniels, D.E., The effects of learning, forgetting, and relearning on decision rule performance in multiproject scheduling, *Dec. Sci.*, 30, 47–82, 1999.

Badiru, A.B., Computational survey of univariate and bivariate learning curve models, *IEEE Trans. Eng. Manage.*, 39, 176–188, 1992.

Badiru, A.B., Multifactor learning and forgetting models for productivity and performance analysis, *Int. J. Hum. Factor. Man.*, 4, 37–54, 1994.

Badiru, A.B., Multivariate analysis of the effect of learning and forgetting on product quality, *Int. J. Prod. Res.*, 33, 777–794, 1995a.

Badiru, A.B., Incorporating learning curve effects into critical resource diagramming, *Project Manage. J.*, 26, 38–45, 1995b.

Bailey, C.D., Forgetting and the learning curve: a laboratory study, *Manage. Sci.*, 35, 340–352, 1989.

Bailey, C.D. and Gupta, S., Judgement in learning-curve forecasting: a laboratory study, *J. Forecasting*, 18, 39–57, 1999.

Bailey, C.D. and McIntyre, E.V., The relationship between fit and prediction for alternative forms of learning curves and relearning curves, *IIE Trans.*, 29, 487–495, 1997.

Bailey, C.D. and McIntyre, E.V., Using parameter prediction models to forecast post-interruption learning, *IIE Trans.*, 35, 1077–1090, 2003.

Baloff, N., Startups in machine-intensive production systems, *Ind. Eng.*, 17, 25–32, 1966.

Baloff, N., Estimating the parameters of the startup model — an empirical approach, *Ind. Eng.*, 18, 248–253, 1967.

Baloff, N., Startup management, *IEEE Trans. Eng. Manage.*, EM-17, 132–141, 1970.

Belkaoui, A.R., *The Learning Curve: A Management Accounting Tool*, Quorum Books, Westport, CT, 1986.

Bevis, F.W., Finniear, C., and Towill, D.R., Prediction of operator performance during learning of repetitive tasks, *Int. J. Prod. Res.*, 8, 293–305, 1970.

Biskup, D., Single-machine scheduling with learning considerations, *Eur. J. Oper. Res.*, 115, 173–178, 1999.

Boone, T. and Ganeshan, R., The effect of information technology on learning in professional service organizations, *J. Oper. Manage.*, 19, 485–495, 2001.

Buck, J., Tanchoco, J., and Sweet, A., Parameter estimation methods for discrete exponential learning curve, *AIIE Trans.*, 8, 185–194, 1976.

Bukchin, J., Luquer, R., and Shtub, A., Learning in tele-operations, *IIE Trans.*, 34, 245–252, 2002.

Carlson, J.G., Cubic learning curve: precision tool for labor estimating, *Man. Eng. Manage.*, 71, 22–25, 1973.

Carlson, J.G. and Rowe, R.J., How much does forgetting cost? *Ind. Eng.*, 8, 40–47, 1976.

Chand, S., Lot sizes and setup frequency with learning and process quality, *Eur. J. Oper. Res.*, 42, 190–202, 1989.

Cherrington, J.E. and Towill, D.R., Learning performance of an industrial long cycle time group task, *Int. J. Prod. Res.*, 18, 411–425, 1980.

Chiu, H.N., Chen, HM, and Weng, L.C., Deterministic time-varying demand lot-sizing models with learning and forgetting in set-ups and production, *Prod. Oper. Manage.*, 12, 120–127, 2003.

Cochran, E.B., Dynamics of work standards, *Man. Eng. Manage.*, 70, 28–31, 1973.

Conway, R. and Schultz, A., The manufacturing progress function, *Ind. Eng.*, 10, 39–53, 1959.

Corlett, N. and Morcombe, V.J., Straightening out the learning curves, *Pers. Manage.*, 2, 14–19, 1970.

Crossman, E.R.F.W., A theory of acquisition of speed skill, *Ergonomics*, 2, 153–166, 1959.

Cunningham, J.A., Using the learning curve as a management tool, *IEEE Spectrum*, 17, 43–48, 1980.

Dar-El, E.M., *Human Learning: From Learning Curves to Learning Organizations*, Kluwer Academic Publishers, Boston, 2000.

Dar-El, E.M., Ayas, K., and Gilad, I., A dual-phase model for the individual learning process in industrial tasks, *IIE Trans.*, 27, 265–271, 1995a.

Dar-El, E.M., Ayas, K., and Gilad, I., Predicting performance times for long cycle time tasks, *IIE Trans.*, 27, 272–281, 1995b.

Dar-El, E.M. and Rubinovitz, J., Using learning theory in assembly lines for new products, *Int. J. Prod. Econ.*, 25, 103–109, 1991.

Darr, E.D., Argote, L., and Epple, D., The acquisition, transfer, and depreciation of knowledge in service organizations: productivity in franchises, *Manage. Sci.*, 41, 1750–1762, 1995.

DeJong, J.R., The effect of increased skills on cycle time and its consequences for time standards, *Ergonomics*, 1, 51–60, 1957.

De Jong, J.R., The effects of increasing skill and methods-time measurement, *Time & Motion Study*, 10, 17, 1961.

Deming, W.E., *Quality, Productivity, and Competitive Position*, MIT Center for Advanced Engineering, Cambridge, MA, 1982.

Dutton, J.M. and Thomas, A., Treating progress functions as a managerial opportunity, *Acad. Manage. Rev.*, 9, 235–247, 1984.

Elmaghraby, S.E., Economic manufacturing quantities under conditions of learning and forgetting (EMQ/LaF), *Prod. Plann. Control*, 1, 196–208, 1990.

Epple, D., Argote, L., and Devadas, R., Organizational learning curves: a method for investigating intra-plant transfer of knowledge acquired through learning by doing, *Org. Sci.*, 2, 58–70, 1991.

Epple, D., Argote, L., and Murphy, K., An empirical investigation of the microstructure of knowledge acquisition and transfer through learning by doing, *Oper. Res.*, 44, 77–86, 1996.

Ernst, C.M., The interaction between cost-management and learning for major surgical procedures — lessons from asymmetric information, *Health Econ.*, 12, 199–215, 2003.

Eynan, A. and Li, C-L., Lot-splitting decisions and learning effects, *IIE Trans.*, 29, 139–146, 1997.

Fine, C.H., Quality improvement and learning in productive systems, *Manage. Sci.*, 32, 1302–1315, 1986.

Franceschini, O. and Galetto, M., Composition laws for learning curves of industrial manufacturing processes, *Int. J. Prod. Res.*, 41, 1431–1447, 2003.

Garvin, D.A., *Managing Quality*, Free Press, New York, 1988.

Globerson, S., The influence of job related variables on the predictability power of three learning curve models, *AIIE Trans.*, 12, 64–69, 1980.

Globerson, S. and Levin, N., Incorporating forgetting into learning curves, *Int. J. Oper. Prod. Manage.*, 7, 80–94, 1987.

Globerson, S. and Levin, N., A learning curve model for an equivalent number of units, *IIE Trans.*, 27, 716–721, 1995.

Globerson, S., Levin, N., and Shtub, A., The impact of breaks on forgetting when performing a repetitive task, *IIE Trans.*, 21, 376–381, 1989.

Globerson, S. and Millen, R., Determining learning curves in group technology settings, *Int. J. Prod. Res.*, 27, 1653–1664, 1989.

Globerson, S., Nahumi, A., and Ellis, S., Rate of forgetting for motor and cognitive tasks, *Int. J. Cognitive Ergon.*, 2, 181–191, 1998.

Globerson, S. and Seidmann, A., The effects of imposed learning curves on performance improvements, *IIE Trans.*, 20, 317–323, 1988.

Glover, J.H., Manufacturing progress functions: an alternative model and its comparison with existing functions, *Int. J. Prod. Res.*, 4, 279–300, 1966.

Goel, S.N. and Bucknell R.H., Learning curves that work, *Ind. Eng.*, 4, 28–31, 1972.

Goldman, J. and Hart, L.W., Jr., Information theory and industrial learning, *J Ind. Eng.*, 16, 306–313, 1965.

Hackett, E.A., Application of a set of learning curve models to repetitive tasks, *Radio Electron. Eng.*, 53, 25–32, 1983.

Hammer, K.F., An Analytical Study of Learning Curves as a Means of Relating Labor Standards, MS thesis, Cornell University, NewYork, 1954.

Hancock, W.M., The prediction of learning rates for manual operations, *Ind. Eng.*, 18, 42–47, 1967.

Hatch, N.W. and Mowery, D.C., Process innovation and learning by doing in semiconductor manufacturing, *Manage. Sci.*, 44, 1461–1477, 1998.

Heathcote, A., Brown, S., and Mewhort, D.J.K., Repealing the power law: the case for an exponential law of practice, *Psychonomic Bull. Rev.*, 7, 185–207, 2000.

Hergenhahn, B.R., *An Introduction to Theories of Learning*, 3rd ed., Prentice-Hall Inc., Englewood Cliffs, NJ, 1988.

Hewitt, D., Sprague, K., Yearout, R., Lisnerski, D., and Sparks, C., The effects of unequal relearning rates on estimating forgetting parameters associated with performance curves. *Int. J. Ind. Ergon.*, 10, 217–224, 1992.

Hirsch, W.Z., Manufacturing progress function, *Rev. Econ. Stat.*, 34, 143–155, 1952.

Hirsch, W.Z., Firm progress function, *Econometrica*, 24, 136–143, 1956.

Hirschmann, W.B., The learning curve, *Chem. Eng.*, 71, 95–100, 1964.

Hoffman, T.R., Effect of prior experience on learning curve parameters, *J. Ind. Eng.*, 19, 412–413, 1968.

Hottenstein, M.P. and Bowman, S.A., Cross-training and worker flexibility: a review of DRC system research, *J. Technol. Manage. Res.*, 9, 157–174, 1998.

Hovland, C.I., Human learning and retention, in *Handbook of Experimental Psychology*, Stevens, S.S., Ed., Wiley, New York, 1951, chap. 17.

Jaber, M.Y. and Abboud, N.E., The impact of random machine unavailability on inventory policies in a continuous improvement environment, *Prod. Plann. Cont.*, 12, 754–763, 2001.

Jaber, M.Y. and Bonney, M., Production breaks and the learning curve: the forgetting phenomena, *Appl. Math. Model.*, 20, 162–169, 1996.

Jaber, M.Y. and Bonney, M.C., A comparative study of learning curves with forgetting, *Appl. Math. Model.*, 21, 523–531, 1997.

Jaber, M.Y. and Bonney, M., The economic manufacture/order quantity (EMQ/EOQ) and the learning curve: past, present, and future, *Int. J. Prod. Econ.*, 59, 93–102, 1999.

Jaber, M.Y. and Bonney, M., Lot sizing with learning and forgetting in set-ups and in product quality, *Int. J. Prod. Econ.*, 83, 95–111, 2003.

Jaber, M.Y. and Guiffrida, A.L., Learning curves for processes generating defects requiring reworks, *Eur. J. Oper. Res.*, 159, 663–672, 2004.

Jaber, M.Y. and Kher, H.V., The two-phased learning-forgetting model, *Int. J. Prod. Econ.*, 76, 229–242, 2002.

Jaber, M.Y. and Kher, H.V., Variant versus invariant time to total forgetting: the learn-forget curve model revisited. *Comput. Ind. Eng.*, 46, 697–705, 2004.

Jaber, M.Y., Kher, H.V., and Davis, D., Countering forgetting through training and deployment, *Int. J. Prod. Econ.*, 85, 33–46, 2003.

Jaber, M.Y. and Sikström, S., A note on: an empirical comparison of forgetting models, *IEEE Trans. Eng. Manage.*, 51, 233–234, 2004a.

Jaber, M.Y. and Sikström, S., A numerical comparison of three potential learning and forgetting models, *Int. J. Prod. Econ.*, 92, 281–294, 2004b.

Kannan, V.R. and Jensen, J.B., Learning and labour assignment in a dual resource constrained cellular shop, *Int. J. Prod. Res.*, 42, 1455–1470, 2004.

Kapp, K.M., Transforming your manufacturing organization into a learning organization, *Hosp. Mater. Manage. Q.*, 20, 46–54, 1999.

Keachie, E.C. and Fontana, R.J., Production lot sizing under a learning effect, *Manage. Sci.*, 13, B102–B108, 1966.

Kher, H.V., Malhotra, M.K., Philipoom, P.R., and Fry, T.D., Modeling simultaneous worker learning and forgetting in dual constrained systems, *Eur. J. Oper. Res.*, 115, 158–172, 1999.

Kilbridge, M.D., Predetermined learning curves for clerical operations, *Ind. Eng.*, 10, 203–209, 1959.

Koulamas, C., Quality improvement through product redesign and the learning curve, *OMEGA: Int. J. Manage. Sci.*, 20, 161–168, 1992.

Lam, K.C., Lee, D., and Hu, T., Understanding the effect of the learning–forgetting phenomenon to duration of projects construction, *Int. J. Proj. Manage.*, 19, 411–420, 2001.

Lapré, M.A. and Van Wassenhove L.N., Managing learning curves in factories by creating and transferring knowledge, *California Manage. Rev.*, 46, 53–71, 2003.

Lapré, M.A., Mukherjee, A.S., and Van Wassenhove, L.N., Behind the learning curve: linking learning activities to waste reduction, *Manage. Sci.*, 46, 597–611, 2000.

Lerch, J.F. and Buck, J.R., An Exponential Learning Curve Experiment, The Human Factor Society 19th Annual Meeting, October, Dallas-Texas, 1975.

Levy, F.K., Adaptation in the production process, *Manage. Sci.*, 11, B136–B154, 1965.

Li, G. and Rajagopalan, S., The impact of quality on learning, *J. Oper. Manage.*, 15, 181–191, 1997.

Li, G. and Rajagopalan, S., A learning curve model with knowledge depreciation, *Eur. J. Oper. Res.*, 105, 143–154, 1998a.

Li, G. and Rajagopalan, S., Process improvement, quality, and learning effects, *Manage. Sci.*, 44, 1517–1532, 1998b.

Lieberman, M.B., The learning curve, diffusion, and competitive strategy, *Strateg. Manage. J.*, 8, 441–452, 1987.

Macher, J.T. and Mowery, D.C., Managing learning by doing: an empirical study in semiconductor manufacturing, *J. Prod. Innov. Manage.*, 20, 391–410, 2003.

Mazur, J.E. and Hastie, R., Learning as accumulation: a re-examination of the learning curve, *Psychol. Bull.*, 85, 1256–1274, 1978.

Mazzola, J.B. and McCardle, K.F., The stochastic learning curve: optimal production in the presence of learning-curve uncertainty, *Oper. Res.*, 45, 440–450, 1997.

McCampbell, E.W. and McQueen, C.W., Cost estimating from the learning curve, *Aeronaut. Dig.*, 73, 36, 1956.

McCreery, J.K. and Krajewski, L.J., Improving performance using workforce flexibility in an assembly environment with learning and forgetting effects, *Int. J. Prod. Res.*, 37, 2031–2058, (1999).

McIntyre, E., Cost-volume-profit analysis adjusted for learning, *Manage. Sci.*, 24, 149–160, 1977.

McKenna, S.P. and Glendon, A.I., Occupational first aid training: decay in cardiopulmonary resuscitation (CPR) skills, *J. Occup. Psychol.*, 58, 109–117, 1985.

Nadler, G. and Smith, W.D., The manufacturing progress functions for types of processes, *Int. J. Prod. Res.*, 2, 115–135, 1963.

Naim, M., Learning Curve Models for Predicting Performance of Industrial Systems, Ph.D. Thesis, School of Electrical and Systems Engineering, University of Wales, Cardiff, U.K., 1993.

Nelson, R.T., Labor and machine limited production systems, *Manage. Sci.*, 13, 648–671, 1967.

Nembhard, D.A., The effect of task complexity and experience on learning and forgetting: a field study, *Hum. Factor.*, 42, 272–286, 2000.

Nembhard, D.A. and Osothsilp, N., An empirical comparison of forgetting models, *IEEE Trans. Eng. Manage.*, 48, 283–291, 2001.

Nembhard, D.A. and Osothsilp, N., Task complexity effects on between-individual learning/forgetting variability, *Int. J. Ind. Ergon.*, 29, 297–306, 2002.

Nembhard, D.A. and Uzumeri, M.V., Experiential learning and forgetting for manual and cognitive tasks, *Int. J. Ind. Ergon.*, 25, 315–326, 2000.

Pegels, C., On startup or learning curves: an expanded view, *AIIE Trans.*, 1, 216–222, 1969.

Reis, D., Learning curves in food services, *J. Oper. Res. Soc.*, 42, 623–629, 1991.

Sinclair, G., Purchasing and the learning curve: a case study of a specialty chemical business unit, *J. Supply Chain Manage.*, 35, 44–49, 1999.

Schilling, M.A., Vidal, P., Ployhart, R.E., and Marangoni, A., Learning by doing something else: variation, relatedness, and the learning curve, *Manage. Sci.*, 49, 39–56, 2003.

Shafer, M.S., Nembhard, D.A., and Uzumeri, M.V., The effects of worker learning, forgetting, and heterogeneity on assembly line productivity, *Manage. Sci.*, 47, 1639–1653, 2001.

Shtub, A., Scheduling of programs with repetitive projects, *Proj. Manage. J.*, 22, 49–53, 1991.

Shtub, A., Levin, N., and Globerson, S., Learning and forgetting industrial tasks: an experimental model, *Int. J. Hum. Factor. Man.*, 3, 293–305, 1993.

Sikström, S., Power function forgetting curves as an emergent property of biologically plausible neural networks model, *Int. J. Psychol.*, 34, 460–464, 1999.

Sikström, S. and Jaber, M.Y., The Power Integration Diffusion (PID) model for production breaks, *J. Exp. Psychol. Appl.*, 8, 118–126, 2002.

Smunt, T.L., Log-linear and non-log-linear learning curve models for production research and cost estimation, *Int. J. Prod. Res.*, 37, 3901–3911, 1999.

Sparks, C. and Yearout, R., The impact of visual display units used for highly cognitive tasks on learning curve models, *Comput. Ind. Eng.*, 19, 351–355, 1990.

Steedman, I., Some improvement curve theory, *Int. J. Prod. Res.*, 8, 189–205, 1970.

Steven, G.J., The learning curve: from aircraft to spacecraft? *Financ. Manage.*, 77, 64–65, 1999.

Summers, E.L. and Welsch, G.A.E., How learning curve models can be applied to profit planning, *Manage. Serv.*, 7, 45–50, 1970.

Tapiero, C.S., Production learning and quality control, *IIE Trans.*, 19, 362–370, 1987.

Teplitz, C.J. and Amor, J.P., Improving CPM's accuracy using learning curves, *Proj. Manage. J.*, 24, 15–19, 1993.

Terwiesch, C. and Bohn, R., Learning and process improvement during production ramp-up, *Int. J. Prod. Econ.*, 70, 1–19, 2001.

Thomopoulos, N.T. and Lehman, M., The mixed-model learning curve, *AIIE Trans.*, 1, 127–132, 1969.

Thorndike, E.L., Animal intelligence: an experimental study of the associative process in animals, *Psychol. Rev.: Monogr. Suppl.*, 2, 1–109, 1898.

Thurstone, L.L., The learning curve equation, *Psychol. Monogr.*, 26, 1–51, 1919.

Towill, D.R., Transfer functions and learning curves, *Ergonomics*, 19, 623–638, 1976.

Towill, D.R., How complex a learning curve model need we use? *Radio Electron. Eng.*, 52, 331–338, 1982.

Treleven, M., A review of the dual resource constrained system research, *IIE Trans.*, 21, 279–287, 1989.

Uzumeri, M. and Nembhard, D.A., A population of learners: a new way to measure organizational learning, *J. Oper. Manage.*, 16, 515–528, 1998.

Vits, J. and Gelders, L., Performance improvement theory, *Int. J. Prod. Econ.*, 77, 285–298, 2002.

Waldman, J.D., Yourstone, S.A., and Smith, H.L., Learning curves in health care, *Health Care Manage. Rev.*, 28, 41–54, 2003.

Waterworth, C.J., Relearning the learning curve: a review of the deviation and applications of learning-curve theory, *Proj. Manage. J.*, 31, 24–31, 2000.

Wertmann, L., Putting learning curves to work, *Tool Eng.*, 43, 99–102, 1959.

Wiltshire, H.C., The variation of cycle times with repetition for manual tasks, *Ergonomics*, 10, 331–347, 1967.

Wisner, J.D., A study of US machine shops with just-in-time customers, *Int. J. Oper. Prod. Manage.*, 16, 62–76, 1996.

Wisner, J.D. and Siferd, S.P., A survey of US manufacturing practices in make-to-order machine shops, *Prod. Invent. Manage. J.*, 36, 1–6, 1995.

Wixted, J.T. and Ebbesen, E.B., On the form of forgetting, *Psychol. Sci.*, 2, 409–415, 1991.

Wixted, J.T. and Ebbesen, E.B., Genuine power curves in forgetting: a quantitative analysis of individual subject forgetting functions, *Mem. Cogn.*, 25, 731–739, 1997.

Wortham, A.W. and Mayyasi, A.M., Learning considerations with economic order quantity, *AIIE Trans.*, 4, 69–71, 1972.

Wright, T., Factors affecting the cost of airplanes, *J. Aeronaut. Sci.*, 3, 122–128, 1936.

Yelle, L.E., The learning curve: historical review and comprehensive survey, *Dec. Sci.*, 10, 302–328, 1979.

Yelle, L.E., Industrial life cycles and learning curves, *Ind. Market. Manage.*, 9, 311–318, 1980.

Zangwill, W.I. and Kantor, P.B., Toward a theory of continuous improvement and learning curves, *Manage. Sci.*, 44, 910–920, 1998.

Zangwill, W.I. and Kantor, P.B., The learning curve: a new perspective, *Int. Trans. Oper. Res.*, 7, 595–607, 2000.

31

Industrial Engineering Applications in the Construction Industry

Lincoln H. Forbes*
Department of Facilities Design and
Standards, Miami-Dade County
Public Schools

31.1 Introduction

The purpose of this chapter is to provide an overview of the possible applications of industrial engineering (IE) techniques in construction. Due to space restrictions and the large number of techniques available, a very limited selection of examples is presented. Readers are encouraged to do further reading of the sources provided in the references.

The Construction Industry has traditionally been one of the largest industries in the United States. As reported by the Bureau of Labor Statistics (BLS), U.S. Department of Labor, the value of construction put in place in 2003 was $916 billion, representing 8.0% of the gross domestic product. The industry employed approximately 6.9 million people in 2003. By its very nature, construction activity in the United States has not been subjected to the trend toward outsourcing that has plagued both the manufacturing and service industries. The BLS report titled "*State of Construction 2002–2012*" forecasts that 58.4% of U.S. jobs will be construction-related at the end of that decade. Yet, although other industries have blazed a trail to higher levels of quality and performance, the majority of construction work is based on antiquated techniques.

The potential for savings and productivity improvement is immense. Studies have pointed to typical losses in construction projects in the range of 30%; were this projected to the nation's annual total, over

*Also acts as an Adjunct Professor in the College of Engineering, Florida International University, Miami, FL 33199, USA.

200 billion may be wasted in a variety of ways. Mistakes, rework, poor communication, and poor workmanship are part of an ongoing litany of deficiencies that seem to be accepted as being a natural part of construction activity. Safety is a major national concern. Construction has an abysmally poor safety record, worse than virtually all other industries.

31.1.1 Categories of Construction

In order to understand how IE techniques can be applied to the construction industry, it is helpful to understand that environment; it is truly diverse, so much so that its participants have found it easy to rely on such clichés as "the industry is like no other," " no two projects are alike," to maintain the status quo in which long-established management traditions are seen as an arcane art that others cannot understand fully.

The BLS refers to three major headings: General Building Contractors SIC Code 15, Heavy Construction (except building) SIC Code 16, and Special Trade Contractors SIC 17. These are further subdivided into 11 SIC Code headings that include:

- Commercial building construction: offices, shopping malls
- Institutional construction: hospitals, schools, universities, prisons, etc.
- Residential: housing construction, including manufactured housing
- Industrial: warehouses, factories, and process plants
- Infrastructure: road and highway construction, bridges, dams, etc.

Who are the parties involved in construction?

- Owners, who originate the need for projects and determine the locations and purpose of facilities.
- Designers — they are usually architects or engineers (electrical, mechanical, civil/structural), who interpret the owner's wishes into drawings and specifications that may be used to guide facility construction. In the design-build (DB) process, they may be part of the construction team.
- Constructors — they are contractors and subcontractors who provide the workforce, materials, equipment and tools, and provide leadership and management to implement the drawings and specifications to furnish a completed facility.
- Construction trades, represented by unions.
- Consumer advocates and building owners.
- The legal industry.
- Developers.
- Major suppliers.
- Code enforcement professionals.
- Financial institutions — banks, construction financial organizations.
- Safety professionals.

31.1.2 Construction Delivery Methods

Several methods are available for carrying out construction projects. Design-bid-build (DBB) is the most traditional method of project delivery. Typically, a project owner engages a design organization to conduct planning, programing, and preliminary and detailed design of facilities. The final design and specifications are used to solicit bids from contractors. A contractor is hired with a binding contract based on the owner's drawings and specifications. Because of the linear nature of this process, several years may elapse between project conceptualization and final completion.

Design-build involves a contractor and designer working as a combined organization to provide both design and construction services. The owner engages a design professional to do a limited amount of preliminary project planning, schematic design, cost, and schedule proposals. DB firms subsequently compete for a contract based on the owner's preliminary information. The selected DB may commence construction while completing the final design. This concurrent engineering approach significantly reduces the duration of each project.

Engineer–procure–construct (EPC) contracts are similar to design build; this type of delivery involves a single organization providing engineering, procurement, and construction. It is most appropriate for engineering-based projects such as construction of manufacturing facilities or large municipal projects.

Construction management (CM) involves coordination and management by a CM firm of design, construction activities. The owner may elect to pay a fee for these services. CM at risk, on the other hand, involves the assumption of risk by the contractor for carrying out the construction through its own forces. Other types of delivery systems may be based on a combination of the foregoing systems. Overall, the methods have advantages and disadvantages that are best identified through systematic analysis.

31.2 Industrial Engineering Applications

There are several areas in the construction industry where IE techniques may be applied. The techniques are as follows:

- Ergonomics/human factors
- Value engineering
- Work measurement
- The learning curve
- Quality management (QM)
- Productivity management
- Continuous improvement
- ISO 9000
- Cycle time analysis
- Lean methods
- Supply-chain management (SCM)
- Automation/robotics
- Radio frequency identification (RFID)
- Safety management
- Systems integration
- Simulation
- Quality function deployment
- Facilities layout
- Operations research and statistical applications
- Sustainable construction

31.2.1 Ergonomics/Human Factors

The study and redesign of construction workspace using traditional and modern IE tools could increase efficiency and minimize on-the-job injuries and worker health impacts. Unlike factories, construction workspace constantly changes in geometry, size, location and type of material, location of work, location of material handling equipment and other tools, etc. These create new and challenging research opportunities. In addition, significant environmental impacts result from construction-related activities. Safety engineering approaches and industrial ecology tools such as life-cycle analysis may be developed to define and measure the impacts of different designs for workspaces and constructions.

Construction workers use a wide assortment of tools and equipment to perform construction tasks. Especially in cases where such aids are used for prolonged periods of time, workers' effectiveness and capacity to work with high levels of concentration, ergonomics are a major concern. Workers cannot be expected to "build in" quality in constructed facilities if they are subjected to awkward positions and excessive physical stress caused by tools and equipment that are difficult to use.

The significance of ergonomics in the construction environment is evident from a study conducted by the Associated General Contractors (AGC) of California to examine ergonomics-related costs. Their findings are:

- Related Workers Comp Insurance claims had increased by up to 40% for many construction companies.
- Financial returns due to ergonomic business strategy — 80% of the companies that had incorporated ergonomics-based methods reported improvements.
- Of 24 companies that measured for productivity, 100% reported improvements in cases where ergonomics-related concerns were addressed.

31.2.1.1 Tool and Equipment Design

Much research has yet to be done in the design of construction-oriented tools and equipment. The factors that may cause fatigue include weight, size, vibration, and operating temperature. Work-related musculoskeletal disorders (WRMSDs) generally include strains, sprains, soft tissue, and nerve injuries; they are cumulative trauma disorders and repetitive motion injuries. The construction workers who are at highest risk for these disorders are carpenters, plumbers, drywall installers, roofers, electricians, structural metal workers, carpet layers, tile setters, plasterers, and machine operators.

The top five contributory risk factors are as follows:

Working in a specific posture for prolonged periods, bending or rotating the trunk awkwardly, working in cramped or awkward positions, working after sustaining an injury, and handling heavy materials or equipment.

The use of a shovel is a very typical example of the labor-intensive material handling activities that are routinely carried out on construction projects. This activity requires workers to bend over, apply force to a shovel in different planes, and rotate the trunk in a flexed position. Such movements impose biomechanical stress which may impose cumulative trauma risk. Freivalds (1986) studied the work physiology of shoveling tasks and identified the shovel design parameters that would increase task efficiency. Friedvald's two-phase experimental study addressed the following parameters:

- The size and shape of the shovel blade
- The lift angle
- Shovel contours — hollow and closed-back design
- Handle length
- Energy expenditure
- Perceived exertion
- Low-back compressive forces

The recommended shovel design is as follows:

- A lift angle of approximately 32°
- A hollow-back construction to reduce weight
- A long tapered handle
- A solid socket for strength in heavy-duty uses
- A large, square point blade for shoveling
- A round, point blade for digging, with a built-in step for digging in hard soil.

31.2.1.2 Ergonomics Applications in Structural Ironwork

The BLS reports that construction trade workers experience higher rates of musculoskeletal injuries and disorders than workers in other industries: 7.9 cases per 100 equivalent workers as compared with the industry average of 5.7 per 100 (Bureau of Labor Statistics, 2001). In overall injuries, construction workers registered 7.8 vs. the industry average of 5.4. Observations by Holstrom et al. (1993), Guo et al. (1995), Kisner and Fosbroke (1994), and others point to a lack of studies in ergonomics, presumably because of high task variability, irregular work periods, changing work environments, and the transient nature of construction trades. As pointed out by Forde and Buchholz (2004), each construction trade and task represents a unique situation; the identification and application of prevention measures, tools and work conditions is best derived from trade and task-specific studies. This approach is the most likely to minimize the incidence of construction trades' WRMSDs.

By way of illustration, Forde and Buchholz (2004) studied construction ironworkers to identify mitigating measures in that group. Construction ironwork refers to outdoor work (not shop fabrication) as four specialties — the erection of structural steel (structural ironwork [SIW]), placement of reinforcing bars (rebars) (reinforcing ironwork [RIW]), ornamental ironwork (OIW), and machinery moving and rigging (MMRIW).

Previous studies determined that construction ironwork involves lifting, carrying and manipulating of heavy loads, maintaining awkward postures in cramped quarters, working with arms overhead for extended periods, using heavy, vibrating pneumatic tools, and extensive outdoor exposure in temperature and weather extremes.

Forde and Buchholz (2004) made the following observations and recommendations on the various categories of ironwork:

- *Machinery moving/rigging*. The erection of equipment such as a crane involves the pushing and pulling of large and heavy segments, and lining them up for bolting together. During an 8-h shift, this activity was observed to require 1.3 h of significant whole-body exertion. Workers in this scenario are most susceptible to overexertion of the back, legs, and shoulders.
- *Ornamental ironwork*. This work was observed to require arms to be above the shoulder level 21% of the time. Trunk flexion or twisting and side bending were observed 23% of the time.

These percentages indicate a high risk of overexertion of the involved muscle groups. Industrial engineers should review the work methods to increase the amount of preassembly at workbench height.

- *Reinforcing ironwork*. The preparation of reinforcement cages and tying of rebars were seen to cause nonneutral trunk postures up to 50% of the time. The handling of heavy loads (50 lb or greater) was observed to occur for 1.9 h of an 8-h shift, representing significant long-term risk. A 2004 study by Forde and Buchholz identified a need to improve the design of hand tools used for securing rebars. Such redesign would reduce nonneutral hand/wrist postures such as flexion, extension, and radial and ulnar deviation. These postures put construction workers at risk of repetitive motion injuries.

31.2.1.3 Auxiliary Handling Devices

A number of research studies have shown that construction workers have suffered back, leg, and shoulder injuries because of overexertion resulting from stooped postures, performing manual tasks above shoulder level, and the lifting of heavy objects. Such overexertion and injuries reduce worker productivity and may negatively affect the timeliness and profitability of construction projects. The use of auxiliary handling devices may reduce the degree of overexertion experienced by construction workers, and enhance productivity. Sillanpaa et al. (1999) studied the following five auxiliary devices:

- Carpet wheels
- A lifting strap for drain pipes
- A portable cutting bench for molding
- A portable storage rack
- A portable cutting bench for rebars.

The survey subjects utilized these devices to carry out typical construction tasks, such as carrying rolls of carpet, mounting drain pipes, cut pieces of molding, and fashioned rebars. The results of the study were mostly positive but mixed, pointing to the need for further research. The auxiliary devices were found to reduce the muscular load of some subjects, but others experienced an increased load because of differences in anthropometric dimensions, work modes, and level of work experience.

31.2.1.4 Drywall Hanging Methods

Drywall lifting and hanging are extensively conducted in both residential and commercial building construction; drywall board has become the standard for interior wall panels. It is the standard for surfacing residential ceilings. Workers are required to handle heavy and bulky drywall sheets and assume and maintain awkward postures in the course of performing installation work. These activities often cause muscle fatigue and lead to a loss of balance; studies have identified drywall lifting and hanging tasks as causing more fall-related injuries than any other tasks. Pan et al. (2000) studied 60 construction workers to identify the methods resulting in the least postural stability during drywall lifting and hanging tasks.

The subjects' instability was measured using a piezoelectric-type force platform. Subjects' propensity for loss of balance was described by two postural-sway variables (sway length and sway area) and three instability indices (PSB, SAR, and WRTI). The study was a randomized repeated design with lifting and hanging methods for lifting and hanging randomly assigned to the subjects. ANOVA indicated that the respective lifting and hanging methods had significant effects on two postural-sway variables and the three postural instability indices.

The recommended methods were:

- Lifting drywall sheets horizontally with both hands positioned on the top of the drywall causes the least postural sway and instability.
- Hanging drywall horizontally on ceilings produces less postural sway and instability than vertically.

31.2.2 Value Engineering

Value engineering (VE) is a proven technique for identifying alternative approaches to satisfying project requirements, while simultaneously lowering costs. It is a process of relating the functions, the quality, and the costs of a project in determining optimal solutions (Dell'Isola, 1988). In the construction environment it involves an organized multidisciplined team effort to analyze the functions of systems, equipment, facilities, processes, products, and services to achieve essential functions at the lowest possible cost consistent with the customer's requirements while improving performance and quality requirements. The multidisciplined approach of the IE is well suited to driving and facilitating the VE process. The IE can be especially valuable in facilitating a multidisciplinary group of design and construction professionals in brainstorming, generating ideas, and in conducting life-cycle analysis for the comparison of alternatives.

Some client organizations, such as government agencies, share the savings derived from VE with the contractor; the ratio varies with the respective type of contract.

Private contractors are generally highly motivated to develop improvements to tasks or projects, because of the financial benefit of lowering their costs; lower costs translate to higher profits.

There are several examples of savings in construction value engineering:

The U.S. Army Corps of Engineers has been using VE principles since 1964; in 2001, the Corps saved $90.78 million in its Civil works programs and has also realized at least $20 for each $1 spent on VE.$421 million in life-cycle cost was saved on a criminal court complex in New York City using VE. At the Bayou Bonfouca project in Louisiana, capital savings of $200,000 were obtained, and operations and maintenance costs were even greater at $4.4 million over a 2-year period.

Value engineering was successfully applied in a project at the Port of San Diego General Services Facilities building. The new structure comprised 45,200 sq. ft of administrative offices and maintenance shops at a cost of $8.9 million. A VE consultant was hired for the project. The building cost was reduced by 10%. In addition, the VE application placed a high priority on energy efficiency. The design was modified to emphasize the use of natural convection ventilation in shop areas vs. forced air, and specialized lighting/controls were selected to reduce energy consumption. Consequently, energy costs were reduced by 10%.

Kubal (1994) points out that while VE is beneficial during the design stages of a project, it can be most effective during the preconstruction phase because it facilitates both product and process improvements. Therefore, it should be perceived not just as a cost reduction exercise, but as a means of improving the entire construction process.

Design for manufacture and assembly (DFMA) techniques may be used to supplement VE activities; DFMA involves the review of designs to identify the optimal choice of materials, component design, fabrication, and assembly for the most cost-effective and functional solution. DFMA is carried out with the participation of a multidisciplinary team — whereas in manufacturing environments the team includes manufacturing engineers, shop floor mechanics, suppliers' representatives and specialists in maintainability and reliability studies, construction projects would include building design engineers, architects, contractors, and maintenance personnel.

A typical VE project may involve seven phases:

Team selection.   A VE team leader supervises a number of team members. These individuals should preferably be construction professionals who are generalists and specialists; the team leader should seek out flexible individuals who are willing to participate in a group activity. The team members should be trained in the VE process.

Information gathering.   Team members gather information on both technical and cost issues relating to a project, using available documents; the team VE leader assembles the information and shares it with the entire team.

Brainstorming.   This phase involves creative thinking to identify alternatives for carrying out a project. Experienced team members may recommend innovative approaches for conducting a project. The brainstorming phase is expected to generate many ideas without judgment. The original design is the point of reference for the alternatives that are generated.

Evaluation of alternatives.   Each alternative is reviewed carefully to determine its feasibility. Cost benefit or life-cycle cost analysis may be conducted in order to rank the possible solutions in order of importance. This ranking may be based on cost, and also on, ease of implementation.

Recommendation of alternatives.   The team leader reports on all the alternatives to the team, then selects the most appropriate ones for the client/owner. The savings derived may be in the range of 5–30% of initial project cost estimates.

Implementation.   The contractor implements the selected alternatives and the savings are divided between the owner and the contractor. The method of division is generally dictated by the form of contract. In the case of U.S. Government contracts, for example, Federal Acquisition Regulations (FAR) advocate the use of VE to reduce project costs. It also prescribes the types of savings and sharing for each type of project; the ratios used for various contract types.

As described by Adrian and Adrian (1995), the VE process matches the worth and cost of building elements; aesthetically pleasing features should not represent a significantly higher percentage of building cost than those attributes that the owner considers most valuable. For example, it is not uncommon for facilities to be built with brass hardware and marble floors, yet lack adequate service access to HVAC equipment. The VE technique is most effective when applied to the design phases of a project, when the influence on cost is greatest.

Optimization of projects with VE: Typical factors to consider in optimizing construction projects are:

- The intended purposes and functions for a project/facility.
- A clear understanding of the owner/client's needs.
- The perceived value to users and aesthetic appeal.
- Architectural systems and finishes and the specified conditions for their operation.
- Structural systems and materials — to maintain the integrity of a project/facility under all design conditions.
- Electrical, lighting, and communications systems — adequate and reliable operation is required.
- HVAC, plumbing, gas, and other systems to maintain a comfortable environment for users.
- Fire protection systems for detection and fire-fighting, adequate means of egress in case of emergencies.
- The constructability of a facility — the proposed construction methods and the projected time frames.
- The maintainability of a facility, the maintenance requirements, and the replacement cycle for components (HVAC, lighting devices, flooring materials, etc.).
- The expected return on investment for the owner/client.

In applying the VE process to building systems and components, the following steps may be used:

1. Identify functions
2. Estimate the value of each function
3. List the components

4. Determine component costs
5. Identify component functions
6. Calculate the cost per function
7. Evaluate and modify the proposed design

A VE team is staffed by knowledgeable individuals — designers, maintenance staff, etc., who understand the consequences of their decisions. They are also trained in the VE process, and participate in steps 1–7 given above.

31.2.3 Work Measurement

Work measurement techniques can help to increase construction productivity. Whereas standard work times are often used by the industry, these standards need to be reviewed and updated. Industrial engineers can tailor these standards to specific projects to reflect the logistics of the work site and also adjust the standards to represent methods improvement. The more accurate the information that is available on work standards, the better construction managers can conduct the preplanning of projects and exert greater control over the costs and schedules of these projects. Many construction standards need to be reengineered to reflect the use of technology in work processes. Methods time measurement (MTM) can be used to develop engineered standards.

Methods time measurement is based on the concept that a method must first be developed, elemental steps defined, and standard times developed. The standard must be based on the average times necessary for trained experienced workers to perform tasks of prescribed quality levels, based on acceptable trade practices. This approach is most practical with repetitive tasks.

In the MTM system, operations are subdivided into tasks; tasks are further reduced to individual body movements such as reaching, grasping, applying pressure, positioning, turning, and disengaging. Other movements include eye travel and focus and body, leg, or foot motions. Each body movement is subdivided into individual actions, such as reaching 2 in., grasp, apply pressure, turn, etc. Each action is assigned a standard time stated in time measurement units (TMU).

Examples of Methods — Time Measurement Application Data (All Times Include a 15% Allowance)

Activity	TMU		
Reach 2^2	4		
Grasp (simple)	2		
Turn	6		
Regrasp	6	ITMU	= 0.00001 H
Look (eye time)	10		= 0.006 min
Leg motion	10		= 0.036 sec
Kneel on one knee	35		
Arise	35		

In applying the MTM system (or any other standardized measurement system) it cannot be overemphasized that an appropriate method must first be established that can be successfully applied by the average, trained worker at definitive quality levels. The effect of the learning curve should also be considered when establishing work standards to ensure that repetition does not render the task times excessively long.

31.2.4 The Learning Curve

A learning curve is the phenomenon demonstrated by the progressive reduction in the time taken by an individual, or by a team to perform a task or a set of tasks on a repetitive basis. The individuals performing the task or project become more proficient with each repetition; the observed improvement serves as a motivator and a learning tool resulting in successively shorter performance times.

The learning curve is represented by an equation of the form

$$T_n = T_1(n)^{**}(-a)$$

where T_n is the time for the nth cycle, T_1 the time for the first cycle, n the number of cycles, and a a constant representing the learning rate. This equation produces a hyperbolic curve.

In order to determine the learning rate of a given activity, time study may be applied to a worker who is performing the task. For example, masons installing concrete blocks to form a wall would be timed as they perform successive iterations of the process.

The learning curve can be applied to construction projects. It can be highly relevant in repetitious projects such as housing construction, but the success of this application requires the IE to understand that interruptions to the construction process limit its use. Examples of such interruptions include prolonged shutdowns and Christmas holidays.

Also, construction tasks are often varied and nonrepetitive, so the IE has to apply the concept very judiciously. On-site managers who understand the learning curve rates for different types of tasks can improve work performance by selecting alternative work methods, especially with less experienced crafts persons.

Oglesby et al. (1989) identified three distinct phases: (1) when construction crews are familiarizing themselves with a process; (2) when a routine is learned so that coordination is improved; and (3) a deliberate and continuing effort to improve with successive iterations of the process.

Oglesby et al. (1989) estimated that learning curves for construction typically fall in the 70–90% range.

The curve below represents a project involving the installation of a number of generator units. The expert's estimate for carrying out this work was 11,000 man-hours. The contractor's bid was lower, i.e., 7200 h per unit. It is unlikely that a bid based on the expert's estimate would have been successful. The use of the learning curve allowed the contractor to complete the project at an even lower level of man-hours, i.e., 5900 h per unit. By using the benefit of the learning curve, the contractor was able to reduce labor hours by $1200 \times 8 = 9{,}600$ h over eight installations. This savings could translate directly to an increased profit margin. This profit margin is represented by the difference in area under the bid estimate line and the 8-unit average line (Figure 31.1).

31.2.4.1 Example — Learning Curve Calculations

A construction crew is carrying out a repetitive task. The first cycle takes the crew 5 h to complete. The third cycle takes the crew 4 h. Learning rate can be calculated by

$$i = 1 \text{ (first cycle)}$$
$$j = 3 \text{ (third cycle)}$$
$$r = 5 \text{ h}$$
$$s = 4 \text{ h}$$

$$r/s = (j/i)^n$$

$$5/4 = (3/1)^n$$
$$1.25 = 3^n$$
$$n \log 3 = \log 1.25$$
$$n = \log 1.25/\log 3 = = 0.203$$
$$\text{Learning rate} = 2^n = 1/2^n = 1/1.151 = 0.868$$
$$\text{Learning rate} = 86.8\%$$

How long should it take to complete the fourth cycle of the task?

$$i = 1$$
$$j = 4$$
$$r = 5$$
$$s = ?$$

From above, learning rate = 86.8%, $n = 0.203$

$$r/s = (j/i)^n$$
$$5/s = (4)0.203 = 1.3205$$
$$s = 5/1.3205 = 3.79 \text{ h}$$

The fourth cycle takes 3.79 h.

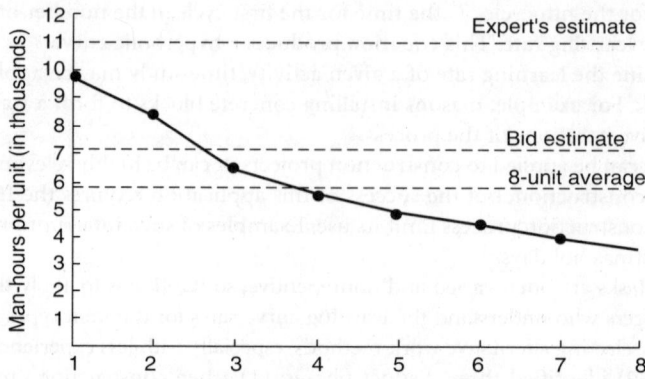

FIGURE 31.1 Progression of learning curve.

31.2.5 Quality Management

Total quality is an approach of doing business that attempts to maximize the competitiveness of an organization through the continual improvement of the quality of its products, services, people, processes, and environments (Goetsch and Davis, 2000).

Historically, the Japanese were among the first to apply quality improvement approaches in construction on a large scale, although they did not embrace this concept until the oil crisis of 1973. Prior to this, they thought that the construction industry was inappropriate for the application of total quality control (TQC), because of the inherent variability in projects and the difficulty in defining "acceptable quality." Takenaka Komuten Company, the sixth largest in Japan, had their formerly impeccable safety and quality image tarnished by the failure of a sheet piling system in Okinawa, in 1975, and embarked on a quality control (QC) program. They were followed by Shimizu Construction Company, the second largest in Japan, that established a QC program in 1976, and by Kajima Corporation, the third largest, in 1978. Subsequently, several U.S. companies have adopted TQC programs and the more familiar total quality management (TQM) programs used by U.S. manufacturers.

In 1992, the Construction Industry Institute (CII) published Guidelines for Implementing Total Quality Management in the Engineering and Construction Industry. Their research studies confirm that TQM has resulted in improved customer satisfaction, reduced cycle times, documented cost savings, and more satisfied and productive workforces (Burati and Oswald, 1993).

31.2.5.1 Benefits of TQM

The application of TQM principles can benefit design and construction organizations in many ways. These include

- Survival in an increasingly competitive world
- Improved levels of customer service
- Reduced project durations and costs
- Improvement of the overall quality and safety of facilities
- Better utilization of employees' skills/talents and increased quality orientation
- Increased profitability

31.2.5.2 Foundations of TQM

Total quality management is based on the total quality concept, which involves everyone in an organization in an integrated effort toward improved performance at each level (Goetsch and Davis, 2003).

It integrates fundamental management techniques, improvement efforts, in a disciplined approach toward continual process improvement. Total quality has the following characteristics: it is driven by an organizational strategy and unity of purpose, an internal and external customer focus, obsession with quality, scientifically based decision making and problem solving, continuous process improvement, long-term commitment, teamwork, employee involvement and empowerment, and education and training.

While total quality approaches have been highly beneficial to the manufacturing and service industries, they have had limited application in the construction environment. The construction industry has been heavily steeped in the traditional ways of executing projects and its constituents — designers and constructors, have been reluctant to make a necessary cultural and behavioral change to adopt total quality approaches.

Top management and senior management are generally preoccupied with short term, project by project profitability, and not with long-term quality-based strategies.

Although organizations have adopted a wide variety of quality improvement programs, these programs are based on the concepts advocated by the total quality pioneers. The most highly acknowledged pioneers are W. Edwards Deming, Joseph M. Juran, and Philip B. Crosby. Armand V. Feigenbaum and Japanese experts Kaoru Ishikawa and Shigeo Shingo were also major contributors to the quality improvement philosophy.

Deming has emerged as the influential and durable proponent of QM in the United States and is best known for the Deming cycle, his 14 points, and the seven deadly diseases.

The 14 points are summarized as:

1. Develop a program of constancy in purpose
2. Adopt this new program and philosophy
3. Stop depending on inspection to achieve quality — build in quality from the start
4. Stop awarding contracts on the basis of low bids
5. Improve continuously and forever the system of production and service
6. Institute training on the job
7. Institute leadership
8. Drive out fear so everyone may work efficiently
9. Eliminate barriers between departments so that people can work as a team
10. Eliminate slogans, targets, and targets for the workforce — they create adversarial relationships
11. Eliminate quotas and management by objectives
12. Remove barriers that rob people of pride of workmanship
13. Establish rigorous programs of education and self-improvement
14. Make the transformation everyone's job.

Juran is known for several quality contributions:

- Three basic steps to progress
- Ten steps to quality improvement
- The quality trilogy

Ishikawa is credited with the development/adaptation of seven quality tools:

- Pareto charts
- Cause and effect diagrams
- Scatter diagrams
- Check sheets
- The histogram
- Stratification
- Control charts

31.2.5.3 Obstacles to TQM

There are many obstacles to the application of TQM in the construction environment, and industrial engineers can help the industry to overcome these concerns:

(1) Measuring results is difficult (Shriener et al., 1995), whereas Deming (1991) advocate that measurement is a critical element in quality improvement efforts. The concept of construction performance does not emphasize productivity and quality initiatives. The work of many researchers

has revealed an industry tendency to measure performance in terms of the following: completion on time, completion within budget, and meeting construction codes. Very little attention has been directed to owner satisfaction as a performance measure.

(2) *The industry has a crisis orientation.* Significant changes have been sparked primarily by catastrophes of one kind or another. Major revisions were made in U.S. engineering codes after the failure of a structure in the Kansas City Hyatt Regency Hotel. Hurricane Andrew devastated Dade County, Florida, in August 1992, resulting in a major scrutiny of building codes and their enforcement. It is probable that with sufficient attention to quality at the front end, more building failures might be avoidable.

(3) *Poor communication.* Communication tends to be via the contract. Essentially, the designer is paid to produce a design expressed in the form of specifications and drawings. The contractor is expected to use these as a means of communication, and produce the completed facility. This communication often does not work as well as it should. Cross-functional communication must include subcontractors and suppliers to solve quality problems.

(4) There are large gaps between expectations and results as perceived by construction owners. Symbolically,

$$\text{Value } (V) = \text{Results } (R) - \text{Expectations } (E)$$

Consequently, since expectations often outweigh the results, construction owners feel that they receive less value than they should. Forbes (1999) quantified the "gaps" or dissonance zones between the three parties to construction, i.e., owners, designers, and contractors in health care facilities projects. In the area of owner satisfaction factors for example, public owners and designers differed on 7 of 9 criteria, owners and contractors differed on 5 of 9 criteria, while designers and contractors disagreed on the relative importance of 2 criteria.

(5) *A focus on inspection, not workmanship.* Code enforcement representatives of government agencies carry out construction inspections. Their role is to inspect critical aspects of the construction process by limited inspections on a number of items including reinforcing elements and concrete samples, but not workmanship.

(6) *The growing emergence of subcontracting.* The subcontractors are often priced in a manner that does not reflect the contract with the owner — even if the owner pays a high price, the subcontractor may still have to work with inadequate budgets, often compromising quality as a result. Deming's fourth point cautions against awarding contracts based on price tags alone.

(7) A culture of slow adoption of innovation — small contractors often lack the expertise or financial resources to adopt technological advances — adoption is inhibited further by fear and uncertainty. Roofing contractors, for example, tend to use the same time-honored methods to ensure that supplies and equipment are on site each day. Items that are frequently forgotten are delivered by expediters, contributing to waste in the industry.

(8) The training needed often does not get to the decision-makers in the construction industry. Construction management programs around the country have been providing higher levels of training for managers; however, this training has not reached the ultimate decision-makers in the industry. Efforts to enhance quality and productivity are likely to be frustrated under this scenario.

(9) Owners have not specifically demanded productivity and quality. There is a general lack of productivity/quality awareness in the industry among all parties, including owners. Owners have come to accept industry pricing — they have not been able to influence the productivity of the industry — prices have simply become higher on a per unit basis. By contrast, manufacturing activities have become cheaper over time on a per unit basis.

(10) Architect/engineer (AE) contracts are said to be unclear with respect to professional standards of performance, often leading to unmet expectations. Construction owners feel that typical A/E contracts protect designers at the owner's expense. For example, prevailing contract language relieves designers of any role in the case of a lawsuit or arbitration between an owner and contractor. An outgrowth of this is the practice of "substantial completion," where a job is usable but has 5% of

the remaining work in the form of a "punch list." An owner often has a very difficult time in persuading a contractor to finish that work.

(11) Few large companies, and virtually no small companies have implemented the concept of a quality or productivity manager — cost-cutting trends have resulted in such a position being viewed as an unjustifiable luxury.

(12) There is little, if any, benchmarking — many manufacturers and service organizations have become preeminent by adopting the best practices of benchmarked organizations. Construction has done very little of this due to distrust, fear of losing competitive advantage, but more likely, simply by being anachronistic.

31.2.5.3.1 Quality Management Systems

The Malcolm Baldrige Quality Award criteria provide an excellent framework for a construction organization's QM system; these criteria embody many of the concepts advocated by the quality pioneers — Deming, Juran, Crosby. Past winners of the Baldrige Award have proven to be been world-class organizations. Industrial engineers can assist construction organizations to improve quality and productivity by applying the Malcolm Baldrige criteria to their business model.

The Baldrige Award Criteria are based on a framework of core values for quality improvement comprised of seven critical areas:

(1) Leadership
(2) Customer and market focus
(3) Strategic quality planning
(4) Information and analysis
(5) Human resource development
(6) Process management
(7) Operational results

Other industry-recognized QM systems include the ISO9000: 2000 standards.

31.2.5.3.2 Industry Awards

The National Association of Home Builders created a National Housing Quality Program in 1993 to promote quality improvement in that industry. The National Housing Quality Award was developed based on the Malcolm Baldrige Award.

31.2.6 Productivity Management

By definition, productivity is measured as the ratio of outputs to inputs; it may be represented as the constant-in-place value divided by inputs such as the dollar value of material and labor. In the construction environment, productivity measurements may be used to evaluate the effectiveness of using supervision, labor, equipment, materials, etc. to produce a building or structure at the lowest feasible cost.

Mali (1978) combines the terms productivity, effectiveness, and efficiency as follows:

$$\text{Productivity index} = \frac{\text{Output obtained}}{\text{Input supplied}}$$

$$= \frac{\text{Performance achived}}{\text{Resources consumed}} \tag{31.1}$$

$$= \frac{\text{Effectiveness}}{\text{Efficiency}}$$

Therefore, productivity is the combination of effectiveness and efficiency. To increase productivity, the ratio(s) mentioned in Equation (31.1) must increase. This can be achieved by increasing the output, reducing the input or permitting changes in both such that the rate of increase in output is greater than that for input.

An increase in productivity can be achieved in five ways as follows:

(i) Reduced costs: $\dfrac{\text{output at same level}}{\text{input decreasing}}$

(ii) Managed growth: $\dfrac{\text{output increasing}}{\text{input increasing (slower)}}$

(iii) Reengineering: $\dfrac{\text{output increasing}}{\text{input constant}}$

(iv) Paring-down: $\dfrac{\text{output down}}{\text{input down (faster)}}$

(v) Effective working: $\dfrac{\text{output increasing}}{\text{input decreasing}}$

31.2.6.1 Total Productivity

Total productivity (TP) is the ratio of output to all inputs. All input resources are factored in this principle. Tracking the productivity changes that occur in different time periods is the most useful application of TP. Sumanth (1984) points to the limitations of partial productivity measures, which are measured by the ratio of output to one class of input such as labor productivity. Such measures if used alone can be misleading, do not have the ability to explain overall cost increases, and tend to shift blame to the wrong areas of management control.

Total productivity may be defined as

$$TP = \frac{\text{Total sales or value of work}}{\substack{\text{Labor cost } (M_1) + \text{Materials cost } (M_2) + \text{Machinery cost } (M_3) + \text{Money cost } (M_4) \\ + \text{ Management cost } (M_5) + \text{Technology cost } (M_6)}}$$

or

$$TP = \frac{T(s)}{M_1 + M_2 + M_3 + M_5 + M_6} \tag{31.2}$$

Since

$$P_i = \frac{T(s)}{M_i}$$

$$P_t = \frac{1}{1/P_1 + 1/P_2 + 1/P_3 + 1/P_4 + 1/P_5 + 1/P_6} \tag{31.3}$$

The above-mentioned factors are expressed as constant dollars (or other currency) for a reference period. To increase TP, it is necessary to determine which partial productivity factor (P_i) has the greatest short- and long-term potential effect on TP.

As pointed out by Oglesby et al. (1989), traditional construction project management tools do not address productivity; they include schedule slippages and cost overruns. Forbes and Golomski (2001) observed that the construction industry as a whole measures performance in terms of completion on time, completion within budget, and meeting construction codes.

Construction organizations (designers and constructors) would benefit significantly by establishing formal productivity and quality improvement programs that build on the knowledge gained from the measurement approaches that have been discussed above.

Industrial engineers can support such organizations in setting up productivity and quality improvement programs and providing ongoing measurement, which is critical to the process of continuous improvement.

Construction productivity is a major concern, especially when compared to other industries. As reported by the U.S. Department of Commerce, construction productivity has been rising at a much slower rate than other industries; between 1990 and 2000 it rose by approximately 0.8% compared to more than 2% for all U.S. industries. Construction costs have been increasing at the same time. Raw materials such as steel, staples have been rising, especially in the face of escalating global demand. Labor costs are a major component of most construction projects — in the vicinity of 40%, yet on many construction sites a large percentage of the daily labor hours are unproductive.

Activity sampling studies have shown that the working portion of activities generally occupies 40 to 60%, and by the same token 40 to 60% of labor hours are unproductive. There are many reasons for lost time — poor communications, waiting on assignments, waiting on resources, double material handling, rework, accidents, late or inaccurate job status reports, lack of supervision, etc. One third of these losses reflect issues that are within management's control. Construction profitability is directly linked to labor productivity. Industry-wide studies suggest that most construction projects yield net profits of 2 to 3% of the total project cost.

A hypothetical example:

Contract price	$10,000,000
Labor cost (40%)	$4,000,000
Other costs, overheads, etc.	$5,700,000
Net profit	$300,000

Assuming a 5% reduction in labor cost due to productivity improvement,

$$\text{savings in labor cost} = \$4,000,000 \times 0.05 = \$200,000$$
$$\text{revised net profit} = \$300,000 + \$200,000 = \$500,000$$

Hence, a 5% improvement in labor productivity can improve profitability by 66.7%.

Similarly, the value of lost labor hours due to management inefficiencies

$$= \$4,000,000 \times 1/3 = \$1,333,333$$

A 50% reduction in management-based losses would save $1,333,333/3 = $666,667

$$\text{revised net profit} = \$300,000 + \$666,667 = \$966,667$$

A 50% improvement in labor deployment would improve profits by

$$967 \times 100\%/300 = 322\%$$

In summary, IEs can have a major impact on construction productivity and profitability by helping management to improve its decision making and the logistics of the labor force. Further gains can be derived by addressing other construction processes and SCM issues.

31.2.7 Continuous Improvement

31.2.7.1 An Example from the Modular Housing Industry

This example is excerpted with permission from research performed by the Housing Constructability Lab (HCL) at the University of Central Florida, as described by Elshennawy et al. (2002).

In response to requests from modular housing manufacturers, the HCL research team undertook a research effort to:

- Benchmark quality systems used by modular manufacturers, other homebuilders and parallel industries
- Identify current best practices
- Develop recommendations for a quality system for use in a typical modular factory

The purpose of this undertaking was to align users of the model with the National Housing Quality Award and improve the modular home building industry beyond the minimal requirements dictated by the Department of Housing and Urban Development (HUD). The model was also intended to provide the foundation of a quality system to serve as a tool for managing, planning, and measuring performance.

31.2.7.2 Benchmarking

The purpose of benchmarking was to compare an organization with the industry and quickly identify "best practices" that could be adopted to facilitate process improvements at a much faster rate than would be possible with continuous improvement approaches, that tend to work incrementally. The exercise involved:

- Benchmarking against five modular manufacturers that had existing QM systems
- Identifying similar production processes in the yacht industry and benchmarking against two manufacturers that had been recognized for quality
- Visiting the site of a recent National Housing Quality Award winner to review quality practices
- Reviewing the practices of Malcolm Baldrige Quality Award winners, i.e., the best companies nationally

The benchmarking study yielded the following information:

Modular homebuilders thought quality was important, but depended on inspections to assure quality instead of "building in" quality. On the other hand quality leaders exhibited the following best practices:

- A mission to satisfy both external and internal customers
- Measuring the satisfaction levels of all customers
- Continuous process improvement
- Employee/team member empowerment — create ownership of improvements
- Training
- Recognition and rewards for outstanding performance
- Active, involved leadership

31.2.7.3 Quality Improvement Concepts

A mission statement should exemplify an organization's focus on exceeding customers' expectations with products of high value and quality, and motivating/empowering its employees.

Leadership involvement — Baldrige winners all shared the common attribute of having leaders which were committed to quality and which demonstrated it by example through their daily activities. Such leaders were active participants or members of quality councils that were set up to promote quality endeavors. They guided and led quality improvement efforts, established and reviewed performance measures, kept quality as a major topic of all meetings or ongoing reviews.

Measurement — Measurement is critical to ensuring that an organization is meeting its goals and expectations. These goals/expectations should be delineated in a strategic plan developed with the participation of all stakeholders to ensure that they are an integral part of the quality endeavor. For quality performance to be achieved there must be specific, meaningful measures of performance based on key drivers. The strategic plan should also delineate the responsibilities of specific individuals and assign time frames for accomplishment.

Key drivers — customer satisfaction, operational performance, financial performance, team member satisfaction, and community service — are indicated in Figure 31.2.

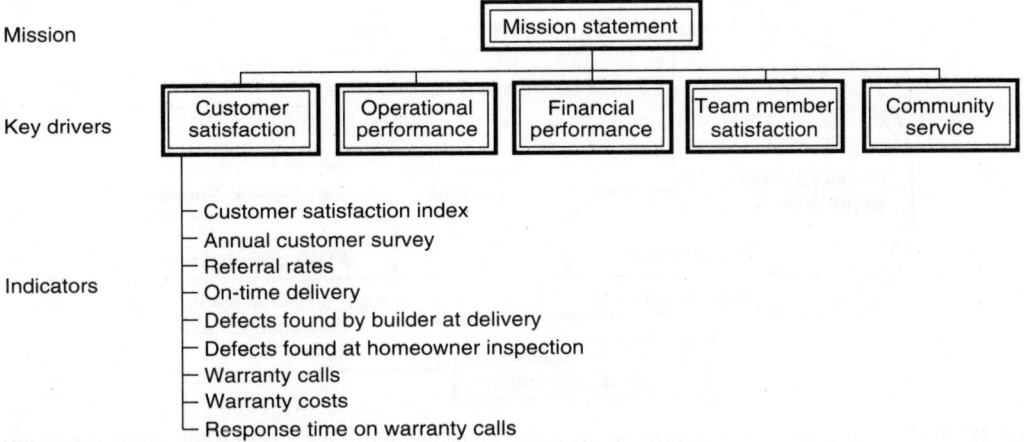

FIGURE 31.2 Typical measurement hierarchy for a modular manufacturer with indicators for customer satisfaction. (With permission from the Housing Constructability Lab, University of Central Florida.)

By way of an example, the indicators of customer satisfaction are customer satisfaction index, referral rates, percentage on-time delivery, defects detected by the builder at delivery, defects deter-mined by homeowner inspection, warranty call rates, warranty costs, and response time on warranty calls.

The measurement of operational performance may include labor efficiency, production costs, quality costs including warranty costs, and rework costs. Financial performance may include return on assets, tracking of fixed and variable costs, profit margins, etc.

Community service activities reflect the organization's citizenship, which are integral to its values, as reflected in the leadership component of the Malcolm Baldrige criteria.

Continuous improvement — The HCL study utilized the PDCA cycle to promote a culture of continuous improvement in the modular homebuilding industry. A modified PDCA cycle includes both incremental improvement and breakthrough improvement, as would be derived from a benchmarking endeavor. The model emphasizes feedback from several stakeholders and the use of performance measurements. Feedback is obtained from homebuyers about product quality as well as service and warranty call responsiveness. Production workers, field installers and other workers provide feedback about potential improvements in materials, design, production, and erection of homes. Cross-functional teams identify and implement improvement opportunities such as:

- Reducing defects and associated costs
- Improving responsiveness in handling customer complaints, service/warranty calls
- Improved homes with higher perceived value
- Improved operational productivity/effectiveness

Employee empowerment — The HCL model focused on empowering employees for the success of the organization, using a number of Deming's 14 points (as described in Section 31.2.5):

- Point 6: Institute training
- Point 8: Drive out fear
- Point 10: Eliminate exhortations for employees and the workforce
- Point 13: Encourage education and self-improvement for everyone (Figure 31.3).

31.2.7.4 Training and Education

The review of best practices of Baldrige winners pointed to the importance of quality training and education — an understanding of quality principles such as statistical process control (SPC), teambuilding, and empowerment is a prerequisite for a capable and motivated workforce.

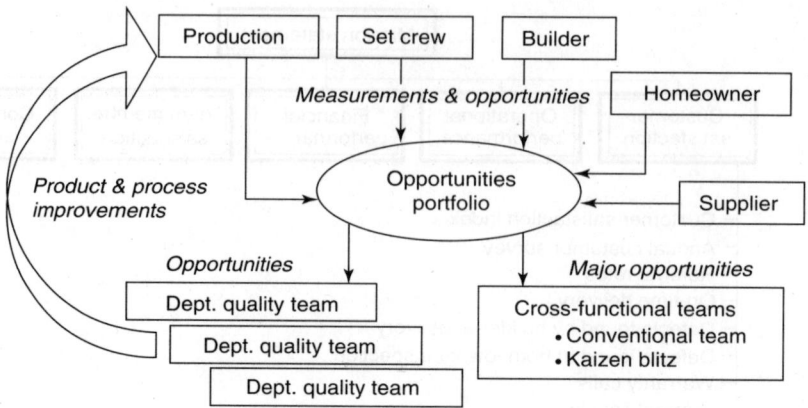

FIGURE 31.3 Continuous improvement in modular homebuilding. (With permission from the Housing Constructability Lab, University of Central Florida.)

31.2.7.5 Recognition and Rewards

Recognition and reward systems were seen to be a major factor in motivating employees to display a commitment to quality and to seek to create a superior organization. Examples include — at the company level — bonuses for meeting or exceeding targets for customer satisfaction or defect rates. At the team level, team recognition is effective for rewarding sustained high performance.

31.2.7.6 Continuous Construction Improvement in Institutional Buildings

The process of post occupancy evaluation (POE) provides a structured and systematic method for learning lessons from past construction projects. Preiser et al. (1988) define POE as the process of evaluating buildings in a systematic and rigorous manner after they have been built and occupied for some time. Although one may intuitively suspect that a facility succeeds (or fails) in serving its users well, one cannot fully appreciate and measure to what extent this occurs without conducting a POE. The quality of constructed facilities is important to its users. Perkins et al. (1992) observed that people experience more satisfaction within their environment if it is kept in an aesthetically pleasing manner. Preiser et al. (1988) noted that "spatial attributes, the sequence, location, relationships, size, and detail of a facility's spaces have been shown to affect occupant behavior." A POE involves the measurement of the functioning of a facility as compared with its purpose as defined in a formal program, and by the objectives of the architect/designer. The POE requires that a systematic research methodology be used to compare specifically the expectations of the client/owner with the effects of the facility on its users. The results of the POE can identify the extent to which the design intent has been met — this feedback can also help to identify "best practices" that can be used to improve future designs.

POE accomplishes the following:

- It measures the functioning of a facility in use compared with the goals of the formal program and the goals of the architect/designer and other specialists.
- It determines how well a facility meets its intended purpose.
- It compares the expectations of the client/owner with the effects of the facility on its users.
- It identifies changes, if any, that can improve future facilities.
- It allows all involved parties to learn from the past.

31.2.7.7 Categories of POE: Historical, Comparative, Longitudinal, Quasi-Experimental

Ex-ante evaluation involves an analysis of facility performance "before the fact." It may be informal, as conducted by some architects, in visioning how a facility may be used. On the other hand, simulation methods evaluate the utility of a facility in terms of travel distances, etc., as well as perceptual criteria.

There are several categories of POE. They include:

- *Historical* — studying the facility in retrospect to determine if actions taken during the design/construction process have been effective.
- *Comparative* — contrasting two situations such as two similar facilities after one has been specifically changed.
- *Longitudinal* — taking baseline measurements before changes are made. Changes are then initiated and differences attributed to them.
- *Quasi-experimental* — using statistical approaches to compare experimental and control situations.
- *"Post mortems"* — a revisitation of the design and construction processes themselves provides critical process-related lessons.

31.2.7.8   Conduct of the POE

A POE survey instrument is developed to address the factors of economy, function, and performance in the project/facility being reviewed. It uses a five-point Likert scale ranging from "very dissatisfied" to "very satisfied," with a midpoint indicating "neither satisfied nor dissatisfied." One version of the survey is developed for facility users (who are familiar with a facility) and another for design, construction, operations, and maintenance personnel. This latter group may represent a design/technical team that can address a very broad range of issues related not only to the owner's intent, but also to long-term performance of a facility.

31.2.7.9   Procedures

- Invitations are sent to the team members for two distinctly different evaluation meetings: (1) a facility user meeting, and (2) a design/technical team meeting. The second group is especially critical as it is the best source of remedial action information.
- The facilitator conducts the evaluation meetings. The design/technical team is instructed on the completion of the survey documents.
- A general discussion is held on the background and conduct of the project. The owner, decision-makers, designers, and project staff provide background on the inception of the project and its subsequent conduct.
- The POE participants walk through the entire facility in small groups and complete the surveys individually.
- Following the site visit, the data from the design/technical team's surveys are collated and analyzed.
- The scored and narrative responses for both users' and design/technical surveys are combined with historical building performance information. Maintenance records are checked for operating costs, breakdowns, and malfunctions.
- A comprehensive POE report is prepared. The survey findings are reconciled with current standards and specifications to develop meaningful recommendations.
- The report is distributed to all involved parties.

31.2.7.10   Implementing Continuous Improvement with the POE

Once a POE is completed and documented, it is critical that additional steps be taken to apply the lessons learned to the design and construction of future facilities. These steps represent the stages in the continuous improvement cycle shown in Figure 31.1, and are required for systematic improvements to occur. A meeting is conducted with selected decision-makers, in which the survey findings and recommendations are presented. Approved changes are subsequently made in the appropriate reference documents, so that the specifications and construction procedures for future projects can have enforceable requirements to utilize identified best practices. These documents may include master specifications, design criteria, and construction procedures manuals (Figure 31.4).

As indicated in the foregoing figure, the post occupancy process serves as the "Study/Check" phase of the PDSA cycle. The recommended changes are implemented in the "Act" phase.

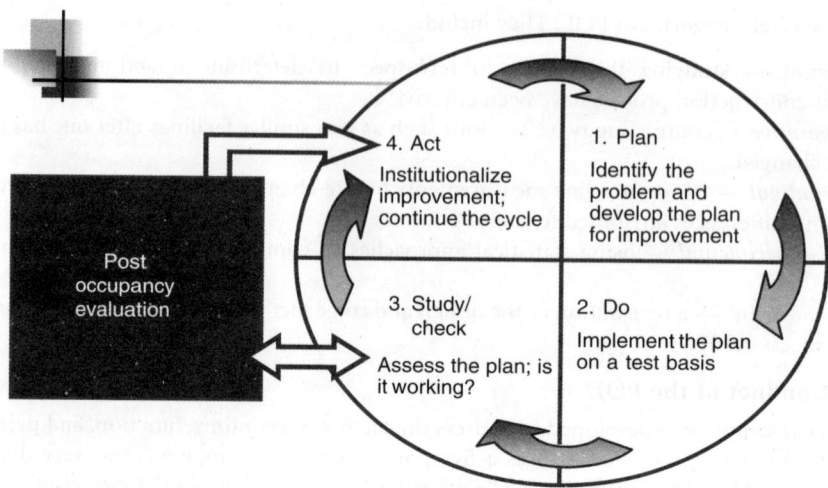

FIGURE 31.4   The Deming cycle (PDSA cycle) and postoccupancy evaluation. (Reprinted with permission from Forbes, *Proceedings of the IIE Annual Conference*, Orlando, FL, 2002.)

31.2.7.11   Quality Score Calculations

The following equation provides for the determination of a "quality score" for each surveyed facility or project. The questionnaire responses are combined to obtain mean ratings (R) for each question. A weight (W) is assigned to reflect the importance of a particular objective. For example, under the heading of the objective "function," respondents could be asked to rate the adequacy of a specific office space. The same question, when asked of different spaces may, have different weights applied.

The composite quality score would be based on the following factors: rating R and relationship W_{jk} of question j to objective k,

$$\text{quality scores } (QS_k) = \frac{\sum_{j=1}^{n} R_j W_{jk}}{\sum_{j=1}^{n} W_{jk}}$$

Industrial engineers can further enhance the POE process through life-cycle analysis to assess building/facility performance on a wider scale that includes not only user satisfaction with everyday utilization, but also issues of operating cost, durability, and reliability. User satisfaction questionnaires provide valuable information on the perceptions of the client (construction owner) with respect to the finished product and, very importantly, the nature of the associated processes. A typical questionnaire would address contractor performance measures relating to such issues as timeliness, responsiveness, communication, empathy, cost, and quality.

Overall, the delivery of design and construction services can be significantly enhanced through the use of "lessons learned" and "best practices." The information collected through POE application should also address how to design processes to improve design quality, cycle time, transfer of learning from past projects, and provide effective performance measurement systems for understanding, aligning and improving performance at all levels. The quasi-experimental type of POE should be investigated for application where two or more facilities may be compared with each other.

31.2.8   ISO9000

Although the ISO9000 series of standards are being adopted by a few design and construction organizations, the impact on industry-wide quality standards is not clear. The standards are voluntary; ISO's mission does not include verifying that organizations conform with the intent of the respective standards.

Conformance assessment is a matter for suppliers and clients or of regulatory bodies if ISO standards are included in public legislation. Hong Kong and Singapore, for example, have been on the leading edge of adopting ISO standards in construction. However, quality is not always internally motivated because it seen as a business requirement, i.e., a license to bid for projects in those countries. In the United States, the majority of construction work is done for private owners — they do not generally require ISO compliance even though a few have begun to do so.

ISO9000 series standards have been applied to major European Community projects; General Motors, Boeing Commercial Airplane Group, and several U.S. Federal Government departments have incorporated ISO9000 standards in their construction specifications, and have begun to require that designers/builders have ISO9000 certification. ISO9000 is an effective control system to reduce labor inefficiencies and waste. Certification enables optimal quality levels, improving market standing. Many organizations pursue certification to satisfy expectations of certain customers that are certified (e.g., the automotive industry.) Acceptance in the global marketplace is extensive.

Obstacles to acceptance are perceived high cost and extensive documentation. Employee resistance to change is also a major problem.

31.2.9 Cycle Time Analysis

Cycle time is the elapsed time from the start of a process until it is completed. It is usually desirable for cycle times to be reduced as this enables construction equipment and resources to be used to provide increased outputs cost effectively. The application of process improvement methods can identify "best methods" for carrying out repetitive work activities that are a major component of construction projects. By optimizing such work cycles, construction costs (and durations) can be reduced; profitability is directly enhanced by the resulting improvements in resource utilization.

The total cycle time of a system $t_s = \sum_{j=1}^{n(s)} t_{sj}$, where t_{sj} is the service time for server s performing action j and $n(s)$ the total number of actions or segments in activity s

Time to service one customer,

$$T_s = [C_t/C_s] t_s$$

where C_s is the capacity of the servicing activity and C_t the capacity of the item being serviced.

The following example models an excavation project in which a backhoe is used to excavate a trench for the installation of a water line (Griffis, F.H., Farr, John V., Morris, M. D., 2000, *Construction Planning for Engineers*. Copyright, McGraw-Hill Companies, Inc.). Dump trucks are used to haul the excavated material 2 miles from the site. The backhoe is 1 yd^3 capacity, the dump truck 10 yd^3 and the trench is 6 ft deep and 4 ft wide.

The cycle time for loading one truck can be determined as follows:

Excavation cycle:

Dig	30 sec
Swing	15
Dump	15
Swing	15
Total	75

$T_s = 75$ sec/cycle

Swell factor of soil = 25%
$C_t = 10$ loose yd/1.25 = 8 bank yd^3
Loader capacity = $C_s = 1$ loose yd/1.25 = 0.8 bank yd.

1. The time to service the truck = $T_s = [C_t/C_s]$. $t_s = 8$ bank yd/0.8 bank yd (75 sec/cycle) = 750 sec/cycle.
2. The service rate μ is the reciprocal of the service time = $60/T_s$.
3. The production rate $N_s = C_t \mu$.

If the activity involves servicing a customer, the cycle time of the serviced activity is $n(s)$.

The total cycle time of a system $T_{sj} = \sum_{j=1} Et_{st}$. The arrival rate of the serviced activity per customer $\lambda = 60/T_t$ (where T is in min). The productive time per hour is calculated using the factor θ.

The production rate $N_t = \theta \times C_t\lambda$.

Example (continued) to calculate backhoe output and completion time.

$$\mu = 1/T_s$$
$$= 60 \text{ sec/min}/750 \text{ sec/truck} = 0.08 \text{ truck/min}$$

Production rate $N_s = C_t \times \mu = (60 \text{ min/h})(8 \text{ bank yd}^3)(0.8 \text{ truck/min}) = 38.4 \text{ bank yd}^3$

Quantity of material to be removed $Q = 6 \times 4 \times 6000/27 = 5334 \text{ bank yd}^3$.

$$T_s = Q/N_s = 5{,}334/38.4 = 139 \text{ h}$$

Calculation of truck output based on 50 productive min/h (assume average speed of 20 mph):

Hauling (2 miles)	6 min
Return trip	6 min
Dumping, turning, acceleration	2 min
Loading	12.5 min
T_t	26.5 min/trip

$$\lambda = 60/26.5 = 2.264 \text{ trips/h}$$

$$N_t = \theta \times C_t \times \lambda = 0.833 \times 8 \times 2.264 = 15.1 \text{ bank yd}^3/\text{h}$$

The total time required $(T_t) = Q/N_t = 5334/15.1 = 353$ h or 51 seven-hour days. This time frame is the optimal time calculated for this activity.

31.2.10   Lean Methods

Lean construction is a new way to design and build facilities. Lean theory, principles, and techniques, jointly provide the foundation for a new form of project management. Lean construction uses production management techniques to make significant improvements, particularly on complex, uncertain, and quick projects.

There are five Lean principles according to Womack and Jones (1996):

1. *Value*: identify the value of the project, the customer's needs, and the agents involved at all stages from inception to the delivery process.
2. *Value stream*: by mapping the whole value stream, establishing cooperation between the agents or participants, and identifying and eliminating the waste, we improve the construction process.
3. *Flow*: business, job site, and supply flows depend of the value stream analysis and their own nature.

   • *Business flow*: related with the information of the project (specifications, contracts, plans, etc.).
   • *Job site flow*: involved the activities and the way they have to be done.
   • *Supply flow*: referred to materials involved in the project. This is similar to any other supply chain.

4. *Pull*: state the demand of the project. Under the Lean transformation, the efforts of all participants are to stabilize pulls during the construction process.
5. *Perfection*: develop work instructions and procedures, and establish quality controls.

31.2.10.1   Concurrent Engineering

Prevailing project management methods utilize a sequential process that results in wasteful iterations, even when constructability reviews are included. Concurrent engineering entails the simultaneous design of the facility and its production process. Concurrent engineering helps to define, create, and deliver value to the customer throughout the life of the project.

31.2.11 Construction Supply-Chain Management

Supply-chain management has saved hundreds of millions of dollars in manufacturing while improving customer service by taking a systems view of production activities of autonomous units. It is proposed that similar savings could be derived in construction, especially since subcontractor and supplier production account for the largest percentage of project costs. Studies have indicated project cost increases of up to 10% because by poor supply-chain practices.

The term supply chain encompasses all the activities that lead to having an end user provided with a product or service — the chain is comparable with a network that provides a conduit for flows in both directions, such as materials, information, funds, paper, and people. The main elements of SCM are information flow, order fulfillment, and product development with faster response times, less waste, more effective information flow, and smaller amounts of inventory. Studies by Bertelsen (1993) indicated project cost increases of up to 10% because of poor supply-chain design. Supply-chain management analyzes the impact of facility design on the construction process and enables superior project planning and management, avoiding the fragmented approach of other methods. Through SCM, all parties are kept aware of commitments, schedules, and expedites — all work as a virtual corporation that can source, produce, and deliver products with minimal lead-time and expense. Supply-chain management application needs to be tailored to the conditions in the geographic area and environment in which projects are executed.

The implementation of SCM in construction is challenging. The industry trend toward subcontracting has resulted in specialization and fragmentation; each subcontractor tends to act on its own interest and the relationships between the various parties often becomes adversarial. The sharing of information is critical for both SCM and Lean construction. Central to their business model is the importance of making and keeping commitments. A general contractor (GC), for example, should be able to extract a reliable commitment from an air conditioning subcontractor that a chiller has been ordered from a specified supplier. Using an online information system, it should be possible for the GC and the owner's project manager to verify the supplier's delivery schedule, although pricing information could be kept confidential. The dimensions and capacities of the chiller could be verified long before delivery to ensure that it will be delivered and installed in seamless agreement with the construction schedule. This system allows a "just in time" approach to be used, so that the chiller is not delivered too far in advance, when on-site storage requirements may become costly and inconvenient. By the same token late deliveries would be avoided. Unavoidable deviations from the construction schedule would be shared with all stakeholders, and needed adjustments made.

Supply-chain systems facilitate the sharing of resources, staff and expertise, problem solving, improved economic performance, and increased innovative capacity (Miller et al., 2001). Supply-chain systems also facilitate the tracking of work performance, resource utilization and provide feedback on workforce productivity. This feedback is indispensable for enabling continuous improvement efforts and corrective actions to be taken on a timely basis.

31.2.11.1 An Example from the Homebuilding Industry

A multiyear study of homebuilding operations found that houses under construction sit idle over 50% of the available work time. This represents over $68 billion in working capital; homebuyers who wait an average of 5 months for a new home.

IEs can couple advanced scheduling techniques with information technology to plan and manage better the complex supply chain that provides building materials and subcontractor services to homebuilders and enable the use of such modern techniques as Lean construction.

The timely delivery and coordination of supplies and crews have a significant impact on the completion time, quality, and expense of construction projects. If supplies are delivered too far in advance they are subject to quality degradation from several sources including weather and on-site traffic. In addition, late delivery of supplies results in costly project delays. Industrial engineering models may be developed to evaluate the proper coordination between project tasks and supply/crew availability. Also, statistical QC tools can be developed for use on the construction site by supervisors and workers with minimal education.

31.2.12   Automation/Robotics

Automation is beginning to emerge as an important method for addressing the negative aspects of construction; then industry is associated with the "three Ds," i.e., many of its tasks are perceived as Dirty, Dangerous, and Dull. Automated systems have been developed to perform such tasks, but they have yet to obtain wide acceptance. One obstacle to acceptance is a relatively low cost of labor in many countries, including the United States. In Japan, the aging of the workforce and the high cost of labor have facilitated the partial adoption of automated systems.

The Robotics Industry Association defines a robot as a reprogrammable multifunctional manipulator designed to move material, parts, tools, and specialized devices for the performance of a variety of tasks.

A study by Slaughter (1997) reviewed 85 robotics/automation technologies used in the construction environment. In keeping with a worldwide pattern, over two-thirds of the technologies in the sample had originated in Japan, while the remainder were distributed between the countries of Europe and the United States, with one each from Israel and Australia.

Industrial robot manipulators are devices that can control both position and movement, and can utilize tools to perform a variety of complex tasks with great precision. They are capable of interfacing via a range of communications devices and are capable of force control and visual serving.

Warzawski and Navon (1998) point out that the construction industry faces a number of problems that may favor the application of robotics and automation: labor efficiency and productivity are low, quality levels are low, construction safety and accident rates are a major concern, skilled workers are in increasingly short supply.

On the other hand, the low cost of labor in some countries such as Portugal limits the viability of large investments to facilitate construction robot applications (Pires and Pereira, 2003).

Gambao and Balaguer (2002) point out that construction automation is low relative to the state of technology. Research in automation and robotics falls into two groups — civil infrastructure and house building. Examples of civil infrastructure projects have been carried out in the European Union (EU) include the EU Computer Integrated Road Construction project of 1997 to 1999. Road paving robots were developed to operate autonomously, using GPS technology for navigation purposes. Automated systems have been developed for compaction of asphalt in road construction.

Japanese companies have been active in several robotics/automation applications — in tunnel construction, excavation machines are equipped with sensor-based navigation devices such as gyrocompasses, lasers, level gauges, and inclinometers. Shield-type tunnels are constructed with automatic drive systems; bolt-tightening robots install the tunnel segments. Tunneling through mountains is facilitated via concrete spraying by a shotcrete machine. The Japanese have also developed automatic/semi-automatic systems for bridge and dam construction — a column to column welding robot has been used for column field welding. A robotic bridge maintenance system has been developed at North Carolina State Univresity based on a truck-mounted inspection robot with four degrees of freedom.

Japanese companies have been very active in residential/commercial building applications — the SMART system was used in the 1990s to construct buildings of 30 stories and higher. The system comprises a robotic factory on the top floor (see www.cv.ic.ac.uk/futurehome). It elevates the construction plant floor-by-floor as the building is erected.

Robots have been developed for interior-finishing applications, such a mobile floor finishing and interior painting, where close tolerances have to be maintained. The KIST floor robotic system involves the network-based actions of a fleet of robots to compact and control the thickness of flooring concrete. Painting robots developed at Technion are used for coating interior building surfaces.

Pires and Periera (2003) give several reasons why it is difficult and often impractical to utilize robots on construction sites: construction sites are unique in nature with varied topography. They involve the conduct of many simultaneous tasks; they represent a hostile environment with dust, debris, and uneven surfaces. In the majority of instances, each building is unique; hence there is little repetition in the construction process. Construction sites are inherently dynamic — several tasks are interlinked, but compete for the same resources. The performance of tasks generally requires complex motions in several planes and moving from one location to another, literally to bring materials and labor to the building in process. The unstructured nature of building sites would make it difficult for robots to sense the environment, interpret the data and carry out the necessary complex tasks. The irregular terrain would also be an obstacle to the robot's mobility.

Pires and Pereira list a number of attributes that robots would need for the construction environment:

Locomotion — the ability to navigate obstacles, climb ladders, and traverse open areas. For robot operations to be feasible, the site would have to be provided with guidance systems, predefined routes, or reference points that are recognizable by robots.

Vision — robots would need artificial vision to recognize and interpret a wide variety of elements involved in construction sites.

Adaptability to hostile environments — construction robots would need to be weather proof, resistant to heavy falling objects as well as withstanding falls from heights. They would need to maintain precision of movement and manipulation when subjected to vibration dust, and abrasive/corrosive agents.

Capacity to handle a wide range of materials — construction materials cover a wide range of sizes, shapes, weights, configurations, and textures. Heavy loads may include beams and precast panels. Fragile loads may include glass, ceramic tiles, and bathroom fixtures.

Pires and Pereira (2003) express the view that robotic systems are immediately adaptable to construction activities in a factory setting, such as in the manufacture of prefabricated building components.

31.2.12.1   Layered Fabrication Technology — Contour Crafting

Khoshnevis (2004) describe the development of the contour crafting (CC) system that promises to take automation in construction from the component level to the fabrication of entire structures. Khoshnevis points out that automation has seen limited application due to the lack of suitable technologies for large-scale products, limitations in building materials and design approaches, economic viability, and managerial issues. Khoshnevis et al. (2001) describes CC as a superior layer fabrication technology as it produces better surface quality, higher fabrication speed, and provides a wider choice of materials.

Contour crafting is achieved through the use of computer control of troweling tools to create smooth and accurate planar and free-form surfaces. It applies computer control to the traditional practice of industrial model building with surface shaping knives; it also combines this technology with an extrusion process and a filling process to build an object core. As shown in the diagram, a material feed barrel supplies the material to a nozzle for layering, while a top and a side trowel shape the deposited material incrementally.

The CC technology is extended to large structures, such as complete houses, as indicated in Figure 31.5. Khoshnevis describes a gantry system which carries a nozzle on two parallel lanes mounted at the construction site. As shown in the figure, the CC machine moves along the parallel lanes and laterally between the gantry supports as the nozzle deposits the ceramic mix. The process is very effective for building the exteriors of building systems, which can subsequently be filled with concrete. The process is also very effective with adobe-type structures that include domes and vaults in their configuration. These designs are typical of CalEarth (www.calearth.org).

Conventional designs can be constructed by incorporating other devices with the gantry system such as a picking and positioning arm; the system can, in a single run, produce a single house or a colony of houses, all of different designs.

Automated reinforcement — the picking and positioning arm complements the extrusion nozzle and trowel system by incorporating modular reinforcement components, which are imbedded between the layers of walls built by CC. The steel reinforcement may be supplied by a module feeder, which may be combined with the concrete filler feeder. This system can simultaneously position internal reinforcement and create a wide variety of smoothly finished exterior surfaces.

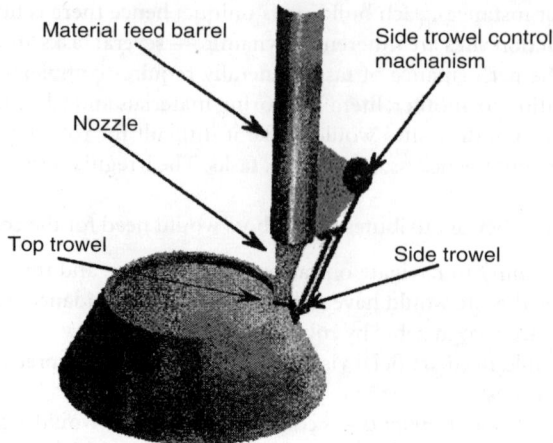

FIGURE 31.5   Contour crafting process. (With permission from B. Khoshnevis, University of Southern California.)

31.2.12.2   Benefits of Contour Crafting

According to Khoshnevis (2004), the CC system has many e-design flexibility benefits: the system makes possible the construction of architecturally complex shapes that are difficult to realize with conventional construction.

31.2.12.3   Material Flexibility

A variety of materials may be used for external materials as well as fillers between surfaces. The system makes it possible to combine materials that normally interact just before they are deposited. This contrasts with the use of concrete, which has a very limited time window for use before it becomes unworkable.

"Smart materials," such as carbon-filled concrete, may be configured to create floor and wall heating elements with specified electric resistance. Similarly, strain sensors may be incorporated into the construction. Nonmetallic materials such as glass or carbon fibers may be extruded to create fiber-reinforced plastics (FRP) in one step (Figure 31.6).

Posttensioning can be used with the CC system — ducts can be built into the structure and metal or FRP lines can be threaded through and tensioned to provide the necessary structural strength.

31.2.12.4   Minimal Waste

Whereas the construction of a typical single family generates 3–7 tons of waste material, CC is an additive process that produces very little or no material waste. Also, the process may be electrically driven and consequently produce few emissions.

31.2.12.5   Simplified Building Utility Systems

Utility conduits defined in a CAD system can be constructed in the field through CC, as material is deposited layer by layer. In the case of plumbing, wall layers are installed and conduit chases are created; lengths of piping are inserted that have joints, which have been pretreated with solder and provided with heating elements. As the height of a wall is built further, the robotics system uses robotic grippers to add lengths of pipe. Each length of pipe is placed in the respective coupling of its predecessor, and a heater ring is used to melt the solder, bonding the pipes together to form a pressure tight joint. The components to be installed may be prearranged in a tray or magazine for easy manipulation by the robotics system.

In the case of electrical and communications wiring, the conductors may be imbedded in insulating material and designed to interconnect in modular fashion. As done in plumbing systems, the electrical modules may be inserted into conduits fabricated in the walls of a structure. This technology requires the use of specialized robotic grippers working in conjunction with a delivery tray or magazine.

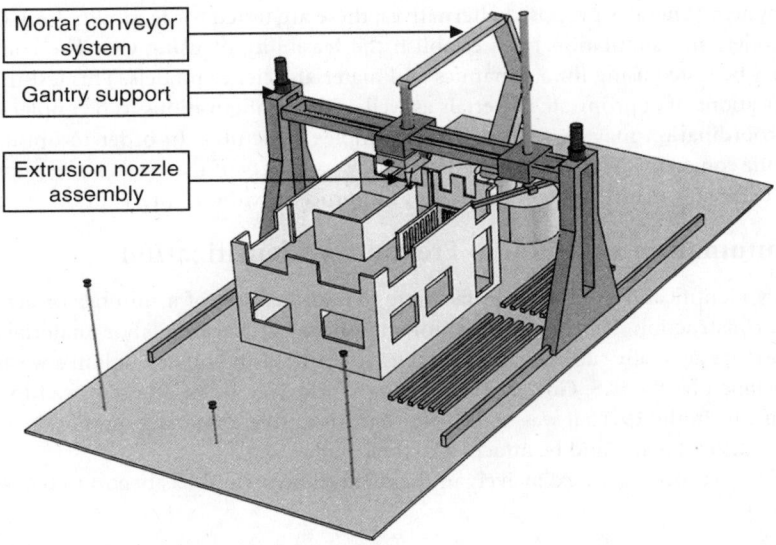

FIGURE 31.6 Construction of conventional building using counter crafting. (With permission from B. Khoshnevis.)

31.2.12.6 Automated Trades

In addition to the CC technology, several skilled trades activities can be integrated with the gantry hardware.

Tiling of walls and floors can be accomplished by having the CC equipment deposit the adhesive material to the respective surfaces. A robotic arm can retrieve the tiles from a stack and place them in the locations where the adhesive has been applied. The arm may be installed on the system that bears the CC nozzle assembly.

Painting may be carried out with a mechanism attached to a robotics manipulator; this mechanism may be a spray nozzle or an ink-jet-based system that can paint very complex patterns.

31.2.12.7 Mobile Robotics Approach

Khoshnevis explains that the gantry robot system has significant limitations — the gantry has to be large enough to accommodate the finished structure within its operating envelope, which results in a large structure. It also requires extensive site preparation. In contrast, a system of multiple, mobile robots has several advantages — it is easier to transport and set up, and several robots may work on the same building simultaneously.

The mobile robot may be equipped with material tanks, material delivery pumps; its end effector would be provided with a CC nozzle. Mobile robots may be used to build supportless structures such as domes and vaults. In the case of planar roofs it is preferable to incorporate roof beams in the design. The erection process would involve having two robots lift the beams at either end and place them on the structure.

The delivery of materials to the roof is challenging — mobile robots may be positioned inside a structure and move materials over the roof beams, in succession as they are placed over each beam. After all beams are mounted, the robots would work from the exterior of the structure.

The NIST RoboCrane system offers a special application — it may be attached to a conventional crane and hoisted overhead where it can manipulate structural members with much great precision than the master crane. The RoboCrane may be provided not only with a gripper for beams, but also may be equipped with a material tank and a CC nozzle for delivering materials to the roof.

31.2.12.8 Information Technology Systems

As described by Khoshnevis several support activities are required to make CC and other automated systems feasible.

A planning system generates proposed alternatives; these are tested for feasibility for CC applications. Engineering models and simulation tools establish the feasibility of using CC. The construction of a vaulted roof may be tested using fluid dynamics and materials science models. This test may identify the required specifications of appropriate materials as well as the configurations to be implemented.

Multirobot coordination may be required with complex structures. In order to optimize the use of automation in the construction environment that involves a variety of materials and equipment, logistics planning is necessary to ensure that work activities can proceed without interruption.

31.2.13 Automation with Radio Frequency Identification

Radio frequency identification (RFID) can facilitate the automation of a number of activities that are associated with construction. It provides an automatic means of tracking labor, materials, and maintenance; these areas are generally subject to inaccuracies in the field and often result in a waste of resources. It was first developed by the U.S. Government during World War II, based on a need to track items in harsh environments. In the 1970s it was used by the agro-industry to manage animals by means of small battery-powered devices that could be attached to their bodies.

More recently, it has been used extensively in the retail industry and has begun to be used in the construction industry.

31.2.13.1 RFID Components

An RFID system comprises a tag that contains an integrated circuit programed with a unique ID number, and a read-write unit. It contains a small antenna that transmits data by means of radio waves. The read-write unit also contains an integrated circuit and an antenna. The read-write unit receives information from the tag and stores it. The RFID system has a distinct advantage over bar code systems in that it can operate in a dirty environment that is also moist and windy.

The system is based on three types of tags.

> Read-only tags have information permanently encoded by the manufacturer, and cannot be altered.
> Write-one, read-many (WORM) tags permit data to be added but do not allow the identification to be changed.
> Read-write tags may be erased and rewritten as required.

Radio frequency identification systems may be passive or active. In the case of passive systems the read-write unit generates radio waves that induce energy in the tag, up to a distance of 6 ft. Active systems include tags that have an internal battery and can be written on at distances up to 100 ft.

Radio frequency identification has great potential for construction applications, and IEs can be a major force in adapting the technology. It may be used to track the materials used on a project, providing real-time accountability for materials expenditures. It can be a critical element in a supply chain, ensuring that materials are requested/ordered to replace inventories that are kept to a minimum. Equipment items would be clearly identified through RFID, reducing errors in selection and installation.

The receipt of items on-site would allow electronic verification, reducing the need for labor-intensive checking.

Worker productivity can be tracked on repetitious tasks involving the installation of framing systems, windows, doors, and other modular components. The modular housing industry would be highly adaptable to the use of RFID, as tasks are more predictable than on most building sites.

31.2.14 Safety Management

The construction industry has an unsafe reputation earned by its high occupational accidents rates during the past years (Bureau of Labor Statistics, 2002). In the year 2000 alone, construction accounted for 21% of all deaths and 11% of all disabling injuries/illnesses in the private industry (Bureau of Labor Statistics, 2002). But no industry has a good record, some are just worse than others. It is often claimed that the nature of construction work makes accidents inevitable, but that is an unacceptable position to take.

Most of the accidents that occur on the job site can be prevented, it is just the matter of priorities for the contractors to reach this goal. Safety programs can help prevent accidents with injuries and deaths at the job site, and they not only pay back with economic incentives, such as higher production rates, lower workers' compensation premiums, etc., but also with a good reputation and motivation of the labor force, etc. The decision of implementing and running a good safety program rests solely in the hands of the companies' top management. They are the only ones responsible for designing and implementing procedures to reduce accidents rates.

Safety improvement begins with the systems-thinking approach, using a methodology that is similar to the approach of TQM. Existing safety management systems can be significantly improved with the adoption of core TQM principles and procedures.

Weinstein (1997) recommends a number of applications of TQM principles that are helpful in occupational or work safety. According to the scope of this research, selected TQM concepts and techniques have been identified which are to be integrated with the existing safety management system.

31.2.15 Systems Integration

A major criticism of the construction industry is that it has become extremely fragmented, with all the parties acting in their own self-interest. The concepts of SCM and Lean construction require close collaboration between the parties; hence there are significant benefits to be derived by systems integration. Industrial engineers can help in overcoming the barriers to this process. Effective communication is a critical prerequisite for a successful project. Some of the hurdles to effective communication include:

- Team members may be geographically dispersed
- Timely information delivery and response
- Document revision control
- Cost control

A web-based collaboration tool allows information to be posted online and viewed by anyone with a login/password and Internet connection. The tool acts as a central point of documentation, allowing team members to always access the most current project information. The tool is highly customizable to meet project and customer needs and can be used to manage most aspects of the project communication, including:

- Design development and drawing revisions
- Construction or design RFP, bids, awards
- Construction budget
- RFI (requests for information)
- Change control
- Punch list/audit/closeout

Through time, the IEs using the tool can improve it by adding automation to the process.

31.2.16 Simulation

Simulation may be a physical or numerical representation of a system, i.e., a group of entities that interact in accordance with defined rules of behavior. Although physical simulation has been well utilized to test engineering systems such as using wind tunnels to represent wind load conditions, numerical simulation has not been used as extensively with the construction process itself. Given the high cost of labor and related lost time, numerical simulation can effect significant economies in the construction environment, especially as an aid to preplanning. Traditionally, experienced estimators visualize how each stage of a project will be executed, and plan the demand for work crews, materials and equipment, completion forecasts, and costs. This deterministic approach has severe limitations, as many variables interact in real time. Simulation, on the other hand, includes the randomness that is likely to occur in each of many activities in a complex project. A major factor in the success of simulation is the representation of activity times with appropriate mathematical distributions.

Simulations may be either static or dynamic; static simulation represents the status of a system at any point in time, whereas dynamic simulation shows changes over time. Monte Carlo simulation is the former type; it is based on input variables represented by distributions. These distributions, in turn, are linked by decision rules and are used to determine the distributions of outcome variables of interest. In Monte Carlo simulation, each activity is assumed to be a random variable that can be represented by a known probability distribution. The duration of the activity is expected to follow the defined probability distribution instead of being a point estimate value. The simulation process uses random numbers to assign this duration during each iteration.

Systems may be classified as discrete or continuous. Discrete systems change at specific points in time, while continuous systems are constantly changing with the passage of time.

31.2.16.1 Advantages and Disadvantages

Advantages: The most important concepts can be represented in the simulation model. Changes in the input variables can be tested very quickly and efficiently.

Disadvantages: The development of the model can be time consuming and costly. It may be difficult to verify and validate the model.

Data collection and analysis can provide suitable approximations to these distributions.

31.2.16.2 Example 1

Precast concrete piles are taken from the end of a production line into a storage area, from where they are collected by a transporter (having a maximum capacity of 18 piles). The transporter(s) call after a day's production has finished. The transporters' arrival is randomly distributed as given in Table 31.1. The daily production figures were collected, over a period of 100 days, and are given in Table 31.1.

The number of days elapsed between successive transporter arrivals is given in Table 31.2.

In the above data, days before next transporter arrival equals 0, meaning that two transporters call on the same day.

The storage area can only hold 20 piles. Extra piles have to be stored at a specially arranged place at a cost of $50 for each pile per day (only charged for when used). Alternatively a new storage space (capacity 20 piles also) can be rented for $300 per day. Assume that the storage area is initially empty and that the transporter called on the first day; simulate this system over a period of 10 days. Should the firm rent the new space by using the simulation result?

The following random numbers were generated in sequence for the simulation:

For number of piles produced	12	57	85	78	36	9	60	73	57	86
For days lapsed	29	73	45	58	95					

31.2.16.3 Solution

Step 1
Tabulate the production figures along with their given probability and calculate the cumulative probability and then assign the random numbers, as shown in Table 31.3.

Step 2
Tabulate the number of days between successive transporter arrivals and assign RN (Table 31.4).

Step 3
Simulate a period of 10 days:

(i)

Day	1	2	3	4	5	6	7	8	9	10
RN	12	57	85	78	36	9	60	73	57	86
No. of piles produced	10	12	13	13	11	9	12	12	12	13

(ii)

Day	1	1 + 1=2	2 + 3 = 5	5 + 1 = 6	6 + 3 = 9
RN	29	73	45	58	95
Days elapsed	1	3	1	3	4

Step 4
Evaluate the simulated result, assuming that the firm has not rented the extra space:

At the end of			Storage (piles)		No. of Transporter	
		Before Transporter			After	Arrivals
	Arrival		Extra Cost			
Day 1	10		—		—	1
Day 2	12		—		—	1
Day 3	13		—		—	0
Day 4	13 + 13 = 26		6×$50 = $300		—	0
Day 5	26 + 11 = 37		—		19	1
Day 6	19 + 9 = 28		—		10	1
Day 7	10 + 12 = 22		2×$50 = $100		—	0
Day 8	22 + 12 = 34		14×$50 = $700		—	0
Day 9	34 + 12 = 46		8×$50 = $400		28	1
Day 10	28 + 13=41		21 × $50 = $1050			0
			Σ = $2550			

Step 5
Make comparison: If the firm has rented the extra space, then the amount of rent is $300×10=$3000. Since $2550 < $3000, therefore the firm should not rent the extra space.

TABLE 31.1 Production History in the Past 100 Days

No. of piles produced	9	10	11	12	13	14
No. of days at this production rate	10	18	29	21	12	10

TABLE 31.2 Interval (Days) of Transporter Arrival

Days before next transporter arrival	0	1	2	3	4
Frequency of occurrence	32	22	10	20	26

TABLE 31.3 Assign Random Numbers for the Number of Piles Produced

No. of piles produced	No. of days at this production rate	Probability	Cumulative probability	RN	No. of piles produced
9	10	0.10	0.10	00–09	9
10	18	0.18	0.28	10–27	10
11	29	0.29	0.57	28–56	11
12	21	0.21	0.78	57–77	12
13	12	0.12	0.90	78–89	13
14	10	0.10	1.00	90–99	14
Total	100	1.00			

TABLE 31.4   Assign Random Numbers for Intervals (Days) of Transporter Arrival

Days before next transporter arrival	Frequency of occurrence	Probability	Cumulative probability	RN
0	32	0.29	0.29	00–28
1	22	0.20	0.49	29–48
2	10	0.09	0.58	49–57
3	20	0.18	0.76	58–75
4	26	0.24	1.00	76–99
Total	110	1.00		

31.2.17   Quality Function Deployment (QFD)

Quality function deployment may be applied to the design and planning of construction projects. As described by Bossert (1991), QFD provides a systematic method of quantifying users' needs and reflects these needs in the features of the respective products/services. In this case, the built facilities should more closely meet users' needs than is derived through current design practices which rely on practitioners' experience.

A QFD model may be used to represent six basic project management areas: project scope (functional requirement), budget costing, scheduling, land requirements, technical and safety requirements, and statutory and environmental requirements. Data from two projects of different type, nature, and scale are fed into the model for testing. QFD can enhance the project planning according to Ahmed et al. (2003) in the following ways:

1. Quality function deployment serves as a road map for navigating the planning process and always keeps track of customer requirements and satisfaction. This actually helps in eliminating human inefficiency (Figure 31.7).
2. The process of building a QFD matrix can be a good communication facilitator that helps break through the communication barriers between client and the designer and among members of the design team.
3. Quality function deployment can be an excellent tool for evaluating project alternatives, balancing conflicting project requirements, and establishing measurable project performance targets.
4. Quality function deployment can be used as a quick sensitivity test when project requirements change.

With use as a project-planning tool, QFD can bring benefits and enhancements to civil engineering capital project planning. Some research topics suggested for further study are streamlining the QFD process, computer-aided QFD applications, evaluation of the cost and benefits of using QFD, use of QFD in detailed design, and how to integrate QFD with total project QM systems (Figure 31.8).

31.2.18   Facilities Layout

Construction professionals often carry out the laying out of construction sites as an intuitive process; the result is a less than optimal solution. There are many variables in the construction environment such as geographic location, weather conditions, and the type and configuration of the materials and equipment and materials involved. The approach taken to laying out sites can have a major impact on project performance. Factors to consider include:

- Utilizing manpower and space effectively
- Minimizing delays, backtracking, and multiple handling
- Maintaining flexibility
- Providing good site hygiene and ease of maintenance
- Promoting safety
- Promoting high worker morale.

Industrial engineering techniques can be used to optimize the use of each construction site, thereby improving productivity and profitability.

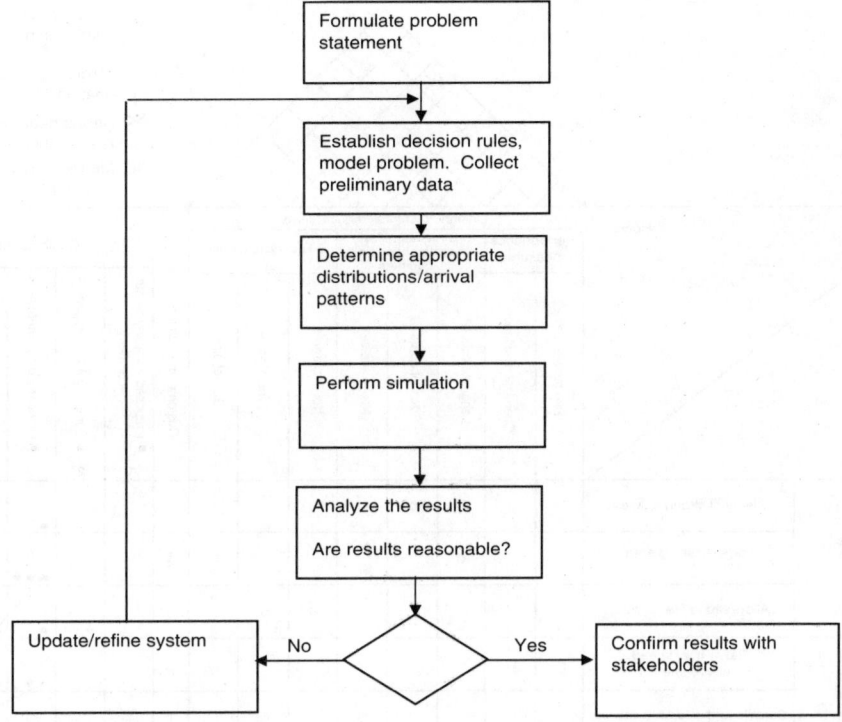

FIGURE 31.7   The simulation process.

An example of a typical site problem is how to position tower cranes for moving materials in the optimal manner. The crane boom needs to swing through a given angle at a particular velocity, and the trolley must travel along the boom for a given distance. Operations research techniques can be used to solve these problems.

31.2.19   Operations Research and Statistical Applications

Modeling the logistics of construction present interesting challenges which include crew scheduling under uncertainty (use of union vs. nonunion workers, regular time and overtime assignments, multishift and rolling shift, etc.). Many areas such as supply chain, transportation, material dispatching and inventory planning and control may be modeled and solved by operation research tools. The current tools for dealing with uncertainty in project management are overly simplistic and are dependent on unrealistic assumptions, such as the independence of completion times for each activity. More robust methods to analyze the distribution of completion times for large-scale construction projects are needed. Furthermore many issues related to automation, such as robot path planning, collision avoidance, image and signal processing, etc. would require OR modeling.

31.2.20   Sustainable Construction

Industrial engineers can play a key role in improving the quality, timeliness, efficiency, variety, and sustainability of homebuilding. A recent study of new home quality suggests that homebuilders are struggling to maintain high quality as they respond to the unprecedented demand for new homes. Industrial engineers can play a leadership role in quality improvement efforts, helping to define housing quality goals and developing the QM systems and tools necessary to achieve these goals.

Construction waste also puts additional pressure on the environmental sustainability of housing. For example, 36.4% of total U.S. primary energy consumption comes from construction. It also accounts

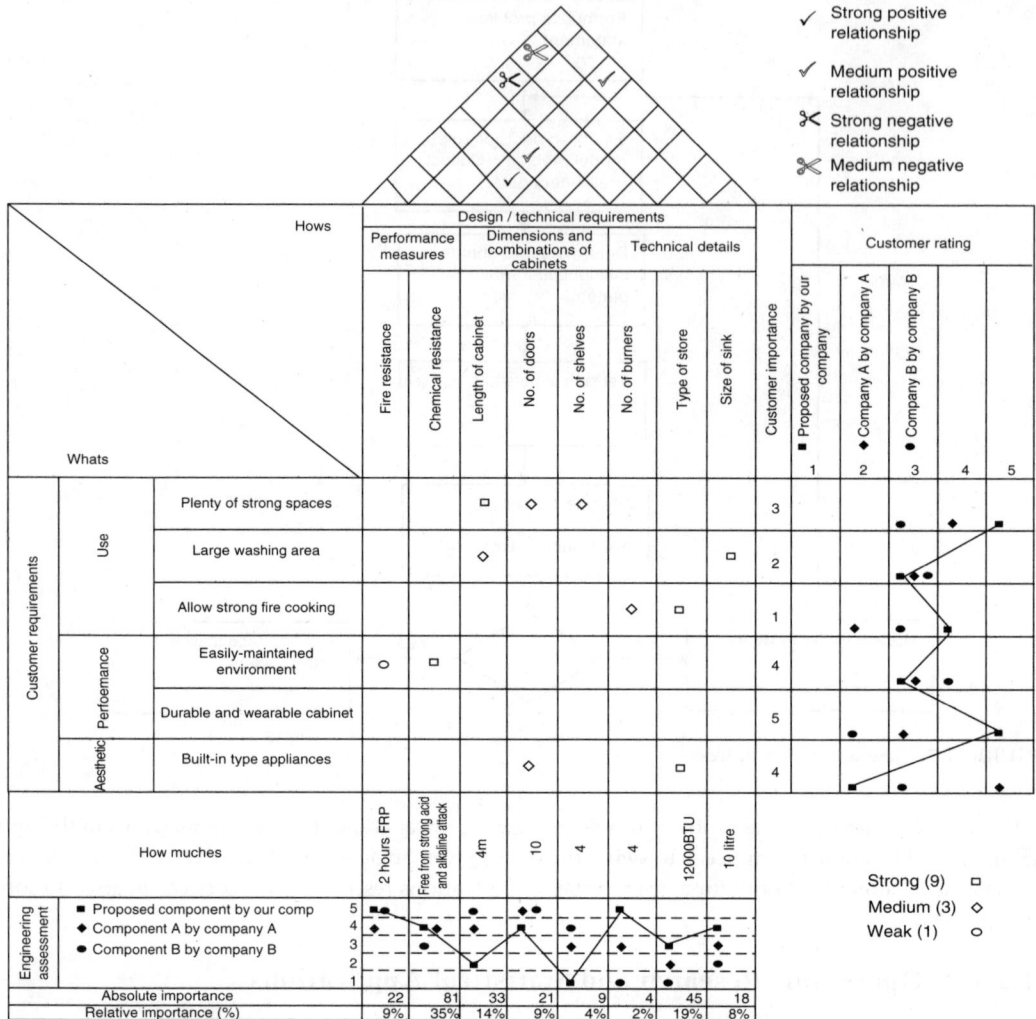

FIGURE 31.8 A simplified example of the house of quality for a kitchen cabinet. (Courtesy of CITC2000 Conference Proceedings: Cheung et al., 2002.)

for36% of total U.S. CO_2 emissions, 30% of total U.S. global warming gases such as methane, nitric oxide, and hydro fluorocarbons, and 60% of total U.S. ozone depleting substances. Construction waste is a major problem; each residential building accounts for 3–7 tons of waste. Nationally, the United States produces 136 million tons of construction waste, but only 20 to 30% of this is recycled or reused. Industrial engineers can do much to reduce waste, optimizing material purchases and developing construction practices that increase the recycling of material waste.

Acknowledgments

The author wishes to acknowledge Dr. Behrokh Khoshnevis (University of Southern California) for special contributions on Contour Crafting, Dr. Michael A. Mullens (University of Central Florida) for Continuous Improvement in Modular Housing, and Dr. Syed M. Ahmed (Florida International University) for Simulation.

References

Adrian, J. and Adrian, D., *Total Productivity and Quality Management in Construction*, Stipes Publishing, LLC, Champaign, IL, 1995.

Ahmed, S.M., Sang, L.P., and Torbica, Z.M. Use of quality function deployment in civil engineering capital project planning, *J. Constr. Eng. Manage. ASCE*, 358–368, 2003.

Bertelsen, S., *Construction Logistics I and II, Materials Management in the Construction Process (in Danish)*, Boligministeriet, Bygge-og, Boligstyrelsen, Kobenhavn, Denmark, 1993.

Bossert, J.L., *Quality Function Deployment: A Practitioner's Approach*, ASQC Quality Press, Milwaukee, WI, 1991.

Burati, J.L., Jr. and Oswald, T.H., Implementing TQM in engineering and construction, *J. Manage. Eng.*, ASCE, 9, 456–470, 1993.

Cheung, K.T., Cheung, S.O., Drew, D., Value optimisation of building components in the design process, *First International Conference on Construction in the 21st Century (CITC2002) — Challenges and Opportunities in Management and Technology*, Miami, FL, USA, 2002, pp. 311–318.

Dell'Isola, A.J., *Value Engineering in the Construction Industry*, Smith Hinchman and Grylls, Washington, DC, 1988.

Deming, W.E., *Out of the Crisis*, Massachusetts Institute of Technology, Center for Advanced Engineering Study, Cambridge, MA, 1991.

Elshennawy, A., Mullens, M., and Nahmens, I., *Quality improvement in the modular housing industry, Industrial Engineering Research '02 Conference Proceedings*, Orlando, May, 2002.

Forbes, L., An Engineering-Management-Based Investigation of Owner Satisfaction, Quality and Performance Variables in Health Care Facilities Construction, Dissertation, University of Miami, 1999.

Forbes, L., Continuous learning through quality-based post occupancy evaluation, *Proceedings of the IIE Annual Conference*, Orlando, FL, 2002.

Forbes, L. and Golomski, W., A contemporary approach to construction quality improvement, in *The Best on Quality*, Vol. 12, Sinha, M.N., Ed., IAQ Book Series, ASQ Quality Press, Milwaukee, WI, 2001, pp. 185–199.

Forde, M. and Buchholz, B., Task content and physical ergonomic risk construction ironwork, *Int. J. Ind. Ergon.*, 34, 319–333, 2004.

Freivalds, A., Ergonomics of shovelling and shovel design — an experimental study, *Ergonomics*, Elsevier Engineering Information Inc., 1986, pp. 19–30.

Gambao, E. and Balaguer, C., Robotics and automation in construction, *IEEE Robotics Automat. Mag.*, 2002.

Goetsch, D. and Davis, S., *Quality Management: Introduction to Total Quality Management for Production, Processing, and Services*, Prentice-Hall, New York, 2003.

Griffis, F.H., Farr, J.V., and Morris, M.D., *Construction Planning for Engineers*, McGraw-Hill, New York, 2000.

Guo, H.R., Tanaka, S., Cameron, L.L., Seligman, P., Behrens, J., Wild, D., and Putz-Anderson, V., Back pain among workers in the United States: national estimates and workers at high risk, *Am. J. Ind. Med.*, 28, 591–592, 1995.

Holstrom, E., Lindell, J., and Moritz, U., Healthy lower backs in the construction industry in Sweden. *Work Stress*, 7, 259–271, 1993.

Khoshnevis, B., Automated construction by contour crafting-related robotics and information technologies, *Automat. Constr.*, 13, 5–19, 2004.

Khoshnevis, B., Russell, R., Kwon, H., and Bukkapatnam, S., Contour crafting — a layered fabrication technique, *IEEE Robotics Automat. Mag.*, (Special Issue) 8, 2001.

Kisner, S. and Fosbroke, D., Injury hazards in the construction industry, *J. Occupat. Med.*, 36, 137–143, 1994.

Kubal, M.T., *Engineered Quality in Construction*, McGraw-Hill, New York, 1994.

Mali, P., *Improving Total Productivity: MBO Strategies for Business, Government, and Not-for profit Organizations*, Wiley, New York, 1978.

Miller, C., Packham, G., and Thomas, B., Harmonization and Lean construction: acknowledging the role of the small contracting firm, Working Paper 15, Welsh Enterprise Institute, University of Glamorgan Business School, 2001.

Oglesby, C., Parker, H.W., and Howell, G.A., *Productivity Improvement in Construction*, McGraw Hill, New York, 1989.

Pan, C.S, Chou, S., Long, D., Zwiener, J., and Skidmore, P., Postural stability during simulated drywall lifting and hanging tasks, *Proceedings of the XIV Triennial Congress of International Ergonomics Association, and the 44th Annual meeting of the Human Factors and Ergonomics Association, Ergonomics for the New Millennium*, 2000, pp. 679–682.

Perkins, D.D., Meeks, JW., and Taylor, R.B., The physical environment of street blocks and resident perceptions of crime and disorder: Implications for theory and measurement, *J. Environ. Psychol.*, 12, 21–34, 1992.

Pires, N. and Pereira, T.D., Robotics and the construction industry, in *System-based Vision for Strategic and Creative Design, Second International Conference on Structural and Construction Engineering, University of Rome, 'La Sapienzia"*, Italy, September 23–26, Bontempi F., Ed., Swets & Zeitlinger, Lisse, ISBN 90 5809 599 1, 2003.

Preiser, W.F.E., Rabinowitz, H.Z., and White, E.T., *Post Occupancy Evaluation*, Van Nostrand Reinhold, New York, 1988.

Shriener, A. and McManamy, Total quality management struggles into a low orbit, *Eng. New Rec.*, 24–28, 1995.

Sillanpaa, J., Lappalainen, J., Kaukiainen, A., Viljanen, M., and Laippala, P., Decreasing the physical workload of construction work with the use of four auxiliary handling devices, *Int. J. Ind. Ergon.*, 211–222, 1999.

Slaughter, E.S., Characteristics of existing construction automation and robotics technologies, *Automat. Constr.*, 6, 109–120, 1997.

State of Construction 2002–2012, Bureau of Labor Statistics, Department of Labor, www.constructioneducation.com.

Sumanth, D.J., *Productivity Engineering and Management*, McGraw-Hill, New York, 1984.

Warzawski, A. and Navon, R., Implementation of robotics in buildings: current status and future prospects, *J. Constr. Eng. Manage.*, 121, 488–455, 1998 (www.cv.ic.ac.uk/futurehome).

Womack, J.P. and Jones, D.T., *Lean Thinking*, Simon and Schuster, New York, NY, 1996.

Bibliography

Acharya, P., Pfrommer, C., and Zirbel, C., Think value engineering, *J. Manage. Eng.*, 13–17, 1995.

Ahmed, S., Forbes, L., and Esquivia, J., *Using the Expected Monetary Value — EMV Criteria for Decision-Making to Implement a Safety Program in Construction Companies*, Rome, 2004.

Davis, R., Creating high performance project teams, *Proceedings, Construction Industry Institute Conference*, Austin, TX, September 26–28, 1999.

Jaselskis, E. and El-Misalami, T., Implementing radio-frequency identification in the construction process, *J. Constr. Eng. Manag.*, 2003.

Marans, R.W. and Spreckelmeyer, K.F., Measuring overall architectural quality: a component of building evaluation, *Environ. Behav.*, 14, 652–670, 1982.

Obrien, W., London, K., and Vrijhoef, R., Construction supply chain modeling: a research review and interdisciplinary research agenda, *Tenth International Group for Lean Construction Conference*, Gramado, Brazil, August, 2002.

Omigbodun, A., Value engineering and optimal building projects, *J. Architect. Eng.*, 2001.

Sommerkamp, J., The Deming approach to construction safety management, *Am. Soc. Saf. Eng.*, 35–37, 1994.

Van Wagenberg, A.F., Post occupancy of a general hospital, Building for People in Hospitals, European Foundation of Living and Working Conditions, Loughlinstown House, Shankill, County Dublin, Ireland, 1990, pp. 155–170.

Biography

Dr. Lincoln H. Forbes, P.E. obtained his Ph.D. from the University of Miami in 1999. His area of study was the improvement of quality and performance, including health care facilities design and construction. Previously, Lincoln obtained both MBA and M.S. in Industrial Engineering degrees at the University of Miami. He earned a B.Sc. in Electrical Engineering at the University of the West Indies.

A Registered Professional Engineer in the State of Florida, Lincoln is a Supervisor in the Division of Facilities Design and Standards at Miami-Dade County Public Schools, Miami, FL. He oversees POE activities and provides research and investigation support on construction quality issues including systems, methods, and materials. He has also held positions in areas such as in-house construction, construction quality control, project warranty services, and paint program quality management.

Lincoln is currently President of the Construction Division of the Institute of Industrial Engineers (IIE). He has previously served the Institute as Director of the Government Division and President of the Miami Chapter. He is also a member of ASQ and The American Society for Healthcare Engineering (ASHE). He is an adjunct professor with the Industrial and Systems Engineering Department at Florida International University, Miami, FL, specializing in Quality and Performance Improvement. He has served as an Examiner for the Florida Sterling Council Quality Award, and as a columnist on Construction Quality for the ASCE publication "Leadership in Management and Engineering." Interested parties may contact him by e-mail at conqualrsh@aol.com and www.iieconstruction.org

32

Scheduling of Production and Service Systems

Bobbie Leon Foote
U.S. Military Academy

32.1 Definition of the Scheduling Decision

The basic function of scheduling is the decision to start a job on a given process and predict when it will finish. This decision is determined by the objective desired, the data known, and the constraints. The basic data required to make a mathematical decision are the process plan, the time of operation projected for each process, the quality criteria, the bill of materials (which may determine start dates if materials are not immediately available), the due date, if any, and the objectives that must be met, if any. A process plan is a list of the operations needed to complete the job and any sequence constraints. It is sometimes thought that there is always a strict precedence requirement, but that is not true. Many times one can paint or cut to shape in either order. A quality requirement may impose a precedence, but this is to be determined. The basic plan can be listed as cut, punch, trim, and smooth (emery wheel or other). This list is accompanied by instructions as to how to execute each process (usually in the form of a drawing or pictures) and standards of quality to be achieved; as well, there will be an estimate of the time to set up, perform the process, do quality checks, record required information for the information and tracking systems, and move. A decision is required when there are two or more jobs at a process and the decision has to be made as to which job to start next. Some jobs require only a single machine. The process could be welding, creating a photo from film, x-ray, creating an IC board, or performing surgery. The study of single machines is very important because this theory forms the basis of many heuristics for more complex manufacturing systems.

32.2 Single Machines

We are assuming here a job that only requires one process, and that the time to process is given as a deterministic number or a random value from a given distribution. The standard objectives are (1) to Min $\bar{F}$ the average flow time, (2) Min Max lateness L_{max}, (3) Min n_t, the number of tardy jobs, and combinations of earliness and lateness. The flow time F is the time to completion from the time the job is ready. If the job is ready at time equal to 0, flow time and completion time are the same. The function (finish time–due date) is the basic calculation. If the value is positive the job is late, if it is negative the job is early. The number of tardy jobs is the number of late jobs. Finish time is simply computed in the deterministic case as the time of start + processing time. There are surprisingly few objectives that have an optimal policy. This fact implies that for most systems we must rely on good heuristics. The following is a list of objectives and the policy that optimizes the objective. The list is not exclusive, as there are many types of problems with unusual assumptions.

32.2.1 Deterministic Processing Time

The objective is followed by the policy that achieves that objective:

1. Min $\bar{F}$: Schedule job with shortest processing time first (SPT). Break ties randomly or based on a second objective criterion. The basic idea is this: if the shortest job is worked first, that is the fastest a job can come out. Every job's completion time is the smallest possible (Figure 32.1).
2. Min Max lateness (L_{max}): Schedule the job with the earliest due first (EDD). Break ties randomly or based on a second objective criterion (Figure 32.2).
3. Min n_t: Execute Moore's algorithm. This algorithm is explained to illustrate a basic sequencing algorithm. Some notation is required (Figure 32.3).

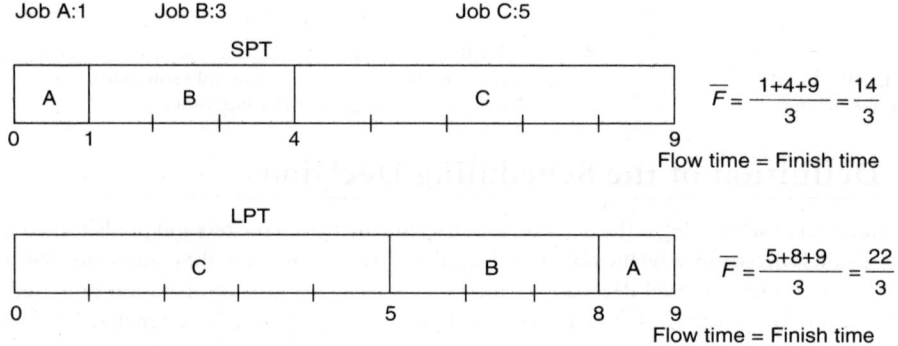

FIGURE 32.1 Optimality of SPT for min $\bar{F}$ (single machine Gantt charts comparing optimal with nonoptimal).

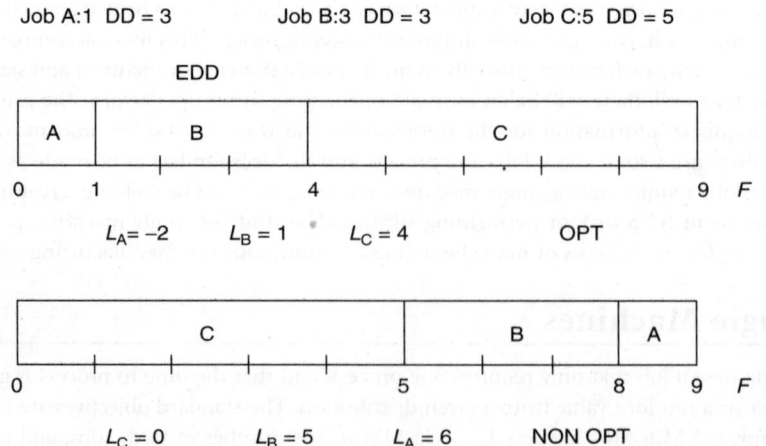

FIGURE 32.2 Optimality of EDD to Min Max lateness (single machine problem).

Job A: 6   DD = 6   Job B: 3   DD = 7   Job C: 2   DD = 8

Step 1: Schedule by EDD

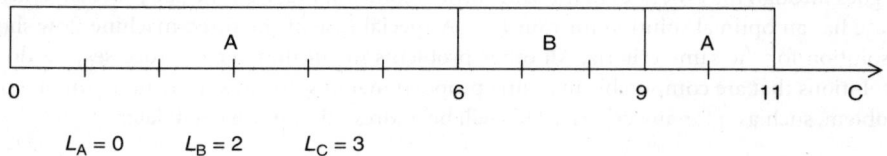

$L_A = 0$ $L_B = 2$ $L_C = 3$

STEP 2: B is first tardy job
Reject A since $P_A = 6 > P_B = 3$

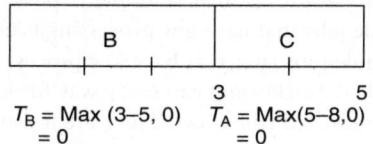

$T_B = \text{Max } (3–5, 0)$ $T_A = \text{Max}(5–8,0)$
$= 0$ $= 0$

Step 3: No further iterations needed.
All jobs on time.

Final schedule

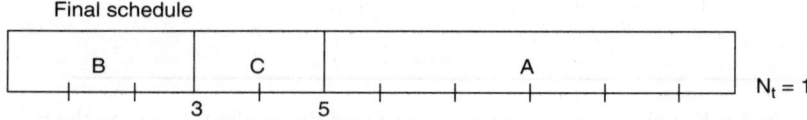

FIGURE 32.3   Use of Moore's algorithm.

32.2.1.1 Notation

Classification of problems: n/m/A/B, where *n* is the number of jobs, *m* the number of machines, *A* the flow pattern (blank when *m* = 1, otherwise *F* for flow-shop, *P* for permutation job shop (jobs in same order on all machines), *G* the general job shop (where jobs can have different priorities at different process plan order), *B* the performance measure (objective). *C* the completion time of a job and is equal to the ready time plus sum of waiting and processing times of jobs ahead of it and *D* the due date of the job

L (lateness of job) $= C - D$
T (tardiness of job) $= \text{Max}(L,0)$
E (earliness of job) $= \text{Max}(-L,0)$
P_i (processing time of job i) $n_t =$ number of tardy jobs

Moore's algorithm solves the $n/1/ /n_t$ problem. The idea of the algorithm is simple. Schedule by EDD to min max tardiness. Then find the first tardy job and look at it and all the jobs in front of it. The problem job is the one with the largest processing time. It keeps other jobs from finishing; remove it. It could be a job that is not tardy. Now repeat the process. When no tardy jobs are found in the set of jobs that remain, schedule them in EDD order and place the rejected jobs at the end in any order.

Figure 32.3 illustrates Moore's algorithm and basic commonsense ideas. For other single-machine problems that can be solved optimally, see French (1982) or any current text on scheduling (a fruitful path is to web-search Pinedo or Baker).

32.3   Flow Shops

More than one-third of production systems are flow shops. This means that all jobs have a processing plan that goes through the processes in the same order (some may take 0 time at a process). Only the two-process case has an optimal solution for min F_{max}. A special case of the three-machine flow shop has an optimal solution for the same criteria. All other problems for m(number of processes)>2 do not have optimal solutions that are computable in a time proportional to some quadratic function of a parameter of the problem, such as n for any criteria. This will be addressed in more detail later.

32.3.1   Two-Process Flow Shops

It has ben proved for flow shops that only schedules that schedule jobs through each process (machine) in the same sequence need be considered. This does not mean that deviations from this form will not sometimes get a criterion value that is optimal. When $m = 2$, an optimal solution is possible using Johnson's algorithm. The idea of the algorithm is to schedule jobs that have low processing times on the first machine and thus get them out of the way. Then schedule jobs later that have low processing times on the last machine, so that when they get on they are finished quickly and can make way for jobs coming up. In this way, the maximum flow time of all jobs is minimized (the time all jobs are finished).

Job	Time on Machine A	Time on Machine B
1	8	12
2	3	7
3	10	2
4	5	5
5	11	4

The algorithm is easy. Look at all the processing times and find the smallest. If it is on the first machine, schedule it as soon as possible and if it is on the second machine, schedule it as late as possible. Then remove this job's data and repeat. The solution here has five positions. The following shows the sequence construction step by step. Schedule is represented by S.

$$S = [\,,\,,\,,3]$$

$$S = [2,\,,\,,\,,3]$$

$$S = [2,\,,\,,5,3]$$

$$S = [2,4,\,,5,3] \text{ or } [2,\,,4,5,3]$$

Job 1 now goes in either sequence in the only position left: $S = [2,4,1,5,3]$ or $S = [2,1,4,5,3]$ (See Figure 32.4 below).

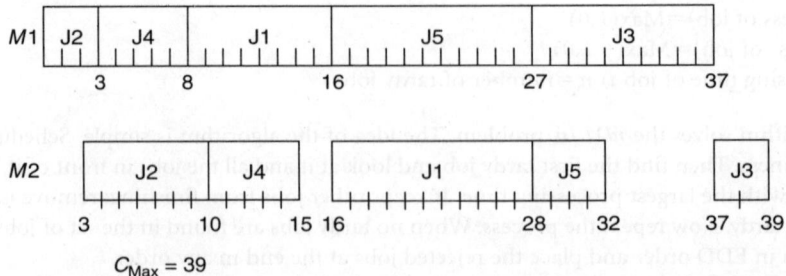

FIGURE 32.4   Gantt chart of an optimal schedule. For the example in two C_{max} problem schedule, using Johnson's algorithm.

32.3.2   Heuristics for General Flow Shops for Some Selected Criteria

$$n/m/F/F_{max}$$

The Campbell–Dudek heuristic makes sense and gives good solutions. There are better heuristics if one is not solving by hand. The idea is common sense. Assume we have six machines. If we find a job that gets through the six machines quickly, schedule it first so it gets out of the way. If there is a job that gets through the first five quickly, schedule it as soon as you can. This follows up to a job that gets off the first machine quickly. Alternatively, if a job gets off the last machine quickly, schedule it last, or if it gets off the last two quickly, or the last three, and so on. This leads to solving five constructed two-machine flow shop problems (surrogates), finding the optimal sequences, and then picking the one with the best maximum flow. A sixth sequence can be tested by sequencing in order of the least total (on all machines) processing time.

Here is a five-job, four-machine problem to illustrate the surrogate problems. There will be m – 1 or three of them.

Job	m1	m2	m3	m4	Total
1	5	7	4	11	27
2	2	3	6	7	18
3	6	10	1	3	20
4	7	4	2	4	17
5	1	1	1	2	5

$S=[5,4,2,3,1]$ when looking at total processing time

Three Surrogate Problems
Surrogate one: times on first and last machines only (solution by Johnson's algorithm: [5,2,3,4,1])

Job	m1′	m2″
1	5	11
2	2	7
3	6	3
4	7	4
5	1	2

Surrogate two: times on first two and last two machines (solution by Johnson's algorithm: [5,2,1,4,3])

Job	m1′	m2′
1	12	15
2	5	13
3	16	4
4	11	6
5	2	3

Surrogate three: times on first three machines and last three machines (solution by Johnson's algorithm: [5,4,3,2,1])

Job	m1′	m2′
1	16	22
2	11	16
3	17	14
4	13	10
5	3	4

These four sequences are then Gantt-charted, and the one with the lowest F max is used. If we want to min average flow time, then these would also be good solutions to try. If we want to min max lateness, then try EDD, and also try these sequences. We could also put job 1 last and schedule by EDD as a commonsense approach. Later we will talk about computer heuristics for these types of problems.

32.3.2.1 The Concept of Fit in Flow Shops

Consider a case where job (i) precedes job (j), with a six-machine flow shop. Let their processing times be (i): [5 4 2 6 8 10] and (j): [3 2 5 8 9 15]. In the Gantt chart (see Figure 32.5) you will see that job (i) is always ready to work when job (j) is finished on a machine. No idle time is created (see Figure 32.5). If computer-processing time is available, then a matrix of job fit can be created to be used in heuristics.

Fit is computed with the following notation: t_{ij} is the processing time of job (i) on machine (j). Then let job (j) precede job (k):

$$\text{Fit} = \sum_{s=2}^{s=m} \max[t_{js} - t_{k,s-1}, 0]$$

Heuristics will try to make job (j) ahead of job (k) if Fit is the largest positive found, i.e., [...k,j,...]. By removing the max operator, one can get a measure of the overall fit or one can count the number of positive terms and use this to pick a pair that should be in sequence. If there are six machines, pick out all the pairs that have a count of 5, then 4, then 3 and so on, and try to make a common sense Gantt chart. Consider the following data: J1: [4 5 4 7], J2: [5 6 9 5], J3: [6 5 6 10], J4: [3 8 7 6], J5: [4 4 6 11]. The largest number of fits is 3.

4→3 3→2 2→1 1→5 have three positive terms in the fit formula. Hence an optimal sequence is S=[4 3 2 1 5] (Figure 32.6).

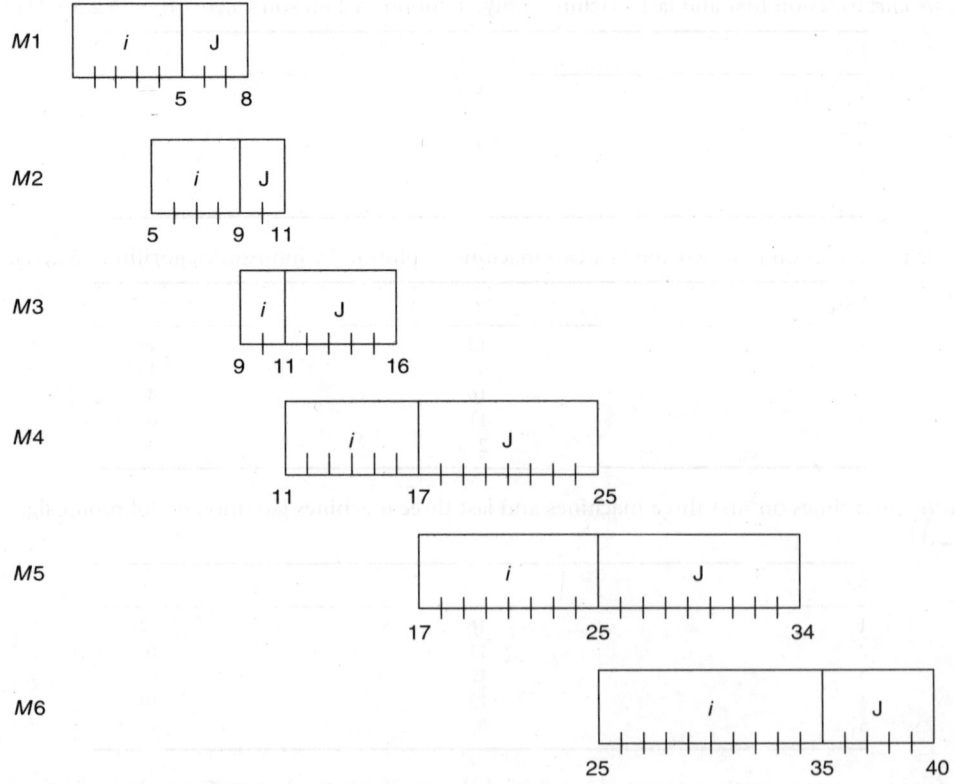

FIGURE 32.5 Job *J* fits behind job *i*.

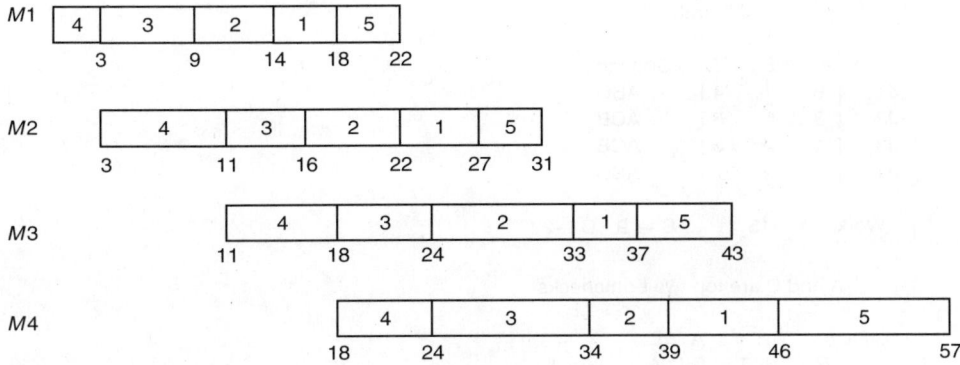

FIGURE 32.6   An optimal sequence since we have perfect Fit.

32.4   Job Shops

Job shops are notoriously tough (see later discussion on NP-complete problems). In job shops, a job may have processing plans that have a variety of sequences. For large job shops some simulation studies have shown that job shops are dominated by their bottlenecks. A bottleneck is the machine that has the largest total work to do when all jobs are completed.

In practice, it has been effective to schedule the jobs on each machine by using the sequence determined by solving a one-machine problem with the bottleneck as the machine. For large job shops, jobs are constantly entering. Thus, the structure constantly changes. This means that at the beginning of the period, each machine is treated as a single machine, and single machine theory is used for scheduling. At the beginning of the next period, a new set of priorities is determined. The determination of period length is a matter still under research. A period length of a multiple of total average job-processing time is a good start. A multiple such as 10 is a good middle ground to try. Perform some experiments and try going up if you have a lot of setup time and down if setup time is minimal.

Some other approaches are to take the top two bottlenecks and schedule the jobs as a two-machine flow shop, using only their times on the two bottlenecks. If you are using min average flow time, use flow shop heuristics designed for that purpose. If you are using min max flow time, use Johnson's algorithm. If a job does not go through the two bottlenecks, schedule it later, if you can. In the real world sometimes jobs can be worked faster than that predicted. Most schedulers try to find such jobs. They then look for small processing time jobs and insert them. Schedulers also look for jobs that have remaining processes that are empty and pull them out if average flow time is an issue.

It is never good to operate at 100% capacity. Quality goes down under this kind of pressure. Studies have shown that around 90% is good. At times, random variation will put the shop at 100% even then. But a little gap is good for preventive maintenance and for training. Most maintenance will be off shift if unplanned, but off shift can be expensive. Figure 32.7 is an example using Johnson's algorithm on the two bottlenecks and then scheduling the three machines with that sequence.

32.4.1   Scheduling and Systems Analysis

If output from one subsystem goes to another, then there is a timing issue and hence a due date issue. Basic queuing issues come into play. If service centers are in a flow mode, then the jobs should be in SPT order. This gets them out of the early systems quickly and allows the other jobs get going. If work has a large variance, then low variance jobs go first. Simulation software analysis is a must for large systems. The biggest aid is total system knowledge. If a truck coming in has an accident and this is known by the warehouse, some healing action can be taken. Some borrowing from other trucks might happen. Work that will not be completed due to the accident can be put off. Top-level

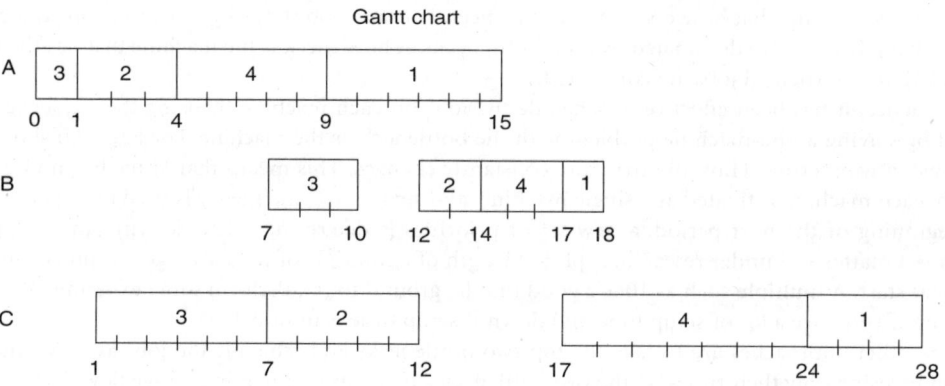

FIGURE 32.7   A heuristic approach to schedule a job shop. Compare with other guesses and pick lowest C_{max}.

knowledge of demand from the primary data is a must. Inferring it from orders is a major mistake. This subsystem knowledge allows coordination.

32.4.2   Control

What does control mean for a system? If a policy is simulated, there will be a mean and standard deviation of the queue length. When a queue exceeds the mean plus two standard deviations something is likely wrong. So, what questions should you ask? Are the processing times different from predictions? Is a large amount of maintenance going on? Do we have critical material shortages? Have we instituted just-in-time scheduling when the variance in system parameters is too high? If your trucks go completely through urban areas, you may not be able to arrive in a just-in-time window.

It is now possible with computers to predict a range of conditions that are normal. Readers should refer to the literature for current information, journals, and books.

32.4.3   Solvability Problems

We now know of many areas that have no solution. The normal distribution function does not have a closed-form integral. It must be computed numerically. Most scheduling problems are such that to guarantee optimality, total enumeration or implied total enumeration (branch and bound) must be used. However, it has been shown over and over that heuristics can get optimal solutions or solutions that are only off by 1 or 2%. When data quality is considered, this will be satisfactory. In linear programming, any point in the vicinity of a basic solution will be near optimal and, in fact, due to round-off errors in computing, the basis may even be better than the basis presented by the computation. Problems that require total enumeration to prove optimality are interesting, but in practice are solved and the solutions are good. Practically speaking, problems such as $n/m/F/n_t$ are what are called NP complete. They must have implied total enumeration to prove the answer is optimal. However, in practice, one can obtain good solutions. Moreover, for some problems, the optimal solution is obvious (for example, in flow shops where all jobs have unit processing times on all machines for the objective of min average flow time).

32.4.4 Bootstrapping

With the advent of modern computers and Monte Carlo simulation packages, it is possible to obtain statistical predictions of systems. By using heuristics and genetic algorithms to generate policies to simulate, a good statistical prediction can be made to set control standards with only three or four replications. Bootstrapping is the technique in which actual data are used for lead times, processing times, etc. These data can be arranged as a histogram from which values can be drawn randomly to determine how long a job will be processed, how long it will take to repair a machine, and so on. This uses the best evidence, the data, and has no chance of making an error in distribution assumption. Good data collection systems are a must. Distribution assumptions do have their virtues, in that we know repair times can occur that have not previously been observed. Running a Monte Carlo simulation both ways can throw much light on the matter.

32.4.5 Practical Issues

In both build-to-stock and build-to-order policies, the unexpected may happen. Machines break down, trucks get caught in traffic delays, materials are found to be defective, and workers do not show up. The plan and schedule must be repaired. A basic plan to illustrate repair is to assign workers to the bottleneck one at a time, then find the new bottleneck and assign the next worker there. Repairing, regaining feasibility, and extending the plan when new work arrives requires software if the system is large ("large" meaning that hand computation is simply not doable). An illustration of repair occurs when some jobs are not ready in an $n/m/G$ or F/C_{max} problem at time 0. One then goes ahead and solves, assuming all ready times are 0. If the solution requires a job to start before it is ready, repair must be done. These needs form the criteria for selecting commercial software from an available list. The software must be algorithmically correct and computationally efficient with good structures for search and information retrieval. Professional magazines such as *IE solutions* provide lists of vendors to assess.

References

Baker, Ken http://www.pearson.ch/pageid/34/artikel/28138PH/PrenticeHall/0130281387/Scheduling Theory Algorithms.aspx, 1995. (also self publishes)

French, S., *Sequencing and Scheduling*, Wiley, New York, 1982.

Pinedo, M.L., http://www.amazon.co.uk/exec/obidos/ASIN/0963974610/tenericbusine-21/026-2125298-8551636, Prentice Hall, 2002.

33

The Application of Industrial Engineering to Marketing Management

Tzong-Ru Lee
National Chung-Hsing University

Shing-Chi Chang
National Taiwan University of Science and Technology

33.1 Introduction

The concept behind industrial engineering is to regulate, predict, and evaluate the performance of systems. Such systems may comprise humans, materials, equipment, information, and power resources, all of which can be examined by applying the special knowledge and technology of mathematics, natural science, and social science, along with the theorems and methods of engineering analysis and design. Industrial engineering may be broadly applied in both industry and academia.

Before discussing the application of industrial engineering to marketing management, we first need to define "marketing" and "marketing management."

Marketing is closely related to individual life, economic activity, and organization operation, and it focuses on making deals, satisfying customer demand, and achieving organization goals. Marketing management comprises the process of analyzing, planning, implementing, and controlling market strategies; creativity, product, and service must be taken into consideration.

The function of marketing management is to consider, implement, and solve all types of marketing problems to help an individual or organization achieve a given goal. The process includes (1) analyzing marketing opportunities, (2) choosing a target market, (3) designing a marketing strategy, (4) planning a marketing policy, and (5) managing the marketing endeavor. A marketing strategy enables an enterprise to achieve its marketing goal in the target market, and the devices used to do this are called the "marketing mix." McCarthy uses the term "4P" — which stands for product, price, place, and promotion — to refer to the marketing mix.

33.2 Four Cases

We offer four cases to demonstrate the applicability of industrial engineering to 4P in order to help the reader to understand more readily the application of industrial engineering to marketing management.

33.2.1   Case 1: Application of Industrial Engineering to Product Strategy

Production is the most important tool that enterprises use to satisfy customer demand. Production strategy is an important marketing strategy and it comprises brand, packaging, service, product line, and product combination. This case takes the product combination of agricultural enterprises as the research subject to demonstrate the application of industrial engineering to product strategy. This research uses data-mining to find the product structure of agricultural enterprises and endeavors in order to adjust their product structure and strengthen their competitiveness.

33.2.1.1   What is Data-Mining Technology?

Berson et al. (2000) consider data mining to be the extraction of patterns in an autodetection database; Han and Kamber (2001) refer to data mining as extracting or digging out useful knowledge from a large amount of data. Thus, data mining involves the extraction of related patterns to extract or dig out useful data from a large amount of data.

33.2.1.2   Knowledge Discovery in Database

Before dealing with data mining, we have to understand what knowledge discovery in database (KDD) is. Knowledge discovery in database is the whole process of sifting out useful knowledge from a large amount of data. The KDD operator takes the following steps to rapidly and correctly extract the knowledge he needs:

1. *Data cleaning*: removing interference and abnormal information
2. *Data integrating*
3. *Data selecting*
4. *Data switching*: switching data to the right position and mode for convenientuse
5. *Data mining*: detecting the related patterns in the data
6. *Pattern evaluating*: analyzing and explaining the needed data
7. *Knowledge presenting*: presenting the results to users correctly

Data mining is used widely. For example, we can analyze the trading records collected by the point of sales (POS) system to learn about the consumer's purchase habits; then we can put together the products that consumers often buy at the same time so as to increase purchase convenience and purchase interest. Customer relationship management (CRM) is another application; by analyzing consumer behavior, we can find out what customers like or dislike, their purchase policy, and the socioeconomic variables affecting their purchasing decisions, which can then provide a point of references for policy makers.

According to the association rules of data mining, when support is higher than minimum support and when confidence is higher than minimum confidence, the group concept is used. Han and Kamber (2001) point out that the threshold value is set on the basis of user demand. This study sets the threshold value according to the number of groups.

33.2.1.3   Research Methodology

Each agricultural enterprise produces its own products and makes different products at different times. Some agricultural enterprises make only one kind of agricultural product. This study endeavors to find the product combination data of the agricultural enterprises using data mining and to make suggestions on an optimal product combination. The steps of data mining are as follows:

1. *Collecting data*: Product data are collected from 100 agricultural enterprises.
2. *Integrating and transforming data*: All product data are processed by Excel and transformed into a matrix expressed by 0 and 1: "1" denotes the product the enterprise produces, and "0" denotes the product the enterprise lacks.

33.2.1.4   Results

This study uses Polyanalyst 4.5 for data mining and conducts association analysis using market basket analysis software. Support represents the probability of an event happening, and confidence represents the probability of another event happening under the condition that part of that event has already happened.

TABLE 33.1   Results of the Product Combination Association of the Agricultural Enterprises

Group Number	Association Group	Support (%)	Confidence (%)
1	Oranges → K153 or persimmons	13	72.22
	K153 → oranges	13	100.00
	Persimmons → oranges	13	100.00
2	Grapes → K153	6	100.00
	K153 → grapes	6	100.00
3	Bamboo shoot → K100 or plums	6	100.00
	K100 → bamboo shoot plums	6	66.67
	→ bamboo shoot	6	66.67
4	Roses → K170	5	71.43

We set support at 3% and confidence at 65% to analyze the product combination. The results are shown in Table 33.1. There are four groups, and each of them consists of 100 large-scale productions. The information data are as follows:

1. *Group 1*: Of the enterprises in this group, 13% grow oranges or persimmons; 72.22% of the enterprises that grow oranges also grow persimmons simultaneously, and 100% of the enterprises that grow persimmons also grow oranges simultaneously (K153 is the control item for the enterprises growing oranges only).
2. *Group 2*: Only 6% of the enterprises grow grapes.
3. *Group 3*: Only 6% of the enterprises grow bamboo shoots or plums; all of the enterprises (i.e., 100%) that grow bamboo shoots grow plums simultaneously, and 66.67% of the enterprises that grow plums grow oranges simultaneously. Confidence of K100 to bamboo shoots is 66.7%, which means that 33.33% of the enterprises that grow bamboo shoots grow other crops simultaneously (K100 is the control item for the enterprises growing bamboo shoots only).
4. *Group 4*: Only 5% of the enterprises grow roses; 71.43% grow roses only, while the other 28.57% also grow other crops.

The above information reveals that some crops have similar growth environments, so the crops of the 100 agricultural enterprises are associated with each other. For example, oranges and persimmons probably have very similar growth environments, so they have a strong association. When the orange-planting acreage is too large and may result in overproduction during the subsequent year, the government can suggest that the enterprises that grow oranges and persimmons at the same time replaces some orange planting with persimmon planting so as to prevent the overproduction of oranges.

33.2.1.5   Conclusion

It is a helpful reference for the agricultural department of the government in crop diversion to establish a database for the products of each agricultural enterprise and analyze the associations among them.

33.2.2   Case 2: Application of Industrial Engineering to Price Strategy

In a marketing mix, price has a close relationship with product design, marketing channels, and promotion because while setting the price, market demand, competition, consumer behavior, and government regulations must be taken into account. This case takes the price prediction of the wholesale market as the research subject to demonstrate the application of industrial engineering to price strategy.

33.2.2.1   Case Preface

In Taiwan, computer auction skills are implemented in four wholesale flower markets, viz., Taipei, Taichung, Changhua, and Tainan. The computer auction quantity of cut flowers in those four markets is about 90% of the total auction quantity in Taiwan. The income of a wholesale flower market comes from the auction charge, which is positively correlated to the total auction value. If wholesale flower markets want to raise their total auction value, they must make an effort to increase the auction price or the auction quantity.

The objective of this case is to analyze the relationship between the auction time per box and the auction price of cut flowers in the four wholesale flower markets in Taiwan. From the results, we formulate a mathematical model to determine the optimal auction time per box of cut flowers at the highest auction price.

33.2.2.2 The Auction Process in the Wholesale Flower Market

The auction unit for cut flowers is the "box," because the flowers are auctioned one box at a time. Before the auction, the cut flowers are packed in boxes and moved by conveyers or carts in sequence. The auction staff will auction the cut flowers box by box. They must orally describe the quality of the cut flowers, predict a deal auction price, and set an initial price by observing the "buying atmosphere" (i.e., competition among the buyers). They usually set the initial price 20% higher than the deal auction price. When the computer auction begins, the initial price set by the auction staff keeps falling at certain intervals until the first buyer presses a control button (the equipment used in computer auctions) to show that he or she wants to buy. The buyer's information will then be shown on the information board. At this moment, the computer will stop the price from falling, and the number of cut flowers bought by the first buyer will be deducted from the total quantity. The auction process for this box of flowers is completed (Chen, 1997). The data from the Taipei Wholesale Flower Market show that the average auction time for a box of cut flowers is 5 sec.

33.2.2.3 Research Approach

We use cut flower auction data from the four wholesale flower markets in Taiwan collected on January 9, 2000, which, according to the auction staff, is representative of the typical auction quantity, the types of cut flowers, and the variation in auction time. The information used in this research contains auction time, price, and quantity per box for each type of cut flower. The probability density function (p.d.f.) for the auction time per box for different types of cut flowers is then determined.

In terms of auction quantity and auction price, there are 30 types of cut flowers auctioned. Owing to insufficient quantity of flowers, 11 types are excluded and the remaining 19 types are our research subjects. They are: Anthurium (An), Chrysanthemum (Ch), Spray Chrysanthemum (SC), Gerbera (Ge), Gludiolus (Gl), Bird of Paradise (Bp), Butterfly Lily (Bl), Dancing Lady (DL), Rose (R), Casablanca (Cb), Longiflorum (Lf), Eustoma (Eu), Baby's Breath (Bb), Lucky Bamboo (LB), Oncidium (On), Celosia plumosa (Cp), Solidago altissima (Sa), Gypsophila paniculata (Gp), and Dendrobium (Do).

Stat:fit software is used to determine whether the p.d.f. for the auction time per box is the same. We estimate the auction time per box for each type of cut flower using the p.d.f. format and their parameters, and find that the auction times per box for all the 19 types do not follow the same p.d.f. at the Taipei Wholesale Flower Market. We also notice that the p.d.f. for auction time for each type of cut flower is not the same in the other three markets.

There are many factors affecting the auction time p.d.f. for each type of cut flower, such as quality and type of flower and buyer demand. The results show that the type of cut flower affects the auction time. Thus, we discuss the relationship between auction time and auction price per box with the factor "types." We perform a regression analysis considering the different types of cut flowers. We set up 19 regression models for auction time and auction price per box based on their types and compute them using SPSS software.

33.2.2.4 Research Results

After concluding the regression relationship for the auction time and the price per box, we set up an integer linear programming model M1 and solve it using CPLEX software. In model M1, we attempt to determine "the optimal auction time for each main type of cut flower at its highest auction price."

The objective function in model M1 is to maximize the auction price for all varieties of cut flowers at the four wholesale flower markets net. A_{mfn}, B_{mfn}, C_{mfn}, and D_{mfn} be the coefficients in the regression models. The optimal auction time per box multiplied by the number of boxes should not exceed the total auction time for each auction line at each wholesale flower market.

We use the average and standard deviation of the auction time for each main type of cut flower to determine the upper and lower limits of auction time and to exclude extreme values. We test the auction time per box on the basis of one, double, or triple standard deviation. Within one standard variation, the estimated auction time per box is closest to the real auction time. Therefore, we set the auction time per box to be within one standard deviation of the auction time, as shown in Equation (33.3).

(M1)

Maximize

$$\sum_{m=1}^{4} \sum_{f=1}^{19} \sum_{n=1}^{3} \sum_{S=U_{mfn}}^{V_{mfn}} (A_{mfn} + B_{mfn}S + C_{mfn}S^2 + D_{mfn}S^3)X_{mfnS} \tag{33.1}$$

subject to

$$\sum_{f=1}^{19} \sum_{S=U_{mfn}}^{V_{mfn}} (Q_{mfn} \times S \times X_{mfnS}) \le T_{mn}, \quad m=1,2,3,4; \ n=1,2,3 \tag{33.2}$$

$$U_{mfn} \le \sum_{S=U_{mfn}}^{V_{mfn}} (S \times X_{mfnS}) \le V_{mfn}, \quad m=1,2,3,4; \ n=1,2,3; \ f=1,2,3,\ldots,19 \tag{33.3}$$

$$\sum_{S=U_{mfn}}^{V_{mfn}} X_{mfnS} \le 1, \quad m=1,2,3,4; \ n=1,2,3; \ f=1,2,3,\ldots,19 \tag{33.4}$$

X_{mfnS} is binary

where

m: indicator of the wholesale flower markets, $m=1,2,3,4$
f: indicator of the types of cut flowers, $f=1,2,\ldots,19$
n: indicator of the auction lines, $n=1,2,3$
S: auction time for a box of cut flowers, $S=U_{mfn}, U_{mfn}+1,\ldots,V_{mfn}$

Decision variables:

$$X_{mfnS} := \begin{cases} 1, & \text{represents the auction time for a box of } f \text{ type cut flowers at } n \\ & \qquad\qquad\qquad \text{auction line in } m \text{ wholesale flow} \\ 0, & \text{otherwise} \end{cases}$$

S: auction time for a box of cut flowers (unit: sec)

Parameters:

A_{mfn}: the intercept for the regression model between auction time and price for a box of f-type cut flowers at n auction line in m wholesale flower market

B_{mfn}: the first-degree coefficient of the regression model between auction time and price for a box of f-type cut flowers at n auction line in m wholesale flower market

C_{mfn}: the second-degree coefficient of the regression model between auction time and price for a box of f-type cut flowers at n auction line in m wholesale flower market

D_{mfn}: the third-degree coefficient of the regression model between auction time and price for a box of f-type cut flowers at n auction line in m wholesale flower market

Q_{mfn}: total number of boxes of f-type cut flowers auctioned at n auction line in m wholesale flower market

T_{mn}: total auction time for n auction line in m wholesale flower market (unit: sec)

U_{mfn}: lower limit of auction time for f-type cut flowers at n auction line in m wholesale flower market

V_{mfn}: upper limit of auction time for f-type cut flowers at n auction line in m wholesale flower market

CPLEX software is used to calculate the optimal auction time per box of each type of cut flower when its auction price is the highest. The results are shown in Table 33.2. The same type of cut flower has different

TABLE 33.2　The Optimal Auction Time Per Box for Each Type of Cut Flowers

Types of Cut Flowers	Taipei		Taichung		Changhua		Tainan	
	Maximum Auction Price (NT$)	Optimal Auction Time (Sec)	Maximum Auction Price (NT$)	Optimal Auction Time (Sec)	Maximum Auction Price (NT$)	Optimal Auction Time (Sec)	Maximum Auction Price (NT$)	Optimal Auction Time (Sec)
An	9	8	6	3	7	4	7	4
Ch	44	18	36	5	37	7	38	5
SC	42	14	32	1	36	10	36	10
Ge	21	5	25	7	27	13	22	5
Gl	67	8	52	1	54	2	72	16
Bp	25	9	20	3	18	3	22	7
Bl	22	19	22	9	15	13	23	3
On	50	6	52	8	51	7	58	14
Cp	35	6	42	15	37	6	38	4
Sa	19	10	19	10	19	13	19	4
Gp	94	15	65	2	72	16	70	3
DL	65	11	66	12	67	8	62	2
R	77	6	65	3	79	13	66	8
Cb	100	9	103	12	101	10	95	3
Lf	53	13	48	4	45	1	49	6
Eu	92	13	95	7	89	4	96	12
Bb	159	15	112	4	123	10	106	1
LB	48	13	42	6	38	5	52	9
Do	74	8	68	4	82	16	65	10

optimal auction times and a maximal auction price if it is auctioned at different wholesale flower markets. The farmers will choose the wholesale flower market with the maximal auction price for each type of cut flowers and sell cut flowers in that market.

On the basis of the information in Table 33.2, the auction staff at the four wholesale flowers markets are provided with a benchmark number to adjust the auction time per box, to increase the auction price, and to benefit the farmers.

33.2.2.5 Conclusion

For auction operations in the wholesale flower market, both buyers and auction staff must consider many factors, such as supply, demand, and quality of commodity, and they must make a decision within a very short time. In this research, "auction time" is an index to represent the relationship between all factors to predict the auction price. As a result, the cost and time to gather data can be saved and easily applied to other kinds of wholesale markets, e.g., wholesale fruit and vegetable markets. We believe that the results from this research will also benefit agricultural industries in other countries.

33.2.3   Case 3: Application of Industrial Engineering to Place Strategy

A marketing channel is the place, institution, or individual that gains product ownership or assists in transferring product ownership when certain products or services are delivered to consumers. The most common traditional marketing channels are substantial stores. Over time, with rapid economic development and increasing personal income, consumers gradually change their consumption habits. To be pioneers of the trend, enterprises must adapt their sales mode to improve their competitiveness and market share. Thus, nonstore marketing emerges.

This case takes the e-shop, a product of nonstore marketing, to demonstrate the application of industrial engineering to place strategy.

33.2.3.1   Case Preface

Internet trading is overwhelmingly popular in the global retail market. With the commercialization of the Internet availability of convenient interfaces and low website system operation cost, e-shops have boomed in a short period of time and have become acceptable and popular retail channels. In e-shops, shopkeepers have no direct contact with customers, and during the sales process, consumers have to actively search for the product they need. Thus, in building an e-shop, the enterprise should pay attention to features of the Internet such as cross-region, all-time, interaction, and abundance, so as to give consumers diverse product information and stimulate their desire to purchase. In recent years, agricultural products have become accessible in several e-stores. However, agricultural products are perishable and not easy to standardize, so consumers are very concerned about the function of e-store and the information they provide about agricultural products regarding quality assurance and low risk of consumption.

This study focuses on the B2C e-store mode from the consumer's viewpoint and categorizes the functions of the agricultural e-stores into "information flow," "physical flow," and "cash flow." We consider product information in "information flow" as an example. Using gray relation analysis, we discuss consumers' concerns about product information. Functions in other dimensions can also be found in this way and can serve as a reference for agricultural e-shops owners in designing their e-shops.

33.2.3.2   Basic Concept of E-Commerce

E-commerce is the exchange of products, services, and information through the Internet (Turban et al., 2000). NIST (1999) defines e-commerce as (1) the use of electronic communication to perform activities such as stocking, trading, advertising, channels, and paying; (2) any commercial trading method based on digital transmission; and (3) any electronic commercial trading service. Among them, electronic business, the virtual store, and the e-store belong to stores in virtual space; customers are able to shop through the Internet, and can obtain service at a lower cost than they would at substantial or physical stores (Yesil, 1997).

Basically, e-commerce is classified into four categories: business to business (B2B), business to customer (B2C), customer to customer (C2C), and customer to business (C2B). E-stores focus on B2C. Also, each deal through e-commerce involves "physical flow" (distribution), "cash flow" (account transferring), and "information flow" (added-value and transmission of information). Physical flow is the process of transporting products to the consumer; cash flow is the process of transferring money, including payments; and information flow is the process of obtaining product information, information supply, promotion, etc.

33.2.3.3 What Is Gray Relation Analysis?

In 1982, Deng Julong, Professor at Chung Hua Science and Engineering University, proposed the gray system theory. According to Julong, human beings still have limited, obscure, and uncertain knowledge of all the ecological systems in nature. Since the ecological systems provide incomplete messages to human beings, these messages are represented by the color gray. Gray, which is between black and white, here signifies uncertainty, so a system with incomplete messages is called a gray system. On the contrary, a white system represents a system with clear factors, relationships, structures, and operation guidelines (Hsiao, 2000). A gray system is characterized by (1) unclear system factors (parameters), (2) unclear relationship between factors, (3) unclear system structure, and (4) unclear guidelines for system operation. Generally speaking, traditional probability statistical methods use probability statistical value to calculate the regularity of a random process. In a gray system, we assume that the variables in any random process are the changing gray quanta within a certain range and time. A random process in a gray system is called a "gray process," and as long as there are three or four more original data, it can be converted into a gray model (Shen et al., 1998). The central theme of the gray theory is to discuss how to research a system with a limited quantity of message and further to form a complete view of the system. In a word, the theory is to perform relation analysis, model establishment under the uncertainty and incompleteness of the system model, and discuss the system by means of prediction and policy-making (Wu et al., 1996).

33.2.3.4 Research Methodology

Our study considers B2C agricultural e-shops as the subject and discusses how much emphasis consumers have on each function and the quality of agricultural e-shops.

33.2.3.5 Research Steps

Step 1: On the basis of previous studies about factors affecting consumers' purchases online, our research collects functions of e-shops that influence the purchase desire of consumers, and through search engines such as Yahoo, Yam, and Sina, finds domestic and foreign agricultural e-shops. We analyze all the functions and services.

Step 2: Having collected and classified related functions that influence the purchase desire of consumers, we find out the research variables of our study.

Step 3: By systematic random sampling, 300 consumers from Taipei, Taichung, and Kaohsung, the three most densely populated areas in Taiwan, are interviewed using questionnaires. The results will help to relate the demand and desire of consumers to the functions of agricultural e-shops.

Step 4: Data from the questionnaires are computed by gray relation analysis, and the functions of agricultural e-shops that consumers consider extremely important are ranked.

33.2.3.6 Research Variables

The research variables of this study are dimensions derived from online consumer purchasing based on previous studies, and we arrange these dimensions to get the related functions of e-shops that influence the purchase desire of consumers. At the same time, we select 15 globally well-known agricultural e-shops (e.g., 1-800-flowers.com, applesonline.com, buytomatoesonline.com, Florida Citrus Ranch, ubox.org.tw) and analyze their functions and services. We choose "information flow," "physical flow," and "cash flow," the three main dimensions of e-commerce, as the basis of classifying functions. Table 33.3 shows the functions of each dimension. In addition, this study uses the Likert scale to evaluate the degree of concern of the interviewees with regard to the functions and subfunctions of each dimension.

TABLE 33.3 Research Variables

Dimension	Function	Subfunction
Information flow	1. Product information	(1) Product price, (2) product name, (3) classified packing level, (4) product warranty, (5) product picture, (6) marketing price of deals, (7) nutrition facts, (8) standard for selection, (9) storage guidelines, (10) product source, (11) product search
	2. Ordering information	(1) Purchase procedure, (2) order inquiry, (3) order quantity and amount, (4) gift packing service
	3. Consulting service	(1) Customer consulting service, (2) on-line consulting, (3) FAQ, (4) discussion forum, (5) complaint line, (6) important information for leaguers
	4. Member service	(1) Registration with charges, (2) registration without charges, (3) request for basic personal files, (4) personalized promotion, (5) reserved deals for members only, (6) decision-assistant system
	5. Website linkage	(1) Connection to domestic websites, (2) connection to foreign websites
	6. Related information	(1) Latest news, (2) promotion of popular seasonal products, (3) introduction to web site, (4) product information, (5) industry introduction, (6) product recommendation, (7) travel and recreation, (8) information on cooking recipes
Physical flow		(1) Account of product pick-up place, (2) account of product pick-up period, (3) product-delivery methods, (4) product follow-up, (5) product-returning mechanism, (6) account of delivery process
Cash flow		(1) Account of charging, (2) multiple payments, (3) safe payment mechanism, (4) sound refund mechanism

TABLE 33.4 Gray Relation Analysis of Product Information

Function	Subfunction	Gray Relation Degree	Gray Relation Order
Product information	Product search	0.7902	1
	Product warranty	0.7493	2
	Nutrition facts	0.7479	3
	Product price	0.7293	4
	Product picture	0.7221	5
	Storage guidelines	0.7200	6
	Product source	0.7057	7
	Classified packing level	0.6951	8
	Standard for selection	0.6876	9
	Marketing price of deals	0.6643	10
	Product name	0.6479	11

33.2.3.7 Analysis

Considering product information in information flow as an example, we employ gray relation analysis and discuss the relation degree between each dimension and its functions as well as subfunctions in agricultural e-shops, and rank the importance of each function to consumers. The steps of gray relation analysis are: (1) normalizing the original data; (2) computing the gray relation coefficient; (3) computing the gray relation degree; (4) and permuting the gray relation.

33.2.3.8 Results

By using gray relation analysis, we compute the gray analysis degree and rank the gray relation order of 11 subfunctions, and rank the importance of each function to consumers. The results are shown in Table 33.4. The analysis shows that product search is most essential, followed by product warranty and nutrition facts.

33.2.3.9 Conclusion

This study demonstrates broad guidelines for developing agricultural e-shops. The analysis model of this study is recommended for those who want to develop agricultural e-shops in their own countries, taking into consideration the three flows (cash flow, physical flow, and information flow) from the consumer's viewpoint. The subdivisions of each of the three flows must be based on the properties of the agricultural products, consumer behavior, and the current e-shop trend in one's own country so as to develop the most appropriate agricultural e-shop mode.

33.2.4 Case 4: Application of Industrial Engineering to "Promotion" Strategy

Common promotion tools include advertisement, sales promotion, public relations, face-to-face promotion, and direct marketing. The purpose of promotion is to communicate with consumers or the public hope in the of increasing their acceptance of the products and directly or indirectly complete the trading. Therefore, even if a company has high-quality products, attractive prices, or effective marketing channels, they will be of no use without an effective promotion strategy to attract consumers. This case takes the supermarket of a Farmers' Association as the subject in order to demonstrate the application of industrial engineering to promotion strategy.

33.2.4.1 Case Preface

In recent years, the emergence of large-scale shopping malls has strongly threatened the survival of traditional supermarkets. This study designs a questionnaire employing an "industry attractiveness–business strength matrix" to help traditional supermarkets discern their most advantageous promotion strategies or improvements so as to upgrade their overall operation performance.

33.2.4.2 What Is the "Industry Attractiveness/Business Strength Matrix"?

Hedley (1977) of Boston Councilor Group (BCG) combines the relative competition position of businesses with the industry growth rate to divide businesses into four types: Problem businesses, Star businesses, Cow businesses, and Dog businesses; he also establishes strategies with the measurement of market growth rate and relative market share. However, the General Electric (GE) approach is more thorough than the BCG approach. The GE approach assumes that other than market growth rate and relative market share, there are other factors to evaluate an existing or future strategy. All the factors are categorized into two types: industry attractiveness and business strength; the former includes market scale, market growth rate, profit, competition condition, demand technology, inflation impact level, power source dependence, impact on the environment, and social/political/legal restrictions; the latter includes market share and market growth, product quality, brand goodwill, sales network, promotion efficiency, production energy, unit cost, raw material supply, R&D performance, and quality and quantity of management personnel. Then, all the factors are put into the matrix analysis to evaluate the feasibility of a strategy business in the whole enterprise (Chen, 1987). However, as the external environment of the enterprise is always changing, influential factors for evaluating industry attractiveness and business strength have to be modified constantly. In any case, it is possible to use the two factors in the BCG model, i.e., market growth rate and relative market share, in the GE model, i.e., the industry attractiveness/business strength matrix analysis.

The industry attractiveness/business strength matrix analyzes two dimensions: industry attractiveness and business strength, and the influential factors are shown in Table 33.5. Graph 33.1 shows different strategies for different industry attractiveness and business strength.

33.2.4.3 Questionnaire Design

1. *Content of questionnaire:*

We have designed five policies — issuing membership cards, conforming to fresh-food stocking time, providing cooked food service, holding sales at specific times, and increasing the number of unique products — each of which is analyzed with the industry attractiveness/business strength matrix. Table 33.6 shows that each dimension has five key success factors, and an extra factor, "others," is included. The managers of each supermarket score the current strength of the supermarket according to the 11 key success factors, and conclude with the final strength of the two dimensions. In this way, we can find the relative position of each policy from the matrix.

2. *Transforming questionnaire scores:*

In the industry attractiveness/business strength matrix, "high" means the total score range is 11–15, "middle" is 6–10, and "low" is 1–5 (Graph 33.2). The current strengths of each policy regarding each factor of

TABLE 33.5 Influential Factors for Industry Attractiveness–Business Strength

Industry Attractiveness	Business Strength
Existing competition condition	Cost advantage
Bargaining ability to supplier and customer	Promotion of efficiency
Demand technology	Raw material supply
Profit	Market share
Market growth rate	Manpower

Source: This study.

Business strength \ Industry attractiveness	High	Middle	Low
High	Enter into investment	Selective growth	Continue to cultivate
Middle	Selective growth	Continue to cultivate	Calculate profit + terminate investment
Low	Continue to cultivate	Calculate profit + terminate investment	Retreat from investment

GRAPH 33.1. Industry attractiveness/business strength matrix.

TABLE 33.6 Key Success Factors

Industry attractiveness	1. Existing competition condition
	2. Bargaining ability to supplier and customer
	3. Demand technology
	4. Profit
	5. Market growth rate
Business strength	6. Cost advantage
	7. Promotion efficiency
	8. Raw material supply
	9. Market share
	10. Manpower
	11. Others

the supermarkets are transformed into scores; that is, "favorable," "neutral," and "unfavorable" are transformed into 3, 2, and 1, respectively. The total scores of each dimension are computed to find their corresponding positions in the industry attractiveness/business strength matrix. "Enter into investment" means that the current operation of the supermarket matches the 11 key success factors, so it is very likely that investment will be successful. "Selective growth" means that this policy can be invested and cultivated, but since some of the key success factors fail to match the current condition, it is necessary to selectively strengthen the weaker factors. "Continue to cultivate" means that the current operation of the supermarket fails to match the key success factors in some aspects, and further investment is held back. After evaluation by high-level managers, if there is a chance to invest successfully, they can choose "Continue to cultivate or no interference." "Cultivate profit+terminate investment" means that once there is investment in the policy, the investment should be completed once profits are made because there may be losses rather than profits with further investment. "Retreat from investment" means that the policy

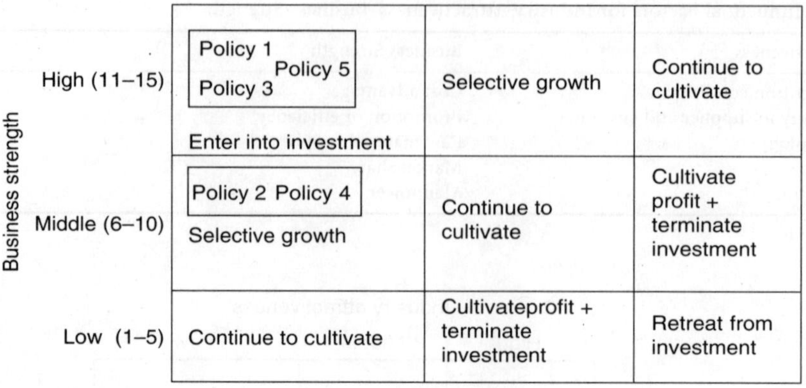

GRAPH 33.2.   The corresponding positions of policies of the supermarket of Farmers' Association in the strategy matrix.

fails to match the current condition of the supermarket, and there is no way to retrieve capital with further investment, so the investment must be retreated from or withdrawn immediately.

33.2.4.4   Analysis Results

This study takes the supermarket of a Farmers' Association as the subject, and transforms the three items "favorable," "neutral," and "unfavorable" in the questionnaires into scores of 3, 2 and 1, respectively. After totaling the scores of each dimension, we find that policies 1, 3, and 5 belong to "enter into investment," whereas policies 2 and 4 belong to "selective growth." The corresponding positions of each policy in the strategy matrix are shown in Graph 33.2.

33.2.4.5   Conclusion

This study uses the industry attractiveness/business strength matrix to help traditional supermarkets pinpoint the most advantageous promotion strategies or improvements so as to upgrade the overall operation performance.

References

Berson, A., Smith, S., and Thearling, K., *Building Data Mining Applications for CRM*, Translated by Ye, L.-C., McGraw-Hill, Taipei, 2000.

Chen, R.S., The introduction to auction system — a flower case, *The Management and Automatic of Wholesale Agricultural Products Market*, Taiwan Agricultural Mechanization Research & Development Center, Taipei, 1997, pp. 28–42.

Chen, X.-Z., *Strategy Management*, Taipei Publisher, Taipei, 1987.

Han, J. and Kamber, M., Eds., *Data Mining: Concepts and Techniques*, Academic Press, San Diego, CA, 2001.

Hedley, B., Strategy and the business portfolio, *Long Range Planning*, 10(1), Feb 1977, pp. 9–15.

Hsiao, S.W., *The Theory and the Application of Quantitative Decision*, Course Handouts, National Cheng Kung University, Taiwan, 2000. (Cited in Chen, C.N. and Ting, S.C., A study using the grey system theory to evaluate the importance of various service quality factors, *Int. J. Qual. Reliabil. Manage.*, 19, 838–861, 2002. NIST, *E-Commerce: The Future is Now*, Nation Institute of Standard and Technology, 1999.)

Shen, J.C., Shi, T.H., and Wang, S.C., An evaluation study using grey relational analysis in travel risks, *Proceedings of the Conference on The Grey System Theory and its Applications*, Taiwan, 1998, pp. 185–193. (Cited in Chen, C.N. and Ting, S.C., A study using the grey system theory to evaluate the importance of various service quality factors, *Int. J. Qual. Reliabil. Manage.*, 19, 838–861, 2002.)

Turban, E., Jae, L., David King, H., and Michael, C., *Electronic Commerce: A Managerial Perspective*, NJ: Prentice Hall, Upper Saddle River, 2000.

Yesil, M., *Creating the Virtual Store*, Wiley, New York, 1997.

Wu, H.-H., Deng, J.-L., and Wen, K.-L., *Introduction to Grey Analysis*, Kao-Li Publisher, Taipei, 1996.

34

A Management Model for Planning Change Based on the Integration of Lean and Six Sigma

Rupy Sawhney
University of Tennessee

Ike C. Ehie
Kansas State University

34.1   Introduction

For the past few decades, efforts have been made and various approaches have been developed to improve the competitiveness of industry. Lean and Six Sigma are the two most popular approaches that address efficiency and quality, respectively, within an organization. Many organizations have selected either Lean or Six Sigma as the primary brand to promote and implement change according to their focus of strategy. Even when organizations have adopted both Lean and Six Sigma, this is normally done from a perspective of implementing two complementary yet distinctly different programs. This is evidenced by many organizations that have functional and Six Sigma silos. There are often several programs that are simultaneously being implemented in the organization that may overlap or at least impact each other. The current mode is to manage each program independently through the appropriate department or function.

Is it possible that both Lean and Six Sigma promise efficiency and quality to organizations, but that each is limited by its scope in its efficiency and quality of delivering the concepts? A core customer requirement of an industry is a quality product that is competitively priced and delivered on time. At an operational level, this requirement translates to both the elimination of waste and the reduction of variation. Neither Lean nor Six Sigma can alone fulfill these operational requirements. This implies that both Lean and Six Sigma are required to meet the customer expectations. Practice has indicated that there is a great deal of synergy because both can positively impact the organizational processes. In essence, they both serve the various needs of the customers. Therefore, the successful implementation of Lean will enhance the performance of Six Sigma and vice versa.

Should Lean and Six Sigma be implemented in an organization as two separate initiatives? Such an approach taxes the organization on several different fronts. First, there is the probability that higher number of resources will be required due to duplication of effort. Second, a higher level of coordination effort may be required to prevent the two groups from being counterproductive to each other. Third, the implementation will be limited in scope to the group heading the effort. Fourth, there may be greater potential for internal strife due to turf protection. These drawbacks can be overcome by effective models to manage multiple overlapping programs, such as Lean and Six Sigma, whose purpose is to provide the structure for better management in an organization. An alternative approach is to integrate the concepts of Lean and Six Sigma into a singular approach.

The objective of this chapter is to introduce a model that allows management to plan change based on integrating Lean and Six Sigma. The focus of this chapter is to guide management systematically through issues that must be addressed before any implementation. It is this phase that is typically taken for granted and therefore the root cause of failures in implementation. In Section 34.2, we review the basics of Lean and Six Sigma. In Section 34.3, we provide a conceptual framework for integrating Lean and Six Sigma, and in Section 34.4, we present a management model to plan changes based on Lean and Six Sigma.

34.2   Basics of Lean and Six Sigma

34.2.1   Lean: Developing the Systems Perspective

The most popular definition of Lean is that Lean is a system to eliminate waste. However, organizations differ in how they define waste. A practical definition of Lean is that a Lean system reduces the time in the system of a product while meeting the flexibility requirements associated with the product mix. Therefore time in system becomes a dominant metric for measuring the efficiency of a production process. Organizations with a proper focus on time in system will align with customer expectations as outlined below:

1. The less time a product stays in the system, the lower the cost to the organization, hence the less a customer has to pay to buy it.
2. The less time a product stays in the system, the greater the organization's ability to meet customer delivery requirements.
3. The less time a product stays in the system, the smaller the probability of operational problems.

A more practical way to utilize time in system is to break it down into four components: process time, setup time, move time, and wait time, as illustrated in Figure 34.1. This perspective is based on the fact that at any given time, a product is being processed, waiting for setup, being moved, or simply waiting in the system. One primary value of this perspective is the framework that provides a view of the entire system. Each of these components is further explained below:

1. *Process time.* This category of time in system is the only value-added element. It is the core essence of the process as it is the only time component that is converting raw material into the final product. It represents the total time a product is being processed by the equipment or operator. It is calculated by summing all processing times as defined by the route sheet.

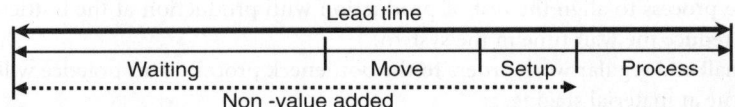

FIGURE 34.1 Decomposing time in system.

2. *Setup time.* This category of time in system is nonvalue-added. It is a key category in Lean and a major area for the performance improvement, since many Lean systems focus on high variety production systems. Setup time represents the downtime required at a station to accommodate changes in equipment setups, tooling, procedures, material, and others to produce a product different than the current. The calculation is based on summing all setup time, given that every station in the route sheet will require a setup.

3. *Move time.* This category of time in system is nonvalue-added. It represents the time a product or service is being transferred from one station to another. It is of interest because it provides insight into the physical layout of the plant and the logistics of moving product within this layout. The calculation is based on summing all the times in which the product is being moved from one station to another.

4. *Wait time.* This category of time in system is nonvalue-added. It represents the total amount of time a product or service is waiting within the system. This category is of particular interest as it provides a direct insight into the efficiency of the system. The greater the time a product waits in the system, the greater the inefficiency in the system, such as imbalanced lines, unavailability of materials, poor quality, equipment downtime. It is calculated by subtracting the sum of the total process, setup and move times from the time in system.

The above categorization of time in system allows multidimensional focus points for process improvements. Lean can be implemented with a focus on process time, setup time, move time, or wait time. The choice is dependent upon specific immediate circumstances. Since they are interrelated no matter where one initiates the efforts, one will have to address eventually all the categories of time in system. If an organization has no preference, a focus on wait time will be prudent because it is typically the largest component of time in system.

34.2.2 Envisioning Time in System Reduction

One concept that will allow planners to visualize how to impact all four components of time in the system is based on the exercise of developing the future state of processes via value stream mapping (VSM) (Rother, 1999). The following highlights major points of developing the future state and their impacts on the four components of time in system:

1. Ensure that the stations are balanced and below the TAKT time. This can be achieved by designing processes, equipment acquisition, combining work elements, distributing work elements, and other actions to reduce process time and wait time.

2. Develop continuous flow to produce one piece at a time, with each item passed immediately from one process to another. This can be achieved by cellular manufacturing to combine operations wherever possible to reduce move time and wait time.

3. Utilize supermarket flow concept to control production where continuous flow is not possible. This coordination between the customer process and the supplier process based on consumer process demand will again have an impact on wait time and (possibly) on move time. Produce all products regularly at the bottleneck/constraining process. This implies the need for the bottleneck process to produce smaller lots of each product more frequently. This demands a reduction in setup time. The major factor here is to reduce setup time and wait time.

4. Design the process to align the rest of production with production at the bottleneck. This alignment will reduce the wait time in the system.
5. Release small and regular work orders to the bottleneck process. This practice will have an impact on wait time at material staging.

34.2.3 Six Sigma: Providing Depth of Analysis to the Systems Perspective

The value of Six Sigma is presented by discussing a simple three-station process illustrated in Figure 34.2. This representation is another format for the four components of time in system. Lean focuses on system performance based on evaluating data such as average processing times for each station. For example, looking at scenario 1, a VSM-based evaluation would conclude that the constraining station is station 2 and, therefore, the product will be coming off the line every 10 min. Most organizations define their processes in this manner. This would be correct if there were no variation in the process. However, variation is part of every process. The focus on the average processing time becomes less significant as process time variation increases with respect to the average processing time. For example, looking at scenario 2, the bottleneck is still obviously station 2 because the variation is small compared with the average processing time at each station. In scenario 3, it is ambiguous which station is the bottleneck at any time.

How can one truly analyze the system without considering variation? Variation stems from a variety of sources: natural variations, variations due to measurement systems, process-based variation, and variation within parts. Lean does not provide the necessary methodologies or the tools to consider variation. The focus of Six Sigma is to provide methodologies and analytical tools to identify, analyze, and consequently reduce variation. Some of the commonly used Six Sigma tools are statistical process control (SPC), measurement system analysis (MSA), failure mode and effect analysis (FMEA), root cause analysis (RCA), design of experiments (DOE), and analysis of variance (ANOVA).

34.2.4 Envisioning a Systematic Approach to Variation Reduction

The implementation of Six Sigma can be described by the DMAIC approach (define, measure, analyze, improve, and control) (Thomas, 2003). Figure 34.3 presents this systematic approach to variation reduction. An explanation of the figure follows.

The goal is to decrease defects per million opportunities (DPMO). The first phase is the "define" phase, in which goals and target completion dates are established. Further, resources required for the

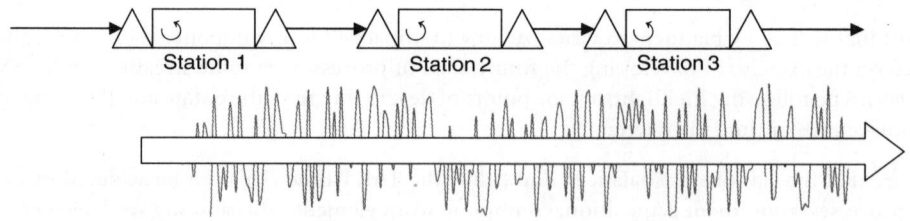

Scenarios		Station 1	Station 2	Station 3
1	Mean (min)	8	10	9
	Std. dev. (min)	N/A	N/A	N/A
2	Mean (min)	8	10	9
	Std. dev. (min)	0.2	0.2	0.2
3	Mean (min)	8	10	9
	Std. dev. (min)	1	3	2

FIGURE 34.2 Production process.

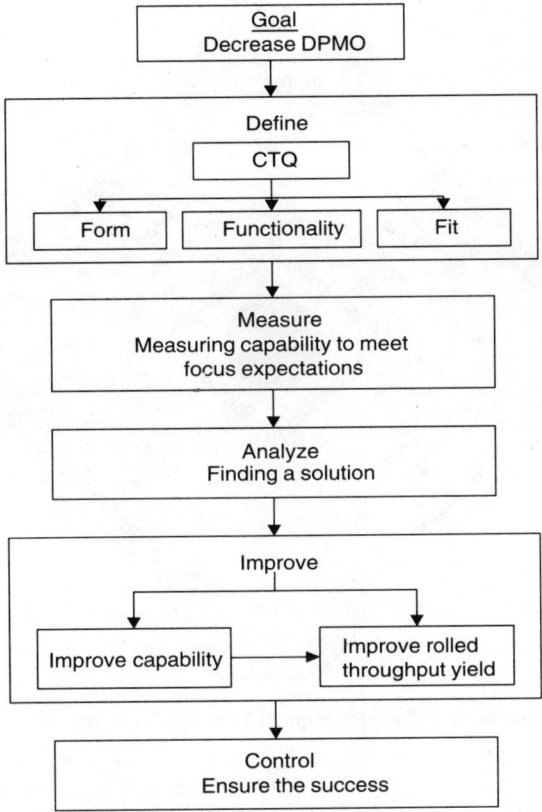

FIGURE 34.3   Six Sigma implementation process.

change and management support are obtained. During the "measure" phase, critical to quality (CTQ) measures and associated measurement systems are designed. Data are collected based on statistical concepts to ensure validity of results. During the "analyze" phase, statistical tools are utilized to identify root cause problems via better process data. This helps in developing an improvement strategy. The "improve" phase selects solutions, designs implementation plans based on proposed solutions, and ensures that support functions are available for the implementation plan. In the "control" phase, processes are standardized and monitored.

34.3   Conceptual Framework for the Integration of Lean and Six Sigma

Lean and Six Sigma cannot be taken out of two packages and simply thrown into a single toolbox. There is a need for a conceptual framework that allows a logical and systematic integration of the two business systems. Figure 34.4 presents such a conceptual framework. The key concepts associated with this conceptual framework are discussed below.

34.3.1   Change Is a Prerequisite

Change is a prerequisite for the survival of all organizations in the global market. Companies that are successful in repositioning themselves in the market are the ones that have well-defined processes to ensure

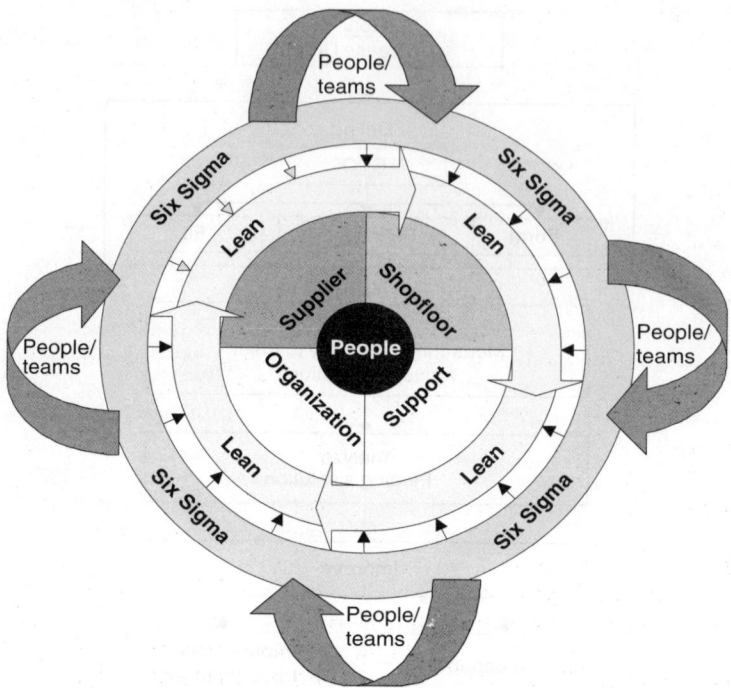

FIGURE 34.4   Conceptual framework of the integration of Lean and Six Sigma.

change. This does not imply that these processes cannot accommodate circumstances, but that there must exist a common starting point or a default.

34.3.2 Human Resources Are the Key

This conceptual framework hinges upon the development of a culture that embraces change. The human assets of an organization are its key asset. The development of such a culture is dependent upon a well-articulated vision, an explicit process, the skills to enable the workforce, and an organization that encourages the desired behavior.

34.3.3 Supporting the Defining and Critical Process

This conceptual framework requires human resources to seek opportunities for improvement throughout the supply chain, but in a systematic manner. There are four explicit phases within this conceptual framework. First and foremost, the focus should be on the design of the critical and defining process within each organization. For example, the critical and defining process for manufacturers is the manufacturing process. In this case, the first task is to design the manufacturing process based on capacity, capability, demand and growth strategies. All other activities are based on developing processes that support the critical and defining process. The logic within the conceptual framework is to move systematically away from the epicenter (critical and defining) process. The second phase involves designing operational support functions such as maintenance, warehousing to support the critical and defining process. There are many organizations where these processes actually constrain the manufacturing process. For example, production is often down because maintenance has not been able to maintain a piece of equipment. The third phase involves the design of organizational support functions such as production planning, scheduling, human resources to support efficiently the critical and defining process. The fourth phase involves extending the improvements to the supply chain.

34.3.4   Focus on Both Time in System and Variation

Each of the four phases above has further been divided into two components. The effort in each phase is to first reduce the time required to provide a product or service. This is followed by an effort to reduce the variations associated with the process. This requires that all employees have a skill set based on both Lean and Six Sigma.

34.4   Overview of Management Model for Planning Change

The conceptual framework presented in Section 34.3 can help guide management to evaluate first the design of the shop floor production system based on the requirement of the customers. If this evaluation leads to the conclusion that the shop floor is appropriately designed, the management moves to the next domain. This does not imply that the shop floor has to be perfect before any other domain is addressed but rather that the focus should be on the domain that is constraining the organization's ability to meet customer requirements. The ability to make the necessary changes proposed by the conceptual model is explicitly based on developing a people-oriented culture. The first step in developing such a culture is to plan the change properly. The six-phase management model for planning change based on both Lean and Six Sigma is presented as Appendix A. The essence of each phase is defined below.

Phase 1: Selling the Idea

One of the most important factors for successful change is selling the idea of change within the organization. Acceptance of change is the essential first step to developing the necessary culture.

Phase 2: Performance Measurement System

1. *Management must validate metrics currently in use.* Wherever appropriate, a new set of performance metrics must be articulated that is aligned with the objectives of the organization. It is recommended that a "change index" be developed utilizing critical performance metrics to highlight and communicate the change within the organization.
2. *Assessment.* Management must benchmark its processes in order to provide a clear understanding of the current state. The benchmarking must include an evaluation of the status of critical performance metrics. The benchmarking should go beyond the operations and include the domains identified in the conceptual model.
3. *Communication.* This phase involves planning the mechanisms to communicate change within the organization. It is critical for the entire organization to communicate with all employees on strategy, progress and results so as to develop the necessary culture. The specific communications that must be considered are defined in three categories:
    1. communicate the strategic plans associated with change;
    2. communicate the current initiatives and results associated with change;
    3. communicate changes in critical plant metrics.

Phase 3: Training

This phase concentrates on identifying the training requirements for all levels within the organization. Universal training is crucial to creating a common understanding and a common language within the organization. However, specific training must also be provided for various constituents based on their anticipated new responsibilities. Training must also be provided for new employees. It is important that a process be defined to provide training to all new employees if the desired changes are to be sustained.

Phase 4: Implementation

Implementation must be broken down into steps in a manner that is logical and palatable to the organization. For this reason this phase is delineated into three categories:

1. *Concept demonstration.* It is critical that concepts be implemented via selected projects that have high visibility and high impact. It is critical to demonstrate the concept and create excitement for change.
2. *Focused implementation.* The first comprehensive implementation strategy should focus on a single product family that is critical to the organization.
3. *Implementation proliferation.* This phase spreads the effort to other product families. Each repetition becomes easier as the process becomes familiar, and certain system-based issues have already been resolved as the efforts move from one product family to another.

It should be noted that it is important to sustain the change through the entire implementation process. It should start from the initiation phase and become part of the organizational culture.

34.4.1 Selling the Idea

Selling the idea of change to those who will implement it is critical. Yet, this concept is not part of the planning process. If human assets are the foundation for any change, then selling should occur throughout the organization and focus on each individual in the organization. At times a single message that is relevant to all employees can resonate with the entire population. For example, a message built on an immediate crisis in an organization facing severe consequences will resonate with everyone. However, in most organizations, a single generic message will not be an effective means of selling change.

It must be recognized that resistance to change is not based upon the need for change. Individuals are inherently aware of an organizational need for change. Neither is the resistance based upon the validity of the concepts. Both Lean and Six Sigma are proven concepts. Which employee would deny the logic to have a product more quickly available to customers or to reduce uncertainty associated with production? The resistance is based upon the perception of how the change is going to impact an individual or group of individuals. The impact of Lean and Six Sigma implementation, and its relevance to each employee, are not perceived the same way by all individuals within an organization. Therefore, the key to selling the idea is to communicate the need for change in a manner that is relevant to the target group of employees.

Each organization must define the various target groups. As a starting point, the organization may consider the following categories: executive management, facility management, functional management, staff, supervisors, and shop floor personnel. The magnitude of perceived relevance is based on how well the message meets the following requirements:

1. Presents the need for change.
2. Presents the approach selected for change.
3. Conveys awareness of the impact of change on the targeted group.
4. Categorizes and prioritizes concerns.
5. Addresses each concern and provides suggestions.
6. Provides assurance of organizational support.
7. Identifies mechanisms to allow employee feedback.

34.4.2 Performance Measurement System

Selling the idea of change communicates the organization's expectations. The expectations can only be achieved if the culture's behavior in the organization supports the expectations. Management has the explicit ability to align the behavior of personnel with the expectations, through the proper selection and application of performance metrics. It is common knowledge today that without the correct metrics and measurement methods it is difficult to sustain change. Industries have implemented Lean and Six Sigma to obtain initial positive results only to see it fade because the culture required to sustain change did not exist.

Initiatives introduced by management will provide results only as long as the management sets a high priority on the initiative. Unfortunately, management usually does not have the luxury to focus on a single initiative and therefore is unable to sustain change. However, management has the responsibility to implement a process that transfers the responsibility for sustaining change throughout the organizational structure. This top-down approach must link all organizational design to align behavior with expectations. It is critical that the approach be top-down to reduce potential conflicts. For example, an organization must be sure not to create situations in which shop floor personnel cannot sustain change because their supervisors have a conflicting set of interests.

The following is a guideline that can be utilized in developing the performance measurement system:

1. First, review the current system in a structured manner. Table 34.1 presents one alternative that will guide management through an exercise. The objective of the exercise is to highlight all the metrics that are utilized to evaluate the various levels in the organization. These metrics are then grouped in five categories: The first category of metrics directly encourages behavior to reduce time in system. The second category directly encourages behavior to reduce variation. The third category of metrics indirectly supports behavior to reduce time in system or variation. The fourth category of metrics has no impact on either the time in system or variation. The fifth category of metrics is the one that conflicts with the desired behavior. The completion of Table 34.1 will provide management with the following:

   (i) Insight into the current behavior at each level in the organization
   (ii) Insight to level of modification required at each level in the organization
   (iii) Insight into ensuring consistency across the reporting structure

   Design the performance measurement system based on the results of the previous step. Table 34.2 presents a list of metrics for consideration.

2. Develop an aggregate change index that combines the relevant metrics identified in the previous step into a single message. This change index communicates in a simple but powerful way the importance and status of change within the organization. The change index validates the efforts at the organization as the index illustrates improvement within the organization. The improvements are calculated based on current performance as compared with a base time period. This base time period is progressively redefined based on an agreed-upon time frame.
3. Plan reporting that ensures that the performance metrics are utilized as a basis to operate the business.

TABLE 34.1   Lean Six Sigma Performance Measurement System Review and Alignment

Objective	Concepts	Organizational Levels						
	Key Principles	Executive Management	Level 2	Facility Management	Level 3	Level 4	Shop Floor
Reduce time in system	Process Move Setup Wait			On-time delivery			
Reduce variation	Nature Measurement Process Part			Cost of quality			
Other supporting metrics				Days to produce			
Other neutral metrics				None			
Other conflicting metrics				Absorption rate			

TABLE 34.2   List of Lean Six Sigma Metrics

Efficiency
  Lead time
    Shop lead time
    Office lead time
  Cycle time
  Value-added time/shop lead time
  Nonvalue-added time/shop lead time
    Setup time/shop lead time
    Movement time/shop lead time
    Wait time/shop lead time
Inventory
  Inventory turnovers
  Inventory level
    Raw material
    Work in progress
    Finished goods
Delivery
  On-time delivery
    Percentage of on-time delivery given date by customers
    Percentage of on-time delivery given date by production
Quality
  Cost of quality
  Cost of yield
  Internal failure
  Scrap rate
  Rework rate
  External failure
    Returned product
  Appraisal
    Inspection
    Audits
    Inspections
    Prevention

34.4.3   Assessment

A historical review of manufacturing organizations implementing change reveals one common thread: inconsistency in achieving expected results. Two primary drivers of this inconsistency are the presence or lack of an explicit and well-articulated strategy for change and the ability or inability of the organization to align human behavior with the requirements of an organization. Alignment of human behavior is based on the skill sets and support perceived from the organizational infrastructure. Management must do a self-assessment prior to any implementation to provide data and the necessary analysis to develop a customized strategy for change. A comprehensive assessment must meet the following criteria. It should be able to

1. link performance metrics to the operational performance and the level of Lean and Six Sigma implementation;
2. ascertain current opportunities on the shop floor, support functions and the supply chain;
3. understand the organizational concerns and the needs for change;
4. consider personnel skills and availability;
5. utilize the above information to develop a customized plan for implementation.

Table 34.3 presents the logic behind the model and provides an assessment tool that meets the criteria established above. This model is based on four basic assessment pillars: operational assessment, metric assessment, personnel assessment, and organizational assessment.

TABLE 34.3   Assessment Logic

Assessment Type	Focus	Issues	Anticipated Outcome		
Operational	Shop floor	Time in system Variation	Shop floor constraints	Operational concerns	Customized plan for change
	Shop floor support	Time in system Variation	Shop support constraints		
	Office support sigma	Time in system Variation	Office support constraints		
	Supply chain	Time in system Variation	Supply chain performance		
Metric	Delivery	On-time delivery Days to produce Inventory levels Setup time	Root causes for delivery concerns	Root cause of concerns	Customized plan for change
	Quality	Cost of quality First article yield Schedule deviation	Root causes for quality concerns		
Personnel	Skills	Technical Lean and six sigma Facilitation Project management	Personnel ability to support change	Personnel support to correct concerns	
	Culture	Personnel stress levels	Current perceived personnel concerns		
Organization	Support	Management commitment Clear vision and plan Clear expectations Organization structure Alignment with rewards	Actual support for change	Company structure support to correct concerns	
	Resources	Availability of personnel Availability of capital Availability of data Availability of tools	Actual resources for change		

Figure 34.5 details the assessment flow, which transfers the current state status to a vision of the future state. Both quantitative and qualitative tools are utilized to conduct the assessment, including VSM, process stream mapping (PSM), surveys, interviews, and others. Each of the assessment types indicated in this assessment model is explained in detail in the following sections.

34.4.3.1   Operational Assessment

The first assessment pillar is the operational assessment. The purpose of this pillar is to evaluate the operational ability of the system to meet customer expectations. Operational ability can be profiled into the system capacity and capability. This is achieved by evaluating the time in system for a product and the associated variation in the process.

The facility is toured and all production and office processes are mapped for a given product family. Capability studies are performed on key areas identified in the VSM. VSM and PSM (for office environment) analyze the time in system and therefore provide the basic framework for highlighting system constraints based on an analysis of each product family. While VSM is extremely popular, its value is limited as it focuses on time in system and does not account for the impact of variation within the system.

Simulation modeling is the ideal tool to perform an operational assessment to both understand the current state and design the future state by considering the various components of time in system, variation and their interrelationships. Simulation modeling is suitable because it has the ability to replicate processes on a computer for the purposes of experimentation, without disrupting the actual process. Also,

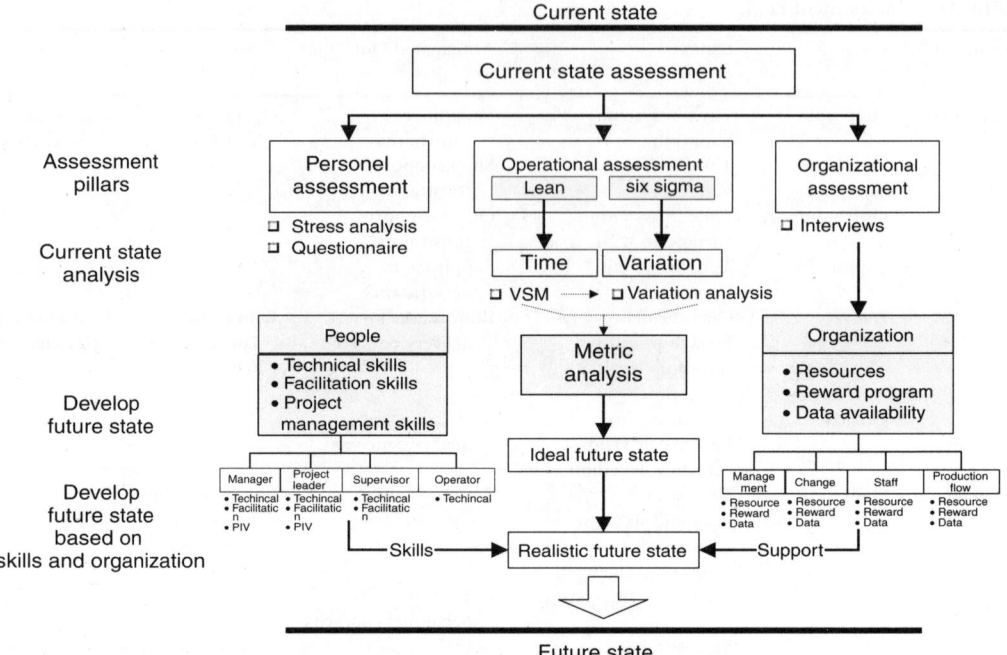

FIGURE 34.5 Assessment model flow for the implementation of Lean and Six Sigma.

the format of a VSM is easily transferred to a simulation model. Simulation has the ability to consider variation in analyzing and designing processes, and its results are presented in terms of the selected performance metrics.

Once the VSM is prepared and simulation modeling is done, an interview is conducted with most functions in the organization. The interviews are utilized for four primary purposes:

1. To verify the VSM
2. To verify the capability
3. To verify the metric analysis
4. To complete an operational survey that provides a baseline ranking of the operations within the organization

34.4.3.2 Metric Assessment

A metric analysis involves tracking basic performance metrics and the root cause for the current levels of the Lean and Six Sigma index. The metrics for measuring operational performance of an organization may be categorized into efficiency, quality, on-time delivery, and inventory. The purpose of metric analysis is to develop a baseline and subsequently analyze these metrics in order to identify areas of improvement within an organization. The operational assessment and metric analysis together yield an ideal future state. This has to be modified based on the results from the personnel and organizational assessments to yield a realistic future state to aim for.

34.4.3.3 Personnel Assessment

The personnel assessment is based on an analysis referred to as a stress analysis. The stress analysis is based on two questionnaires: a personnel stress questionnaire and a skill questionnaire. The stress analysis links the two questionnaires to provide management insight into the change in employee stress due to the introduction, implementation, and refinement phases of Lean and Six Sigma.

34.4.3.3.1   Stress Analysis

The purpose of stress analysis is to evaluate the current stress level of employees and the impact of the implementation of change on the stress level. The assessment may contain a questionnaire to be answered by all the employees, and the result will give the stress level of the entire organization (Hossain, 2004).

34.4.3.3.2   Skill Analysis

The purpose of personnel analysis is to develop a mechanism that allows management to baseline the personnel to determine the personnel requirements so that the system requirements are aligned with the personnel requirements (Sawhney & Chason, 2005). The matrix for the skill analysis of personnel is given in Table 34.4.

34.4.3.3.3   Personnel Analysis

Knowledge: This category covers knowledge that the worker carries within himself (unlike work instructions posted at work stations). While classroom, computer, and web-based training are the major media for providing workers with information, activities addressing worker knowledge may include newsletters, bulletin boards, shift meetings, and other postings.

Capacity: The worker's fundamental ability to do work. Ability is acquired during the recruitment and selection process. Aids to improve capacity also fall in this category, for example, work schedules, shift rotations, human ergonomics, color coding.

Motive: This category focuses on matching the worker's motive and interest in the work with the realities of the work situation. For example, if a worker who is interested in a fast-paced work environment is assigned to electronics assembly work, the match is not motivating and will result in poor performance.

The horizontal categories represent a conceptual approach to implementing Lean and Six Sigma in an organization. It is based on the concept that there is a sequential set of events: planning, creating the fundamental work environment for improvement, designing efficient flow through the shop floor, designing support systems to actually support production, reducing the variation in all designed processes, and developing systems that sustain improvement. The Lean implementation scheme is defined by the following sequence of events:

Planning: This category is associated with initiating the implementation. The first activity is the development of a production policy, which includes baselining the process, benchmarking (internally and externally) the process, developing metrics, developing targets for metrics, and developing a strategy for improvement. The second activity is to develop a culture within the organization, which includes a reporting structure, resource allocation, and team development. The third activity is developing a training and communication strategy.

Workplace: This category is associated with modifying the workplace and reducing variation in the manner that work is being performed to a point at which one explicitly understands the flow through the workplace. The activities include ensuring that (1) the workplace is safe and organized, (2) the work is performed according to standard operating procedures, (3) the work can be managed by visual controls, and (4) the processes are mistake-proofed to reduce chances of errors.

Flow: This category is associated with developing the basis to flow product through the process in a smooth manner. The activities include reduction of setup time, designing the process to match with takt time, line balancing, small lot production, cellular design, and implementing pull systems.

Support: This category is associated with designing the functional support systems to support actually the production systems. The activities include ERP systems, maintenance, supplier development, production control, and others.

TABLE 34.4   Personnel Skill Analysis in Lean Six Sigma Implementation

Human	Behavior	Planning	Workplace	Flow	Support	Consistency	Sustainability
People	Knowledge	Cell 1	Cell 2	Cell 3	Cell 4	Cell 5	Cell 6
	Capacity	Cell 7	Cell 8	Cell 9	Cell 10	Cell 11	Cell 12
	Motives	Cell 13	Cell 14	Cell 15	Cell 16	Cell 17	Cell 18

Consistency: This category is associated with reducing variation from design production and support processes. Activities include defining variation, measuring variation, analyzing variation, improving processes by reducing variation, and controlling processes.

Sustainability: This category is associated with sustaining improvements and a continuous improvement effort.

Each cell in the matrix can provide detailed information as to how an organization has addressed personnel requirements within a specific phase of the implementation scheme. Each row and column can provide information about the organization's ability to implement the change strategy. A consistently low evaluation in a given row would indicate an organizational weakness in the implementation. For example, the organizational weakness could manifest itself in the form of information, resources, incentives, personnel skills, and others. A consistently low evaluation in a column could indicate that the organization's strategy in implementing the change may be incorrect. For example, is it possible that manufacturers have attempted to implement a manufacturing cell without providing the necessary fundamentals to ensure the success of the cell?

34.4.3.4 Organizational Assessment

The fourth pillar is the organizational assessment. Components of the interviews performed in the operational assessment are utilized to baseline the organizational support. The organizational infrastructure is evaluated in terms of data availability, resource availability, and level of recognition to encourage Lean and Six Sigma implementation. If the skill sets or the organizational infrastructure is not available, it is the responsibility of the assessment to make management aware of the shortcomings. Further, if the decision of the management is not to increase the resource level, then the desired future state has to be modified to a more realistic future state.

34.4.3.4.1 Interviews

Interviews are a common approach for understanding the organization before implementation of a change. It is important to include all functions within the organizations when conducting the interviews. The interviews should be designed in such a way that employees are encouraged to present potential improvements in their functional areas.

34.4.3.4.2 Support Analysis

The purpose of support analysis is to determine the preparation of resources within an organization for the implementation. The matrix for the support analysis of organization is given in Table 34.5.

34.4.3.4.3 Organizational Analysis

Data: This category includes frequent feedback about the performance, clear expression of performance expectations, clear guidance for performance, and timely information necessary for decision-making that supports performance.

Instruments: In other words, resources, the means for accomplishing a desired performance. Many strategies and tools for achieving improvement come under this category, for example, well-designed tools, availability of tools at the point of use, well-designed and appropriately lighted work areas.

Incentives: Means of performance improvement in the form of financial and nonfinancial incentives. Financial incentives may include bonuses; pay for production and such other monetary incentives; and nonfinancial incentives could include awards, promotion, and special assignment.

TABLE 34.5 Organizational Support Analysis in Lean Six Sigma Implementation

Human		Behavior	Planning	Workplace	Flow	Support	Consistency	Sustainability
Organization		Data	Cell 1	Cell 2	Cell 3	Cell 4	Cell 5	Cell 6
		Instruments	Cell 7	Cell 8	Cell 9	Cell 10	Cell 11	Cell 12
		Incentives	Cell 13	Cell 14	Cell 15	Cell 16	Cell 17	Cell 18

The six Lean Six Sigma principles that constitute the horizontal axis of the matrix in Table 34.5 are described in the previous section.

34.4.3.4.4 Communication

Communication within an organization that is proposing change is critical to the success of implementing change. There are two major objectives associated with a communication strategy. First, communication strategy should provide visibility to all employees. Second, the organization selects a unique subset of communication mechanisms based on resources, constraints, and current culture. This alignment is significant as it defines the essence of an organization's ability to sustain the necessary level of communication to promote change. Such a communication strategy cannot be an implicit understanding but should be part of an explicit strategy. The strategy must include the following major information components: communicate the strategic plans associated with change, the current initiatives and results associated with change, and the changes in critical plant metrics. Organizations normally adopt a combination of the following: plant-wide meetings, operational meetings, organization-wide newsletters, e-mails, web pages, and display boards.

A checklist for developing a communication strategy is proposed below to ensure consistency within the different business units and facilities.

Communicate the Strategy for Change

1. Are there specific plans and mechanisms to communicate the final strategy to all management? It is assumed that management is heavily involved with developing the operational strategy. However, it is beneficial to communicate the final strategy to everyone at the same time to provide consistency of understanding. Further, the communication should provide guidelines and expectations for management communication to the rest of the facility.
2. Is there a benefit to having an organization-wide meeting to present the operational strategy to all associates? Is there a plan to provide documentation of the strategy to associates?
3. Is there a plan for facility managers and specialist to present the strategy to subgroups to discuss the impact and expectations to each work area?
4. Is there a plan to develop a web site to communicate the strategy?
5. Is there a plan to develop a board to communicate the strategy?
6. Is there a plan to provide a feedback mechanism to address associated concerns?

Communicate Current Initiatives and Results Associated with Change

1. Are there plans to recognize initiative success?
2. Are there planned meetings in which current initiatives are presented?

   - Is there a plan to develop a web site to communicate initiatives?
   - Are there plans to develop a best practices database?
   - Is there a plan to develop boards to commnicate the initiatives?

Communicate Changes in Critical Plant Metrics

1. Are there plans to recognize initiative metrics?
2. Are there planned meetings in which current metrics are presented?
3. Is there a plan to develop a web site to communicate progresses?
4. Are there plans to develop a best practices database?
5. Is there a plan to develop boards to communicate the progresses?

34.4.4 Training

In order for Lean Six Sigma to be understood and implemented by everyone concerned, a training program should be conducted for the entire organization. This training program should be started from the initiation phase of the implementation and designed in such a way that it involves the entire organization, enhances the techniques and skills for change, and makes the idea of change part of the organizational culture.

34.4.4.1 Design a Training Strategy

In order for training to achieve the predetermined goal, the following need to be considered when designing the training program:

General goal. There is a definite desire to enhance the implementation of Lean Six Sigma within the entire organization. The training program is designed to prepare key front-line leaders to facilitate critical projects and for all employees to apply daily concepts that support the Lean Six Sigma initiatives. Focus will be on building skill and practice in areas including group facilitation, problem-solving and decision-making, teambuilding, and project and meeting management.

Audience/participants. The target audience for this training is primarily manufacturing engineers, engineering management, production supervisors and managers, project leaders, or anyone responsible for leading or participating in similar projects. Parts of the program may also be offered to plant employees involved in implementation of the projects.

Certification. It is critical that portions of this training be mandatory for all supervisors and other individuals responsible for implementing Lean Six Sigma. A certificate of completion should be given to participants who have met the requirements of the program, including attaining an appropriate level of knowledge from each session and the ability to apply that knowledge to their work area. It is recommended that the management should establish a timeframe requiring all appropriate people working in the facility to be certified according to their respective job responsibilities. Certifications may require annual refresher training.

Training structure. There are three primary categories of competency that are important for individuals managing and facilitating change as defined below:

1. *Technical competence.* The change agents must not only have a clear understanding of Lean Six Sigma concepts but must also have the ability to translate those concepts into systematic actions in the workplace.
2. *Facilitation competence.* The change agents must have the ability to understand the workforce and develop a strategy by which to lead, facilitate, and communicate change.
3. *Project management competence.* The change agents must have the detailed ability to manage projects to a successful conclusion.

Training methodology. The in-class sessions should consist of informal lecture, small group interaction, video or audiotapes, paper/pencil instrumentation, simulations and applicable action planning, and pre- and posttests. On-line programs consist of pre- and posttests, online presentations, and simulations. Each segment of training is scheduled in such a way as to allow multiple exposures to key concepts and the opportunity for each participant to apply the learning principles to actual projects.

Training effectiveness measurement. Each training session has predetermined key behavioral objectives that are evaluated through testing or observation. It should be noted that this measurement indicates how the training affects knowledge and not necessarily business results. To measure how the training ultimately impacts the strategic initiatives, such as productivity, turn rates, scrap, and customer complaints, key measures should be base-lined and tracked for progress. Additional sources of measurement include employee surveys, feedback from the Lean Six Sigma Steering council, and percentage of employees certified. Each business leader is recommended to identify key metrics that will serve as an indicator of success of the program. Also, participants at the end of each class complete a training evaluation.

34.4.4.2 Designing a Training Program

Table 34.6 presents the training requirements for Lean Six Sigma *specialists* (site program champions), *project leaders* (anyone responsible for leading a designated improvement project), *supervisors* (not

TABLE 34.6   Sample of Training Design for the Implementation of Lean Six Sigma

Category	Topics	Instructor(s)	Hours	Specialist	Project Leaders	Supervisors	Employee
Technical	Lean principles			M	R		
Technical	Six sigma principles			M	R		
Technical	Developing lean six sigma strategy			M	M	R	
Technical	Applying lean six sigma for leaders			M	M	M	
Technical	Applying lean six sigma for employees						M/R
Facilitation	Introduction to teamwork			M	M	M	M
Facilitation	Workstyle assessment			M	M	M	R
Facilitation	Interpersonal/ communication skills			M	M	M	
Facilitation	Conflict resolution			M	M	M	R
Facilitation	Meeting management/ facilitation			M	M	M	R
Facilitation	Presentational skills			R	M	R	
Project management	Project management fundamentals			M	M	R	
Project management	Planning a project			M	M	R	
Project management	Scheduling and budget			M	M	R	
Project management	Controlling and closing			M	M	R	

M, mandatory; R, recommended.

responsible for projects but for the daily application of Lean Six Sigma principles), and *employees* (shop floor employees, operators).

34.4.5   Implementation

34.4.5.1   Demonstration of Concept

Industry needs an approach that is able to obtain immediate improvements, provide motivation for the implementation process, and explore other opportunities for the organization. This phase concentrates on illustrating the value of integrating Lean and Six Sigma to achieve quick results utilizing the concepts of Kaizen/Blitz events. Kaizen/Blitz is a focused approach that applies Lean Six Sigma techniques only to selected areas for quicker results. The focus for Kaizen/Blitz events is the areas that have direct impact on the organizational performance. This implies that the projects selected will have components associated with reducing time in system as well as reducing variation. The projects must be strategically selected for content, probability of success, involvement of informal leaders and personnel, and project location to provide the greatest impression and visibility. Such a demonstration is critical after the training phase and it will provide several benefits. The first benefit is that results are achieved over very short periods of time. The second benefit is that the demonstration will quench the general concern that the initiatives introduced will not be implemented at all or not implemented properly. The third benefit is the ability to provide practical training to complement the previous training. In essence, the personnel involved are being taught how to apply the concepts to their work areas. The fourth benefit is that it provides a referable example for implementation in the rest of the process. The fifth benefit is that it provides the organization an opportunity to address constrained and problem areas. The sixth benefit is that it allows the management the ability to judge the level and source of resistance and develop a response. It is inevitable that resistance to change appears as one proposes change within people's areas.

34.4.5.2 Focused Implementation

34.4.5.2.1 Product Family Selection

The organization must select a single product family that is significant either in terms of sales or growth. The objective of the focused implementation phase is to allow an organization to coordinate its resources in implementing changes along the entire value stream for the given product family. The advantages of this approach are as follows:

1. The focus on product families provides a direct link between Lean and Six Sigma initiatives and commercial initiative within an organization. Operational improvements are made for product families that have the greatest impact on the organization.
2. This approach is more efficient in achieving bottom-line results than the traditional approach. The traditional approach emphasizes a systematic implementation of tools as compared to the focus on value stream and bottom-line improvements.
3. A focus on the entire value stream intensifies the requirement for cross-functional cooperation. Improvements to the entire value stream will also lead to greater visibility of Lean and Six Sigma within the organization.
4. The approach provides organizations with a single, focused, and well-understood target that is better aligned with the current resources at the organization.
5. This approach induces organizations to utilize quickly the comprehensive set of Lean Six Sigma tools. Such requirements will intensify the skill levels and knowledge within the organization.
6. The approach allows resolution of systematic issues for one product family to translate to similar problems for other product families. Therefore, analysis of subsequent product families may be less intensive.

34.4.5.2.2 Process for Strategy Development

The primary tool utilized in developing the strategy is the A-3, a methodology whose name refers to a common size of European paper that is divided into four quadrants, as illustrated in Appendix B. The first quadrant is utilized to present a business case study and associated baseline metrics. The second quadrant is utilized to present the current state value stream map. The third quadrant presents the future state value stream map and the associated hypothesis. Finally, the fourth quadrant provides an implementation plan highlighting the projects and the logistics associated with the projects. The fourth quadrant also presents the anticipated results for critical metrics.

A hierarchy of A-3s can be utilized to articulate the strategy for the organization. High-level A-3s present the strategy at the corporate or business unit level. The focus of this A-3 level is to develop a high-level strategy for the selected product family over the entire value stream. The implementation plan for the high-level A-3 defines the specific projects, person responsible for the project, team members for the project, stakeholders, facilitators available to the project, and start date and end date for projects. A-3s provide projected improvements to the selected product families as measured by Lean and Six Sigma index. Similarly, lower level A-3s are developed as needed. For example projects at each facility can develop A-3s to support the business or corporate level A-3s. Further, lower level A-3s can be developed to support facility level initiatives. The development of this hierarchy ensures consistency of effort throughout the organization and an inherent infrastructure for responsibility and accountability.

34.4.5.2.3 Reviewing the Strategy

It is recommended that a core team be selected to review the strategies for each business unit and each facility. There are specific reasons for this suggestion. There is little consistency in the process adopted between business units to prepare the necessary strategy. The range can vary from one unit involving numerous people over multiple days, with long discussions and input, to plans created by one or a few individuals in a short time. The cumulative impact of the level of participation in each facility and the amount of time dedicated to this critical task can culminate in inconsistent quality. A set of criteria has been developed to allow organizations to review the strategies submitted based on an explicit standardized process. The following are the recommended criteria and the associated rating scheme utilized for the reviews:

Strategic business impact (SBI). This category attempts to ensure that the product family selected aligns with the strategic goals of the business, as follows:

1. No significant benefit
2. Increasing capacity without significant cost impact
3. Increasing capacity with significant cost impact
4. Increasing market share with existing products
5. Increasing market share with existing products and new product introduction.

Participation level in preparing strategy (PLPS): This category attempts to define the level of participation in developing the strategy. The more comprehensive the level of participation the more robust the plan, and the greater the "buy in" to the proposed strategy.

1. No participation
2. Single individual without management responsibilities developed strategy
3. Single individual with management responsibilities developed strategy
4. Small core group with management involvement developed strategy
5. Facility-wide participation in developing strategy

Opportunities over the entire value stream (OEVS): This category evaluates the proposal to ensure that the opportunities are considered over the entire value stream. This will ensure that the focus is beyond simply manufacturing, as some of the greatest opportunities exist outside of manufacturing.

1. Strategy not focused on value stream
2. Strategy focused only on components of the manufacturing process
3. Strategy focused only on the entire manufacturing process
4. Strategy focused on the internal value stream including support functions
5. Strategy focused on entire supply chain

Lean Six Sigma implementation level (LSSIL): The attempt in this category is to ensure that the concept of Lean and Six Sigma is utilized in a manner to ensure that waste is eliminated from the system in the most efficient manner. Further, the proposal is reviewed to ensure that there is a focus on process improvements rather than equipment acquisition.

1. The strategy is based on maintaining status quo
2. The strategy is primarily based on equipment acquisition
3. The strategy is based upon utilizing subset of Lean Six Sigma concepts
4. The strategy is based upon improvements utilizing all Lean Six Sigma concepts
5. The strategy focuses on maximizing impact via project selection

Alignment with proposed results (APR): The attempt in this category is to ensure that the projected results are aligned with resources and necessary skill sets.

1. Resources not identified
2. Inadequate resource allocation
3. Adequate resources but with limited skill sets and no facilitators identified
4. Adequate resources with skill sets but with no facilitators identified
5. Adequate resources with skill sets and facilitators identified

Participation level in implementing (PLI): The attempt in this category is to ensure that all levels of the facility are involved with Lean Six Sigma strategy. One focus is to ensure that shop floor personnel are part of the implementation strategy.

1. No participation identified
2. Participation primarily includes Lean Six Sigma
3. Participation primarily includes Lean Six Sigma and management
4. Participation primarily includes Lean Six Sigma, management, and support
5. Participation primarily includes Lean Six Sigma, management, support, and shop

Responsibility and accountability for implementation (RAI): The attempt in this category is to ensure that the plan contains an explicit structure for accountability and responsibility.

1. No responsibility and accountability plans
2. Plans provides no structure for accountability and responsibility
3. Responsibility and accountability for stakeholders only
4. Responsibility and accountability for stakeholders and facilitators
5. Plans have structure with single point accountability and responsibility

Table 34.7 illustrates how one can provide feedback to the unit that has developed the strategy. The table must accompany a summary of the strengths and weaknesses associated with each category of the strategy review. Examples and other facilities that have done outstanding jobs must be presented to the unit. Further, the name of a facilitator that can assist must be provided to the unit. The following is a sample summary attached to the review table.

Summary: A single (at most a couple of) individual(s) developed the plan. The effort of developing the plan was disjointed. The plan is focused primarily on components of the manufacturing process only utilizing the concepts of cells and the CONWIP system. The plan addresses manufacturing-based opportunities that will increase capacity and impact cost. Resources are identified, but they have limited skills sets to accomplish the tasks within the proposal. Availability of resources is suspect because of time constraints. Retraining is required if all levels including shop floor personnel are to be involved. There is no local facilitator identified and the details of individual accountability are incomplete.

34.4.5.3 Responsibility and Accountability

The above process explicitly recommends a structure that enhances the accountability and responsibility for implementation. This can be accomplished by increasing the participation and involvement of all levels of an organization with Lean Six Sigma. This increased participation will increase the visibility of Lean Six Sigma within the organization. The structure for responsibility and accountability starts with the stakeholders who are responsible for the overall implementation of the plan and moves down to individuals who are responsible for each initiative and further down to each task within an initiative. Further, facilitators and team members provide the necessary resources and support to ensure successful implementation. The following recommendations are presented for consideration:

1. A council led by the president of the organization reviews the progress of each facility toward achieving the projected results. Any changes to the projected plan can be made only at the quarterly meetings. These changes are noted and explicitly marked. The individuals accountable to the president are the stakeholders of each facility and the corporate facilitator.
2. The stakeholders have the responsibility to ensure the validity of the developed strategy for their facility. Further they must ensure that the organizational and resource requirements are available for implementation. When assistance is required, they are to work with the corporate facilitator.

TABLE 34.7   Review of Strategy

Business Unit	Review Score	Review Criteria						
		SBI	PLPS	OEVS	LSSIL	APR	PLI	RAI
1	2.3	2.5	2	2	2	1.5	4	2
2	2.8	3.5	2	2.5	3	2	3	3.5
3	3.5	3	4	3.5	3	3.5	3.5	4
4	2.9	2.5	2.5	2.5	3	3.5	3	3
5	2.2	3	2	1.5	1.5	3.0	2.5	1.5
6	2.7	2	3	2	2	3.5	3	3.5
7	2.7	2	3	2	2	3.5	3	3.5

3. The corporate facilitator works with the stakeholders to provide them with assistance. Further, the corporate facilitator works with business facilitators to insure timely transfer of technology. Appropriate mechanisms for technology transfer and communications are developed if required. Lastly, the corporate facilitator provides leadership and technical expertise to individual initiatives.

4. The business facilitators have a responsibility to the stakeholders to ensure that associated teams are performing. The business facilitators work with the corporate facilitator on various issues, including transfer of technology across facilities. The business facilitators also provide the necessary tools and skills to each team to ensure their success.

5. The team leaders on the high-level A-3s are responsible for the business facilitators to meet the requirements of the given initiative.

6. The team members are accountable to the team leaders to provide the necessary resources to ensure successful completion.

7. The task leaders on the lower level A-3s are accountable to the team leaders to ensure the tasks in an initiative are completed successfully.

8. The task members are accountable to the task leaders to provide the necessary resources to ensure successful completion.

34.4.5.4 Implementation Proliferation

Management must develop a plan on how to translate the knowledge from the previous phase to incorporate other product families.

34.4.6 Sustaining

The six-phase approach is an explicit articulation of Lean Six Sigma strategy that combines an efficient and practical emphasis for developing a culture that sustains and builds upon the Lean Six Sigma. Figure 34.6 provides a simplified visual for the strategy, given the understanding that each phase is not necessarily sequential and that there are overlaps and parallel processing of various phases. The visualization of the model is critical, as it presents several key assumptions of the model. First, the model assumes that there will be a timetable for the organization to go through the six phases of the implementation plan. Each

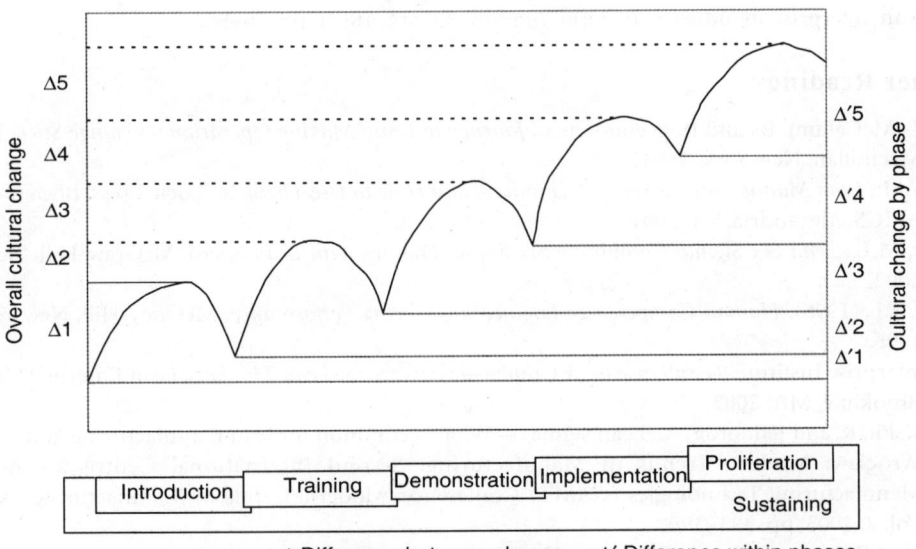

FIGURE 34.6   Lean Six Sigma strategy implementation.

successive phase will lead to a greater magnitude of cultural change required for Lean Six Sigma implementation. Second, the implementation strategy provides for the reality that there will be some level of backsliding within each phase after some level of success. A success is indicated when the ending point of cultural change is higher than the beginning point within each phase. The third assumption is that any implementation strategy requires explicit points to rejuvenate the commitment to Lean Six Sigma. The model provides such an opportunity at the beginning of each phase. This renewal ensures that the culture moves toward one sought by the Lean Six Sigma.

34.5   Conclusion

The integration of Lean and Six Sigma provides great advantages in increasing the competitiveness of an organization by reducing the lead time while ensuring that the quality of product meets the customer's requirements. This chapter provides a human-based approach for managing the integration and implementation of Lean and Six Sigma. The six-phase process will help the organizations improve their abilities to produce products in a timely manner and consequently increase their competitiveness in today's market.

Acknowledgment

We acknowledge, with high appreciation, the efforts of Yanzhen "Lee" Li, a PhD student in Industrial Engineering at University of Tennessee, Knoxville. Without his contributions and work, this chapter would not have taken shape.

References

Hossain, N.N., Utilizing employee stress to establish guidelines for managing personnel during Lean transition, *Unpublished Masters Thesis*, University of Tennessee, Knoxville, 2004.

Pyzdek, T., *Six SIGMA Handbook*, McGraw-Hill Professional, New York, 2003.

Rother, M., *Learning to See: Value Stream Mapping to Add Value and Eliminate Muda*, Lean Enterprise Institute, Brookline, MA, 1999.

Sawhney, R. and Chason, S., Human behavior based exploratory model for successful implementation of lean enterprise in industry, *Perform. Improve. Quart.*, 18(2), pp. 76–96.

Further Reading

Drew, J., McCallum, B., and Roggenhofer, S., *Journey to Lean: Making Operational Change Stick*, Palgrave Macmillan, New York, 2004.

Feld, W.M., *Lean Manufacturing Tools, Techniques, and How to Use Them*, St. Lucie Press, Boca Raton, FL, APICS, Alexandria, VA, 2001.

George, M.L., *Lean Six Sigma: Combining Six Sigma Quality with Lean Speed*, McGraw-Hill, New York, 2002.

Gilbert, T.F. (1996). *Human Competence: Engineering Worthy Performance*, McGraw-Hill, New York, NY, 1996.

Lean Enterprise Institute, *Lean Lexicon: A Graphical Glossary for Lean Thinkers*, Lean Enterprise Institute, Brookline, MA, 2003.

Torczewski., K. and Jednorog, A., Lean sigma — What is common for lean manufacturing and six sigma, Wroclaw: Modern Trends in Manufacturing: Second International Centre for Advanced Manufacturing Technologies (CAMT) Conference: Modern Trends in Manufacturing, February, Vol. 3, 2003, pp. 393–399.

Womack, J.P., Jones, D.T., and Roos, D., *The Machine that Changed the World*, Rawson Associates, New York, NY, 1990.

Womack, J.P. and Jones, D.T., *Lean Thinking: Banish Waste and Create Wealth in your Corporation*, Simon & Schuster, New York, NY, 1996.

Appendix A   Lean Six Sigma Implementation Process.

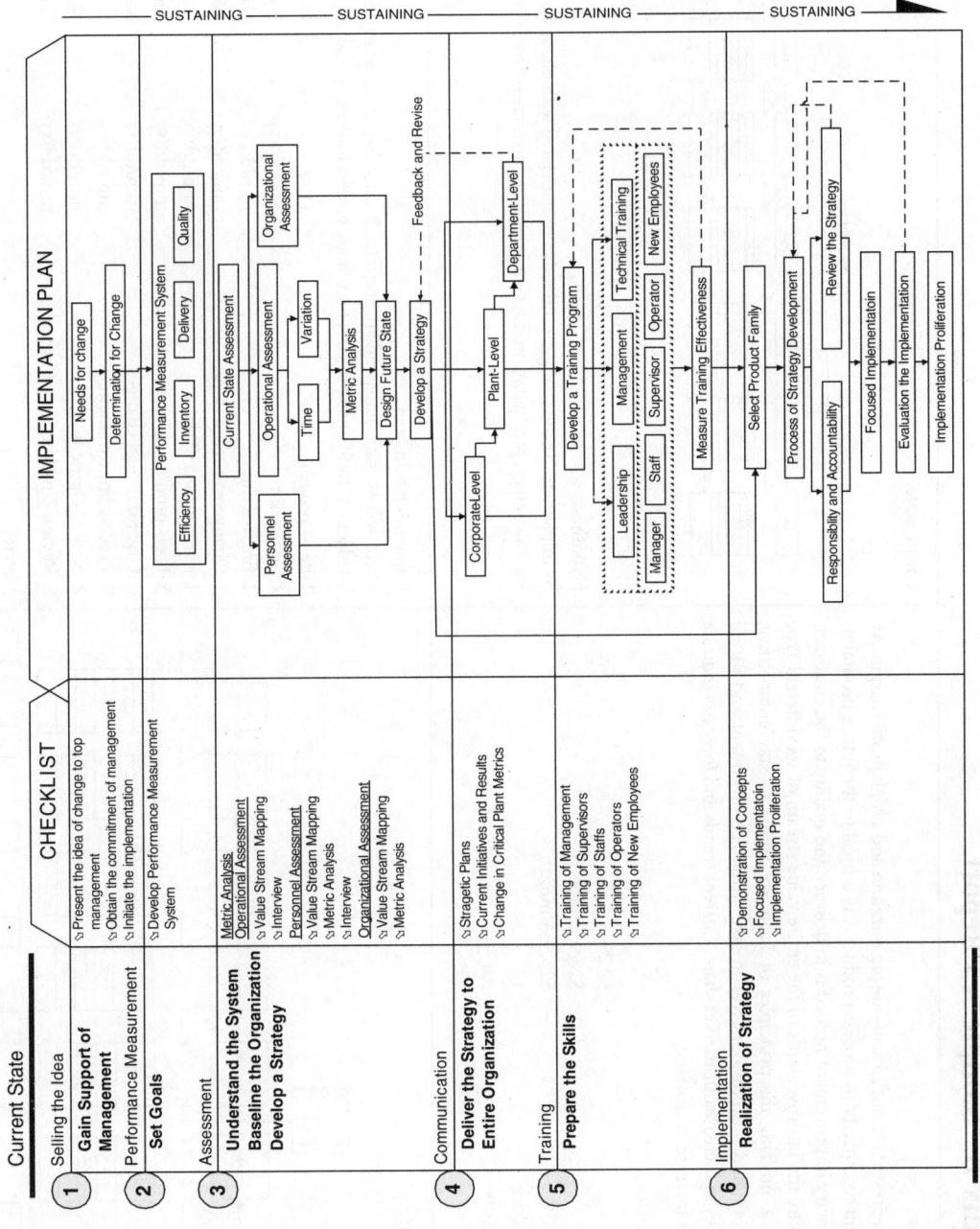

Appendix B   Sample A3 Report.

Business case

Product A represents about 25% of one organization and a high profit margin. An improvement to capacity to process product A is critical to the future profitability. The primary improvement needed is in delivery and quality to the customer. We will ensure the improvements to the entire value stream of six different type of product A utilizing the principles of Lean Six Sigma to stay competitive, improve profit, and position ourselves to take advantage of present and future opportunities to increase our market share. Current trends and forecasts indicate a significant increase in demand.

Baseline metrics

OTD	69.7%
COQ	$28,952/month
COY	$26,556/month
Days to produce	12
Inventory turns	3.6
Margin %	35%

Current state

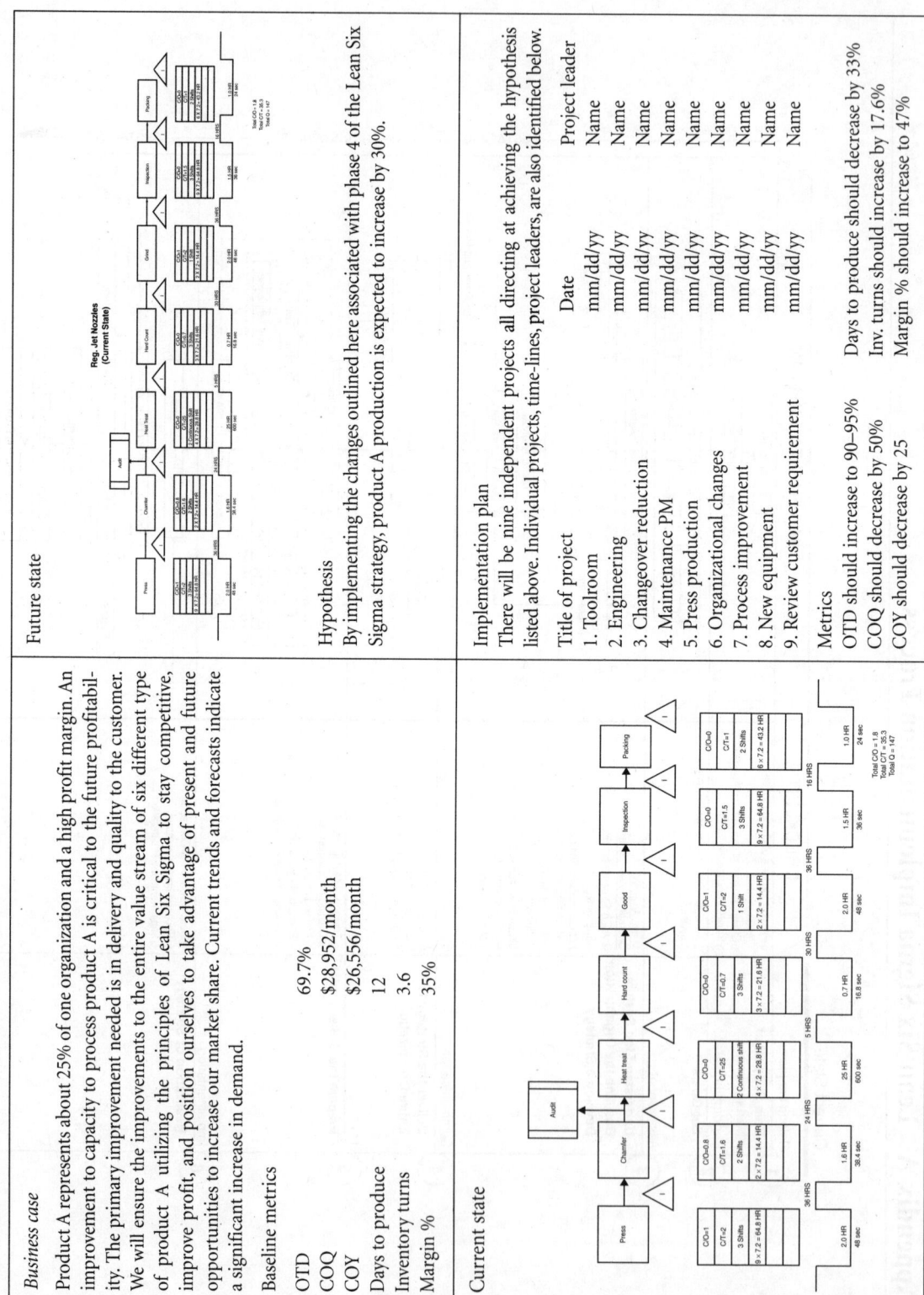

Future state

Hypothesis

By implementing the changes outlined here associated with phase 4 of the Lean Six Sigma strategy, product A production is expected to increase by 30%.

Implementation plan

There will be nine independent projects all directing at achieving the hypothesis listed above. Individual projects, time-lines, project leaders, are also identified below.

Title of project	Date	Project leader
1. Toolroom	mm/dd/yy	Name
2. Engineering	mm/dd/yy	Name
3. Changeover reduction	mm/dd/yy	Name
4. Maintenance PM	mm/dd/yy	Name
5. Press production	mm/dd/yy	Name
6. Organizational changes	mm/dd/yy	Name
7. Process improvement	mm/dd/yy	Name
8. New equipment	mm/dd/yy	Name
9. Review customer requirement	mm/dd/yy	Name

Metrics

OTD should increase to 90–95%   Days to produce should decrease by 33%
COQ should decrease by 50%   Inv. turns should increase by 17.6%
COY should decrease by 25   Margin % should increase to 47%

35

Critical Resource Diagramming: A Tool for Resource Utilization Analysis

Adedeji B. Badiru
University of Tennessee

35.1 Introduction

Basic critical path method (CPM) and program evaluation and review technique (PERT) approaches assume unlimited resource availability in project network analysis. In realistic projects, both the time and resource requirements of activities should be considered in developing network schedules. Projects are subject to three major constraints: time limitations, resource constraints, and performance requirements. Since these constraints are difficult to satisfy simultaneously, trade-offs must be made. The smaller the resource base, the longer the project schedule. Resource allocation facilitates the transition of a project from one state to another state. Given that the progress of a project is in an initial state defined as S_i and a future state is defined as S_f, then the following three possible changes can occur.

1. Further progress may be achieved in moving from the initial state to the future state (i.e., $S_f > S_i$).
2. Progress may be stagnant between the initial state and the future state (i.e., $S_f = S_i$).
3. Progress may regress from the initial state to the future state (i.e., $S_f < S_i$).

Resource allocation strategies must be developed to determine which is the next desired state of the project, when the next state is expected to be reached, and how to move toward that next state. Resource availability and criticality will determine how activities should be assigned to resources to facilitate progress of a project from one state to another. Graphical tools can provide guidance for resource allocation strategies.

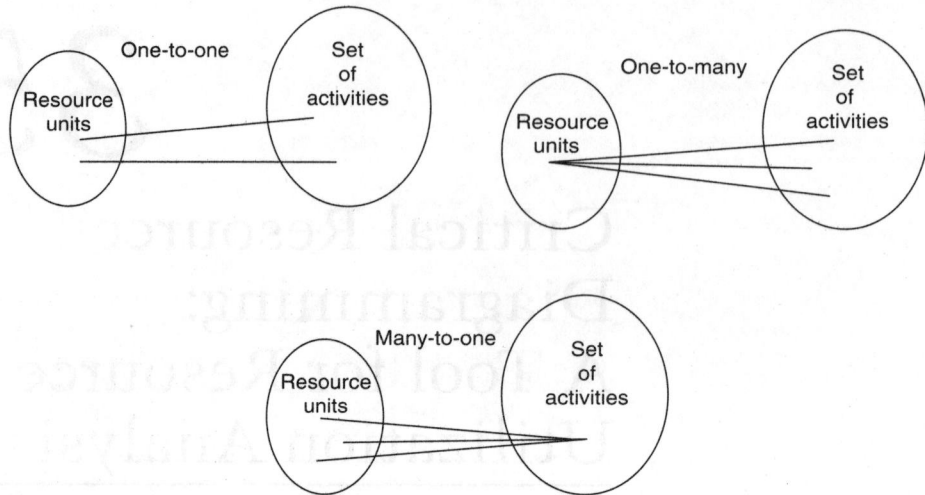

FIGURE 35.1   Activity–resource assignment options.

Critical path method, PERT, and precedence diagramming method are examples of graphical tools based on activity scheduling. Unfortunately, similar tools are not available for resource scheduling. There is a need for simple tools for resource allocation planning, scheduling, tracking, and control.

Figure 35.1 shows three resource loading options. In the first case, activity–resource assignments are made on a one-to-one basis. In the second case, one resource is assigned to multiple activities. There are also cases where a single activity may be assigned to more than one resource unit. The specific strategy used will depend on the prevailing resource constraints.

The importance of resource considerations in project management is emphasized by the following relationship:

$$RESOURCE \rightarrow ACTIVITY \rightarrow PRODUCT \rightarrow PROJECT \rightarrow ORGANIZATION$$

Resources are required for activities, activities produce products, products constitute projects, and projects make up organizations. Thus, resource management can be viewed as a basic component of the management of any organization.

It is logical to expect different resource types to exhibit different levels of criticality in a resource allocation problem. For example, some resources may be very expensive, some may possess special skills, and some may have very limited supply. The relative importance of different resource types should be considered when carrying out resource allocation in activity scheduling. The critical resource diagram helps in representing resource criticality.

35.2   Resource Profiling

Resource profiling involves the development of graphical representations to convey information about resource availability, utilization, and assignment. Resource loading and resource leveling graphs are two popular tools for profiling resources. Resource idleness graph and critical resource diagram are two additional tools that can effectively convey resource information.

35.2.1   Resource Loading

Resource loading refers to the allocation of resources to work elements in a project network. A resource loading graph presents a graphical representation of resource allocation over time. Figure 35.2 shows an

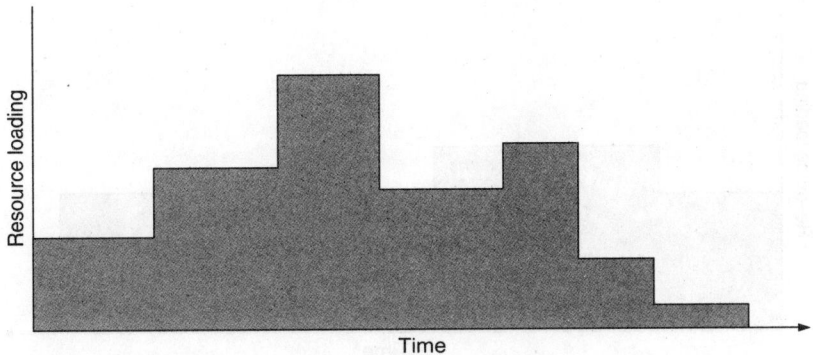

FIGURE 35.2   Resource loading graph.

TABLE 35.1   Example of Resource Availability Database

Resource ID	Brief Description	Special Skills	When Available	Duration of Availability	How Many
Type 1	Technician	Electronics	5 Aug 1993	2 months	15
Type 2	Programer	Database	12 Dec 1993	Indefinite	2
Type 3	Engineer	Design	Immediate	2 years	27
⋮	⋮	⋮	⋮	⋮	⋮
Type $n–1$	Operators	Data entry	Always	Indefinite	10
Type n	Accountant	Contract laws	2 Sept 1993	6 months	1

example of a resource loading graph. A resource loading graph may be drawn for the different resources types involved in a project.

The graph provides information useful for resource planning and budgeting purposes. In addition to resource units committed to activities, the graph may also be drawn for other tangible and intangible resources of an organization. For example, a variation of the graph may be used to present information about the depletion rate of the budget available for a project. If drawn for multiple resources, it can help in identifying potential areas of resource conflicts. For situations where a single resource unit is assigned to multiple tasks, a variation of the resource loading graph can be developed to show the level of load (responsibilities) assigned to the resource over time. Table 35.1 shows an example of a resource availability database for drawing a resource loading graph.

35.2.2   Resource Leveling

Resource leveling refers to the process of reducing the period-to-period fluctuation in a resource loading graph. If resource fluctuations are beyond acceptable limits, actions are taken to move activities or resources around in order to level out the resource loading graph. Proper resource planning will facilitate a reasonably stable level of the work force. Advantages of resource leveling include simplified resource tracking and control, lower cost or resource management, and improved opportunity for learning. Acceptable resource leveling is typically achieved at the expense of longer project duration or higher project cost. Figure 35.3 shows a somewhat leveled resource loading.

It should be noted that not all of the resource fluctuations in a loading graph can be eliminated. Resource leveling attempts to minimize fluctuations in resource loading by shifting activities within their available slacks. One heuristic procedure for leveling resources, known as the Burgess's method, is based on the technique of minimizing the sum of the squares of the resource requirements in each period over the duration of the project.

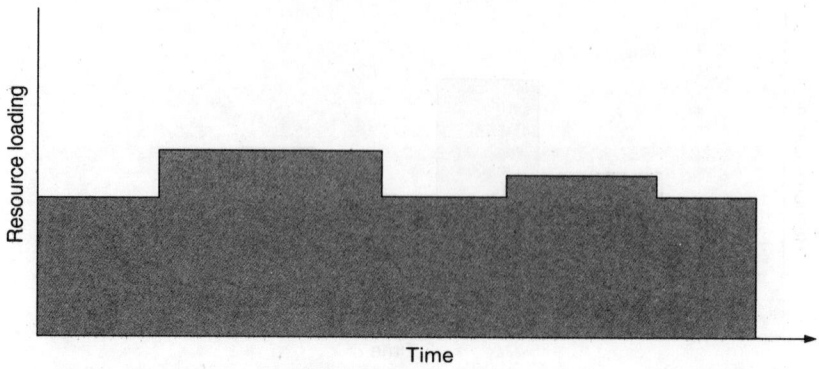

FIGURE 35.3   A somewhat leveled resource loading.

35.2.3   Resource Idleness

A resource idleness graph is similar to a resource loading graph except that it is drawn for the number of unallocated resource units over time. The area covered by the resource idleness graph may be used as a measure of the effectiveness of the scheduling strategy employed for a project. Suppose two scheduling strategies yield the same project duration and a measure of the resource utilization under each strategy is desired as a means to compare the strategies. Figure 35.4 shows two hypothetical resource idleness graphs for the alternate strategies. The areas are computed as follows:

$$\text{Area A} = 6(5) + 10(5) + 7(8) + 15(6) + 5(16)$$

$$= 306 \text{ resource units time}$$

$$\text{Area B} = 5(6) + 10(9) + 3(5) + 6(5) + 3(3) = 12(12)$$

$$= 318 \text{ resource units time}$$

As can be seen from the preceding sections, resource management is a complex task, which is subject to several limiting factors, including the following:

- Resource interdependencies
- Conflicting resource priorities
- Mutual exclusivity of resources
- Limitations on resource substitutions
- Variable levels of resource availability
- Limitations on partial resource allocation
- Limitations on duration of availability.

Several tools have been presented in the literature for managing resource utilization.

The critical resource diagramming (CRD) is introduced as a tool for resource management. Figure 35.5 shows an example of a critical resource diagram for a project that requires six different resource types. The example illustrates how CRD can be used to develop strategies for assigning activities to resources and vice versa. Each node identification, RES *j*, refers to a task responsibility for resource type *j*. In a CRD, a node is used to represent each resource unit. The interrelationships between resource units are indicated by arrows. The arrows are referred to as *resource-relationship (R-R) arrows*. For example, if the job of resource 1 must precede the job of resource 2, then an arrow is drawn from the node for resource 1 to the node for resource 2.

Task durations are included in a CRD to provide further details about resource relationships. Unlike activity diagrams, a resource unit may appear at more than one location in a CRD provided that there are

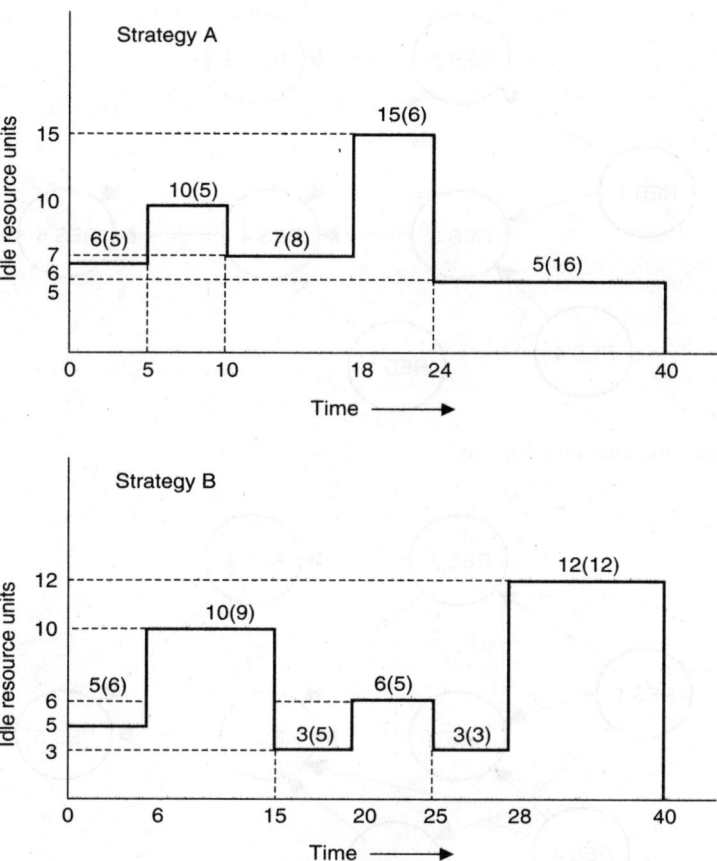

FIGURE 35.4   Resource idleness graphs for resource allocation strategies.

no time or task conflicts. Such multiple locations indicate the number of different jobs for which the resource is responsible. This information may be useful for task distribution and resource leveling purposes. In Figure 35.6, resource type 1 (RES 1) and resource type 4 (RES 4) appear at two different nodes, indicating that each is responsible for two different jobs within the same work scenario.

However, appropriate precedence constraints may be attached to the nodes associated with the same resource unit if the resource cannot perform more than one task at the same time. This is illustrated in Figure 35.6.

In an application context, CRD can be used to evaluate the utilization of tools, operators, and machines in a manufacturing system. Effective allocation of these resources will improve their utilization levels. If tools that are required at several work sites are not properly managed, bottleneck problems may develop. Operators may then have to sit idle waiting for tools to become available, or an expensive tool may have to sit unused while waiting for an operator. If work cannot begin until all required tools and operators are available, then other tools and operators that are ready to work may be rendered idle while waiting for the bottleneck resources. A CRD analysis can help identify when and where resource interdependencies occur so that appropriate reallocation actions may be taken. When there are several operators, any one operator that performs his/her job late will hold up everyone else.

35.2.3.1 CRD Network Analysis

The same forward and backward computations used in CPM are applicable to a CRD diagram. However, the interpretation of the critical path may be different since a single resource may appear at multiple

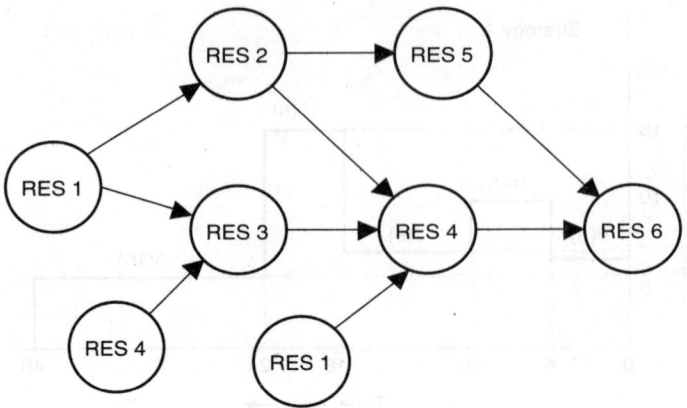

FIGURE 35.5   Basic critical resource diagram.

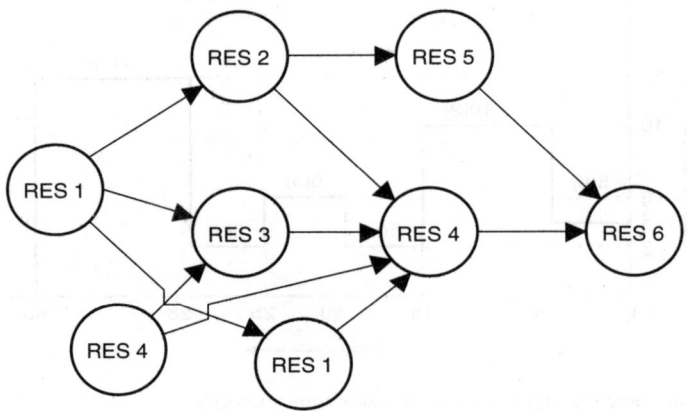

FIGURE 35.6   CRD with singular resource precedence constraint.

nodes. Figure 35.7 presents a computational analysis of the CRD network in Figure 35.5. Task durations (days) are given below the resource identifications. Earliest and latest times are computed and appended to each resource node in the same manner as in CPM analysis. RES 1, RES 2, RES 5, and RES 6 form the critical resource path. These resources have no slack times with respect to the completion of the given project. Note that only one of the two tasks of RES 1 is on the critical resource path.

Thus, RES 1 has a slack time for performing one job, while it has no slack time for performing the other. None of the two tasks of RES 4 is on the critical resource path. For RES 3, the task duration is specified as zero. Despite this favorable task duration, RES 3 may turn out to be a bottleneck resource. RES 3 may be a senior manager whose task is that of signing a work order. But if he/she is not available to sign at the appropriate time, then the tasks of several other resources may be adversely affected. A major benefit of a CRD is that both the senior- and lower-level resources can be included in the resource planning network.

A *bottleneck* resource node is defined as a node at which two or more arrows merge. In Figure 35.7, RES 3, RES 4, and RES 6 have bottleneck resource nodes. The tasks to which bottleneck resources are assigned should be expedited in order to avoid delaying dependent resources. A *dependent* resource node is a node whose job depends on the job of immediate preceding nodes. A *critically dependent* resource node is defined as a node on the critical resource path at which several arrows merge. In Figure 35.7, RES 6 is both a critically dependent resource node and a bottleneck resource node. As a scheduling heuristic, it is recommended that activities that require bottleneck resources be scheduled as early as possible. A *burst* resource node is defined as a resource node from which two or more arrows emanate.

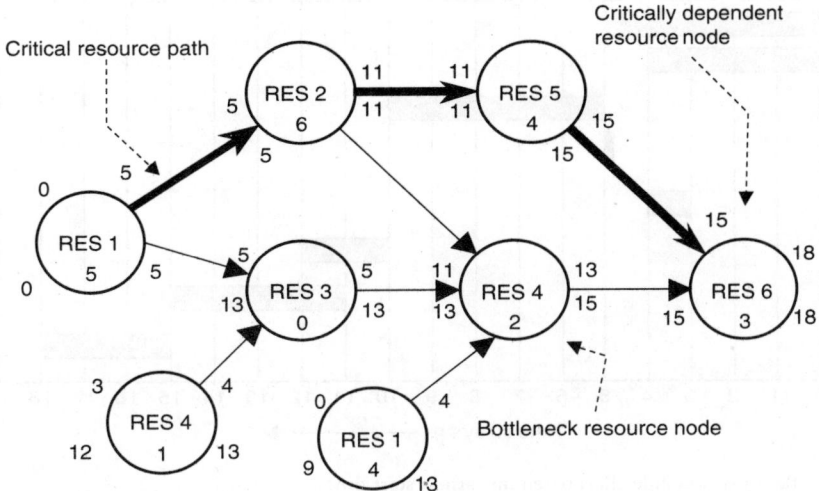

FIGURE 35.7 CRD network analysis.

Like bottleneck resource nodes, burst resource nodes should be expedited since their delay will affect several following resource nodes.

35.2.4 Resource Schedule Chart

The critical resource diagram has the advantage that it can be used to model partial assignment of resource units across multiple tasks in single or multiple projects. A companion chart for this purpose is the resource schedule (RS) chart. Figure 35.8 shows an example of an RS chart based on the earliest times computed in Figure 35.7. A horizontal bar is drawn for each resource unit or resource type. The starting point and the length of each resource bar indicate the interval of work for the resource. Note that the two jobs of RES 1 overlap over a 4-day time period. By comparison, the two jobs of RES 4 are separated by a period of 6 days. If RES 4 is not to be idle over those 6 days, "fill-in" tasks must be assigned to it. For resource jobs that overlap, care must be taken to ensure that the resources do not need the same tools (e.g., equipment, computers, and lathe) at the same time. If a resource unit is found to have several jobs overlapping over an extensive period of time, then a task reassignment may be necessary to offer some relief for the resource. The RS chart is useful for a graphical representation of the utilization of resources. Although similar information can be obtained from a conventional resource loading graph, the RS chart gives a clearer picture of where and when resource commitments overlap. It also shows areas where multiple resources are working concurrently.

35.3 Resource Work Rate Analysis

When resources work concurrently at different work rates, the amount of work accomplished by each may be computed. The critical resource diagram and the RS chart provide information to identify when, where, and which resources work concurrently. The general relationship between work, work rate, and time can be expressed as

$$w = rt$$

where w is the amount of actual work accomplished (expressed in appropriate units), such as miles of road completed, lines of computer code typed, gallons of oil spill cleaned, units of widgets produced, and surface area painted; r the rate at which the work is accomplished; and t the total time required to accomplish the work.

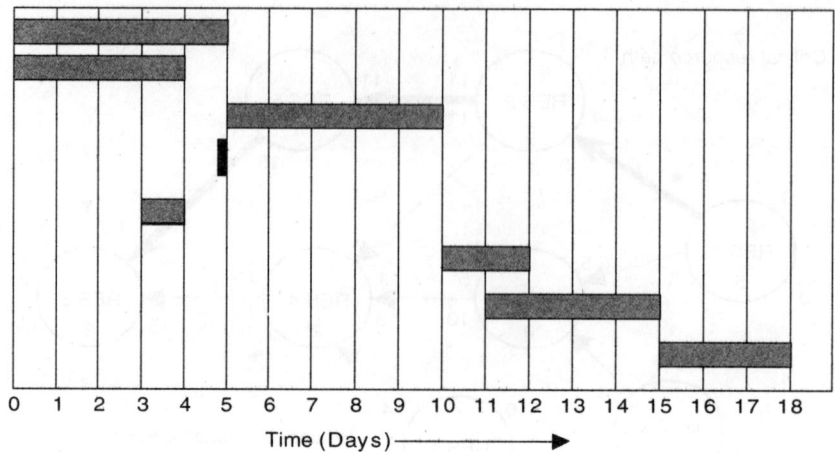

FIGURE 35.8   Resource schedule chart based on earliest start times.

It should be noted that work rate can change due to the effects of learning curves. In the discussions that follow, it is assumed that work rates remain constant for at least the duration of the work being analyzed.

Work is defined as a physical measure of accomplishment with uniform destiny (i.e., homogeneous). For example, a computer-programming task may be said to be homogeneous if one line of computer code is as complex and desirable as any other line of code in the program. Similarly, cleaning one gallon of oil spill is as good as cleaning any other gallon of oil spill within the same work environment. The production of one unit of a product is identical to the production of any other unit of the product. If uniform work density cannot be assumed for the particular work being analyzed, then the relationship presented above will need to be modified. If the total work to be accomplished is defined as one whole unit, then the tabulated relationship below will be applicable for the case of a single resource performing the work:

Resource	Work Rate	Time	Work Done
Machine A	$1/x$	t	1.0

$1/x$ is the amount of work accomplished per unit time. For a single resource to perform the whole unit of work, we must have the following:

$$(1/x)(t) = 1.0$$

which means that magnitude of x must be equal to the magnitude of t. For example, if Machine A is to complete one work unit in 30 min, it must work at the rate of 1/30 of work per unit time. If the magnitude of x is greater than the magnitude of t, then only a fraction of the required work will be performed. The information about the proportion of work completed may be useful for resource planning and productivity measurement purposes. In the case of multiple resources performing the work simultaneously, the work relationship is as follows:

Resource Type i	Work Rate r_i	Time t_i	Work Done w_i
RES 1	r_1	t_1	$(r_1)(t_1)$
RES 2	r_2	t_2	$(r_2)(t_2)$
$\vdots$	$\vdots$	$\vdots$	$\vdots$
RES n	r_n	t_n	$(r_n)(t_n)$
		Total	1.0

For multiple resources, we have the following expression:

$$\sum_{i=1}^{n} r_i t_i = 1.0$$

where n is the number of different resource types, r_i the work rate of resource type i, and t_i the work time of resource type i.

The expression indicates that even though the multiple resources may work at different rates, the sum of the total work they accomplished together must be equal to the required whole unit. For partial completion of work, the expression becomes

$$\sum_{i=1}^{n} r_i t_i = p$$

where p is the proportion of the required work actually completed. Suppose that RES 1, working alone, can complete a job in 50 min. After RES 1 has been working on the job for 10 min, RES 2 was assigned to help RES 1 in completing the job. Both resources working together finished the remaining work in 15 min. It is desirable to determine the work rate of RES 2.

The amount of work to be done is 1.0 whole unit. The work rate of RES 1 is 1/50 of work per unit time. Therefore, the amount of work completed by RES 1 in the 10 min when it worked alone is $(1/50)(10) =$ 1/5 of the required work. This may also be expressed in terms of percent completion or earned value by using C/SCSC (cost–schedule control systems criteria). The remaining work to be done is 4/5 of the total work. The two resources working together for 15 min yield the following results:

Resource Type i	Work Rate r_i	Time t_i	Work Done w_i
RES 1	1/50	15	15/50
RES 2	r_2	15	$15(r_2)$
		Total	4/5

Thus, we have $15/50 + 15(r_2) = 4/5$, which yields $r_2 = 1/30$ for the work rate of RES 2. This means that RES 2, working alone, could perform the job in 30 min. In this example, it is assumed that both resources produce identical quality of work. If quality levels are not identical for multiple resources, then the work rates may be adjusted to account for the different quality levels or a quality factor may be introduced into the analysis. The relative costs of the different resource types needed to perform the required work may be incorporated into the analysis as shown below:

Resource i	Work Rate r_i	Time t_i	Work Done w	Pay Rate p_i	Total Cost C_i
Machine A	r_1	t_1	$(r_1)(t_1)$	p_1	C_1
Machine B	r_2	t_2	$(r_2)(t_2)$	p_2	C_2
$\vdots$	$\vdots$	$\vdots$	$\vdots$	$\vdots$	$\vdots$
Machine n	r_n	t_n	$(r_n)(t_n)$	p_n	C_n
		Total	1.0		Budget

As another example, suppose that the work rate of RES 1 is such that it can perform a certain task in 30 days. It is desirable to add RES 2 to the task so that the completion time of the task could be reduced. The work rate of RES 2 is such that it can perform the same task alone in 22 days. If RES 1 has already worked 12 days on the task before RES 2 comes in, find the completion time of the task. It is assumed that RES 1 starts the task at time 0.

As usual, the amount of work to be done is 1.0 whole unit (i.e., the full task). The work rate of RES 1 is 1/30 of the task per unit time and the work rate of RES 2 is 1/22 of the task per unit time. The amount of work completed by RES 1 in the 12 days it worked alone is $(1/30)(12) = 2/5$ (or 40%) of the required work. Therefore, the remaining work to be done is 3/5 (or 60%) of the full task. Let T be the time for which both resources work together. The two resources working together to complete the task yield the following:

Resource Type i	Work Rate r_i	Time t_i	Work Done w_i
RES 1	1/30	T	$T/30$
RES 2	1/22	T	$T/22$
		Total	3/5

Thus, we have $T/30 + T/22 = 3/5$, which yields $T = 7.62$ days. Consequently, the completion time of the task is $(12 + T) = 19.62$ days from time 0. The results of this example are summarized in the RS charts in Figure 35.9. It is assumed that both resources produce identical quality of work and that the respective work rates remain consistent. As mentioned earlier, the respective costs of the different types may be incorporated into the work rate analysis.

35.3.1 Resource Assignment Problem

Operations research techniques are frequently used to enhance resource allocation decisions. One common resource allocation tool is the resource assignment algorithm. This algorithm can be used to enhance resource allocation decisions. Suppose that there are n tasks, which must be performed by n workers. The cost of worker i performing task j is c_{ij}. It is desirable to assign workers to the tasks in a fashion that minimizes the cost of completing the tasks. This problem scenario is referred to as the *assignment problem*. The technique for finding the optimal solution to the problem is called the *assignment method*. Like the transportation method, the assignment method is an iterative procedure that arrives at the optimal solution by improving on a trial solution at each stage of the procedure.

Critical path method and PERT can be used in controlling projects to ensure that the project will be completed on time. As was mentioned previously, these two techniques do not consider the assignment of resources to the tasks that make up a project. The *assignment method* can be used to achieve an optimal assignment of resources to specific tasks in a project. Although the assignment method is cost-based, task duration can be incorporated into the modeling in terms of time–cost relationships; of course, task precedence requirements and other scheduling of the tasks. The objective is to minimize the total cost. Thus, the formulation of the assignment problem is as follows:

Let

$$x_{ij} = \begin{cases} 1 & \text{if worker } i \text{ is assigned to task } j, \quad i, j = 1, 2, \ldots, n \\ 0 & \text{if worker } i \text{ is not assigned to task } j \end{cases}$$

$$c_{ij} = \text{cost of worker } i \text{ performing task } j$$

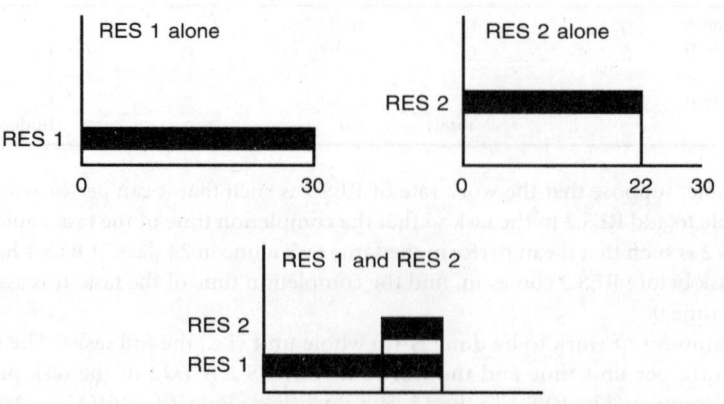

FIGURE 35.9   Resource schedule charts for RES 1 and RES 2.

$$\text{minimize} \quad z = \sum_{i=1}^{n} \sum_{j=1}^{n} c_{ij} x_{ij}$$

$$\text{subject to} \quad \sum_{j=1}^{n} x_{ij} = 1, \quad i = 1, 2, \ldots, n$$

$$\sum_{i=1}^{n} x_{ij} = 1, \quad j = 1, 2, \ldots, n$$

$$x_{ij} \geq 0, \quad i, j = 1, 2, \ldots, n$$

It can be seen that the above formulation is a transportation problem with $m = n$ and all supplies and demands are equal to 1. Note that we have used the nonnegativity constraint, $x_{ij} \geq 0$, instead of the integer constraint, $x_{ij} = 0$ or 1. However, the solution of the model will still be integer-valued. Hence, the assignment problem is a special case of the transportation problem with $m=n$, $S_i=1$ (supplies), and $D_i=1$ (demands). Conversely, the transportation problem can also be viewed as a special case of the assignment problem. A transportation problem can be modeled as an assignment problem and vice versa. The basic requirements of an assignment problem are as follows:

1. There must be two or more tasks to be completed.
2. There must be two or more resources that can be assigned to the tasks.
3. The cost of using any of the resources to perform any of the tasks must be known.
4. Each resource is to be assigned to one and only one task.

If the number of tasks to be performed is greater than the number of workers available, we will need to add *dummy workers* to balance the problem. Similarly, if the number of workers is greater than the number of tasks, we will need to add *dummy tasks* to balance the problem. If there is no problem of overlapping, a worker's time may be split into segments so that the worker can be assigned more than one task. In this case, each segment of the worker's time will be modeled as a separate resource in the assignment problem. Thus, the assignment problem can be extended to consider partial allocation of resource units to multiple tasks.

Although the assignment problem can be formulated for and solved by the simplex method or the transportation method, a more efficient algorithm is available specifically for the assignment problem. The method, known as the *Hungarian method*, is a simple iterative technique. Details of the assignment problem and its solution techniques can be found in operations research texts. As an example, suppose that five workers are to be assigned to five tasks on the basis of the cost matrix presented in Table 35.2. Task 3 is a machine-controlled task with a fixed cost of $800 regardless of which worker it is assigned to. Using the assignment, we obtain the optimal solution presented in Table 35.3, which indicates the following:

$$x_{15} = 1, \quad x_{23} = 1, \quad x_{31} = 1, \quad x_{44} = 1, \quad \text{and} \quad x_{52} = 1$$

Thus, the minimum total cost is given by

$$TC = c_{15} + c_{23} + c_{31} + c_{44} + c_{52} = \$(400 + 800 + 300 + 400 + 350) = \$2250$$

The technique of work rate analysis can be used to determine the cost elements that go into an assignment problem. The solution of the assignment problem can then be combined with the technique of CRD. This combination of tools and techniques can help in enhancing resource management decisions.

TABLE 35.2 Cost Matrix for Resource Assignment Problem

Worker	Task 1	Task 2	Task 3	Task 4	Task 5
1	300	200	800	500	400
2	500	700	800	1250	700
3	300	900	800	1000	600
4	400	300	800	400	400
5	700	350	800	700	900

TABLE 35.3   Solution to Resource Assignment Problem

Worker	Task 1	Task 2	Task 3	Task 4	Task 5
1	0	0	0	0	1
2	0	0	1	0	0
3	1	0	0	0	0
4	0	0	0	1	0
5	0	1	0	0	0

35.3.2   Line Balancing

Line balancing involves adjusting tasks at workstations so that it takes about the same amount of time to complete the work at each station. Most production facilities involve an integrated collection of workstations. Line balancing helps in controlling the output rates in continuous production systems. As work is completed on a product at one station, it is passed on to the next for further processing. Cycle time is the time the product spends at each workstation. The cycle time is dependent on the expected output from the line. The cycle time can be calculated on the basis of the production rate. A balanced line refers to the equality of output of each successive workstation on the assembly line. The maximum output of the line is controlled by its slowest operation. Perfect balance exists when each workstation requires the same amount of time to perform its assigned tasks and there is no idle time at any of the workstations. Because of bottleneck operations and different work rates, perfect balance is rarely achieved.

The CRD approach can be effective in line-balancing analysis. The work rate table can identify specific work rates that may need to be adjusted to achieve line balance within a specified margin. The *margin*, in this case, is defined as the time deviation from perfect balance. The following definitions are important for line-balancing analysis:

1. *Workstation.* The immediate work environment of an operator, which offers all the tools necessary to perform the tasks assigned to the operator.
2. *Work element.* This is the smallest unit of work that cannot be divided between two or more operators without having schedule or resource conflict.
3. *Operation.* A set of work elements assigned to a single workstation.
4. *Cycle time.* The time that one unit of the product spends at each workstation.
5. *Balance delay.* The total idle time in the line as a result of unequal division of work between workstations.

Line balancing involves grouping work elements into stations so as to minimize the number of stations based on a desired output rate or cycle time. A precedence diagram is often used to determine the logical basis for grouping work elements. The critical resource diagram can be used for this purpose. Since there will be a large number of possible ways to group work elements, the analysis must use precedence constraints to eliminate the infeasible groupings.

A conveyor system is the simplest type of line balancing. The conveyor moves work elements past workstations at a constant rate. Each station is expected to perform its task within the time period allowed by the conveyor speed. A sign of imbalance occurs when work piles up in a particular station because more units of the product arrive at the station before the preceding ones are dispatched. In this case, the next operator in the line will experience some idle time while waiting for work to arrive from the preceding station.

The choice of cycle time depends on the desired output rate. The minimum number of workstations is the total work element duration divided by the cycle time. Fractional results are rounded up to the next higher integer value. This is calculated as

$$n = \frac{\sum_{i=1}^{k} t_i}{C}$$

where k is the number of work elements, t_i the processing time of work element i, and C the cycle time.

In most cases, the theoretical minimum number of workstations will be impossible to achieve because of physical constraints in the production line and work element times that are not compatible. It should be noted that the theoretical minimum number of stations is not necessarily devoid of idle times. A balance with the theoretical minimum number of workstations will contain the least total idle time. The maximum possible efficiency with the theoretical minimum number of stations is represented as

$$f_{max} = \frac{\sum_{i=1}^{k} t_i}{nC}$$

For cases where the theoretical minimum number of stations cannot be achieved, the actual efficiency is computed by substituting the actual number of stations, m, for n in the above equation. Since actual efficiency, f_a, will be less than or equal to f_{max}, the analyst would attempt to increase efficiency toward the maximum value by rearranging work elements. Several mathematical and heuristic methods are used for investigating the rearrangements. These include linear programming, dynamic programming, computer simulation, trial-and-error, and the ranked positional weight technique.

35.4   Human Resource Management

Human resources make projects successful. Human resources are distinguished from other resources because of the ability to learn, adapt to new project situations, and set goals. Human resources, technology resources, and management resources must coexist to pursue project goals. Managing human resources involves placing the right people with the right skills in the right jobs in the right environment. Good human resource management motivates workers to perform better. Both individual and organizational improvements are needed to improve the overall quality by enriching jobs with the following strategies:

- Specify project goals in unambiguous terms
- Encourage and reward creativity on the job
- Eliminate mundane job control processes
- Increase accountability and responsibility for project results
- Define jobs in terms of manageable work packages that help identify line of responsibility
- Grant formal authority to make decisions at the task level
- Create advancement opportunities in each job
- Give challenging assignments that enable a worker to demonstrate his/her skill
- Encourage upward (vertical) communication of ideas
- Provide training and tools needed to get job done
- Maintain a stable management team

Several management approaches are used to manage human resources. Some of these approaches are formulated as direct responses to the cultural, social, family, or religious needs of workers. Examples of these approaches are:

- Flextime
- Religious holidays
- Half-time employment

These approaches can have a combination of several advantages. Some of the advantages are for the employer, while some are for the workers. The advantages are:

- Low cost
- Cost savings on personnel benefits
- Higher employee productivity

- Less absenteeism
- Less work stress
- Better family/domestic situation, which may have positive effects on productivity

Retraining work force is important for automation projects. Continuing education programs should be developed to retrain people who are only qualified to do jobs that do not require skilled manpower. The retraining will create a ready pool of human resource that can help boost manufacturing output and competitiveness. Management stability is needed to encourage workers to adapt to the changes in industry. If management changes too often, workers may not develop a sense of commitment to the policies of management.

The major resource in any organization, both technical and nontechnical, is manpower. People are the overriding factor in any project life cycle. In automated operations, the role played by even if people are involved can be very significant. Such operations invariably require the services of technical people with special managerial and professional needs. The hi-tech manager in such situations would need special skills in order to discharge the managerial duties effectively. The manager must have auto-management skills that relate to the following:

- Managing self
- Being managed
- Managing others

Many of the managers who supervise technical people rise to the managerial posts from technical positions. Consequently, they often lack the managerial competence needed for the higher offices. In some cases, technical professionals are promoted to managerial levels and then transferred to administrative posts in functional areas different from their areas of technical competence. The poor managerial performance of these technical managers is not necessarily a reflection of poor managerial competence, but rather an indication of the lack of knowledge of the work elements in their surrogate function. Any technical training without some management exposure is, in effect, an incomplete education. Technical professionals should be trained for the eventualities of their professions.

In the transition from the technical to the management level, an individual's attention would shift from detail to overview, specific to general, and technical to administrative. Since most managerial positions are earned based on qualifications (except in aristocratic and autocratic systems), it is important to train technical professionals for possible administrative jobs. It is the responsibilities of the individual and the training institution to map out career goals and paths and institute specific education aimed at the realization of those goals. One such path is outlined below:

1. *Technical Professional.* This is an individual with practical and technical training or experience in a given field, such as industrial engineering. The individual must keep a current awareness in his/her area of specialization through continuing education courses, seminars, conferences, and so on. The mentor program, which is now used in many large organizations, can be effectively utilized at this stage of the career ladder.
2. *Project manager.* This is an individual who is assigned the direct responsibility of supervising a given project through the phases of planning, organizing, scheduling, monitoring, and control. The managerial assignment may be limited to just a specific project. At the conclusion of the project, the individual returns to his/her regular technical duties. However, his/her performance on the project may help identify him/her as a suitable candidate for permanent managerial assignment later on.
3. *Group manager.* This is an individual who is assigned direct responsibility to plan, organize, and direct the activities of a group of people with a specific responsibility—for example, a computer data security advisory committee. This is an ongoing responsibility that may repeatedly require the managerial skills of the individual.
4. *Director.* An individual who oversees a particular function of the organization. For example, a marketing director has the responsibility of developing and implementing the strategy for getting the organization's products to the right market, at the right time, at the appropriate price, and in

the proper quantity. This is a critical responsibility that may directly affect the survival of the organization. Only the individuals who have successfully proven themselves at the earlier career stages get the opportunity to advance to the director's level.

5. *Administrative manager.* This is an individual who oversees the administrative functions and staff of the organization. His/her responsibilities cut across several functional areas. He/she must have proven his/her managerial skills and diversity in previous assignment.

The above is just one of the several possible paths that can be charted for a technical professional as he/she gradually makes the transition from the technical ranks to the management level. To function effectively, a manager must acquire nontechnical background in various subjects. His/her experience, attitude, personality, and training will determine his/her managerial style. His/her appreciation of the human and professional needs of his/her subordinates will substantially enhance his/her managerial performance. Examples of subject areas in which a manager or an aspiring manager should get training include:

1. Project management

   (a) *Scheduling and budgeting.* Knowledge of project planning, organizing, scheduling, monitoring, and controlling under resource and budget restrictions.
   (b) *Supervision.* Skill in planning, directing, and controlling the activities of subordinates.
   (c) *Communication.* Skill of relating to others both within and outside the organization. This includes written and oral communication skills.

2. Personal and personnel management

   (a) *Professional development.* Leadership roles played by participating in professional societies and peer recognition acquired through professional services.
   (b) *Personnel development.* Skills needed to foster cooperation and encouragement of staff with respect to success, growth, and career advancement.
   (c) *Performance evaluation.* Development of techniques for measuring, evaluating, and improving employee performance.
   (d) *Time management.* Ability to prioritize and delegate activities as appropriate to maximize accomplishments within given time periods.

3. Operations management

   (a) *Marketing.* Skills useful for winning new business for the organization or preserving existing market shares.
   (b) *Negotiating.* Skills for moderating personnel issues, representing the organization in external negotiations, or administering company policies.
   (c) *Estimating and budgeting.* Skills needed to develop reasonable cost estimates for company activities and the assignment of adequate resources to operations.
   (d) *Cash flow analysis.* An appreciation for the time value of money, manipulations of equity and borrowed capitals, stable balance between revenues and expenditures, and maximization of returns on investments.
   (e) *Decision analysis.* Ability to choose the direction of work by analyzing feasible alternatives.

A technical manager can develop the above skills through formal college courses, seminars, workshops, short courses, professional conferences, or in-plant company training. Several companies appreciate the need for these skills and are willing to bear the cost of furnishing their employees with the means of acquiring the skills. Many of the companies have custom formal courses, which they contract out to colleges to teach for their employees. This is a unique opportunity for technical professionals to acquire managerial skills needed to move up the corporate ladder.

Technical people have special needs. Unfortunately, some of these needs are often not recognized by peers, superiors, or subordinates. Inexperienced managers are particularly prone to the mistake of not distinguishing between technical and nontechnical professional needs. In order to perform more effectively, a manager must be administratively adaptive. He/she must understand the unique expectations of technical professionals in terms of professional preservation, professional peers, work content, hierarchy of needs, and the technical competence or background of their managers.

35.4.1 Professional Preservation

Professional preservation refers to the desire of a technical professional to preserve his/her identification with a particular job function. In many situations, the preservation is not possible due to a lack of manpower to fill specific job slots. It is common to find people trained in one technical field holding assignments in other fields. An incompatible job function can easily become the basis for insubordination, egotism, and rebellious attitudes. While it is realized that in any job environment there will sometimes be the need to work outside one's profession, every effort should be made to match the surrogate profession as close as possible. This is primarily the responsibility of the human resources manager.

After a personnel team has been selected in the best possible manner, a critical study of the job assignments should be made. Even between two dissimilar professions, there may be specific job functions that are compatible. These should be identified and used in the process of personnel assignment. In fact, the mapping of job functions needed for an operation can serve as the basis for selecting a project team. In order to preserve the professional background of technical workers, their individualism must be understood. In most technical training programs, the professional is taught how to operate in the following ways:

1. Make decisions based on the assumption of certainty of information
2. Develop abstract models to study the problem being addressed
3. Work on tasks or assignments individually
4. Quantify outcomes
5. Pay attention to exacting details
6. Think autonomously
7. Generate creative insights to problems
8. Analyze systems operatability rather than profitability

However, in the business environment, not all of the above characteristics are desirable or even possible. For example, many business decisions are made with incomplete data. In many situations, it is unprofitable to expend the time and efforts to seek perfect data. As another example, many operating procedures are guided by company policies rather than creative choices of employees. An effective manager should be able to spot cases where a technical employee may be given room to practice his/her professional training. The job design should be such that the employee can address problems in a manner compatible with his/her professional training.

35.4.2 Professional Peers

In addition to having professionally compatible job functions, technical people like to have other project team members to whom they can relate technically. A project team consisting of members from diversely unrelated technical fields can be a source of miscommunication, imposition, or introversion. The lack of a professional associate on the same project can cause a technical person to exhibit one or more of the following attitudes:

1. Withdraw into a shell, and contribute very little to the project by holding back ideas that he/she feels the other project members cannot appreciate.
2. Exhibit technical snobbery, and hold the impression that only he/she has the know-how for certain problems.
3. Straddle the fence on critical issues, and develop no strong conviction for project decisions.

Providing an avenue for a technical "buddy system" to operate in an organization can be very instrumental in ensuring congeniality of personnel teams and in facilitating the eventual success of project endeavors. The manager in conjunction with the selection committee (if one is used) must carefully consider the mix of the personnel team on a given project. If it is not possible or desirable to have more than one person from the same technical area on the project, an effort should be made to provide as good a mix as possible. It is undesirable to have several people from the same department taking issues against the views of a lone project member from a rival department. Whether it is realized or not, and whether it is admitted or not, there is a keen sense of rivalry among technical fields. Even within the same field, there are subtle rivalries between specific functions. It is important not to let these differences carry over to a project environment.

35.4.3   Work Content

With the advent of new technology, the elements of a project task will need to be designed to take advantage of new developments. Technical professionals have a sense of achievement relative to their expected job functions. They will not be satisfied with mundane project assignments that will bring forth their technical competence. They prefer to claim contribution mostly where technical contribution can be identified. The project manager will need to ensure that the technical people of a project have assignments for which their background is really needed. It will be counterproductive to select a technical professional for a project mainly on the basis of personality. An objective selection and appropriate assignment of tasks will alleviate potential motivational problems that could develop later in the project.

35.4.4   Hierarchy of Needs

Recalling Maslow's hierarchy of needs, the needs of a technical professional should be more critically analyzed. Being professionals, technical people are more likely to be higher up in the needs hierarchy. Most of their basic necessities for a good life would have been met. Their prevailing needs will tend to involve esteem and self-actualization. As a result, by serving on a project team, a technical professional may have expectations that cannot usually be quantified in monetary terms. This is in contrast to nontechnical people who may look forward for overtime pay or other monetary gains that may result from being on the project. Technical professionals will generally look forward to one or several of the following opportunities.

1. *Professional growth and advancement.* Professional growth is a primary pursuit of most technical people. For example, a computer professional has to be frequently exposed to challenging situations that introduce new technology developments and enable him to keep abreast of his/her field. Even occasional drifts from the field may lead to the fear of not keeping up and being left behind. The project environment must be reassuring to the technical people with regard to the opportunities for professional growth in terms of developing new skills and abilities.

2. *Technical freedom.* Technical freedom, to the extent permissible within the organization, is essential for the full utilization of a technical background. A technical professional will expect to have the liberty of determining how best the objective of his/her assignment can be accomplished. One should never impose a work method on a technical professional with the assurance that "this is the way it has always been done and will continue to be done!" If the worker's creative input to the project effort is not needed, then there is no need having him/her on the team in the first place.

3. *Respect for personal qualities.* Technical people have profound personal feelings despite the mechanical or abstract nature of their job functions. They will expect to be respected for their personal qualities. In spite of frequently operating in professional isolation, they do engage in interpersonal activities. They want their nontechnical views and ideas to be recognized and evaluated based on merit. They do not want to be viewed as "all technical." An appreciation for their personal qualities gives them the sense of belonging and helps them to become productive members of a project team.

4. *Respect for professional qualification.* A professional qualification usually takes several years to achieve and is not likely to be compromised by any technical professional. Technical professionals cherish the attention they receive due to their technical background. They expect certain preferential treatments. They like to make meaningful contributions to the decision process. They take

approval of their technical approaches for granted. They believe they are on a project because they are qualified to be there. The project manager should recognize these situations and avoid the bias of viewing the technical person as being conceited.

5. *Increased recognition.* Increased recognition is expected as a by-product of a project effort. The technical professional, consciously or subconsciously, views his/her participation in a project as a means of satisfying one of his/her higher-level needs. He/she expects to be praised for the success of his/her efforts. He/she looks forward to being invited for subsequent technical endeavors. He/she savors hearing the importance of his/her contribution being related to his/her peers. Without going to the extreme, the project manager can ensure the realization of the above needs through careful comments.

6. *New and rewarding professional relationship.* New and rewarding professional relationships can serve as a bonus for a project effort. Most technical developments result from joint efforts of people who share closely allied interests. Professional allies are most easily found through project groups. A true technical professional will expect to meet new people with whom he/she can exchange views, ideas, and information later on. The project atmosphere should, as a result, be designed to be conducive to professional interactions.

35.4.5　Quality of Leadership

The professional background of the project leader should be such that he/she commands the respect of technical subordinates. The leader must be reasonably conversant with the base technologies involved in the project. He/she must be able to converse intelligently on the terminologies of the project topic and be able to convey the project ideas to upper management. This serves to give him/her technical credibility. If technical credibility is lacking, the technical professionals on the project might view him/her as an ineffective leader. They will consider it impossible to serve under a manager to whom they cannot relate technically.

In addition to technical credibility, the manager must also possess administrative credibility. There are routine administrative matters that are needed to ensure a smooth progress for the project. Technical professionals will prefer to have those administrative issues successfully resolved by the project leader so that they can concentrate their efforts on the technical aspects. The essential elements of managing a group of technical professionals involve identifying the unique characteristics and needs of the group and then developing the means of satisfying those unique needs.

Recognizing the peculiar characteristics of technical professionals is one of the first steps in simplifying project management functions. The nature of manufacturing and automation projects calls for the involvement of technical human resources. Every manager must appreciate the fact that the cooperation or the lack of cooperation from technical professionals can have a significant effect on the overall management process. The success of a project can be enhanced or impeded by the management style utilized.

35.4.6　Work Simplification

Work simplification is the systematic investigation and analysis of planned and existing work systems and methods for the purpose of developing easier, quicker, less fatiguing, and more economic ways of generating high-quality goods and services. Work simplification facilitates the content of workers, which invariably leads to better performance. Consideration must be given to improving the product or service, raw materials and supplies, the sequence of operations, tools, work place, equipment, and hand and body motions. Work simplification analysis helps in defining, analyzing, and documenting work methods.

35.5　Conclusion

The CRD and RS charts are simple extensions of very familiar management tools. They are simple to use, and they convey resource information quickly. They can be used to complement existing resource management tools. CRD can be modified for specific resource planning, scheduling, and control needs.

36

Integrating Six Sigma and Lean Manufacturing for Process Improvement: A Case Study

Ike C. Ehie
Kansas State University

Rupy Sawhney
University of Tennessee

36.1 Introduction

Faced with an ever increasingly competitive market, manufacturing firms must constantly make changes to improve their operations. There are numerous continuous improvement (CI) strategies available. This reports on the combination of Six Sigma and Lean manufacturing to improve production system performance. Six Sigma uses statistical techniques to understand, measure, and reduce process variation with the primary goal of achieving improvement in both quality and cost. Lean manufacturing focuses on eliminating nonvalue-added activities in a process with a goal of eliminating waste, reducing process cycle time, improving on-time delivery and reducing cost. While these two CI strategies differ in the aspects of implementation procedure, they are not exclusive of each other. This chapter proposes an integrated framework that offers a balanced combination of the speed and variation reduction power of both Lean and Six Sigma to achieve distinct competitive advantage. The integrated approach is illustrated using a case study of an actual manufacturing company that is situated in the Midwestern United States. The goal is to demonstrate the value of combining Six Sigma and Lean manufacturing to make process improvement.

36.2 Six Sigma

The philosophy of process improvement has been around since the early 1900 through the work of Frederick Taylor. Therefore the concept of Six Sigma, although not particularly new, encompasses many well-tested and generally accepted practices (Levinson and Rerick, 2002). The organization develops and deploys a set of best practices and standardizes these procedures throughout the company. Six Sigma, first popularized by Motorola in 1990, is a business process improvement methodology in which sigma represents a statistical measure of variability in the process.

Six Sigma is basically a business process improvement methodology in which sigma represents a statistical measure of variability in the process. Under conditions of normality the Six Sigma represents two parts per billion; however, considering the long-term process drift of 1.5σ espoused by Motorola in early 1990s. A process operating at Six Sigma will produce 3.4 parts per million which is equivalent to 4.5σ away from the mean (Harry and Schroeder, 2000). The Six Sigma strategy involves the use of statistical tools within a structured methodology for gaining knowledge needed to achieve better, faster, and less expensive products and services than the competition (Breyfogle, 1999). A Six Sigma initiative in a company is designed to change the culture through breakthrough improvements by focusing on thinking out-of-the-box in order to achieve aggressive stretch goals. When deployed appropriately, Six Sigma can infuse intellectual capital into a company and produce unprecedented knowledge gains that would translate into shared bottom line results (Kiemele et al., 1997). Six Sigma is becoming a breakthrough philosophy among the world's leading corporations because it has proven itself by generating substantial business returns.

The implementation of Six Sigma strategy involves a series of steps specifically designed to facilitate a process of CI. The strategy takes the key manufacturing, engineering, and transactional processes of entire process through the five transformational phases (Plotkin, 1999). These are define, measure, analyze, improve, and control and they are popularly referred to as the DMAIC process.

The primary objective of the five-step process is to recognize critical customer requirements, identify and validate the improvement opportunity, and upgrade the business processes. Harry and Schroeder extend DMAIC to an eight-step process: recognize, define, measure, analyze, improve, control, standardize, and integrate. These activities rely on the standard plan-do-check-act improvement cycle and they roughly parallel the Ford Motor Company Eight Discipline (8D) Team-oriented problem solving (TOPS). The 8D-TOPS approach is a quality management tool. It acts as a vehicle for a cross-functional team to articulate thoughts and provides scientific certitude to details of problems and long-lasting solutions. Eight discipline heightens awareness of team members and often results in several eye-opening revelations by staff, who are not necessarily managers. Whereas Six Sigma focuses on data and process variables, the 8D-TOPS uses cross-functional teams, looks for root causes, and implements and test permanent corrections or improvements.

A large number of companies have boosted their profitability, increased market share, and improved customer satisfaction through the implementation of Six Sigma (Harry, 1998). These companies embarked on Six Sigma become of the need to improve their processes and provide value to their customers. In all, these companies have seen positive impact of Six Sigma on their bottom line. Six Sigma is management methodology driven by data. Six Sigma focuses on projects that will produce measurable business results.

36.3 Lean Manufacturing

Manufacturing improvement efforts can be sustained by the continuous search for waste. In most organizations, waste is rife and easy to detect, however, finding ways to eliminate the waste could be difficult. Ohno (1978) identified seven sources of waste as being from producing defects, transportation, inventory, overproduction, waiting time, processing, and motion. Lean manufacturing is aimed at the elimination of waste in every area of production including customer relations, product design, supplier networks and factory management. Its goal is to incorporate less human effort, less inventory, less time to develop products, and less space to become highly responsive to customer demand while producing top quality products in the most efficient and economical manner possible. The relentless pursuit of the elimination of waste is main concept behind lean manufacturing system. Lean manufacturing is defined as a systematic approach to identifying and eliminating waste (nonvalue-added activities) through CI by flowing the product at the pull of the customer in pursuit of perfection (NIST Lean Network). Lean means adding value by eliminating waste, being responsive to change, focusing on quality, and enhancing the effectiveness of the workforce (Source: Lean Aerospace Initiative, MIT). Lean manufacturing focuses on eliminating nonvalue-added activities in a process with goal of reducing process cycle times, improving on-time delivery performance and reducing cost. Lean thinking or principles can be applied in service industry as well as in manufacturing industry to reduce lead time, improve quality and productivity by eliminating wastes in the system. There are seven elements of lean manufacturing and these are listed below (Nicholas, 1998):

1. Small-lot production
2. Setup-time reduction
3. Maintaining and improving equipment
4. Pull production systems
5. Focused factories and group technology
6. Workcells and cellular manufacturing
7. Standard operations

36.4   Lean-Sigma Approach

Lean Sigma is a systematic approach to identify waste and other nonvalue-added activities through CI in all processes. Combining the speed and power of both Lean manufacturing and Six Sigma can give a company a unique competitive position in the market. When one compares these two techniques, one would naturally ask the following questions: how do these two techniques complement each other? Can they work together, or do they exclude each other? While different vocabularies may be used, both methods actually are very consistent with focusing on identifying key variables, designing critical measures, improving key processes, changing current systems to support improvement, and monitoring the results of improvement. The Lean focus is primarily on eliminating waste, whereas Six Sigma's focus is primarily on reducing variability. This is not an "either/or" proposition since they complement each other and indeed share common tools. The appropriate methodology and tools depend entirely upon the improvement opportunity at hand. Six Sigma provides specific statistical tools and engineering techniques for implementing changes while lean manufacturing serves as a framework for waste reduction and CI.

Overall, when the integrated framework is applied to improve a specific process, these two techniques seem to complement each other. The integration is made by combining the waste reduction of Lean manufacturing and the statistical approach of Six Sigma. Generally, firms that apply Lean-Sigma adopt a global perspective in identifying the constraints and examining necessary changes that would impact the entire system. Lean-Sigma brings in the perspectives of customer needs, performance measures, and engineering and statistical tools during the stages of constraints identification and exploitation. Table 36.1 provides a comparison between Six Sigma and Lean manufacturing on a DMAIC framework.

36.5   The Case

Dana Corporation is a leader in the development of modular systems technology and a worldwide resource for engineering, research, and development. The company is one of the world's largest independent suppliers of vehicular components to automobile manufacturers. It employs nearly 80,000 people worldwide and has sales reaching $10 billion. The company's quest for CI and their dedication for furthering vehicular component design have provided the opportunity to bring to market one of the most advanced and comprehensive ranges of purpose-built products available today. The success of the company has been

TABLE 36.1 Comparison between Six Sigma and Lean Manufacturing

Improvement Process	Six Sigma	Lean Manufacturing
Define	Identify customer needs and project suitable for Six-Sigma effort (voice of the customer)	Select project family and identify the key areas of concern using value-stream mapping
Measure	Determine what and how to measure the performance of the selected process	Map the current state of the process and establish performance improvement measures
Analyze	Understand and determine the variables that create quality variations	Analyze the value-stream map and separate waste from value-added activities
Improve	Identify means to remove causes of defects and modify the process	Invent the future value-stream map and eliminate source of waste
Control	Maintain the improvement	Achieve and sustain the future value-stream map. Provide visual control to sustain the improvement

based on a long history of listening to customers. Only by consistently meeting customer needs has the company been able to develop an extensive range of product options that suit each market segment. The extensive years of experience combined with a massive sustained investment in cutting edge technology has enabled the Dana Corporation to design, develop, and manufacture vehicular components that meet customers needs more precisely than ever before.

One of the divisions of Dana Corporation, Axle facility, manufactures a variety of axle products and related components for internal operations for original equipment vehicle manufacturers including DaimlerChrysler, Ford, Isuzu, and Land Rover. The Axle division incorporates advanced technological manufacturing systems and engineering methods to provide design in a very cost-effective manner. Each new component and axle system is put through rigorous testing, such as fatigue, metallurgical, dynamometer, and chassis dynamometer/quiet chamber testing. Axle uses the latest and the most effective manufacturing technology to product quality products.

The business case for this project was initiated because of the eroding sales revenue, which went down by 23% in 2000, while fixed expenses went up by as much as 22% within the same year. Management was faced with either shutting down the plant or eliminating the nonvalue-added processes to increase capacity without incurring new capital expenditures. After receiving training in Six Sigma and Lean manufacturing, managers decided to combine both methodologies to guide their improvement effort. As shown in Figure 36.1, the basic DMAIC framework was adopted and complemented with some elements of Lean manufacturing. We term this approach Lean-Sigma. Whereas Six Sigma unleashes the power of statistical process analysis, Lean manufacturing focuses the company on waste reduction and process improvement. With careful study and planning, an eight-member project team was formed. The project team was composed of the plant manager, the controller, two Six Sigma certified employees, and four operators from the plant.

The team was charged with the responsibility of seeking process improvement that would result in a minimum of $175,000 savings per year. This was the minimum standard established by the plant for any major process improvement project. The team started by reviewing the process map to determine possible bottlenecks in the process. The process mapping is shown in Figure 36.2. Extensive interviews were conducted, and an in-depth observation of the processes was undertaken to identify probable causes of inefficiencies in the system. Many new ideas were generated while breaking the problem down, so the team considered breaking the project further down to subprojects. After the extensive investigation, the cutting process was singled out as the likely bottleneck operation. The cutter grind showed up quite often when tracking downtime on the cutting machines. The team found that many cutting heads sent to the cutting machines do not reach maximum operating life. The heads becomes dull and the cutter grind finds it difficult to keep up with production and in numerous occasions, the cutting machines were idle waiting for the blades to be ground.

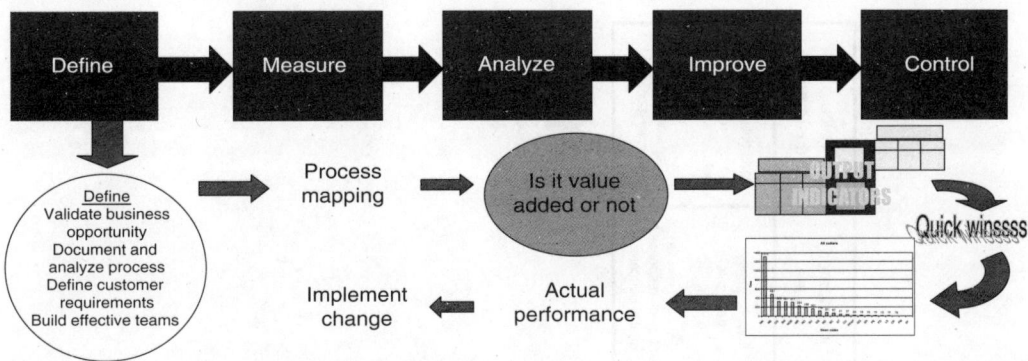

FIGURE 36.1   Lean-Sigma improvement process.

Process: Gear cutting				Process owner: Operator	
Suppliers (providers of the required resources)	Inputs (resources required by the process)	Process (top level description of activity)	Outputs (deliverables from the process)	Customers (anyone who receives a deliverable from the process)	Requirements (critical and measurable customer expectations of the process)
Blanking department	Gear blanks within tolerances	Cut gears to customer expectations	Consistency in tolerances	Heat treat	On time delivery
				Lapping department	Quality
Cutter grind	Cutter bodies within tolerances		Quality tolerances maintained		
	Process capability			End customer	
	Maint. support				

FIGURE 36.2   Process mapping of the gear-cutting process.

According to Lean-Sigma approach, the first phase of improvement involves identifying the process to be improved based on the voice of the business and voice of the customer. The voice of the business involves establishing the critical business requirement. In this case, it involves improving the efficiency and machine utilization of the cutting operation. The voice of the customer entails reviewing critical customer requirements. For this case the review lead to ensuring that the number of defects shipped to the customer is substantially reduced. As shown in Figure 36.3, by not meeting both requirements the company was having difficulty satisfying its customer and meeting the revenue goals established for the company.

With the knowledge of Lean Sigma, the project team leaders decided to choose the improvement project that would lead to local improvement in global benefits. The management team made a thorough evaluation of the plant processes from the aspect of customer satisfaction and throughput. Traditional unit cost reduction and operations productivity increase were used as the output measures of the improvement effort. The impact of the improvement on overall quality of axle and system throughput was also considered as an output indicator.

The process flow map is shown in Figure 36.4. Using the value-stream mapping methodology, the project team identified the operation that was the cause of the large amount of defects in the plant. These

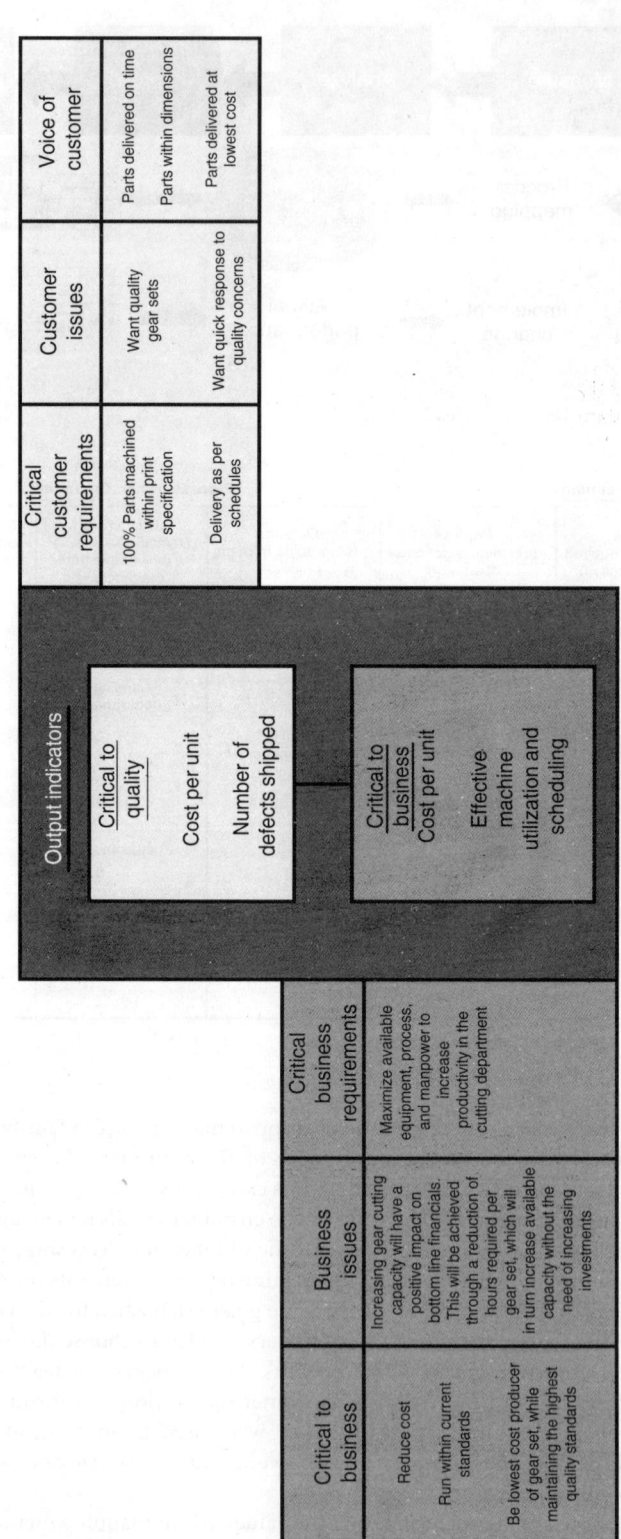

FIGURE 36.3　Integrating voice of business/customer in Lean Sigma.

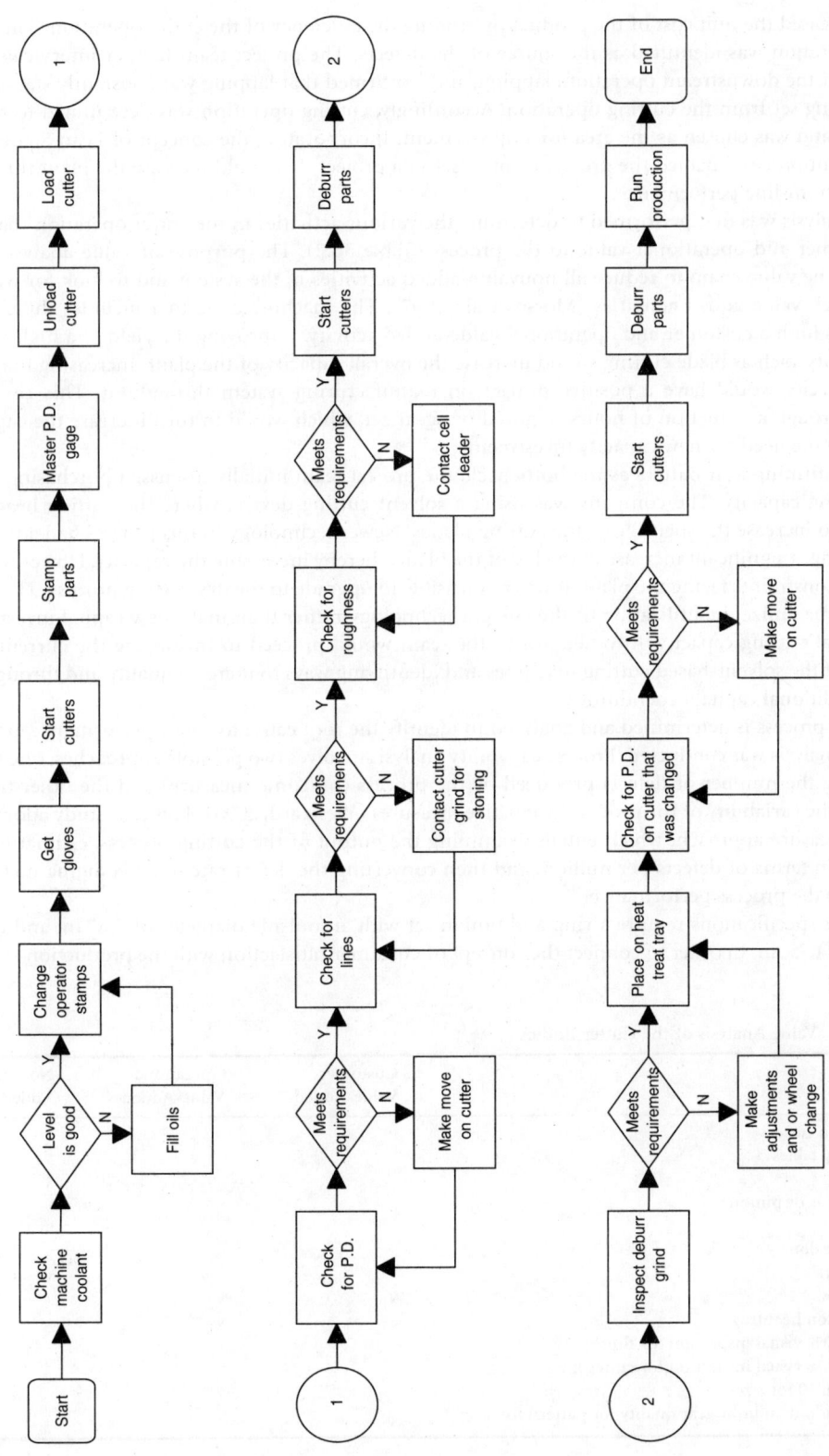

FIGURE 36.4 Value-stream mapping of the cutter body process.

defects increased the unit cost of the product by eroding the efficiency of the entire operation. The gear-cutting operation was identified as the source of the defects. The project team further interviewed the operators of the downstream operation, lapping, and confirmed that lapping was constantly starved for complete ring set from the cutting operation. Accordingly, cutting operation was determined to be the bottleneck and was chosen as the area for improvement. Incorporating the concept of Lean Sigma into improvement process enabled the project team to select a project that could increase the plant through-put and bottom-line performance.

Value analysis was first performed to determine the various activities in the cutter operation that add both customer and operational value to the process (Table 36.2). The purpose of value analysis is to streamline the value chain to reduce all nonvalue-added activities in the system and to look for ways to enhance high value-added activities (Morse et al., 1997). The machine cycle that includes the cutting operation is both a customer and operational value-added activity. Improving the yield of a high-value-added activity such as blade cutting would increase the overall capacity of the plant. Increasing the gear-cutting capacity would have a positive impact on manufacturing system throughput. This could be achieved through a reduction of hours required per gear set, which would in turn increase the capacity and remove the need for new capacity investments.

After confirming gear cutters as the bottleneck, the project team initially discussed purchasing additional cutting capacity. The company was using a solvent-cutting device, where the cutting head was lubricated to increase the shelf life of the cutting blades. Newer technology in this process had advanced to dry cutting, a significant increase in the life of the blade, thereby increasing the capacity. However, with the capital constraints facing the plant, it was not feasible to upgrade to the dry cutting process. The team decided to maximize the utilization of the current technology rather than make new capital investment in additional cutting capacity. In other words, the team would proceed to investigate the current per-formance of the solvent-based cutting machines and identifying ways to increase quality and throughput without additional capital expenditures.

After the process is determined and analyzed to identify the root cause for poor performance, process capability analysis was conducted. Process capability analysis involves two possible approaches, one based on counting the number of defects produced by the process (outcome measure) and the other tied to measuring the variability of the process (predictive measure) (Westgard, 2003). This case study adopts the outcome measure approach, which entails examining the output of the cutting process, estimating the defect rate in terms of defects per million, and then converting the defect rate to a Six Sigma metric to characterize the process performance.

Customer specifications require a ring and pinion set with an outside diameter of 7.67 in. and inside diameter of 4.75 in. In order to connect the concept of customer satisfaction with the production system,

TABLE 36.2 Value Analysis of the Cutter Bodies

Activity	Customer Value-Added	Operational Value-Added	Nonvalue-Added
Move blanks to cutters			√
Load stamping tables			√
Stamp blanks			√
Remove cut gear or pinion			√
Load blank			√
Close machine door			√
Push cycle start			√
Machine cycles	√	√	
Place cut gear on heat tray		√	
Gage parts 100% visual inspection for finish		√	
Gage parts 100% visual inspection for cutter lines		√	
Gage parts 1 in 10 for size		√	
Gage parts random audit by gear quality for pattern for size		√	

the company decided to convert the defective rate into a "customer specification index (CSI)". CSI was computed based on the ratio of "the number of products produced that meet customer specifications" and "peak capacity", defined as "the maximum output rate a process can achieve under ideal conditions" (Krajewski and Ritzman, 2002). The company believes that the CSI effectively communicates how well a process meets customer specifications, and it provides more useful feedback to the production system. For instance, a 65% of CSI in cutter operation was measured over a 2-month period, which indicates that only 65% of peak capacity was utilized to meet customer needs. In other words, 35% of machine capacity was either wasted (due to setup, wait for material, maintenance, or breakdown) or produced items that failed to meet customer specifications. The 65% of CSI was then used as a baseline to measure the level of improvement made by this project. The goal (performance outcome) established for this process was 80% of CSI, which was considered to be the standard for world-class practice.

Once the measure and process capability for cutting operations were defined, the project team proceeded to analyze the root cause of poor CSI performance. The project team interviewed the operators and found that due to lack of proper lubrication of the blades, many cutting heads did not attain their maximum life. Moreover, as the dull blades were removed for resharpening, cutter grinders became idle and thus failed to keep up with the production schedule. Apparently, the dullness of the blades caused substantial downtime at the cutters. Furthermore, a dull blade also resulted in many defect-prone items including rough finish along the cutter lines and machine crash. In summary, the root cause of poor CSI was found to be blade inefficiency, since it caused machine downtime and defective products.

The team decided that any improvement made must increase the blade efficiency, reduce the cutter downtime, and subsequently improve the CSI. As part of the capability analysis, the project team concluded that they must design a process that would have equal flow and pressure for the blades to cut the drive and the coast sides of the product. Based on the operator's input and the observation of the current process, the project team found that most coolant was deflected from the blades during the process of cutting the coast side of the ring. Insufficient coolant reached the coastline of the blades, rendering them dull, thus reducing the efficiency of the blade and increasing the amount of defects that resulted from the dull blades. The solution was to acquire a locally manufactured muffler pipe that discharged an equal flow of coolants to all faces of the cutting operation. Because pressure and flow were important to the life of the cutter body, two actions were taken to insure that filters were changed on a timely basis. The first action was to install coolant lines that would feed coolants to the coast side of the cut. This change increased the shelf lives and the yield of the blades. Furthermore, the blades were no longer necessary to be reground, which dramatically reduce the downtime of the cutters. The performance of the new coolant feed lines is presented in Table 36.3. There was a substantial improvement in productivity from 106.5% in period 1 to 134.6% in period 5. The second action was to test a number of blades to determine those that would perform close to the dry cut process. The blade chosen was outsourced through a local supplier who agreed to carry a stock of 1,000 blades to support the company's production lines. This allowed the company to order the blades in smaller lots, reducing the cost of stocking the blades in-house.

TABLE 36.3   Productivity Improvement in the Cutter Body Operation

	Base	Target	Week 1	Week 2	Week 3	Week 4	Week 5
Unit	38,131	38,131	42,305	52,687	52,553	44,739	51,685
FTE's	62.0	58.0	60.4	60.1	61.3	57.1	58.4
Units/FTE	615	657	700	877	857	884	885
Productivity improvement (%)		6.9	6.5	25.3	−2.3	8.5	12.9
Productivity improvement actual vs. target (%)			106.5	133.3	130.4	130.2	134.6

Once the new processes are in place, the goal shifts to supporting the change and monitoring the processes to make sure the improvements are sustained (Plotkin, 1999). The phase is designed to document and monitor the improved process conditions via a statistical process control methodology. In this study, the operators were trained on appropriate statistical tools to help monitor the new process. Control charts of the cutter surface were maintained by the technicians, and the charts enabled the cutters and the project team to monitor the process and assess the extent of the improvement made. A greater portion of employee training was held on the job, because the company retained the same experienced technicians who were running the machines before the process improvement. Finally, a crucial aspect of the CI process was the buy-in of the technicians/operators. The company had implemented a gain-sharing plan in which the employees share benefits from any improvements made. In this project, many employees received additional pay from the savings resulting from the implementation of the Lean-Sigma improvement. The additional employee-sharing benefits reduced most of the resistance to any changes resulting from process improvement, such as the use of process control charts by the operators to control the process.

Figure 36.5 shows an X-bar chart of the CSI prior to and after the improvement. This project had clearly improved the process capability of cutting operation by raising its CSI from 65% to approximately 85%. Meanwhile, the yield increased from 106% the first week of implementation to an average of more than 130% the next 4 weeks after implementation (Table 36.3).

This improvement project was initiated in late 2002 and completed in March 2003. Once the cutting operation was improved and became efficient, it would no longer be the bottleneck of the production system. The project team continued to analyze the manufacturing performance and customer needs to detect potential new constraints, which would be the target of the new improvement project.

36.6   Conclusions

Many studies have investigated the integration of various CI techniques, such as TOC and JIT (Cook, 1994; Rahman, 1998), TOC and TQM (Ronen and Pass, 1994; Lepore and Cohen, 1999), TOC and Reengineering (Gritzuk et al., 1999), Six Sigma and TOC (Ehie and Sheu, 2005). This study adds to the CI literature and practice by demonstrating the value of a method of combining lean manufacturing and Six Sigma. Specifically, we proposed an integrated Lean-Six Sigma framework and applied this framework to an axle manufacturing company to improve its gear-cutting operation. Under this framework, Six Sigma focused on reducing process variation, thus reducing the amount of defective products while Lean manufacturing focused on reducing waste in the system. The results of the case study indicate that the company

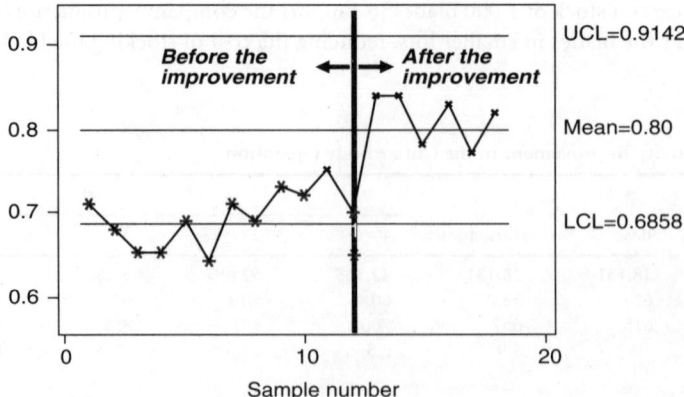

FIGURE 36.5   X-bar chart for CSI. (1) The control limit 80% was the target established by the project team since it was considered to be world-class practice. (2) The upper and lower limits were determined based on three-sigma variations.

benefited tremendously from its emphasis on global improvement guided by the integrated Lean-Sigma approach. Managers were able to select a CI project that had greater impact on bottom line performance. On the other hand, Six Sigma provided various statistical tools and engineering techniques for defining the specific process to be improved, analyzing the root causes, and designing actions for making improvement. After the implementation of the change, the Lean-Six Sigma framework ensured that the new change was supported and substantiated by proper employee training and continuous monitoring. Communication for obtaining buy-in was made to reduce resistance to changes from employees. Ultimately, the gear-cutting project improved customer satisfaction index, increased throughput and quality of the cutting operation, and reduced the inventory level of blades.

The implementation of this CI project took 4 months. It was estimated that the project resulted in a total saving of $200,000 per year, exceeding the average $170,000 savings for a typical Six Sigma project at the plant. The system-wide "soft" savings from the project exceeded $1.5 million. The integrated Lean/Six Sigma approach was clearly a success.

Overall, the integration of Six Sigma and Lean manufacturing provides managers with an excellent platform for two CI techniques to work together and create great synergy. Companies often implement more than one CI technique, and each CI technique is usually led by a different project team and teams rarely communicated with each other. Consequently, they do not collaborate with each other, but actually compete with each other for corporate resource and top management support. An integration model such as the one described herein provides a platform for different CI techniques to interact and work together. It enables mangers to understand that various CI approaches are not exclusive to each other. The case study presents an evidence of the value of an integrated CI framework.

References

Breyfogle, F., III, *Implementing Six Sigma: Smarter Solution Using Statistical Methods*, Wiley, New York, 1999.

Cook, D., A simulation comparison of traditional, JIT, and TOC manufacturing systems in a flow shop with bottlenecks, *Prod. Invent. Manage. J.*, 35, 73–78, 1994.

Ehie, Ike C. and Sheu, C., Integrating Six-Sigma and theory of constraints for continuous improvement: A case study, *The International Journal of Manufacturing Technology Management*, 16 (5), 542–553, 2005.

Harry, M.J., Six-Sigma: A breakthrough strategy for profitability, *Qual. Prog.*, 1998, 60–64.

Harry, M. and Schreoder, R., *Six Sigma: The Breakthrough Management Strategy Revolutionizing the World's Top Corporation*, Currency, New York, 2000.

Kiemele, M.J., Schmidt, S.R., and Berdine, R.J., *Basic Statistical Tool for Continuous Improvement*, Air Academy Press, Colorado Springs, CO, 1997.

Krajewski, L. and Ritzman, L., *Operations Management: Strategies and Analysis*, Prentice-Hall, Englewood Cliffs, NJ, 2002.

Lepore, D. and Cohen, O., *Deming and Goldratt – The Theory of Constraints and the Systems of Profound Knowledge*, The North River Press, Croton-on-Hudson, New York, 1999.

Levinson, W.A. and Rerick, R.A., *Lean Enterprise: A Synergistic Approach to Minimizing Waste*, ASQ Quality Press, Milwaukee, WI, USA, 2002.

Gritzuk, M., Janis, W.S., Blundon, E.G., and Myers, J., Multiple benefits of reengineering, *Journal of American Water Works Association*, 91 (11), 48–55, November 1999.

Morse, M., Selders, D. McIntyre, K., and Hruska, J., Quest for Quality: A Common Sense Approach to Business Operating Systems, Quality Insight Column, 1997 (http://www.bnp.com/industrial_heating/archive/quality7-97.html).

Nicholas, J.M., *Competitive Manufacturing Management: Continuous Improvement, Lean Production, and Customer-Focused Quality*, Irwin_McGraw-Hill Publishers, New York, NY, 1998.

Ohno, T., *Toyota Production System – Beyond Management of Large-Scale Production*, Diamond Publishing, Tokyo, Japan, 1978.

Plotkin, H., Six Sigma: What it is and How to Use it, Harvard Management Update: A Newsletter from Harvard Business School Publishing, Reprint # u9906c, 1999.

Rahman, S., Theory of constraints: A review of the philosophy and its applications, *Int. J. Oper. Prod. Manage.*, 18, 336–356, 1998.

Ronen, B. and Pass, S., Focused management: a business oriented approach to total quality management, *Ind. Manage.*, May/June, 9–12, 1994.

Westgard, J.O., Six Sigma basics: Outcome measurement of process performance accessed at: http://www.westgard.com/lesson66.html, September 29, 2003.

Appendix

Systems Conversion Factors and Formulae

Numbers and Prefixes

yotta (10^{24})	1 000 000 000 000 000 000 000 000
zetta (10^{21})	1 000 000 000 000 000 000 000
exa (10^{18})	1 000 000 000 000 000 000
peta (10^{15})	1 000 000 000 000 000
tera (10^{12})	1 000 000 000 000
giga (10^{9})	1 000 000 000
mega (10^{6})	1 000 000
kilo (10^{3})	1 000
hecto (10^{2})	100
deca (10^{1})	10
deci (10^{-1})	0.1
centi (10^{-2})	0.01
milli (10^{-3})	0.001
micro (10^{-6})	0.000 001
nano (10^{-9})	0.000 000 001
pico (10^{-12})	0.000 000 000 001
femto (10^{-15})	0.000 000 000 000 001
atto (10^{-18})	0.000 000 000 000 000 001
zepto (10^{-21})	0.000 000 000 000 000 000 001
yacto (10^{-24})	0.000 000 000 000 000 000 000 001
Stringo (10^{-35})	0.000 000 000 000 000 000 000 000 000 000 000 01

Conversion Factors

Area

Multiply	By	To Obtain
acres	43,560	sq feet
	4,047	sq meters
	4,840	sq yards
	0.405	hectare
sq centimeter	0.155	sq inches
sq feet	144	sq inches
	0.09290	sq meters
	0.1111	sq yards
sq inches	645.16	sq millimeters
sq kilometers	0.3861	sq miles
sq meters	10.764	sq feet
	1.196	sq yards
sq miles	640	acres
	2.590	sq kilometers

Volume

Multiply	By	To Obtain
acre-foot	1233.5	cubic meters
cubic centimeter	0.06102	cubic inches
cubic feet	1728	cubic inches
	7.480	gallons (U.S.)
	0.02832	cubic meters
	0.03704	cubic yards
liter	1.057	liquid quarts
	0.908	dry quarts
	61.024	cubic inches
gallons (U.S.)	231	cubic inches
	3.7854	liters
	4	quarts
	0.833	British gallons
	128	U.S. fluid ounces
quarts (U.S.)	0.9463	liters

Energy, heat power

Multiply	By	To Obtain
BTU	1,055.9	joules
	0.2520	kg-calories
watt-hour	3,600	joules
	3.409	BTU
HP (electric)	746	watts
BTU/second	1,055.9	watts
watt-second	1.00	joules

Mass

Multiply	By	To Obtain
carat	0.200	cubic grams
grams	0.03527	ounces
kilograms	2.2046	pounds
ounces	28.350	grams
pound	16	ounces
	453.6	grams
stone (U.K.)	6.35	kilograms
	14	pounds
ton (net)	907.2	kilograms
	2000	pounds
	0.893	gross ton
	0.907	metric ton
ton (gross)	2240	pounds
	1.12	net tons
	1.016	metric tons
tonne (metric)	2,204.623	pounds
	0.984	gross pound
	1000	kilograms

Temperature

Conversion Formulas

Celsius to Kelvin	$K = C + 273.15$
Celsius to Fahrenheit	$F = (9/5)C + 32$
Fahrenheit to Celsius	$C = (5/9)(F - 32)$
Fahrenheit to Kelvin	$K = (5/9)(F + 459.67)$
Fahrenheit to Rankin	$R = F + 459.67$
Rankin to Kelvin	$K = (5/9)R$

Velocity

Multiply	By	To Obtain
feet/minute	5.080	mm/second
feet/second	0.3048	meters/second
inches/second	0.0254	meters/second
km/hour	0.6214	miles/hour
meters/second	3.2808	feet/second
	2.237	miles/hour
miles/hour	88.0	feet/minute
	0.44704	meters/second
	1.6093	km/hour
	0.8684	knots
knot	1.151	miles/hour

Pressure

Multiply	By	To Obtain
atmospheres	1.01325	bars
	33.90	feet of water
	29.92	inches of mercury
	760.0	millimeter of mercury
bar	75.01	centimeter of mercury
	14.50	pounds/square inch
dyne/sq centimeter	0.1	Newtons/square meter
Newtons/square centimeter	1.450	pounds/square inch
pounds/square inch	0.06805	atmospheres
	2.036	inches of mercury
	27.708	inches of water
	68.948	millibars
	51.72	mm of mercury

Constants

Speed of light	2.997925×10^{10} cm/sec
	983.6×10^{6} ft/sec
	186,284 miles/sec
Velocity of sound	340.3 m/sec
	1116 ft/sec
Gravity (acceleration)	9.80665 m/sec^2
	32.174 ft/sec^2
	386.089 in./sec^2

Distance

Multiply	By	To Obtain
angstrom	10^{-10}	meters
feet	0.30480	meters
	12	inches
inches	25.40	millimeters
	0.02540	meters
	0.08333	feet
kilometers	3280.8	feet
	0.6214	miles
	1094	yards
meters	39.370	inches
	3.2808	feet
	1.094	yards
miles	5280	feet
	1.6093	kilometers
	0.8694	nautical miles
millimeters	0.03937	inches
nautical miles	6076	feet
	1.852	kilometers
yards	0.9144	meters
	3	feet
	36	inches

Physical Science Equations

$$D = \frac{m}{V}$$

D　density

m　mass,　　$\dfrac{\text{g}}{\text{cm}^3} = \dfrac{\text{kg}}{\text{m}^3}$

V　volume

$$P = \frac{W}{t}$$

P　power,　　W (=watts)

W　work,　　J

t　time,　　sec

$$d = vt$$

d　distance,　m

v　velocity,　m/sec

t　time,　　sec

$$KE = \frac{1}{2} mv^2$$

KE　kinetic energy

m　mass,　　kg

v　velocity,　m/sec

$$a = \frac{vf - vi}{t}$$

a　acceleration,　m/sec²

vf　final velocity,　m/sec

vi　initial velocity,　m/sec

t　time,　　sec

$$Fe = \frac{kQ_1 Q_2}{d^2}$$

Fe　electric force,　N

k　Coulomb's constant,　$k = 9 \times 10^9 \dfrac{\text{Nm}^2}{\text{c}^2}$

$Q1, Q2$　electrical charges,　C

d　separation distance,　m

$$d = vit + \tfrac{1}{2} at^2$$

d　distance,　　m

vi　initial velocity,　m/sec

t　time,　　sec

a　acceleration,　m/sec²

$$V = \frac{W}{Q}$$

V　electrical potential difference V(=volts)

W　work done,　　J

Q　electic charge flowing,　C

$$F = ma$$

F　net force,　　N(=Newton)

m　mass,　　kg

a　acceleration,　m/sec²

$$I = \frac{Q}{t}$$

I　electric current ampères

Q　electic charge moving,　C

t　time,　　sec

$$Fg = \frac{Gm_1 m_2}{d^2}$$

Fg　force of gravity,　N

G　universal gravitational constant,

　　　$G = 6.67 \times 10^{-11} \dfrac{\text{N}-\text{m}^2}{\text{kg}}$

m_1, m_2　masses of the two objects,　kg

d　separation distance, m

$$W = VIt$$

W　electical energy,　J

V　voltage,　　V

I　current,　　A

t　time,　　sec

$p = mv$	p	momentum,	kg m/sec	$P = VI$	P	power,	W
	m	mass			V	voltage,	V
	v	velocity			I	current,	A
	W	work,	J($=$joules)		H	heat energy,	J
$W = Fd$	F	force,	N	$H = cm\Delta T$	m	mass,	kg
	d	distance,	m		ΔT	change in temperature,	°C
					c	specific heat,	J/Kg °C

Units of Measurement

English System		Metric System			
1 foot (ft)	12 inches (in.); $1'=12''$	mm	millimeter	0.001 m
1 yard (yd)	3 feet	cm	centimeter	0.01 m
1 mile (mi)	1,760 yards	dm	decimeter	0.1 m
1 sq foot	144 sq inches	m	meter	1 m
1 sq yard	9 sq feet	dam	decameter	10 m
1 acre	4840 sq yards=43560 ft^2	hm	hectometer	100 m
1 sq mile	640 acres	km	kilometer	1000 m

Note: Prefixes also apply to l (liter) and g (gram).

Common Units used with the International System

Units of Measurement	Abbreviation	Relation	Units of Measurements	Abbreviation	Relation
meter	m	length	degree Celsius	°C	temperature
hectare	ha	area	Kelvin	K	thermodynamic temperature
tonne	t	mass	Pascal	Pa	pressure, stress
kilogram	kg	mass	joule	J	energy, work
nautical mile	M	distance (navigation)	Newton	N	force
knot	kn	speed (navigation)	watt	W	power, radiant flux
liter	L	volume or capacity	ampere	A	electric current
second	s	time	volt	V	electric potential
hertz	Hz	frequency	ohm	Ω	electric resistance
candela	cd	luminous intensity	coulomb	C	electric charge

Kitchen Measurements

A pinch	1/8 tsp. or less
3 tsp	1 tbsp.
2 tbsp	1/8 c.
4 tbsp	1/4 c.
16 tbsp	1 c.
5 tbsp. + 1 tsp	1/3 c.
4 oz	1/2 c.
8oz	1 c.
16 oz	1 lbs.
1 oz	2 tbsp. fat or liquid
1 c. of liquid	1/2 pt.
2 c	1 pt.
2 pt	1 qt.
4 c. of liquid	1 qt.
4 qts	1 gallon
8 qts	1 peck (such as apples, pears, etc.)
1 jigger	$1\frac{1}{2}$ fl. oz.
1 jigger	3 tbsp.

Index